SOCIOLOGY
Your Compass for a New World

SOCIOLOGY
Your Compass for a New World

Robert J. Brym
University of Toronto

John Lie
University of Michigan

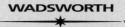

WADSWORTH
TM
THOMSON LEARNING

Australia • Canada • Mexico • Singapore • Spain
United Kingdom • United States

WADSWORTH

✦ ™

THOMSON LEARNING

Sociology Editor: Eve Howard
Senior Executive Editor: Sabra Horne
Development Editor: Lisa Hensley
Assistant Editor: Analie Barnett
Marketing Manager: Matthew Wright
Advertising Project Manager: Linda Yip
Project Manager, Editorial Production: Angela Williams Urquhart
Print/Media Buyer: Nancy Panziera
Permissions Editor: Charlotte Thomas

Production Service: Graphic World
Text Designer: Van Mua
Photo Researcher: Cheri Throop
Copy Editor: Dennis Webb
Illustrator: Graphic World
Cover Designer: Van Mua
Cover Images: Van Mua
Cover Printer: Transcontinental Printing Inc.
Compositor: Graphic World
Printer: Transcontinental Printing Inc.

Printed in Canada

2 3 4 5 6 7 05 04 03 02

For more information about our products, contact us at:
Thomson Learning Academic Resource Center
1-800-423-0563
For permission to use material from this text,
contact us by:
Phone: 1-800-730-2214
Fax: 1-800-730-2215
Web: http://www.thomsonrights.com

ISBN 0-15-507212-9
Library of Congress Control Number: 2002100040

Asia
Thomson Learning
60 Albert Street, #15-01
Albert Complex
Singapore 189969

Australia
Nelson Thomson Learning
102 Dodds Street
South Melbourne, Victoria 3205
Australia

Canada
Nelson Thomson Learning
1120 Birchmount Road
Toronto, Ontario M1K 5G4
Canada

Europe/Middle East/Africa
Thomson Learning
Berkshire House
168-173 High Holborn
London WC1 V7AA
United Kingdom

Latin America
Thomson Learning
Seneca, 53
Colonia Polanco
11560 Mexico D.F.
Mexico

Spain
Paraninfo Thomson Learning
Calle/Magallanes, 25
28015 Madrid, Spain

DEDICATION

Many authors seem to be afflicted with stoic family members who gladly allow them to spend endless hours buried in their work. We suffer no such misfortune. The members of our families have demanded that we focus on what really matters in life. We think that focus has made this a better book. We are deeply grateful to Rhonda Lenton, Shira Brym, Talia Lenton-Brym, Ariella Lenton-Brym, Charis Thompson, Thomas Cussins, Jessica Cussins, and Charlotte Lie. We dedicate this book to them with thanks and love.

Robert J. Brym (pronounced "brim") is an internationally known scholar. He studied in Israel and Canada and received his PhD from the University of Toronto, where he is now on faculty and where he especially enjoys teaching introductory sociology to 1,500 students every year. He has won numerous awards for his teaching and scholarly work, including four Dean's Excellence Awards, the Oswald Hall Award for Undergraduate Teaching Excellence, and the Outstanding Contribution Award and the Distinguished Service Award of the Canadian Sociology and Anthropology Association. His main areas of research are political sociology, race and ethnic relations, and sociology of culture. His major works include *Intellectuals and Politics* (London and Boston: Allen & Unwin, 1980); *From Culture to Power* (Toronto: Oxford University Press, 1989); *The Jews of Moscow, Kiev, and Minsk* (New York: New York University Press, 1994); and *New Society* (Toronto: Harcourt Canada, 2001), one of Canada's best-selling introductory sociology textbooks, now in its third edition. In 2001, he was co-investigator for the world's first large-scale survey of Internet dating, sponsored by MSN. His works have been translated into several languages. From 1992–97, Robert served as editor of *Current Sociology*, the journal of the International Sociological Association, and he is now co-editor of *East European Jewish Affairs*, published in London. He is currently involved in two research projects. With sociologists at the Institute of Sociology, Russian Academy of Sciences, Moscow, he is conducting a study of recruitment and mobility in the Russian civil service. As part of an international team of more than 100 sociologists and political scientists centered at the University of Michigan, he is also analyzing the results of the 2000 World Values Survey of 86 countries, including the United States.

John Lie (pronounced "lee") was born in South Korea, grew up in Japan and Hawaii, and attended Harvard University. Currently Professor of Sociology at the University of Michigan, he has taught at the University of Hawaii at Manoa, the University of California at Berkeley, the University of Illinois at Urbana–Champaign, and Harvard University in the United States, as well as universities in Japan, South Korea, Taiwan, and New Zealand. He served as Chair of the Department of Sociology at the University of Illinois, Urbana–Champaign. His main research interests are comparative macrosociology and comparative race and ethnic relations. John's major publications include *Blue Dreams: Korean Americans and the Los Angeles Riots* (Cambridge MA: Harvard University Press, 1995), *Han Unbound: The Political Economy of South Korea* (Stanford CA: Stanford University Press, 1998), and *Multiethnic Japan* (Cambridge MA: Harvard University Press, 2001). He has taught introductory sociology classes ranging in size from 3 to more than 700 students in several countries and hopes that this book will stimulate your sociological imagination.

BRIEF CONTENTS

CONTENTS

◾

PART I

Foundations, 1

PART III

Inequality, 171

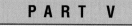

PREFACE

Why is Sociology a Compass for a New World?

Soon after European explorers arrived in North and South America, they started calling the twin continents the "New World." Everything was different here. A native population perhaps a hundredth the size of Europe's occupied a territory more than four times larger. The New World was unimaginably rich in resources. European rulers saw that by controlling it they could increase their power and importance. Christians recognized new possibilities for spreading their religion. Explorers discerned fresh opportunities for rewarding adventures. A wave of excitement swelled as word spread of the New World's vast potential and challenges.

Today, it is easy for us to appreciate that wave of excitement. For we too have reached the frontiers of a New World. And we are also full of anticipation. Our New World is one of virtually instant long-distance communication, global economies and cultures, weakening nation-states, and technological advances that often make the daily news seem like reports from a distant planet. In a fundamental way, the world is not the same place it was just 50 years ago. Orbiting telescopes that peer to the fringes of the universe, human genetic code laid bare like a road map, fiber optic cable that carries a trillion bits of information per second, and spacecraft that transport robots to Mars help to make this a New World.

Five hundred years ago, the early European explorers of North and South America set themselves the preliminary task of mapping the contours of the New World. We set ourselves a similar task here. Their frontiers were physical. Ours are social. Their maps were geographical. Ours are sociological. But in terms of functionality, our maps are much like theirs. All maps allow us to find our place in the world and see ourselves in the context of larger forces. As C. Wright Mills wrote, *sociological* maps allow us to "grasp the interplay of [people] and society, of biography and history" (Mills, 1959: 4). This book shows you how to draw sociological maps so you can see your place in the world, figure out how to navigate through it, and perhaps discover how to improve it. It is your sociological compass.

We are not as naive as the early European explorers. Where they saw only hope and bright horizons, minimizing the significance of the violence required to conquer the people of the New World, our anticipation is mixed with dread. Scientific breakthroughs are announced almost daily, but the global environment has never been in worse shape, and AIDS is now the leading cause of death in Africa. Marriages and nations unexpectedly break up and then reconstitute themselves in new and unanticipated forms. We celebrate the advances made by women and racial minorities only to find that some people oppose their progress, sometimes violently. Waves of people suddenly migrate between continents, establishing cooperation but also conflict between previously separated groups. New technologies make work more interesting and creative for some, offering unprecedented opportunities to get rich and become famous. They also make jobs more onerous and routine for others. The standard of living goes up for many people but stagnates for many more. The world's mood and its political and economic outlook can be buoyant one day and uncertain the next. As Tom Brokaw wrote in the aftermath of the terrorist attacks of September 11, 2001, "change is coming . . . in forms we cannot foresee" (Brokaw, 2001).

Is it any wonder that, amid all this contradictory news, good and bad, uncertainty about the future prevails? We wrote this book to show undergraduate college students that sociology can help them make sense of their lives, however uncertain they may appear to be. Moreover, we show that sociology can be a liberating practical activity, not just an abstract intellectual exercise. By revealing the opportunities and constraints you face, sociology can help teach you who you are and what you can become in this particular social and historical context. We cannot know what the future will bring, but we can at least know the choices we confront and the likely consequences of our actions. From this point of view, sociology can help us create the best possible future. That has always been sociology's principal justification, and so it should be today.

Unique Features

We have tried to keep sociology's main purpose and relevance front and center in this book. As a result, *Sociology: Your Compass for a New World* differs from other major introductory sociology textbooks in six ways:

1. ***Drawing connections between one's self and the social world.*** To varying degrees, all introductory sociology textbooks try to show students how their personal experiences are connected to the larger social world. However, we employ two devices to make these connections clearer than in other textbooks. First, we illustrate key sociological ideas by using fresh examples that resonate deeply with student interests and experiences. For example, to show how radical subcultures often become commercialized, we analyze the development of rap music. To demonstrate how functionalists study religion, we discuss the Super Bowl. To illustrate the use of two- and three-variable tables, we study the most recent General Social Survey data on the number of sex partners Americans had in the past year. To characterize contemporary processes of urbanization, we consider the growth of theme parks. To demonstrate the reach of globalization processes, we trace the use of American slang among Japanese teenagers. To portray various aspects of the mass media, we investigate the growth of the World Wide Web and its convergence with television and telephony. We think these examples speak directly to today's students about important sociological ideas in terms they understand, thus making the connection between self and society clear.

   We also developed several pedagogical features to draw the connection between students' experiences and the larger social world. *"Where Do You Fit In?"* is a question we ask in every chapter. In this feature, we repeatedly challenge students to consider how and why their own lives conform to, or deviate from, various patterns of social relations and actions. *"It's Your Choice"* is a feature of each chapter that sets out public policy alternatives on a range of pressing social issues. It teaches students that sociology can be a matter of the most urgent practical importance. Students also learn they can have a say in the development of public policy. *"Sociology at the Movies"* takes a universal and popular element of contemporary culture and renders it sociologically relevant. We provide brief reviews of movies, most of them recent releases, and highlight the sociological insights they contain.

2. ***What to think versus how to think.*** All textbooks teach students both *what* to think about a subject and *how* to think about it from a particular disciplinary perspective. In our judgment, however, introductory sociology textbooks usually place too much stress on the "what" and not enough on the "how." The result: They sometimes read more like encyclopedias than enticements to look at the world in a new way. We have tipped the balance in the other direction. To be sure, *Sociology: Your Compass for a New World* contains definitions and literature reviews. It features standard pedagogical aids such as a list of "Chapter Aims" at the beginning of each chapter and a "Summary," a "Glossary," a list of "Suggested Readings," a list of "Recommended Web Sites," and a set of "Questions to Consider" at the end of each chapter. However, we devote more space than other authors to showing how sociologists think. We typically relate an anecdote to highlight an issue's importance, present contending interpretations of the issue, and then adduce data to judge the merits of the various interpretations. We do not just refer to tables and graphs, we analyze them. When evidence warrants, we reject theories and endorse others. Thus, many sections of the book read more like a simplified journal article than an encyclopedia. If all this sounds just like what sociologists do professionally, then we have achieved our aim: to present a less antiseptic, more realistic, and therefore intrinsically exciting account of how sociologists practice their craft. Said differently, one of the strengths of this book is that it does not present sociology as a set of immutable truths carved in stone tablets. Instead, it shows how sociologists actually go about the business of solving sociological puzzles.

3. ***Objectivity versus subjectivity.*** Sociologists since Max Weber have understood that sociologists—indeed, all scientists—are members of society, and that their thinking and research are influenced by the social and historical context in which they work. Yet a recent article in the *American Sociological Review* shows that introductory sociology textbooks present a stylized and not very sociological view of the research process. Textbooks tend to emphasize sociology's objectivity and the hypothetico-deductive method of reasoning, for the most part ignoring the more subjective factors that go into the research mix (Lynch and Bogen, 1997). We think this emphasis is a pedagogical error. In our own teaching, we have found that drawing the connection between objectivity and subjectivity in sociological research makes the discipline more appealing to students. It shows how research issues are connected to the lives of real flesh-and-blood women and men, and how sociology is related to students' existential concerns. Therefore, in each chapter of *Sociology: Your Compass for a New World*, we feature a "Personal Anecdote" that explains how certain sociological issues first arose in our own minds. We also place the ideas of important sociological figures in social and historical context. We show how sociological methodologies serve as a reality check, but we also make it clear that socially grounded personal concerns often lead sociologists to decide which aspects of reality are worth checking on in the first place. We believe *Sociology: Your Compass for a New World* is unique in presenting a realistic and balanced account of the role of objectivity and subjectivity in the research process.

4. ***Diversity and a global perspective.*** It is gratifying to see how much less parochial American introductory sociology textbooks are today than they were just twenty years ago. Contemporary textbooks highlight gender and race issues. They broaden the student's understanding of the world by comparing the United States with other societies. They show how global processes affect local issues and how local issues affect global processes. *Sociology: Your Compass for a New World* is no different in this regard. We have made diversity and globalization prominent themes of this book. We incorporate a "Global Perspective" feature in each chapter. We make frequent and effective use of cross-national comparisons between the United States and countries as diverse as India and Sweden. And we remain sensitive to gender and race issues throughout. This has been easy for us because we are members of racial

and ethnic minority groups. We are multilingual. We have lived in other countries for extended periods. And we have published widely in four countries other than the United States. Robert Brym specializes in the study of Russia and Canada, while John Lie's research focuses on South Korea and Japan. As you will see in the following pages, our backgrounds have enabled us to bring greater depth to issues of diversity and globalization than other textbooks.

5. *Currency.* Every book bears the imprint of its time. It is significant, therefore, that the first editions of all major American introductory sociology textbooks were published about fifteen years ago. Circa 1988, just over 10% of Americans owned PCs. The World Wide Web did not exist. Genetic engineering was in its infancy. The USSR was a major world power. Nobody could imagine teenage boys committing mass murder at school with semi-automatic weapons. *Sociology: Your Compass for a New World* is one of the first American introductory sociology textbooks of the 21st century, and it is the most up-to-date. This is reflected in the currency of our illustrations and references. For instance, we do not just recommend a few Web sites at the end of each chapter, as is usual in other introductory sociology textbooks. Instead, Web resources form an integral part of this book; fully one-sixth of our citations are to materials on the Web. The currency of this book is also reflected in our "Technology Bytes" feature, which emphasizes the benefits and disadvantages of science and technology as they apply to each of sociology's major sub-fields. It is reflected in the fact that we devote entire chapters to "The Mass Media" and "Technology and the Global Environment." And it is reflected in the book's theoretical structure.

It made sense in the 1980s to simplify the sociological universe for introductory students by claiming that three main theoretical perspectives—functionalism, symbolic interactionism, and conflict theory—pervade all areas of the discipline. However, that approach is no longer adequate. Functionalism is less influential than it once was. Feminism is an important theoretical perspective in its own right. Conflict theory and symbolic interactionism have become internally differentiated. For example, there is no longer a single conflict theory of politics but at least three important variants. Highly influential new theoretical perspectives, such as postmodernism and social constructionism have emerged, and not all of them fit neatly into the old categories. *Sociology: Your Compass for a New World* incorporates not just the latest research findings in sociology but also recent theoretical innovations that are given insufficient attention in other major textbooks.

6. *Organization.* Traditionally, introductory sociology textbooks come in two sizes. For semester courses, there is the 21–24 chapter volume, roughly 700 pages in length. For courses taught over a quarter, there is the 14- or 15-chapter "essentials" version, about 450 pages long. We have used both types of books in our classrooms, and we have discussed the question of textbook length with other instructors. We concluded that the 700-page tome is more than many instructors and students can handle in a semester, while the stripped-down 450-page version covers the field too superficially. We decided, therefore, to write *Sociology: Your Compass for a New World* in a concise style, combining some topics and eliminating some nonessential elements one finds in other textbooks. We believe many instructors will find the roughly 600 pages of *Sociology: Your Compass for a New World* crisp and to the point, without sacrificing any essential elements of the full-length introductory sociology textbook. Some instructors of semester courses may want to flesh out some of the themes covered here by adopting one of Wadsworth's companion readers, such as *Sociological Odyssey: Contemporary Readings in Sociology,* by Adler & Adler; *Classic Readings in Sociology,* by Howard; or *Understanding Society: An Introductory Reader,* by Andersen, Logio, & Taylor. Some instructors of courses taught over a quarter may decide not to teach two or three chapters of the book. But we believe *Sociology: Your Compass for a New World* is well suited for use in both semester- and quarter-length courses, overcoming frequently voiced problems associated with both longer and shorter works of its type.

Ancillaries

A full range of high quality ancillaries has been prepared to help instructors and students get the most out of *Sociology: Your Compass for a New World*.

Supplements for Instructors

✦ *Instructor's Manual.* The Instructor's Manual has been prepared by Jenifer Kunz, PhD, of West Texas A&M University. It contains lecture outlines, supplemental lecture material, suggested activities for students, and Internet and InfoTrac College Edition exercises. There is a table of contents for the *CNN Today* Sociology Video Series. Concise user guides for InfoTrac College Edition and Web Tutor are included as appendices.

✦ *Test Bank.* The Test Bank has been prepared by Arthur Jipson, PhD, of the University of Dayton. It consists of 75–100 multiple-choice questions and 15–20 true/false questions per chapter, all with rejoinders and page references. This supplement also includes 10–20 short answer and 5–10 extended essay questions per chapter.

✦ *ExamView Computerized Test Bank.* Create, deliver, and customize tests and study guides (both print and online) in minutes with this easy-to-use assessment and tutorial system. *ExamView* offers both a *Quick Test Wizard* and an *Online Test Wizard* that guide you step-by-step through the process of creating tests. The test appears on screen exactly as it will print or display online. Using *ExamView's* complete word processing capabilities, you can enter an unlimited number of new questions or edit existing questions.

✦ *InfoTrac College Edition.* Ignite discussions or augment your lectures with the latest developments in sociology and societal change. Create your own course reader by selecting articles or by using the search keywords provided at the end of each chapter. *InfoTrac College Edition* (available as a free option with this text) gives you and your students four months of free access to an easy-to-use online database of reliable, full-length articles (not abstracts) from hundreds of top academic journals and popular sources. Among the journals available twenty-four hours a day, seven days a week are *American Journal of Sociology*, *Social Forces*, *Social Research*, and *Sociology*. Contact your Wadsworth/Thomson Learning representative for more information. *InfoTrac College Edition* is available only to North American college and university students. Journals are subject to change.

Classroom Presentation Tools for the Instructor

✦ *PowerPoint series.* Robert Brym wrote and designed the PowerPoint series that accompanies the book. Occasionally using animation to add variety and enhance visual appeal, the PowerPoint series avoids cartoons and other graphical elements that distract students and trivialize the subject matter. Instead, it clearly outlines each chapter of the text and reproduces all of the textbook's Figures. Instructors can access supplementary graphics that allow them to develop themes introduced in the text. The PowerPoint series is also available as a set of acetates.

✦ *Wadsworth's Introduction to Sociology Transparency Acetates.* A set of four-color acetates is available to help prepare lecture presentations.

✦ *Multimedia Manager for Sociology: A Microsoft PowerPoint Link Tool.* The easy way to great multimedia lectures! This one-stop digital library and presentation tool helps you assemble, edit, and present custom lectures. The CD-ROM brings together art (figures, tables, photos) from the text itself, pre-assembled PowerPoint lecture slides, and CNN video. You can use the materials as they are or add your own materials for a truly customized lecture presentation. Available Fall 2002.

✦ *CNN Today Sociology Video Series, Volumes I-V.* The *CNN Today* Sociology Video Series is an exclusive series jointly created by Wadsworth and *CNN* for the introduction to sociology course. Each video in the series consists of approximately

45 minutes of footage originally broadcast on *CNN* within the last several years and selected specifically to illustrate important sociological concepts. The videos are broken into short, two- to seven-minute segments that are perfect for classroom use as lecture launchers or to illustrate key sociological concepts. An annotated table of contents accompanies each video, with descriptions of the segments and suggestions for their possible use within the course.

◆ ***Wadsworth Sociology Video Library.*** Qualified adopters can also select from an extensive selection of videos from *Films for the Humanities and Sciences*. Contact your Wadsworth/Thomson Learning representative for more information.

◆ ***Demonstrating Sociology: ShowCase Presentation Software.*** This is a software package for instructors that allows them to analyze data live in front of a classroom. It is a powerful yet easy-to-use statistical analysis package that enables professors to show students how sociologists ask and answer questions using sociological theory. A resource book accompanies it with detailed "scripts" for using ShowCase in class. (This software is for Windows users with CD-ROM capability.)

Supplements for Students

◆ ***Study Guide.*** The Study Guide was prepared by Michael Perez, PhD, of California State University—Fullerton. Each chapter of the guide includes learning objectives, chapter outlines, key terms, Internet resources, and multiple-choice and true/false questions with rejoinders and page references, as well as short answer and essay questions to enhance student understanding.

◆ ***Understanding Society: An Introductory Reader*** by Margaret L. Anderson, Kim Logio, and Howard F. Taylor is an anthology of contemporary and classic articles that can be used to supplement this text. Articles reflect themes such as classical sociological theory, contemporary sociological research, diversity in society, globalization, and the application of the sociological perspective. The articles have been selected to be highly accessible to undergraduate readers, thus further engaging them in sociological analysis.

◆ ***Classic Readings in Sociology: Second Edition,*** edited by Eve Howard, is an inexpensive alternative to a full-sized reader. This choice collection contains a series of both classic and contemporary articles written by key sociologists. Excerpts include C. Wright Mills's "The Promise of Sociology," Erving Goffman's "Presentation of Self in Everyday Life," Max Weber's "Characteristics of Bureaucracy," and Jonathan Kozol's "Savage Inequalities."

◆ ***Sociological Odyssey: Contemporary Readings in Sociology*** by Patricia A. Adler and Peter Adler is a collection of articles that is contemporary, based on current research, and demonstrates the new sociological issues in the world today. Articles are highly readable and based on the everyday concerns that influence students' lives.

◆ ***Data Maps Supplement on CD-ROM.*** Prepared by Robert Brym, this ancillary contains 48 global and United States maps based mainly on data from 1998–2001. It is designed to help students "see" how social forces shape the world around them and stimulate discussion inside and outside the classroom. Each map is tied directly to the book's content and is accompanied by a brief description, a set of critical thinking questions, a link to Infotrac College Edition, and suggested Infotrac search terms. The data on which the maps are based are included in an Excel spreadsheet, so students can create their own customized maps. Web resources are also provided. This free CD-ROM is straightforward and easy to use—a must-have supplement for Introductory Sociology students. Instructors may also use the CD in class. Topics include:

- Cultural Exports as Percent of Total Cultural Trade, World, 1997
- Percent Women Among Adult HIV/AIDS Cases, World, 1997
- Days of Parental Leave, World, 1998
- Ethnic Groups, World, 2001
- Death Penalty, World, 2001

- Hate Crimes per 100,000 Population, U.S.A., 2000
- Right-to-Work States, U.S.A., 2001
- Destinations of U.S. Air Travelers, 200

✦ *2001 Researching Sociology on the Internet Guide.* Prepared by D.R. Wilson, Houston Baptist University, and David L. Carlson, Texas A&M University, this useful guide is designed to assist sociology students in all of their needs when doing research on the Internet. Part One contains general information needed to get started and answers questions about security, the type of material available on the Internet, the information that is reliable and the sites that are not, the best ways to find research, and the best links to take students where they want to go. Part Two looks at each main subfield in sociology and refers students to sites with the most enlightening research. Specific drawbacks and issues to watch out for in each of the areas are noted, as well as specific resources and subjects that are well represented in the online world.

✦ *Discovering Sociology: An Introduction Using MicroCase ExplorIt.* Prepared by Steven Barkan, University of Maine, *Discovering Sociology* is a software-based workbook that allows students to explore dozens of sociological topics and issues, using data from the United States and around the world. The workbook is accompanied by the ExplorIt software and data sets. An updated Second Edition will be available Fall 2002.

✦ *Doing Sociology: A Global Perspective, Fourth Edition.* This software and workbook package, written by Rodney Stark, shows students what it takes to do real sociological research, using the same data and techniques used by professional researchers. The step-by-step approach in the workbook includes explanations of basic research concepts and methods, expanded exercises, and suggestions for independent research projects, effectively guiding students through the research process and offering them a real sense of what sociologists do. (IBM-compatible only, Windows 95 or later). The workbook is accompanied by an instructor's manual.

✦ *Web Tutor.* Web Tutor is a content-rich, Web-based teaching and learning tool that helps students succeed by taking the course beyond classroom boundaries to an anytime, anywhere environment. Web Tutor is rich with study and mastery tools, communication tools, and course content. Professors can use Web Tutor to provide virtual office hours, post their syllabi, set up threaded discussions, track student progress with the quizzing material, and more. For students, WebTutor offers real-time access to a full array of study tools, including flashcards (with audio), practice quizzes, interactive games, and highly relevant Web links. Web Tutor also provides rich communication tools including a course calendar, asynchronous discussion, "real time" chat, and an integrated e-mail system. Web Tutor is available on both WebCT and Blackboard. Please ask your Wadsworth/Thomson Learning representative for more information.

Web Site to Accompany *Sociology: Your Compass for a New World:*

✦ *Companion Web Site.* In terms of substance, visual appeal, and interactivity, the book's Web site contains an unusually rich collection of materials. Alone among introductory sociology Web sites, this one was written by one of the textbook authors, Robert Brym. All of the materials on the site are closely tied to themes introduced in the text. For each chapter of the book, the site features three online tests, one online research project, one interactive exercise, and four carefully chosen Web links. A separate set of online tests is provided for each of the interactive exercises. In addition, the site contains more than 100 lectures and interviews on a wide variety of sociological topics, many of them featuring leading sociologists. Students hear the lectures and interviews using the free RealPlayer plug-in. A "Research Center" offers students easy access to a full range of official statistics. Instructors can download the Instructor's Manual and the PowerPoint series directly from the site.

✦ ***Virtual Society: The Wadsworth Sociology Resource Center.*** The Wadsworth Sociology Resource Center contains a wealth of information and useful tools for both instructors and students. After logging on to http://sociology.wadsworth.com, click on Brym and Lie, *Sociology: Your Compass for a New World*. Proceed to the student Resources Section to find flashcards, links to key sociological sites, and online quizzes for each chapter. Also here you will find three special features of the site.

First, the *Virtual Tours for Introductory Sociology*, prepared by Robert Wood of Rutgers University, Camden, provide a hands-on learning experience using the Internet to study the core topics covered in the introductory sociology course. The user is directed through a series of key Web sites relating to chapter topics. Using pull-down menus and short answers, the tours are fully interactive. Students may e-mail responses to their instructor for credit. The tours are also available upon request in hard copy to provide ease of use.

Second, *MicroCase Online*, prepared by Matt Bahr of the American Religion Data Archive, Purdue University, allows students to analyze real-world data using the powerful MicroCase software and professorial data sets. A series of exercises for each chapter of the text is provided. These Web-based exercises offer activities designed around key pieces of current social science research. Basic univariate statistics, mapping, and cross-tabulations give students a chance to discover sociology by using the same data and methods used by professional social science researchers. All of the exercises rely on the same high-quality data available to professional researchers and include variables from sources such as the 2000 General Social Survey (GSS), the World Values Survey, United States Census data, the Federal Election Commission, and others—including data on crime, education, religion, and political participation.

Third, *Student Guide to InfoTrac College Edition for Sociology*, prepared by Michele Adams of the University of California, Riverside, consists of exercises based on 23 core subjects vital to the study of sociology. These exercises use InfoTrac College Edition's huge database of articles. The exercises help students to narrow down the search of articles related to each subject and ask questions that enable them to see the ideas more clearly and pique their interest.

Acknowledgements

Anyone who has gone sailing knows that when you embark on a long voyage you need more than a compass. Among other things, you need a helm operator blessed with a strong sense of direction and an intimate knowledge of likely dangers. You need at least one crew member who knows all the ropes and can use them to keep things intact and in their proper place. And you need sturdy hands to raise and lower the sails. On the four-year voyage to complete this book, our crew demonstrated all these skills. Our Acquisitions Editors— Brenda Weeks, Lin Marshall, Bryan Leake, and Eve Howard (*de facto*)—saw this book's promise from the outset, understood clearly the direction we had to take to develop its potential, and on several occasions steered us clear of threatening shoals. We still marvel at how Michele Tomiak, our Production Editor, was able to keep the many parts of this project in their proper order and prevent the whole thing from flying apart at the seams even in stormy weather. But more than anyone else it was Lisa Hensley, our Developmental Editor, who made this book sail. She knew just when to trim the jib and when to hoist the mainsail. We are deeply grateful to her and to all the members of our crew for a successful voyage.

This book would have been of far inferior quality if the following people had not generously shared their knowledge with us, offered painstaking criticisms of chapter drafts, and given us emotional support:

Dean Behrens, University of Toronto

Nachman Ben-Yehuda, Hebrew University of Jerusalem

Steve Berkowitz, University of Vermont

Monica Boyd, Florida State University and University of Toronto

Clem Brooks, University of Indiana

Harvey Choldin, University of Illinois, Urbana-Champaign

Dan Clawson, University of Massachusetts, Amherst

Weizhen Dong, University of Toronto

John Fox, McMaster University, Canada

Rosemary Gartner, University of Toronto

Michael Goldman, University of Illinois, Urbana-Champaign

Randy Hodson, Ohio State University

David Hopping, University of Illinois, Urbana-Champaign

Elizabeth Jenner, University of Illinois, Urbana-Champaign

Baruch Kimmerling, Hebrew University of Jerusalem

Larisa Kosova, Russian Center for Public Opinion Research, Moscow

Rhonda Lenton, McMaster University, Canada

Vladimir Magun, Institute of Sociology, Russian Academy of Science, Moscow

Jeff Manza, Northwestern University

John Myles, Florida State University and University of Toronto

Gregg Olsen, University of Manitoba, Canada

William Outhwaite, University of Sussex, UK

Jim Richardson, University of New Brunswick, Canada

Michael Shalev, Hebrew University of Jerusalem

Hira Singh, York University, Canada

Murray Straus, University of New Hampshire

Shelley Ungar, University of Toronto

Jack Veugelers, University of Toronto

We are also grateful to the following colleagues who reviewed the manuscript and provided a wealth of helpful suggestions:

Robert Kettlitz, Hastings College

Gregg Lee Carter, Bryant College

Robert Graham, Lee University

Ron Wohlstein, Eastern Illinois University

Phyllis Gorman, Ohio State University

Hadley Klug, University of Wisconsin, Whitewater

Christopher Mele, State University of New York, Buffalo

Arthur Jipson, University of Dayton

Jan Fiola, Moorhead State University

Lisa Slattery Rashotte, University of North Carolina, Charlotte

William Canack, Middle Tennessee State University

Michael Perez, California State University, Fullerton

Jenifer Kunz, West Texas A&M University

Ron Hammond, Utah Valley State College

Matthew Smith-Lahrman, Dixie College

George Stine, Millersville University

Dale Lund, University of Utah Gerontology Center

Karen Connor, Drake University
Peter Adler, University of Denver
Katheryn Dietrich, Texas A&M University
Emily Ignacio, Loyola University of Chicago
Pelgy Vaz, Ft. Hays State University
Steven Lybrand, University of St. Thomas
Anne Baird, Morehouse University
Douglas Constance, Sam Houston State University
Luis Salinas, University of Houston
Gershon Shafir, University of California, La Jolla
Joel Snell, Kirkwood Community College
Tim Britton, Lenoir Community College
Michael Goslin, Tallahassee Community College
Steve Kroll-Smith, University of New Orleans
Stephen Couch, Pennsylvania State University, Schuylkill
Billye Nipper, Redlands Community College
Terry Reuther, Anoka-Ramsey Community College
Martin Orr, Boise State University
Elizabeth Meyer, Pennsylvania College of Technology
Anna Wall Scott, Parkland Community College
Sarah F. Anderson, Northern Virginia Community College
Ian Lapp, Monmouth University
J. Russell Willis, Grambling State University
Juanita Firestone, University of Texas, San Antonio
Hence Parson, Hutchinson Community College
Kent Sandstrom, University of Northern Iowa
Deborah Abowitz, Bucknell University
Ione DeOllos, Ball State University
Gary Hampe, University of Wyoming
William Smith, Georgia Southern University

Robert J. Brym
John Lie

I

FOUNDATIONS

IN THIS CHAPTER, YOU WILL LEARN THAT:

✦ The causes of human behavior lie partly in the patterns of social relations that surround and permeate us.

✦ Sociology is the systematic study of human behavior in social context.

✦ Sociologists examine the connection between social relations and personal troubles.

✦ Sociologists are often motivated to do research by the desire to improve people's lives. At the same time, sociologists adopt scientific methods to test their ideas.

✦ Sociology originated at the time of the Industrial Revolution. The founders of sociology diagnosed the massive social transformations of their day. They also suggested ways of overcoming social problems created by the Industrial Revolution.

✦ Today's Postindustrial Revolution similarly challenges us. Sociology clarifies the scope, direction, and significance of social change. It also suggests ways of dealing with the social problems created by the Postindustrial Revolution.

✦ At the personal level, sociology can help to clarify the opportunities and constraints you face. It suggests what you can become in today's social and historical context.

C H A P T E R

1

A SOCIOLOGICAL COMPASS

INTRODUCTION

Why Robert Brym Decided *Not* to Study Sociology

"When I started college at the age of 18," says Robert Brym, "I was bewildered by the wide variety of courses I could choose from. Having now taught sociology for more than 20 years and met a few thousand undergraduates, I am quite sure most students today feel as I did then.

"One source of confusion for me was uncertainty about why I was in college in the first place. Like you, I knew higher education could improve one's chance of finding good work. But, like most students, I also had a sense that higher education is supposed to provide something more than just the training necessary to embark on a career that is interesting and pays well. Several high school teachers and guidance counselors had told me that college was also supposed to 'broaden my horizons' and teach me to 'think critically.' I wasn't sure what they meant, but they made it sound interesting enough to make me want to know more. Thus, I decided in my first year to take mainly 'practical' courses that might prepare me for a law degree (economics, political science, and psychology). I also enrolled in a couple of other courses to indulge my 'intellectual' side (philosophy, drama). One thing I knew for sure. I didn't want to study sociology.

"Sociology, I came to believe, was thin soup with uncertain ingredients. When I asked a sophomore what sociology is, he told me it deals mainly with why people are unequal— why some are rich and others poor, some powerful and others weak. Coming as I did from a poor immigrant family in an economically depressed region, it appeared that sociology could teach me something about my own life. But it also seemed a lot like what I imagined economics and political science to be about. What, then, was unique about sociology? My growing sense that sociology had nothing special to offer was confirmed when another sophomore told me that sociologists try to describe the ideal society and figure out how to make the world a better place. That description appealed to my youthful sense of the world's injustice. However, it also sounded a lot like philosophy. A junior explained that sociology analyzes how and why people assume different roles in their lives. She made sociology appear similar to drama. Finally, one student reported that in her sociology class she was learning why people commit suicide, homicide, and other deviant acts. That seemed like abnormal psychology to me. I concluded that sociology had no distinct flavor all its own. Accordingly, I decided to forego it for tastier courses.

A Change of Mind

"Despite the opinion I'd formed, I found myself taking no fewer than four sociology courses a year after starting college. That revolution in my life was due in part to the pull of an extraordinary professor I happened to meet just before I began my sophomore year. He set me thinking in an altogether new way about what I could and should do with my life. He exploded some of my deepest beliefs. He started me thinking sociologically.

"Specifically, he first put Yorick's dilemma to me. Yorick is a character—sort of—in *Hamlet.* Toward the end of the play, Hamlet finds two gravediggers at work. They unearth the remains of the former court jester, Yorick, who used to amuse Hamlet and carry him around on his back when Hamlet was a child. Holding high his old friend's skull, Hamlet reflects on what we must all come to. Even the remains of Alexander the Great, he says, turn to dust.

"This incident implies Yorick's dilemma and, indeed, the dilemma of all thinking people. Life is finite. If we wish to make the most of it we must figure out how best to live. That is no easy task. It requires study, reflection, and the selection of values and goals. Ideally, higher education is supposed to supply students with just that opportunity. Finally, I was beginning to understand what I could expect from college apart from job training.

"The professor I met also convinced me that sociology in particular could open up a new and superior way of comprehending my world. Specifically, he said, it could clarify my place in society, how I might best maneuver through it, and perhaps even how I might contribute to improving it, however modestly. Before beginning my study of sociology, I had always taken for granted that things happen in the world—and to me—because physical and emotional forces cause them. Famine, I thought, is caused by drought, war by territorial greed, economic success by hard work, marriage by love, suicide by bottomless depression, rape by depraved lust. But now, this professor repeatedly threw evidence in my face that contradicted my easy formulas. If drought causes famine, why have so many famines occurred in perfectly normal weather conditions or involved some groups hoarding or destroying food so others would starve? If hard work causes prosperity, why are so many hard workers poor? If love causes marriage, why are so many families sites of violence against women and children? And so the questions multiplied.

"As if it were not enough that the professor's sociological evidence upset many of my assumptions about the way the world worked, he also challenged me to understand sociology's unique way of explaining social life. He defined **sociology** as the systematic study of human behavior in social context. He explained that *social* causes are distinct from physical and emotional causes. Understanding social causes can help clarify otherwise inexplicable features of famine, marriage, and so forth. In public school, my teachers taught me that people are free to do what they want with their lives. However, my new professor taught me that the organization of the social world opens some opportunities and closes others, thus constraining our freedom and helping to make us what we are. By examining the operation of these powerful social forces, he said, sociology can help us to know ourselves, our capabilities and limitations. I was hooked. And so, of course, I hope you will be too."

The Goals of This Chapter

In this chapter we aim to achieve three goals:

1. We first illustrate the power of sociology to dispel foggy assumptions and help us see the operation of the social world more clearly. To that end, we examine a phenomenon that at first glance appears to be solely the outcome of breakdowns in individual functioning: suicide. We show that, in fact, social relations powerfully influence suicide rates. This exercise introduces you to what is unique about the sociological perspective.

2. We show that, from its origins, sociological research has been motivated by a desire to improve the social world. Thus, sociology is not just a dry, academic exercise but a means of charting a better course for society. At the same time, however, sociologists adopt scientific methods to test their ideas, thus increasing their validity. We illustrate these points by briefly analyzing the work of the founders of the discipline.

3. We suggest that sociology can help you come to grips with your century, just as it helped the founders of sociology deal with theirs. Today we are witnessing massive and disorienting social changes. Entire countries are becoming unglued. Women are demanding equality with men in all spheres of life. People's wants are increasingly governed by the mass media. Computers are radically altering the way people work and entertain themselves. There are proportionately fewer good jobs to go around. Violence surrounds us. Environmental ruin threatens us all. As was the case a hundred years ago, sociologists today try to understand social phenomena and suggest credible ways of improving their societies. By promising to make sociology relevant to you, this chapter should be viewed as an open invitation to participate in sociology's challenge.

But first things first. Before showing how sociology can help you understand and improve your world, we briefly examine the problem of suicide. That will help illustrate how the sociological perspective can clarify and sometimes overturn commonsense beliefs.

THE SOCIOLOGICAL PERSPECTIVE

By analyzing suicide sociologically, you can put to a tough test our claim that sociology takes a unique, surprising, and enlightening perspective on social events. After all, suicide appears to be the supremely antisocial and nonsocial act. It is condemned by nearly everyone in society. It is typically committed in private, far from the public's intrusive glare. It is rare. In 1998, there were fewer than 11 suicides for every 100,000 Americans (Centers for Disease Control and Prevention, 1999a: 28). And when you think about why people commit such acts, you are likely to focus on their individual states of mind rather than on the state of society. In other words, what usually interests us are the aspects of specific individuals' lives that caused them to become depressed or angry enough to kill themselves. We usually don't think about the patterns of social relations that might encourage such actions in general. If sociology can reveal the hidden social causes of such an apparently antisocial and nonsocial phenomenon, there must be something to it!

The Sociological Explanation of Suicide

At the end of the 19th century, French sociologist Émile Durkheim, one of the pioneers of the discipline, demonstrated that suicide is more than just an individual act of desperation resulting from psychological disorder, as people commonly believed at the time (Durkheim, 1951 [1897]). Suicide rates, he showed, are strongly influenced by social forces.

Durkheim made his case by examining the association between rates of suicide and rates of psychological disorder for different groups. The idea that psychological disorder causes suicide would be supported, he reasoned, only if suicide rates are high where rates of psychological disorder are high, and low where rates of psychological disorder are low. However, his analysis of European government statistics and hospital records revealed nothing of the kind. He discovered there were slightly more women than men in insane asylums. Yet there were four male suicides for every female suicide. Jews had the highest rate of psychological disorder among the major religious groups in France. However, they also had the lowest suicide rate. Psychological disorders occurred most frequently when a person reached maturity. Suicide rates, though, increased steadily with advancing age.

Émile Durkheim (1858–1917) was the first professor of sociology in France and is often considered to be the first modern sociologist. Particularly in *The Rules of the Sociological Method* (1895) and *Suicide* (1897), he argued that human behavior is shaped by "social facts," or the social context in which people are embedded. In Durkheim's view, social facts define the constraints and opportunities within which people must act. Durkheim was also keenly interested in the conditions that promote social order in "primitive" and modern societies, and he explored this problem in depth in such works as *The Division of Labor in Society* (1893) and *The Elementary Forms of the Religious Life* (1912).

"Pacific." Alex Colville. 1967.

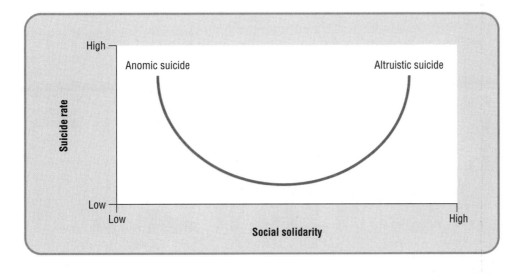

✦ **FIGURE 1.1** ✦

Durkheim's Theory of Suicide
Durkheim argued that, as the level of social solidarity increases, the suicide rate declines. Then, beyond a certain point, it starts to rise. Hence the U-shaped curve in this graph. Durkheim called suicides that occur in very high-solidarity settings *altruistic*. They occur when norms very tightly govern behavior. For example, when soldiers knowingly give up their lives to protect members of their unit, they commit altruistic suicide out of a deep sense of patriotism and comradeship. In contrast, suicide that occurs in very low-solidarity settings is *anomic*, said Durkheim. That is, it results from vaguely defined norms governing behavior.

So rates of suicide and psychological disorder did not rise and fall together. What then accounts for variations in suicide rates? Durkheim argued that suicide rates vary due to differences in the degree of **social solidarity** in different groups. According to Durkheim, the more a group's members share beliefs and values, and the more frequently and intensely they interact, the more social solidarity there is in the group. In turn, the higher the level of social solidarity, the more firmly anchored individuals are to the social world and the less likely they are to take their own life if adversity strikes. In other words, Durkheim expected groups with a high degree of solidarity to have lower suicide rates than groups with a low degree of solidarity—at least up to a certain point (see Figure 1.1).

To support his argument, Durkheim showed that married adults are half as likely as unmarried adults are to commit suicide. That is because marriage usually creates social ties and a sort of moral cement that bind the individual to society. Similarly, women are less likely to commit suicide than men are. Why? Women are generally more involved in the intimate social relations of family life. Jews, Durkheim wrote, are less likely to commit suicide than Christians are. The reason? Centuries of persecution have turned them into a group that is more defensive and socially more tightly knit. And the elderly are more prone than the young and the middle-aged to take their own lives in the face of misfortune. That is because they are most likely to live alone, to have lost a spouse, and to lack a job and a wide network of friends. In general, Durkheim wrote, "suicide varies with the degree of integration of the social groups of which the individual forms a part" (Durkheim, 1951 [1897]: 209). Note that his generalization tells us nothing about why any particular individual may take his or her own life. That is a question for psychology. But it does tell us that a person's likelihood of committing suicide decreases with the degree to which he or she is anchored in society. And it says something surprising and uniquely sociological about how and why suicide rates vary from group to group.

Strong social bonds decrease the probability that a person will commit suicide if adversity strikes.

✦ **FIGURE 1.2** ✦

Suicide Rate by Sex and Age Cohort, United States, 1997 (per 100,000 people)

SOURCE: Centers for Disease Control and Prevention (1999c).

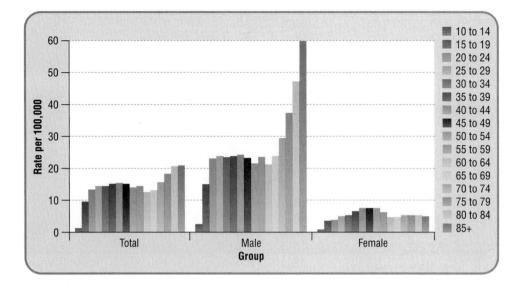

Durkheim's theory is not just a historical curiosity. It also sheds light on suicide here and now. We noted above that about 11 Americans out of every 100,000 take their own lives each year. However, as the cluster of bars at the far left of Figure 1.2 shows, the suicide rate varies with age, just as it did a century ago in France. The elderly are most likely to commit suicide because they are the age group least firmly rooted in society. Moreover, among the elderly, suicide is most common among the divorced and widowed (National Center for Injury Prevention and Control, 2000). Figure 1.2 also shows that suicide rates differ for men and women. As in France in the late 1800s, men are on average less involved than women are in childcare and other intricacies of family life. And, as in Durkheim's France, men are about four times more likely than women are to commit suicide. Research also shows that parts of the United States with high rates of church membership have low suicide rates while areas with high divorce rates have high suicide rates (Breault, 1986). This finding, too, is consistent with Durkheim's theory.

One thing Figure 1.2 does not show is that suicide among young people has become more common over the past half century. For example, from 1952 to 1995, the suicide rate for men aged 65 and older hardly changed at all. For men between the ages of 15 and 24, however, the suicide rate nearly tripled (National Center for Injury Prevention and Control, 2000). Why do you think this happened? Are there ways in which family life, the work world, religion, and other areas of life have changed to weaken young people's ties to society? Can *you* explain increased youth suicide sociologically?

From Personal Troubles to Social Structures

You have known for a long time that you live in a society. Yet until now, you may not have fully appreciated that society also lives in you. That is, patterns of social relations affect your innermost thoughts and feelings, influence your actions, and thus help shape who you are. As we have seen, one such pattern of social relations is the level of social solidarity characteristic of the various groups to which you belong.

Sociologists call stable patterns of social relations **social structures.** One of the sociologist's main tasks is to identify and explain the connection between people's personal troubles and the social structures in which they are embedded. This is harder work than it may at first seem. In everyday life, we usually see things from our own point of view. Our experiences appear unique to each of us. If we think about them at all, social structures may appear remote and impersonal. To see how social structures operate inside us, we require sociological training.

An important step in broadening one's sociological awareness involves recognizing that three levels of social structure surround and permeate us. Think of these structures as concentric circles radiating out from you:

1. **Microstructures** are patterns of intimate social relations. They are formed during face-to-face interaction. Families, friendship circles, and work associations are all examples of microstructures.

   Understanding the operation of microstructures can be useful. Let's say you're looking for a job. You might think you'd do best to ask as many close friends and relatives as possible for leads and contacts. However, sociological research shows that people you know well are likely to know many of the same people. After asking a couple of close connections for help landing a job, you'd therefore do best to ask more remote acquaintances for leads and contacts. People to whom you are *weakly* connected (and who are weakly connected among themselves) are more likely to know *different* groups of people. Therefore, they will give you more information about job possibilities and ensure that word about your job search spreads farther. You're more likely to find a job faster if you understand "the strength of weak ties" in microstructural settings (Granovetter, 1973).

2. **Macrostructures** are patterns of social relations that lie outside and above your circle of intimates and acquaintances. Macrostructures include class relations, bureaucracies, and **patriarchy,** the traditional system of economic and political inequality between women and men in most societies (for exceptions, see Chapter 9, "Sexuality and Gender," and Chapter 12, "Families").[1]

   Understanding the operation of macrostructures can also be useful. Consider, for example, one aspect of patriarchy. Most married women who work full-time in the paid labor force do more housework, childcare, and care for the elderly than their husbands. Governments and businesses support this arrangement insofar as they give little assistance to families in the form of nurseries, after-school programs for children, senior homes, and so forth. Yet the unequal division of work in the household is a major source of dissatisfaction with marriage, especially in families that can't afford to buy these services privately. Thus, sociological research shows that where spouses share domestic responsibilities equally, they are happier with their marriages and less likely to divorce (Hochschild with Machung, 1989). When a marriage is in danger of dissolving, it is common for partners to blame themselves and each other for their troubles. However, it should now be clear that forces other than incompatible personalities often put stresses on families. Understanding how the macrostructure of patriarchy crops up in everyday life, and doing something to change that structure, can thus help people lead happier lives.

3. The third level of society that surrounds and permeates us is composed of **global structures.** International organizations, patterns of worldwide travel and communication, and the economic relations between countries are examples of global structures. Global structures are increasingly important as inexpensive travel and communication allow all parts of the world to become interconnected culturally, economically, and politically.

   Understanding the operation of global structures can be useful too. For instance, many people are concerned about the world's poor. They donate money to charities to help with famine relief. Some people also approve of the American government giving foreign aid to poor countries. However, many of these same people do not appreciate that charity and foreign aid alone do not seem able to end world poverty. That is because charity and foreign aid have been unable to overcome the structure of social relations between countries that have created and sustain global inequality.

   Let us linger on this point for a moment. As we will see in Chapter 16 ("Population, Urbanization, and Development"), Britain, France, and other imperial powers locked some countries into poverty when they colonized them between the 17th and 19th centuries. In the 20th century, the poor (or "developing") countries borrowed money from these same rich countries and Western banks to pay for airports, roads, harbors, sanitation systems, basic health care, and so forth. Today, poor countries pay far more to rich countries and Western banks in interest on those loans than they

---

[1]Some sociologists also distinguish "mesostructures," social relations that link microstructures and macrostructures.

✦ **FIGURE 1.3** ✦

Foreign Aid, Debt, and Interest Payments of Developing Countries, 1992 and 1997 (in $ United States billions)

SOURCE: World Bank (1999a; 1999b).

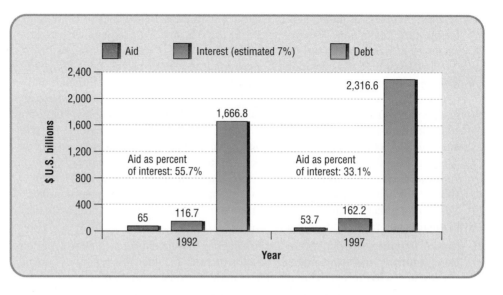

C. Wright Mills (1916–62) argued that the sociological imagination is a unique way of thinking. It allows people to see how their actions and potential are affected by the social and historical context in which they find themselves. Mills employed the sociological imagination effectively in his most important works. For example, *The Power Elite* (1956) is a study of the several hundred men who occupy the "command posts" of the United States' major institutions. It suggests that economic, political, and military power is highly concentrated in United States society, which is therefore less of a democracy than we are often led to believe. The implication of Mills's study is that to make our society more democratic, power must be more evenly distributed among the citizenry.

receive in aid and charity (see Figure 1.3). Thus, it seems that relying exclusively on foreign aid and charity can do little to help solve the problem of world poverty. Understanding how the global structure of international relations created and helps maintain global inequality suggests new policy priorities for helping the world's poor. One such priority might involve campaigning for the cancellation of foreign debt in compensation for past injustices.

As these examples illustrate, personal problems are connected to social structures at the micro, macro, and global levels. Whether the personal problem involves finding a job, keeping a marriage intact, or figuring out a way to act justly to end world poverty, social-structural considerations broaden our understanding of the problem and suggest appropriate courses of action.

The Sociological Imagination

Nearly half a century ago, the great American sociologist C. Wright Mills (1959) called the ability to see the connection between personal troubles and social structures the **sociological imagination.** He emphasized the difficulty of developing this quality of mind (see Box 1.1). His language is sexist by today's standards, but his argument is as true and inspiring today as it was in the 1950s:

> When a society becomes industrialized, a peasant becomes a worker; a feudal lord is liquidated or becomes a businessman. When classes rise or fall, a man is employed or unemployed; when the rate of investment goes up or down, a man takes new heart or goes broke. When war happens, an insurance salesman becomes a rocket launcher; a store clerk, a radar man; a wife lives alone; a child grows up without a father. Neither the life of an individual nor the history of a society can be understood without understanding both.
>
> Yet men do not usually define the troubles they endure in terms of historical change. . . . The well-being they enjoy, they do not usually impute to the big ups and downs of the society in which they live. Seldom aware of the intricate connection between the patterns of their own lives and the course of world history, ordinary men do not usually know what this connection means for the kind of men they are becoming and for the kind of history-making in which they might take part. They do not possess the quality of mind essential to grasp the interplay of men and society, of biography and history, of self and world. They cannot cope with their personal troubles in such a way as to control the structural transformations that usually lie behind them.
>
> What they need . . . is a quality of mind that will help them to [see] . . . what is going on in the world and . . . what may be happening within themselves. It is this quality . . . that . . . may be called the sociological imagination. (Mills, 1959: 3–4)

BOX 1.1
SOCIOLOGY AT THE MOVIES

Forrest Gump let life happen to him without ever understanding the larger social and historical forces impinging on him. His was a life entirely devoid of the sociological imagination.

FORREST GUMP (1994)

Forrest Gump is a cinematic tour of the United States from the 1950s to the 1980s. The movie stars Tom Hanks as Forrest Gump, a man with an IQ of 75. Although Forrest Gump has limited intelligence and little education, he leads a remarkable life. He is a football hero in high school, wins the Congressional Medal of Honor during the Vietnam War, and manages to become a millionaire by buying stock in a start-up company called Apple Computers. In between, he teaches Elvis to swivel his hips, sits next to John Lennon in a television talk show, and meets many United States presidents. In short, he leads a charmed life.

In contrast, the girl Forrest Gump falls in love with in elementary school—and whom he stays in love with for the rest of his life—becomes a hippie. She preaches flower power and peace while Forrest is fighting in Vietnam. Eventually, she becomes a drug addict, a stripper, and HIV positive.

Between these two lives, one charmed, the other damaged, *Forrest Gump* manages to capture many important moments in United States history over a 40-year period. However, Forrest, because he is so limited intellectually, can't make much sense of his life or the world around him.

The sociological imagination urges us to connect our biography with history and social structure—to make sense of our lives against a larger historical and social background and to act in light of our understanding. The message of the sociological imagination is similar to that of the classical Greek philosopher, Socrates: "The unexamined life is not worth living." In contrast, Forrest just lets things happen and doesn't think much about them. Remarkable things happen to him but he never fully appreciates or understands the significance of the world he lives in.

What do you think about these two conflicting views? Would you like things to happen to you without really understanding what's going on? Or is it better to live an examined life? Had Forrest been equipped with the sociological imagination, would he have gotten more out of life? Have you ever tried to put events in your own life in the context of history and social structure? Did the exercise help you make sense of your life? Did it in any way lead to a life more worth living? Is the sociological imagination a worthy goal?

Although movies are just entertainment to many people, they often achieve by different means what the sociological imagination aims for. Therefore, in each chapter of this book, we review a movie to shed light on topics of sociological importance.

The sociological imagination is a recent addition to the human repertoire. It is only about as old as the United States. True, in ancient and medieval times, some philosophers wrote about society. However, their thinking was not sociological. They believed God and nature controlled society. They spent much of their time sketching blueprints for the ideal society and urging people to follow those blueprints. And they relied on speculation rather than evidence to reach conclusions about how society works.

The sociological imagination was born when three modern revolutions pushed people to think about society in an entirely new way. First, the **Scientific Revolution** began about 1550. It encouraged the view that sound conclusions about the workings of society must be based on solid evidence, not just speculation. Second, the **Democratic Revolution** began about 1750. It suggested that people are responsible for organizing society and that human intervention can therefore solve social problems. Third, the **Industrial Revolution** began about 1780. It created a host of new and serious social problems that attracted the attention of many social thinkers. Let us briefly consider these three sources of the sociological imagination.

The Scientific Revolution began in Europe around 1550. Scientists proposed new theories about the structure of the universe and developed new methods to collect evidence so they could test those theories. Shown here is an astrolabe used by Copernicus to solve problems relating to the position of the sun, the planets, and the stars.

Origins of the Sociological Imagination

The Scientific Revolution

It is said that a group of medieval monks once wanted to know how many angels could dance on the head of a pin. They consulted ancient books in Hebrew, Greek, and Latin. They thought long and hard. They employed all their debating skills to argue the issue. They did not, however, resolve the dispute. That is because they never considered inspecting the head of a pin and counting. Any such suggestion would have been considered heresy. We, in contrast, would call it the beginning of a scientific approach to the subject.

People often link the Scientific Revolution to specific ideas, such as Copernicus's theory that the earth revolves around the sun and Newton's laws of motion. However, science is less a collection of ideas than a method of inquiry. For instance, in 1609, Galileo pointed his newly invented telescope at the heavens, made some careful observations, and showed that his observations fit Copernicus's theory. This is the core of the scientific method: using evidence to make a case for a particular point of view. By the mid-1600s, some philosophers, such as Descartes in France and Hobbes in England, were calling for a science of society. When sociology emerged as a distinct discipline in the 19th century, commitment to the scientific method was one firm pillar of the sociological imagination.

The Democratic Revolution

The second pillar of the sociological imagination is the realization that people control society and can change it. Four hundred years ago, most Europeans thought otherwise. For them, God ordained the social order.

Consider the English engraving reproduced in Figure 1.4. It shows how most educated Europeans pictured the universe in Shakespeare's time. Note the cloud at the top of the circle. The Hebrew name of God is inscribed on it. God's hand extends from the cloud. It holds a chain, which is attached to a woman representing Nature. Nature also holds a chain in her hand. It is connected to "the ape of Nature," representing humankind. The symbolism is clear: God and his intermediary, Nature, control human action. Note also that the engraving arranges everything in a linked hierarchy. The hierarchy includes the mineral, vegetable, and animal kingdoms, the elements, heavenly objects, angels, and so forth. Each level of the hierarchy corresponds to and controls some aspect of the level below it. For

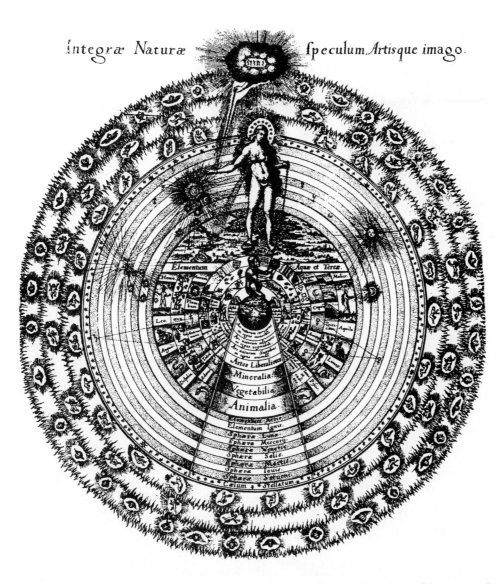

♦ **FIGURE 1.4** ♦

Robert Fludd. *Utriusque Cosmi Maioris Scilicet et Minoris Metaphysica, Physica Atqve Technica Historia.* **1617–19. (Oppenheim, Germany. Johan-Theodori de Bry.)** This engraving shows how educated Europeans viewed the universe in Shakespeare's time, nearly 400 years ago.

example, people believed Archangels regulate the movements of the planet Mercury and the movements of Mercury affect human commerce. Similarly, in the medieval view, God ordained a hierarchy of people. The richest people were seen as the closest to God and therefore deserving great privilege. Supposedly, kings and queens ruled because God wanted them to (Tillyard, 1943).

The American Revolution (1775–83) and the French Revolution (1789–99) helped to undermine these ideas. These democratic political upheavals showed that society could experience massive change in a short period. They proved that people could replace unsatisfactory rulers. And they suggested that *people* control society. The implications for social thought were profound. For if it were possible to change society by human intervention, then a science of society could play a big role. The new science could help people figure out ways of overcoming various social problems, improving the welfare of all citizens, and finding the most effective way to reach given goals. Much of the justification for sociology as a science arose out of the democratic revolutions that shook Europe and North America.

The Industrial Revolution

The third pillar of the sociological imagination was the Industrial Revolution. It began in England about 1780. Due to the growth of industry, masses of people moved from countryside to city, worked agonizingly long hours in crowded and dangerous mines and factories, lost faith in their religions, confronted faceless bureaucracies, and reacted to the

Eugene Delacroix. *Liberty Leading the People, July 28, 1830.* The democratic forces unleashed by the French Revolution suggested that people are responsible for organizing society and that human intervention can therefore solve social problems. As such, democracy was a foundation stone of sociology.

Diego Rivera. *Detroit Industry, North Wall.* 1932–33. Fresco (detail). Copyright 1997, The Detroit Institute of Arts. The so-called Second Industrial Revolution began in the early 20th century. Wealthy entrepreneurs formed large companies. Steel became a basic industrial material. Oil and electricity fueled much industrial production. At the same time, Henry Ford's assembly lines and other mass-production technologies transformed the workplace.

filth and poverty of their existence by means of strikes, crime, revolution, and war. Scholars had never seen a sociological laboratory like this. The Scientific Revolution suggested that a science of society is possible. The Democratic Revolution suggested that people can intervene to improve society (see Box 1.2). The Industrial Revolution now presented social thinkers with a host of pressing social problems crying out for solution. They responded by giving birth to the sociological imagination.

SOCIOLOGICAL THEORIES

The French social thinker Auguste Comte coined the term *sociology* in 1838 (Comte, 1975). Comte tried to place the study of society on scientific foundations. He wanted to understand the social world as it is, not as he or anyone else imagined it should be. Yet

BOX 1.2
IT'S YOUR CHOICE

IS THE MONICA LEWINSKY AFFAIR A PERSONAL TROUBLE OR A PUBLIC ISSUE?

In the second term of Bill Clinton's presidency, the United States was mired in a debate over the President's affair with Monica Lewinsky. Much of the debate centered on whether President Clinton's actions were just a big personal mistake or whether they had important implications for public life as well.

From a sociological point of view, the Lewinsky affair was as much a public issue as the President's personal trouble. At least three issues made the entanglement a matter of public concern.

First, the Lewinsky affair violated a consensus about what constitutes appropriate behavior between managers and employees. Before the 1970s, sexual relations between managers and employees were considered a private matter. Reports of executive-suite dalliances were routinely brushed under the boardroom rug. Since then, however, feminist sociologists have shown that the problem is widespread, and they have succeeded in increasing the public's awareness of the problem. Some people have argued that sex cannot be consensual when there is a big power imbalance between partners. There must always be an element of coercion in such cases, they say. Others note that office romances routinely upset fellow workers who think that the boss favors his or her lover. Still others point out that affairs gone sour often fuel harassment lawsuits. Due to growing public opposition to workplace romances between bosses and their employees, the Supreme Court defined sexual harassment as sexual discrimination in 1986. Today, most large businesses, the military, government organizations, and colleges have policies banning or restricting relations between managers and at least direct subordinates. Thus, the Lewinsky affair was not just a private matter because it openly violated public opinion (Kasindorf, Armour, and Stone, 1998).

The Lewinsky affair was also a public issue because it undermined confidence in a public leader. Many people look up to the President as an authority figure and a role model. By having an affair with a direct subordinate and then denying it, Clinton arguably served as a poor example to the many people who look up to him—especially young people. The Lewinsky affair caused many Americans to question the President's integrity and his judgment. In the view of some people, the affair and the denials damaged not just the authority of the President but also the authority of the presidency.

Finally, and paradoxically, the Lewinsky affair was a public issue in that it raised questions about the citizen's right to privacy. In 1890, Supreme Court Justice Louis Brandeis argued that Americans deserve "the right to be left alone." He viewed the right to privacy as a foundation stone of any free society. Thus, the protection of privacy is itself a public issue, especially in an era when the use of electronic surveillance techniques in the form of computer databases, hidden cameras, telephones taps, and so forth are becoming more widespread. From this point of view, President Clinton had as much right to privacy as any other citizen of the United States. Arguably, if he was denied that right, one of our society's basic freedoms was denied.

Sociology cannot help you decide whether President Clinton deserved to be impeached. That is a political question. However, by emphasizing the public dimensions of the Lewinsky affair, sociology helps us see the President's personal troubles in the larger context of public policy, or the laws and regulations passed by organizations and governments. Even our most intimate social relations are shaped in part by public policy. Therefore, we review a public policy debate in each chapter of this book. It's good exercise for the sociological imagination, and it will help you gain more control over the forces that shape your life.

there was a tension in his work. For although Comte was eager to adopt the scientific method in his study of society, he was a conservative thinker, motivated by strong opposition to rapid change in French society. This was evident in his writings. Comte witnessed the democratic forces unleashed by the French Revolution, the early industrialization of society, and the rapid growth of cities. What he saw shocked and angered him. Rapid social change was destroying much of what he valued, especially respect for traditional authority. He therefore urged slow change and the preservation of all that was traditional in social life. Thus, scientific methods of research *and* a vision of the ideal society were evident in sociology at its origins.

The same sort of tension exists in the work of the most important early figures in the history of sociology, Karl Marx, Émile Durkheim, and Max Weber. These three men lived in the century between 1820 and 1920. They witnessed various phases of Europe's wrenching transition to industrial capitalism. They wanted to explain the great transformation of Europe and suggest ways of improving people's lives. Like Comte, they were all committed to the scientific method of research. However, they also wanted to chart a better course for their societies. The ideas they developed are therefore not just diagnostic tools from which we can still learn much, but also, like many sociological ideas, prescriptions for combating social ills.

The tension between analysis and ideal, diagnosis and prescription, is evident throughout sociology. This becomes clear if we distinguish three important terms: theories, research, and values.

✦ Sociological ideas are usually expressed in the form of theories. **Theories** are tentative explanations of some aspect of social life. They state how and why certain facts are related. For example, in his theory of suicide, Durkheim related facts about suicide rates to facts about social solidarity. This enabled him to explain suicide as a function of social solidarity. In our broad definition, even a hunch qualifies as a theory if it suggests how and why certain facts are related.

✦ *After* sociologists formulate theories, they can conduct research. **Research** is the process of carefully observing social reality, often to "test" a theory or assess its validity. For example, Durkheim collected suicide statistics from various government agencies to see whether the data supported or contradicted his theory. It is because research can call the validity of a theory into question that theories are only *tentative* explanations. We discuss the research process in detail in Chapter 2 ("Research Methods").

✦ *Before* sociologists can formulate a theory, however, they must make certain judgments. For example, they must decide which problems are worth studying. They must make certain assumptions about how the parts of society fit together. If they are going to recommend ways of improving the operation of some aspect of society, they must even have an opinion about what the ideal society ought to look like. As we will soon see, these issues are shaped in large measure by sociologists' values. **Values** are ideas about what is right and wrong. Inevitably, values help sociologists formulate and favor certain theories over others (Edel, 1965; Kuhn, 1970 [1962]). So sociological theories may be modified and even rejected due to research, but they are often motivated by sociologists' values.

Durkheim, Marx, and Weber initiated three of the major theoretical traditions in sociology—functionalism, conflict theory, and symbolic interactionism. A fourth perspective, feminism, arose in recent decades to correct some deficiencies of the three long-established traditions. It will become clear as you read this book that there are many more theories than just these four. However, because these four traditions have been especially influential in the development of sociology, we present a thumbnail sketch of each one here at the beginning.

Functionalism

Durkheim's theory of suicide is an early example of what sociologists now call **functionalism.** Functionalist theories incorporate four features:

1. They stress that human behavior is governed by stable patterns of social relations, or social structures. For example, Durkheim emphasized how suicide rates are influenced by patterns of social solidarity. Usually the social structures analyzed by functionalists are macrostructures.

2. Functionalist theories show how social structures maintain or undermine social stability. Typically, Durkheim analyzed how the growth of industries and cities in 19th-century Europe lowered the level of social solidarity and contributed to social instability. One aspect of instability, said Durkheim, is a higher suicide rate. Another is frequent strikes by workers.

3. Functionalist theories emphasize that social structures are based mainly on shared values. Thus, when Durkheim wrote about social solidarity, he sometimes meant the frequency and intensity of social interaction, but more often he thought of social solidarity as a kind of moral cement that binds people together.

4. Functionalism suggests that reestablishing equilibrium can best solve most social problems. Thus, Durkheim said social stability could be restored in late 19th-century Europe by creating new associations of employers and workers that would lower workers' expectations about what they could expect out of life. If, said Durkheim, more people could agree on wanting less, social solidarity would rise and there would be fewer strikes, less suicide, and so on. Functionalism, then, was a conservative response to widespread social unrest in late 19th-century France. A more liberal or radical response would have been to argue that if people are expressing discontent because they are getting less out of life than they expect, discontent can be lowered by figuring out ways for them to get more out of life.

Although functionalist thinking influenced American sociology at the end of the 19th century, it was only during the country's greatest economic crisis ever, the Great Depression of 1929–39, that functionalism took deep root here (Russett, 1966). With 30% of the labor force unemployed and labor unrest reaching unprecedented levels by 1934, it is not entirely surprising that sociologists with a conservative frame of mind were attracted to a theory that focused on how social equilibrium could be restored. Functionalist theory remained popular in the United States for about 30 years. It experienced a minor revival in the early 1990s but never regained the dominance it enjoyed from the 1930s to the early 1960s.

Sociologist Talcott Parsons was the foremost American proponent of functionalism. Parsons is best known for identifying how various institutions must work to ensure the smooth operation of society as a whole. He argued that society is well integrated and in equilibrium when the family successfully raises new generations, the military successfully defends society against external threats, schools are able to teach students the skills and values they need to function as productive adults, and religions create a shared moral code among people (Parsons, 1951).

Parsons was criticized for exaggerating the degree to which members of society share common values and social institutions contribute to social harmony. This led North America's other leading functionalist, Robert Merton, to propose that social structures may have different consequences for different groups of people. Merton noted that some of those consequences may be disruptive or **dysfunctional** (Merton, 1968 [1949]). Moreover, said Merton, while some functions are **manifest** (intended and easily observed), others are **latent** (unintended and less obvious). For instance, a manifest function of schools is to transmit skills from one generation to the next. A latent function of schools is to encourage the development of a separate youth culture that often conflicts with parents' values (Coleman, 1961; Hersch, 1998).

Conflict Theory

The second major theoretical tradition in sociology emphasizes the centrality of conflict in social life. **Conflict theory** incorporates these features:

1. It generally focuses on large, macrolevel structures, such as relations between classes.

2. Conflict theory shows how major patterns of inequality in society produce social stability in some circumstances and social change in others.

3. Conflict theory stresses how members of privileged groups try to maintain their advantages while subordinate groups struggle to increase theirs. From this point of view, social conditions at a given time are the expression of an ongoing power struggle between privileged and subordinate groups.

4. Conflict theory typically leads to the suggestion that eliminating privilege will lower the level of conflict and increase total human welfare.

Conflict theory originated in the work of German social thinker Karl Marx. A generation before Durkheim, Marx observed the destitution and discontent produced by the

Robert Merton (1910–) made functionalism a more flexible theory from the late 1930s to the 1950s. In *Social Theory and Social Structure* (1949), he proposed that social structures are not always functional. They may be dysfunctional for some people. Moreover, not all functions are manifest; some are latent, according to Merton. Merton also made major contributions to the sociology of science, notably in *On the Shoulders of Giants* (1956), a study of creativity, tradition, plagiarism, the transmission of knowledge, and the concept of progress.

Karl Marx (1818–83) was a revolutionary thinker whose ideas affected not just the growth of sociology but the course of world history. He held that major sociohistorical changes are the result of conflict between society's main social classes. In his major work, *Capital* (1867–94), Marx argued that capitalism would produce such misery and collective strength among workers that they would eventually take state power and create a classless society in which production would be based on human need rather than profit.

Max Weber (1864–1920) was Germany's greatest sociologist. He profoundly influenced the development of the discipline internationally. Engaged in a lifelong "debate with Marx's ghost," Weber held that economic circumstances alone do not explain the rise of capitalism. As he showed in *The Protestant Ethic and the Spirit of Capitalism* (1904–5), independent developments in the religious realm had unintended, beneficial consequences for capitalist development in some parts of Europe. He also argued that capitalism would not necessarily give way to socialism. Instead, he regarded the growth of bureaucracy and the overall "rationalization" of life as the defining characteristics of the modern age. These themes were developed in *Economy and Society* (1922).

Industrial Revolution and proposed a sweeping argument about the way societies develop (Marx, 1904 [1859]; Marx and Engels, 1972 [1848]). Marx's theory was radically different from Durkheim's. Class conflict lies at the center of his ideas.

Marx argued that owners of industry are eager to improve the way work is organized and to adopt new tools, machines, and production methods. These innovations allow them to produce more efficiently, earn higher profits, and drive inefficient competitors out of business. However, the drive for profits also causes capitalists to concentrate workers in larger and larger establishments, keep wages as low as possible, and invest as little as possible in improving working conditions. Thus, said Marx, in factory and in mine, a large and growing class of poor workers comes to oppose a small and shrinking class of wealthy owners.

Marx felt that workers would ultimately become aware of belonging to the same exploited class. Their sense of "class consciousness," he wrote, would encourage the growth of trade unions and labor parties. These organizations would eventually seek to put an end to private ownership of property, replacing it with a system in which everyone shared property and wealth. This was the "communist" society envisaged by Marx.

Marx's predictions about the inevitable collapse of capitalism are now largely discredited. Max Weber, a German sociologist who wrote his major works a generation after Marx, was among the first to point out some of the flaws in Marx's argument (Weber, 1946). Weber noted the rapid growth of the "service" sector of the economy with its many nonmanual workers and professionals. He argued that many members of these occupational groups stabilize society because they enjoy higher status and income than manual workers in the manufacturing sector. In addition, Weber showed that class conflict is not the only driving force of history. In his view, politics and religion are also important sources of historical change (see below). Other writers pointed out that Marx did not understand how investment in technology would make it possible for workers to toil fewer hours under less oppressive conditions. Nor did he foresee that higher wages, better working conditions, and welfare state benefits would pacify manual workers. We see, then, that many of the particulars of Marx's theory were called into question by Weber and other sociologists.

Nonetheless, Marx's insights about the fundamental importance of conflict in social life were influential. They still are. For example, in the United States, W. E. B. Du Bois was an early advocate of conflict theory. For a man writing at the end of the 19th century, he had a remarkably liberal and even radical frame of mind. The first African American to receive a Ph.D. from Harvard, Du Bois conducted pioneering studies of race. He was also a founder of the National Association for the Advancement of Colored People (NAACP). His best-known work is *The Philadelphia Negro*. This book is based on the first major sociological research project conducted in the United States. Du Bois showed that poverty and other social problems faced by African Americans are not due to some "natural" inferiority (which was widely believed at the time) but to white prejudice (Du Bois, 1967 [1899]). He believed the elimination of white prejudice would reduce racial conflict and create more equality between blacks and whites. Du Bois was also critical of economically successful African Americans. He criticized them for failing to help less fortunate blacks and segregating themselves from the African-American community to win acceptance among whites.

Du Bois was a pioneer in American conflict theory, particularly as it applies to race and ethnic relations. While conflict theory had some advocates in the United States after Du Bois, it did not really flower in this country until the 1960s, a decade that was rocked by growing labor unrest, antiwar protests, the rise of the black power movement, and the first stirrings of feminism. Strikes, demonstrations, and riots were almost daily occurrences in the 1960s and early 1970s, and it seemed self-evident to many sociologists of that generation that conflict between classes, nations, races, and generations is the very essence of society. Many of today's leading sociologists attended graduate school in the 1960s and 1970s. They were strongly influenced by the spirit of the times. As you will see throughout this book, they have made important contributions to conflict theory during their professional careers.

Symbolic Interactionism

Above we noted that Weber criticized Marx's interpretation of the development of capitalism. Among other things, Weber argued that early capitalist development was caused not just by favorable *economic* circumstances. In addition, he said, certain *religious* beliefs facilitated robust capitalist growth. In particular, 16th- and 17th-century Protestants believed their religious doubts could be reduced, and a state of grace assured, if they worked diligently and lived modestly. Weber called this belief **the Protestant ethic.** He believed it had an unintended effect: people who adhered to the Protestant ethic saved and invested more than others. Thus, according to Weber, capitalism developed most robustly where the Protestant ethic took hold. He concluded that capitalism did not develop due to the operation of economic forces alone, as Marx argued. Instead, it depended partly on the religious meaning individuals attached to their work (Weber 1958 [1904–5]).

The idea that subjective meanings must be analyzed in any complete sociological analysis was only one of Weber's contributions to early sociological theory. Weber was also an important conflict theorist, as you will learn. At present, however, it is enough to note that his emphasis on subjective meanings found rich soil in the United States in the late 19th and early 20th centuries. For here was an idea that resonated deeply with the individualism of American culture. A century ago, it was widely believed that individual talent and initiative could achieve just about anything in this land of opportunity. Small wonder, then, that much of early American sociology focused on the individual or, more precisely, on the connection between the individual and the larger society. For example, George Herbert Mead at the University of Chicago was the driving force behind the study of how individual identity is formed in the course of interaction with other people. We discuss his contribution in Chapter 4 ("Socialization"). Here we note only that the work of Mead and his colleagues gave birth to symbolic interactionism, a distinctively American theoretical tradition that continues to be a major force in American sociology today.

Functionalist and conflict theories assume that people's group memberships—whether they are rich or poor, male or female, black or white—determine their behavior. This can sometimes make people seem like balls on a pool table. They get knocked around and can't choose their own destinations. We know from our everyday experience, however, that people are not like that. You often make choices, sometimes difficult ones. You sometimes change your mind. Moreover, two people with similar group memberships may react differently to similar social circumstances. That is because they interpret those circumstances differently.

Recognizing these issues, some sociologists focus on the subjective side of social life. They work in the symbolic interactionist tradition. **Symbolic interactionism** incorporates these features:

1. It focuses on face-to-face communication, or interaction in microlevel social settings. This distinguishes it from both functionalist and conflict theories.

2. Symbolic interactionism emphasizes that an adequate explanation of social behavior requires understanding the subjective meanings people attach to their social circumstances.

3. Symbolic interactionism stresses that people help to create their social circumstances and do not merely react to them.

4. By focusing on the subjective meanings people create in small social settings, symbolic interactionists validate unpopular and nonofficial viewpoints. This increases our understanding and tolerance of people who may be different from us.

To understand symbolic interactionism better, let us briefly return to the problem of suicide. If a police officer discovers a dead person at the wheel of a car that has run into a tree, it may be difficult to establish with certainty whether the death was accidental or suicidal. Interviewing friends and relatives to discover the driver's state of mind just before the crash may help rule out the possibility of suicide. But, as this example illustrates,

W. E. B. Du Bois (1868–1963), Harvard's first African-American Ph.D., was a pioneer in the study of race in the United States and a founder of the NAACP. His *The Philadelphia Negro* (1899) is a classic that went against the grain of much contemporary social thought. In his view, social inequality and discrimination—between blacks and whites, and between successful and less successful blacks—were the main sources of problems faced by the African-American community. He argued that only a decline in inequality and prejudice would solve the problems of the African-American community.

understanding the intention or motive of the actor is critical to understanding the meaning of a social action and explaining it. A state of mind must be interpreted, usually by a coroner, before the dead body becomes a suicide statistic (Douglas, 1967).

For surviving family and friends, suicide is always painful and sometimes embarrassing. Insurance policies often deny payments to beneficiaries in the case of suicide. As a result, coroners are inclined to classify deaths as accidental whenever such an interpretation is plausible. Being human, they want to minimize a family's suffering after such a horrible event. Sociologists believe that, for this reason, suicide rates according to official statistics are about one-third lower than actual suicide rates.

The study of the subjective side of social life reveals many such inconsistencies. It helps us go beyond the official picture, deepening our understanding of how society works, and supplementing the insights gained from macrolevel analysis. Moreover, by stressing the importance and validity of subjective meanings, symbolic interactionists also increase tolerance for minority and deviant viewpoints.

Feminist Theory

Few women figured prominently in the early history of sociology. The strict demands placed on them by the 19th-century family and the lack of opportunity for women in the larger society prevented them from getting a higher education and making major contributions to the discipline. The women who did make their mark on the discipline in its early years tended to have unusual biographies. These exceptional people introduced gender issues that were largely ignored by Marx, Durkheim, Weber, Mead, and other early sociologists. Appreciation for the sociological contribution of these pioneer women has grown in recent years as concern with gender issues has come to form a substantial part of the modern sociological enterprise.

For example, Harriet Martineau is often called the first woman sociologist. Born in England to a prosperous family in 1802, she never married. She was able to support herself comfortably from her journalistic writings. Martineau translated Comte into English. She wrote one of the first books on research methods. She undertook critical studies of slavery and factory laws and of gender inequality. She was a leading advocate of voting rights and higher education for women, as well as gender equality in the family. As such, Martineau was one of the first feminists (Martineau, 1985).

Despite its encouraging beginnings, feminist thinking had little impact on sociology until the mid-1960s. It was then that the rise of the modern women's movement drew attention to the many remaining inequalities between women and men. Since then, feminist theory has had such a big influence on sociology it may fairly be regarded as sociology's fourth major theoretical tradition. There are several variants of modern feminism (see Chapter 9, "Sexuality and Gender"). However, the various strands of **feminist theory** share the following features:

1. Feminist theory focuses on various aspects of patriarchy, the system of male domination in society. Patriarchy, feminists contend, is at least as important as class inequality in determining a person's opportunities in life, and perhaps more so.

2. Feminist theory holds that male domination and female subordination are determined not by biological necessity but by structures of power and social convention. From their point of view, women are subordinate to men only because men enjoy more legal, economic, political, and cultural rights.

3. Feminist theory examines the operation of patriarchy in both micro and macro settings.

4. Feminist theory contends that existing patterns of gender inequality can and should be changed for the benefit of all members of society. The main sources of gender inequality include differences in the way boys and girls are brought up, barriers to equal opportunity in education, paid work, and politics, and the unequal division of domestic responsibilities between women and men.

Harriet Martineau (1802–76) was the first woman sociologist. Unlike most women of her time and place, she was able to live the life of a scholar because she came from a wealthy family and never married or had children. She translated Comte into English and conducted studies on research methods, slavery, factory laws, and gender inequality. As a leading advocate of voting rights and higher education for women, as well as gender equality in the family, Martineau was one of the first feminists.

Theoretical Tradition	Main Level of Analysis	Main Focus	Main Question
Functionalist	Macro	Values	How do the institutions of society contribute to social stability and instability?
Conflict	Macro	Inequality	How do privileged groups seek to maintain their advantages and subordinate groups seek to increase theirs, often causing social change in the process?
Symbolic interactionist	Micro	Meaning	How do individuals communicate so as to make their social settings meaningful?
Feminist	Micro and macro	Patriarchy	Which social structures and interaction processes maintain male dominance and female subordination?

✦ **TABLE 1.1** ✦
Four Theoretical Traditions in Sociology

The theoretical traditions outlined above are summarized in Table 1.1. As you will see in the following pages, sociologists in the United States and elsewhere have applied them to all of the discipline's branches. They have elaborated and refined each of them. Some sociologists work exclusively within one tradition. Others conduct research that borrows from more than one tradition. But all sociologists are deeply indebted to the founders of the discipline. Standing on the shoulders of giants, we are able to see farther.

THEIR REVOLUTION AND OURS

In the 19th century, the founders of sociology devoted their lives to solving the great sociological puzzle of their time, the causes and consequences of the Industrial Revolution. However, the ideas that stirred them did not spring fully grown from their minds. Rather, their social experiences helped to shape their ideas. There is an important lesson to be learned here. In general, sociological ideas are influenced by the social settings in which they emerge.

This lesson immediately suggests two important questions. First, what are the great sociological puzzles of *our* time? Second, how are today's sociologists responding to the challenges presented by the social settings in which *they* live? We devote the rest of the book to answering these questions in depth. In the remainder of this chapter, we offer an outline of what you can expect to learn. To provide a context for this outline, we first say a few words about how the Industrial Revolution of the 19th century was transformed into the Postindustrial Revolution of our day.

The Industrial Revolution

The Industrial Revolution began in Britain in the 1780s. It involved the application of science and technology to industrial processes, the creation of factories, and the formation of a large class of **blue-collar workers.** Their jobs were manual, dirty, physically demanding, and often required little training. Within about a century, the Industrial Revolution had taken firm root throughout Western Europe, North America, and Japan. A century after that, industry had begun implanting itself in most of the rest of the world.

As noted in our discussion of Marx, the industrial working class protested long workdays, low pay, and dangerous working conditions. Workers went out on strike, formed unions, and joined political parties. Their protests forced governments to tax citizens and provide at least minimal protection against ill health, unemployment, and poverty. Working-class protests also forced employers to limit the length of the workweek to 40 hours, improve working conditions, and raise wages. Employers were still able to increase their profits, however, by making the organization of work more efficient and introducing new technologies.

Collecting taxes, administering social services, and investing heavily in technological change required the growth of government and business offices, hospitals, schools, colleges, and research laboratories. Thus, alongside the old manufacturing sector of the economy, the new service sector was born. Its employees became known as white-collar workers. Highly trained professionals stood at the peak of the service sector. Secretaries and clerks were positioned near its base. By 1960, more than half of all people working in the paid labor force of the world's rich countries were in nonmanual occupations.

Sociologists call this most recent transformation of human society the **Postindustrial Revolution.** Specifically, the Postindustrial Revolution refers to the technology-driven shift from manufacturing to service industries and the consequences of that shift for nearly all human activities (Bell, 1976; Toffler, 1990). The causes and consequences of postindustrialism form the great sociological puzzle of our time. Since much of this book analyzes postindustrialism and its effects, it will prove useful if we now offer a guided tour to *Sociology: Your Compass for a New World.*

Postindustrialism: Opportunities and Pitfalls

In 1998, *Wired* magazine, the exuberant voice of North American computer culture, published a special fifth anniversary issue. It was devoted to analyzing the state of the planet. One contributor wrote that "[t]he life of *Wired* coincides with the best five years humanity has ever experienced." The world is not only free of large-scale conflict. In addition, "we are squarely in the midst of the most amazing upsurge of knowledge and wealth ever seen on Earth. And that trend is—for the first time in human history—irreversible" (Simon, 1998: 66). A few months earlier, in a special issue of the *New York Times Magazine* devoted to technology, one staff writer gushed:

> Individuals are acquiring more control over their lives, their minds and their bodies, even their genes, thanks to the transformations in medicine, communications, transportation and industry. At the same time, these technologies are providing social benefits and undoing some of the damage of the past. Technology helps to conserve natural resources and diminish pollution . . . The Information Revolution, besides enabling us to visit Mars at will, is fostering peaceful cooperation on Earth by decentralizing power. Political tyrants and demagogic warmongers are losing control now that their subjects have tools to communicate directly with one another. People are using the tools to do their jobs without leaving their families. They're forming new communities in cyberspace and forming new bonds with their neighbors in real space. Technology has the potential to increase individual freedom and strengthen community. . . . (Tierney, 1997: 46–7)

Most sociologists are less starry-eyed than these journalists. Sociologists agree that postindustrialism promises many exciting opportunities to enhance the quality of life. However, they also see many social-structural barriers to the realization of that promise. For most sociologists, the Postindustrial Revolution is so far only half a revolution. And it is uncertain whether the second half will turn out as well as the optimistic writers at *Wired* and the *New York Times Magazine* imagine.

The main unresolved issues confronting the postindustrial era may be sketched in the form of a compass—a sociological compass (see Figure 1.5). Each axis of the compass contrasts a postindustrial ideal (equality of opportunity, freedom) with its opposite (inequality of opportunity, constraint):

Web Interactive Exercises
Will Spiritual Robots Replace Humans by 2100?

1. *Equality versus inequality of opportunity.* Optimists forecast that postindustrialism will provide not just more opportunity to find creative, interesting, challenging, and rewarding work. In addition, they say, the new era will generate more "equality of opportunity," that is, a better chance for *all* people to get a good education, influence government policy, and find good jobs.

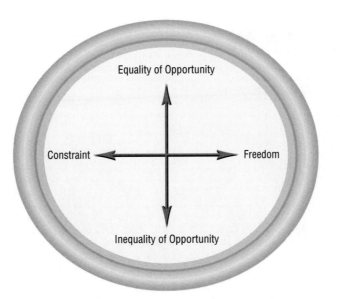

You will certainly find evidence to support these optimistic claims in the following pages. For example, Chapters 7 ("Stratification: United States and Global Perspectives") and 10 ("Work and the Economy") show that the average standard of living and the number of good jobs are increasing in postindustrial societies such as the United States. In Chapters 9 ("Sexuality and Gender") and 13 ("Religion and Education") you will learn about the rapid strides women in particular are making in the economy, the education system, and other institutions. Chapter 8 ("Race and Ethnicity") shows that postindustrial societies like the United States are characterized by a decline in discrimination against members of ethnic and racial minorities. Chapter 11 ("Politics") demonstrates that democracy is spreading quickly throughout the world.

Yet, as you read this book, it will also become clear that all of these seemingly happy stories have a dark underside. For example, it turns out that the number of routine jobs with low pay and few benefits is growing faster than the number of creative, high-paying jobs. Inequality between the wealthiest and poorest Americans has grown in recent decades, as has inequality between the wealthiest and poorest nations. An enormous opportunity gulf still separates women from men. Racism and discrimination are still very much a part of the world we live in. Our health care system is in crisis just as our population is aging rapidly and most in need of health care (see Chapter 15, "Health, Medicine, and Aging"). Many of the world's new democracies are only superficially democratic, while Americans and citizens of other postindustrial societies are increasingly cynical about the ability of their political systems to respond to their demands and are looking for alternative forms of political expression (see Chapter 17, "Collective Action and Social Movements"). In short, equality of opportunity is an undeniably attractive ideal, but it is far from clear that it is the inevitable outcome of the growth of postindustrial society.

2. *Individual freedom versus individual constraint.* The same may be said of the ideal of freedom. In an earlier era, most people retained their religious, ethnic, racial, and sexual identities for a lifetime, even if they weren't particularly comfortable with them. They often remained in social relationships that made them unhappy. One of the major themes of *Sociology: Your Compass for a New World* is that many people are freer to construct their identities and form social relationships in ways that suit them. To a greater degree than ever before, it is possible to *choose* who you want to

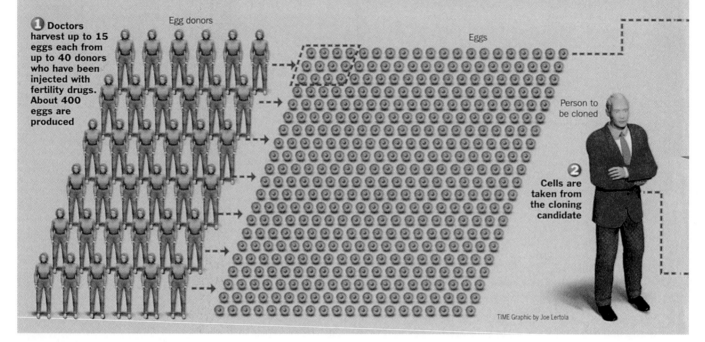

HOW TO CLONE A HUMAN

If it works in humans as it has in other mammals, cloning will be technically possible, but also terribly inefficient and risky.

According to experts, producing a single viable clone will require scores of volunteers to donate eggs and carry embryos—most of which will have major abnormalities and never come to term. The clones that do survive could suffer more subtle problems that might show up well after birth. Here's how it might be done.

① Doctors harvest up to 15 eggs each from up to 40 donors who have been injected with fertility drugs. About 400 eggs are produced

Egg donors

Eggs

Person to be cloned

② Cells are taken from the cloning candidate

TIME Graphic by Joe Lertola

It is widely predicted that the cloning of humans will soon be attempted so that organs can be "harvested," otherwise infertile couples can have children, and so forth. However, as with many other scientific breakthroughs, "progress" in cloning techniques is likely to create serious ethical dilemmas and social problems. Social-structural barriers hinder the full realization of still other scientific advances.

be, who you want to associate with, and how you want to associate with them. The postindustrial era frees people from traditional constraints by encouraging virtually instant global communication, international migration, greater acceptance of sexual diversity and a variety of family forms, the growth of ethnically and racially diverse cities, and so forth. For instance, in the past people often remained in marriages even if they were unhappy. Families often involved a father working in the paid labor force and a mother keeping house and rearing children without pay. Today, people are freer to end unhappy marriages and create family forms that are more suited to their individual needs (see Chapter 12, "Families"). We take up the theme of increasing individual freedom in Chapter 3 ("Culture"), Chapter 4 ("Socialization"), Chapter 8 ("Race and Ethnicity"), Chapter 9 ("Sexuality and Gender"), Chapter 13 ("Religion and Education"), Chapter 14 ("The Mass Media"), and Chapter 16 ("Population, Urbanization, and Development").

Again, however, we must face the less rosy aspects of postindustrialism. In many of the following chapters, we point out how increased freedom is experienced only within certain limits and how social diversity is limited by a strong push to conformity in some spheres of life. For example, we can choose a far wider variety of consumer products than ever before, but consumerism itself increasingly seems a compulsory way of life (Chapter 3, "Culture"). Moreover, it is a way of life that threatens the natural environment (Chapter 18, "Technology and the Global Environment"). Meanwhile, some new technologies, such as surveillance cameras, cause us to modify our behavior and act in more conformist ways (Chapter 6, "Deviance and Crime"). Large, impersonal bureaucracies and standardized products and services dehumanize both staff and customers (Chapter 5, "Interaction and Organization"). The tastes and the profit motive of vast media conglomerates, most of them American owned, govern most of our diverse cultural consumption and arguably threaten the survival of distinctive national cultures (Chapter 14, "The Mass

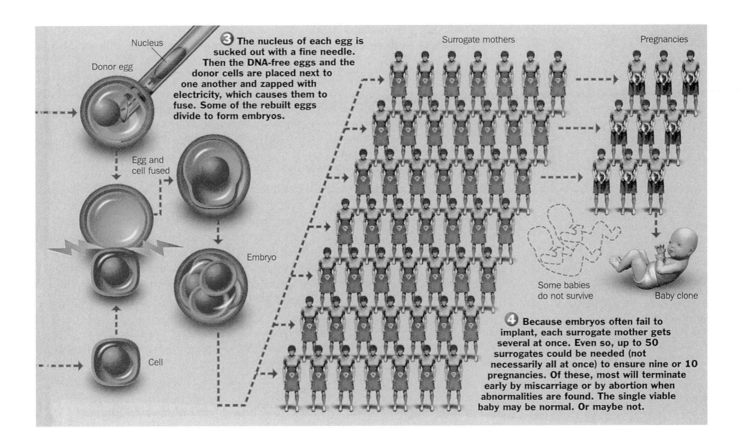

③ The nucleus of each egg is sucked out with a fine needle. Then the DNA-free eggs and the donor cells are placed next to one another and zapped with electricity, which causes them to fuse. Some of the rebuilt eggs divide to form embryos.

Nucleus

Donor egg

Egg and cell fused

Embryo

Cell

Surrogate mothers

Pregnancies

Some babies do not survive

Baby clone

④ Because embryos often fail to implant, each surrogate mother gets several at once. Even so, up to 50 surrogates could be needed (not necessarily all at once) to ensure nine or 10 pregnancies. Of these, most will terminate early by miscarriage or by abortion when abnormalities are found. The single viable baby may be normal. Or maybe not.

Media"). Powerful interests are trying to shore up the traditional nuclear family even though it does not suit some people (Chapter 12, "Families"). As these examples show, the push to uniformity counters the trend toward growing social diversity. Postindustrialism may make us freer in some ways, but it also places new constraints on us.

WHY SOCIOLOGY?

Our overview of themes in *Sociology: Your Compass for a New World* drives home a point made by the renowned British sociologist, Anthony Giddens. We live in an era "suspended between extraordinary opportunity . . . and global catastrophe" (Giddens, 1987: 166). A whole range of environmental issues, profound inequalities in the wealth of nations and of classes, racial and ethnic violence, and unsolved problems in the relations between women and men continue to stare us in the face and profoundly affect the quality of our everyday lives.

Despair and apathy is one possible response to these complex issues. But it is not a response that humans have often favored. If it were our nature to give up hope, we would still be sitting around half-naked in the mud outside a cave.

People are more inclined to look for ways of improving their lives, and this period of human history is full of opportunities to do so. We have, for example, advanced to the point where for the first time we have the means to feed and educate everyone in the world. Similarly, it now seems possible to erode some of the inequalities that have always been the major source of human conflict.

Sociology offers some useful advice on how to achieve these goals. For sociology is more than just an intellectual exercise. It is also an applied science, with practical,

Web Research Projects
The Uses of Sociology

everyday uses. There are perhaps 20,000 people with sociology Ph.D.s in the United States today and hundreds of thousands with sociology M.A.s and B.A.s. People with sociological training teach at various levels from high school to graduate school. They conduct research and give advice in a wide variety of settings. These include local, state, and federal governments, colleges, corporations, management consulting firms, trade unions, social service agencies, international organizations, and private research and testing firms. "Applied" sociologists help various organizations solve immediate problems that impede their smooth operation. "Clinical" sociologists intervene as counselors, environmental impact assessors, conflict mediators, and so forth. They help guide the process of social change (United States Department of Labor, 2000c).

Sociology has benefits even for people who do not work as sociologists. For, although it offers no easy solutions as to how the goal of improving society may be accomplished, it does promise a useful way of understanding our current predicament and seeing possible ways of dealing with it, of leading us a little farther away from the mud outside the cave. You sampled its ability to tie personal troubles to social-structural issues when we discussed suicide. You reviewed the major theoretical perspectives that enable sociologists to connect the personal with the social-structural. When we outlined the half-fulfilled promises of postindustrialism, you saw sociology's ability to provide an understanding of where we are and where we can head.

We frankly admit that the questions we raise in this book are tough to answer. Sharp controversy surrounds them all. But we are sure that if you try to grapple with them you will enhance your understanding of your society's, and your own, possibilities. Sociology can help you figure out where you fit in to society and how you can make society fit you.

SUMMARY

1. Durkheim showed that even apparently nonsocial and antisocial actions are influenced by social structures. Specifically, levels of social solidarity affect suicide rates.

2. Sociologists analyze the connection between personal troubles and social structures.

3. Sociologists analyze the influence of three levels of social structure on human action: microstructures, macrostructures, and global structures.

4. A theory is a tentative explanation of some aspect of social life. It states how and why specific facts are connected. Research is the process of carefully observing social reality to test the validity of a theory.

5. There are four major theoretical traditions in sociology. Functionalism analyzes how social order is supported by macrostructures. The conflict approach analyzes how social inequality is maintained and challenged. Symbolic interactionism analyzes how meaning is created when people communicate in microlevel settings. Feminism focuses on the social sources of patriarchy in both macro and micro settings.

6. The rise of sociology was stimulated by the Scientific, Industrial, and Democratic Revolutions.

7. The Postindustrial Revolution is the technology-driven shift from manufacturing to service industries and the consequences of that shift for virtually all human activities.

8. The causes and consequences of postindustrialism form the great sociological puzzle of our time. The tensions between equality and inequality of opportunity, and between freedom and constraint, are among the chief interests of sociology today.

GLOSSARY

Blue-collar workers have manufacturing-sector jobs that are manual, dirty, physically demanding, and often require little training.

Conflict theory generally focuses on large, macrolevel structures, such as the relations between classes. It shows how major patterns of inequality in society produce social stability in some circumstances and social change in others. It stresses how members of privileged groups try to maintain their advantages while subordinate groups struggle to increase theirs. And it typically leads to the suggestion that eliminating privilege will lower the level of conflict and increase the sum total of human welfare.

The **Democratic Revolution** began about 1750. It involved the citizens of the United States, France, and other countries broadening their participation in government. It also suggested that people organize society, and that human intervention can therefore resolve social problems.

Dysfunctions are effects of social structures that create social instability.

Feminist theory claims that patriarchy is at least as important as class inequality in determining a person's opportunities in life. It holds that male domination and female subordination are determined not by biological necessity but by structures of power and social convention. It examines the operation of patriarchy in both micro and macro settings. And it contends that existing patterns of gender inequality can and should be changed for the benefit of all members of society.

Functionalist theory stresses that human behavior is governed by relatively stable social structures. It underlines how social structures maintain or undermine social stability. It emphasizes that social structures are based mainly on shared values or preferences. And it suggests that reestablishing equilibrium can best solve most social problems.

Global structures are patterns of social relations that lie outside and above the national level. They include international organizations, patterns of worldwide travel and communication, and the economic relations between countries.

The **Industrial Revolution** refers to the rapid economic transformation that began in Britain in the 1780s. It involved the large-scale application of science and technology to industrial processes, the creation of factories, and the formation of a working class. It created a host of new and serious social problems that attracted the attention of many social thinkers.

Latent functions are invisible and unintended effects of social structures.

Macrostructures are overarching patterns of social relations that lie outside and above your circle of intimates and acquaintances. Macrostructures include classes, bureaucracies, and power systems such as patriarchy.

Manifest functions are visible and intended effects of social structures.

Microstructures are the patterns of relatively intimate social relations formed during face-to-face interaction. Families, friendship circles, and work associations are all examples of microstructures.

Patriarchy is the traditional system of economic and political inequality between women and men.

The **Postindustrial Revolution** refers to the technology-driven shift from manufacturing to service industries and the consequences of that shift for virtually all human activities.

The **Protestant ethic** is the 16th- and 17th-century Protestant belief that religious doubts can be reduced, and a state of grace assured, if people work diligently and live ascetically. According to Weber, the Protestant ethic had the unintended effect of increasing savings and investment and thus stimulating capitalist growth.

Research is the process of carefully observing reality to assess the validity of a theory.

The **Scientific Revolution** began in Europe about 1550. It encouraged the view that sound conclusions about the workings of society must be based on solid evidence, not just speculation.

Social structures are stable patterns of social relations.

The **sociological imagination** is the quality of mind that enables one to see the connection between personal troubles and social structures.

Sociology is the systematic study of human behavior in social context.

Social solidarity refers to (1) the degree to which group members share beliefs and values and (2) the intensity and frequency of their interaction.

Symbolic interactionist theory focuses on face-to-face communication, or interaction in microlevel social settings. It emphasizes that an adequate explanation of social behavior requires understanding the subjective meanings people attach to their social circumstances. It stresses that people help to create their social circumstances and do not merely react to them. And, by underscoring the subjective meanings people create in small social settings, it validates unpopular and nonofficial viewpoints. This increases our understanding and tolerance of people who may be different from us.

A **theory** is a tentative explanation of some aspect of social life that states how and why certain facts are related.

Values are ideas about what is right and wrong.

QUESTIONS TO CONSIDER

1. Do you think the promise of freedom and equality will be realized in the 21st century? Why or why not?

2. In this chapter, you learned how variations in the level of social solidarity affect the suicide rate. How do you think variations in social solidarity might affect other areas of social life, such as criminal behavior and political protest?

3. Is a science of society possible? If you agree that such a science is possible, what are its advantages over common sense? What are its limitations?

WEB RESOURCES

Companion Web Site for This Book

http://sociology.wadsworth.com
Begin by clicking on the Student Resources section of the Web site. Choose "Introduction to Sociology" and finally the Brym and Lie book cover. Next, select the chapter you are currently studying from the pull-down menu. From the Student Resources page you will have easy access to InfoTrac College Edition®, MicroCase Online exercises, additional Web links, and many resources to aid you in your study of sociology, including practice tests for each chapter.

Infotrac Search Terms

These search terms are provided to assist you in beginning to conduct research on this topic by visiting http://www.infotrac college. com/wadsworth.

Conflict theory	**Social structure**
Feminism	**Suicide**
Functionalism	**Symbolic interactionism**

Recommended Web Sites

For an inspiring essay on the practice of the sociological craft by one of America's leading sociologists, see Gary T. Marx, "Of Methods and Manners for Aspiring Sociologists: 36 Moral Imperatives," on the World Wide Web at http://web.mit.edu/gtmarx/ www/37moral.html. This article was originally published in *The American Sociologist* (28: 1997) 102–25.

SocioWeb is a comprehensive guide to sociological resources on the World Wide Web at http://www.socioweb.com/markbl/ socioweb.

For descriptions of departments of sociology at universities throughout the world, visit http://www.socioweb.com/markbl/ socioweb/univ.

The American Sociological Association (ASA) is the main professional organization of sociologists in the United States. The ASA Web site is at http://www.asanet.org.

SUGGESTED READINGS

Peter Berger. *Invitation to Sociology*: *A Humanistic Approach* (New York: Doubleday, 1963). A classic, brief introduction to the discipline by one of America's most respected sociologists.

Randall Collins. *Sociological Insight: An Introduction to Non-Obvious Sociology* (New York: Oxford University Press, 1992). A leading American sociologist highlights some of sociology's most surprising insights.

Anthony Giddens. *Sociology: A Brief But Critical Introduction*, 2nd ed. (New York: Harcourt Brace Jovanovich, 1987). A spirited, conflict-based introduction to the discipline by Britain's leading sociologist.

W. Richard Stephens, Jr. *Careers in Sociology*, 2nd ed. (Boston: Allyn and Bacon, 1999). Two dozen revealing portraits of how sociology graduates have used their education to find interesting, secure, and well-paying jobs in a wide range of fields.

Immanuel Wallerstein, ed. "The Heritage of Sociology and the Future of the Social Sciences in the 21st Century," *Current Sociology* (46, 2: 1998). Leading sociologists from around the world assess the state of the discipline and its future.

IN THIS CHAPTER, YOU WILL LEARN THAT:

✦ Scientific ideas differ from common sense and other forms of knowledge. Scientific ideas are assessed in the clear light of systematically collected evidence and public scrutiny.

✦ Sociological research depends not just on the rigorous testing of ideas but also on creative insight. Thus, the objective and subjective phases of inquiry are both important in good research.

✦ The main methods of collecting sociological data include systematic observations of natural social settings, experiments, surveys, and the analysis of existing documents and official statistics.

✦ Each data collection method has characteristic strengths and weaknesses. Each method is appropriate for different kinds of research problems.

CHAPTER

2

RESEARCH METHODS

Science and Experience

OTTFFSSENT

Scientific Versus Nonscientific Thinking

The Research Cycle

Participant Observation

Experiments

Surveys

Analysis of Existing Documents and Official Statistics

The Importance of Being Subjective

SCIENCE AND EXPERIENCE

OTTFFSSENT

"Okay, Mr. Smarty Pants, see if you can figure this one out." That's how Robert Brym's 11-year-old daughter, Talia, greeted him one day when she came home from school. "I wrote some letters of the alphabet on this sheet of paper. They form a pattern. Take a look at the letters and tell me the pattern."

Robert took the sheet of paper from Talia and smiled confidently. "Like most North Americans, I'd had a lot of experience with this sort of puzzle," says Robert. "For example, most IQ and SAT tests ask you to find patterns in sequences of letters, and you learn certain ways of solving these problems. One of the commonest methods is to see if the 'distance' between adjoining letters stays the same or varies predictably. For example, in the sequence ADGJ, there are two missing letters within each adjoining pair. Insert the missing letters and you get the first 10 letters of the alphabet: A(BC)D(EF)G(HI)J.

"This time, however, I was stumped. On the sheet of paper, Talia had written the letters OTTFFSSENT. I tried to use the distance method to solve the problem. Nothing worked. After 10 minutes of head scratching, I gave up."

"The answer's easy," Talia said, clearly pleased at my failure. Spell out the numbers 1 to 10. The first letter of each word—one, two, three, and so forth—spells OTTFFSSENT. Looks like you're not as smart as you thought. See ya'." And with that she bounced off to her room.

"Later that day, it dawned on me that Talia had taught me more than just a puzzle. She had shown me that experience sometimes prevents people from seeing things. My experience in solving letter puzzles by using certain set methods obviously kept me from solving the unusual problem of OTTFFSSENT. Said differently, reality (in this case, a pattern of letters) is not just a thing 'out there' we can learn to perceive 'objectively.' As social scientists have appreciated for over a century, experience helps determine how we perceive reality, including what patterns we see and whether we are able to see patterns at all" (Hughes, 1967: 16).

The fact that experience filters perceptions of reality is the single biggest problem for sociological research. In sociological research, the filtering occurs in four stages (see Figure 2.1). First, as noted in Chapter 1, the real-life experiences and passions of sociologists motivate much research. That is, our *values* often help us decide which problems are worth investigating. These values may reflect the typical outlook of our class, race, gender, region, historical period, and so forth. Second, our values lead us to formulate and adopt favored *theories* for interpreting and explaining those problems. Third, sociologists' interpretations are influenced by *previous research*, which we consult in order to find out what we already know about a subject. And fourth, the *methods* we use to gather data mold our perceptions. The shape of our tools often helps to determine which bits of reality we dig up.

Given that values, theories, previous research, and research methods filter our perceptions, you are right to conclude we can never perceive society in a pure or objective form.[1]

✦ **FIGURE 2.1** ✦
How Research Filters Perception

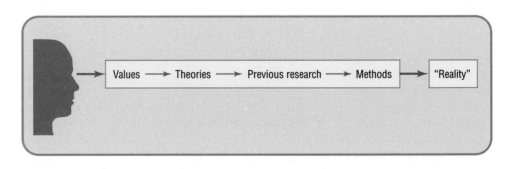

Values → Theories → Previous research → Methods → "Reality"

---

[1]Some scholars think it is possible to examine data without any preconceived notions and then formulate theories on the basis of this examination. However, they seem to form a small minority (Medawar, 1996: 12–32).

What we can do is use techniques of data collection that minimize bias. We can also clearly and publicly describe the filters that influence our perceptions. Doing so enables us to eliminate obvious sources of bias. It also helps others see biases we miss and try to correct for them. The end result is a more accurate perception of reality than is possible by relying exclusively on blind prejudice or common sense.

It is thus clear that a healthy tension pervades all sociological scholarship. On the one hand, researchers generally try to be objective in order to perceive reality as clearly as possible. They follow the rules of scientific method and design data collection techniques to minimize bias. On the other hand, the values and passions that grow out of personal experience are important sources of creativity. As Max Weber said, we choose to study "only those segments of reality which have become significant to us because of their value-relevance" (Weber, 1964 [1949]: 76). So objectivity and subjectivity each play an important role in science, including sociology. Oversimplifying a little, we can say that, while objectivity is a reality check, subjectivity leads us to define which aspects of reality are worth checking on in the first place.

Most of this chapter is about the reality check. It explores how sociologists try to adhere to the rules of scientific method. We first contrast scientific and nonscientific thinking. We next discuss the steps involved in the sociological research process. We then describe the main methods of gathering sociological data and the decisions that have to be made during the research process. In the final section, we return to the role of subjectivity in research.

Scientific Versus Nonscientific Thinking

In science, seeing is believing. In everyday life, believing is seeing. In other words, in everyday life our biases easily influence our observations. This often leads us to draw incorrect conclusions about what we see. In contrast, scientists, including sociologists, develop ways of collecting, observing, and thinking about evidence that minimize their chance of drawing biased conclusions.

On what basis do you decide statements are true in everyday life? In the following list we describe 10 types of nonscientific thinking (Babbie, 2000 [1973]). As you read about each one, ask yourself how frequently you think unscientifically. If you often think unscientifically, this chapter's for you.

1. "Chicken soup helps get rid of a cold. *It worked for my grandparents, and it works for me.*" This statement represents knowledge based on *tradition*. While some traditional knowledge is valid (sugar will rot your teeth), some is not (masturbation will not blind you). Science is required to separate valid from invalid knowledge.

2. "Weak magnets can be used to heal many illnesses. *I read all about it in the newspaper.*" This statement represents knowledge based on *authority*. We often think something is true because we read it in an authoritative source or hear it from an expert. But authoritative sources and experts can be wrong. For example, 19th-century Western physicians commonly "bled" their patients with leeches to draw "poisons" from their bodies. This often did more harm than good. As this example suggests, scientists should always question authority in order to arrive at more valid knowledge.

3. "The car that hit the cyclist was dark brown. I was going for a walk last night when *I saw the accident.*" This statement represents knowledge based on *casual observation*. Unfortunately, we are usually pretty careless observers. That is why good lawyers can often trip up eyewitnesses in courtrooms. Eyewitnesses are rarely certain about what they saw. In general, uncertainty can be reduced by observing in a conscious and deliberate manner and by recording observations. That is just what scientists do.

Perhaps the first major advance in modern medicine took place when doctors stopped using unproven interventions in their treatment of patients. One such intervention involved using leeches to bleed patients, shown here in a medieval drawing.

Even Albert Einstein, often hailed as the most intelligent person of the 20th century, sometimes ignored evidence in favor of pet theories. However, the social institution of science, which makes ideas public and subjects them to careful scrutiny, often overcomes such bias.

4. "If you work hard, you can get ahead. *I know because several of my parents' friends started off poor but are now comfortably middle class.*" This statement represents knowledge based on *overgeneralization*. For instance, if you know a few people who started off poor, worked hard, and became rich you may think any poor person may become rich if he or she works hard enough. You may not know about the more numerous poor people who work hard and remain poor. Scientists, however, sample cases that are representative of entire populations. This enables them to avoid overgeneralization. They also avoid overgeneralization by repeating research. This ensures that they don't draw conclusions from an unusual set of research findings.

5. "I'm right because *I can't think of any contrary cases.*" This statement represents knowledge based on *selective observation*. Sometimes we unconsciously ignore evidence that challenges our firmly held beliefs. Thus, you may actually know some people who work hard but remain poor. However, to maintain your belief that hard work results in wealth, you may keep them out of mind. The scientific requirement that evidence be drawn from representative samples of the population minimizes bias arising from selective observation.

6. "Mr. Smith is poor even though he works hard, but that's because he's disabled. Disabled people are the only *exception to the rule* that if you work hard you can get ahead." This statement represents knowledge based on *qualification*. Qualifications or "exceptions to the rule" are often made in everyday life, and they are made in science, too. The difference is that in everyday life qualifications are easily accepted as valid, while in scientific inquiry they are treated as statements that must be carefully examined in the light of evidence.

7. "The San Francisco Giants won 50% of their baseball games over the last 3 months, but they won 80% of the games they played on Thursdays. *Because it happened so often before,* I bet they'll win next Thursday." This statement represents knowledge based on *illogical reasoning*. In everyday life, we may expect the recurrence of events without reasonable cause, ignoring the fact that rare sequences of events often occur just by chance. For example, it is possible for you to flip a coin 10 times and have it come up heads each time. On average, this will happen once every 1,024 times you flip a coin 10 times. In the absence of any apparent reason for this happening, it is merely coincidental. It is illogical to believe otherwise. Scientists refrain from such illogical reasoning. They also use statistical techniques to distinguish between events that are probably due to chance and those that are not.

8. *"I just can't be wrong."* This statement represents knowledge based on *ego-defense*. Even scientists may be passionately committed to the conclusions they reach in their research because they have invested much time, energy, and money in them. It is

other scientists—more accurately, the whole institution of science, with its commitment to publishing research results and critically scrutinizing findings—that puts strict limits on ego-defense in scientific understanding.

9. *"The matter is settled once and for all."* This statement represents knowledge based on the *premature closure of inquiry.* This involves deciding all the relevant evidence has been gathered on a particular subject. Science, however, is committed to the idea that all theories are only temporarily true. Matters are never settled.

10. *"There must be supernatural forces at work here."* This statement represents knowledge based on *mystification.* When we can find no rational explanation for a phenomenon, we may attribute the phenomenon to forces that cannot be observed or fully understood. Although such forces may exist, scientists remain skeptical. They are committed to discovering observable causes of observable effects.

THE RESEARCH CYCLE

Sociological research seeks to overcome the kind of unscientific thinking described above. It is a cyclical process that involves six steps (see Figure 2.2).

First, the sociologist must *formulate a research question.* A research question must be stated so it can be answered by systematically collecting and analyzing sociological data. Sociological research cannot determine whether God exists or what is the best political system. Answers to such questions require faith more than evidence. Sociological research can determine why some people are more religious than others and which political systems create more opportunities for higher education. Answers to such questions require evidence more than faith.

Second, the *existing research literature must be reviewed.* Researchers must elaborate their research questions in the clear light of what other sociologists have already debated and discovered. Why? Because reading the relevant sociological literature stimulates researchers' sociological imaginations, allows them to refine their initial questions, and prevents duplication of effort.

Selecting a research method is the third step in the research cycle. As we will see in detail below, each data collection method has strengths and weaknesses. Each method is

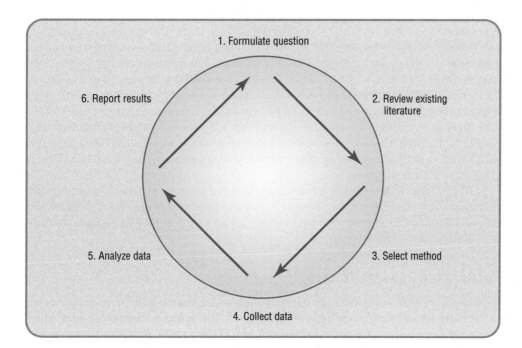

◆ **FIGURE 2.2** ◆
The Research Cycle

Carol Wainio. *We Can Be Certain.* 1982. Research involves taking the plunge from speculation to testing ideas against evidence.

therefore best suited to studying a different kind of problem. When choosing a method, one must keep these strengths and weaknesses in mind. (In the ideal but, unfortunately, infrequent case, several methods are used simultaneously to study the same problem. This can overcome the drawbacks of any single method and increase confidence in one's findings.)

The fourth stage of the research cycle involves *collecting the data* by observing subjects, interviewing them, reading documents produced by or about them, and so forth. Many researchers think this is the most exciting stage of the research cycle because it brings them face-to-face with the puzzling sociological reality that so fascinates them.

Other researchers find the fifth step of the research cycle, *analyzing the data,* the most challenging. During data analysis, you can learn things that nobody ever knew before. This is when data confirm some of your expectations and confound others, requiring you to think creatively about familiar issues, reconsider the relevant theoretical and research literature, and abandon pet ideas.

Of course, research isn't much use to the sociological community, the subjects of the research, or the wider society if researchers don't *publicize the results* in a report, a scientific journal, or a book. This is the research cycle's sixth step. Publication serves another important function too. It allows other sociologists to scrutinize and criticize the research. On this basis, new and more sophisticated research questions can be formulated for the next round of research.

Throughout the research cycle, researchers must be mindful of the need to *respect their subjects' rights.* This means, in the first instance, that researchers must do their subjects no harm. This is the right to safety. Second, research subjects must have the right to decide whether their attitudes and behaviors may be revealed to the public and, if so, in what way. This is the right to privacy. Third, researchers cannot use data in a way that allows them to be traced to a particular subject. This is the subject's right to confidentiality. Fourth, subjects must be told how the information they supply will be used. They must also be allowed to judge the degree of personal risk involved in answering questions. This is the right to informed consent.

Bearing in mind this thumbnail sketch of the research cycle, let us now examine sociology's major research methods. These include the examination of existing documents and official statistics, experiments, surveys, and participant observation. We begin by describing participant-observation research.

Participant Observation

In participant-observation research, the immediate social environment of the people being investigated becomes the sociological "laboratory." The participant-observer goes wherever people meet, from the Italian-American slum (Whyte, 1981 [1943]) to the intensive

Lillian Rubin (University of California at Berkeley) is one of the most talented participant-observation researchers in the United States. One of her most widely acclaimed works is *Families on the Fault Line* (1994), which gives voice to the voiceless by investigating how race, ethnicity, and gender divide the working class. For example, she sensitively captures the ambivalence working-class men feel about their wives working in the paid labor force. She also weaves her interviews into a revealing story about white ethnic pride as a reaction to the economic upheavals of the 1970s and 1980s and the demands of minorities.

care unit of a major hospital (Chambliss, 1996), from the white teenage heavy-metal gang in small-town New Jersey (Gaines, 1990) to the audience of a daytime TV talk show (Grindstaff, 1997).

Sociologists engage in **participant observation** when they attempt to observe a social milieu objectively *and* take part in the activities of the people they are studying (Lofland and Lofland, 1995 [1971]). By participating in the lives of their subjects, researchers are able to see the world from their subjects' point of view. This allows them to achieve a deep and sympathetic understanding of people's beliefs, values, and motives. In addition, participant observation requires that sociologists step back and observe their subjects' milieu from an outsider's point of view. This helps them see their subjects more objectively. In participant-observation research, then, there is a tension between the goals of subjectivity and objectivity. As you will see, however, this is a healthy tension that enhances our understanding of many social settings.

A well-known example of participation observation research is Carl B. Klockars's analysis of the professional "fence," a person who buys and sells stolen goods (Klockars, 1974). Among other things, Klockars wanted to understand how criminals can knowingly hurt people and live with the guilt. Are criminals capable of this because they are "sick" or unfeeling? Klockars came to a different conclusion by examining the case of Vincent Swaggi (a pseudonym).

Swaggi buys cheap stolen goods from thieves and then sells them in his store for a handsome profit. His buying is private and patently criminal. His selling is public and, to his customers, it appears to be legal. Consequently, Swaggi faces the moral dilemma shared by all criminals to varying degrees. He has to reconcile the very different moral codes of the two worlds he straddles, canceling out any feelings of guilt he derives from conventional morality.

"The way I look at it, I'm a businessman," says Swaggi. "Sure I buy hot stuff, but I never stole nothing in my life. Some driver brings me a couple of cartons, though, I ain't gonna turn him away. If I don't buy it, somebody else will. So what's the difference? I might as well make money with him instead of somebody else." Swaggi thus denies responsibility for his actions. He also claims his actions never hurt anyone:

> Did you see the paper yesterday? You figure it out. Last year I musta had $25,000 wortha merchandise from Sears. In this city last year they could'a called it Sears, Roebuck, and Swaggi. Just yesterday I read where Sears just had the biggest year in history, made more money than ever before. Now if I had that much of Sears's stuff can you imagine how much they musta lost all told? Millions, must be millions. And they still had their biggest year ever. . . You think they end up losing when they get clipped? Don't you believe it. They're no different from anybody else. If they don't get it back by takin' it off their taxes, they get it back from insurance. Who knows, maybe they do both.

And if he has done a few bad things in his life, then, says Swaggi, so has everyone else. Besides, he's also done a lot of good. In fact, he believes his virtuous acts more than compensate for the skeletons in his closet. Consider, for example, how he managed to protect one of his suppliers and get him a promotion at the same time:

> I had this guy bringin' me radios. Nice little clock radios, sold for $34.95. He worked in the warehouse. Two a day he'd bring me, an' I'd give him fifteen for the both of 'em. Well, after a while he told me his boss was gettin' suspicious 'cause inventory showed a big shortage. . . . So I ask him if anybody else is takin' much stuff. He says a couple of guys do. I tell him to lay off for a while an' the next time he sees one of the other guys take somethin' to tip off the boss. They'll fire the guy an' clear up the shortage. Well he did an' you know what happened? They made my man assistant shipper. Now once a month I get a carton delivered right to my store with my name on it. Clock radios, percolators, waffle irons, anything I want fifty off wholesale (quoted in Klockars, 1974: 135–61).

Without Klockars's research, we might think that all criminals are able to live with their guilt only because they are pathological or lack empathy for their fellow human beings. But thanks partly to Klockars's research, we know better. We understand that criminals are able to avoid feeling guilty about their actions and get on with their work because they weave a blanket of rationalizations for their criminal activities. These justifications make their illegal activities appear morally acceptable and normal, at least to the criminals themselves. We understand this aspect of criminal activity better because Klockars spent 15 months befriending Swaggi and closely observing him on the job. He interviewed Swaggi for a total of about 400 hours, taking detailed "field notes" most of the time. He then wrote up his descriptions, quotations, and insights in a book that is now considered a minor classic in the sociology of crime and deviance (Klockars, 1974).

Why are observation *and* participation necessary in participant-observation research? Because sociological insight is sharpest when researchers stand both inside and outside the lives of their subjects. Said differently, we see more clearly when we move back and forth between inside and outside.

By immersing themselves in their subjects' world, by learning their language and their culture in depth, insiders are able to experience the world just as their subjects do. Subjectivity can, however, go too far. After all, "natives" are rarely able to see their cultures with much objectivity, and inmates of prisons and mental institutions do not have access to official information about themselves. It is only by regularly standing apart and observing their subjects from the point of view of outsiders that researchers can raise analytical issues and see things their subjects are blind to, or are forbidden from seeing.

Objectivity can also go too far. Observers who try to attain complete objectivity will often not be able to make correct inferences about their subjects' behavior. That is because they cannot fully understand the way their subjects experience the world and cannot ask them about their experiences. Instead, observers who seek complete objectivity must rely only on their own experiences to impute meaning to a social setting. Yet the meaning a situation holds for observers may differ from the meaning it holds for their subjects.

In short, opting for pure observation or pure participation compromises the researcher's ability to see the world sociologically. Instead, participant observation requires the researcher to keep walking a tightrope between the two extremes of objectivity and subjectivity.

It is often difficult for participant-observers to gain access to the groups they wish to study. They must first win the confidence of their subjects, who must feel at ease in the presence of the researcher before they behave naturally. *Reactivity* occurs when the researcher's presence influences the subjects' behavior (Webb, Campbell, Schwartz, and Sechrest, 1966). Reaching a state of nonreactivity requires patience and delicacy on the researcher's part. It took Klockars several months to meet and interview about 60 imprisoned thieves before one of them felt comfortable enough to recommend that he contact Swaggi. Klockars had to demonstrate genuine interest in the thieves' activities and convince them he was no threat to them before they opened up to him. Often, sociologists can minimize reactivity by gaining access to a group in stages. At first, researchers may simply attend a group meeting. After a time, they may start to attend more regularly. Then, when their faces are more familiar, they may strike up a conversation with some of the friendlier group members. Only later will they begin to explain their true motivation for attending.

Klockars and Swaggi are both white men. Their similarity made communication between them easier. In contrast, race, gender, class, and age differences sometimes make it difficult, and occasionally even impossible, for some researchers to study some groups. There are many participant-observation studies in which social differences between sociologists and their subjects were overcome and resulted in excellent research (e.g., Liebow, 1967; Stack, 1974). On the other hand, one can scarcely imagine a sociologist nearing retirement conducting participant-observation research on youth gangs or an African-American sociologist using this research method to study the Ku Klux Klan.

Most participant-observation studies begin as **exploratory research.** This means researchers have at first only a vague sense of what they are looking for, and perhaps no sense at all of what they will discover in the course of their study. They are equipped only with some hunches based on their own experience and their reading of the relevant research literature. They try, however, to treat these hunches as hypotheses. **Hypotheses** are unverified but testable statements about the phenomena that interest researchers. As they immerse themselves in the life of their subjects, their observations constitute sociological data that allow them to reject, accept, or modify their initial hypotheses. Indeed, researchers often purposely seek out observations that enable them to determine the validity and scope of their hypotheses. ("From previous research I know elderly people are generally more religious than young people, and that seems to be true in this community too. But does religiosity vary among people of the same age who are rich, middle-class, working class, and poor? If so, why? If not, why not?") Purposively choosing observations results in the creation of a grounded theory. A *grounded theory* is an explanation of a phenomenon based not on mere speculation but on the controlled scrutiny of one's subjects (Glaser and Straus, 1967).

Methodological Issues

The great advantage of participant observation is that it lets researchers get "inside the minds" of their subjects and discover their view of the world in its full complexity. It is an especially valuable technique when little is known about the group or phenomenon under investigation and the sociologist is interested in constructing a theory about it. But participant observation has drawbacks too. To understand them we must say a few words about measurement in sociology.

When researchers think about the social world, they use mental constructs or concepts such as "race," "class," "gender," and so forth. Concepts that can have more than one value are called **variables.** Height and wealth are variables. Perhaps less obviously, affection and perceived beauty are, too. Just as one can be 5'7" or 6'2", rich or poor, one can be passionately in love with, or indifferent to, the girl next door on the grounds that she is

Web Interactive Exercises
Are IQ and SAT Tests Valid?

✦ **FIGURE 2.3** ✦

Measurement as Target Practice: Validity, Reliability, and Generlizability Compared

Validity, reliability, and generalizability may be explained by drawing an analogy between measuring a variable and firing at a bull's-eye. In case 1, shots (measures) are far apart (not reliable) and far from the bull's-eye (not valid). In case 2, shots are close to one another (reliable) but far from the bull's-eye (not valid). In case 3, shots are close to the bull's-eye (valid) but far from one another (not reliable). In case 4, shots are close to the bull's-eye (valid) and close to one another (reliable). In case 5, we use a second target. Our shots are again close to one another (reliable) and close to the bull's-eye (valid). Because our measures were valid and reliable for both the first and second targets in cases 4 and 5, we conclude our results are generalizable.

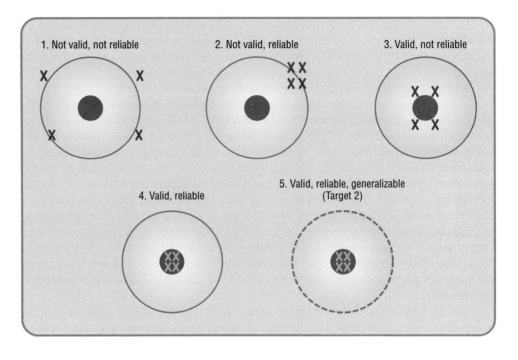

beautiful or plain. In each case, we know we are dealing with a variable because height, wealth, affection, and perceived beauty can take different values.

Once researchers identify the variables that interest them, they must decide which real-world observations correspond to each variable. Should "class," for example, be measured by determining people's annual income? Or should it be measured by determining their accumulated wealth or years of formal education or some combination of these or other indicators of rank? Deciding which observations to link to which variables is known as **operationalization.**

Sociological variables can sometimes be measured by casual observation. It's usually pretty easy to tell if someone is a man or a woman, and participant-observers can learn a great deal more about their subjects through extended discussion and careful observation. When researchers find out how much money their subjects earn, how satisfied they are with their marriages, whether they have ever been the victims of a criminal act, and so forth, they are measuring the values of the sociological variables embedded in their hypotheses.

Typically, researchers must establish criteria for assigning values to variables. At exactly what level of annual income can someone be considered "upper class?" What are the precise characteristics of settlements that allow them to be characterized as "urban?" What features of a person permit us to say she is a "leader?" Answers to such questions all involve measurement decisions.

And there's the rub. In any given research project, participant-observers usually work alone and usually investigate only one group or one type of group. Thus, when we read their research results we must be convinced of three things if we are to accept their findings: (a) We must be confident the findings extend beyond the single case examined; (b) we must be confident their interpretations are accurate; and (c) we must be confident another researcher would interpret things in the same way. Let us examine each of these points in turn (see Figure 2.3):

1. *Would another researcher interpret or measure things in the same way?* This is the problem of **reliability.** If a measurement procedure repeatedly yields consistent results, we consider it reliable. However, in the case of participant observation, there is usually only one person doing the measuring in only one setting. Therefore, there is really no way of knowing whether repeating the procedures would yield consistent results.

2. *Are the researcher's interpretations accurate?* This is the problem of **validity.** If a measurement procedure measures exactly what it is supposed to measure, then it is valid. Whether a measure is reliable has no bearing on its validity. Measuring a person's shoe with a ruler may give us a reliable indicator of that person's shoe size. That is because the ruler repeatedly yields the same results. However, regardless of consistency, shoe size as measured by a ruler is a totally invalid measure of a person's annual income. Similarly, you may think you are measuring annual income by asking people how much they earn. Another interviewer at another time may get exactly the same result when posing the same question. But, despite this reliability, respondents may understate their true income. (A respondent is a person who answers the researcher's questions.) Our measure of annual income may therefore lack validity. Perfectly consistent measures may, in other words, have little truth-value.

Participant-observers have every right to feel they are on solid ground when it comes to the question of validity. If anyone can tell whether respondents are understating their true income, surely it is someone who has spent months or even years getting to know everything about their lifestyle. Still, doubts may creep in if the criteria used by the participant observers to assess the validity of their measures are all *internal* to the settings they are investigating. Our confidence in the validity of researchers' measures increases if we are able to use *external* validation criteria.

Two methods of external validation are commonly used. First, confidence in the validity of researchers' measures increases if their findings are consistent with what we already know from our own experience and the research literature. Second, confidence in the validity of measures increases if they enable us to make useful predictions. For example, researcher A may develop a set of questions that distinguishes people with liberal political opinions from those with conservative political opinions. In another research setting, researcher B may find those same questions distinguish between Democrats and Republicans. Since we know Democrats tend to be liberal and Republicans tend to be conservative, we may conclude that researcher A's measure of liberalism/conservatism has good predictive power. This increases our confidence in the validity of the researchers' questions as measures of political opinion.

3. *Do the research findings apply beyond the specific case examined?* This is the problem of **generalizability,** and it is one of the most serious problems faced by participant-observation studies. For example, Klockars studied just one professional fence in depth. Can we safely conclude his findings are relevant to all professional fences? Do we dare apply his insights to all criminals? Are we foolhardy if we generalize his conclusions to nearly all of us on the grounds that most of us commit deviant acts at one time or another and must deal with feelings of guilt? None of this is clear from Klockars's research, nor are questions of generalizability clearly answered by many participant-observation studies since they are usually studies of single cases.

4. Related to the issue of generalizability is that of **causality,** the analysis of causes and their effects.[2] Information on how widely or narrowly a research finding applies can help us establish the causes of a social phenomenon. For instance, we might want to know how gender, race, class, parental supervision, police surveillance, and other factors shape the type and rate of juvenile delinquency. If so, we require information on types and rates of criminal activity among teenagers with a variety of social characteristics in a variety of social settings. A participant-observation study of crime is unlikely to provide that sort of information. It is more likely to clarify the process by which a specific group of people in a single setting learns to become criminals. Indeed, researchers who conduct participant-observation studies tend not to think in somewhat mechanical, cause-and-effect terms at all. They prefer instead

---

[2]On philosophical grounds, some researchers avoid the terms "cause" and "effect." We use these terms because they are widely accepted and easy to understand. Moreover, we do not want to introduce philosophical complications in an elementary treatment of the subject.

to view their subjects as engaged in a fluid process of social interaction. As a result, participant observation is not the preferred method for discovering the general causes of social phenomena.

In sum, participant observation has both strengths and weaknesses. It is especially useful in exploratory research, constructing grounded theory, creating internally valid measures, and developing a sympathetic understanding of the way people see the world. It is often deficient when it comes to establishing reliability, generalizability, and causality. As you will soon learn, these are precisely the strengths of surveys and (with the exception of generalizability) experiments. Only a small percentage of sociologists conduct experiments. Nonetheless, experiments are important because they set certain standards that other more popular methods try to match. We can show this by discussing experiments concerning the effects of television on real-world violence.

Experiments

In the mid-1960s, about 15 years after commercial TV was introduced in the United States, rates of violent crime began to increase dramatically. Some people were not surprised. The first generation of American children exposed to high levels of TV violence virtually from birth had reached their mid-teens. TV violence, some commentators said, legitimized violence in the real world, making it seem increasingly normal and acceptable. As a result, they concluded, American teenagers in the 1960s and subsequent decades were much more likely than pre-1960s teens to commit violent acts.

Social scientists soon started investigating the connection between TV and real-world violence using experimental methods. An **experiment** is a carefully controlled artificial situation that allows researchers to isolate hypothesized causes and measure their effects precisely (Campbell and Stanley, 1963). It uses a special procedure called **randomization** to create two similar groups. Randomization involves assigning individuals to the groups by chance processes. It then introduces the hypothesized cause to only one of the groups. By comparing the state of the two groups before and after only one of the groups has been exposed to the hypothesized cause, an experiment can determine whether the presumed cause has the predicted effect.

Here is how an experiment on the effects of TV violence on aggressive behavior might work:

When children fight at home, an adult is often present to intervene. By repeatedly separating the children and not *sanctioning* their aggressive behavior, the adult can teach them that fighting is unacceptable. In contrast, experiments on the effect of television on aggressive behavior lack validity, in part because they may sanction violence and may even encourage it.

1. *Selection of subjects.* Researchers advertise in local newspapers for parents willing to allow their children to act as research subjects. Fifty children are selected for the experiment.

2. *Random assignment of subjects to experimental and control groups.* At random, each child draws a number from 1 to 50 from a box. The number is recorded and then returned to the box. The researchers assign children who draw odd numbers to the **experimental group.** This is the group that will be exposed to a violent TV program during the experiment. They assign children who draw even numbers to the **control group.** This is the group that will not be exposed to a violent TV program during the experiment.

    Note that randomization and repetition make the experimental and control groups similar. That is, by assigning subjects to the two groups using a chance process, and repeating the experiment many times, researchers ensure that the experimental and control groups are likely to have the same proportion of boys and girls, members of different races, children highly motivated to participate in the study, and so forth. Random assignment eliminates bias by allowing a chance process and only a chance process to decide which group each child is assigned to.

3. *Measurement of dependent variable in experimental and control groups.* Researchers put small groups of children in a room and give them toys to play with. They observe the children through a one-way mirror, rating each child in terms of the aggressiveness of his or her play. This is the child's pretest score on the **dependent variable,** aggressive behavior. The dependent variable is the effect in any cause-and-effect relationship.

4. *Introduction of independent variable to experimental group.* The researchers show children in the experimental group an hour-long TV show in which many violent and aggressive acts take place. They do not show the film to children in the control group. In this experiment, the violent TV show is the **independent variable.** The independent variable is the presumed cause in any cause-and-effect relationship.

5. *Remeasurement of dependent variable in experimental and control groups.* Immediately after the children see the TV show, the researchers again observe the children in both groups at play. Each child's play is given a second aggressiveness rating— the posttest score.

6. *Assessment of experimental effect.* Posttest minus pretest scores are calculated for both the experimental and control groups. If the posttest minus pretest score for the experimental group is significantly greater than the posttest minus pretest score for the control group, the researchers conclude the independent variable (watching violent TV) has a significant effect on the dependent variable (aggressive behavior). This conclusion is warranted because the introduction of the independent variable is the only difference between the experimental and control groups.

As this example shows, an experiment is a precision instrument for isolating the single cause of theoretical interest and measuring its effect in an exact and repeatable way. But high reliability and the ability to establish causality come at a steep price. Cynics sometimes say experimental sociology allows researchers to know more and more about less and less. Many sociologists argue that experiments are highly artificial situations. They believe that removing people from their natural social settings usually lowers the validity of one's findings.

These misgivings are evident in experimental studies of the effects of TV violence (Felson, 1996). Experiments show that watching violent TV usually increases violent behavior in the short term. However, in the real world, violent behavior usually means attempting to physically harm another person. Shouting, hitting a doll, or kicking a toy is just not the same thing. In fact, such acts may enable children to relieve frustrations in a fantasy world, thus lowering their chance of acting violently in the real world. Moreover, in a laboratory situation, aggressive behavior may be encouraged because it is legitimized.

Simply showing a violent TV program may suggest to subjects how the experimenter expects them to behave during the experiment. Subjects who are influenced by the prestige of the researcher and the scientific nature of the experiment compound this problem. They will try to do what is expected of them in order not to appear poorly adjusted. (Changing people's behavior by making them aware they are being studied is known as the *Hawthorne effect.* It is so named because researchers at the Western Electric Company's Hawthorne factory in the 1930s claimed to find that workers' productivity increased no matter how they changed their work environment. Productivity increased, they said, just because the researchers were paying attention to the workers.[3]) Finally, aggressive behavior is not punished or controlled in the laboratory setting as it is in the real world. If a boy watching *Power Rangers* stands up and delivers a karate kick to his younger brother, a parent is likely to take action to prevent a recurrence. This teaches the boy not to engage in such aggressive behavior. This doesn't happen in the lab, where the lack of disciplinary control may facilitate unrealistically high levels of aggression.

In an effort to overcome the validity problem and still retain many of the benefits of experimental design, some sociologists have conducted experiments in natural settings. In such experiments, researchers forego strict randomization of subjects. They compare groups that are already quite similar. They either introduce the independent variable themselves (this is called a *field experiment*) or observe what happens when the independent variable is introduced to one of the groups in the normal course of social life (this is called a *natural experiment*).

Some field experiments on media effects compare boys in institutionalized settings. The researchers expose half the boys to violent TV programming. Measures of aggressiveness taken before and after the introduction of violent programming allow researchers to calculate its effect on behavior. One reanalysis of 28 such studies yielded mixed results. While 16 of the field experiments (57%) suggested that subjects engage in more aggression following exposure to violent films, 12 (43%) did not (Wood, Wong, and Chachere, 1991).

Natural experiments have compared rates of aggressive behavior in towns with and without TV service, but their results are inconclusive too. They are also muddied by the fact that there are substantial differences between the towns apart from the presence or absence of TV service. It is therefore unclear whether differences in child aggressiveness are due to media effects.

Because of the validity problems noted above, it has not been convincingly demonstrated that TV violence generally encourages violent behavior. The sociological consensus is that TV violence probably does have an effect on a small percentage of viewers, but the effect is not large (Felson, 1996: 123). The extent of the effect is unclear partly because the experimental method makes it difficult to generalize from the specific groups studied to the entire population. The subjects of an experiment on media effects may be white, middle-class people from a college town in the Midwest who read newspaper ads and are in a position to take a day off to participate in the experiment. This is hardly a representative group of Americans. But experimentalists are rarely concerned that their subjects are representative of an entire population. As we will now see, one of the strong points of surveys is that they allow us to make safer generalizations.

Surveys

Sampling

Surveys are part of the fabric of everyday life in America. You see surveys in action when a major television network conducts a poll to discover the percentage of Americans who approve of the President's performance, when someone phones to ask about your taste in breakfast cereal, and when advice columnist Ann Landers asks her readers, "If you had to

---

[3]Subsequent analysis questioned the existence of a productivity effect in the Hawthorne study (Franke and Kaul, 1978). However, the general principle derived from the Hawthorne study—that social science researchers can influence their subjects—is now widely accepted (Webb, Campbell, Schwartz, and Sechrest, 1966).

Researchers collect information using surveys by asking people in a representative sample a set of identical questions. People interviewed on a downtown street corner do *not* constitute a representative sample of American adults. That is because the sample does not include people who live outside the urban core, underestimates the number of elderly and disabled people, does not take into account regional diversity, and so forth.

do it over again, would you have children?" In every survey, people are asked questions about their knowledge, attitudes, or behavior, either in a face-to-face or telephone interview or in a paper-and-pencil format.

Remarkably, Ann Landers found that fully 70% of parents would not have children if they could make the choice again. She ran a shocking headline saying so. Should we have confidence in her finding? Hardly. As the letters from her readers indicated, many of the people who answered her question were angry with their children. All 10,000 respondents felt at least strongly enough about the issue to take the trouble to mail in their replies at their own expense. Like all survey researchers, Ann Landers aimed to study part of a group—a **sample**—in order to learn about the whole group—the **population** (in this case, all American parents). The trouble is, she got replies from a *voluntary response sample,* a group of people who chose *themselves* in response to a general appeal. People who choose themselves are unlikely to be representative of the population of interest. In contrast, a *representative sample* is a group of people chosen so their characteristics closely match those of the population of interest. The difference in the quality of knowledge we can derive from the two types of samples cannot be overstated. Thus, a few months after Ann Landers conducted her poll, a scientific survey based on a representative sample found that 91% of American parents *would* have children again (Moore, 1995: 178).

How can survey researchers draw a representative sample? You might think that setting yourself up in a public place like a shopping mall and asking willing passers-by to answer some questions would work. However, this sort of *convenience sample,* which chooses the people who are easiest to reach, is also highly unlikely to be representative. Most people who go to malls earn above-average income. Moreover, a larger proportion of homemakers, retired people, and teenagers visit malls than can be found in the American population as a whole. Convenience samples are almost always unrepresentative.

To draw a representative sample, respondents cannot select themselves, as in the Ann Landers case. Nor can the researcher choose respondents, as in the mall example. Instead, respondents must be chosen at random, and an individual's chance of being chosen must be known and greater than zero. A sample with these characteristics is known as a **probability sample** (see Box 2.1)

To draw a probability sample, you first need a *sampling frame.* This is a list of all the people in the population of interest. You also need a randomizing method. This is a way of ensuring every person in the sampling frame has a known and nonzero chance of being selected.

Up-to-date membership lists of organizations are useful sampling frames if you want to survey members of organizations. But if you want to investigate, say, the religious beliefs of Americans, then the membership lists of places of worship are inadequate. That is because many Americans do not belong to such institutions. In such cases, you might turn to another frequently used sampling frame, the telephone directory. The telephone directory is now available for the entire country on CD-ROM. However, even the telephone directory lacks the names and addresses of some poor and homeless people (who don't have phones) and some rich people (who have unlisted phone numbers). Computer programs

BOX 2.1
SOCIOLOGY AT THE MOVIES

The Blair Witch Project is frightening partly because it lacks a sociological perspective.

THE BLAIR WITCH PROJECT (1999)

In sociology, some big surveys cost a few million dollars. They employ hundreds of people as interviewers, data analysts, project managers, and so forth. They use computers and sophisticated software to analyze data. The typical Hollywood movie costs 10 times more than even big sociology research projects. It employs many more people and uses much more sophisticated technology for special effects.

You might think there is little room for small-scale work in either sociological research or movie making, but that isn't so. The *Blair Witch Project,* directed by Daniel Myrick and Eduardo Sanchez, cost only $35,000 to make. It was a surprise hit in the summer of 1999, earning $50 million in its first week of national release. Besides raising hopes for all low-budget projects everywhere, the movie can also teach us something about research methods.

The *Blair Witch Project* begins as a research project. Three people trek to Burkittsville, Maryland, to find out about the local witch legend. Like good researchers, they interview local people. Some respondents dismiss the legend. Others provide tantalizing hints that the witch really exists. The results of this research are inconclusive, so the three investigators decide to hike into the woods, hoping to discover for themselves whether the witch exists.

Unfortunately, they overestimate their skills as hikers and campers. They lose their map, soon get lost, and proceed to get on each other's nerves. Eventually, one of them disappears after a big argument with the other two. (We don't know why he disappears—either because the Blair witch got him or because he was so angry with his coinvestigators). The two remaining people then stumble upon an old, vacant house. Exploring the house, they hear odd noises. In

the end, the camera falls to the floor. The video footage ends. Whether the two characters were attacked by the Blair witch or their angry coinvestigator is unclear.

The power of the movie—the reason that it was so frightening to so many people—derives from the fact that it *lacks* a sociological viewpoint. After the three investigators conduct their interviews, anything resembling sociological research stops. Subsequently, our only source of knowledge is the camera held by the investigators. We are rarely given a panoramic view or a sense of context to improve our understanding of what is happening to the three people. The narrow perspectives of the three people certainly provide a sense of being there. Members of the audience feel they are seeing things just like the three investigators do, facing the unknown terror of the Blair witch, who is nowhere to be seen. But if the members of the audience were able to draw on other sources of information about the Blair witch or the angry coinvestigator who left in a huff, if they were able to see things from a broader perspective than is afforded by the individual viewpoints of the three investigators, if they were able to make sense of the larger context of events, some of the

terror might subside. For example, if the two remaining people had better evidence that the Blair witch was real, they might not have entered the house. If they knew their traveling companion was a deeply disturbed young man with violent tendencies, they might have taken steps to protect themselves rather than leaving themselves open to assault. As the old saying goes, it's better to face the devil you know than the devil you don't know.

Sociologically speaking, *The Blair Witch Project* is unsatisfying because it doesn't escape the narrow, individual points of view provided by the video camera. In contrast, sociological research tries to get beyond individual points of view. For instance, by taking random samples rather than convenience samples in opinion surveys, researchers ensure that their data accurately reflect opinion in the population from which the samples are drawn. Similarly, by comparing experimental and control groups, researchers eliminate the possibility that variables other than the independent variable of interest are responsible for observed differences between the two groups. By using these and other research methods, sociologists avoid getting lost in the woods.

Homeless people may be interested in public policy, but public policy will ignore them if they are not counted in the census.

are available that dial residential phone numbers at random, including unlisted numbers. However, that still leaves some American households that will be excluded from any survey relying on the telephone directory as a sampling frame.

As the example of the telephone directory shows, few sampling frames are perfect. Even one of the largest and most expensive surveys in the world, the United States census, missed an estimated 2.1% of the population in 1990. Although minority groups composed only about a quarter of the United States population, half of the undercounted population was composed of members of minority groups because most of the undercount was in poor sections of big cities. For African Americans, the undercount was 4.8%. The figure for Native Americans was 5.0%. Some 5.2% of Hispanic Americans were not counted in the 1990 census (Anderson and Feinberg, 2000: 90; see Box 2.2). Nevertheless, researchers maximize the accuracy of their generalizations by using the least biased sampling frames available and adjusting their analyses and conclusions to take account of known sampling bias.

Once a sampling frame has been chosen or created, individuals must be selected by a chance process. One way to do this is by picking, say, the 10th person on your list and then every 20th (or 30th, or 100th) person after that, depending on how many people you need in your sample. A second method is to assign the number 1 to the first person in the sampling frame, the number 2 to the next person, and so on. Then, you create a separate list of random numbers by using a computer or consulting a table of random numbers, which you can find at the back of almost any elementary statistics book. Your list of random numbers should have as many entries as the number of people you want in your sample. The individuals whose assigned numbers correspond to the list of random numbers are the people in your sample.

How many respondents do you need in a sample? That depends on how much inaccuracy you are willing to tolerate. Large samples give more precise results than small samples. For most sociological purposes, however, a random sample of 1,500 people will give acceptably accurate results, even if the population of interest is the entire adult population of the United States. More precisely, if you draw 20 random samples of 1,500 individuals each, 19 of them will be accurate within 2.5%. Imagine, for example, that 50% of a random sample of 1,500 respondents say they believe the President is doing a good job. We can be reasonably confident that only 1 in 20 random samples of that size will yield results below 47.5% or above 52.5%. This leads us to conclude that the actual percentage of people in the *population* who think the President is doing a good job is probably between 47.5% and 52.5%. When we read that a finding is **statistically significant,** it usually means we can expect similar findings in 19 out of 20 samples of the same size. Said differently, researchers in the social sciences are conventionally prepared to tolerate a 5% chance that the characteristics of a population are actually different from the characteristics of their sample (1/20 = 5%).

BOX 2.2
IT'S YOUR CHOICE

THE POLITICS OF
THE UNITED STATES CENSUS

When John Lie was teaching at the University of Oregon in the early 1990s, several Spanish-speaking sociologists he knew were busy counting the number of Hispanic (or Latino) migrant agricultural workers in the state's rural areas. Why is it important for the government to pay sociologists to count Spanish-speaking migrant workers or, for that matter, other residents of the United States?

Counting the number of Americans may seem a simple matter. Most people don't have trouble counting the number of people in a classroom, so what's the big deal about conducting a national census? Well, imagine counting the number of people at a rock concert or a major sports event. Not only would it take a long time to count them one by one, but the crowd is constantly in motion, making it still harder to count accurately. If it's difficult to count thousands of people in one place, you can appreciate how hard it is to count nearly 300 million Americans. At a given time, many Americans are on the move. They may be traveling or living abroad. They may be driving or flying within the United States, camping in the Sierras or stuck in an elevator in New York. It is little short of a wonder, then, that the United States Bureau of the Census succeeded in counting about 98.5% of the nation's population in 2000.

Why do we need to know how many Americans there are? The census is important because many important decisions are made based on population figures. For example, the number of congressional seats is decided on the basis of population figures. So is the amount of money each state government receives from the federal government. Decisions about everything from school budgets to highway construction rely on census counts. Thus, if census figures are lower than the actual population in a particular area, this can be a serious liability for the people living there.

Census undercounting is especially problematic in the case of racial and ethnic minorities. For example, some recent Hispanic-American immigrants may not get counted because they have difficulty communicating with census takers who don't speak Spanish. Other Hispanic-Americans may worry that census takers are undercover police agents looking for undocumented migrants. Due to these problems, the Spanish-speaking sociologists working in Oregon in the early 1990s were trying to arrive at a more accurate count of the Hispanic-American population in the state.

Because so many political decisions are based on the census count, the United States Bureau of the Census faces a lot of political controversy. For example, some people think it doesn't matter much if about 1.5% of the United States population (over 4 million people in 2000) aren't counted. They believe it is a waste of money to create a more accurate census. Others say that a more accurate census is important because the groups that are undercounted are, in effect, the victims of discrimination. Thus, a 2001 United States Census Monitoring Board study found that 31 states, including the District of Columbia, would lose $4.1 billion in federal funding due to the census undercount. One of the biggest losers: New York City (Scott, 2001).

Among those who think a more accurate census is necessary, controversy exists about how best to deal with undercounting. Democrats favor using sampling techniques to estimate the number and characteristics of the undercounted. They cite the findings of a blue-ribbon panel appointed by the National Academy of Sciences in the 1990s. The panel determined that sampling could produce more accurate results than the current census. However, on January 25, 1999, the Supreme Court rejected a federal plan to supplement the census with a sample. Republicans oppose sampling, fearing it could be manipulated to give desired results. Instead, some Republicans have suggested that census information be collected in as many as 33 languages, including English Braille. They have also suggested that more money be spent on marketing and outreach to increase the response rate (Anderson, 1999; Anderson and Fienberg, 2000; Choldin 1994; Democratic National Committee, 1998; "Supreme Court...," 1999).

Aside from Hispanic Americans, what other groups of people are particularly susceptible to census undercounting in your opinion? If you were in charge of the United States Bureau of the Census, what steps would you take to ensure a more accurate count of the population? Do you think the characteristics of the census takers in the field, particularly their race and ethnicity, affect the census count? If so, how? Do you think that supplementing the census with a sample survey could increase the accuracy of the census? Does sampling introduce more risk of political manipulation than a straight count?

In sum, probability sampling enables us to conduct surveys that permit us to generalize from a part (the sample) to the whole (the population) within known margins of error. Now let us consider the validity of survey data.

Survey Questions and Validity

There are three main ways to conduct a survey. Sometimes, a *self-administered questionnaire* is used. For example, a form containing questions and permitted responses may be mailed to the respondent and returned to the researcher through the mail system. The main advantage of this method is that it is relatively inexpensive. It also has drawbacks. For one thing, it sometimes results in unacceptably low *response rates*. The response rate is the number of people who answer the questionnaire divided by the number of people asked to do so, expressed as a percentage. Moreover, if you use mail questionnaires, an interviewer is not present to explain problematic questions and response options to the respondent. *Face-to-face interviews* are therefore generally preferred over mail questionnaires. In this type of survey, questions and allowable responses are presented to the respondent by the interviewer during a meeting. However, training interviewers and sending them around to conduct interviews is very expensive. That is why *telephone interviews* have become increasingly popular over the past two or three decades. They can elicit relatively high response rates and are relatively cheap to administer.

Questionnaires can contain two types of questions. A *closed-ended question* provides the respondent with a list of permitted answers. Each answer is given a numerical code so the data can later be easily input into a computer for statistical analysis. *Open-ended questions* allow respondents to answer questions in their own words. They are particularly useful in exploratory research, where the researcher does not have enough knowledge to create a meaningful and complete list of possible answers. Open-ended questions are more time-consuming to analyze than closed-ended questions, although computer programs for analyzing text make the task much easier.

Researchers want the answers elicited by surveys to be valid, to actually measure what they are supposed to. To maximize validity, researchers must guard against several dangers. We have already considered one threat to validity in survey research: *undercounting* some categories of the population due to an imperfect sampling frame. Even if an individual is contacted, however, he or she may refuse to participate in the survey. This is the second threat to validity in survey research: *nonresponse*. If nonrespondents differ from respondents in ways that are relevant to the research topic, then the conclusions one draws from the survey may be in jeopardy. For instance, some alcoholics may not want to participate in a survey on alcohol consumption because they regard the topic as sensitive. If so, a measure of the rate of alcohol consumption taken from the sample would not be an accurate reflection of the rate of alcohol consumption in the population. Actual alcohol consumption in the population would be higher than the rate reflected in the sample.

Survey researchers pay careful attention to nonresponse. They try to discover whether nonrespondents differ systematically from respondents so they can take this into account before drawing conclusions from their sample. They must also take special measures to ensure that the response rate remains acceptably high—generally, 70% or more of people contacted. Proven tactics ensure a high response rate. Researchers can notify potential respondents about the survey in advance. They can remind them to complete and mail in survey forms. They can get universities and other prestigious institutions to sponsor the survey. They can stress the practical and scientific value of the research. And they can give people small rewards, such as a dollar or two, for participating.

If respondents do not answer questions completely accurately, then a third threat to validity is present: *response bias*. The survey may focus on sensitive, unpopular, or illegal behavior. As a result, some respondents may not be willing to answer questions frankly. The interviewer's attitude, gender, or race may suggest that some responses are preferred rather than others. This can elicit biased responses. Some of these problems can be

overcome by carefully selecting and training interviewers and closely supervising their work. Response bias on questions about sensitive, unpopular, or illegal behavior can be minimized by having such questions answered in private. For example, the General Social Survey (GSS) is a nationwide survey that has been conducted by the National Opinion Research Center at the University of Chicago every year since 1972. It is one of the most important ongoing surveys in the United States, and we will refer to GSS data many times in the following chapters. Every year, the GSS measures the opinions, social characteristics, and behaviors of a representative sample of 1,500 or more American adults. Interviews are conducted face-to-face in people's households. Since 1988, the GSS has asked questions about how many sexual partners the respondent has had in the past year, the relation of those sex partners to the respondent, and the gender of the sex partners. But rather than having the interviewer ask these questions, almost certainly causing response bias, the respondent is given a card containing the questions. He or she completes the card in private, places it in an envelope provided by the interviewer, seals the envelope, and is assured that the interviewer will not read the card. Researchers believe this procedure minimizes response bias (Smith, 1992).

Fourth, validity may be compromised due to *wording effects*. That is, the way questions are phrased or ordered can influence and invalidate responses. Experienced survey researchers have turned questionnaire construction into a respected craft. Increasingly, they refine the lessons learned from experience with evidence from field experiments. These experiments divide samples into two or more randomly chosen subsamples. Different question wording or ordering is then administered to the people in each subsample so that wording effects can be measured. Detected problems can then be resolved in future research.

Both experience and field experiments suggest that survey questions must be specific and simple. They should be expressed in plain, everyday language. They should be phrased neutrally, never leading the respondent to a particular answer and never using inflammatory terms. Because people's memories are often faulty, questions are more likely to elicit valid responses if they focus on important, singular, current events rather than less salient, multiple, past events. Breaking these rules lowers the validity of survey findings (Converse and Presser, 1986; Ornstein, 1998).

Finally, one can increase the validity of self-administered surveys by avoiding what might be called the "Palm Beach effect." You may recall the controversy that resulted from the poorly designed ballot in Palm Beach County, Florida, in the 2000 presidential election. Some people who thought they were voting for the Democratic ticket actually voted for the tiny Reform party. The mix-up occurred because the Democratic candidates for President and Vice-President were listed second on the ballot but the hole that had to be punched to select them was listed third (see Figure 2.4). The presidential race was extremely close in 2000, and controversy concerning the Palm Beach ballot (among other issues) resulted in a delay in declaring the winner of the election. In general, to avoid the Palm Beach effect and ensure the validity of responses, one must be careful to design self-administered questionnaires so that response categories are clear and unambiguous.

Causality

A survey is not an ideal instrument for conducting exploratory research. It cannot provide the kind of deep and sympathetic understanding one gains from participant observation. On the other hand, surveys do produce results from which we can confidently generalize. If properly crafted, they provide valid measures of many sociologically important variables. Because they allow the same questions to be asked repeatedly, surveys enable researchers to establish the reliability of measures with relative ease. And finally, as we will now see, survey data are useful for discovering relationships among variables, including cause-and-effect relationships.

Recall how causality is established in experiments. Randomly assigning subjects to experimental and control groups makes the two groups similar. Exposing only the experimental group to an independent variable lets the researcher say the independent variable alone is probably responsible for any measured effect. That conclusion is warranted because the effects of irrelevant variables have been removed by randomization. In

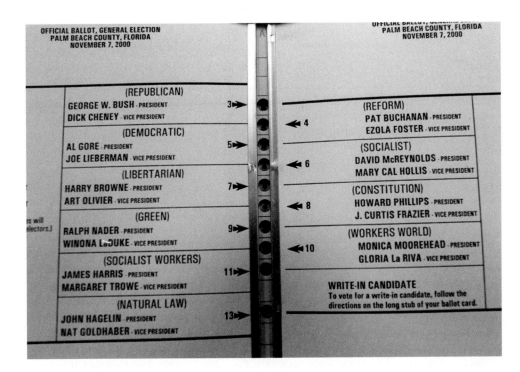

✦ **FIGURE 2.4** ✦
The Butterfly Ballot Used in the 2000 Presidential Election in Palm Beach County, Florida
Because response categories were ambiguous, controversy erupted over voters' real intentions. The lesson for survey research is that one must be careful to design self-administered questionnaires so that response categories are clear.

SOURCE: Official Ballot (2000).

surveys, too, the effects of independent variables can be measured. However, the effects of irrelevant variables are removed not by randomization but by manipulating the survey data.

One of the most useful tools for manipulating survey data is the **contingency table.** A contingency table is a cross-classification of cases by at least two variables that allows you to see how, if at all, the variables are associated. This might sound complex, but it's really not. It's as simple as the corners of your classroom.

Let's say we want to test the hypothesis that the number of hours one spends watching TV per week (the independent variable) increases the frequency of one's violent behavior (the dependent variable). We can test this hypothesis by first asking the students in your class who watch TV more than 10 hours a week to stand by the right wall and the other students to stand by the left wall. We can then ask the students who committed at least one act of physical violence against another person in the past year to move to the front of the room and the others to move to the back. This procedure would, in effect, create a contingency table in the four corners of your classroom. The students would be simultaneously classified (or "cross-classified") by how much TV they watch and their physical aggressiveness (see Figure 2.5).

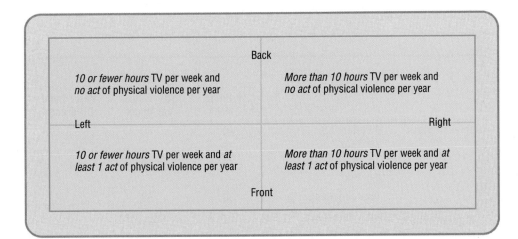

✦ **FIGURE 2.5** ✦
Turning a Classroom Into a Contingency Table

◆ TABLE 2.1 ◆

TV Viewing by Aggressiveness (in percent)

In order to interpret a table, you must pay careful attention to the way it is percentaged, that is, exactly what it is that adds up to 100%. This table says that 52% *of all students who watch TV less than 10 hours a week* commit zero violent acts per year, while 46% *of all students who watch TV 10 or more hours a week* commit zero violent acts per year. You know this because each category of the variable "TV Viewing" equals 100%. (The actual number of students in each category of the variable "TV Viewing" is given in the row labeled "Total Frequency.")

		TV Viewing		
		<10 Hours per Week	10+ Hours per Week	Percentage Difference
Aggressiveness	**0 Violent Acts per Year**	52	46	6
	1+ Violent Act(s) per Year	48	54	6
	Total Frequency (n)	130	70	
	Total Percent	100	100	

There is an **association** between two variables if the value of one variable changes with the value of the other. For example, if the percentage of students who committed an act of physical violence in the past year is higher among frequent TV viewers than among infrequent TV viewers, there is an association between the two variables. The greater the percentage difference between frequent and infrequent TV viewers, the stronger the association. In Table 2.1, for instance, the difference is 6%.

The existence of such an association does not by itself prove that watching TV causes physical aggression. The association may exist for other reasons. For example, it may be that men are more aggressive than women due to the way they are brought up. They may just happen to watch more TV than women do, too.

We can test the hypothesis that watching TV increases physical aggression by creating a second contingency table. Continuing with our classroom example, we can ask all the women to leave the room. In effect, this breaks our original contingency in two, allowing us to examine the association between TV viewing and aggressiveness within a category of a third variable, gender. The third variable, gender, acts as a **control variable.** This means we have manipulated the data to remove the effect of gender from the original association.

Table 2.2 shows TV viewing by aggressiveness for men only. It says that 40% of men who watched TV infrequently and 40% of men who watch TV frequently committed no acts of physical violence in the past year. The percentage difference between these two groups of men is zero. Said differently, once we remove the effect of gender by means of statistical control, there is no longer an association between the independent and dependent variables (TV viewing and physical violence). This obliges us to conclude that the original association in Table 2.1 is **spurious** or accidental. In our example, then, watching TV in and of itself does not seem to cause physical violence. Instead, the association between watching TV and committing acts of physical violence is due to the fact that men happen to watch more TV and are more physically aggressive than women.

In general, to conclude that the association between an independent and a dependent variable is nonspurious or causal, three conditions must hold:

◆ TABLE 2.2 ◆

TV Viewing by Aggressiveness, Men Only (in percent)

		TV Viewing		
		<10 Hours per Week	10+ Hours per Week	Percentage Difference
Aggressiveness	**0 Violent Acts per Year**	40	40	0
	1+ Violent Act(s) per Year	60	60	0
	Total Frequency (n)	50	50	
	Total Percent	100	100	

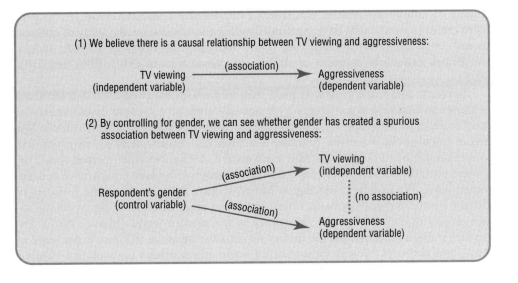

1. There must be an association between the two variables.
2. The presumed cause must occur before the presumed effect.
3. When a control variable is introduced, the original association must not disappear.

If an initial association disappears once a control variable is introduced, the association is spurious. If an initial association stays the same after we introduce a control variable, then we tentatively conclude that the association is causal. We say "tentatively" because there may be other variables that are responsible for the association. If we control for these other variables and find that the association persists, then we will have greater confidence that the association is causal. In an experiment, all extraneous variables are eliminated by randomization. In the analysis of survey data, the best we can hope for is the elimination, by means of statistical control, of those variables that might plausibly explain the original association. This leaves us with a genuine causal association.

Our argument is illustrated in Figure 2.6. The top half of Figure 2.6 shows the original association between watching TV and physical aggression. The arrow indicates the existence of a presumed causal relationship. The bottom half of Figure 2.6 shows what happens when we control for gender: the original association disappears, as suggested by the broken line. The arrows in the bottom half of Figure 2.6 show the actual relationships among the variables.

The analysis of survey data involves more than just searching for nonspurious associations. Many interesting and unexpected things can happen when a two-variable association is elaborated by controlling for a third variable (Hirschi and Selvin, 1972). The original association may remain unchanged. It may strengthen. It may weaken. It may disappear or weaken in only some categories of the control variable. It may even change direction entirely. Data analysis is therefore full of surprises, and accounting for the outcomes of statistical control requires a lot of creative theoretical thinking.

Analysis of Existing Documents and Official Statistics

Apart from participant observation, experiments, and surveys, there is a fourth important sociological research method: the analysis of existing documents and official statistics. What do existing documents and official statistics have in common? They are created by people other than the researcher for purposes other than sociological research.

The three types of existing documents that sociologists have mined most widely and deeply are diaries, newspapers, and published historical works. For example, one of the early classics of American sociology, a study of Polish immigrants, is based on a close

Charles Tilly (Columbia University) is one of the most prolific and respected sociologists in the world. He specializes in the study of large-scale social change and its relation to contentious politics in western Europe, using existing documents such as newspapers, administrative reports, and secondary historical works as sources of evidence. His pathbreaking work has helped to reorient the study of state formation and social movements. Major works include *The Rebellious Century, 1830–1930*, with Louise Tilly and Richard Tilly (1975); *From Mobilization to Revolution* (1978); *Big Structures, Large Processes, Huge Comparisons* (1985); *The Contentious French* (1986); and *Roads from Past to Future* (1997).

reading of immigrants' diaries and letters (Thomas and Znaniecki, 1958 [1918–20]). More recently, sociologists have made outstanding contributions to the study of political protest by systematically classifying 19th- and early 20th-century French, Italian, and British newspaper accounts of strikes and demonstrations (Tilly, Tilly, and Tilly, 1975).

In recent decades, sociologists have tried to discover the conditions that led some countries to dictatorship and others to democracy, some to economic development and other to underdevelopment, some to become thoroughly globalized, others to remain less tied to global social processes. In trying to answer such broad questions, they have had to rely on published histories as their main source of data. No other method would allow the breadth of coverage and depth of analysis required for such comparative and historical work. For example, Barrington Moore spent a decade reading the histories of Britain, France, Russia, Germany, China, India, and other countries to figure out the social origins of dictatorship and democracy in the modern world (Moore, 1967). Immanuel Wallerstein canvassed the history of virtually the entire world to make sense of why some countries became industrialized while others remain undeveloped (Wallerstein, 1974–89). What distinguishes this type of research from purely historical work is the kinds of questions posed by the researchers. Moore and Wallerstein asked the same kinds of big, theoretical questions (and used the same kinds of research methods) as Marx and Weber. They have inspired a generation of younger sociologists to adopt a similar approach. Comparative-historical research is therefore one of the growth areas of the discipline.

Census data, police crime reports, and records of key life events are perhaps the most frequently used sources of official statistics. The first United States census was taken in 1790, and censuses have been conducted at regular intervals since then. The modern census tallies the number of United States residents and classifies them by place of residence, race, ethnic origin, occupation, age, and hundreds of other variables. The FBI publishes an annual *Uniform Crime Report* that tallies the number of crimes in the United States and classifies them by location and type of crime; the age, sex, and race of offenders and victims; and other variables. And the Centers for Disease Control and Prevention regularly publish "vital statistics" reports on births, deaths, marriages, and divorces by sex, race, age, and so forth.

Existing documents and official statistics have four main advantages over other types of data. First, they can save the researcher time and money. That is because they are usually available at no cost in libraries or on the World Wide Web. (See the "Suggested Readings" at the end of the chapter for useful Web sites containing official statistics.) Second, official statistics usually cover entire populations and are collected using rigorous and uniform methods, thus yielding high-quality data. Third, existing documents and official statistics are especially useful for historical analysis. The analysis of data from these sources is the only sociological method that does not require live subjects. Fourth, since the method does not require live subjects, reactivity is not a problem; the researcher's presence does not influence the subjects' behavior.[4]

However, existing documents and official statistics share one big disadvantage. These data sources are not created with the researchers' needs in mind. They often contain biases that reflect the interests of the individuals and organizations that created them. Therefore, they may be less than ideal for research purposes and must always be treated cautiously. For instance, if police enforcement officials decide to patrol minority-group neighborhoods more than majority-group neighborhoods, their action may result in an increase in

[4]When researchers finish analyzing survey data, they typically deposit computer-readable files of the data in an archive. This allows other researchers to conduct secondary analyses of survey data years later. Such data are widely used. They are not collected by government departments, but they have all of the advantages of official statistics listed above, although they are based on samples rather than populations. The largest social science data archive is at the University of Michigan's Inter-University Consortium for Political and Social Research (ICPSR). The ICPSR Web site, at http://www.icpsr.umich.edu, allows visitors to conduct elementary data analysis online.

the number of apprehended minority-group criminals over a given period. However, the increase may not be due to a rise in the underlying crime rate. It may be due to the administrative decision of the officials to increase patrols in certain neighborhoods. It follows that official crime statistics are not ideal measures of crime rates, especially for certain types of crime (see Chapter 6, "Deviance and Crime").

To illustrate further the potential bias of official statistics, consider how researchers used to compare the well-being of Americans and people living in other countries. For years, they used a measure called Gross Domestic Product Per Capita (GDPpc). GDPpc is the total dollar value of goods and services produced in a country in a year divided by the number of people in the country. It was a convenient measure because all governments regularly published GDPpc figures.

Researchers were aware of a flaw in GDPpc. The cost of living varies from one country to the next. A dollar can buy you a cup of coffee in many American restaurants, but the same cup of coffee will cost you $4 in a Japanese restaurant. GDPpc looks at how many dollars you have, not at what the dollars can buy. Therefore, researchers were happy when governments started publishing an official statistic called Purchasing Power Parity (PPP). It takes the cost of goods and services in each country into account.

Significantly, however, both PPP and GDPpc ignore two serious problems. First, it is possible for GDPpc and PPP to go up while most people in a society are worse off. This is just what happened in the United States in the 1980s. The richest people in the country earned all of the newly created wealth while the incomes of most Americans fell. Any measure of well-being that ignores the *distribution* of well-being in society is biased toward measuring the well-being of the well-to-do. Second, in some countries the gap in well-being between women and men is greater than in others. A country like Kuwait ranks quite high on GDPpc and PPP. However, women benefit far less than men do from that country's prosperity. A measure of well-being that ignores the gender gap is biased toward measuring the well-being of men.

This story has a happy ending. Realizing the biases in official statistics like GDPpc and PPP, social scientists at the United Nations created two new measures of well-being in the mid-1990s. First, the Human Development Index (HDI) combines PPP with a measure of average life expectancy and average level of education. The reasoning of the UN social scientists is that people living in countries that distribute well-being more equitably will live longer and be better educated. Second, the Gender Empowerment Measure (GEM) combines the percent of parliamentary seats, good jobs, and earned income controlled by women.

Table 2.3 lists the countries ranked first through fifth on all four measures of well-being we have mentioned. As you can see, the list of the top five countries differs for each measure. There is no "best" measure. Each measure has its own bias, and researchers have to be sensitive to these biases, as they must whenever they use official statistics.

✦ **TABLE 2.3** ✦
Rank of Countries by Four Measures of Well-Being

SOURCE: Adapted from United Nations (1999a; 1999b; 1999c).

Countries Ranked . . .	1st	2nd	3rd	4th	5th
Measure of Well-Being					
Gross Domestic Product Per Capita (1997)	Luxembourg	Bermuda	Switzerland, Liechtenstein	Norway	Japan
Purchasing Power Parity (1995)	Luxembourg	Brunei	United States	Switzerland	Hong Kong
Human Development Index (1995)	Canada	France	Norway	United States	Iceland
Gender Empowerment Measure (1998)	Norway	Sweden	Denmark	Finland	New Zealand

THE IMPORTANCE OF BEING SUBJECTIVE

In the following chapters, we show how participant observation, experiments, surveys, and the analysis of existing documents and official statistics are used in sociological research. You are well equipped for the journey. You should by now have a pretty good idea of the basic methodological issues that confront any sociological research project. You should also know the strengths and weaknesses of some of the most widely used data collection techniques.

Our synopsis of sociology's "reality check" should not obscure the fact that sociological research questions often spring from real-life experiences and the pressing concerns of the day. But prior to sociological analysis, we rarely see things as they are. We see them as *we* are. Then, a sort of waltz begins. Subjectivity leads, objectivity follows. When the dance is finished, we see things more accurately.

Feminism provides a prime example of this process. Here is a *political* movement of people and ideas that, over the past 35 years, has helped to shape the sociological *research* agenda. The division of labor in the household, violence against women, the effects of child-rearing responsibilities on women's careers, the social barriers to women's participation in politics and the armed forces, and many other related concerns were sociological "nonissues" before the rise of the modern feminist movement. Sociologists did not study these problems. Effectively, they did not exist for the sociological community (although they did of course exist for women). But subjectivity led. Feminism as a political movement brought these and many other concerns to the attention of the American public. Objectivity followed. Large parts of the sociological community began doing rigorous research on feminist-inspired issues and greatly refined our knowledge about them.

The entire sociological perspective began to shift as a growing number of scholars abandoned gender-biased research (Eichler, 1988; Tavris, 1992). Thus, *male-centeredness,* or approaching sociological problems from an exclusively male perspective, is now less common than it used to be. For instance, it is less likely in 2003 than in 1973 that a sociologist would study work but ignore unpaid housework as one type of labor. Similarly, *overgeneralization,* or using data on one sex to draw conclusions about all people, is now generally frowned upon. Today, for example, few researchers would be inclined to make claims about the social factors influencing health based on a sample of men only. In addition, *gender-blindness,* or excluding gender as an independent variable, is becoming less common. Thus, 30 years ago, many researchers failed to notice that the experiences of elderly men and women often differ radically because women tend to live longer and are poorer than men on average. That sort of mistake is less common today. Finally, applying a *double standard,* or assuming that women and men should necessarily be assessed on the basis of different criteria, is now viewed as problematic by many sociologists. Increasingly, for example, we understand there is nothing inevitable about husbands being the only breadwinners and wives being the only nurturers.

As these advances in sociological thinking show, and as has often been the case in the history of the discipline, objective sociological knowledge has been enhanced as a result of subjective experiences. And so the waltz continues. As in *Alice in Wonderland,* the question now is, "Will you, won't you, will you, won't you, join the dance?"

SUMMARY

1. The aim of science is to arrive at knowledge that is less subjective than other ways of knowing. A degree of objectivity is achieved by testing ideas against systematically collected data and leaving research open to public scrutiny.

2. The subjective side of the research enterprise is no less important than the objective side. Creativity and the motivation

to study new problems from new perspectives arise from individual passions and interests.

3. Certain methodological issues have to be addressed in any research project to maximize its scientific value. These issues include reliability (consistency in measurement), validity (precision in measurement), generalizability (assessing the applicability of findings beyond the case studied), and

causality (assessing cause-and-effect relations among variables).

4. One of the main sociological methods is participant observation, which involves carefully observing people's face-to-face interactions and actually participating in their lives over a long period of time. Participant observation is particularly useful for exploratory research, constructing grounded theory, and validating measures on the basis of internal criteria. Issues of external validity, reliability, generalizability, and causality make participant observation less useful for other research purposes.

5. An experiment is a carefully controlled artificial situation that allows researchers to isolate hypothesized causes and measure their effects by randomizing the allocation of subjects to experimental and control groups and exposing only the experimental group to an independent variable. Experiments get high marks for reliability and their analysis of causality, but issues of validity and generalizability make them less than ideal for many research purposes.

6. In a survey, people are asked questions about their knowledge, attitudes, or behavior, either in a face-to-face or telephone interview or in a paper-and-pencil format. Surveys rank high on reliability and validity as long as researchers train interviewers well, phrase questions carefully, and take special measures to ensure high response rates. Generalizability is achieved through probability sampling, statistical control, and the analysis of causality by means of data manipulation.

7. Existing documents and official statistics are inexpensive and convenient sources of high-quality data. However, they must be used cautiously since they often reflect the biases of the individuals and organizations that create them rather than the interests of the researcher.

GLOSSARY

An **association** exists between two variables if the value of one variable changes with the value of the other.

Causality means that a change in the independent variable (x) produces a change in the dependent variable (y). In analyzing survey data, we establish causality by demonstrating that (a) there is an association between x and y; (b) x precedes y; and (c) the introduction of a causally prior control variable does not result in the original association disappearing.

A **contingency table** is a cross-classification of cases by at least two variables that allows you to see how, if at all, the variables are associated.

A **control group** in an experiment is the group that is not exposed to the independent variable.

A **control variable** is a variable whose influence is removed from the association between an independent and a dependent variable.

A **dependent variable** is the presumed effect in a cause-and-effect relationship.

An **experiment** is a carefully controlled artificial situation that allows researchers to isolate hypothesized causes and measure their effects precisely.

An **experimental group** in an experiment is the group that is exposed to the independent variable.

Exploratory research is an attempt to describe, understand, and develop theory about a social phenomenon in the absence of much previous research on the subject.

Generalizability exists when research findings apply beyond the specific case examined.

A **hypothesis** is an unverified but testable statement about the relationship between two or more variables.

An **independent variable** is the presumed cause in a cause-and-effect relationship.

Operationalization is the procedure by which researchers establish criteria for assigning values to variables.

Participant observation involves carefully observing people's face-to-face interactions and actually participating in their lives over a long period of time, thus achieving a deep and sympathetic understanding of what motivates them to act in the way they do.

A **population** is the entire group about which the researcher wishes to generalize.

In a **probability sample,** the units have a known and nonzero chance of being selected.

Random means "by chance"—for example, having an equal and nonzero probability of being sampled. Randomization involves assigning individuals to groups by chance processes.

Reliability is the degree to which a measurement procedure yields consistent results.

A **sample** is the part of the population of research interest that is selected for analysis.

A **spurious association** exists between an independent and a dependent variable when the introduction of a causally prior control variable makes the initial association to disappear.

Statistical significance exists when a finding is unlikely to occur by chance, usually in 19 out of every 20 samples of the same size.

In a **survey,** people are asked questions about their knowledge, attitudes, or behavior, either in a face-to-face or telephone interview or in a paper-and-pencil format.

Validity is the degree to which a measure actually measures what it is intended to measure.

A **variable** is a concept that can take on more than one value.

QUESTIONS TO CONSIDER

1. What is the connection between objectivity and subjectivity in sociological research?

2. What criteria do sociologists apply to select one method of data collection over another?

3. What are the methodological strengths and weaknesses of various methods of data collection?

WEB RESOURCES

Companion Web Site for This Book

http://sociology.wadsworth.com

Begin by clicking on the Student Resources section of the Web site. Choose "Introduction to Sociology" and finally the Brym and Lie book cover. Next, select the chapter you are currently studying from the pull-down menu. From the Student Resources page you will have easy access to InfoTrac College Edition®, MicroCase Online exercises, additional Web links, and many resources to aid you in your study of sociology, including practice tests for each chapter.

Infotrac Search Terms

These search terms are provided to assist you in beginning to conduct research on this topic by visiting http://www.infotraccollege.com/wadsworth.

Census

Historical sociology

Participant observation

Sociology experiment

Sociological survey

Recommended Web Sites

Bill Trochim at Cornell University has put together a comprehensive and impressive sociological research methods course on the World Wide Web. Visit it at http://trochim.human.cornell.edu.

For a comprehensive listing of Web sites devoted to qualitative research, go to http://www.nova.edu/ssss/QR/web.html.

"Statistics Every Writer Should Know" is an exceptionally clear presentation of basic statistics on the World Wide Web at http://www.robertniles.com/stats.

The World Wide Web contains many rich sources of official statistics. In preparing this book we relied heavily on data from the Web sites of the United States Bureau of the Census (at http://www.census.gov), the United States Bureau of Labor Statistics (at http://stats.bls.gov), the National Center for Health Statistics (at http://www.cdc.gov/nchs/fastats/fastats.htm), the FBI's Uniform Crime Reports (at http://www.fbi.gov/ucr.htm), and the United Nations (at http://www.un.org).

SUGGESTED READINGS

Earl Babbie. *The Practice of Social Research,* rev. ed. of 9th ed. (Belmont, CA: Wadsworth, 2000 [1973]). This book is generally considered to be the best single-volume introduction to sociological research methods. Babbie is a real craftsman, and he explains in detail how to use all the tools in the sociologist's kit.

William H. Frey with Cheryl L. First. *Investigating Change in American Society: Exploring Social Trends with United States Census Data and StudentChip* (Belmont, CA: Wadsworth, 1997). This book shows you how to get your hands good and dirty analyzing data. In a clear and engaging way, Frey and First ask serious research questions and show you how to answer them using recent United States census

data and a simple statistics program, both of which are provided on diskette.

Carol Tavris. *The Mismeasure of Woman* (New York: Simon & Schuster, 1992). A path-breaking and provocative work by a leading academic feminist. In a hard-hitting but balanced style, Tavris shows how social science research has systematically produced biased findings by considering men the normal standard against which everyone should be judged and measured.

William Foote Whyte. *Street Corner Society: The Social Structure of an Italian Slum,* 3rd revised and expanded ed. (Chicago: University of Chicago Press, 1981 [1943]). This is perhaps the most famous and frequently cited participant-observation study of all time. It is a must read for all aspiring sociologists.

APPENDIX

FOUR STATISTICS YOU SHOULD KNOW

In this book we sometimes report the results of sociological research in statistical form. You need to know four basic statistics to understand this material:

1. The *mean* (or arithmetic average). Imagine we know the height and annual income of the first nine people who entered your sociology classroom today. The height and income data are arranged in Table 2.4. From Table 2.4 you can calculate the mean by summing the values for each student or *case* and dividing by the number of cases. For example, the nine students are a total of 609 inches tall. Dividing 609 by 9, we get the mean height—67.7 inches.

2. The *median*. The mean can be deceiving when some cases have exceptionally high or low values. For example, in Table 2.4, the mean income is $37,667, but because one lucky fellow has an income of $200,000, the mean is higher than the income of seven of the nine students. It is therefore a poor measure of the center of the income distribution. The median is a better measure. If you order the data from the lowest to the highest income, the median is the value of the case at the midpoint. The median income in our example is $15,000. Four students earn more than that, four earn less. (Note: If there is an even number of cases, the midpoint is the average of the middle two values.)

3. *Correlation*. We have seen how valuable contingency tables are for analyzing relationships among variables. However, for variables that can assume many values, such as height and income, contingency tables become impracticably large. In such cases, sociologists prefer to analyze relationships among variables using *scatterplots*. Markers in the body of the graph indicate the score of each case on both the independent and dependent variables. The pattern formed by the markers is inspected visually and through the use of statistics. The strength of the association between the two variables is measured by a statistic called the *correlation coefficient* (signified as "r"). The value of r can vary from −1.0 to 1.0. If the markers are scattered around a straight, upward sloping trend line, r takes a positive value. A positive r suggests that, as the value of one variable increases, so does the value of the other (see Figure 2.7, scatterplot 1). If the markers are scattered around a straight, downward sloping trend line, r takes a negative value. A negative r suggests that, as

Web Research Projects
Online Data Analysis

Student	Height (inches)	Income ($ thousands)
1	67	5
2	65	8
3	60	9
4	64	12
6	68	15
7	70	20
8	69	30
5	72	40
9	74	200

✦ **TABLE 2.4** ✦
The Height and Annual Income of Nine Students

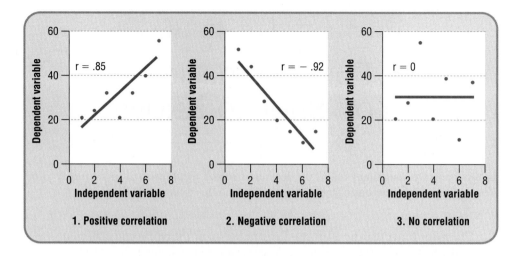

1. Positive correlation   2. Negative correlation   3. No correlation

the value of one variable increases, the value of the other decreases (see Figure 2.7, scatterplot 2). Whether positive or negative, the magnitude (or absolute value) of r decreases the more widely scattered the markers are from the line. If the degree of scatter is very high, r = 0. That is, there is no association between the variables (see Figure 2.7, scatterplot 3). However, a low r or an r of zero may derive from a relationship between the two variables that does not look like a straight line. It may look like a curve. As a result, it is always necessary to inspect scatterplots visually and not just rely on statistics like r to interpret the data.

4. A *rate* lets you compare the values of a variable among groups of different size. For example, let's say 1,000 women got married last year in a city of 100,000 people and 2,000 women got married in a city of 300,000 people. If you want to compare the likelihood of women getting married in the two cities, you have to divide the number of women who got married in each city by the total number of women in each city. Since 1,000 / 100,000 = .01 or 1%, and 2,000 / 300,000 = .00666 or .67%, we can say that the *rate* of women marrying is higher in the first city even though fewer women got married there last year. Note that rates are often expressed in percentage terms. In general, dividing the number of times an event occurs (e.g., a woman getting married) by the total number of people to whom the event could occur in principle (e.g., the number of women in a city) will give you the rate at which an event occurs.

II

BASIC SOCIAL PROCESSES

IN THIS CHAPTER, YOU WILL LEARN THAT:

✦ Culture is the sum of shared ideas, practices, and material objects that people create to adapt to, and thrive in, their environments.

✦ Humans have thrived in their environments because of their unique ability to think abstractly, cooperate with one another, and make tools.

✦ Although sociologists recognize that biology sets broad human limits and potentials, most sociologists do not believe that specific human behaviors and social arrangements are biologically determined.

✦ In some respects, the development of culture makes people freer. For example, culture has become more diversified and consensus has declined in many areas of life, allowing people more choice in how they live.

✦ In other respects, the development of culture puts limits on who we can become. For example, the culture of buying consumer goods has become a virtually compulsory national pastime. Increasingly, therefore, people define themselves by the goods they purchase.

3

CULTURE

CULTURE AS PROBLEM SOLVING

If you follow professional baseball, you probably know that Nomar Garciaparra, the star shortstop of the Boston Red Sox, can take 10 seconds to repeatedly pull up his batting gloves and kick the dirt with the toes of his cleats before he swings the bat. He believes this routine brings him luck. Garciaparra has other superstitious practices as well. For example, he never changes his cap. And although his name is really Anthony, he adopted "Nomar," his father's name spelled backwards, for good luck.

Garciaparra's nervous prebatting dance, as well as his other superstitious practices, make some people chuckle. But they put Garciaparra at ease. They certainly didn't hurt his league-leading .357 batting average in 1999. As Garciaparra says: "I have some superstitions, definitely, and they're always going to be there. I think a lot of people have them in baseball . . . [It] definitely helps because it gets you in the mind set" ("Garciaparra Explains his Superstitions," 2000; see Figure 3.1).

Like soldiers going off to battle, college students about to write final exams, and other people in high-stress situations, athletes invent practices to help them stop worrying and focus on the job at hand. Some wear a lucky piece of jewelry or item of clothing. Others say special words or a quick prayer. Still others cross themselves. And then there are those who engage in more elaborate rituals. For example, two sociologists interviewed 300 college students about their superstitious practices before final exams. One student felt she would do well only if she ate a sausage and two eggs sunny-side up on the morning of each exam. She had to place the sausage vertically on the left side of her plate and the eggs to the right of the sausage so they formed the "100" percent she was aiming for (Albas and Albas, 1989). Of course, the ritual had more direct influence on her cholesterol level than on her grades. Yet indirectly it may have had the desired effect. To the degree it helped to relieve her anxiety and relax her, she may have done better on her exams.

When some people say "culture," they refer to opera, ballet, art, and fine literature. For sociologists, however, this definition is too narrow.[1] Sociologists define **culture** broadly as all the ideas, practices, and material objects that people create to deal with real-life

✦ **FIGURE 3.1** ✦

Culture Can Solve Practical Problems

Nomar Garciaparra creates a little culture . . . and then knocks one out of Fenway Park.

[1]Sociologists call opera, etc., **high culture** to distinguish it from **popular** or **mass culture.** While high culture is consumed mainly by upper classes, popular or mass culture is consumed by all classes.

problems. For example, when Nomar Garciaparra developed the practice of pulling at his gloves and the college student invented the ritual of preparing for exams by eating sausage and eggs arranged just so, they were creating culture in the sociological sense. These practices helped Garciaparra and the student deal with the real-life problem of high anxiety.

Similarly, tools help people solve the problem of how to plant crops and build houses. Religion helps people face the problem of death and how to give meaning to life. Tools and religion are also elements of culture because they, too, help people solve real-life problems. Note, however, that religion, technology, and many other elements of culture differ from the superstitions of Garciaparra and the college student in one important respect. Superstitions are often unique to the individuals who create them. In contrast, religion and technology are widely shared. They are even passed on from one generation to the next. How does cultural sharing take place? By means of communication and learning. Thus, shared culture is *socially* transmitted. We conclude that culture is composed of the socially transmitted ideas, practices, and material objects that enable people to adapt to, and thrive in, their environments.

The Origins of Culture

You can appreciate the importance of culture for human survival by considering the predicament of early humans about 100,000 years ago. They lived in harsh natural environments. They had poor physical endowments, being slower runners and weaker fighters than many other animals. Yet, despite these disadvantages, they survived. More than that: they prospered and came to dominate nature. This was possible largely because they were the smartest creatures around. Their sophisticated brains enabled them to create cultural survival kits of enormous complexity and flexibility. These cultural survival kits contained three main tools. Each tool was a uniquely human talent. Each gave rise to a different element of culture.

The first tool in the human cultural survival kit was **abstraction,** the capacity to create general ideas, or ways of thinking that are not linked to particular instances. **Symbols,** for example, are one important type of idea. They are things that carry particular meanings. Languages, mathematical notations, and signs are all sets of symbols. Symbols allow us to classify experience and generalize from it. For example, we recognize that we can sit on many objects but that only some of those objects have four legs, a back, and space for one person. We distinguish the latter from other objects by giving them a name: "chairs." By the time a baby reaches the end of her first year, she has heard that word repeatedly and understands it refers to a certain class of objects. True, a few chimpanzees have been taught how to make some signs with their hands. In this way, they have learned a few dozen words and how to string together some simple phrases. Yet even these extraordinarily intelligent animals cannot learn any rules of grammar, teach other chimps what they know, or advance much beyond the vocabulary of a 2-year-old human (Pinker, 1994). Abstraction beyond the most rudimentary level is a uniquely human capacity. The ability to abstract enables humans to learn and transmit knowledge in a way no other animal can.

Cooperation is the second main tool in the human cultural survival kit. It is the capacity to create a complex social life. This is accomplished by establishing **norms,** or generally accepted ways of doing things. When we raise children and build schools we are cooperating to reproduce and advance the human race. When we create communities and industries, we are cooperating by pooling resources and encouraging people to acquire specialized skills. This enables them to accomplish things that no person could possibly do on his or her own. An enormous variety of social arrangements and institutions, ranging from healthcare systems to forms of religious worship to political parties, demonstrates the advanced human capacity to cooperate and follow norms. Of course, there is also plenty of war, crime, and revolution in the world. However, even when people engage in conflict they must cooperate and respect norms or fail to achieve their survival aims. The bank robber who is left stranded by his getaway man will be caught; the navy captain whose sailors mutiny will lose the battle.

By acquiring specialized skills, people are able to accomplish things that no person could possibly do on his or her own.

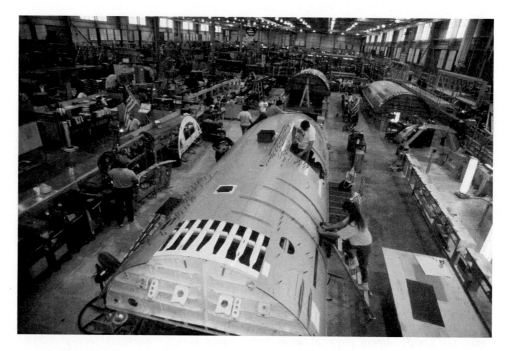

Production is the third main tool in the human cultural survival kit. It involves making and using tools and techniques that improve our ability to take what we want from nature. Such tools and techniques are known as **material culture.** Of course, all animals take from nature in order to subsist, and an ape may sometimes use a rock to break another object. But only humans are sufficiently intelligent and dexterous to *make* tools and use them to produce everything from food to computers. Understood in this sense, production is a uniquely human activity.

Table 3.1 illustrates each of the basic human capacities and their cultural offshoots with respect to three types of human activity: medicine, law, and religion. It shows, for all three types of activity, how abstraction, cooperation, and production give rise to specific kinds of ideas, norms, and elements of material culture. In medicine, theoretical ideas about the way our bodies work are evaluated using norms about how to test theories experimentally. Experimentation, in turn, results in the production of new medicines and therapies. These are part of material culture. In law, values, or shared ideas about what is right and wrong, are embodied in a legal code, or norms defining illegal behavior and punishments for breaking the law. The application of the law requires the creation of courts and jails, which are also part of material culture. Religious folklore—traditional ideas about how the universe was created, the meaning of life, and so forth—is expressed in religious customs regarding how to worship and how to treat fellow human beings. Religious folklore and customs can give rise to material culture that includes churches, their

✦ **TABLE 3.1** ✦
The Building Blocks of Culture

SOURCE: Adapted from Bierstedt (1963).

Human Capacities			
	Abstraction	**Cooperation**	**Production**
	↓	↓	↓
	Ideas	**Norms**	**Material Culture**
Elements of Culture			
		Cultural Activities	
Medicine	Theories	Experiments	Treatments
Law	Values	Laws	Courts, jails
Religion	Religious folklore	Religious customs	Church art, architecture

associated art and architecture, and so forth. As these examples suggest, then, the capacity for abstraction, cooperation, and production are evident in all spheres of culture.

In concluding this discussion of the origins of culture, we must note that people are usually rewarded when they follow cultural guidelines and punished when they do not. These rewards and punishments aimed at ensuring conformity are known as **sanctions.** Taken together they are called the system of **social control.** Rewards (or positive sanctions) include everything from praise and encouragement to money and power. Punishments (or negative sanctions) range from avoidance and contempt to physical violence and arrest. Punishment is more severe for the violation of core norms. These are norms that people feel are essential for the survival of their group or their society. Sociologist William Graham Sumner (1940 [1907]) called such core norms "mores" (the Latin word for "customs," pronounced MORE-ays). Punishment is less severe for the violation of less important norms (which Sumner called "folkways").

Despite efforts to control them, people often reject elements of existing culture and create new elements of culture. Reasons for this are discussed later in this chapter and in Chapter 6 ("Deviance and Crime"), Chapter 14 ("The Mass Media"), and Chapter 17 ("Collective Action and Social Movements"). Here it is enough to say that, just as social control is needed to ensure stable patterns of interaction, so resistance to social control is needed to ensure cultural innovation and social renewal. Stable but vibrant societies are able to find a balance between social control and cultural innovation.

CULTURE AND BIOLOGY

We have seen how the human capacity for abstraction, cooperation, and production enables us to create culture and makes us distinctively human. This capacity is built on a solid biological foundation. For example, without supple vocal chords we could not speak. Without the ability to grasp small objects we could not make tools. Without complex brains we could not even conceive of sophisticated social institutions such as colleges and armies. Biology, as every sociologist recognizes, sets broad human limits and potentials, including the potential to create culture.

However, some students of human behavior who are trained as biologists or psychologists go a step farther. Practitioners of what originated as "sociobiology" and is now commonly known as "evolutionary psychology" claim that genes—chemical units that carry traits from parents to children—account not just for physical characteristics but also for specific behaviors and social practices (Wilson, 1975). From their point of view, genes determine not just whether our eyes are blue or brown, but also whether we are law-abiding, whether we are sexually faithful to our partners, and just about every other aspect of our social behavior. This kind of argument has become increasingly popular since the early 1970s. The overwhelming majority of sociologists disagrees with it. It is therefore worth devoting a few paragraphs to the misconceptions of evolutionary psychology (see also Chapter 9, "Sexuality and Gender").

Evolutionary psychology's starting point is Charles Darwin's theory of evolution. Darwin (1859) observed wide variations in the physical characteristics of members of each species. For example, some deer can run quickly. Others run slower. The coloring of some frogs lets them blend perfectly into their surroundings. The coloring of others frogs does not camouflage them as well. Some tigers are more ferocious than others. Because of such variations, some members of each species—the quicker deer, the better-camouflaged frog, the more ferocious tiger—are more likely to survive. In general, the species members who are best adapted to their environments (or "fittest") are most likely to live long enough to have offspring. Therefore, concluded Darwin, the species characteristics that endure are those that increase the survival chances of the species.

Contemporary evolutionary psychologists make similar arguments about human behavior and social arrangements. Typically, *they first identify a supposedly universal human behavioral trait.* For example, they claim that men are more likely than women to want many sexual partners.

They next offer an explanation as to why this behavior increases survival chances. Thus, to continue with our example, they account for supposedly universal male promiscuity and female fidelity as follows. Every time a man ejaculates, he produces hundreds of millions of sperm, and he can achieve this feat from puberty until old age. In contrast, a woman typically releases fewer than 400 mature eggs from her ovaries over her entire lifetime—one egg per month between puberty and menopause in periods when she is not pregnant. From these observed sex differences, evolutionary psychologists jump to the assertion that men and women develop different "reproductive strategies" to increase the chance they will reproduce their genes. Specifically, because a woman produces few eggs, she improves her chance of reproducing her genes if she has a mate who stays around to help and protect her during those few occasions when she is pregnant, gives birth, and nurses a small infant. Because a man's sperm is so plentiful, he improves his chance of reproducing his genes if he tries to impregnate as many women as possible. In short, women's desire for a single mate and men's desire for many sexual partners is simply the way men and women play out the game of survival of the fittest. Even male rapists, writes one evolutionary psychologist, may just be "doing the best they can to maximize their [reproductive] fitness" (Barash, 1981: 55).

The final part of the evolutionary psychologists' argument is that the behavior in question cannot be changed. Once metal pathways are stamped into a computer's circuit boards they determine how electrical current can flow. In much the same way, the characteristics that maximize the survival chances of a species supposedly get encoded or "hardwired" in our genes. It follows that what exists is necessary.

Most sociologists and many biologists and psychologists are critical of the reasoning of evolutionary psychologists. In the first place, *many behaviors discussed by evolutionary psychologists are not universal and some are not even that common.* Consider male promiscuity. Is it true that men are promiscuous and that women are not? The data tell a different story. As Table 3.2 shows, in 1998 only a small minority of adult American men (20%) claimed they had more than one sex partner in the previous year. The figure for adult American women is lower, but not dramatically so (11%). Moreover, if we consider married adults only, the figures fall to 4% for men and 2% for women, a small and statistically insignificant difference (see Table 3.3). True, some groups of men are more promiscuous than others. For example, 41% of unmarried American men claim to have had more than one sex partner in the previous year. The proportion is higher still for unmarried young men. You can exercise your sociological imagination to explain why certain *social* arrangements such as the institution of marriage and male youth culture account for variations in promiscuity. For present purposes, however, the important point is that the evolutionary

◆ **TABLE 3.2** ◆

Number of Sex Partners by Respondent's Sex, United States, 1996 (in percent)

SOURCE: National Opinion Research Center (1999).

Number of Sex Partners	Respondent's Sex	
	Male	Female
0 or 1	80	89
More than 1	20	11
Total	100	100
n	870	1,008

◆ **TABLE 3.3** ◆

Number of Sex Partners by Respondent's Sex, United States, 1996, Married Respondents Only (in percent)

SOURCE: National Opinion Research Center (1999).

Number of Sex Partners	Respondent's Sex	
	Male	Female
0 or 1	96	98
More than 1	4	2
Total	100	100
n	487	605

psychologists' claim that men *in general* are highly promiscuous, and much more promiscuous than women, is false. So are many of their other claims about so-called behavioral constants or universals.

The second big problem with evolutionary psychology is that one of its key arguments—that specific behaviors and social arrangements are associated with specific genes—has never been verified. Researchers *have* identified gene mutations associated with many diseases, including more than 20 types of hereditary cancer (Fearon, 1997). Most people are optimistic that these discoveries will lead to new medical treatments. But what is true for diseases is not true for behaviors and social arrangements. No convincing evidence supports the view that specific behaviors and social arrangements are associated with specific genes. Therefore, when it comes to supporting their key argument, evolutionary psychologists have little to stand on apart from a fragile string of maybes and possibilities: "[W]e *may* have to open our minds and admit the *possibility* that our need to maximize our [reproductive] fitness *may* be whispering somewhere deep within us and that, *know it or not,* most of the time we are heeding these whisperings" (Barash, 1981: 31; our emphasis). Maybe. Then again, maybe not.

Finally, even if researchers eventually discover an association between particular genes and particular behaviors, it would be wrong to conclude that variations among people are due just to their genes. Why? Because, in the words of R. C. Lewontin, one of the world's leading biologists, "variations among individuals within species are a unique consequence of both genes *and environment* in a constant interaction . . . [and] random variation in growth and division of cells during development" (Lewontin, 1991: 26–7; our emphasis). Genes *never* develop without environmental influence. The genes of a human embryo, for example, are profoundly affected by whether the mother consumes the recommended daily dosage of calcium or nearly overdoses daily on crack cocaine. And what the mother consumes is, in turn, determined by many social factors. Even if one inherits a mutant cancer gene, the chance of developing cancer is strongly influenced by diet, exercise, tobacco consumption, and factors associated with occupational and environmental pollution. Some cancers are more heritable than others, but even the most heritable cancers seem to be much more strongly influenced by environmental than genetic factors (Fearon, 1997; Hoover, 2000; Kevles, 1999; Lichtenstein, Holm, Verkasalo, Iliadou, Kaprio, Koskenvuo, Pukkala, Skytthe, and Hemminki, 2000; Remennick, 1998). It follows that the pattern of your life is not entirely hardwired by your genes (see Figure 3.2). Changes in social environment do produce physical and, to an even greater degree, behavioral change. However, to figure out the effects of the social environment on human behavior we have to abandon the premises of evolutionary psychology and develop specifically sociological skills for analyzing the effects of social structure and culture. We begin that task by first considering how it is possible to observe culture in an unbiased fashion.

✦ **FIGURE 3.2** ✦

A Genetic Misconception

When scientists announced they had finished sequencing the human genome on June 26, 2000, some people thought all human characteristics could be read from the human genetic "map." They cannot. The functions of most genes are still unknown. Moreover, because genes mutate randomly and interact with environmental (including social) conditions, the correspondence between genetic function and behavioral outcome is highly uncertain.

SOURCE: "Human Genome . . . (2000)."

CULTURE FROM THE INSIDE AND THE OUTSIDE

Web Interactive Exercises
How Is Social Inequality
Justified in Our Culture?

"I was once introduced to an interesting woman at a party and began a conversation with her that started agreeably," says Robert Brym. "Within ten minutes, however, I found myself on the other side of the room, my back pressed hard against the wall, trying to figure out how I could politely end our interaction. I wasn't immediately aware of the reason for my discomfort. Only after I told the woman I had to make an important phone call and had left the room did I realize the source of the problem: she had invaded my culturally-defined comfort zone. Research shows the average North American prefers to stand 30 to 36 inches away from strangers or acquaintances when they are engaged in face-to-face interaction (Hall, 1959: 158–80). But this woman had recently arrived from her home in a part of the Middle East where the culturally-defined comfort zone is generally smaller. She stood only about two feet away from me as we spoke. Without thinking about it, I retreated half a step. Without thinking about it, she advanced half a step. And soon we had waltzed across the faculty club lounge, completely unaware of what we were doing, until I had no more room to retreat and had to concoct a means of escape."

As this example shows, culture, despite its central importance in human life, is often invisible. That is, people tend to take their own culture for granted; it usually seems so sensible and natural they rarely think about it. In contrast, people are often startled when confronted by cultures other than their own. That is, the ideas, norms and techniques of other cultures frequently seem odd, irrational, and even inferior.

Judging another culture exclusively by the standards of one's own is known as **ethnocentrism.** Ethnocentrism impairs sociological analysis (see Box 3.1). This can be illustrated by a practice that seems bizarre to many Westerners: cow worship among Hindu peasants in India.

Hindu peasants refuse to slaughter cattle and eat beef because, for them, the cow is a religious symbol of life. Pinup calendars throughout rural India portray beautiful women with the bodies of fat, white cows, milk jetting out of each teat. Cows are permitted to wander the streets, defecate on the sidewalks, and stop to chew their cud in busy intersections or on railroad tracks, causing traffic to come to a complete halt. In Madras, police stations maintain fields where stray cows that have fallen ill can graze and be nursed back to health.

Many Westerners find the Indian practice of cow worship bizarre. However, cow worship performs a number of useful economic functions and is in that sense entirely rational. By viewing cow worship exclusively as an outsider (or, for that matter, exclusively as an insider), we fail to see its rational core.

BOX 3.1
SOCIOLOGY AT THE MOVIES

THE JOY LUCK CLUB (1993)

In China in the 1930s and 1940s, arranged marriage was common. Often, you wouldn't meet the person with whom you were going to spend the rest of your life until the wedding ceremony. One of the women in *The Joy Luck Club* (1993) prays hard that her husband will be "not too old." After the ceremony, in the married couple's bedroom, she discovers that her husband is a 10-year-old boy. "Maybe I prayed too hard!" she says.

Every week, four Chinese-American women meet to play mah-jongg and gossip about their lives. This is the Joy Luck Club. The movie revolves around the lives of the four women and their four daughters. The narrator—June, played by Ming-Na Wen—and the three other Chinese-American daughters love, but do not understand, their Chinese-born mothers. The daughters grew up in San Francisco. They assimilated into American culture. So they find their mothers' Chinese ways strange. To be sure, their mothers find their daughters strange, too, not really understanding their American ways. Cultural differences between mothers and daughters inevitably result in conflict.

The movie begins with a farewell party for June, who is about to set off to China to meet her half-sisters. This is something that had bothered June for years; she just cannot understand how her mother could have abandoned her infants. In the course of the movie, she comes to understand the hardships that her mother—as well as her "aunties"—had

The Joy Luck Club (1993).

to endure in wartime China of the 1940s. June learns that her mother left her children because she was convinced she would die and thought it would be bad luck for her children to have a dead mother.

In the course of the movie, we come to understand with June the hardships experienced by all four mothers. For example, one of them was the fourth wife of a rich man. When she had a son, the second wife snatched him away from her. However, the mothers do not tell their tragic stories to their daughters. They keep silent about their past lives. As a result, their daughters don't really understand their mothers. Conversely, the older women find the choices their American daughters make, the things they do, the men they marry, equally incomprehensible. We thus see that even within one family there may exist two cultures, each of which is ethnocentric.

The problems of the mothers and daughters in *The Joy Luck Club* are not peculiar to Chinese-Americans, nor are they restricted to immigrant communities. In modern and postmodern societies, parents *typically* grow up in a culture that differs from that of their children. Because of the universality of cultural conflict between generations, *The Joy Luck Club* appeals to a wide audience.

If you're like most people, your culture seems natural to you, while the culture of your parents seems somewhat alien. But, reflecting on the sociological lessons of *The Joy Luck Club,* why do you think your culture differs from that of your parents? How do you think you could achieve a better understanding of your parent's culture? How could they achieve a better understanding of yours? In general, how can you avoid ethnocentrism?

The government even runs old-age homes for cows where dry and decrepit cattle are kept free of charge. All this seems utterly inscrutable to most Westerners, for it takes place amid poverty and hunger that could presumably be alleviated if only the peasants would slaughter their "useless" cattle for food instead of squandering scarce resources feeding and protecting them.

According to anthropologist Marvin Harris, however, ethnocentrism misleads many Western observers (Harris, 1974: 3–32). Cow worship, it turns out, is an economically rational practice in rural India. For one thing, Indian peasants can't afford tractors, so cows are needed to give birth to oxen, which are in high demand for plowing. For another, the cows produce hundreds of millions of pounds of recoverable manure, about half of which is used as fertilizer and half as a cooking fuel. With oil, coal, and wood in short supply, and with the peasants unable to afford chemical fertilizers, cow dung is, well, a godsend. What is more, cows in India don't cost much to maintain since they eat mostly food that isn't fit for human consumption. And they represent an important source of protein and a livelihood for members of low-ranking castes, who have the right to dispose of the bodies of dead cattle. These "untouchables" eat beef and form the workforce of India's large leather craft industry. The protection of cows by means of cow worship is thus a perfectly sensible and highly efficient economic practice. It only seems irrational when judged by the standards of Western agribusiness.

We can draw much the same lesson from the case of cow worship in India as from Robert Brym's hurried exit from the faculty club lounge. Culture is most clearly visible from the margins, as it were. We see its contours most sharply if we are neither too deeply immersed in it (as Robert was during his faculty club conversation) nor too much removed from it (as many Western observers are when they analyze cow worship in India). Said differently, if you refrain from taking your own culture for granted and judging other cultures by the standards of your own, you will have taken important first steps towards developing a sociological understanding of culture.

THE TWO FACES OF CULTURE

Culture has two faces. First, culture provides us with an opportunity to exercise our *freedom*. We create elements of culture in our everyday life to solve practical problems and express our needs, hopes, joys, and fears.

However, creating culture is just like any other act of construction in that we need raw materials to get the job done. The raw materials for the culture we create consist of cultural elements that either existed before we were born or other people created since our birth. We may put these elements together in ways that produce something genuinely new. But there is no other well to drink from, so existing culture puts limits on what we can think and do. In that sense, culture *constrains* us. This is culture's second face.

Because culture can be seen both as an opportunity for freedom and as a source of constraint, we examine both faces of culture below. We begin with the view that culture is an opportunity for freedom. We first establish that people are not just passive recipients but active producers and interpreters of culture. Next, we show that the range of cultural choices available to us has never been greater because we live in a society that is characterized by unparalleled cultural diversity. We then show how globalization processes contribute to the diversification of culture and broaden the range of cultural choices open to us. We argue that this has led to the emergence of a new, "postmodern" era of culture. After developing the idea that culture is a source of freedom, we turn to culture's flip side as a source of social constraint.

Culture as Freedom

Cultural Production

Until the 1960s, many sociologists argued that culture is simply a "reflection" of society. Using the language of Chapter 2, we can say they regarded culture as a dependent variable. Television, for example, became a household necessity after World War II. Sociologists noted that its spread depended on the existence of an affluent and technologically advanced society. Moreover, they said, the programming content of television revealed much about the concerns and aspirations of people in post–World War II society. As a part of both material and symbolic culture, then, television was said to reflect the society from which it emerged.

More recently, sociologists have emphasized culture as an *independent* variable. Increasingly, they stress that people do not just accept culture passively. That is, we are hardly inert and empty vessels into which society pours a defined assortment of beliefs, symbols, and values. Instead, we actively produce and interpret culture, creatively fashioning it to suit our own needs.

British literary critic Richard Hoggart (1958) and social historian E. P. Thompson (1968) wrote pioneering works emphasizing how people produce and interpret culture. Hoggart and Thompson showed how working-class people shape the cultural *milieux* in which they live. For instance, religious ideas and secular reading materials may be created for members of the working class by people in higher class positions—"from the outside," as it were. What then happens, according to Hoggard and Thompson, is that members of the working class make sense of these elements of culture on their own terms. In general, audiences always change ideas to make them meaningful to themselves. This line of thought was developed by sociologist Stuart Hall (1980) and his colleagues, who showed how people mold culture to fit their sense of self. It gave rise to the field of "cultural studies," which overlaps the sociology of culture (Griswold, 1992; Long, 1997; Wolff, 1999). Later in this chapter and again in Chapter 14 ("The Mass Media"), we take up some of the themes introduced by Hoggart, Thompson, and Hall.

Cultural Diversity

The fact that people actively produce and interpret culture means that, to a degree, we are at liberty to choose how culture influences us. We are increasingly able to exercise that ability because there is more to choose from. Like most societies in the world, American society is undergoing rapid cultural diversification, partly due to a high rate of immigration.

The proportion of the population that consists of immigrants is higher in the United States than in all but three other countries (Israel, Australia, and Canada). Moreover, ethnically and racially, the United States is a more heterogeneous society than at any point in its history. According to the United States Census Bureau, more than 28% of the United States population was nonwhite or Hispanic in 2000. Over the next 50 years, Hispanic- and Asian-American groups are projected to grow more than 200%. African- and Native-American groups are projected to grow 60–70%. Meanwhile, the white non-Hispanic group will grow less than 6%. It will begin to shrink after 2030. Sometime around 2060, white non-Hispanics will form a *minority* of the United States population (United States Bureau of the Census, 2000g).

The cultural diversification of American society is evident in all aspects of life, from the growing popularity of Latino music to the increasing influence of Asian design in clothing and architecture to the ever broadening international assortment of foods consumed by most Americans. Marriage between people of different ethnic groups is widespread, and marriage between people of different races is increasingly common. For example, about half of Asian Americans and a tenth of African Americans now marry outside their racial group (Stanfield, 1997).

At the political level, however, cultural diversity has become a source of conflict. This is nowhere more evident than in the debates that have surfaced in recent years concerning curricula in the American educational system.

Until recent decades, the American educational system stressed the common elements of American culture, history, and society. Students learned the story of how European settlers overcame great odds, prospered, and forged a unified nation out of diverse ethnic and racial elements. School curricula typically neglected the contributions of non-whites and non-Europeans to America's historical, literary, artistic, and scientific development. Moreover, students learned little about the less savory aspects of American history, many of which involved the use of force to create a strict racial hierarchy that persists until today, albeit in modified form (see Chapter 8, "Race and Ethnicity"). History books did not deny that African Americans were enslaved and that force was used to wrest territory from Native Americans and Mexicans. They did, however, make it seem as if these unfortunate events were part of the American past with few implications for the present. The history of the United States was presented as a history of progress involving the *elimination* of racial privilege.

For the past several decades, advocates of **multiculturalism** have argued that school and college curricula should present a more balanced picture of American history, culture, and society—one that better reflects the country's ethnic and racial diversity in the past and

The United States continues to diversify culturally.

its growing ethnic and racial diversity today (Ball, Berkowitz, and Mzamane, 1998). A multicultural approach to education highlights the achievements of non-whites and non-Europeans in American society. It gives more recognition to the way European settlers came to dominate non-white and non-European communities. It stresses how racial domination resulted in persistent social inequalities. And it encourages elementary-level instruction in Spanish in California, Texas, New Mexico, Arizona, and Florida, where a substantial minority of people speak Spanish at home.[2]

Most critics of multiculturalism do not argue against teaching cultural diversity. What they fear is that multiculturalism is being taken too far (Glazer, 1997; Schlesinger, 1991; Stotsky, 1999; see Box 3.2). Specifically, they say multiculturalism has three negative consequences:

✦ Critics believe that multicultural education hurts minority students by forcing them to spend too much time on noncore subjects. To get ahead in the world, they say, one needs to be skilled in English and math. By taking time away from these subjects, multicultural education impedes the success of minority group members in the work world. (Multiculturalists counter that minority students develop pride and self-esteem from a curriculum that stresses cultural diversity. They argue that this helps minority students get ahead in the work world.)

✦ Critics also believe that multicultural education causes political disunity and results in more interethnic and interracial conflict. Therefore, they want schools and colleges to stress the common elements of the national experience and highlight Europe's contribution to American culture. (Multiculturalists reply that political

---

[2]About one in seven Americans over the age of five speaks a language other than English at home. Of these people, more than half speak Spanish. Most Spanish speakers live in the states listed above.

unity and interethnic and interracial harmony simply maintain inequality in American society. Conflict, they say, while unfortunate, is often necessary to achieve equality between majority and minority groups.)

✦ Finally, critics of multiculturalism complain that it encourages the growth of **cultural relativism.** Cultural relativism is the opposite of ethnocentrism. It is the belief that all cultures and all cultural practices have equal value. The trouble with this view is that some cultures oppose the most deeply held values of most Americans. Other cultures promote practices that most Americans consider inhumane. Should we respect racist and antidemocratic cultures, such as the apartheid regime that existed in South Africa from 1948 until 1992? How about female circumcision, which is still widely practiced in Somalia, Sudan, and Egypt? Or the Australian aboriginal practice of driving spears through the limbs of criminals (Garkawe, 1995)? Critics argue that by promoting cultural relativism, multiculturalism encourages respect for practices that are abhorrent to most Americans. (Multiculturalists reply that cultural relativism need not be taken to such an extreme. *Moderate* cultural relativism encourages tolerance, and it should be promoted.)

Clearly, multiculturalism is a complex and emotional issue requiring much additional research and debate. It is worth pondering here, however, because it says something important about the state of American culture today and, more generally, about how world culture has developed since our remote ancestors lived in tribes. Even 50 years ago, the American ideal was to create one new culture out of many—*E Pluribus Unum*. Today, multiculturalism stands for the opposite—creating many cultures out of one. Nor is this shift unique to the United States. In general, as we will now see, cultures tend to become more heterogeneous over time, with important consequences for everyday life.

From Diversity to Globalization

In preliterate or tribal societies, cultural beliefs and practices are virtually the same for all group members. For example, many tribal societies organize **rites of passage.** These are cultural ceremonies that mark the transition from one stage of life to another (e.g., baptisms, confirmations, weddings) or from life to death (e.g., funerals). These religious rituals involve elaborate body painting, carefully orchestrated chants and movements, and so forth. They are conducted in public. No variation from prescribed practice is allowed. Culture is homogeneous (Durkheim, 1976 [1915]).

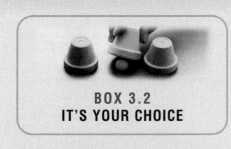

**BOX 3.2
IT'S YOUR CHOICE**

MULTICULTURALISM

In 1994, the school board of Lake County, Florida, passed a resolution stating: "Instruction shall include and instill in our students an appreciation of our American heritage and culture such as: our republican form of government, capitalism, a free-enterprise system, patriotism, strong family values, freedom of religion and other basic values that are superior to other foreign or historic cultures" (quoted in Glazer, 1997: 1).

In the 1950s, the Lake County resolution would have been common sense. Half a century ago, most Americans believed in the superiority of American culture and could talk about core American values as if they were self-evident. In the 1990s, however, a heated debate about multiculturalism erupted. It was sometimes described as a "culture war." Thus, a 1991 Florida law encouraged students to "eliminate personal and national ethnocentrism so that they understand that a specific culture is not intrinsically superior or inferior to another" (quoted in Glazer, 1997: 2). The Lake County resolution was largely a reaction to that law. Many people opposed the Lake Country resolution. One critic claimed to be embarrassed: "We've become sort of a laughing stock." A history teacher stated: "We regard American culture as very diverse, and we're not sure what values they see as American culture" (quoted in Glazer, 1997: 3).

What values do you think of as American? If you were a member of the Lake County school board, which values would you want schools to transmit to their students? In your opinion, should the content of school curricula depend on the ethnic and racial mix of the students in the region or the ethnic type of students in the country as a whole, or should it ignore such questions entirely? Do you think that multiculturalist school policies strengthen or weaken the United States? Why do you hold that opinion?

In contrast, preindustrial Western Europe and North America were rocked by artistic, religious, scientific, and political forces that fragmented culture. The Renaissance, the Protestant Reformation, the Scientific Revolution, the French and American Revolutions—between the 14th and 18th centuries, all of these movements involved people questioning old ways of seeing and doing things. Science placed skepticism about established authority at the very heart of its method. Political revolution proved there was nothing ordained about who should rule and how they should do so. Religious dissent ensured that the Catholic Church would no longer be the supreme interpreter of God's will in the eyes of all Christians. Authority and truth became divided as never before.

Cultural fragmentation picked up steam during industrialization, as the variety of occupational roles grew and new political and intellectual movements crystallized. Its pace is quickening again today in the postindustrial era. This is due to **globalization,** the process by which formerly separate economies, states, and cultures are being tied together.

The roots of globalization are many. International trade and investment are expanding. Even a business as "American" as McDonald's now reaps 60% of its profits from outside the United States, and its international operations are expected to grow at four times the rate of its United States outlets (Commins, 1997). At the same time, members of different ethnic and racial groups are migrating and coming into sustained contact with one another. A growing number of people date, court, and marry across religious, ethnic, and racial lines. Influential "transnational" organizations such as the International Monetary Fund, the World Bank, the European Union, Greenpeace, and Amnesty International are multiplying. Relatively inexpensive international travel and communication make contacts between people from diverse cultures routine. The mass media make Arnold Schwarzenegger and *Survivor* nearly as well known in Warsaw as in Wichita. MTV brings rock music to the world via MTV Latino, MTV Brazil, MTV Europe, MTV Asia, MTV Japan, MTV Mandarin, and MTV India (Hanke, 1998). Globalization, in short, destroys political, economic, and cultural isolation, bringing people together in what Canadian media analyst Marshall McLuhan (1964) called a "global village." As a result of globalization, people are less obliged to accept the culture into which they are born and freer to combine elements of culture from a wide variety of historical periods and geographical settings. Globalization is a schoolboy in Bombay, India, listening to Bob Marley on his MP3 player as he rushes to slip into his Levis, wolf down a bowl of Kellogg's Basmati Flakes, and say good-bye to his parents in Hindi because he's late for his English-language school (see Figure 3.3).

A good indicator of the influence and extent of globalization is the spread of English since 1600. In 1600, English was the mother tongue of between 4 and 7 million people. Not even all people in England spoke it. Today, 750 million to 1 billion people speak English worldwide, over half as a second language. With the exception of the many varieties of Chinese, English is the most widespread language on earth. Over half the world's technical and scientific periodicals are written in English, as are three quarters of the world's letters, telexes, and telegrams, and 80 percent of the nonnumerical data stored in the world's computers. English is the official language of the Olympics, of the Miss Universe contest, of navigation in the air and on the seas, and of the World Council of Churches.

English is dominant because Britain and the United States have been the world's most powerful and influential countries—economically, militarily, and culturally—for 200 years. (Someone once defined "language" as a dialect backed up by an army.) In recent decades, the global spread of capitalism, the popularity of Hollywood movies and American TV shows, and widespread access to instant communication via telephone and the Internet have increased the reach of the English language (see Figure 3.4). There are now more speakers of excellent English in India than in Britain, and when a construction company jointly owned by German, French, and Italian interests undertakes a building project in Spain, the language of business is English (McCrum, Cran, and MacNeil, 1992).

Even in Japan, where relatively few people speak the language, English words are commonly used. For example, when you learn to open a computer file's *ai-kon* (icon) you are told to *daburu-kurikku* (double-click) the *mausu* (mouse). In view of the extensive use of English words in Japan, *The Japanese Times,* one of Tokyo's four English daily newspapers, ran a story a few years ago noting the pressures of globalization and suggesting it might be time for Japan to switch to English. True, the Health and Welfare Ministry

✦ **FIGURE 3.3** ✦
**The Effect of Globalization
on Corn Flakes**
The idea of globalization first gained
prominence in marketing strategies
in the 1970s. In the 1980s, companies like Coca-Cola and McDonald's
expanded into non-Western countries to find new markets. Today,
Kellogg's markets products in more
than 160 countries. Basmati Flakes
cereal was first produced by the
Kellogg's plant in Tajola, India,
in 1992.

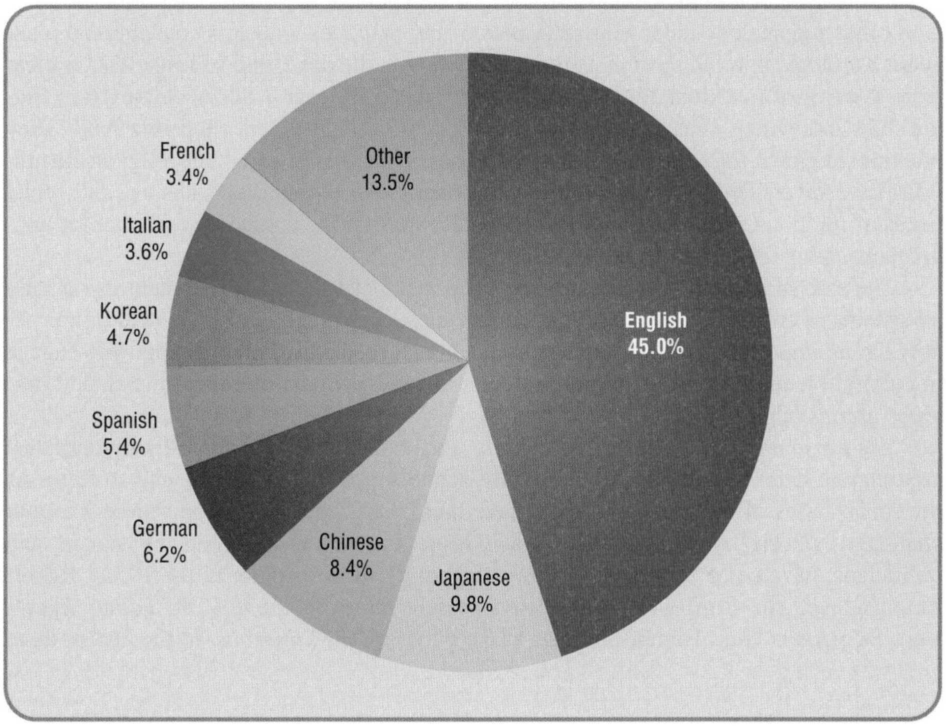

✦ **FIGURE 3.4** ✦
**Internet Usage by Language
Group, June 2001**

SOURCE: Global Reach (2001).

Pie chart (Figure 3.4): English 45.0%, Other 13.5%, French 3.4%, Italian 3.6%, Korean 4.7%, Spanish 5.4%, German 6.2%, Chinese 8.4%, Japanese 9.8%

banned excessive use of English in its documents a couple of years ago. But, as one Japanese newspaper noted, given the popularity of English words, it's doubtful there will be much *foro-uppu* (follow-up).

For Japanese teenagers, English is certainly considered very cool. A 15-year-old girl, wearing her trademark *roozu sokusu* (loose socks), might greet a friend sporting new sunglasses with a spirited *chekaraccho* (Check it out, Joe). If she likes the shades, she might say they're *cho beri gu* (ultra-good) and invite her friend *deniru* (to go to a Denny's restaurant) or *hageru* (to go to a Häagen-Dazs ice cream outlet). Of course, the girl might also *disu* (diss, or show disrespect toward) her friend. She might come right out and inform him that the new shades look *cho beri ba* (ultra-bad) or *cho beri bu* (ultra-blue, depressing or ultra-ugly). If so, the situation that develops could be a little *denjarasu* (dangerous). Terms of affection, such as *wonchu* (I want you), might not be exchanged. The budding relationship might go nowhere. Nonetheless, we can be pretty sure that Japanese teenagers' use of English slang will intensify under the pressures of globalization (Kristof, 1997).

Postmodernism

Some sociologists think so much cultural fragmentation and reconfiguration has taken place in the last few decades that a new term is needed to characterize the culture of our times: **postmodernism.** Scholars often characterize the last half of the 19th century and the first half of the 20th century as the era of modernity. During this hundred-year period, belief in the inevitability of progress, respect for authority, and consensus around core values characterized much of Western culture. In contrast, postmodern culture involves an eclectic mixing of elements from different times and places, the erosion of authority, and the decline of consensus around core values. Let us consider each of these aspects of postmodernism in turn.

An eclectic mixing of elements from different times and places. In the postmodern era, it is easier to create individualized belief systems and practices by blending facets of different cultures and historical periods. Consider religion. In the United States today, there are many more ways of worshiping than there used to be. The latest edition of the *Encyclopedia of American Religions* lists more than 2,100 religious groups, and one can easily construct a personalized religion involving, say, belief in the divinity of Jesus *and* yoga (Melton, 1996 [1978]). In the words of one journalist: "In an age when we trust ourselves to assemble our own investment portfolios and cancer therapies, why not our religious beliefs?" (Creedon, 1998). Nor are religious beliefs and practices drawn just from conventional sources. Even fundamentalist Christians who believe the Bible is the literal word of God often supplement Judeo-Christian beliefs and practices with less conventional ideas about astrology, psychic powers, communication with the dead, and so forth. This is clear from a series of questions that were asked in the 1989 General Social Survey (see Figure 3.5). Individuals thus draw on religions much like consumers shop in a mall. They practice religion *à la carte.* Meanwhile, churches, synagogues, and other religious institutions have diversified their menus in order to appeal to the spiritual, leisure, and social needs of religious consumers and retain their loyalties in the competitive market for congregants and parishioners (Finke and Stark, 1992).

The mix-and-match approach we see when it comes to religion is evident in virtually all spheres of culture. Purists may scoff at this sort of cultural blending. However, it probably has an important positive social consequence. It seems likely that people who engage in cultural blending are usually more tolerant and appreciative of ethnic, racial, and religious groups other than their own.

The erosion of authority. Half a century ago, Americans were more likely than they are today to defer to authority in the family, schools, politics, medicine, and so forth. As the social bases of authority and truth have multiplied, however, we are more likely to challenge authority. Authorities once widely respected, including parents, physicians, and politicians, have come to be held in lower regard by many people. In the 1950s, Robert Young played the firm, wise, and always-present father in the TV hit, *Father Knows Best.* Fifty years later, Homer Simpson plays a fool in *The Simpsons.* In the 1950s, three

✦ FIGURE 3.5 ✦
Unconventional Beliefs Among Christian Fundamentalists, United States, 1989 (in percent; n = 312)

"How often have you had any of the following experiences: Felt in touch with someone when they were far away from you ('ESP')? Felt as though you were really in touch with someone who had died ('Spirits')? Seen events that happened at a great distance as they were happening ('Visions')?" Responses are shown for respondents who said they are Protestant or Catholic and who believe the Bible is the literal word of God.

SOURCE: National Opinion Research Center (1999).

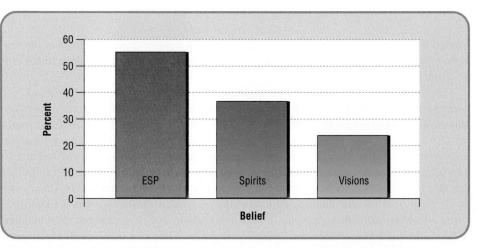

A hallmark of postmodernism is the combining of cultural elements from different times and places. Architect I. M. Pei unleashed a storm of protest when his 72-foot glass pyramid became an entrance to the Louvre in Paris. It created a postmodern nightmare in the eyes of some critics.

quarters of Americans expressed confidence in the federal government's ability to do what is right "just about always" or "most of the time." Fifty years later, the figure stood at just one third (see Figure 3.6). The rise of Homer Simpson and the decline of confidence in government both reflect the society-wide erosion of traditional authority (Nevitte, 1996).

The decline of consensus around core values. Half a century ago, people's values remained quite stable over the course of their adult lives and many values were widely accepted. Today, value shifts are more rapid and consensus has broken down on many issues. For example, in the middle of the 20th century, the great majority of adults remained loyal to one political party from one election to the next. By the third quarter of the 20th century, however, specific issues and personalities had eclipsed party loyalty as the driving forces of American politics (Nie, Verba, and Petrocik, 1976). Today, people are more likely to vote for different parties in succeeding elections than they were in 1950.

The decline of consensus may also be illustrated by considering the fate of Big Historical Projects. For most of the past 200 years, consensus throughout the world was built around Big Historical Projects. Various political and social movements convinced people they could take history into their own hands and create a glorious future just by signing

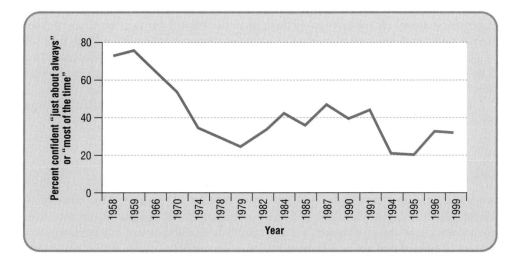

✦ FIGURE 3.6 ✦
Confidence in Washington, 1958–1999 (in percent)
"How much of the time do you think you can trust the government in Washington to do what is right—just about always, most of the time, or only some of the time?" Polls show that confidence in Washington increased after the terrorist attacks of September 11, 2001. However, it is unclear how long the increase will last.

SOURCE: United States Information Agency (1998–99: Vol. 1, 46; Vol. 2, 42).

✦ **FIGURE 3.7** ✦

**Does Science Benefit Human-
ity? (in percent)**

"In the long run, do you think the
scientific advances we are making
will help or harm humankind?"

SOURCE: *World Values Survey* (1994).

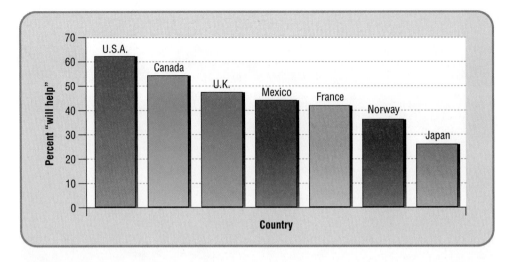

✦ **FIGURE 3.7** ✦

**Does Science Benefit Human-
ity? (in percent)**

"In the long run, do you think the
scientific advances we are making
will help or harm humankind?"

SOURCE: *World Values Survey* (1994).

up. German Nazism was a Big Historical Project. Its followers expected the Reich to enjoy 1,000 years of power. Communism was an even bigger Big Historical Project, mobilizing hundreds of millions of people for a future that promised to end inequality and injustice for all time. However, the biggest and most successful Big Historical Project was not so much a social movement as a powerful idea—the belief that progress is inevitable, that life will always improve, due mainly to the spread of democracy and scientific innovation.

The 20th century was unkind to Big Historical Projects. Russian communism lasted 74 years. German Nazism endured a mere 12. And the idea of progress fell on hard times as a hundred million soldiers and civilians died in wars, the forward march of democracy took wrong turns into fascism, communism, and regimes based on religious fanaticism, and pollution due to urbanization and industrialization threatened the planet. In the postmodern era, more and more people recognize that apparent progress, including scientific advances, often have negative consequences (Scott, 1998; see Figure 3.7). As the poet e e cummings once wrote, nothing recedes like progress.

Postmodernism has many parents, teachers, politicians, religious leaders, and not a few university professors worried. Given the eclectic mixing of cultural elements from different times and places, the erosion of authority, and the decline of consensus around core values, how can we make binding decisions? How can we govern? How can we teach children and adolescents the difference between right and wrong? How can we transmit accepted literary tastes and artistic standards from one generation to the next? These are the kinds of issues that plague people in positions of authority today.

Although their concerns are legitimate, many authorities seem not to have considered the other side of the coin. The postmodern condition, as we have described it above, empowers ordinary people and makes them more responsible for their own fate. It frees people to adopt religious, ethnic, and other identities they are comfortable with, as opposed to identities imposed on them by others. It makes them more tolerant of difference. That is no small matter in a world torn by group conflict. And the postmodern attitude encourages healthy skepticism about rosy and naive scientific and political promises.

Thus, the news about postmodern culture is not all bad. However, as you will now see, it's not all good either.

Culture as Constraint

We noted above that culture has two faces. One we labeled "freedom," the other "constraint." Diversity, globalization, and postmodernism are all aspects of the new freedoms that culture allows us today. We now turn to an examination of two contemporary aspects of culture that act as constraining forces on our lives: rationalization and consumerism.

Rationalization

In 14th-century Europe, an upsurge in demand for textiles caused loom owners to look for ways of increasing productivity. To that end, they imposed longer hours on loom workers. They also turned to a new technology for assistance: the mechanical clock. They installed public clocks in town squares. The clocks, known as *Werkglocken* ("work clocks") in German, signaled the beginning of the workday, the timing of meals, and quitting time.

Workers were accustomed to enjoying many holidays and a fairly flexible and vague work schedule regulated only approximately by the seasons and the rising and setting of the sun. The regimentation imposed by the work clocks made life harder. So the workers staged uprisings to silence the clocks. But to no avail. City officials sided with the employers and imposed fines for ignoring the *Werkglocken.* Harsher penalties, including death, were imposed on anyone trying to use the clocks' bells to signal a revolt (Thompson, 1967).

Now, more than 600 years later, many people are, in effect, slaves of the *Werkglock.* This is especially true of big-city North American couples who are employed full-time in the paid labor force and have preteen children. For them, life often seems an endless round of waking up at 6:30 a.m., getting everyone washed and dressed, preparing the kids' lunches, getting them out the door in time for the school bus or the car pool, driving to work through rush-hour traffic, facing the speedup at work that resulted from the recent downsizing, driving back home through rush-hour traffic, preparing dinner, taking the kids to their soccer game, returning home to clean up the dishes and help with homework, getting the kids washed, brushed, and into bed, and (if you haven't brought some office work home) grabbing an hour of TV before collapsing, exhausted, for $6\frac{1}{2}$ hours before the story repeats itself. Life is less hectic for residents of small towns, unmarried people, couples without small children, retirees, and the unemployed. But the lives of most people are so packed with activities that time must be carefully regulated, each moment precisely parceled out so that we may tick off item after item from an ever growing list of tasks that needs to be completed on schedule (Schor, 1992).

After more than 600 years of conditioning, it is unusual for people to rebel against the clock in the town square anymore. In fact, we now wear a watch on our wrist without giving it a second thought, as it were. This signifies that we have accepted and internalized the regime of the *Werkglock.* Allowing clocks to precisely regulate our activities seems the most natural thing in the world—which is a pretty good sign that the internalized *Werkglock* is, in fact, a product of culture.

Is the precise regulation of time rational? It certainly is rational as a means of ensuring the goal of efficiency. Minding the clock maximizes how much work you get done in a day. The regulation of time makes it possible for trains to run on schedule and university classes to begin punctually. But is minding the clock rational as an end in itself? For many people, it is not. They complain that the precise regulation of time has gotten out of hand. Life has simply become too hectic for many people to enjoy. In this sense, a *rational*

Have we come to depend too heavily on the *Werkglock?* Harold Lloyd in "Safety Last" (1923).

◆ **FIGURE 3.8** ◆

The Rationalization of Chinese Script

Reprinted here are the Chinese characters for "listening" *(t'ing)* in traditional Chinese script (left) and simplified, modern script (right). Each character is composed of several word-symbols. In classical script, listening is depicted as a process involving the eyes, the ears, and the heart. It implies that listening demands the utmost empathy and involves the whole person. In contrast, modern script depicts listening as something that merely involves one person speaking and the other "weighing" speech. Modern Chinese script has been rationalized. Has empathy been lost in the process?

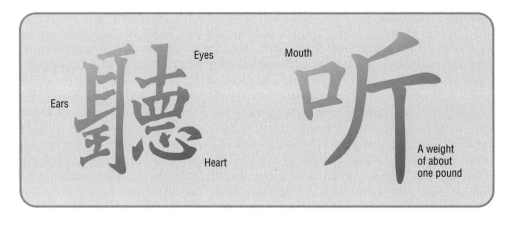

means (the *Werkglock*) has been applied to a *given goal* (maximizing work) but has led to an *irrational end* (a hectic life).

This, in a nutshell, is Max Weber's thesis about the rationalization process. **Rationalization,** in Weber's usage, means: (a) the application of the most efficient means to achieve given goals and (b) the unintended, negative consequences of doing so. Weber claimed that rationality of means has crept into all spheres of life, leading to unintended consequences that dehumanize and constrain us (see Figure 3.8).

Weber believed that the rationalization process is exemplified by bureaucracies—large, impersonal organizations composed of many clearly defined positions arranged in a hierarchy (see Chapter 5, "Interaction and Organization"). Modern bureaucracies, Weber said, are increasingly influential organizations. The factory, the government office, the military, the system of higher education, and the institutions of science are all bureaucratically organized. Yet bureaucracies are composed of nonelected officials. Consequently, they concentrate power and threaten democracy. Moreover, bureaucracies discourage officeholders from considering what the goals of their organization ought to be. Bureaucrats are asked only to determine the best way of achieving the goals defined by their superiors. Officeholders thus lose their spontaneity, their inventiveness, and all opportunity to act heroically. It is "horrible to think," wrote Weber, "that the world could one day be filled with nothing but those little cogs, little men clinging to little jobs and striving towards bigger ones . . ." (quoted in Mayer, 1944: 127).

Sociologist George Ritzer argues that, just as the modern bureaucracy epitomized the rationalization process for Weber at the turn of the 20th century, the McDonald's restaurant is the epitome of rationalization today (Ritzer, 1993; 1996). Instead of adapting institutions to the needs of people, says Ritzer, people must increasingly adapt to the needs of "McDonaldization." We are dehumanized in the process (Leidner, 1993; Reiter, 1991).

As Ritzer shows, McDonald's has lunch down to a science. The meat and vegetables used to prepare your meal must meet minimum standards of quality and freshness. Each food item contains identical ingredients. Each portion is carefully weighed and cooked according to a uniform and precisely timed process. McDonald's executives have carefully thought through every aspect of your lunch. They have turned its preparation into a model of rationality. With the goal of making profits, they have optimized food preparation to make it as fast and as cheap as possible.

Unfortunately, however, the rationalization of lunch dehumanizes both staff and customers. For instance, meals are prepared by nonunionized, uniformed workers who receive minimum wage. They must execute their tasks quickly and within specific time limits. To boost sales, they must smile as they recite fixed scripts ("Would you like some fries with your burger?"). Nearly half of all McDonald's employees are so dissatisfied with their work they quit after a year or less. To deal with this problem, McDonald's is now field-testing vending machines that will be used to replace staff and boost sales. Anyone for an e-burger? ("McDonald's Testing E-burgers," 1999).

Max Weber likened the modern era to an "iron cage." Sociology promises to teach us both the dimensions of the cage and the possibilities for release.

Meanwhile, customers are expected to spend as little time as possible eating the food—hence the drive-through window, chairs designed to be comfortable for only about 20 minutes, and small express outlets in subways and department stores where customers eat standing up or on the run. Customers are also expected to eat unhealthy food. A Big Mac, small fries, medium Coke, and an apple Danish can bring you up to 67% of your recommended daily calorie intake and 88% of your recommended daily fat intake (calculated from McDonald's, 1999). That is why physicians and nutritionists regularly decry the popularity of fast food.

Nonetheless, powerful forces make the Big Mac popular. On the demand side, fast food fits the rushed lifestyle of many individuals and families in the more affluent countries of the world and the growing middle class in the developing countries. (More than half of McDonald's sales are now outside North America). On the supply side, one can make big profits by turning meal preparation into a mass production industry. Motivated

"McDonaldization" is a global phenomenon, as this busy McDonald's restaurant in Beijing, China, suggests.

by these forces, rationality of means (turning lunch into a science) results in irrationality of ends (dehumanizing staff and customers).

As the examples of the *Werkglock,* bureaucracy, and McDonald's show, rationalization enables us to do just about everything more efficiently, but at a steep cost. Because it is so widespread, rationalization is one of the most constraining aspects of culture today. In Weber's view, it makes life in the modern world akin to living inside an "iron cage."

The second constraining aspect of culture that we wish to examine is consumerism. **Consumerism** is the tendency to define ourselves in terms of the goods and services we purchase.

Consumerism

In 1998, apparel sales in North America were lagging. As a result, the GAP launched a new ad campaign to help revitalize sales. The company hired Hollywood talent to create a slick and highly effective series of TV spots for khaki pants. According to the promotional material for the ad campaign, the purpose of the ads was to "reinvent khakis," that is, to stimulate demand for the pants. In *Khakis rock,* "skateboarders and in-line skaters dance, glide, and fly to music by the Crystal Method." In *Khakis groove,* "hip-hop dancers throw radical moves to the funky beat of Bill Mason." In *Khakis swing,* "two couples break away from a crowd to demonstrate swing techniques to the vintage sounds of Louis Prima" (Gap.com, 1999).

About 55 seconds of each ad featured the dancers. During the last 5 seconds, the words "GAP khakis" appeared on the screen. The GAP followed a similar approach in its 2000 ad campaign, inspired by the 1957 musical, *West Side Story.* The 30-second spots replaced the play's warring street gangs, the Jets and the Sharks, with fashion factions of their own, the Khakis and the Jeans. Again, most of the ad was devoted to the riveting dance number. The pants were mentioned for only a few seconds at the end.[3]

As the imbalance between stylish come-on and mere information suggests, the people who created the ads understood well that it was really the appeal of the dancers that would sell the pants. They knew that to stimulate demand for their product, they had to associate the khakis with desirable properties such as youth, good health, coolness, popularity, beauty, and sex. As an advertising executive said in the 1940s: "It's not the steak we sell. It's the sizzle."

Because advertising stimulates sales, there is a tendency for business to spend more on advertising over time (see Figure 3.9). Because advertising is widespread, most people unquestioningly accept it as part of their lives. In fact, many people have *become* ads.

Web Research Projects
Anti-Consumerism

✦ **FIGURE 3.9** ✦
Advertising as Percent of Gross Domestic Product (GDP), United States, 1975–98

SOURCE: Television Bureau of Advertising (2000).

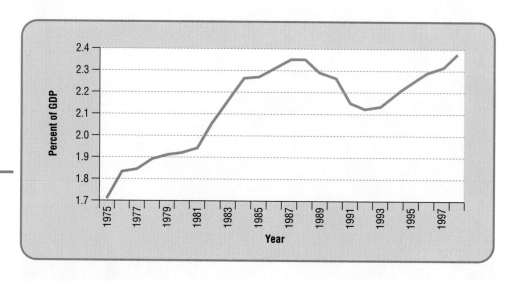

---

[3]To view the Gap ads on the World Wide Web, go to http://www.AdCritic.com and search for "Gap."

Thus, when your father was a child and quickly threw on a shirt, allowing a label to hang out, your grandmother might admonish him to "tuck in that label." In contrast, many people today proudly display consumer labels as marks of status and identity. Advertisers teach us to associate the words "Gucci" and "Nike" with different kinds of people, and when people display these labels on their clothes they are telling us something about the kind of people they are. Advertising becomes us.

Where do you fit in? If you don't display labels on your clothes, what does your reluctance to do so tell people about who you are? If you do display labels on your clothes, which ones are they? What do the labels tell others about who you are? Do your friends display clothing labels similar to yours? How about people you dislike? Do you think it is reasonable to conclude that clothing labels are cultural artifacts that increase the solidarity of social groups and segregate them from other groups?

The rationalization process, when applied to the production of goods and services, enables us to produce more efficiently, to have more of just about everything than our parents did. But it is consumerism, the tendency to define our selves in terms of the goods we purchase, that ensures all the goods we produce will be bought. Of course, we have lots of choice. We can select from dozens of styles of running shoes, cars, toothpaste, and all the rest. We can also choose to buy items that help to define us as members of a particular **subculture,** adherents of a set of distinctive values, norms, and practices within a larger culture. But, regardless of individual tastes and inclinations, nearly all of us have one thing in common: we tend to be good consumers. We are motivated by advertising, which is based on the accurate insight that people will likely be considered cultural outcasts if they fail to conform to stylish trends. By creating those trends, advertisers push us to buy even if we must incur large debts to do so (Schor, 1999). That is why North Americans' "shop-till-you-drop" lifestyle prompted French sociologist Jean Baudrillard to remark pointedly that even what is best in America is compulsory (Baudrillard, 1988 [1986]).

As is the case for rationalization, consumerism has unintended, negative consequences. For example, our culture of excessive consumption causes environmental degradation. We discuss this problem in detail in Chapter 18 ("Technology and the Global Environment"). In addition, consumerism is remarkably effective in taming expressions of freedom and individualism, including acts of dissent and rebellion. That is, deviations from mainstream culture often lose their power to drive change and get turned simply into means of making money. This is nowhere more apparent than in the recent history of hip-hop, a case to which we now turn.

What is being sold here? The pants or the attitude?

Taming Subcultural Revolt: The Case of Hip-Hop

Music can sometimes act as a kind of social cement. Reflecting the traditions, frustrations, and ambitions of the communities that create it, music can help otherwise isolated voices sing in unison. It can help individuals shape a collective identity. Sometimes, music can even inspire people to engage in concerted political action (Mattern, 1998).

Under some circumstances, however, music can have the opposite effect. It can individualize feelings of collective unrest and thereby moderate dissent. This occurs, for example, when music that originates as an act of rebellion is turned into a mass-marketed commodity. Music that develops in opposition to the mainstream typically loses its edge when it is transformed into something that can be bought and sold on a wide scale. By commodifying dissent and broadening its appeal to a large and socially heterogeneous audience, consumer culture renders it mainstream. The way it accomplishes this remarkable feat is well illustrated by the musical genre called hip-hop.[4]

Hip-hop originated in the social conditions of African-American inner-city youth in the 1970s and 1980s. During those decades, manufacturing industries left the cities for suburban or foreign locales where land values were lower and labor was less expensive. Unemployment among African-American youth rose to more than 40%. Middle-class blacks left the inner city for the suburbs. This robbed the remaining young people of successful role models they could emulate. The out-migration also eroded the taxing capacity of municipal governments, leading to a decline in public services. Meanwhile, the American public elected conservative governments at the state and federal levels. The conservative political climate led to cuts in funding for social services and education, thus deepening the destitution of ghetto life (Wilson, 1987).

Understandably, many young African Americans grew angrier as the conditions of their existence worsened. With few legitimate prospects for advancement, some turned to crime and, in particular, to the drug trade.

In the late 1970s, cocaine was expensive and demand for the drug was flat. So, in the early 1980s, Colombia's Medellin drug cartel introduced a less expensive form of cocaine called rock or crack. Crack offered a quick and intense high. It was highly addictive. It gave people a temporary escape from hopelessness. As a result, it was soon widely used in the inner city. Turf wars spread as gangs tried to outgun each other for control of the local traffic. The sale and use of crack became so widespread it corroded the inner-city African-American community (Davis, 1990).

The conditions described above gave rise to a new musical form: hip-hop. Stridently at odds with the values and tastes of both whites and middle-class African Americans, hip-hop described and glorified the mean streets of the inner city while holding the police, the mass media, and other pillars of society in utter contempt. Furthermore, hip-hop tried to offend middle-class sensibilities, black and white, by using highly offensive language.

In 1988, more than a decade after its first stirrings, hip-hop reached its political high point with the release of the CD, *It Takes a Nation to Hold Us Back* by Chuck D and Public Enemy. In "Don't Believe the Hype," Chuck D accused the mass media of maliciously distributing lies. In "Black Steel in the Hour of Chaos," he charged the FBI and the CIA with assassinating the two great leaders of the African-American community in the 1960s, Martin Luther King, Jr. and Malcolm X. In "Party for Your Right to Fight" he blamed the federal government for organizing the fall of the Black Panthers, the radical black nationalist party of the 60s. Here, it seemed, was an angry expression of subcultural revolt that could not be tamed (Best and Kellner, 1999).

However, there were elements in hip-hop that soon transformed it (Bayles, 1994: 341–62; Neal, 1999: 144–8). In the first place, early, radical hip-hop was not written as

---

[4]Scholars and music buffs disagree about the exact difference and degree of overlap between hip-hop and rap. They seem to agree, however, that rap refers to a particular tradition of black rhythmic *lyrics* while hip-hop refers to a particular black *beat* (often jerky and offbeat) mixed with samples of earlier recordings and LP scratches (now largely *passé*). See Mink-Cee (2000). Here we use the terms interchangeably.

dance music. It therefore cut itself off from a large audience. Moreover, hip-hop entered a self-destructive phase with the emergence of Gangsta rap, which extolled the criminal lifestyle, denigrated women, and replaced politics with drugs, guns, and machismo. The release of Ice T's "Cop Killer" in 1992 provoked strong political opposition from Republicans and Democrats, white church groups and black middle-class associations. Time/Warner was forced to withdraw the song from circulation. The sense that hip-hop had reached a dead end, or at least a turning point, grew in 1996, when rapper Tupac Shakur was murdered in the culmination of a feud between two hip-hop record labels (Death Row in Los Angeles and Bad Boy in New York) (Springhall, 1998: 149–51).

If these events made it seem that hip-hop was self-destructing, the police and insurance industries helped to speed up its demise. In 1988, a group called NWA released "F___ the Police," a critique of police violence against black youth. Law enforcement officials in several cities dared the group to perform the song in public, threatening to detain the performers or shut down their shows. Increasingly thereafter, ticket holders at rap concerts were searched for drugs and weapons, and security was tightened. Insurance companies, afraid of violence, substantially raised insurance rates for hip-hop concerts, making them a financial risk. Soon, the number of venues willing to sponsor hip-hop concerts dwindled.

While the developments noted above did much to mute the political force of hip-hop, the seduction of big money did more. As early as 1982, with the release of Grandmaster Flash and the Furious Five's "The Message," hip-hop began to win acclaim from mainstream rock music critics. With the success of Public Enemy in the late 80s, it became clear there was a big audience for hip-hop. Much of that audience was composed of white youths, "who relished . . . the subversive 'otherness' that the music and its purveyors represented" (Neal, 1999: 144). Sensing the opportunity for profit, major media corporations, such as Time/Warner, Sony, CBS/Columbia, and BMG Entertainment signed distribution deals with the small independent recording labels that had formerly been the exclusive distributors of hip-hop CDs. In 1988, *Yo! MTV Raps* debuted. The program brought hip-hop to middle America.

Most hip-hop recording artists proved they were more than eager to forego politics for commerce. For instance, the rap group, WU-Tang Clan, started a line of clothing called WU Wear, and companies as diverse as Tommy Hilfiger, Timberland, Starter, and Versace started to market clothing influenced by ghetto styles with the help of major hip-hop recording artists. By the early 1990s, hip-hop was no longer just a musical form but a commodity with spin-offs. Rebellion had been turned into mass consumption.

No rapper has done a better job of turning rebellion into a commodity than Sean Combs, better known as Puff Daddy (and more recently as "P. Diddy"). At first blush, Puff Daddy seems to promote rebellion. For example, the liner notes for his 1999 hit CD, *Forever,* advertise his magazine, *Notorious,* as follows:

> There is a revolution out there. Anyone can do anything. There are no rules. There are no restrictions. *Notorious* magazine presents provocative profiles of rebels, rulebreakers and mavericks—Notorious people who are changing the world with their unique brand of individuality.
>
> Our goal is to inform and inspire, to educate and elevate the infinite range of individual possibility . . . In essence, *Notorious* is for everyone who wants to live a sexy, daring life—a life that makes a difference. After all, you can't change the world without being a little . . . *Notorious* (Combs, 1999).

Sean "Puffy" Combs

This individualistic brand of dissent appeals to a broad audience, much of it white and middle-class. As his video director, Martin Weitz, observed in an interview for *Elle* magazine, Puff Daddy's market is not the ghetto: "No ghetto kid from Harlem is going to buy Puffy. They think he sold out. It's more like the 16-year-old white girls in the Hamptons, baby!" (quoted in Everett-Green, 1999).

In 1998, *Forbes* magazine ranked Puff Daddy 15th among top-earning entertainment figures, with an income of $53.5 million (Forbes.com, 1998). He is entirely forthright about his apolitical, self-enriching aims. In his 1997 song, "I Got the Power," Puff Daddy referred to himself as "that n____ with the gettin money game plan" (Combs and The Lox, 1997). And in *Forever,* he reminds us: "N____ get money, that's simply the plan."

Hip-hop emerged among poor African-American inner-city youth as a counsel of despair with strong political overtones. Much of it has become an apolitical commodity that increasingly appeals to a white, middle-class audience. The story of how hip-hop was tamed is not unique. Similar stories could be told about the blues that originated in the Mississippi delta at the turn of the 20th century, rock and roll in the 1960s, and British punk music in the 1980s. These and other musical genres are testimony to the capacity of postmodern culture to constrain expressions of freedom, individualism, dissent, and rebellion (Frank and Weiland, 1997). They are compelling illustrations of postmodern culture's second face.

SUMMARY

1. Humans have been able to adapt to their environments because they can create culture. In particular, the ability to create symbols, make tools, and cooperate has enabled humans to thrive.

2. No hard evidence supports the view that specific human behaviors and social arrangements are biologically determined, although biology does set human limits and potentials.

3. We can see the contours of culture most sharply if we are neither too deeply immersed in it nor too much removed from it. Understanding culture requires refraining from taking your own culture for granted and judging other cultures by the standards of your own.

4. Culture has two faces. In some respects, culture provides us with increasing opportunities to exercise our freedom. The growth of multiculturalism, globalization, and postmodernism reflect this tendency. In other respects, culture constrains us, putting limits on what we can become. The growth of rationalization and consumerism reflect this tendency.

5. Advocates of multiculturalism want school and college curricula to reflect the country's growing ethnic and racial diversity. They also want school and college curricula to stress that all cultures have equal value. They believe that multicultural

education will promote self-esteem and economic success among members of racial minorities. Critics fear that multiculturalism results in declining educational standards. They believe that multicultural education causes political disunity and interethnic and interracial conflict. And they argue that it promotes an extreme form of cultural relativism.

6. The globalization of culture has resulted from the growth of international trade and investment, ethnic and racial migration, influential "transnational" organizations, and inexpensive travel and communication.

7. Postmodernism involves an eclectic mixing of elements from different times and places, the decline of authority, and the erosion of consensus around core values.

8. Rationalization involves the application of the most efficient means to achieve given goals and the unintended, negative consequences of doing so. Rationalization is evident in the growth and operation of bureaucracies, in the increasingly regulated use of time, and in many other areas of social life.

9. Consumerism is the tendency to define ourselves in terms of the goods we purchase. Excessive consumption puts limits on who we can become, constrains our capacity to dissent from mainstream culture, and degrades the natural environment.

GLOSSARY

Abstraction is the human capacity to create general ideas, or ways of thinking that are not linked to particular instances. For example, languages, mathematical notations, and signs allow us to classify experience and generalize from it.

Consumerism is the tendency to define ourselves in terms of the goods we purchase.

Cooperation is the human capacity to create a complex social life.

Cultural relativism is the belief that all cultures have equal value.

Culture is the sum of practices, languages, symbols, beliefs, values, ideologies, and material objects that people create to deal with real-life problems. Cultures enable people to adapt to, and thrive in, their environments.

Ethnocentrism is the tendency to judge other cultures exclusively by the standards of one's own.

Globalization is the process by which formerly separate economies, states, and cultures are being tied together.

High culture is culture consumed mainly by upper classes.

Mass culture (*see* Popular culture).

Material culture is composed of the tools and techniques that enable people to get tasks accomplished.

Supporters of **multiculturalism** argue that the curricula of America's public schools and colleges should reflect the country's ethnic and racial diversity and recognize the equality of all cultures.

Norms are generally accepted ways of doing things.

Popular culture (or **mass culture**) is culture consumed by all classes.

Postmodernism is characterized by an eclectic mixing of cultural elements and the erosion of consensus.

Production is the human capacity to make and use tools. It improves our ability to take what we want from nature.

Rationalization is the application of the most efficient means to achieve given goals and the unintended, negative consequences of doing so.

Rites of passage are cultural ceremonies that mark the transition from one stage of life to another (e.g., baptisms, confirmations, weddings) or from life to death (e.g., funerals).

Sanctions are rewards and punishments intended to ensure conformity to cultural guidelines.

The system of **social control** is the sum of sanctions in society by means of which conformity to cultural guidelines is ensured.

A **subculture** is a set of distinctive values, norms, and practices within a larger culture.

A **symbol** is anything that carries a particular meaning, including the components of language, mathematical notations, and signs. Symbols allow us to classify experience and generalize from it.

QUESTIONS TO CONSIDER

1. We imbibe culture but we also create it. What elements of culture have you created? Under what conditions were you prompted to do so? Was your cultural contribution strictly personal or was it shared with others? Why?

2. Select a subcultural practice that seems odd, inexplicable, or irrational to you. By interviewing members of the subcultural group and reading about them, explain how the subcultural practice you chose makes sense to members of the subcultural group.

3. Do you think the freedoms afforded by postmodern culture outweigh the constraints it places on us? Why or why not?

WEB RESOURCES

Companion Web Site for This Book

http://sociology.wadsworth.com

Begin by clicking on the Student Resources section of the Web site. Choose "Introduction to Sociology" and finally the Brym and Lie book cover. Next, select the chapter you are currently studying from the pull-down menu. From the Student Resources page you will have easy access to InfoTrac College Edition®, MicroCase Online exercises, additional Web links, and many resources to aid you in your study of sociology, including practice tests for each chapter.

Infotrac Search Terms

These search terms are provided to assist you in beginning to conduct research on this topic by visiting http://www.infotraccollege.com/wadsworth.

Consumerism	**Multiculturalism**
Culture	**Postmodernism**
Globalization	**Rationalization**

Recommended Web Sites

Bernard Barber "Jihad vs. McWorld," on the World Wide Web at http://www.theatlantic.com/politics/foreign/barberf.htm is a brief, masterful analysis of the forces that are simultaneously making world culture more homogeneous and more heterogeneous. The article was originally published in *The Atlantic Monthly* (March 1992). For the full story, see Bernard Barber *Jihad vs. McWorld: How Globalism and Tribalism are Reshaping the World* (New York: Ballantine Books, 1996).

Adbusters is an organization devoted to analyzing and criticizing consumer culture. Its provocative Web site is at http://adbusters.org.

Sharon Zupko's Cultural Studies Center is our favorite site on the sociology of popular culture. Visit it at http://www.popcultures.com.

The Resource Center for Cyberculture Studies is an organization devoted to studying emerging cultures on the World Wide Web. Its Web site is at http://otal.umd.edu/rccs.

SUGGESTED READINGS

Wendy Griswold. "The Sociology of Culture: Four Good Arguments (And One Bad One)," *Acta Sociologica* (35: 1992) 322–28. Concisely analyzes major issues in the subfield.

R. C. Lewontin. *Biology as Ideology: The Doctrine of DNA* (New York: Harper Collins, 1991). A brilliant short critique of sociobiology, evolutionary psychology, and related ideologies by one of the world's leading scientists.

Mark Anthony Neal. *What the Music Said: Black Popular Music and Black Public Culture* (New York: Routledge, 1999). A fine case study in the sociology of culture. Shows how post–World War II black popular music emerged from the struggle to maintain community in the face of poverty and brutality.

IN THIS CHAPTER, YOU WILL LEARN THAT:

✦ The view that social interaction unleashes human abilities is supported by studies showing that children raised in isolation do not develop normal language and other social skills.

✦ While the socializing influence of the family decreased in the 20th century, the influence of schools, peer groups, and the mass media increased.

✦ People's identities change faster, more often, and more completely than they did just a couple of decades ago; the self has become more plastic.

✦ The main socializing institutions often teach children and adolescents contradictory lessons, making socialization a more confusing and stressful process than it used to be.

✦ Declining parental supervision and guidance, increasing assumption of adult responsibilities by youth, and declining participation in extracurricular activities are transforming the character of childhood and adolescence today.

SOCIALIZATION

SOCIAL ISOLATION AND THE CRYSTALLIZATION OF SELF-IDENTITY

One day in the year 1800, a 10- or 11-year-old boy walked out of the woods in southern France. He was filthy, naked, and unable to speak and had not been toilet trained. After being taken by the police to a local orphanage, he repeatedly tried to escape and refused to wear clothes. No parent ever claimed him. He became known as "the wild boy of Aveyron." A thorough medical examination found no major abnormalities of either a physical or mental nature. Why, then, did the boy seem more animal than human? Apparently because, until he walked out of the woods, he had been raised in isolation from other human beings (Shattuck, 1980).

Similar horrifying reports lead to the same conclusion. Occasionally a child is found locked in an attic or a cellar, where he or she saw another person for only short periods each day to receive food. Like the wild boy of Aveyron, such children rarely develop normally. Typically, they remain disinterested in games. They cannot form intimate social relationships with other people. They develop only the most basic language skills.

Some of these children may suffer from congenitally subnormal intelligence. It is uncertain how much and what type of social contact they had before they were discovered. Some may have been abused. Therefore, their condition may not be due only to social isolation. However, these examples do at least suggest that the ability to learn culture and become human is only a potential. To be actualized, **socialization** must unleash this human potential. Socialization is the process by which people learn their culture—including norms, values, and roles—and become aware of themselves as they interact with others.

More convincing evidence of the importance of socialization in unleashing human potential comes from a study conducted by René Spitz (Spitz, 1945; 1962). Spitz compared children who were being raised in an orphanage with children who, for medical reasons, were being raised in a nursing home. Both institutions were hygienic and provided good food and medical care. However, while their mothers cared for the babies in the nursing home, just six nurses cared for the 45 orphans. The orphans therefore had much less contact with other people. Moreover, from their cribs, the nursing home infants could taste a slice of society. They saw other babies playing and receiving care. They saw mothers, doctors, and nurses talking, cleaning, serving food, and giving medical treatment. In contrast, it was established practice in the orphanage to hang sheets from the cribs to prevent the infants from seeing the activities of the institution. Depriving the infants of social stimuli for most of the day apparently made them less demanding.

Social deprivation had other effects too. Because of the different patterns of child care described above, by the age of 9–12 months the orphans were more susceptible to infections and had a higher death rate than the babies in the nursing home. By the time they were 2–3 years old, all the children from the nursing home were walking and talking, compared to fewer than 8% of the orphans. Normal children begin to play with their own genitals by the end of their first year. Spitz found that the orphans began this sort of play only in their fourth year. He took this as a sign they might have an impaired sexual life when they reached maturity. This had happened to rhesus monkeys raised in isolation. Spitz's natural experiment thus amounts to quite compelling evidence for the importance of childhood socialization in making us fully human. Without childhood socialization, most of our human potential remains unlocked.

The formation of a sense of self continues in adolescence. This is a particularly turbulent period of rapid self-development. Consequently, many people can remember experiences from their youth that helped to crystallize their self-identity. Do you? Robert Brym clearly recalls one such defining moment.

"I can date precisely the pivot of my adolescence," says Robert. "I was in Grade 10. It was December 16th. At 4 p.m. I was a nobody and knew it. Half an hour later, I was walking home from school, delighting in the slight sting of snowflakes melting on my upturned face, knowing I had been swept up in a sea change.

In the 1960s, researchers Harry and Margaret Harlow placed baby rhesus monkeys in various conditions of isolation to witness and study the animals' reactions. Among other things, they discovered that baby monkeys raised with an artificial mother made of wire mesh, a wooden head, and the nipple of a feeding tube for a breast, were later unable to interact normally with other monkeys. However, when the artificial mother was covered with a soft terry cloth, the infant monkeys clung to it in comfort and later revealed less emotional distress. Infant monkeys preferred the cloth mother even when it gave less milk than the wire mother. The Harlows concluded that emotional development requires affectionate cradling.

"About two hundred students sat impatiently in the auditorium that last day of school before the winter vacation. We were waiting for Mr. Garrod, the English teacher who headed the school's drama program, to announce the cast of *West Side Story*. I was hoping for a small speaking part and was not surprised when Mr. Garrod failed to read my name as a chorus member. However, as the list of remaining characters grew shorter, I became despondent. Soon only the leads remained. I knew an unknown kid in Grade 10 couldn't possibly be asked to play Tony, the male lead. Leads were almost always reserved for more experienced Grade 12 students.

"Then the thunderclap. 'Tony,' said Mr. Garrod, 'will be played by Robert Brym.'

"'Who's Robert Brym?' whispered a girl two rows ahead of me. Her friend merely shrugged in reply. If she had asked *me* that question, I might have responded similarly. Like nearly all 15-year-olds, I was deeply involved in the process of figuring out exactly who I was. I had little idea of what I was good at. I was insecure about my social status. I wasn't sure what I believed in. In short, I was a typical teenager. I had only a vaguely defined sense of self.

"A sociologist once wrote that 'the central growth process in adolescence is to define the self through the clarification of experience and to establish self-esteem' (Friedenberg, 1959: 190). From this point of view, playing Tony in *West Side Story* turned out to be the first section of a bridge that led me from adolescence to adulthood. Playing Tony raised my social status in the eyes of my classmates, made me more self-confident, taught me I could be good at something, helped me to begin discovering parts of myself I hadn't known before, and showed me that I could act rather than merely be acted upon. In short, it was through my involvement in the play (and, subsequently, in many other plays throughout high school) that I began to develop a clear sense of who I am."

The crystallization of self-identity during adolescence is just one episode in a life-long process of socialization. To paint a picture of the socialization process in its entirety, we must first review the main theories of how one's sense of self develops during early childhood. We then discuss the operation and relative influence of society's main socializing institutions or "agents of socialization": families, schools, peer groups, and the mass media. In these settings, we learn, among other things, how to control our impulses, think of ourselves as members of different groups, value certain ideals, and perform various roles. (A **role** is the behavior expected of a person occupying a particular position in society.) You will see that these institutions do not always work hand-in-hand to produce happy, well-adjusted adults. They often give mixed messages and are often at odds with each other. That is, they teach children and adolescents different and even contradictory lessons. You will also see that while recent developments give us more freedom to decide who we are, they can make socialization more disorienting than ever before. Finally, in the concluding section of this chapter, we examine how decreasing supervision and guidance by adult family members, increasing assumption of adult responsibilities by youth, and declining participation in extracurricular activities are changing the nature of childhood and adolescence today. Some analysts even say that childhood and adolescence are vanishing before our eyes. Thus, the main theme of this chapter is that the development of one's self-identity is often a difficult and stressful process—and it is becoming more so.

It is during childhood that the contours of one's self are first formed. We therefore begin by discussing the most important social-scientific theories of how the self originates in the first years of life.

THEORIES OF CHILDHOOD SOCIALIZATION

Freud

Socialization begins soon after birth. Infants cry out, driven by elemental needs, and are gratified by food, comfort, or affection. Because their needs are usually satisfied immediately, they do not at first seem able to distinguish themselves from their main caregivers,

Sigmund Freud (1856–1939) was the founder of psychoanalysis. Many issues have been raised about the specifics of his theories. Nevertheless, his main sociological contribution was his insistence that the self emerges during early social interaction and that early childhood experience exerts a lasting impact on personality development.

usually their mothers. However, social interaction soon enables infants to begin developing a self-image or sense of **self**—a set of ideas and attitudes about who they are as independent beings.

Sigmund Freud proposed the first social-scientific interpretation of the process by which the self emerges (Freud, 1962 [1930]; 1973 [1915–17]). Freud was the Austrian founder of psychoanalysis. He referred to the part of the self that demands immediate gratification as the **id.** According to Freud, a self-image begins to emerge as soon as the id's demands are denied. For example, at a certain point, parents usually decide not to feed and comfort a baby every time it wakes up in the middle of the night. The parents' refusal at first incites howls of protest. Eventually, however, the baby learns certain practical lessons from the experience—to eat more before going to bed, sleep for longer periods, and put itself back to sleep if it wakes up. Equally important, the baby begins to sense that its needs differ from those of its parents, that it has an existence independent of others, and that it must somehow balance its needs with the realities of life.

Because of many such lessons in self-control, including toilet training, the child eventually develops a sense of what constitutes appropriate behavior and a moral sense of right and wrong. Soon a personal conscience or, to use Freud's term, a **superego,** crystallizes. The superego is a repository of cultural standards. In addition, the child develops a third component of the self, the **ego.** According to Freud, the ego is a psychological mechanism that, in well-adjusted individuals, balances the conflicting needs of the pleasure-seeking id and the restraining superego.

In Freud's view, the emergence of the superego is a painful and frustrating process. In fact, said Freud, to get on with our daily lives we have to repress memories of denying the id immediate gratification. Repression involves storing traumatic memories in a part of the self that we are not normally aware of: the **unconscious.** Repressed memories influence emotions and actions even after they are stored away. Particularly painful instances of childhood repression may cause psychological problems of various sorts later in life, requiring therapy to correct. However, some repression is the cost of civilization. As Freud said, we cannot live in an orderly society unless we deny the id (Freud, 1962 [1930]).

Researchers have called into question many of the specifics of Freud's argument. Three criticisms stand out:

1. *The connections between early childhood development and adult personality are more complex than Freud assumed.* Freud wrote that when the ego fails to balance the needs of the id and the superego, individuals develop personality disorders. Typically, he said, this occurs if a young child is raised in an overly repressive atmosphere. To avoid later psychiatric problems, Freud and his followers recommended raising young children in a relaxed and permissive environment. Such an environment is characterized by prolonged breast-feeding, nursing on demand, gradual weaning, lenient and late bladder and bowel training, frequent mothering, freedom from restraint and punishment, and so forth. However, sociological research reveals no connection between these aspects of early childhood training and the development of well-adjusted adults (Sewell, 1958). One group of researchers who were influenced by Freud's theories tracked people from infancy to age 32 and made *incorrect* predictions about personality development in two thirds of the cases. They "had failed to anticipate that depth, complexity, problem-solving abilities, and maturity might derive from painful [childhood] experiences" (Coontz, 1992: 228).

2. *Many sociologists criticize Freud for gender bias in his analysis of male and female sexuality.* Freud argued that psychologically normal women are immature and dependent on men because they envy the male sexual organ. Women who are mature and independent he classified as abnormal. We discuss this fallacy in detail in Chapter 9 ("Sexuality and Gender").

3. *Sociologists often criticize Freud for neglecting socialization after childhood.* Freud believed that the human personality is fixed by about the age of 5. However,

BOX 4.1
SOCIOLOGY AT THE MOVIES

Nick Nolte and James Coburn in *Affliction*.

AFFLICTION (1997)

Nearly everyone accepts the importance of socialization in shaping people's personalities. In casual conversation, we talk about the influence of family and friends, neighborhood and school, and other agents of socialization in making us who we are. *Affliction,* starring Nick Nolte, Sissy Spacek, and James Coburn, is a powerful movie about socialization and its legacy. Based on a novel by Russell Banks, it shows that some individuals cannot overcome the impact of early socialization while others can.

James Coburn plays an alcoholic and abusive father. He thinks nothing of beating his wife and his two children. Nick Nolte plays the elder son who is afflicted by his father's curse. When he was a child, he was afraid of his father. As an adult, he tries to overcome his father's influence. Yet he cannot resist the lure of alcohol and violence. He is an unsuccessful police officer, and he

has trouble maintaining the love and respect of the women around him: his ex-wife, his daughter, and his new girlfriend. Although he tries to be caring and responsible, his dependence on alcohol, his quick temper, and his inclination to violence ultimately doom his good intentions. In the end, tragedy befalls him.

The movie is not, however, fatalistic. It does not suggest that childhood socialization casts the adult personality in stone. The movie's narrator, the younger brother, managed to break away from the affliction. The younger brother does not tell the audience

how this came about. However, he does give us a clue. He left his family and community to pursue higher education in the city. That is, he found another life, another set of social influences. Adult socialization in a new social context set him free.

As *Affliction* shows, then, socialization has a big impact on all of us. Childhood socialization is not, however, one's destiny. In making us think about the different paths taken by the two brothers, *Affliction* offers a good case study in the power of childhood socialization—and its limitations.

sociologists have shown that socialization continues throughout the life course (see Box 4.1). We devote much of this chapter to exploring socialization after early childhood.

Despite the shortcomings listed above, the sociological implications of Freud's theory are profound. His main sociological contribution was his insistence that the self emerges during early social interaction and that early childhood experience exerts a lasting impact on personality development. As we will now see, the great American social psychologist, George Herbert Mead, took these ideas in a still more sociological direction.

Mead

A century ago, the American sociologist, Charles Horton Cooley, introduced the idea of the "looking glass self." Cooley wrote that, when we interact with others, they gesture and react to us. Just as we see our physical body reflected in a mirror, so we see our social selves reflected in people's gestures and reactions. In other words, our feelings about who we are depend largely on how we see ourselves judged by others (Cooley, 1902).

George Herbert Mead (1934) took up and developed the idea of the looking glass self. Like Freud, Mead noted that a subjective and impulsive aspect of the self is present from birth. Mead called it simply the **I.** Again like Freud, Mead argued that a repository of

Much socialization takes place informally, with the participants unaware
they are being socialized. These girls
are learning gender roles as they go
to the mall dressed like Britney
Spears.

culturally approved standards emerges as part of the self during social interaction. Mead called this objective, social component of the self the **me.** However, while Freud focused on the denial of the id's impulses as the mechanism that generates the self's objective side, Mead drew attention to the unique human capacity to "take the role of the other" as the source of the me.

Mead understood that human communication involves seeing yourself from the point of view of other people. How, for example, do you interpret your mother's smile? Does it mean "I love you," "I find you humorous," or something else entirely? According to Mead, you can know the answer only if you use your imagination to take your mother's point of view for a moment and see yourself as she sees you. In other words, you must see yourself objectively, as a "me," to understand your mother's communicative act. All human communication depends on being able to take the role of the other, wrote Mead. The self thus emerges from people using symbols such as words and gestures to communicate. It follows that the "me" is not present from birth. It emerges only gradually during social interaction.

Unlike Freud, Mead did not think that the emergence of the self was traumatic. On the contrary, he thought it was fun. Mead saw the self as developing in four stages of role-taking. At first, children learn to use language and other symbols by *imitating* important people in their lives, such as their mother and father. Mead called such people **significant others.** Second, children pretend to *be* other people. That is, they use their imaginations to role-play in games such as "house," "school," and "doctor." Third, by the time they reach the age of about 7, children learn to play complex games requiring that they simultaneously take the role of *several* other people. In baseball, for example, the infielders have to be aware of the expectations of everyone in the infield. A shortstop may catch a line drive. If she wants to make a double play, she must almost instantly be aware that a runner is trying to reach second base and that the person playing second expects her to throw to second. If she hesitates, she probably cannot execute the double play. Once a child can think in this complex way, she can begin the fourth stage in the development of the self. This involves taking the role of what Mead called the **generalized other.** Years of experience may teach an individual that other people, employing the cultural standards of their society, usually regard her as funny or temperamental or intelligent. A person's image of these cultural standards and how they are applied to her is what Mead meant by the generalized other.

Recent Developments

Since Mead, psychologists have continued to study childhood socialization. For example, they have identified the stages in which thinking and moral skills develop from infancy to the late teenage years. Let us briefly consider some of their most important contributions.

The Swiss psychologist Jean Piaget divided the development of thinking (or "cognitive") skills during childhood into four stages (Piaget and Inhelder, 1969). In the first 2 years of life, he wrote, children explore the world only through their 5 senses. Piaget called this the "sensorimotor" stage of cognitive development. At this point in their lives, children's knowledge of the world is limited to what their senses tell them. They cannot think using symbols.

According to Piaget, children begin to think symbolically between the ages of 2 and 7. He called this the "preoperational" stage of cognitive development. Language and imagination blossom during these years. However, children are still unable to think abstractly. Piaget illustrated this by asking a series of 5- and 6-year-olds to inspect two identical glasses of colored water. He then asked them whether the glasses contained the same amount of colored water. All of the children said they did. Next, the children watched Piaget pour the water from one glass into a wide, low beaker and the water from the second glass into a narrow, tall beaker. Obviously, the water level was higher in the second beaker although the volume of water was the same in both containers. Piaget then asked each child whether the two beakers contained the same amount of water. Nearly all of the children said that the second beaker contained more water. Clearly, the abstract concept of volume had no meaning for them.

In contrast, most 7- or 8-year-old children understood that the volume of water is the same in both beakers, despite the different water levels. This suggests that abstract thinking begins at about the age of 7. Moreover, between the ages of 7 and 11, children are able to see the connections between causes and effects in their environment. Piaget called this the "concrete operational" stage of cognitive development. Finally, by about the age of 12, children develop the ability to think more abstractly and critically. This is the beginning of what Piaget called the "formal operational" stage of cognitive development.

Lawrence Kohlberg, an American social psychologist, took Piaget's ideas in a somewhat different direction. He showed how children's *moral* reasoning—their ability to judge right from wrong—also passes through developmental stages (Kohlberg, 1981). Kohlberg argued that young children distinguish right from wrong only on the basis of whether something gratifies their immediate needs. At this stage of moral growth, which Kohlberg labeled the "preconventional" stage, what is "right" is simply what satisfies the young child. For example, from the point of view of a 2-year-old, it is entirely appropriate to grab a cookie from a playmate and eat it. An abstract moral concept like theft has no meaning for the very young child.

Teenagers, in contrast, begin to think about right and wrong in terms of whether specific actions please their parents and teachers and are consistent with cultural norms. This is the "conventional" stage of moral growth in Kohlberg's terminology. At this stage, a child understands that theft is a proscribed act and that getting caught stealing will result in punishment.

Some people never advance beyond conventional morality. Others, however, develop the capacity to think abstractly and critically about moral principles. This is Kohlberg's "postconventional" stage of moral development. At this stage, one may ponder the meaning of such abstract terms as freedom, justice, and equality. One may question whether the laws of one's society or the actions of one's parents, teachers, or other authorities conform to lofty moral principles. For instance, a 19-year-old who believes the settlement of Europeans in North America involved the theft of land from native peoples is thinking in postconventional moral terms. Such an adolescent is applying abstract moral principles independently and is not merely accepting them as interpreted by authorities.

Modern psychology has done much to reveal the cognitive and moral dimensions of childhood development. However, from a sociological point of view, the main problem with this body of research is that it minimizes the extent to which society shapes the way we think. Thus, most psychologists assume that people pass through the same stages of mental development and think in similar ways, regardless of the structure of their society and their position in it. Many sociologists disagree with these assumptions.

A few psychologists do, too. The Belarusian psychologist Lev Vygotsky and the American educational psychologist Carol Gilligan offer the most sociological approaches to thinking about cognitive and moral development, respectively. For Vygotsky, ways of thinking are determined not so much by innate factors as they are by the nature of the social institutions in which individuals grow up. Consider, for example, the contrast between ancient China and Greece. The rice agriculture of ancient southern China required substantial cooperation among neighbors. It was centrally organized in a complex hierarchy within a large state. Harmony and social order were therefore central to ancient Chinese life. Ancient Chinese thinking, in turn, tended to stress the importance of mutual social obligation and consensus rather than debate. Ancient Chinese philosophies focused on the way in which wholes, not analytical categories, caused processes and events. In contrast, the hills and seashores of ancient Greece were suited more to small-scale herding and fishing than large-scale, centrally organized agriculture. Ancient Greek society was less socially complex than that of ancient China, it was politically decentralized, and it gave its citizens more personal freedom. As a result, ancient Greek thinking stressed personal agency. Debate was an integral part of politics. Philosophies tended to be analytical, which means, among other things, that processes and events were viewed as the result of discrete categories rather than whole systems. Markedly different civilizations grew up on these different cognitive foundations; ways of thinking depended less on innate characteristics than on the structure of society (Cole, 1995; Nisbett, Peng, Choi, and Norenzayan, 2001; Vygotsky, 1987).

In a like manner, Gilligan emphasized the sociological foundations of moral development in her studies of American boys and girls. She attributed differences in the moral development of boys and girls to the different cultural standards parents and teachers pass on to them (Gilligan, 1982; Gilligan, Lyons, and Hanmer, 1990; Brown and Gilligan, 1992). For example, Gilligan found that, unlike boys, girls suffer a decline in self-esteem between the ages of 5 and 18. She attributed this to their learning our society's cultural standards over time. Specifically, our society tends to define the ideal woman as eager to please and therefore nonassertive. Most girls learn this lesson as they mature, and their self-esteem suffers as a result. The fact that girls encounter more male and fewer female teachers and other authority figures as they age reinforces this lesson, according to Gilligan.

Influenced more by the approaches of Vygotsky and Gilligan than Piaget and Kohlberg, we now assess the contribution of various agents of socialization to the development of the self. These agents of socialization include families, schools, peer groups, and the mass media. We emphasize differences in socialization between societies, social groups, and historical periods. Our approach to socialization, therefore, is rigorously sociological.

In her research, Carol Gilligan attributes differences in the moral development of boys and girls to the different cultural standards parents and teachers pass on to them. By emphasizing that moral development is socially differentiated and does not follow universal rules, Gilligan has made a major sociological contribution to our understanding of childhood development.

AGENTS OF SOCIALIZATION

Families

Freud and Mead understood well that the family is the most important agent of **primary socialization,** the process of mastering the basic skills required to function in society during childhood. They argued that, for most babies, the family is the world. This is as true today as it was a hundred years ago. The family is well suited to providing the kind of careful, intimate attention required for primary socialization. The family is a small group. Its members are in frequent face-to-face contact. Child abuse and neglect exist, but most parents love their children and are therefore highly motivated to care for them. These characteristics make most families ideal even today for teaching small children everything from language to their place in the world.

The family into which one is born also exerts an *enduring* influence over the course of one's entire life. Consider the long-term effect of the family's religious atmosphere, for instance. We used data from the General Social Survey to examine respondents who, in their youth, had mothers who attended church once a month or more. As adults, 82% of these respondents attended church once a month or more themselves. In contrast, only 18% of them attended church once a month or less.[1] Clearly, the religious atmosphere of the family into which one is born exerts a strong influence on one's religious practice as an adult.

Despite the continuing importance of the family in socialization, things have changed since Freud and Mead wrote their important works in the early 1900s. They did not foresee how the relative influence of various socialization agents would alter during the next century. The influence of some socialization agents increased, while the influence of others—especially the family—declined.

The socialization function of the family was more pronounced a century ago, partly because adult family members were more readily available for child care than they are today. As industry grew across America, families left farming for city work in factories and offices. Especially after the 1950s, many women had to work outside the home for a wage to maintain an adequate standard of living for their families. Fathers, for the most part, did not compensate by spending more time with their children. In fact, because divorce rates have increased, and many fathers have less contact with their children after divorce, children probably see less of their fathers on average now than they did a century ago. As a result of these developments, child care—and therefore child socialization—became a big social problem in the 20th century.

The family is still an important agent of socialization, although its importance has declined since the 19th century.

Schools

For children over the age of 5, the child care problem was partly resolved by the growth of the public school system, which was increasingly responsible for **secondary socialization,** or socialization outside the family after childhood. American industry needed better trained and educated employees. Therefore, by 1918, every state required children to attend school until the age of 16 or the completion of grade 8. By the beginning of the 21st century, more than four fifths of Americans over the age of 25 had graduated from high school and about a quarter had graduated from college. This makes Americans the most highly educated people in the world.

Although schools help to prepare students for the job market, they do not necessarily give them an accurate picture of what the job market requires. In 1992, for example, a nationwide survey highlighted the mismatch between the ambitions of American high school students and the projected needs of the American economy in 2005 (Schneider and Stevenson, 1999: 77–8). The number of high school students wanting to become lawyers and judges was five times the projected number needed. The number who wanted to become writers, artists, entertainers, and athletes was 14 times higher than expected openings in 2005. At the other extreme, in 2005 there will be five times more administrative and clerical jobs than students now interested in such work. And there will be seven times more service jobs than teenagers wanting them. American high school students, it seems safe to say, often have unrealistically high expectations about the kinds of jobs they are likely to get when they finish their education (see Figure 4.1).

Instructing students in academic and vocational subjects is just one part of the school's job. In addition, a **hidden curriculum** teaches students what will be expected of them in the larger society once they graduate. The hidden curriculum teaches them how to be conventionally "good citizens." Most parents approve of this instruction. According to a survey conducted in the United States and the highly industrialized countries of Europe in 1998, the capacity of schools to socialize students is more important to the public than all academic subjects except mathematics (Galper, 1998).

---

[1]See Chapter 13, "Religion and Education," Table 13.4. The data are from 1989 since this is the last year mother's church attendance was measured in the GSS.

◆ **FIGURE 4.1** ◆

Adolescent Job Preferences and Projected Jobs in Paid Labor Force, United States, 2005 (in percent)

SOURCE: Schneider and Stevenson (1999:77).

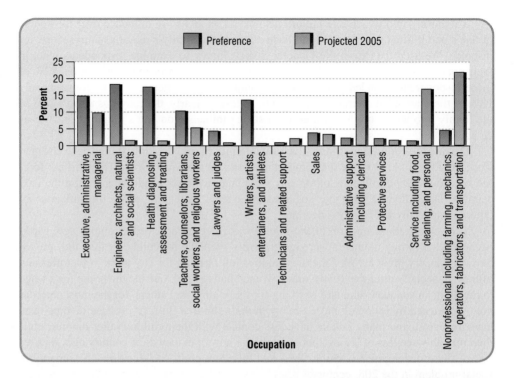

Learning disciplined work habits is an important part of the socialization that takes place in schools.

What is the content of the hidden curriculum? In the family, children tend to be evaluated on the basis of personal and emotional criteria. As students, however, they are led to believe that they are evaluated solely on the basis of their performance on impersonal, standardized tests. They are told that similar criteria will be used to evaluate them in the world of work. The lesson is, of course, only partly true. As you will see in Chapter 8 ("Race and Ethnicity"), Chapter 9 ("Sexuality and Gender"), and Chapter 13 ("Religion and Education"), it is not just performance, but also class, gender, and racial criteria that help determine success in school and in the work world. But the accuracy of the lesson is not the issue here. The important point is that the hidden curriculum has done its job if it convinces students they are judged on the basis of performance alone. Similarly, a successful hidden curriculum teaches students punctuality, respect for authority, the importance of competition in leading to excellent performance, and other conformist behaviors and beliefs that are expected of good citizens, conventionally defined.

Many students from poor and racial minority families reject the hidden curriculum in whole or in part. Their experience, and the experience of their friends, peers, and family members, may make them skeptical about the ability of school to open job opportunities for them. As a result, they rebel against the authority of the school. Expected to be polite and studious, they openly violate rules and neglect their work.

Believing that school does not lead to economic success can act as a **self-fulfilling prophecy,** an expectation that helps cause what it predicts. W. I. Thomas and Dorothy Swaine Thomas had a similar idea in stating what became known as the **Thomas theorem:** "Situations we define as real become real in their consequences" (Thomas, 1966 [1931]: 301). For example, believing that school won't help you get ahead may cause you to do poorly in school, and performing poorly makes it more likely you will wind up near the bottom of the class structure (Willis, 1984 [1977]).

Teachers, for their part, can also develop expectations that turn into self-fulfilling prophecies. In one famous study, two researchers informed the teachers in a primary school that they were going to administer a special test to the pupils to predict intellectual "blooming." In fact, the test was just a standard IQ test. After the test, they told teachers which students they could expect to become high achievers and which they could expect to become low achievers. In fact, the researchers assigned pupils to the two groups at random. At the end of the year, the researchers repeated the IQ test. They found that the students singled out as high achievers scored significantly higher than those singled out as low achievers. Since the only difference between the two groups of students was that teachers expected one group to do well and the other to do poorly, the researchers concluded that teachers' expectations alone influenced students' performance (Rosenthal and Jacobson, 1968). The clear implication of this research is that if a teacher believes that poor or minority children are likely to do poorly in school, chances are they will.

Peer Groups

A second socialization agent whose importance increased in the 20th century is the **peer group.** Peer groups consist of individuals who are not necessarily friends but are about the same age and of similar status. (**Status** refers to a recognized social position that an individual can occupy.) Peer groups help children and adolescents separate from their families

Gender segregation during school-yard play.

◆ **FIGURE 4.2** ◆

Percentage of Americans Age 12–17 Years Who Used Cigarettes, Alcohol, Marijuana, or Cocaine in Month Prior to Survey, 1990–97

SOURCE: United States Department of Health and Human Services (1999:22).

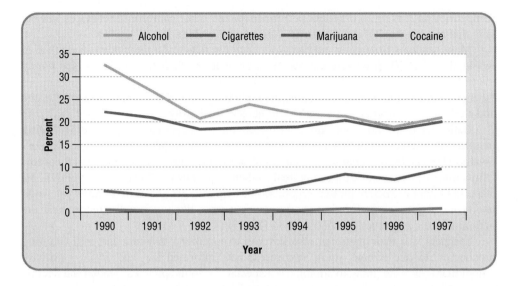

and develop independent sources of identity. They are especially influential over such lifestyle issues as appearance, social activities, and dating. In fact, from middle childhood through adolescence, the peer group is often the dominant socializing agent.

As you probably learned from your own experience, there is often conflict between the values promoted by the family and those promoted by the adolescent peer group. Adolescent peer groups are controlled by youth, and through them young people begin to develop their own identities. They do this by rejecting some parental values, experimenting with new elements of culture, and engaging in various forms of rebellious behavior, including the consumption of alcohol, cigarettes, and drugs (see Figure 4.2). In contrast, families are controlled by parents. They represent the values of childhood. Under these circumstances, such issues as hair and dress styles, music, curfew time; tobacco, drug, and alcohol use; and political views are likely to become points of conflict between the generations.

We should not, however, overstate the significance of adolescent–parent conflict. For one thing, the conflict is usually temporary. Once adolescents mature, the family exerts a more enduring influence on many important issues. Research shows that families have more influence than peer groups over the educational aspirations and the political, social, and religious preferences of adolescents and college students (Davies and Kandel, 1981; Milem, 1998).

A second reason why we should not exaggerate the extent of adolescent–parent discord is that peer groups are not just sources of conflict. They also help to *integrate* young people into the larger society. A recent study of preadolescent children in a small city in the Northwest illustrates the point. Over a period of 8 years, sociologists Patricia and Peter Adler conducted in-depth interviews with school children between the ages of 8 and 11. They lived in a well-to-do community in the Northwest composed of about 80,000 whites and 10,000 Hispanics and other racial minority group members (Adler and Adler, 1998). In each school they visited, they found a system of cliques arranged in a strict hierarchy, much like the arrangement of classes and racial groups in adult society. In schools with a substantial number of Hispanics and other nonwhites, cliques were divided by race. Nonwhite cliques were usually less popular than white cliques. In all schools, the most popular boys were highly successful in competitive and aggressive achievement-oriented activities, especially athletics. The most popular girls came from well-to-do and permissive families. One of the main bases of their popularity was that they had the means and the opportunity to participate in the most interesting social activities, ranging from skiing to late-night parties. Physical attractiveness was also an important basis of girls' popularity. Thus, elementary school peer groups prepared these youngsters for the class and racial inequalities of the adult world and the gender-specific criteria that would often be used to evaluate them as adults, such as competitiveness in the case of boys and attractiveness in the case of girls. (For more on gender socialization, see the discussion of the mass media below and Chapter 9, "Sexuality and Gender.") What we learn from this research is that the function of peer groups is not just to help adolescents form an independent identity by separating them from their families. In addition, peer groups teach young people how to adapt to the ways of the larger society.

The Mass Media

Like the school and the peer group, the mass media have also become increasingly important socializing agents in the 20th century. The mass media include television, radio, movies, videos, CDs, audio tapes, the Internet, newspapers, magazines, and books.

The fastest growing mass medium is the Internet. Worldwide, the number of Internet users jumped sevenfold from 40 million in 1995 to 280 million in 2000 (see Figure 4.3). However, TV viewing consumes more of the average American's time than any other mass medium. In 1992, the A. C. Nielsen Company, which measures audience size, estimated that more than 98% of American households owned a TV. On average, each TV was turned on for 7 hours a day. The University of Maryland's 1993–95 "Americans' Use of Time Project" collected national survey data showing that watching TV is the most time-consuming waking activity for women between the ages of 18 and 24 and the second most time-consuming waking activity for men in the same age group (see Figure 4.4). Survey research shows that American adults watched more TV in the 1970s than in the 1960s, more in the 1980s than in the 1970s, and more in the early 1990s than in the 1980s. Heavy users of TV are concentrated among socially disadvantaged groups, and that trend is intensifying over time (Hao, 1994; Robinson and Bianchi, 1997; see Table 4.1).[2]

Children and adolescents use the mass media for entertainment and stimulation. The mass media also help young people cope with anger, anxiety, and unhappiness. Finally, the cultural materials provided by the mass media help young people construct their identities—for example, by emulating the appearance and behavior of appealing movie stars, rock idols, and sports heroes. In performing these functions, the mass media offer youth much choice. Many Americans have access to scores of radio stations and TV channels, hundreds of magazines, thousands of CD titles, hundreds of thousands of books, and millions of Web sites. Most of us can gain access to hip-hop, heavy metal, or

Web Research Projects
Male Socialization,
Pornography, and Women

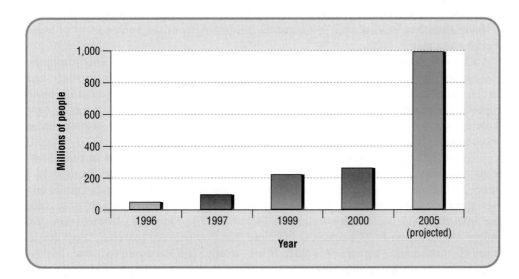

✦ FIGURE 4.3 ✦
**Number of Internet Users,
1996–2005 (projected)**

SOURCE: "Face of the Web . . . " (2000);
"Internet Growth" (1999).

---

[2]Note, however, that the proportion of light TV viewers, who watch television 0–2 hours per day, *increased* from 1989–98. This suggests that it is heavy viewers who account for the rising average number of hours watched.

✦ **FIGURE 4.4** ✦

Top Four Waking Activities of American Women and Men, Age 18–24, 1993–95 (hours per day)

SOURCE: Robinson and Bianchi (1997).

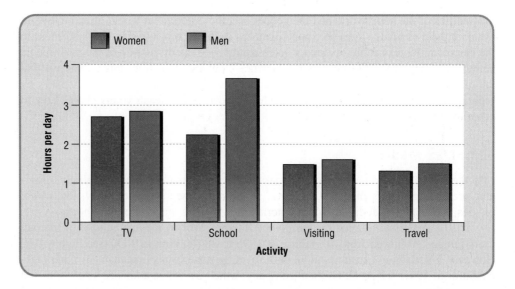

✦ **FIGURE 4.4** ✦

Top Four Waking Activities of American Women and Men, Age 18–24, 1993–95 (hours per day)

SOURCE: Robinson and Bianchi (1997).

✦ **TABLE 4.1** ✦

Hours of TV Viewing per Week by Years of Education, United States, 1998 (n = 2,327)

SOURCE: Computed from National Opinion Research Center (1999).

Hours of TV Viewing/Week	Years of Education				
	0–1	12	13–14	15–16	17–20
0–3	14.6	19.7	26.1	36.6	45.0
2–3	36.3	44.3	50.3	46.5	42.7
4–5	30.0	26.0	16.8	13.6	9.2
6+	19.1	9.9	6.8	3.3	3.1
Total	100.0	100.0	100.0	100.0	100.0
n	397	695	517	456	262

Haydn with equal ease. Thus, while adolescents have little choice over how they are socialized by their family and their school, the very proliferation of the mass media gives them more say over which media messages will influence them. To a degree, the mass media allow adolescents to engage in what sociologist Jeffrey Jensen Arnett (1995) calls **self-socialization,** or choosing socialization influences from the wide variety of mass media offerings.

Although people are to some extent free to choose socialization influences from the mass media, they choose some influences more often than others. Specifically, they tend to choose influences that are more pervasive, fit existing cultural standards, and are made especially appealing by those who control the mass media. We can illustrate this by considering how we learn gender roles from the mass media. **Gender roles** are widely shared expectations about how males and females are supposed to act.

The social construction of gender roles by the mass media begins when small children learn that only a kiss from Snow White's Prince Charming will save her from eternal sleep. It continues in magazines, romance novels, television, advertisements, music, and the Internet. It is big business. For example, Harlequin Enterprises of Toronto dominates the production and sale of romance novels worldwide. The company sells more than 160 million books a year in 23 languages and more than 100 national markets. About one in every six mass-market paperbacks sold in North America is a Harlequin romance. The average romance reader spends $800 a year on the genre. Most readers of Harlequin romances consume between 3 and 20 books a month. A central theme in these romances is the transformation of women's bodies into objects for men's pleasure. In the typical Harlequin romance, men are expected to be the sexual aggressors. They are typically more

Frantically she got up, her eyes flooding with tears, knocking over her chair in her desperate attempt to avoid crying in front of Alex and completely humiliating herself. But as she tried to run to the sanctuary of the bathroom the length of her bathrobe hampered her, and she had only taken a few steps before Alex caught up with her, bodily grabbed hold of her and swung her around to face him, his own face taut with emotion . . .

'Men aren't worth loving . . .'

'No?' Alex asked her huskily.

'No,' Beth repeated firmly, but somehow or other her denial had lost a good deal of its potency. Was that perhaps because of the way Alex was cupping her face, his mouth gently caressing hers, his lips teasing the stubbornly tight line of hers, coaxing it to soften and part. . . .?

As Alex continued to kiss her the most dizzying sweet sensation filled Beth.

She had the most overpowering urge to cling blissfully to Alex and melt into his arms like an old-fashioned Victorian maiden. Behind her closed eyelids she could have sworn there danced sunlit images of tulle and confetti scented with the lilies of a bridal bouquet, and the sound of a triumphant 'Wedding March' swelled and boomed and gold sunbeams formed a circle around her.

Dreamily Beth sighed, and then smiled beneath Alex's kiss, her own lips parting in happy acquiescence to the explorative thrust of his tongue.

Alex was dressed casually, in jeans and a soft shirt. Beneath her fingertips Beth could feel the fabric of that shirt, soft and warm, but the body that lay beneath it felt deliciously firm . . . hard, masculine, an unfamiliar and even forbidden territory that her fingers were suddenly dangerously eager to explore.

✦ **FIGURE 4.5** ✦

A Harlequin Romance

SOURCE: Jordan (1999: 97–8).

experienced and promiscuous than women. Women are expected to desire love before intimacy. They are assumed to be sexually passive, giving only subtle cues to indicate their interest in male overtures. Supposedly lacking the urgent sex drive that preoccupies males, women are often held accountable for moral standards and contraception (e.Harlequin.com, 2000; Jensen, 1984; Grescoe, 1996; see Figure 4.5).

Boys and girls do not passively accept such messages about appropriate gender roles. They often interpret them in unique ways and sometimes resist them. For the most part, however, they try to develop skills that will help them perform gender roles in a conventional way (Eagley and Wood, 1999: 412–13). Of course, conventions change. It is important to note in this regard that what children learn about femininity and masculinity today is less sexist than what they learned just a few generations ago. For example, comparing *Cinderella* and *Snow White* with *Mulan,* we see immediately that children going to Disney movies today are sometimes presented with more assertive and heroic female role models than the passive heroines of the 1930s and 1940s. On the other hand, the amount of change in gender socialization should not be exaggerated. *Cinderella* and *Snow White* are still popular movies. Moreover, for every *Mulan* there is a *Little Mermaid,* a movie that simply modernizes old themes about female passivity and male conquest. In the end, the Little Mermaid's salvation comes through her marriage. The heroic, gutsy, smart, and enterprising leads in nearly all children's movies are still boys (Douglas, 1994: 296–7).

As the learning of gender roles through the mass media suggests, then, not all media influences are created equal. We may be free to choose which media messages influence us, but most people are inclined to choose the messages that are most widespread, most closely aligned with existing cultural standards, and made most enticing by the mass media. In the case of gender roles, these messages are those that support conventional expectations about how males and females are supposed to act.

Not all initiation rites or "rites of passage" involve resocialization, in which powerful socializing agents deliberately cause rapid change in people's values, roles, and self-conception, sometimes against their will. Some rites of passage are a normal part of primary and secondary socialization and merely signify the transition from one status to another. Here, an Italian family celebrates the first communion of a young boy.

Resocialization and Total Institutions

In concluding our discussion of socialization agents, we must underline the importance of **resocialization** in contributing to the lifelong process of social learning. Resocialization takes place when powerful socializing agents deliberately cause rapid change in people's values, roles, and self-conception, sometimes against their will.

You can see resocialization at work in the ceremonies that are staged when someone joins a fraternity, a sorority, the marines, or a religious order. Such a ceremony, or **initiation rite,** signifies the transition of the individual from one group to another and ensures his or her loyalty to the new group. Initiation rites require new recruits to abandon old self-perceptions and assume new identities. Often they are composed of a three-stage ceremony involving: (a) separation from one's old status and identity (ritual rejection), (b) degradation, disorientation, and stress (ritual death), and (c) acceptance of the new group culture and status (ritual rebirth).

Much resocialization takes place in what sociologist Erving Goffman (1961) called **total institutions.** Total institutions are settings where people are isolated from the larger society and under the strict control and constant supervision of a specialized staff. Asylums and prisons are examples of total institutions. Because of the "pressure cooker" atmosphere in such institutions, resocialization in total institutions is often rapid and thorough, even in the absence of initiation rites.

A famous failed experiment illustrates the immense resocializing capacity of total institutions (Haney, Banks, and Zimbardo, 1973; Zimbardo, 1972). In the early 1970s, a group of researchers in Palo Alto, California, created their own mock prison. They paid about two dozen male volunteers to act as guards and inmates. The volunteers were mature, emotionally stable, intelligent college students from middle-class homes in the United States and Canada. None had a criminal record. By the flip of a coin, half the volunteers were designated prisoners, the other half guards. The guards made up their own rules for maintaining law and order in the mock prison. The prisoners were picked up by city police officers in a squad car, searched, handcuffed, fingerprinted, booked at the Palo Alto station house, and taken blindfolded to the mock prison. At the mock prison, each prisoner was stripped, deloused, put into a uniform, given a number, and placed in a cell with two other inmates.

To better understand what it means to be a prisoner or a prison guard, the researchers wanted to observe and record social interaction in the mock prison for 2 weeks. However, they were forced to end the experiment abruptly after only 6 days because what they witnessed frightened them. In less than a week, the prisoners and prison guards could no longer tell the difference between the roles they were playing and their "real" selves. Much of the socialization that these young men had undergone over a period of about 20 years was quickly suspended.

About a third of the guards began to treat the prisoners like despicable animals, taking pleasure in cruelty. Even the guards who were regarded by the prisoners as tough but fair stopped short of interfering in the tyrannical and arbitrary use of power by the most sadistic guards.

All of the prisoners became servile and dehumanized, thinking only about survival, escape, and their growing hatred of the guards. If they were thinking as college students, they could have walked out of the experiment at any time. Some of the prisoners did in fact beg for parole. However, by the fifth day of the experiment they were so programmed to think of themselves as prisoners that they returned docilely to their cells when their request for parole was denied.

The Palo Alto experiment suggests that your sense of self and the roles you play are not as fixed as you may think. Radically alter your social setting and, like the college students in the experiment, your self-conception and patterned behavior are likely to change too. Such change is most evident among people undergoing resocialization in total institutions. However, the sociological eye is able to observe the flexibility of the self in all social settings—a task made easier by the fact that the self has become more flexible over time. We now turn to an examination of the growing flexibility of the self.

THE FLEXIBLE SELF

Older sociology textbooks acknowledge that the development of the self is a lifelong process. They note that when young adults enter a profession and get married they must learn new occupational and family roles. If they marry someone from an ethnic, racial, or religious group other than their own, they are likely to adopt new cultural values or at least modify old ones. Retirement and old age present an entirely new set of challenges. Giving up a job, seeing children leave home and start their own families, losing a spouse and close friends—all these changes later in life require people to think of themselves in new ways, to redefine who they are.

In our judgment, however, older treatments of adult socialization underestimate the plasticity or flexibility of the self (Mortimer and Simmons, 1978). We believe that today, people's identities change faster, more often, and more completely than they did just a couple of decades ago.

One important factor contributing to the growing flexibility of the self is globalization. As we saw in Chapter 3, people are now less obliged to accept the culture into which they are born. Due to globalization, they are freer to combine elements of culture from a wide variety of historical periods and geographical settings.

A second factor increasing our freedom to design our selves is our growing ability to fashion new bodies from old. People have always defined themselves partly in terms of their bodies; your self-conception is influenced by whether you're a man or a woman, tall or short, healthy or ill, conventionally good-looking or plain. But our bodies used to be fixed by nature. People could do nothing to change the fact that they were born with certain features and grew older at a certain rate.

Now, however, you can change your body, and therefore your self-conception, radically and virtually at will—if, that is, you can afford it. Bodybuilding, aerobic exercise, and weight reduction regimes are more popular than ever. Plastic surgery allows people to buy new breasts, noses, lips, eyelids, and hair—and to remove unwanted fat, skin, and hair from various parts of their body. Roughly 2 million plastic surgeries were conducted in the United States in 1996, about 90% of them on women. While the annual number of reconstructive operations hardly changed in the 1990s, the number of cosmetic enhancements increased by more than 150% according to the American Society of Plastic and Reconstructive Surgeons (MacCarthy, 1999: 19, 20). Sex-change operations, while infrequent, are no longer a rarity. Organ transplants are routine. At any given time, about 50,000 Americans are waiting for a replacement organ. There is a brisk illegal international trade in human hearts, lungs, kidneys, livers, and eyes that enables well-to-do people to enhance and extend their lives (Rothman, 1998).

As if all this is were not enough to change how people think of themselves, Dr. Robert J. White of Case Western Reserve University School of Medicine in Cleveland has started to do whole-body transplants. In 1998, he removed the head of a rhesus monkey and connected it by tubes and sutures to the trunk of another monkey. The new entity lived and gained consciousness, although it was paralyzed below the neck because there is no way yet to connect the millions of neurons bridging the brain and the spinal column. Could the same operation be done on a human? No problem, says White. Because the human body is larger and we know more about human than monkey anatomy, the operation would in fact be simpler than it is on monkeys. And, notes White, because research on spinal regeneration is advancing rapidly, it is only a matter of time before it will be possible to create a fully functional human out of one person's head and another's body (Browne, 1998; ABC Evening News, 30 April 1998). We used to think of our selves as congruent with our bodies but the two have now become disjointed. As a result, the formerly simple question—"Who are you?"—has grown complex.

Further complicating the process of identity formation today is the growth of the Internet and its audio-visual component, the World Wide Web. In the 1980s and early 1990s, most observers believed that social interaction by means of computer would involve only the exchange of information between individuals (Wellman et al., 1996). It turns out they

Web Interactive Exercises
Identity and Community in Cyberspace

were wrong. Computer-assisted social interaction profoundly affects how people think of themselves.

Internet users interact socially by exchanging text, images, and sound via e-mail, Internet phone, video conferencing, computer-assisted work groups, and participation in **virtual communities.** Virtual communities have the most radical implications for the way we see ourselves. Virtual communities are associations of people, scattered across the country or the planet, who communicate via computer and modem about subjects of common interest. Some virtual communities cater to people's interest in specialized subjects, such as Latino culture, BMWs, or white-water canoeing. Others are MUDs (multiple user dimensions), computer programs that allow thousands of people to role-play and engage in a sort of collective fantasy. These programs define the aims and rules of the virtual community and the objects and spaces it contains. Users log on to the MUD from their PCs around the world and define their character—their identity—any way they wish. They interact with other users either by exchanging text messages or by having their "avatars" (graphical representations) act and speak for them.

The first MUD was created in 1979 at the University of Essex in England. In June 2001, there were nearly 1,800 MUDs worldwide and probably over a million MUD users ("The MUD Connector," 2001). MUD users form social relationships. They exchange confidences, give advice, share resources, get emotionally involved, and talk sex. Although their true identities are usually concealed, they sometimes decide to meet and interact in real life.

Some people may dismiss all this as yet another computer game played mainly by bored college students, a sort of high-tech version of *Dungeons and Dragons.* The fact is, however, that a large and growing number of people are finding that virtual communities affect their identities in profound ways (Dibbell, 1993). Specifically, because virtual communities allow people to interact using concealed identities, MUD users are free to assume new identities and are encouraged to discover parts of themselves they were formerly unaware of. In virtual communities, shy people can become bold, normally assertive people can become voyeurs, old people can become young, straight people can become gay, and women can become men.

Take Doug, a Midwestern college junior interviewed by sociologist Sherry Turkle. Doug plays four characters distributed across three different MUDs: a seductive woman, a macho cowboy type, a rabbit who wanders its MUD introducing people to each other, and a fourth character "I'd rather not even talk about because my anonymity there is very important to me. Let's just say that I feel like a sexual tourist." Doug often divides his computer screen into separate windows, devoting a couple of windows to MUDs and a couple to other applications. This allows him, in his own words, to

> split my mind . . . I can see myself as being two or three or more. And I just turn on one part of my mind and then another when I go from window to window. I'm in some kind of argument in one window and trying to come on to a girl in a MUD in another, and another window might be running a spreadsheet program or some other technical thing for school . . . And then I'll get a real-time message . . . that's RL [real life] . . . RL is just one more window . . . and it's not usually my best one (quoted in Turkle, 1995: 13).

Turkle (1995: 14) comments:

> [I]n the daily practice of many computer users, windows have become a powerful metaphor for thinking about the self as a multiple, distributed system. The self is no longer simply playing different roles in different settings at different times, something that a person experiences when, for example, she wakes up as a lover, makes breakfast as a mother, and drives to work as a lawyer. The life practice of windows is that of a decentered self that exists in many worlds and plays many roles at the same time . . . MUDs . . . offer parallel identities, parallel lives.

Since Turkle did her research in the first half of the 1990s, other forms of Internet community building have surpassed the importance of MUDs. The most widely used are ICQ and MSN Messenger Service. These are instant messaging programs that allow users to interact in real time either in an unstructured way or focused on certain themes of their own choosing. ICQ and MSN "chat groups" have role-playing capabilities, but they are usually less formally organized than MUDs. Millions of people use ICQ and MSN Messenger Service daily ("ICQ.com," 2000). Regardless of the forum, however, experience on the Internet reinforces our main point. In recent decades, the self has become increasingly flexible, and people are freer than ever to shape their selves as they choose (Brym and Lenton, 2001).

However, this freedom comes at a cost, particularly for young people. In concluding this chapter, we consider some of the socialization challenges American youth faces today. To set the stage for this discussion, we first examine the emergence of "childhood" and "adolescence" as categories of social thought and experience some 400 years ago.

DILEMMAS OF CHILDHOOD AND ADOLESCENT SOCIALIZATION

The Emergence of Childhood and Adolescence

In preindustrial societies, children are thought of as small adults. From a young age, they are expected to conform as much as possible to the norms of the adult world. That is largely because children are put to work as soon as they can contribute to the welfare of their families. Often, this means doing chores by the age of 5 and working full-time by the age of 10 or 12. Marriage, and thus the achievement of full adulthood, is common by the age of 15 or 16.

Until the late 1600s, children in Europe and North America fit this pattern. It was only in the late 1600s that the idea of childhood as a distinct stage of life emerged. At that time, the feeling grew among well-to-do Europeans and North Americans that boys should be allowed to play games and receive an education that would allow them to develop the emotional, physical, and intellectual skills they would need as adults. Girls continued to be treated as "little women" (the title of Louisa May Alcott's 1869 novel) until the 19th century. Most working-class boys didn't enjoy much of a childhood until the 20th century. Thus, it is only in the last century that the idea of childhood as a distinct and prolonged period of life became universal in the West (Ariès, 1962).

The idea of childhood emerged when and where it did due to social necessity and social possibility. Prolonged childhood was *necessary* in societies that required better educated adults to do increasingly complex work. That is because it gave young people a chance to prepare for adult life. Prolonged childhood was *possible* in societies where improved hygiene and nutrition allowed most people to live more than 35 years, the average life span in Europe in the early 1600s. In other words, before the late 1600s, most people didn't live long enough to permit the luxury of childhood. Moreover, there was no social need for a period of extended training and development before the comparatively simple demands of adulthood were thrust upon young people.

In general, wealthier and more complex societies whose populations enjoy a long average life expectancy stretch out the preadult period of life. For example, we saw that in Europe in 1600, most people reached mature adulthood by the age of about 16. In contrast, in countries like the United States today, most people are considered to reach mature adulthood only around the age of 30, by which time they have completed their formal education, married, and "settled down." Once teenagers were relieved of adult responsibilities, a new term had to be coined to describe the teenage years: "adolescence." Subsequently, the term "young adulthood" entered popular usage as an increasingly large number of people in their late teens and 20s delayed marriage to attend college (see Table 4.2).

Although these new terms describing the stages of life were firmly entrenched in North America by the middle of the 20th century, some of the categories of the population

✦ **TABLE 4.2** ✦

The Stages of Life in the United States and Russia

Note: Among industrialized countries, those that are wealthiest and enjoy the longest life expectancy seem to distinguish more *stages* of life, each with distinct needs and features. For example, Russia is among the poorest of industrialized countries, and its average life expectancy is only about 65 years. There, only two age groups are distinguished between birth and the age of 29: children and youth. In contrast, the United States is among the richest of the industrialized countries, and its average life expectancy is about 76 years. Here, six age groups are commonly distinguished between birth and age 29: infants, toddlers, children, preteens, teenagers (or adolescents), and young adults. Some people in the United States also distinguish different groups among the elderly—the "young old" (aged 65–74), the "old" (75–84), and "old old" (85+). See Chapter 15 ("Health, Medicine, and Aging"). As this example suggests, long life and material well-being allow people to invent new terms for more age groups and focus on the special needs of each one.

SOURCE: Adapted from Markowitz (2000); United Nations (1999c); World Health Organization (1996).

United States		Russia	
GDP per capita, 1997: $28,789		GDP per capita, 1996: $3,028	
Average life expectancy, 1996: 76.0 years		Average life expectancy, 1996: 65.0 years	
Infants	0–2	Children (*deti*)	0–16
Toddlers	3–5		
Children	6–10		
Preteens	11–12		
Teenagers (adolescents)	13–17 (or 20)	Youth (*molodezh*)	17–29
Young adults	18 (or 21)–25 (or 30)		
Adults	26 (or 31)–40 (or 45)	Adults (*vzroslie*)	30–44
Middle-aged	41 (or 46)–64	Older adults (*pozhilie*)	45–54 (women), 59 (men)
Elderly	65+	Elderly/pensioners (*starie/pensioneri*)	55 (women), 60 (men)

they were meant to describe soon began to change dramatically. Somewhat excitedly, a number of analysts began to write about the "disappearance" of childhood and adolescence altogether (Friedenberg, 1959; Postman, 1982). While undoubtedly overstating their case, these social scientists identified some of the social forces responsible for the changing character of childhood and adolescence in recent decades. Let us examine these social forces in the concluding section of this chapter.

Problems of Childhood and Adolescent Socialization Today

When you were between the ages of 10 and 17, how often were you at home or with friends but without adult supervision? How often did you have to prepare your own meals or take care of a younger sibling while your parent or parents were at work? How many hours a week did you spend cleaning house? How many hours a week did you have to work at a part-time job to earn spending money and save for college? How many hours a week did you spend on extracurricular activities associated with your school? On TV viewing and other mass media use? If you were like most American preteens and teenagers, many of your waking hours outside of school were spent without adult supervision and/or assuming substantial adult responsibilities such as those listed above. You are unlikely to have spent much time on extracurricular activities associated with your school but quite a lot of time viewing TV and using other mass media.

Declining adult supervision and guidance, increasing mass media and peer group influence, and the increasing assumption of substantial adult responsibilities to the neglect of extracurricular activities have done much to change the socialization patterns of American youth over the past 40 years or so. Let us consider each of these developments in turn.

✦ *Declining adult supervision and guidance.* In her recent 6-year, in-depth study of American adolescence, Patricia Hersch wrote that "in all societies since the beginning of time, adolescents have learned to become adults by observing, imitating and interacting with grown-ups around them" (Hersch, 1998: 20). However, in contemporary America, notes Hersch, adults are increasingly absent from the lives of adolescents. Why? According to Hersch, "American society has left its children behind as the cost of progress in the workplace" (Hersch, 1998: 19). What she means is that more American adults are working longer hours than ever before. Consequently, they have less time to spend with their children than they used to. We examine some reasons for the increasing demands of paid work in Chapter 5 ("Interaction and Organization") and Chapter 10 ("Work and the Economy"). Here, we stress a major consequence for American youth: Young people are increasingly left alone to socialize themselves and build their own community. This community sometimes revolves around high-risk behavior. To be sure, more is involved in high-risk behavior than socialization patterns (see Box 4.2). However, it is not coincidental that the peak hours for juvenile crime are between 3 p.m. and 6 p.m. on weekdays—that is, after school and before most parents return home from work (Hersch, 1998: 362). It is also significant in this connection that girls are less likely to engage in juvenile crime than boys. That is partly because parents tend to supervise and socialize their sons and daughter differently (Hagan, Simpson, and Gillis, 1987). Parents typically exert more control over girls, supervising them more closely and socializing them to avoid risk. These research findings suggest that many of the teenage behaviors commonly regarded as problematic result from declining adult guidance and supervision.

BOX 4.2
IT'S YOUR CHOICE

SOCIALIZATION VERSUS GUN CONTROL

On April 20, 1999, Columbine High School in Littleton, Colorado, was the scene of a mass killing by two students. The shooters, Dylan Klebold and Eric Harris, murdered 13 of their fellow students and then turned their guns on themselves.

After the massacre at Columbine High School, newspapers, magazines, Internet chat rooms, and radio and television talk shows were abuzz with the problem of teenage violence. "What is to be done," people asked? One solution that seems obvious to many people outside of the United States—and to an increasing number of Americans—is to limit the availability of firearms. Their reasoning is simple. All the advanced industrial societies except the United States restrict gun ownership. Only the United States has a serious problem with teenagers shooting one another. Other countries have problems with teenage violence. However, since guns are not readily available, teenage violence does not lead to mass killings in, say, Canada, Australia, Britain, or Japan. According to a 1995 Canadian government report, the rate of homicide us-

Citizens of most developed countries must purchase a license before they can possess firearms and buy ammunition. Licensing allows officials to require that applicants pass a safety course and a background check. This lowers the risk that firearms will be used for illegal purposes.

ing firearms per 100,000 people is 2.2 in Canada, 1.8 in Australia, 1.2 in Japan, 1.3 in Britain, and 9.3 in the United States (Department of Justice, Canada, 1995).

In the United States, however, most political discussions about teenage violence focus on the problem of socialization, not on gun control. Soon after the Littleton tragedy, for example, the House of Representatives passed a "juvenile crime bill." It cast blame on the entertainment industry, especially Hollywood movies, and the decline of "family values." Henry Hyde, an Illinois Republican, complained: "People were misled and disinclined to oppose the powerful entertainment industry" (quoted in Lazare, 1999: 57). Tom DeLay, a Republican congressman from Texas, worried: "We place our children in daycare centers where they learn their socialization skills . . . under the law of the jungle . . . " (quoted in Lazare, 1999: 58). In other words, according to these politicians, teenage massacres result from poor childhood socialization: the corrupting influence of Hollywood movies and declining family values.

Some politicians, including Hyde and DeLay, want to reintroduce Christianity into public schools to help overcome this presumed decay. DeLay thus reported an e-mail message he received. It read: "'Dear God, why didn't you stop the shootings at Columbine?' And God writes, 'Dear student, I would have, but I wasn't allowed in school'" (quoted in Lazare, 1999: 57–8). One consequence of the Littleton massacre was not a gun control bill, but a bill to display the Ten Commandments in public schools.

What do you think? Is the problem of students shooting each other a problem of socialization, lack of gun control, or a combination of both? In answering this question, think about the situation in other countries and refer back to the discussion of media influence in Chapter 2.

✦ *Increasing media influence.* Declining adult supervision and guidance also leaves American youth more susceptible to the influence of the mass media and peer groups. As one parent put it, "[w]hen they hit the teen years it is as if they can't be children anymore. The outside world has invaded the school environment" (quoted in Hersch, 1998: 111). In an earlier era, family, school, church, and community usually taught young people more or less consistent beliefs and values. Now, however, the mass media offer a wide variety of cultural messages, many of which differ from each other and from those taught in school and at home. The result for many adolescents is confusion (Arnett, 1995). Should the 10-year-old girl dress modestly or in a sexually provocative fashion? Should the 14-year-old boy devote more time to attending church, synagogue, temple, or mosque—or to playing electric guitar in the garage? Should you just say no to drugs? The mass media and peer groups often pull young people in different directions from the school and the family, leaving them uncertain about what constitutes appropriate behavior and making the job of growing up more stressful than it used to be.

✦ *Declining extracurricular activities and increasing adult responsibilities.* As the opening anecdote about Robert Brym's involvement in high school drama illustrates, extracurricular activities are important for adolescent personality development. That is because they provide opportunities for students to develop concrete skills and thereby make sense of the world and their place in it. In schools today, academic subjects are too often presented as disconnected bits of knowledge that lack relevance to the student's life. Drama, music, and athletics programs are often better at giving students a framework within which they can develop a strong sense of self; for these are concrete activities with clearly defined rules. By training and playing hard on a football team, mastering electric guitar or acting in plays, you can learn something about your physical, emotional, and social capabilities and limitations, about what you are made of, and what you can and can't do. These are just the sorts of activities adolescents require for healthy self-development.

However, if you're like most young Americans today, you spent fewer hours per week on extracurricular activities associated with school than your parents did when they went to school. Educators estimate that only about a quarter of today's high school students take part in sports, drama, music, and so forth (Hersch, 1998). Many of them are simply too busy with household chores, child care responsibilities, and part-time jobs to enjoy the benefits of school activities outside the classroom.

Some analysts wonder whether the assumption of so many adult responsibilities, the lack of extracurricular activities, declining adult supervision and guidance, and increasing mass media and peer group influence are causing childhood and adolescence to disappear. As early as 1959, one sociologist spoke of "the vanishing adolescent" in American society (Friedenberg, 1959). More recently, another commentator remarked: "I think that we who were small in the early sixties were perhaps the last generation of Americans who actually had a childhood, in the . . . sense of . . . a space distinct in roles and customs from the world of adults, oriented around children's own needs and culture rather than around the needs and culture of adults" (Wolf, 1997: 13; see also Postman, 1982). Childhood and adolescence became universal categories of social thought and experience in the 20th century. Under the impact of the social forces discussed above, however, the experience and meaning of childhood and adolescence now seem to be changing radically.

SUMMARY

1. Studies show that children raised in isolation do not develop normally. This corroborates the view that social interaction unleashes human potential.

2. Freud developed the first social-scientific theory of how the self develops. He called the part of the self that demands immediate gratification the id. He argued that a self-image

begins to emerge when the id's demands are denied. Because of many lessons in self-control, a child eventually develops a sense of what constitutes appropriate behavior, a moral sense of right and wrong, and a personal conscience or superego. The superego is a repository of cultural standards. A third component of the self, the ego, then develops. In psychologically healthy individuals, the ego balances the demands of the id and superego.

3. Like Freud, Mead noted that a subjective and impulsive aspect of the self is present from birth. He called it the "I." Mead also argued that a repository of culturally approved standards emerges as part of the self during social interaction. Mead called it the "me." However, Mead drew attention to the unique human capacity to "take the role of the other" as the source of the me. At first the child imitates and then pretends to be his or her significant others. Next, the child learns to play complex games in which he or she must understand several roles simultaneously. Finally, a person's image of cul-tural standards and how they are applied to him or her stimu-lates the growth of what Mead called "the generalized other."

4. In the 20th century, the increasing socializing influence of schools, peer groups, and the mass media was matched by the decreasing socializing influence of the family.

5. People's self-conceptions are subject to more flux now than they were even a few decades ago. Cultural globalization, medical advances, and computer-assisted communication are among the factors that have made the self more plastic.

6. Decreasing parental supervision and guidance, the increasing assumption of substantial adult responsibilities by children and adolescents, declining participation in extracurricular ac-tivities, and increased mass media and peer group influence are causing changes in the character and experience of child-hood and adolescence. According to some analysts, childhood and adolescence as they were known in the first half of the twentieth century are disappearing.

GLOSSARY

The **ego,** according to Freud, is a psychological mechanism that balances the conflicting needs of the pleasure-seeking id and the restraining superego.

Gender roles are widely shared expectations about how males and females are supposed to act.

The **generalized other,** according to Mead, is a person's image of cultural standards and how they apply to him or her.

A **hidden curriculum** teaches students what will be expected of them as conventionally good citizens once they leave school.

The **I,** according to Mead, is the subjective and impulsive aspect of the self that is present from birth.

The **id,** according to Freud, is the part of the self that demands immediate gratification.

An **initiation rite** is a ritual that signifies the transition of the individual from one group to another and ensures his or her loyalty to the new group.

The **me,** according to Mead, is the objective component of the self that emerges as people communicate symbolically and learn to take the role of the other.

One's **peer group** is composed of people who are about the same age and of similar status as the individual. The peer group acts as an agent of socialization.

Primary socialization is the process of acquiring the basic skills needed to function in society during childhood. Primary socialization usually takes place in a family.

Resocialization occurs when powerful socializing agents deliberately cause rapid change in one's values, roles, and self-conception, sometimes against one's will.

A **role** is the behavior expected of a person occupying a particular position in society.

Secondary socialization is socialization outside the family after childhood.

The **self** consists of your ideas and attitudes about who you are.

A **self-fulfilling prophecy** is an expectation that helps bring about what it predicts.

Self-socialization involves choosing socialization influences from the wide variety of mass media offerings.

Significant others are people who play important roles in the early socialization experiences of children.

Socialization is the process by which people learn their culture—including norms, values, and roles—and become aware of themselves as they interact with others.

Status refers to a recognized social position that an individual can occupy.

The **superego,** according to Freud, is a part of the self that acts as a repository of cultural standards.

The **Thomas theorem** states: "Situations we define as real become real in their consequences."

Total institutions are settings where people are isolated from the larger society and under the strict control and constant supervision of a specialized staff.

The **unconscious,** according to Freud, is the part of the self containing repressed memories that we are not normally aware of.

A **virtual community** is an association of people, scattered across the country, continent, or planet, who communicate via computer and modem about a subject of common interest.

QUESTIONS TO CONSIDER

1. Do you think of yourself in a fundamentally different way from the way your parents (or other close relatives or friends at least 20 years older than you) thought of themselves when they were your age? Interview your parents, relatives, or friends to find out. Pay particular attention to the way in which the forces of globalization may have altered self-conceptions over time.

2. Have you ever participated in an initiation rite in college, the military, or in a religious organization? If so, describe the ritual rejection, ritual death, and ritual rebirth that made up the rite. Do you think that the rite increased your identification with the group you were joining? Did it increase the sense of solidarity—the "we-feeling"—of group members?

3. List the contradictory lessons that different agents of socialization taught you as an adolescent. How have you resolved these contradictory lessons? If you have not, how do you intend to do so?

WEB RESOURCES

Companion Web Site for This Book

http://sociology.wadsworth.com

Begin by clicking on the Student Resources section of the Web site. Choose "Introduction to Sociology" and finally the Brym and Lie book cover. Next, select the chapter you are currently studying from the pull-down menu. From the Student Resources page you will have easy access to InfoTrac College Edition®, MicroCase Online exercises, additional Web links, and many resources to aid you in your study of sociology, including practice tests for each chapter.

Infotrac Search Terms

These search terms are provided to assist you in beginning to conduct research on this topic by visiting http://www.infotraccollege.com/wadsworth.

Hidden curriculum   Primary socialization
Initiation rite   Secondary socialization
Peer group

Recommended Web Sites

A useful summary of major ideas in the sociological study of socialization can be found at http://www.nwmissouri.edu/nwcourses/martin/general/socialization/168108.html.

For the socialization experiences that characterize different generations, go to a major search engine on the Web, such as Yahoo at http://www.yahoo.com, and search for "teenagers," "generation X," "baby boomers," "the elderly," etc.

Initiation rites (or rites of passage) are conveniently summarized in the online version of the *Encarta* encyclopedia. Go to the Encarta search engine at http://encarta.msn.com and search for "rites of passage."

SUGGESTED READINGS

Patricia A. Adler and Peter Adler. *Peer Power: Preadolescent Culture and Identity* (New Brunswick, NJ: Rutgers University Press, 1998). The best sociological study of the role of peer groups in preadolescent socialization.

Sigmund Freud. *Civilization and Its Discontents.* James Strachey, trans. (New York: W. W. Norton, 1962 [1930]). Freud's classic explanation of how there can be no civilization without repression.

Patricia Hersch. *A Tribe Apart: A Journey Into the Heart of American Adolescence* (New York: Ballantine Books, 1998). An insightful portrait of American adolescents today.

IN THIS CHAPTER, YOU WILL LEARN THAT:

✦ People interact for a variety of reasons. One important motivation is to gain valued resources. However, if one party competes too eagerly for valued resources, preventing others from benefiting, interaction breaks down.

✦ Competition is only one basis for interaction. Other bases include domination, which depends on fear, and cooperation, which depends on trust.

✦ For people to interact, they must understand the basic values and norms held by others. They must also accept their own and others' roles and statuses. To a degree, people work out these elements of interaction during interaction itself.

✦ Bureaucracies are large, impersonal organizations that operate with varying degrees of efficiency.

✦ Efficient bureaucracies are those that keep hierarchy to a minimum, distribute decision making to all levels of the bureaucracy, and keep lines of communication open between different units of the bureaucracy.

INTERACTION AND ORGANIZATION

THE STRUCTURE OF SOCIAL INTERACTION

In 1941, the large stone and glass train station was one of the proudest structures in Smolensk, a provincial capital of about 100,000 people on Russia's western border. Always bustling, it was especially busy on the morning of June 28. For besides the usual passengers and well-wishers, hundreds of Soviet Red Army soldiers were nervously talking, smoking, writing hurried letters to their loved ones, and sleeping fitfully on the station floor waiting for their train. Nazi troops had invaded the nearby city of Minsk in Belarus a couple of days before. The Soviet soldiers were being positioned to defend Russia against the inevitable German onslaught.

Robert Brym's father, then in his 20s, had been standing in line for nearly 2 hours to buy food when he noticed flares arching over the station. Within seconds, Stuka bombers, the pride of the German air force, swept down, releasing their bombs just before pulling out of their dive. Inside the station, shards of glass, blocks of stone, and mounds of earth fell indiscriminately on sleeping soldiers and nursing mothers alike. Everyone panicked. People trampled over one another to get out. In minutes, the train station was rubble.

Nearly 2 years earlier, Robert's father had managed to escape Poland when the Nazis invaded his hometown near Warsaw. Now, he was on the run again. By the time the Nazis occupied Smolensk a few weeks after their dive-bombers destroyed its train station, Robert's father was deep in the Russian interior serving in a workers' battalion attached to the Soviet Red Army.

"My father was one of 300,000 Polish Jews who fled eastward into Russia before the Nazi genocide machine could reach them," says Robert. "The remaining 3 million Polish Jews were killed in various ways. Some died in battle. Many more, like my father's mother and younger siblings, were rounded up like diseased cattle and shot. However, most of Poland's Jews wound up in the concentration camps. Those deemed unfit were shipped to the gas chambers. Those declared able to work were turned into slaves until they could work no more. Then they, too, met their fate. A mere 9% of Poland's 3.3 million Jews survived World War II.

"Two questions always perplexed my father about the war. How was it possible for many thousands of ordinary Germans—products of what he regarded as the most advanced civilization on earth—to systematically murder millions of defenseless and innocent Jews, Roma ("Gypsies"), homosexuals, and mentally disabled people in the death camps? And why did the innocents often march to the gas chambers without protest rather than always making it as difficult as possible for the Nazis to carry out their vile plans?"

To answer these questions adequately, we must borrow ideas from the sociological study of interaction and organizations. Consider first the question of how ordinary German citizens could commit the crime of the century. The conventional, nonsociological answer is that many Nazis were evil, sadistic, or deluded enough to think that Jews and other undesirables threatened the existence of the German people. Therefore, in the Nazi mind, the innocents had to be killed. This is the answer of the 1993 movie, *Schindler's List*, and other accounts (Goldhagen, 1996).

Yet it is far from the whole story. Sociologists emphasize two other factors:

1. *Structures of authority tend to render people obedient.* Most people find it difficult to disobey authorities because they fear ridicule, ostracism, and punishment. This was strikingly demonstrated in an experiment conducted by social psychologist Stanley Milgram (1974). Milgram informed his experimental subjects they were taking part in a study on punishment and learning. He brought each subject to a room where a man was strapped to a chair. An electrode was attached to the man's wrist. The experimental subject sat in front of a console. It contained 30 switches with labels ranging from "15 volts" to "450 volts" in 15-volt increments. Labels ranging from "SLIGHT SHOCK" to "DANGER: SEVERE SHOCK" were pasted below the switches. The experimental subjects were told to administer a 15-volt shock for the man's first wrong answer and then increase the voltage each time he made an error. The man

German industrialist Oskar Schindler (Liam Neeson, center), searches for his plant manager Itzhak Stern among a trainload of Polish Jews about to be deported to Auschwitz-Birkenau in *Schindler's List*. The movie turns the history of Nazism into a morality play, a struggle between good and evil forces. It does not probe into the sociological roots of good and evil.

strapped in the chair was in fact an actor. He did not actually receive a shock. However, as the experimental subject increased the current, the actor began to writhe, shouting for mercy and begging to be released. If the experimental subjects grew reluctant to administer more current, Milgram assured them that the man strapped in the chair would be just fine and insisted that the success of the experiment depended on the subject's obedience. The subjects were, however, free to abort the experiment at any time. Remarkably, 71% of experimental subjects were prepared to administer shocks of 285 volts or more even though the switches at that level were labeled "IN-TENSE SHOCK," "EXTREME INTENSITY SHOCK" and "DANGER: SEVERE SHOCK" and despite the fact that the actor appeared to be in great distress at this level of current.

Milgram's experiment teaches us that as soon as we are introduced to a structure of authority, we are inclined to obey those in power. This is the case even if the authority structure is brand new and highly artificial, even if we are free to walk away from it with no penalty, even if we think that by remaining in its grip we are inflicting terrible pain on another human being. In this context, the actions and inactions of German citizens in World War II become more understandable if no more forgivable.[1]

2. *Bureaucracies are highly effective structures of authority*. A second reason why the Nazi genocide machine was so effective is that it was bureaucratically organized. As Max Weber (1978a) defined the term, a **bureaucracy** is a large, impersonal organization composed of many clearly defined positions arranged in a hierarchy. A bureaucracy has a permanent, salaried staff of qualified experts and written goals, rules, and procedures. Staff members always try to find ways of running their organization more efficiently. "Efficiency" means achieving the bureaucracy's goals at the least cost. The goal of the Nazi genocide machine was to kill Jews and other undesirables. To achieve that goal with maximum efficiency, the job was broken into many small tasks. Most officials performed only one function: checking train schedules, organizing entertainment for camp guards, maintaining supplies of Zyklon B

---

[1]Significantly, a study of Christians who helped Jews in World War II Europe found that their heroism was not related to their educational attainment, political orientation, religious background, or even their attitudes toward Jews. What the rescuers had in common was that they were not well integrated in society. That is, for one reason or another, they were poorly socialized and therefore less inclined than nonheroes to respect authority (Tec, 1986).

✦ **FIGURE 5.1** ✦

Obedience to Authority Increases With Separation From the Negative Effects of One's Actions

Milgram's experiment supports the view that separating people from the negative effects of their actions increases the likelihood of compliance. When subject and actor were in the same room and the subject was told to force the actor's hand onto the electrode, 30% of subjects administered the maximum 450-volt shock. When subject and actor were merely in the same room, 40% of subjects administered the maximum shock. When subject and actor were in different rooms but the subject could see and hear the actor, 62.5% of subjects administered the maximum shock. When subject and actor were in different rooms and the actor could be seen but not heard, 65% of subjects administered the maximum shock.

SOURCE: Milgram (1974).

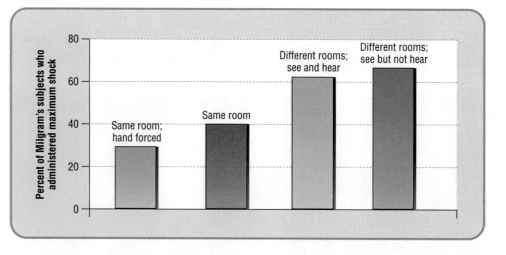

gas, removing ashes from the crematoria, and so forth. The full horror of what was happening eluded many officials, or at least could be conveniently ignored as they concentrated on their jobs, most of them far removed from the gas chambers and death camps in occupied Poland (Figure 5.1). Many factors account for variations in Jewish victimization rates across Europe during World War II (Fein, 1979; Marrus, 1987). One was bureaucratic organization. Not coincidentally, the proportion of Jews killed was highest not in the Nazi-controlled countries where the hatred of Jews was most intense (e.g., Romania), but in countries where the Nazi bureaucracy was best organized (e.g., Holland) (Arendt, 1977 [1963]; Bauman, 1991 [1989]; Hilberg, 1961; Sofsky, 1997 [1993]).

In short, the sociological reply to the first question posed by Robert's father is that it was not just blind hatred but the social organization of authority and, in particular, its bureaucratic structure that made it possible for the Nazis to kill innocent people so ruthlessly.

And why did the innocents only sometimes resist their oppressors rather than always fighting them tooth and nail? The short sociological answer is that they had few resources to fight with. Not only did they lack the most rudimentary weapons. In addition, they had been systematically stripped of their strength, their health, their dignity, and their courage when they were shaven, deloused, starved, intimidated, worked half to death, and beaten by the camp guards. Jews with resources did fight the Nazis. They were the second most highly decorated ethnic group in the Soviet Red Army (next to the Russians themselves). They fought heroically in the underground resistance movement. They organized uprisings against the Nazis in Warsaw and other places (Dawidowicz, 1975). Only in the camps, where Jews lacked nearly all means of resistance, did their interaction with the Nazis tend to take the form of fearful compliance.

The kinds of issues raised above lie at the heart of the sociological study of interaction and organization. How is social interaction maintained? What are the most efficient forms of social organization? In one form or another, these questions have concerned social thinkers for centuries. They are our chief focus here.

How is social interaction maintained? **Social interaction** involves people communicating face-to-face, acting and reacting in relation to other people. Social interaction is structured around three elements. First, each person engaged in social interaction tends to adhere to specific *norms* or generally accepted ways of doing things. Second, each person acts according to the demands of a particular *role* or set of expected behaviors. Third, each person assumes a certain *status* or recognized position in the interaction. But if norms, roles, and statuses are the building blocks of social interaction, what is the cement that holds them together? In other words, why do people maintain stable patterns of

In primary groups such as families, norms, roles, and statuses are agreed upon but are not set down in writing. Social interaction creates strong emotional ties. It extends over a long period. It involves a wide range of activities. It results in group members knowing one another well.

interaction in the first place? This is the most fundamental sociological question one can ask, for it is really a question about how society is at all possible. As we will see, there are three main ways of maintaining social interaction and thereby cementing society: by means of domination (as in the concentration camps), by means of competition, and by means of cooperation. In the first half of the chapter, we investigate each of these modes of interaction.

What are the most efficient forms of social organization? Face-to-face social interaction is the microstructural "raw material" out of which larger and more enduring social structures are built (see Chapter 1, "A Sociological Compass," on the distinction between micro- and macrostructures). For instance, some face-to-face interaction gives rise to **social groups,** clusters of people who identify with one another and adhere to defined norms, roles, and statuses. Social groups are usually distinguished from **social categories,** or people who share a similar status but do *not* identify with one another.

There are many kinds of social groups. However, sociologists make a basic distinction between primary and secondary groups. In **primary groups,** norms, roles, and statuses are agreed upon but are not set down in writing. Social interaction creates strong emotional ties. It extends over a long period. It involves a wide range of activities. It results in group members knowing one another well. The family is the most important primary group.

Secondary groups are larger and more impersonal than primary groups. Compared to primary groups, social interaction in secondary groups creates weaker emotional ties. It extends over a shorter period. It involves a narrow range of activities. It results in most group members having at most a passing acquaintance with one another. Your sociology class is an example of a secondary group.

Another example of a secondary group is bureaucracy, which we defined above. Bureaucracies are widespread because they are often more efficient than other kinds of secondary groups—that is, they achieve similar goals at a lower cost. But, despite its success, is bureaucracy always the most efficient type of secondary group? We devote much of the second half of the chapter to answering that question.

Before analyzing the operation of bureaucratic macrostructures, however, we must first examine the microstructural foundations of social organization in general. Specifically, how do domination, competition, and cooperation maintain social interaction? We begin to answer this question by dissecting the competitive character of much everyday conversation.

Web Interactive Exercises
Does the Internet Isolate
People Socially?

THREE MODES OF SOCIAL INTERACTION

Interaction as Competition and Exchange

Have you ever been in a conversation where you can't get a word in edgewise? If you're like most people, this is bound to happen to you from time to time. The longer this kind of one-sided conversation persists, the more neglected you feel. You may make increasingly less subtle attempts to turn the conversation your way. But if you fail, you may decide to end the interaction altogether. If this experience repeats itself—if the person you're talking to monopolizes conversations repeatedly—you're likely to want to avoid getting into conversations with him or her in the future. Maintaining interaction (and maintaining a relationship) requires that the need of both parties for attention be met.

Most people don't consistently try to monopolize conversations. If they did, there wouldn't be much talk in the world. However, a remarkably large part of all conversations involves a subtle competition for attention. Consider the following snippet of dinner conversation:

> John: "I'm feeling really starved."
> Mary: "Oh, I just ate."
> John: "Well, I'm feeling really starved."
> Mary: "When was the last time you ate?"

Sociologist Charles Derber recorded this conversation (Derber, 1979: 24). John starts by saying how hungry he is. Attention is on him. Mary replies she's not hungry. Attention shifts to her. John insists he's hungry, shifting attention back to him. Mary finally allows the conversation to focus on John by asking him when he last ate. John thus "wins" the competition for attention.

Derber recorded 1,500 conversations in family homes, workplaces, restaurants, classrooms, dormitories, and therapy groups. He concluded that Americans usually try to turn conversations toward themselves. They usually do so in ways that go unnoticed. Nonetheless, says Derber, the typical conversation is a covert competition for attention. In Derber's words, there exists

> a set of extremely common conversational practices which show an unresponsiveness to other's topics and involve turning them into one's own. Because of norms prohibiting blatantly egocentric behavior, these practices are often exquisitely subtle . . . Although conversationalists are free to introduce topics about themselves, they are expected to maintain an appearance of genuine interest in those about others in a conversation. A delicate face-saving system requires that people refrain from openly disregarding others' concerns and keep expressions of disinterest from becoming visible (Derber, 1979: 23).

You can observe the competition for attention yourself. Tape-record a couple of minutes of conversation in your dorm, home, or workplace. Then play back the tape. Evaluate each statement in the conversation. Does the statement try to change who is the subject of the conversation? Or does it say something about the *other* conversationalist(s) or ask them about what they said? How does not responding, or merely saying "uh-huh" in response, operate to shift attention? Are other conversational techniques especially effective in shifting attention? Who "wins" the conversation? What is the winner's gender, race, and class position? Is the winner popular or unpopular? Do you think there is generally a connection between the person's status in the group and his or her ability to win? (Hint: Sociological research shows that men interrupt conversations more than women and more often become the focus of attention [Tannen, 1994a; 1994b].) You might even want to record yourself in conversation. Where do you fit in?

Derber is careful to point out that conversations are not winner-take-all competitions. Unless both people in a two-person conversation receive some attention, the interaction is likely to cease. It follows that conversation typically involves the *exchange* of attention. Furthermore, attention is only one valued resource that people trade during social interaction. Other media of exchange include pleasure, approval, prestige, information, and money.

The idea that social interaction involves trade in valued resources is the central insight of **exchange theory** (Blau, 1964; Homans, 1961). A variant of this approach is **rational choice theory** (Coleman, 1990; Hechter, 1987). Rational choice theory focuses less on the resources being exchanged than the way interacting people weigh the benefits and costs of interaction. According to rational choice theory, interacting people always try to maximize benefits and minimize costs. Businesspeople want to keep their expenses to a minimum so they can keep their profits as high as possible. Similarly, everyone wants to gain the most from their interactions—socially, emotionally, and economically—while paying the least.

Undoubtedly, one can explain many types of social interaction in terms of exchange and rational choice theories. However, some types of interaction cannot be explained in these terms. For example, people often act in ways they consider fair or just, even if this does not maximize their personal gain (Frank, 1988; Gamson, Fireman, and Rytina, 1982). Some people even engage in altruistic or heroic acts from which they gain nothing at all. They do so although altruism and heroism can sometimes place them at considerable risk.

Consider a woman who hears a drowning man cry "help" and decides to risk her life to save him (Lewontin, 1991: 73–4). Some analysts assert that such a hero is willing to save the drowning man because she thinks the favor may be returned in the future. To us, this seems far-fetched. In the first place, there is close to zero probability that today's rescuer will be drowning someday and that the man who is drowning today will be present to save her. Second, a man who can't swim well enough to save himself is just about the last person on earth you'd want to try to rescue you if the need arose. Third, heroes typically report they decide to act in an instant, before there is a chance to weigh any costs and benefits at all. Heroes respond to cries for help based on emotion (which, physiologists tell us, takes 1/125 of a second to register in the brain), not calculation (which takes seconds or even minutes).

When people behave fairly or altruistically, they are interacting with others based on *norms* they have learned—norms that say they should act justly and help people in need, even if substantial costs are attached. Such norms are for the most part ignored by exchange and rational choice theorists. Exchange and rational choice theorists assume that most of the norms governing social interaction are like the "norm of reciprocity," which states that you should try to do for others what they try to do for you, because if you don't, then others will stop doing things for you (Homans, 1950). But social life is richer than this narrow view suggests. Interaction is not all selfishness.

Moreover, as you will now see, we cannot assume what people want. That is because norms (as well as roles and statuses) are not presented to us fully formed. Nor do we mechanically accept norms when they are presented to us. Instead, we constantly negotiate and modify norms—as well as roles and statuses—as we interact with others. Let us now explore this theme by considering the ingenious ways in which people manage the impressions they give to others during social interaction.

Web Research Projects
Conversation Analysis

Interaction as Impression Management

> *"The best way of impressing [advisers] with your competence is asking questions you know the answer to. Because if they ever put it back on you, "Well what do you think?" then you can tell them what you think and you'd give a very intelligent answer because you knew it. You didn't ask it to find out information. You ask it to impress people."*

> —A THIRD-YEAR MEDICAL STUDENT

Soon after they enter medical school, students become adept at **impression management.** That is, they learn how to manipulate the way they present themselves so they can appear in the best possible light and be judged competent by their teachers and patients. As Jack

Impression management involves manipulating the way you present yourself so that others will view you in the best possible light. It is especially important for politicians to be adept at impression management since their success depends heavily on voters' opinions of them.

Haas and William Shaffir (1987) show in their study of professional socialization, students adopt a new, medical vocabulary and wear a white lab coat to set themselves off from patients. They try to model their behavior after that of doctors who have authority over them. They may ask questions they know the answer to so they can impress their teachers. When dealing with patients, they may hide their ignorance under medical jargon to maintain their authority. By engaging in these and related practices, medical students reduce the distance between their premedical-school selves and the role of doctor. By the time they finish medical school, they have reduced the distance so much that they no longer see any difference at all between who they are and the role of doctor. They come to take for granted a fact they once had to socially construct—the fact that they are doctors (Haas and Shaffir, 1987: 53–83).

Haas and Shaffir's study is an application of symbolic interactionism, a theoretical approach introduced in Chapter 1. Symbolic interactionists regard people as active, creative, and self-reflective. Whereas exchange theorists assume what people want, symbolic interactionists argue that people create meanings and desires in the course of social interaction. According to Herbert Blumer (1969), symbolic interactionism is based on three principles. First, "human beings act toward things on the basis of the meaning which these things have for them." Second, "the meaning of a thing" emerges from the process of social interaction. Third, "the use of meanings by the actors occurs through a process of interpretation" (Blumer, 1969: 2; see also Berger and Luckmann, 1966; Strauss, 1993; Wiley, 1994).

While there are several distinct approaches to symbolic interactionism (Denzin, 1992), probably the most widely applied approach is **dramaturgical analysis.** As first developed by sociologist Erving Goffman (1959 [1956]), dramaturgical analysis takes literally Shakespeare's line from *As You Like It*: "All the world's a stage and all the men and women merely players."

From Goffman's point of view, we are constantly engaged in role-playing. This is most clearly evident when we are "front stage," that is, in public settings that require the use of props, set gestures, and memorized lines. A server in a restaurant, for example, must dress in a uniform, smile, and recite fixed lines. ("How are you? My name is Sam and I'm your server today. May I get you a drink before you order your meal?") When the server goes "backstage," he or she can relax from the front stage performance and discuss it with fellow actors ("Those kids at table six are driving me nuts!"). Thus, we often distinguish between our public roles and our "true" selves. When we do so, we experience what Goffman calls **role distance.** Note, however, that even backstage we engage in role-playing and impression management. It's just that we are less likely to be aware of it. For instance, in the kitchen, a server may try to present herself in the best possible light in order to impress another server so she can eventually ask him out for a date. Thus, the implication of dramaturgical analysis is that there is no single self, just the ensemble of roles we play in various social contexts.

By emphasizing how social reality is constructed in the course of interaction, symbolic interactionists downplay the importance of norms and understandings that precede any given interaction. **Ethnomethodology** tries to correct this shortcoming. Ethnomethodology is the study of the methods ordinary people use, often unconsciously, to make sense of what others do and say. Ethnomethodologists stress that everyday interactions could not take place without *pre-existing* shared norms and understandings. To illustrate this point, Harold Garfinkel (1967: 44) got one of his students to interpret a casual greeting in an unexpected way:

Acquaintance: [waving cheerily] How are you?
Student: How am I in regard to what? My health, my finances, my schoolwork, my peace of mind, my . . . ?
Acquaintance: [red in the face and suddenly out of control] Look! I was just trying to be polite. Frankly, I don't give a damn how you are.

As this example shows, social interaction requires tacit agreement between the actors about what is normal and expected. Without shared norms and understandings, there can be no sustained interaction; people are likely to get upset and end an interaction when the assumptions underlying the stability and meaning of daily life are violated.

Assuming the existence of shared norms and understandings, let us now inquire briefly into the way people communicate in face-to-face interaction. This may seem a trivial issue. However, as you will soon see, having a conversation is actually a wonder of intricate complexity. Even today's most advanced supercomputer cannot conduct a natural-sounding conversation with a person (Kurzweil, 1999: 61, 91).

Verbal and Nonverbal Communication

Fifty years ago, an article appeared in the British newspaper, *News Chronicle*, trumpeting the invention of an electronic translating device at the University of London. According to the article, "[a]s fast as [a user] could type the words in, say, French, the equivalent in Hungarian or Russian would issue forth on the tape" (quoted in Silberman, 2000: 225). The report was an exaggeration, to put it mildly. It soon became a standing joke that if you ask a computer to translate "The spirit is willing, but the flesh is weak" into Russian, the output will read "The vodka is good, but the steak is lousy." Today, we are closer to high-quality machine translation than we were in the 1950s. However, a practical Universal Translator exists only on *Star Trek*.

The main problem with computerized translation systems is that computers find it difficult to make sense of the *social and cultural context* in which language is used. The same words may mean different things in different settings, so computers, lacking contextual cues, routinely botch translations. For this reason, metaphors are notoriously problematic for computers. The following machine translation, which contains both literal and metaphorical text, illustrates this point:

English original:

Babel Fish is a computerized translation system that is available on the World Wide Web (at http://babel.altavista.com/translate.dyn). You can type a passage in a window and receive a nearly instant translation in one of four languages. Simple, literal language is translated fairly accurately. But when understanding requires an appreciation of social context, as most of our everyday speech does, the computer can quickly get you into a pickle. What a drag!

Machine translation from English to Spanish:

El pescado de Babel es un sistema automatizado de la traducción que está disponible en el World Wide Web (en http://babel.altavista.com/translate.dyn). Usted puede pulsar un paso en un Window y recibir una traducción casi inmediata en uno de cuatro lenguajes. El lenguaje simple, literal se traduce bastante exactamente. Pero cuando la comprensión requiere un aprecio del contexto social, como la mayoría de nuestro discurso diario, el ordenador puede conseguirle rápidamente en una salmuera. Una qué fricción!

Machine translation from Spanish back to English:

The fish of Babel is an automated system of the translation that is available in the World Wide Web (in http://babel.altavista.com/translate.dyn). You can press a passage in a Window and receive an almost immediate translation in one of four languages. The simple, literal language is translated rather exactly. But when the understanding requires an esteem of the social context, like most of our daily speech, the computer can obtain to him in a brine quickly. One what friction!

Despite the complexity involved in accurate translation, human beings are much better at it than computers. Why is this so? A hint comes from computers themselves. Machine translation works best when applications are restricted to a single social context—say, weather forecasting or oil exploration. In such cases, specialized vocabularies and meanings specific to the context of interest can be built into the program. Ambiguity is thus reduced and computers can "understand" the meaning of words well enough to translate them reasonably accurately. Similarly, humans must be able to reduce ambiguity and make sense of words to become good translators. They do so by learning the nuances of meaning in different cultural and social contexts over an extended period of time.

Mastery of one's own language happens the same way. People are able to understand one another not just because they are able to learn words—computers can do that well enough—but because they can learn the social and cultural contexts that give words meaning. They are greatly assisted in that task by *nonverbal* cues.

Let us linger for a moment on the question of how nonverbal cues enhance meaning. Sociologists, anthropologists, and psychologists have identified numerous nonverbal means of communication that establish context and meaning. The most important types of nonverbal communication involve the use of facial expressions, gestures, body language, and status cues.

Facial Expressions, Gestures, and Body Language

The April 2000 issue of *Cosmopolitan* magazine featured an article advising female readers on "how to reduce otherwise evolved men to drooling, panting fools." Basing his analysis on the work of several psychologists, the author of the article first urges readers to "[d]elete the old-school seductress image (smoky eyes, red lips, brazen stare) from your consciousness." Then, he writes, you must "[u]pload a new inner temptress who's equal parts good girl and wild child." This involves several steps, including the following: (1) Establish eye contact by playing sexual peek-a-boo. Gaze at him, look away, peek again, etc. By interrupting the intensity of your gaze, you heighten his anticipation of the next glance. The trick is to hold his gaze long enough to rouse his interest yet briefly enough to make him want more. Three seconds of gazing followed by 5 seconds of looking away seems to be the ideal. (2) Sit down with your legs crossed to emphasize their shapeliness. Your toes should be pointed toward the man who interests you and should reach inside the 3-foot "territorial bubble" that defines his personal space. (3) Speak quietly. The more softly you speak, the more intently he must listen. Speaking just above a whisper will grab his full attention and force him to remain fixed on you. (4) Invade his personal space and enter his "intimate zone" by finding an excuse to touch him. Picking a piece of lint off his jacket and then leaning in to tell him in a whisper what you've done ought to do the trick. Then you can tell him how much you like his cologne. (5) Raise your arm to flip your hair. The gesture subliminally beckons him forward. (6) Finally, smile—and when you do, tilt your head to reveal your neck because he'll find it exciting (Willardt, 2000). If things progress, another article in the same issue of *Cosmopolitan* explains how you can read his body language to tell whether he's lying (Dutton, 2000).

Whatever we may think of the soundness of *Cosmopolitan*'s advice or the image of women and men it tries to reinforce, this example drives home the point that social

✦ **FIGURE 5.2** ✦
Among other things, body language communicates the degree to which people conform to gender roles, or widely shared expectations about how males or females are supposed to act. In these photos, which postures suggest power and aggressiveness? Which suggest pleasant compliance? Which are "appropriate" to the sex of the person?

interaction typically involves a complex mix of verbal and nonverbal messages. The face alone is capable of more than 1,000 distinct expressions reflecting the whole range of human emotion. Arm movements, hand gestures, posture, and other aspects of body language send many more messages to one's audience (Birdwhistell, 1970; Wood, 1999 [1996]; see Figure 5.2).

Despite the wide variety of facial expressions in the human repertoire, most researchers believed until recently that the facial expressions of six emotions are similar across cultures. These six emotions are happiness, sadness, anger, disgust, fear, and surprise (Ekman, 1978). A smile, it was believed, looks and means the same to advertising executives in Manhattan and members of an isolated tribe in Papua New Guinea. Researchers concluded that the facial expressions that express these basic emotions are reflexes rather than learned responses.

Since the mid-1990s, however, some researchers have questioned whether a universally recognized set of facial expressions reflects basic human emotions. Among other things, critics have argued that "facial expressions are not the readout of emotions but displays that serve social motives and are mostly determined by the presence of an audience" (Fernandez-Dols, Sanchez, Carrera, and Ruiz-Belda, 1997: 163; Harrigan and Tiang, 1997). From this point of view, a smile will reflect pleasure if it serves a person's interest to present a smiling face to his or her audience. On the other hand, a person may be motivated to conceal anxiety by smiling or to conceal pleasure by suppressing a smile.

Some people are better at deception than others. Most people find it hard to deceive others because facial expressions are hard to control. If you've ever tried to stop yourself from blushing you will know exactly what we mean. Sensitive analysts of human affairs—not just sociologists trained in the fine points of symbolic interaction, but police detectives, lawyers, and other specialists in deception—can often see through phony performances. They know that a crooked smile, a smile that lasts too long, or a smile that fades too quickly may suggest that something fishy is going on beneath the superficial level of impression management. Still, smooth operators can fool experts. Moreover, experts can be mistaken. Crooked smiles and the like may be the result of innocent nervousness, not deception.

No gestures or body postures mean the same thing in all societies and all cultures. In our society, people point with an outstretched hand and an extended finger. However, people raised in other cultures tip their head or use their chin or eyes to point out something. We nod our heads "yes" and shake "no," but others nod "no" and shake "yes."

Finally, we must note that in all societies people communicate by manipulating the space that separates them from others (Hall, 1959; 1966). This is well illustrated in our *Cosmopolitan* example, where women are urged to invade a man's "personal space" and "intimate zone" to arouse his interest. Sociologists commonly distinguish four zones that surround us. The size of these zones varies from one society to the next. In North America, an intimate zone extends about 18 inches from the body. It is restricted to people with whom we want sustained, intimate physical contact. A personal zone extends from about 18 inches to 4 feet away. It is reserved for friends and acquaintances. We tolerate only a little physical intimacy from such people. The social zone is situated in the area roughly 4 to 12 feet away from us. Apart from a handshake, no physical contact is permitted from people we restrict to this zone. The public zone starts around 12 feet from our bodies. It is used to distinguish a performer or a speaker from an audience.

Status Cues

Aside from facial expressions, gestures, and body language, a second type of nonverbal communication takes place by means of **status cues,** or visual indicators of other people's social position. Goffman (1959 [1956]) observed that when individuals come into contact, they typically try to acquire information that will help them define the situation and make interaction easier. This is accomplished in part by attending to status cues. Elijah Anderson (1990) developed this idea by studying the way African Americans and European Americans interact on the street in two adjacent urban neighborhoods. Members of both groups visually inspect strangers before concluding that they are not dangerous. They make assumptions about others on the basis of skin color, age, gender, companions, clothing, jewelry, and the objects they carry with them. They evaluate the movements of strangers, the time of day, and other factors to establish how dangerous they might be. In general, children pass inspection easily. White women and men are treated with greater

Stereotypes are rigid views of how members of various groups act, regardless of whether individual group members really behave that way. Stereotypes create social barriers that impair interaction or prevent it altogether. Which stereotypes about Japanese people are reinforced by this American World War II poster?

Louseous Japanicas

The first serious outbreak of this lice epidemic was officially noted on December 7, 1941, at Honolulu, T. H. To the Marine Corps, especially trained in combating this type of pestilence, was assigned the gigantic task of extermination. Extensive experiments on Guadalcanal, Tarawa, and Saipan have shown that this louse inhabits coral atolls in the South Pacific, particularly pill boxes, palm trees, caves, swamps and jungles.

Flame throwers, mortars, grenades and bayonets have proven to be an effective remedy. But before a complete cure may be effected the origin of the plague, the breeding grounds around the Tokyo area, must be completely annihilated.

caution, but not as much caution as black women and black men. Urban dwellers are most suspicious of black male teenagers. People are most likely to interact verbally with individuals who are perceived as the safest.

While status cues may be useful in helping people define the situation and thus greasing the wheels of social interaction, they also pose a social danger. For status cues can quickly degenerate into **stereotypes,** or rigid views of how members of various groups act, regardless of whether individual group members really behave that way. Stereotypes create social barriers that impair interaction or prevent it altogether. For instance, police officers in some states routinely stop young black male drivers without cause to check for proper licensing, possession of illegal goods, etc. In this case, a social cue has become a stereotype that guides police policy. Young black males, the great majority of whom never commit an illegal act, view the police practice as harassment. Racial stereotyping therefore helps to perpetuate the sometimes poor relations between the African-American community and law enforcement officials.

As these examples show, then, face-to-face interaction may at first glance appear to be straightforward and unproblematic. Most of the time it is. However, underlying the taken-for-granted surface of human communication is a wide range of cultural assumptions, unconscious understandings, and nonverbal cues that make interaction possible.

Interaction and Power

So far we have made four main points:

1. One of the most important forces that cements social interaction is the competitive exchange of valued resources. People communicate to the degree they get something valuable out of the interaction. Simultaneously, however, they must engage in a careful balancing act. For if they compete too avidly and prevent others from getting much out of the social interaction, communication will break down. This is exchange and rational choice theory in a nutshell.

2. Nobody hands values, norms, roles, and statuses to us fully formed. Nor do we accept them mechanically. We mold them to suit us as we interact with others. For example, we constantly engage in impression management so others will see the roles we perform in the best possible light. This is a major argument of symbolic interactionism and its most popular variant, dramaturgical analysis.

3. Norms do not emerge entirely spontaneously during social interaction either. In general form, they exist before any given interaction takes place. Indeed, sustained interaction would be impossible without preexisting shared understandings. This is the core argument of ethnomethodology.

4. Nonverbal mechanisms of communication greatly facilitate social interaction. These mechanisms include facial expressions, hand gestures, body language, and status cues.

We now want to highlight a fifth point that has so far been lurking in the background of our discussion. When people interact, their statuses are often arranged in a hierarchy. Those on top enjoy more power than those on the bottom. The degree of inequality strongly affects the character of social interaction between the interacting parties (Bourdieu, 1977; Collins, 1975; 1982; Gamson, Fireman, and Rytina, 1982; Molm, 1997).

Max Weber (1947: 152) defined **power** as "the probability that one actor within a social relationship will be in a position to carry out his [or her] own will despite resistance." We can clearly see how the distribution of power affects interaction by examining male–female interaction. Women are typically socialized to assume subordinate positions in life, men to assume superordinate positions. As we saw in Chapter 4 ("Socialization"), this is evident in the way men usually learn to be aggressive and competitive, women cooperative and supportive (see also Chapter 9, "Sexuality and Gender"). Because of this learning, men often dominate conversations. Thus, conversation analyses conducted by Deborah Tannen show that men are more likely than women to engage in long monologues and interrupt when others are talking. They are also less likely to ask for help or directions because doing so would imply a reduction in their authority. Much male–female conflict results from these differences. A stereotypical case is the lost male driver and the helpful female passenger. The female passenger, seeing that the male driver is lost, suggests that they stop and ask for directions. The male driver doesn't want to ask for directions because he thinks that would make him look incompetent. If both parties remain firm in their positions, an argument is bound to result (Tannen, 1994a; 1994b; see Box 5.1).

To get a better grasp on the role of power in social interaction, it is useful to consider two extreme cases and the case that lies at the midpoint between the extremes (see Figure 5.3 and Table 5.1).[2] **Domination** represents one extreme type of interaction. In social interaction based on domination, nearly all power is concentrated in the hands of people of similar status, while people of different status enjoy almost no power. Guards versus inmates in a concentration camp, and landowners versus slaves on plantations in the antebellum South, were engaged in social interaction based on domination. In extreme cases of domination, subordinates live in a state of near-constant fear.

The other extreme involves interaction based on **cooperation.** Here, power is more or less equally distributed between people of different status. Cooperative interaction is based on feelings of trust. For example, as we will see in Chapter 12 ("Families"), marriages are happier when spouses share housework and child care equitably. Perceived inequity breeds resentment and dissatisfaction. It harms intimacy. It increases the chance that people will have extramarital affairs and divorce. In contrast, a high level of trust between spouses is associated with marital stability and enduring love (Wood, 1999 [1996]).

Between the two extremes of interaction based on domination and interaction based on cooperation is interaction based on **competition.** In this mode of interaction, power is unequally distributed but the degree of inequality is less than in systems of domination. Most of the social interactions analyzed by exchange and rational choice theorists are of this type. If trust is the prototypical emotion of relationships based on cooperation, and

Social interaction is not based entirely on selfishness. Love, for example, is based on trust.
Gustav Klimt. "The Kiss." 1907–08.

---

[2]The following discussion of power and social interaction in different types of organizations is based mainly on Amitai Etzioni's (1975) classic analysis of types of formal organizations and Randall Collins's useful essay on power (Collins, 1982: 60–85).

BOX 5.1
IT'S YOUR CHOICE

ALLOCATING TIME FAIRLY IN CLASS DISCUSSIONS

As Chair of the Department of Sociology at the University of Illinois (Urbana-Champaign), John Lie was used to hearing student complaints. Sometimes they were reasonable. Sometimes they were not. A particularly puzzling complaint came from a self-proclaimed feminist taking a women's studies class. She said: "The professor lets the male students talk in class. They don't seem to have done much of the reading, but the professor insists on letting them say something even when they don't really have anything to say." John later talked to the professor, who claimed she was only trying to let different opinions come out in class.

Policy debates often deal with important issues at the state, national, and international levels. However, they may revolve around everyday social interaction. For instance, as this chapter's discussion of Deborah Tannen's work suggests, gender differences in conversational styles have a big impact on gender inequality. Thus, many professors use class participation to evaluate students. Your grade may depend in part on how often you speak up and whether you have something interesting to say. But Tannen's study suggests that men tend to speak up more often and more forcefully than women. Men are more likely to dominate classroom discussions. Therefore, does the evaluation of class participation in assigning grades unfairly penalize female students? If so, what policies can you recommend that might overcome the problem?

One possibility is to eliminate class participation as a criterion for student evaluation. However, most professors would object to this approach on the grounds that good discussions can demonstrate students' familiarity with course material, sharpen one's ability to reason logically, and enrich every-one's educational experience. A college lacking energetic discussion and debate wouldn't be much of an educational institution.

A second option is to systematically encourage women to participate in classroom discussion. A third option is to allot equal time for women and men or to allot each student equal time. Criticisms of such an approach come readily to mind. Shouldn't time be allocated only to people who have done the reading and have something interesting to say? That is the point made by the woman who complained to John Lie. Encouraging everyone to speak or forcing each student to speak for a certain number of minutes, even if they don't have something interesting to contribute, would probably be boring or frustrating for better prepared students.

As you can see, there is no obvious solution to the question of how time should be allocated in class discussions. In general, the realm of interpersonal interaction and conversation is an extremely difficult area in which to impose rules and policies. So what should your professor do to ensure that class discussion time is allocated fairly?

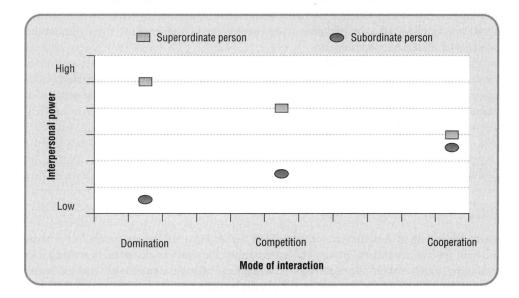

+ FIGURE 5.3 +
Interpersonal Power by Mode of Interaction

+ TABLE 5.1 +
Main Modes of Interaction

Mode of Interaction	Domination	Competition	Cooperation
Level of Inequality	High	Medium	Low
Characteristic Emotion	Fear	Envy	Trust
Efficiency	Low	Medium	High

varying degrees of fear the characteristic emotion of subordinates involved in relationships based on domination, envy is an important emotion in competitive interaction.

Significantly, the mode of interaction in an organization strongly influences its efficiency or productivity, that is, its ability to achieve its goals at the least possible cost. Thus, African-American slaves on plantations in the antebellum South and Jews in Nazi concentration camps were usually regarded as slow and inept workers by their masters (Collins, 1982: 66–9). This characterization was not just a matter of prejudice. Slavery *is* inefficient. That is because, in the final analysis, it is only the fear of coercion that motivates slaves to work. Yet, as psychologists have known for half a century, punishment is a far less effective motivator than reward (Skinner, 1953). Slaves hate their tedious and often backbreaking labor, they get little in exchange for it, and therefore they typically work with less than maximum effort.

In a competitive mode of interaction, subordinates receive more benefits, including prestige and money. Prestige and money are stronger motivators than the threat of coercion. Thus, if bosses pay workers reasonably well, and treat them with respect, they will work more efficiently than slaves even if they don't particularly enjoy their work or identify with the goals of the company. Knowing that they can make more money by working harder, and that their efforts are appreciated, workers will often put in extra effort (Collins, 1982: 63–5).

As Randall Collins and others have shown, however, the most efficient workers are those who enjoy their work and identify with their employer (Collins, 1982: 60–85; Lowe, 2000). Giving workers a bigger say in decision making, encouraging worker creativity, and ensuring that salaries and perks are not too highly skewed in favor of those on top all help to create high worker morale and foster a more cooperative work environment. Company picnics, baseball games, and, in Japan, the singing of company songs before the workday begins all help workers feel they are in harmony with their employer and are playing on the same team. Similarly, while sales meetings and other conferences have an instrumental purpose (the discussion of sales strategies, new products, etc.), they also offer opportunities for friendly social interaction that increase workers' identification with their employer. When workers identify strongly with their employers, they will be willing to undergo self-sacrifice, take the initiative, and give their best creative effort, even without the prospect of increased material gain.

In the next section, we have more to say about competition versus cooperation as modes of interaction. As you will see, modes of interaction influence how efficiently bureaucracies operate. We introduce this theme by first considering the types and sources of bureaucratic *inefficiency*.

BUREAUCRACIES AND NETWORKS

Bureaucratic Inefficiency

At the beginning of this chapter, we noted that Weber regarded bureaucracies as the most efficient type of secondary group. This runs against the grain of common knowledge. In everyday speech, when someone says "bureaucracy," people commonly think of bored clerks sitting in small cubicles spinning out endless trails of "red tape" that create needless waste and frustrate the goals of clients (see Box 5.2). The idea that bureaucracies are efficient may seem very odd.

Real events often reinforce the common view. Consider, for instance, the case of the Challenger space shuttle, which exploded shortly after takeoff on January 28, 1986. All seven crewmembers were killed. The weather was cold, and the flexible "O-rings" that were supposed to seal the sections of the booster rockets had become rigid. This allowed burning gas to leak. The burning gas triggered the explosion. Some engineers at NASA and at the company that manufactured the O-rings knew they wouldn't function properly in cold weather. However, this information did not reach NASA's top bureaucrats:

BOX 5.2
SOCIOLOGY AT THE MOVIES

IKIRU (1952)

Mounds of paperwork, cluttered office desks, long lines of complaining citizens, and indifferent clerks who quietly shuffle papers—this image of bureaucracy can be found in all modern and postmodern societies. Few people are without a story or two of frustrating struggles against one bureaucracy or another. Most movies depict bureaucracy as perpetually mired in red tape and as an impersonal, soulless machine.

Ikiru, directed by the Japanese filmmaker Akira Kurosawa is a profound portrait of the individual versus bureaucracy. Many film critics consider it one of the top 10 films of the 20th century.

At the beginning of the film, the main character, Kanji Watanabe, seems little more than a living corpse. As a minor clerk in a large city bureaucracy, he spends much of the day plodding through documents. He is a true bureaucratic ritualist. He doesn't see and doesn't seem to care about the peo-

In *Ikiru,* Kanji Watanabe, played by Takashi Shamura, finds solace from bureaucracy and the prospect of death by helping to create a small park for children in his neighborhood.

ple whom he is supposed to be serving. He simply shuffles paper.

One day, however, Watanabe learns he is suffering from stomach cancer and has only a year left to live. Without a word, he leaves his job of 30 years. He decides to devote his remaining time to finding meaning in life. But he is alone in the world. His wife is dead. His son is indifferent. His coworkers are strang-

ers. So he decides to go to a bar for the first time in his life and drink himself into oblivion. He finds the experience meaningless.

Then Watanabe spots a pretty young woman from his office. Perhaps she can divert his attention from his looming mortality? In the end, she does, though not in the way Watanabe expected. She inspires him to do something small that will make the world a better place. He hears about a struggle to create a small park for children in his neighborhood. Soon, we find him devoting all of his energy to turning the idea of the park into a reality. Ironically, he spends much of his time battling an uncooperative bureaucracy staffed by uncaring and indifferent officials.

In the end, Watanabe dies. Initially, those who had fought with him vow to go on, to realize the dead man's dream. Soon, however, the rhythm of bureaucratic life resumes. Nearly everyone returns to his or her role as a functionary. As a charismatic leader of a small social movement, the hero briefly made an impact on his society. In the end, however, the wheels of bureaucracy grind on.

Can you think of a situation in which you or someone you know attempted to challenge and reform a bureaucracy? Can change come from within the bureaucracy or does it need to come from outside? Can individuals overcome bureaucratic inertia? Or are we doomed to have the wheels of bureaucracy roll over us?

[The] rigid hierarchy that had arisen at NASA . . . made communication between departments formal and not particularly effective. [In the huge bureaucracy,] most communication was done through memos and reports. Everything was meticulously documented, but critical details tended to get lost in the paperwork blizzard. The result was that the upper-level managers were kept informed about possible problems with the O-rings . . . but they never truly understood the seriousness of the issue (Pool, 1997: 257).

As this tragedy shows, then, bureaucratic inefficiencies can sometimes have tragic consequences.

How can we reconcile the reality of bureaucratic inefficiencies—even tragedies—with Weber's view that bureaucracies are the most efficient type of secondary group? The answer is twofold. First, we must recognize that when Weber wrote about the efficiency of bureaucracy, he was comparing it to older organizational forms. These operated on the basis of either traditional practice ("We do it this way because we've always done it this way") or the charisma of their leaders ("We do it this way because our chief inspires us to do it this way"). Compared to such "traditional" and "charismatic" organizations, bureaucracies *are* generally more efficient. Second, we must recognize that Weber thought bureaucracies could operate efficiently only in the ideal case. He wrote extensively about some of bureaucracy's less admirable aspects in the real world. In other words, he understood that reality is often messier than the ideal case. So should we. In reality, bureaucracies vary in efficiency. Therefore, rather than proclaiming bureaucracy efficient or inefficient, it makes sense to find out what makes bureaucracies work well or poorly. The knowledge can then be applied to improving the operation of bureaucracies.

Traditionally, sociologists have lodged four main criticisms against bureaucracies. First is the problem of **dehumanization.** Rather than treating clients and personnel as people with unique needs, bureaucracies sometimes treat clients as standard cases and personnel as cogs in a giant machine. This frustrates clients and lowers worker morale. Second is the problem of **bureaucratic ritualism** (Merton, 1968 [1949]). Bureaucrats sometimes get so preoccupied with rules and regulations they make it difficult for the organization to fulfill its goals. Third is the problem of **oligarchy,** or "rule of the few" (Michels, 1949 [1911]). Some sociologists have argued there is a tendency in all bureaucracies for power to become increasingly concentrated in the hands of a few people at the top of the organizational pyramid. This is particularly problematic in political organizations because it hinders democracy and renders leaders unaccountable to the public. Fourth is the problem of **bureaucratic inertia.** Bureaucracies are sometimes so large and rigid they lose touch with reality and continue their policies even when their clients' needs change. Like the Titanic, they are so big they find it difficult to shift course and steer clear of dangerous obstacles.

Two main factors underlie bureaucratic inefficiency: size and social structure. Consider size first. There is something to be said for the view that bigger is almost inevitably more problematic. Some of the problems caused by size are evident even when you think about the difference between two- and three-person relationships (known respectively as "dyads" and "triads"). When only two people are involved in a relationship, they may form a strong social bond. If they do, communication is direct and sometimes unproblematic. Once a third person is introduced, however, a secret may be kept, a coalition of two against one may crystallize, and jealousy may result. Thus, triads are usually more unstable and conflict ridden than dyads (Simmel, 1950).

Problems can multiply in groups of more than three people. For example, as Figure 5.4 shows, while there can be only one relationship between two people, there can be three relationships among three people and six relationships among four people. The number of *potential* relationships increases exponentially with the number of people, so there are 325 possible relationships among 25 people and 1,225 possible relationships among 50 people. The possibility of clique formation, rivalries, conflict, and miscommunication rises as quickly as the number of possible social relationships in an organization.

The second factor underlying bureaucratic inefficiency is social structure. Figure 5.5 shows a typical bureaucratic structure. Note it is a hierarchy. The bureaucracy has a head. Below the head are three divisions. Below the divisions are six departments. As you move up the hierarchy, the power of the staff increases. Note also the lines of communication that join the various bureaucratic units. Staff members in the departments report only to their divisions. Staff members in the divisions report only to the head.

Usually, the more levels in a bureaucratic structure, the more difficult communication becomes. That is because people have to communicate indirectly, through department heads, rather than directly with each other. Information may be lost, blocked, reinterpreted, or distorted as it moves up the hierarchy; or there may simply be so much information that the top levels become engulfed in a "paperwork blizzard" that prevents them from clearly seeing the needs of the organization and its clients. Bureaucratic heads may have only a vague and imprecise idea of what is happening "on the ground" (Wilensky, 1967).

Consider also what happens when the lines of communication directly joining divisions or departments are weak or nonexistent. As the lines joining units in Figure 5.5 suggest, department A1 may have information that could help department A2 do its job better but may have to communicate that information indirectly, through the division level. There the information may be lost, blocked, reinterpreted, or distorted. Thus, just as people who have authority may lack information, people who have information may lack the authority to act on it directly (Crozier, 1964 [1963]).

Below, we consider some ways of overcoming bureaucratic inefficiency. As you will see, these typically involve establishing patterns of social relations that flatten the bureaucratic hierarchy and cut across the sort of bureaucratic rigidities illustrated in Figure 5.5.

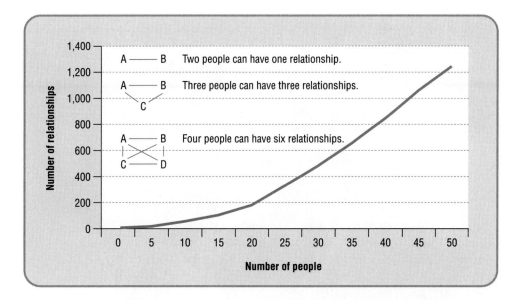

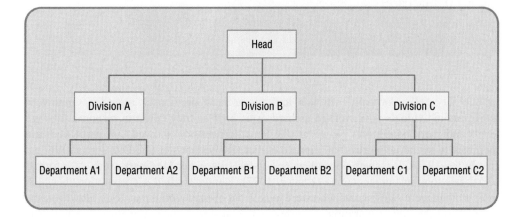

As a useful prelude to this discussion, we first note some shortcomings of Weber's analysis of bureaucracy. Weber tended to ignore both bureaucracy's "informal" side and the role of leadership in influencing bureaucratic performance. Yet, as you will learn, it is precisely by paying attention to such issues that bureaucracies can be made more efficient.

Bureaucracy's Informal Side

Weber was concerned mainly with the formal structure or "chain of command" in a bureaucracy. He paid little attention to the fact that intricate webs of social relations, known as social networks, underlie the chain of command. A **social network** is a bounded set of individuals who are linked by the exchange of material or emotional resources. The patterns of exchange determine the boundaries of the network. Members exchange resources more frequently with each other than with nonmembers. They also think of themselves as network members. Social networks may be formal (defined in writing), but they are often informal (defined only in practice) (Wellman and Berkowitz, 1997 [1988]).

Evidence for the existence of social networks and their importance in the operation of bureaucracies goes back to the 1930s. Officials at the Hawthorne plant of the Western Electric Company near Chicago wanted to see how various aspects of the work environment affected productivity. They sent social scientists in to investigate. Among other things, researchers found workers in one section of the plant had established a norm for daily output. Workers who failed to meet the norm were helped by coworkers until their output increased. Workers who exceeded the norm were chided by coworkers until their

Parents can help their graduating children find jobs by getting them "plugged into" the right social networks. Here, in the 1968 movie *The Graduate,* a friend of the family advises Dustin Hoffman that the future lies in the plastics industry.

productivity fell. Company officials and researchers previously regarded employees merely as individuals who worked as hard or as little as they could in response to wage levels and work conditions. However, the Hawthorne study showed that employees are members of social networks that regulate output (Roethlisberger and Dickson, 1939).

In the 1970s, Rosabeth Moss Kanter conducted another landmark study of informal social relations in bureaucracies (Kanter, 1977). Kanter studied a corporation in which most women were sales agents who were locked out of managerial positions. However, she did not find that the corporation discriminated against women as a matter of policy. She did find a male-only social network whose members shared gossip, went drinking, and told sexist jokes. The cost of being excluded from the network was high: to get good raises and promotions, one had to be accepted as "one of the boys" and be sponsored by a male executive. This was impossible for women. Thus, despite a company policy that did not discriminate against women, an informal network of social relations ensured that the company discriminated against women in practice.

Despite their overt commitment to impersonality and written rules, bureaucracies rely profoundly on informal interaction to get the job done (Barnard, 1938; Blau, 1963 [1955]). This is true even at the highest levels. For example, executives usually decide important matters in face-to-face meetings, not in writing or over the phone. That is because people feel more comfortable in intimate settings, where they can get to know "the whole person." Meeting face-to-face, people can use their verbal and nonverbal interaction skills to gauge other people's trustworthiness. Socializing—talking over dinner, for example—is an important part of any business because the establishment of trust lies at the heart of all social interactions that require cooperation (Gambetta, 1988).

Leadership

Apart from overlooking the role of informal relations in the operation of bureaucracies, Weber also paid insufficient attention to the issue of leadership. Weber thought the operation of a bureaucracy is determined largely by its formal structure. However, sociologists now realize that leadership style also has a bearing on bureaucratic performance (Barnard, 1938; Ridgeway, 1983; Selznick, 1957).

Research shows that the least effective leader is the one who allows subordinates to work things out largely on their own, with almost no direction from above. This is known as *laissez-faire* **leadership,** from the French expression, "let them do." At the other extreme is **authoritarian leadership.** Authoritarian leaders demand strict compliance from subordinates. They are most effective in a crisis such as a war or the emergency room of a hospital. They may earn grudging respect from subordinates for achieving the group's goals in the face of difficult circumstances, but they rarely win popularity contests. **Democratic leadership** offers more guidance than the *laissez-faire* variety but less control than the authoritarian type. Democratic leaders try to include all group members in the decision-making process, taking the best ideas from the group and molding them into a strategy that all can identify with. Except for crisis situations, democratic leadership is usually the most effective leadership style.

In sum, contemporary researchers have modified Weber's characterization of bureaucracy in two main ways. First, they have stressed the importance of informal social networks in shaping bureaucratic operations. Second, they have shown that democratic leaders are most effective in noncrisis situations because they tend to distribute decision-making authority and rewards widely. As you will now see, these are important lessons when it comes to thinking about how to make bureaucracies more efficient.

Overcoming Bureaucratic Inefficiency

In the business world, large bureaucratic organizations sometimes find themselves unable to compete against smaller, innovative firms, particularly in industries that are changing quickly (Burns and Stalker, 1961). That is partly because innovative firms tend to have flatter and more democratic organizational structures, such as the network illustrated in Figure 5.6. Compare the flat network structure in Figure 5.6 with the traditional bureaucratic structure in Figure 5.5. Note that the network structure has fewer levels than the traditional bureaucratic structure. Moreover, in the network structure, lines of communication link all units. In the traditional bureaucratic structure, information flows only upward.

Much evidence suggests that flatter bureaucracies with decentralized decision making and multiple lines of communication produce more satisfied workers, happier clients, and bigger profits (Kanter, 1977; 1983; 1989). Some of this evidence comes from Sweden and Japan. There, beginning in the early 1970s, corporations such as Volvo and Toyota were at the forefront of bureaucratic innovation. They began eliminating middle-management positions. They allowed worker participation in a variety of tasks related to their main functions. They delegated authority to autonomous teams of a dozen or so workers that were allowed to make many decisions themselves. They formed "quality circles" of workers to monitor and correct defects in products and services. As a result, product quality, worker morale, and profitability improved. Today, these ideas have spread well beyond the Swedish and Japanese automobile industry and are evident in such American corporate giants as General Motors, Ford, Boeing, and Caterpillar.

In the 1980s and 1990s, companies outside the manufacturing sector introduced similar bureaucratic reforms, again with positive effects. Consider the case of Bob R., who

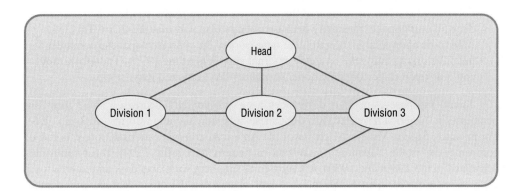

♦ FIGURE 5.6 ♦
Network Structure

works as a field technician for a major southern telephone company. His company created small teams of field technicians who are each responsible for keeping a group of customers happy. The technicians set their own work schedule. They figure out when they need to do preventive maintenance, when they need to conduct repairs, and when it is time to try to sell customers new services. Bob describes his new job as follows:

> I've been at the company for twenty-three years, and I always thought we were overmanaged, overcontrolled, and oversupervised. They treated us like children. We're having a very good time under the new system. They've given us the freedom to work on our own. This is the most intelligent thing this company has done in years. It's fun (quoted in Hammer, 1999: 87).

Managers and workers in many industries have offered similar testimonies to the benefits of more democratic, network-like structures.

Organizational Environments

If flatter organizations are more efficient, why aren't all bureaucracies flatter? Mainly, say sociologists, because of the environment in which they operate. An **organizational environment** is composed of a host of economic, political, and cultural factors that lie outside an organization and affect the way it works (Aldrich, 1979; McKelvey, 1982; Meyer and Scott, 1983; Martin, 1992). Some organizational environments are conducive to the formation of flatter, network-like bureaucracies. Others are not. We can illustrate the effects of organizational environments by discussing two cases that have attracted much attention in recent years: the United States and Japan.

In the 1970s, American business bureaucracies tended to be more hierarchical than their Japanese counterparts. As a result, worker dissatisfaction was high and labor productivity was low in the United States. In Japan, where corporate decision making was more decentralized, worker morale and productivity were high (Dore, 1983; Vogel, 1975). Several aspects of the organizational environment help to explain Japanese–American differences in the 1970s. Specifically:

✦ *Japanese workers were in a position to demand and achieve more decision-making authority than United States workers.* That is because, after World War II, the proportion of Japanese workers in unions increased while the proportion of American workers in unions declined (see Chapter 17, "Collective Action and Social Movements") (Lie, 1998). Unions gave Japanese workers more clout than their American counterparts enjoyed.

✦ *International competition encouraged bureaucratic efficiency in Japan.* Many big Japanese corporations matured in the highly competitive post–World War II international environment. Many big American corporations originated earlier, in an international environment with few competitors. Thus, Japanese corporations had a bigger incentive to develop more efficient organizational structures (Harrison, 1994).

✦ *The availability of external suppliers allowed Japanese firms to remain lean.* Many large American companies matured when external sources of supply were scarce. For example, when IBM entered the computer market in the 1950s, it had to produce all components internally because nobody else was making them. This led IBM to develop a large, hierarchical bureaucracy. In contrast, Japanese computer manufacturers could rely on many external suppliers in the 1970s. Therefore, they could develop flatter organizational structures (Podolny and Page, 1998).

Today, Japanese–American differences have substantially decreased since most big businesses in America have introduced Japanese-style bureaucratic reforms (Tsutsui, 1998). For instance, Silicon Valley, the center of the American computer industry today, is full of companies that fit the "Japanese" organizational pattern (McCarthy, 1999). These companies originated in the 1980s and 1990s, when external suppliers were abundant and international competitiveness was intense. We thus see how changes in the organizational environment help account for convergence between Japanese and American bureaucratic forms.

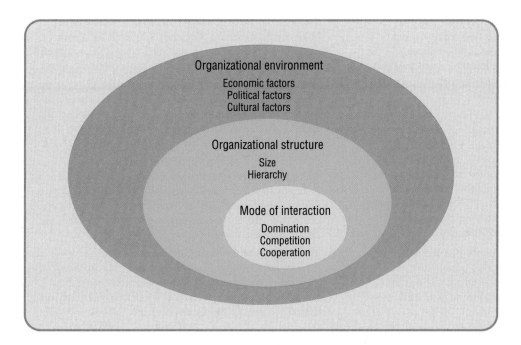

✦ **FIGURE 5.7** ✦
The Structure of Interaction and Organization

The experience of the United States over the past few decades holds out hope for increasing bureaucratic efficiency and the continued growth of employee autonomy and creativity at work. This does not mean that in 20 or 50 or 100 years bureaucracies in Japan and the United States will be alike in all respects. The organizational environment is unpredictable, and sociologists are just beginning to understand its operation. It is therefore anyone's guess how far convergence will continue.

We may summarize our discussion by saying that social interaction and social organization fit together like a set of nested Russian dolls or Chinese boxes (see Figure 5.7). At the microstructural level, face-to-face interaction takes place through verbal and nonverbal communication. Typically, face-to-face interaction involves negotiation and impression management. Interaction may be sustained by means of domination, competition, and cooperation. Sustained interaction may, in turn, give rise to social groups, both primary (such as families) and secondary (such as bureaucracies). Bureaucracies are macrostructural organizations. Their operation is affected by such factors as their size and the degree to which they are hierarchical or "flat." At a higher macrostructural level, forces in the organizational environment influence bureaucracies. We have illustrated the operation of these broad economic, political, and cultural forces here. We discuss them at length in later chapters (see especially Chapter 10, "Work and the Economy," and Chapter 11, "Politics").

SUMMARY

1. According to exchange and rational choice theories, the competitive exchange of valued resources cements social interaction. People communicate to the degree they get something valuable out of the interaction. However, if they compete too avidly and prevent others from getting much out of the social interaction, communication will break down.

2. According to symbolic interactionism and its most popular variant, dramaturgical analysis, values, norms, roles, and statuses are not handed to us fully formed. Nor do we accept them mechanically. We mold them to suit us as we interact with others.

3. According to ethnomethodology, norms do not emerge entirely spontaneously during social interaction. In general

form, they exist before any given interaction takes place. Interaction would be impossible without preexisting shared understandings.

4. Nonverbal mechanisms of communication greatly facilitate social interaction. These mechanisms include facial expressions, hand gestures, body language, and status cues.

5. When people interact, their statuses are often arranged in a hierarchy. Those on top enjoy more power than those on the bottom. The degree of inequality strongly affects the character of social interaction between the interacting parties.

6. The three main modes of interaction are domination (in which inequality between superordinates and subordinates is

high), cooperation (in which inequality is low), and competition (in which inequality is moderate). While varying degrees of fear pervade interaction based on domination, envy is characteristic of competitive interaction and trust suffuses cooperative interaction. Efficiency is highest in cooperative interaction and lowest in interaction based on domination.

7. Bureaucratic inefficiency is evident in dehumanization (treating clients as standard cases and personnel as cogs in a giant machine), bureaucratic ritualism (a preoccupation with rules and regulations that makes it difficult for the organization to fulfill its goals), oligarchy (concentration of power in the hands of a few people), and bureaucratic inertia (rigidity that causes organizations to lose touch with reality and continue their policies even when clients' needs change).

8. In general, inefficiency increases with the size and degree of hierarchy in bureaucracies. By flattening bureaucratic structures, decentralizing decision-making authority, and opening lines of communication between bureaucratic units, efficiency can often be improved.

9. Social networks, or bounded sets of individuals who are linked by the exchange of material or emotional resources, underlie the chain of command in all bureaucracies and affect their operation. Weber ignored this aspect of bureaucracy.

10. Weber also downplayed the importance of leadership in the functioning of bureaucracy. However, research shows that democratic leadership improves the efficiency of bureaucratic operations in noncrisis situations, authoritarian leadership works best in crises, and laissez-faire leadership is the least effective form of leadership in all situations.

11. Although bureaucracies can often be made more efficient by flattening their structures, decentralizing decision-making authority, and opening lines of communication between bureaucratic units, the organizational environment determines the degree to which this is possible.

12. Three aspects of the organizational environment seem to have the greatest bearing on how hierarchical a bureaucracy is. Bureaucracies are less hierarchical where workers are more powerful, competition with other bureaucracies is high, and external sources of supply are available.

GLOSSARY

Authoritarian leaders demand strict compliance from subordinates. They are most effective in a crisis such as a war or the emergency room of a hospital.

Bureaucracies are large, impersonal organizations composed of many clearly defined positions arranged in a hierarchy. Bureaucracies have a permanent, salaried staff of qualified experts and written goals, rules, and procedures. Staff members always try to find ways of running the bureaucracy more efficiently.

Bureaucratic inertia refers to the tendency of large, rigid bureaucracies to continue their policies even when their clients' needs change.

Bureaucratic ritualism involves bureaucrats getting so preoccupied with rules and regulations they make it difficult for the organization to fulfill its goals.

Competition is a mode of interaction in which power is unequally distributed but the degree of inequality is less than in systems of domination. Envy is an important emotion in competitive interactions.

Cooperation is a basis for social interaction in which power is more or less equally distributed between people of different status. The dominant emotion in cooperative interaction is trust.

Dehumanization occurs when bureaucracies treat clients as standard cases and personnel as cogs in a giant machine. This frustrates clients and lowers worker morale.

Democratic leadership offers more guidance than the *laissez-faire* variety but less control than the authoritarian type. Democratic leaders try to include all group members in the decision-making process, taking the best ideas from the group and molding them into a strategy that all can identify with. Outside crisis situations, democratic leadership is usually the most effective leadership style.

Domination is a mode of interaction in which nearly all power is concentrated in the hands of people of similar status. Fear is the dominant emotion in systems of interaction based on domination.

Dramaturgical analysis views social interaction as a sort of play in which people present themselves so as to appear in the best possible light.

Ethnomethodology is the study of how people make sense of what others do and say by adhering to preexisting norms.

Exchange theory holds that social interaction involves trade in valued resources.

Impression management is the process by which people present themselves so they can appear in the best possible light.

***Laissez-faire* leaders** allow subordinates to work things out largely on their own, with almost no direction from above. They are the least effective type of leader.

Oligarchy means "rule of the few." There is a supposed tendency in all bureaucracies for power to become increasingly concentrated in the hands of a few people at the top of the organizational pyramid.

An **organizational environment** is composed of a host of economic, political, cultural, and other factors that lie outside an organization and affect the way it works.

Power is the probability that one actor within a social relationship will be in a position to carry out his own will despite resistance.

In **primary groups,** norms, roles, and statuses are agreed upon but are not set down in writing. Social interaction leads to strong emotional ties. It extends over a long period. It involves a wide range of activities. It results in group members knowing one another well.

Rational choice theory focuses on the way interacting people weigh the benefits and costs of interaction. According to rational choice theory, interacting people always try to maximize benefits and minimize costs.

Role distance is awareness of the difference between our public roles and our "true" selves.

Secondary groups are larger and more impersonal than primary groups. Compared to primary groups, social interaction in secondary groups creates weaker emotional ties. It extends over a shorter period. It involves a narrow range of activities. It results in most group members having at most a passing acquaintance with one another.

A **social category** is composed of people who share a similar status but do not identify with one another.

A **social group** is a cluster of people who identify with one another and adhere to defined norms, roles, and statuses.

Social interaction involves people communicating face-to-face, acting and reacting in relation to other people.

A **social network** is a bounded set of individuals who are linked by the exchange of material or emotional resources. The patterns of exchange determine the boundaries of the network. Members exchange resources more frequently with each other than with nonmembers. They also think of themselves as network members. Social networks may be formal (defined in writing), but they are more often informal (defined only in practice).

Status cues are visual indicators of other people's social position.

Stereotypes are rigid views of how members of various groups act, regardless of whether individual group members really behave that way.

QUESTIONS TO CONSIDER

1. Think about your everyday interactions with friends, professors, and parents. In each case, how do the exchange of valued resources, the dictates of power, and the influence of norms shape your interactions?

2. If you were starting your own business, how would you organize it? Why? Base your answer on theories and research discussed in this chapter.

WEB RESOURCES

Companion Web Site for This Book

http://sociology.wadsworth.com

Begin by clicking on the Student Resources section of the Web site. Choose "Introduction to Sociology" and finally the Brym and Lie book cover. Next, select the chapter you are currently studying from the pull-down menu. From the Student Resources page you will have easy access to InfoTrac College Edition®, MicroCase Online exercises, additional Web links, and many resources to aid you in your study of sociology, including practice tests for each chapter.

Infotrac Search Terms

These search terms are provided to assist you in beginning to conduct research on this topic by visiting http://www.infotraccollege.com/wadsworth.

Bureaucracy
Dramaturgical analysis
Exchange theory
Rational choice theory
Social network

Recommended Web Sites

Erving Goffman's *The Presentation of Self in Everyday Life* (Garden City, NY: Anchor, 1959 [1956]) is a classic in the analysis of social interaction. An excerpt can be found at http://wizard.ucr.edu/~bkaplan/soc/lib/goffself.html.

If you need convincing that social interaction on the Internet can have deep emotional and sociological implications, read Julian Dibbell "A Rape in Cyberspace," on the World Wide Web at http://www.levity.com/julian/bungle.html. This compelling article is especially valuable for showing how social structure emerges in virtual communities. Originally published in *The Village Voice* (21 December 1993) 36–42.

For social interaction on the World Wide Web, visit "The MUD Connector" at http://www.mudconnect.com.

The Society for the Study of Social Interaction is a professional organization of sociologists who study social interaction. Visit their Web site at http://sun.soci.niu.edu/~sssi.

SUGGESTED READINGS

Erving Goffman. *The Presentation of Self in Everyday Life* (Garden City, NY: Anchor, 1959 [1956]). The classic dramaturgical analysis and the model for many later works.

Rosabeth Moss Kanter. *Men and Women of the Corporation* (New York: Basic, 1977). A landmark study of gender in corporate America. Shows how social interaction shapes social organizations and how social organizations shape social interactions.

George Ritzer. "The McDonaldization Thesis: Is Expansion Inevitable?" *International Sociology* (11: 1996) pp. 291–307. Ritzer explains why the world is increasingly being organized on the model of a McDonald's restaurant. He also asks whether McDonaldization is avoidable. His answer: probably not.

IN THIS CHAPTER, YOU WILL LEARN THAT:

✦ Deviance and crime vary among cultures, across history, and from one social context to another.

✦ Rather than being inherent in the characteristics of individuals or actions, deviance and crime are socially defined and constructed. The distribution of power is especially important in the social construction of deviance and crime.

✦ Following dramatic increases in the 1960s and 1970s, crime rates eased in the 1980s and fell in the 1990s. This was due mainly to more effective policing, a declining proportion of young men in the population, and a booming economy.

✦ Statistics show that a disproportionately large number of African Americans are arrested, convicted, and imprisoned. That is due largely to bias in the way crime statistics are collected, the low social standing of the African-American community, and racial discrimination in the criminal justice system.

✦ There are many theories of deviance and crime. Each theory illuminates a different aspect of the process by which people break rules and are defined as deviants and criminals.

✦ As in deviance and crime, conceptions of appropriate punishment vary culturally and historically.

✦ In some respects, modern societies are characterized by less conformity than premodern societies but in other respects they tolerate less deviance.

✦ Imprisonment is one of the main forms of punishment in industrial societies, and in the United States the prison system has grown quickly in the past 30 years and become harsher.

✦ Fear of crime is on the increase, but it is based less on rising crime rates than on manipulation by commercial and political groups that benefit from it.

✦ There are cost-effective and workable alternatives to the regime of punishment currently in place.

CHAPTER 6

DEVIANCE AND CRIME

THE SOCIAL DEFINITION OF DEVIANCE AND CRIME

For serious television viewers, the last week of May 1999 was much like any other. Of the five most watched prime-time network TV programs, three were about crime. Two episodes of *Law and Order* and one of *NYPD Blue* pulled in fully 55.7 million viewers. A long list of other crime-related programs—including *The Practice; Homicide; COPS; Martial Law; Walker, Texas Ranger; Texas Justice;* and *Diagnosis Murder*—attracted many millions more ("Mr. Showbiz," 1999). If one were to add cable, local, daytime, and late-night shows to the list, one might conclude that the United States is a society obsessed with crime.

As one might expect in such a society, punishment is also a big issue. This is evident from the fact that there are now about 2 million people in state and federal prisons and local jails, and their number is increasing by 50,000–80,000 per year. The United States has more people behind bars than any other country on earth. In fact, over 10% more people are behind bars in the United States (2001 population: about 284 million) than in China and India combined (2001 population: about 2.3 billion). State prisons in California alone hold more criminals in their grip than do Japan, Germany, France, Great Britain, the Netherlands, and Singapore combined. As of 2000, the United States had the highest rate of incarceration (the number of people imprisoned per 100,000 population) of any country in the world (Schlosser, 1998; The Sentencing Project, 1997, 2001; United States Bureau of the Census, 2000a, 2001a).

Why is there so much concern with crime in the United States and so much imprisonment? Since we commonly think of criminals as "the bad guys," we might be excused for thinking that the United States simply contains a disproportionately large number of bad people who have broken the law. For the sociologist, however, this is an oversimplification.

Consider the fact that in 1872, Susan B. Anthony, whose image graced the original dollar coin, was arrested and fined. What was her crime? The criminal indictment charged that she "knowingly, wrongfully and unlawfully voted for a representative to the Congress of the United States." Justice Ward Hunt advised the jury: "There is no question for the jury, and the jury should be directed to find a verdict of guilty" (quoted in Flexner, 1975: 170). In the late 1950s and early 1960s, Martin Luther King, Jr., whose birthday we now celebrate as a federal holiday, was repeatedly arrested. What was his crime? He marched in the streets of Birmingham and other Southern cities for African Americans' civil rights, including their right to vote.

Susan B. Anthony and Martin Luther King, Jr., were considered deviant and criminal in their lifetimes. Few Americans in the 1870s thought women should be allowed to vote. Anthony disagreed. In acting on her deviant belief, she committed a crime. Similarly, in the 1950s, most people in the American South believed in white superiority. That belief was expressed in many ways, including so-called Jim Crow laws that prevented many African Americans from voting. Martin Luther King, Jr., and other civil rights movement participants challenged the existing law and were therefore arrested.

Most of you would consider the sexist and racist society of the past, rather than Anthony and King, deviant or criminal. That is because norms and laws have changed dramatically. The 19th Amendment guaranteed women's suffrage in 1920. The Voting Rights Act of 1965 guaranteed voting rights for African Americans. Today, anyone arguing that women or African Americans should not be allowed to vote is considered deviant. Preventing them from voting would result in arrest.

As the examples of Anthony and King suggest, definitions of deviance and crime change over time. For example, homosexuality used to be considered a crime and then a sickness, but an increasing proportion of Americans in the United States and elsewhere now recognize homosexuality as a legitimate sexual orientation (Greenberg, 1988). Similarly, acts that are right and heroic for some people are wrong and treacherous for others. Ask any of the 13,000 law enforcement officials who were brought in to restore order in Los Angeles after the eruption of the 1992 race riot, and they will almost certainly tell you

Martin Luther King Jr. is arrested on September 4, 1958, in Montgomery, Alabama. Considered by many a deviant in his time, King is today hailed by most Americans as a hero. This suggests how the social definition of deviance may change over time.

that the people who engaged in mass looting and violence were all common criminals. Ask a politically sophisticated radical African American, such as Sanyika Shakur (a.k.a. "Monster" Kody Scott), a former leader of the notorious Los Angeles gang, the Crips, and he'll tell you the riot was largely a political reaction to the oppression and powerlessness of inner-city blacks (Shakur, 1993: 381). Sociologists, however, will not jump to hasty conclusions. Instead, they will try to understand how social definitions, social relationships, and social conditions led to the rioting and the labeling of the rioters as criminal by the police.

That is the approach to deviance and crime we take in this chapter. We first discuss how deviance and crime are socially defined. We then analyze crime patterns in the United States—who commits crimes and what accounts for changing crime rates over time. Next we assess the major theories of deviance and crime. Finally, we examine the social determinants of different types of punishment.

Types of Deviance and Crime

Deviance involves breaking a norm. If you were the only man in a college classroom full of women, you probably wouldn't be considered deviant. However, if a man were to use a woman's restroom, we would regard him as deviant. That is because deviance is not merely departure from the statistical average. It implies violating an accepted rule of behavior.

Many deviant acts go unnoticed or are considered so trivial they warrant no punishment. However, people who are observed committing more serious acts of deviance are typically punished, either informally or formally. **Informal punishment** is mild. It may involve raised eyebrows, gossip, ostracism, "shaming," or **stigmatization** (Braithwaite, 1989). When people are stigmatized, they are negatively evaluated because of a marker that distinguishes them from others (Goffman, 1963). One of this book's authors, John Lie, was stigmatized as a young child and often bullied by elementary school classmates because he had a Korean name in a Japanese school. "I was normal in other ways," says John. "I played the same sports and games, watched the same television shows, and looked, dressed, and acted like other Japanese students. However, my one deviation was enough to stigmatize me. It gave license to some of my classmates to beat me up from time to time. I wondered at the time why no rules banned bullying and why no law existed against what I now call racial discrimination. If such a law did exist, my classmates would have been subject to formal punishment, which is more severe than informal punishment." **Formal**

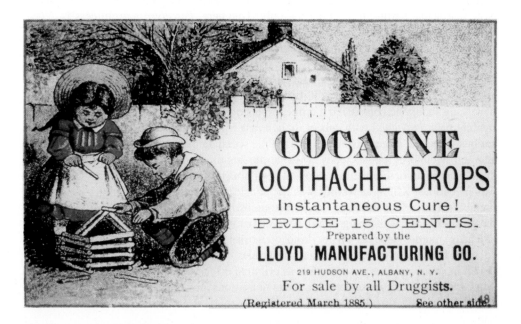

One of the determinants of the seriousness of a deviant act is its *perceived* harmfulness. Perceptions vary historically, however. For instance, until the early part of the 20th century, cocaine was considered a medicine. It was an ingredient of Coca-Cola and toothache drops, and in these forms was commonly given to children.

Transvestites dress in clothing generally considered appropriate to members of the opposite sex. Is transvestitism a social diversion, a social deviation, a conflict crime, or a consensus crime? Why?

punishment results from people breaking laws, which are norms stipulated and enforced by government bodies. For example, criminals may be formally punished by having to serve time in prison or perform community service.

Sociologist John Hagan (1994) usefully classifies various types of deviance and crime along three dimensions (see Figure 6.1). The first dimension is the *severity of the social response*. At one extreme, homicide and other very serious forms of deviance result in the most severe negative reactions, such as life imprisonment or capital punishment. At the other end of the spectrum, some people may do little more than express mild disapproval of slight deviations from a norm, such as wearing a nose ring.

The second dimension of deviance and crime is the *perceived harmfulness* of the deviant or criminal act. While some deviant acts, such as rape, are generally seen as very harmful, others, such as tattooing, are commonly regarded as being of little consequence. Note that actual harmfulness is not the only issue here. *Perceived* harmfulness is. Coca-Cola got its name because, in the early part of the 20th century, it contained a derivative of cocaine. Now cocaine is an illegal drug because people's perceptions of its harmfulness changed.

The third characteristic of deviance is the *degree of public agreement* about whether an act should be considered deviant. For example, people disagree about whether smoking marijuana should be considered a crime, especially since it may have therapeutic value in treating pain associated with cancer. In contrast, virtually everyone agrees that murder is seriously deviant. Note, however, that even the social definition of murder varies over time and across cultures and societies. Thus, at the beginning of the 20th century, Inuit ("Eskimo") communities sometimes allowed newborns to freeze to death. Life in the far north was precarious. Killing newborns was not considered a punishable offense if community members agreed that investing scarce resources in keeping the newborn alive could endanger everyone's well-being. Similarly, whether we classify the death of a miner as an accident or murder depends on the kind of worker safety legislation in existence. Some societies have more stringent worker safety rules than others, and deaths considered accidental in some societies are classified as criminal offenses in others. So we see that, even when it comes to consensus crimes, social definitions are variable.

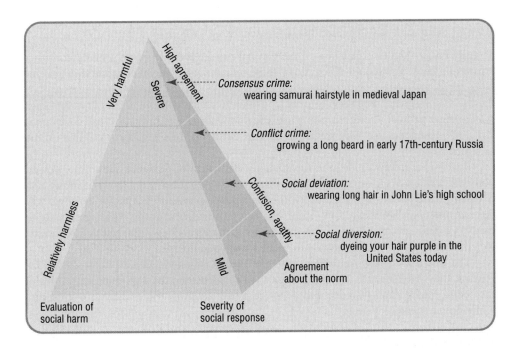

As Figure 6.1 shows, Hagan's analysis allows us to classify four types of deviance and crime:

1. **Social diversions** are minor acts of deviance such as participating in fads and fashions like dyeing one's hair purple. People usually perceive such acts as harmless. They evoke, at most, a mild societal reaction such as amusement or disdain. That is because many people are apathetic or unclear about whether social diversions are in fact deviant.

2. **Social deviations** are more serious acts. Large proportions of people agree they are deviant and somewhat harmful, and they are usually subject to institutional sanction. For example, John Lie's high school in Hawaii had a rule making long hair for boys a fairly serious deviation punishable by a humiliating public haircut.

3. **Conflict crimes** are deviant acts that the state defines as illegal, but the definition is controversial in the wider society. For instance, Tsar Peter the Great of Russia wanted to Westernize and modernize his empire, and he viewed long beards as a sign of backwardness. On September 1, 1698, he imposed a fine on beards in the form of a tax. Many Russians disagreed with his policy. Others agreed that growing long beards harmed Russia because it symbolized Russia's past rather than her future. Due to this disagreement in the wider society about the harmfulness of the practice, wearing a long beard in late 17th-century Russia may be classified as a conflict crime.

4. Finally, **consensus crimes** are widely recognized to be bad in themselves. There is little controversy over their seriousness. The great majority of people agree that such crimes should be met with severe punishment. For instance, in medieval Japan, hairstyle was an important expression of people's status. If you were a peasant and sported the hairstyle of the *samurai* (warrior caste), you could be arrested and even killed because you were seen as calling the entire social order into question.

As these examples show, then, people's conceptions of deviance and crime vary substantially over time and between societies. Under some circumstances, an issue that seems quite trivial to us, such as hairstyle, can even be a matter of life and death.

Power and the Social Construction of Crime and Deviance

To truly understand deviance and crime, you have to study how people socially construct norms and laws. The school of sociological thought known as **social constructionism** emphasizes that various social problems, including crime, are *not* inherent in certain actions themselves. Instead, some people are in a position to create norms and pass laws that stigmatize other people. Therefore, one must study how norms and laws are created (or "constructed") to understand why particular actions get defined as deviant or criminal in the first place.

Power is a crucial element in the social construction of deviance and crime. Power, you will recall from Chapter 5 ("Interaction and Organization"), is "the probability that one actor within a social relationship will be in a position to carry out his [or her] own will despite resistance"(Weber, 1947: 152). An "actor" may be an entire social group. Relatively powerful groups are generally able to create norms and laws that suit their interests. Relatively powerless social groups are usually unable to do so.

The powerless, however, often struggle against stigmatization. If their power increases, they may succeed in their struggle. We can illustrate the importance of power in the social construction of crime and deviance by considering crimes against women and white-collar crime.

Crimes Against Women

We argued above that definitions of crime are usually constructed so as to bestow advantages on the more powerful members of society and disadvantages on the less powerful. As you will learn in detail in Chapter 9 ("Sexuality and Gender"), women are generally less powerful than men in all social institutions. Has the law therefore been biased against women? We believe it has.

Until recently, many types of crimes against women were largely ignored in the United States and most other parts of the world. This was true even for rape. Admittedly, so-called "aggravated rape," involving strangers, was sometimes severely punished. But so-called "simple rape," involving a friend or an acquaintance, was rarely prosecuted. And marital rape was viewed as a contradiction in terms, as if it were logically impossible for a married woman to be raped by her spouse. In her research, Susan Estrich (1987) found that rape law was not taught at American law schools in the 1970s. Law professors, judges, police officers, rapists, and even victims did not think simple rape was "real rape." Similarly, judges, lawyers, and social scientists rarely discussed physical violence against women and sexual harassment until the 1970s. Governments did not collect data on the

The 1873 Comstock Law was meant to stop trade in "obscene literature" and "immoral articles." It was targeted against "dirty books," birth control devices, abortion, and information on sexuality and sexually transmitted diseases—all perfectly legal today. In this 1915 cartoon, Robert Minor satirizes the morality underlying the Comstock Law. The caption reads: "Your honor, this woman gave birth to a naked child!"

topic, and few social scientists showed any interest in what has now become a large and important area of study.

Today, the situation has improved. To be sure, as Diana Scully's (1990) study of convicted rapists shows, rape is still associated with a low rate of prosecution. Rapists often hold women in contempt and do not regard rape as a real crime. Yet efforts by Estrich and others to have all forced sex defined as rape have raised people's awareness of date, acquaintance, and marital rape. Rape is more often prosecuted now than it used to be. The same is true for violence against women and sexual harassment.

Why the change? In part because women's position in the economy, the family, and other social institutions has improved over the past 30 years. Women now have more autonomy in the family, earn more, and enjoy more political influence. They also created a movement for women's rights that heightened concern about crimes disproportionately affecting them. For instance, until very recently, male sexual harassment of female workers was considered normal. Following Catharine MacKinnon's path-breaking work on the subject, however, feminists succeeded in having the social definition of sexual harassment transformed (MacKinnon, 1979). Sexual harassment is now considered a social deviation and, in some circumstances, a crime. Increased public awareness of the extent of sexual harassment has probably made it less common. We thus see how social definitions of crimes against women have changed with a shift in the distribution of power.[1]

White-Collar Crime

White-collar crime refers to illegal acts "committed by a person of respectability and high social status in the course of his [or her] occupation" (Sutherland, 1949: 9). Such crimes include embezzlement, false advertising, tax evasion, insider stock trading, fraud, unfair labor practices, copyright infringement, and conspiracy to fix prices and restrain trade. Sociologists often contrast white-collar crimes with **street crimes.** The latter include arson, burglary, robbery, assault, and other illegal acts. While street crimes are committed disproportionately by people from lower classes, white-collar crime is committed disproportionately by people from middle and upper classes.

Many sociologists think white-collar crime is costlier to society than street crime. Consider that armed robbers netted perhaps $400 million in the 1980s, but the savings and loan scandal, in which bankers mismanaged funds and committed fraud, cost the American public $500 to $600 *billion* during that decade (Brouwer, 1998). Nonetheless, white-collar criminals, including corporations, are prosecuted relatively infrequently. They are convicted even less often. This is true even in extreme cases, when white-collar crimes result in environmental degradation or death due, for example, to the illegal relaxation of safety standards. The police and the FBI routinely pursue burglars, but, typically, many of the guilty parties in the savings and loan scandal of the 1980s were not even charged with a misdemeanor (see Box 6.1).

White-collar crime results in few prosecutions and still fewer convictions for two main reasons. First, much white-collar crime takes place in private and is therefore difficult to detect. For example, corporations may illegally decide to fix prices and divide markets, but executives make these decisions in boardrooms and private clubs that are not generally subject to police surveillance. Second, corporations can afford legal experts, public relations firms, and advertising agencies that advise their clients on how to bend laws, build up their corporate image in the public mind, and influence lawmakers to pass laws "without teeth" (Blumberg, 1989; Clinard and Yeager, 1980; Hagan, 1989; Sherrill, 1997; Sutherland, 1949).

Governments, too, commit serious crimes. However, it is hard to punish political leaders (Chambliss, 1989). Authoritarian governments often call their critics "terrorists" and even torture people who are fighting for democracy, but such governments rarely have to account for their deeds (Herman and O'Sullivan, 1989). Some analysts argue that even the

---

[1]Significantly, black rapists of white women receive much more severe punishments than white rapists of white women (LaFree, 1980). This suggests that race is still an important power factor in the treatment of crime, a subject we have much to say about below.

BOX 6.1
SOCIOLOGY AT THE MOVIES

A CIVIL ACTION (1998)

A Civil Action is based on a true story. John Travolta plays a personal-injury lawyer who hosts a radio show. In response to a call from a woman representing a group of parents in Woburn, Massachusetts, he comes face-to-face with a horrible case of industrial pollution. Many parents in the town have lost children to leukemia. They believe a local factory dumping waste chemicals into the city's water supply caused the disease.

At first, Travolta doesn't want to get involved. He tells the families that a lawsuit makes sense only if the defendant has a lot of money or a large insurance policy. However, on his way home from his meeting with the parents, he gets caught speeding on a freeway. By chance, he notices trucks and railway cars streaming by bearing the logos of W. R. Grace and Beatrice Food. He suddenly realizes that the plant is associated with two large corporations and that he and the parents can make a great deal of money in this case.

White-collar crime doesn't get as much attention as other kinds of crime, but *A Civil Action* casts a spotlight on the problem. In addition, it shows some reasons why prosecuting large corporations is difficult. In the first place, corporations can afford to hire brilliant legal minds, such as the lawyer played by Robert Duvall in *A Civil Action*. Second, because of the difficulty of finding evidence—the frequent absence of a "smoking gun"—proving a case of corporate

John Travolta in *A Civil Action*.

crime is often hard. Third, corporations are in a financial position to offer out-of-court settlements without admitting their guilt. This is what happens in *A Civil Action*. The lawyer played by Travolta finally finds a W. R. Grace worker who admits dumping toxic chemicals. Although the corporation is willing to offer a modest out-of-court settlement, its executives refuse to admit responsibility and apologize for their negligence. Meanwhile, Travolta's firm spends all its resources prosecuting the case, and Travolta finds himself virtually penniless. The movie ends well, however. Travolta manages to get the federal government's Environmental Protection Agency (EPA) interested in the Woburn case, and the EPA lawyers succeed in their appeal.

As you think about white-collar crime, you should contrast it with street crime. You should also consider the way both types of

crime are portrayed in the movies. For example, in *Pulp Fiction* (1994), also starring John Travolta, we get an inside view of the world of organized and street crime, including the drug trade, racketeering, and gambling. The filmmaker introduces the viewer to the character behind each criminal act. The characters are often made to seem likable or at least interesting. Nonetheless, most viewers would insist on stiff penalties for drug trading or murder. In contrast, in *A Civil Action*, we never see the faces of the people responsible for the negligence that led to the deaths of the children. Many members of the public are indifferent to corporate crimes, although they may kill more people than street crimes. Why do you think white-collar and street crimes are portrayed so differently in the movies? Why do you think most members of the public think so differently about the two types of crime?

United States government, in spite of its democratic ideals, sometimes behaves in a manner that may be regarded as criminal. In the late 1980s, for example, while the United States was engaged in a war on drugs, the CIA participated in the drug trade to help arm the right-wing Contra military forces in Nicaragua (Scott and Marshall, 1991).

In sum, white-collar crime is underdetected, underprosecuted, and underconvicted because it is the crime of the powerful and the well-to-do. The social construction of crimes against women has changed over the past 30 years, partly because women have become more powerful. In contrast, the social construction of white-collar crime has changed little since 1970 because upper classes are no less powerful now than they were then.

A rare, successful prosecution of corporate crime was portrayed in the hit film *Erin Brokovich* (2000), which is based on a true story. In the film, Brokovich (Julia Roberts) discovers that Pacific Gas & Electric is illegally dumping cancer-causing chemicals into an unlined pond in Hinkley, California, causing high rates of cancer and other diseases in the area. Against all odds, Brokovich helps her boss try and win a case against the company.

Crime Rates

Some crimes are more common than others, and rates of crime vary over place and time and among different social groups. We will now describe some of these variations. Then we will review the main sociological explanations of crime and deviance.

First, a word about crime statistics. Much crime is not reported to the police. For example, many common assaults go unreported because the assailant is a friend or a relative of the victim. Similarly, many rape victims are reluctant to report the crime because they are afraid they will be humiliated and stigmatized by making it public. Moreover, authorities and the wider public decide which criminal acts to report and which to ignore. If, for instance, the authorities decide to crack down on drugs, more drug-related crimes will be counted, not because there are more drug-related crimes but because more drug criminals are apprehended. Third, many crimes are not incorporated in major crime indexes published by the FBI. Excluded are many so-called **victimless crimes,** such as prostitution and illegal drug use, which involve violations of the law in which no victim steps forward and is identified. Also excluded from the indexes are most white-collar crimes.

Recognizing these difficulties, students of crime often supplement official crime statistics with other sources of information. **Self-report surveys** are especially useful. In such surveys, respondents are asked to report their involvement in criminal activities, either as perpetrators or victims. In the United States, the main source of data on victimization is the National Crime Victimization Survey, conducted by the Department of Justice annually since 1973 and involving a nationwide sample of some 43,000 households (United States Department of Justice, 1999). Among other things, such surveys show about the same rate of serious crime (such as murder and nonnegligent manslaughter) as official statistics but two to three times the rate of less serious crime, such as assault. Figure 6.2 shows some of the results of an international self-report survey conducted in 1996. Among the 11 Western countries studied, the United States is at the high end with respect to violent offenses and household burglary and below average with respect to theft of personal property. Survey data, however, are influenced by people's willingness and ability to discuss criminal experiences frankly. Therefore, indirect measures of crime are sometimes used as well. For instance, sales of syringes are a good index of the use of illegal intravenous drugs. Indirect measures are unavailable for many types of crime, however.

Bearing these caveats in mind, what does the official record show? *Every hour* during 1999, law enforcement agencies in the United States received verifiable reports on an

✦ **FIGURE 6.2** ✦

Percent of Population Victimized Once or More in Preceding 12 Months, by Type of Crime, 1996

Note: Horizontal lines indicate international average for each type of crime.

SOURCE: Besserer (1998).

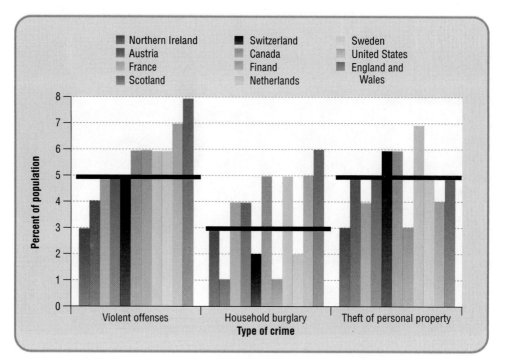

✦ **FIGURE 6.3** ✦

Violent Crime, United States, 1978–1999, Rate per 100,000 Population

Note: 1999 figures for January–June only.

SOURCES: United States Federal Bureau of Investigation (1999a; 1999b).

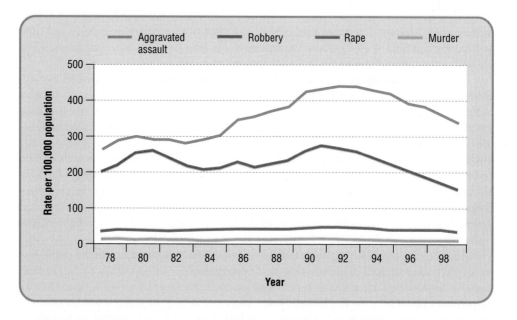

average of nearly 2 murders or nonnegligent manslaughters, 10 rapes, 60 robberies, 106 aggravated assaults, 133 motor vehicle thefts, 240 burglaries, and 720 larceny-thefts (calculated from United States Federal Bureau of Investigation, 2000: 4). Between 1960 and 1992, the United States experienced a roughly 500% increase in the rate of violent crime, including murder and nonnegligent manslaughter, rape, robbery, and aggravated assault. (Remember, the *rate* refers to the number of cases per 100,000 population.) Over the same period, the rate of major property crimes—motor vehicle theft, burglary, and larceny-theft— rose about 150%.

Although these statistics are alarming, we can take comfort from the fact that the long crime wave that began its upswing in the early 1960s and continued to surge in the 1970s eased in the 1980s and fell in the 1990s. The good news is evident in Figures 6.3 and 6.4, which show trends in violent and property crime between 1978 and 1999. Except for aggravated assault, the major crime rates for 1990 were about the same as or lower than the

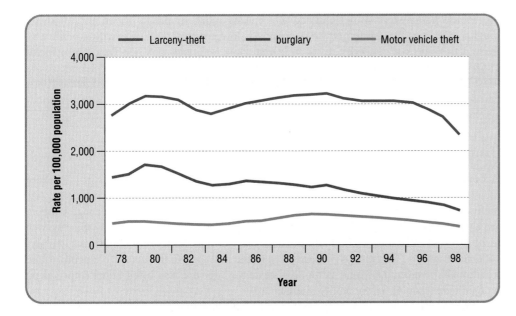

✦ **FIGURE 6.4** ✦
Property Crime, United States, 1978–1999, Rate per 100,000 Population

Note: 1999 figures for January–June only.
SOURCES: United States Federal Bureau of Investigation (1999a; 1999b).

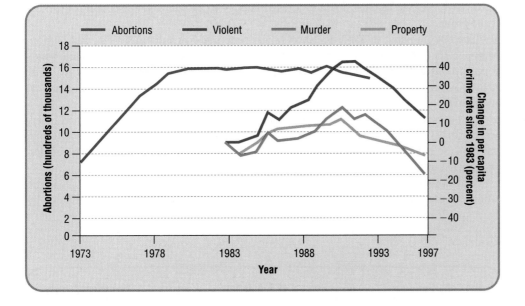

✦ **FIGURE 6.5** ✦
Abortions and Crime, United States, 1973–1997

SOURCE: Holloway (1999:24).

major crime rates for 1980. After about 1990, the rates for all forms of major crime began to fall significantly. The rate of murder and nonnegligent manslaughter, for instance, fell 22% between 1991 and 1999, and the burglary rate fell 41% in the same period.

Four factors are usually mentioned in explaining the decline. First, in the 1990s, governments put more police on the streets and many communities established their own systems of surveillance and patrol. This inhibited street crime. Second, young men are most prone to street crime, but America is aging and the proportion of young men in the population has declined. Third, the economy boomed in the 1990s. Usually, crime rates fluctuate with unemployment rates. When fewer people have jobs, there's more crime. With an unemployment rate below 5% for much of the decade, economic conditions in the United States favored less crime. Finally, and more controversially, some researchers have recently noted that the decline in crime started 19 years after abortion was legalized in the United States. There were proportionately fewer unwanted children in the population beginning in 1992, and unwanted children are more crime prone than wanted children because they tend to receive less parental supervision and guidance (Donahue and Levitt, 2001; Hochstetler and Shover, 1997; Holloway, 1999; LaFree, 1998; Skolnick, 1997; see Figure 6.5).

Please note: We have not claimed that putting more people in prison and imposing tougher penalties for crime help to account for lower crime rates. We will explain why these actions do not generally result in lower crime rates below, when we discuss **social control** (methods of ensuring conformity) and punishment. We will also probe one of the most fascinating questions raised by the FBI statistics: If crime rates steadied in the 1980s and fell in the 1990s, what accounts for the exploding prison population and our widespread and growing fear of crime over the past 20 years?

Criminal Profiles

According to FBI statistics, 78% of all persons arrested in the United States in 1998 were men (United States Federal Bureau of Investigation, 1999a). In the violent crime category, men accounted for 83% of arrests. As in all things, women, and especially teenage women, are catching up, albeit slowly. Men are still six-and-a-half times more likely than women to be arrested. However, with every passing year women compose a slightly bigger percentage of arrests. This change is partly due to the fact that, in the course of socialization, traditional social controls and definitions of femininity are less often being imposed on women (see Chapter 9, "Sexuality and Gender").

Most crime is committed by people who have not reached middle age. As the top panel of Table 6.1 shows, in 1998 Americans between the ages of 10 and 39 accounted for 58.0% of the population but 82.7% of arrests. The 15- to 24-year-old age cohort is the most crime prone.

The bottom panel of Table 6.1 shows that crime also has a distinct racial distribution. Although the United States Bureau of the Census classified 82.5% of the United States population as white in 1998, whites accounted for only 68.0% of arrests in that year. For African Americans, the story is reversed. They accounted for 29.7% of arrests but made up only 12.7% of the population.

Most sociologists agree that the disproportionately high arrest, conviction, and incarceration rates of African Americans are due to three main factors: bias in the way crime statistics are collected, the low class position of blacks in American society, and racial discrimination in the criminal justice system (Hagan, 1994).

The statistical bias is due largely to the absence of data on white-collar crimes in the official crime indexes. Since white-collar crimes are committed disproportionately by

♦ **TABLE 6.1** ♦

Arrests by Age Cohort and Race, United States, 1998

Note: "Other" includes Native American, Alaskan Native, Asian, and Pacific Islander.

SOURCES: Calculated from United States Bureau of the Census (2000g); United States Federal Bureau of Investigation (1999a).

Age Cohort	Percent of Population	Percent of Arrests
Under 10	14.4	0.3
10–19	14.4	27.8
20–29	13.4	30.7
30–39	15.8	23.9
40–49	15.1	12.7
50–59	10.4	3.4
60+	16.5	1.2
Total	100.0	100.0

Racial Group	Percent of Population	Percent of Arrests
White	82.5	68.0
Black	12.7	29.7
Other	4.8	2.3
Total	100.0	100.0

In the 1950s, a sort of "racial profiling" was still commonly applied to disadvantaged members of white ethnic minorities, such as Italian Americans. In the 1957 classic, *12 Angry Men,* the character played by Henry Fonda convinces the members of a jury to overcome their prejudices, examine the facts dispassionately, and allow a disadvantaged minority youth accused of murdering his father to go free.

whites, official crime indexes make it seem as if blacks commit a higher proportion of all crimes than they actually do.

The low class standing of African Americans means they experience twice the unemployment rate of whites, three times the rate of child poverty, and more than three times the rate of single motherhood. All these factors are associated with higher crime rates; the great majority of poor people are law abiding, but poverty and its associated disabilities are associated with elevated crime rates. The effect of poverty on crime rates is much the same for blacks and whites, but the problem worsened for the African-American community in the last quarter of the 20th century. During this period, the United States economy was massively restructured and budgets for welfare and inner-city schools were massively cut. Many manufacturing plants in or near United States inner cities were shut down in the 1970s and 1980s, causing high unemployment among local residents, a disproportionately large number of whom were African Americans. Many young African Americans, with little prospect of getting a decent education and finding meaningful work, turned to crime as a livelihood and a source of prestige and self-esteem (Sampson and Wilson, 1995).

Finally, as Jerome Miller has convincingly shown, the criminal justice system efficiently searches out African-American males for arrest and conviction (Miller, 1996: 48–88). Many white citizens are more zealous in reporting African-American than white offenders. Many police officers are more eager to arrest African Americans than whites. Court officials are less likely to allow African Americans than whites to engage in plea bargaining. Fewer African Americans than whites can afford to pay fines that would prevent them from being jailed. Especially since the onset of the "war against drugs" in the 1980s, African Americans have been targeted, arrested, sentenced, and imprisoned in disproportionate numbers (see Box 6.2). The fact that some 40% of the United States prison population consists of African-American men is not just the result of their criminal activity. Thus, in the mid-1990s the crime rate of African-American men was not much different from their crime rate in 1980, but their imprisonment rate rose more than 300% during that period (Tonry, 1995).

BOX 6.2
IT'S YOUR CHOICE

THE WAR ON DRUGS

Did your high school conduct random drug searches? Did you have to take a Breathalyzer test at your prom? Increasingly, companies are demanding that employees take urine and other tests for drug use. The war on drugs, symbolized by Nancy Reagan's plea to "Just Say No" during the 1980s, thus continues in the United States.

One consequence of the continuing war on drugs is the stiff penalties imposed on drug offenders. For example, if you're caught selling one vial of crack or one bag of heroin, your sentence is 5 to 25 years, depending on what state you're in. New York drug laws are toughest. If you're caught selling 2 or more ounces of heroin or owning 4 ounces of cocaine in New York, then, even as a first-time offender, you receive the

maximum prison sentence: life in prison. The authorities make more than 1.5 million arrests every year for drug-related offenses, including 700,000 for the sale or possession of marijuana (Massing et al., 1999: 11–20).

Despite all these arrests, most people think our drug control policy is ineffective. The United States government spends 18 times more on drug control now than it did in 1980 ($18 billion vs. $1 billion). Eight times as many Americans are in jail today for drug-related offenses (400,000 vs. 50,000 in 1980). Yet there are still an estimated 4 million hardcore drug users in the United States (Massing et al., 1999: 32). What should we do?

Rather than continuing the war on drugs, some sociologists suggest it is time to think of alternative policies. We can, for example, estimate the effectiveness of four major policies on drug control: controlling the drug trade abroad, stopping drugs at the border, arresting drug traders and users, and drug prevention and treatment. In one major government-funded study, "[t]reatment was found to be seven times more cost-effective than law enforcement, ten times more effective than interdiction [stopping drugs at the border], and twenty-three times more effec-

tive than attacking drugs at their source" (quoted in Massing et al., 1999: 14). Yet the United States government spends less than 10 percent of its $18 billion drug-control budget on prevention and treatment. Over two thirds of the money is spent on reducing the supply of drugs (Massing et al., 1999:14).

Another, more radical option, is to seek limited legalization of drugs. Two arguments support this proposal. First, the United States' major foray into the control of substance abuse—the prohibition of alcohol during the 1920s and early 1930s—turned out to be a major fiasco. It led to an increase in the illegal trade in alcohol and the growth of the Mafia. Second, the Netherlands, for example, has succeeded in decriminalizing marijuana use. Even after it became legal, no major increase in the use of marijuana or more serious drugs, such as heroin, took place (Massing et al., 1999: 28–9).

Clearly, the citizens of the United States need to discuss drug policy in a serious way. Just saying "no" and spending most of our drug-control budget on trying to curb the supply of illegal drugs are ineffective policies (Reinarman and Levine, 1999).

EXPLAINING DEVIANCE AND CRIME

Lep: "*I remember your li'l ass used to ride dirt bikes and skateboards, actin' crazy an' shit. Now you want to be a gangster, huh? You wanna hang with real muthaf_____ and tear shit up, huh? . . . Stand up, get your l'il ass up. How old is you now anyway?*"

Kody: "*Eleven, but I'll be twelve in November.*"

—Sanyika Shakur (1993: 8)

"Monster" Scott Kody eagerly joined the notorious gang, the Crips, in South Central Los Angeles in 1975, when he was in grade 6. He was released from Folsom Prison on parole in 1988, at the age of 24. Until about 3 years before his release, he was one of the most ruthless gang leaders in Los Angeles and the California prison system. In 1985, however, he decided to reform. He adopted the name of Sanyika Shakur, became a black nationalist, and began a crusade against gangs. Few people in his position have chosen that path. In Kody's heyday, about 30,000 gang members roamed Los Angeles County. Today there are an estimated 150,000.

What makes the criminal life so attractive to so many young men and women? In general, why do deviance and crime occur at all? Sociologists have proposed dozens of explanations. However, we can group them into two basic types. **Motivational theories** identify the social factors that *drive* people to commit deviance and crime. **Constraint theories** identify the social factors that *impose* deviance and crime (or conventional behavior) on people. Let us briefly examine three examples of each type of theory.

		Institutionalized Means		
		Accept	**Reject**	**Create New**
Cultural Goals	**Accept**	Conformity	Innovation	—
	Reject	Ritualism	Retreatism	—
	Create New	—	—	Rebellion

◆ **TABLE 6.2** ◆
Merton's Strain Theory of Deviance

SOURCE: Adapted from Merton (1938).

Motivational Theories

Strain Theory

You will recall Durkheim's idea that the absence of clear norms—"anomie"—can result in elevated rates of suicide and other forms of deviant behavior (see Chapter 1, "A Sociological Compass"). Robert Merton's **strain theory,** summarized in Table 6.2, extends Durkheim's insight (Merton, 1938). Merton argued that cultures often teach people to value material success. Just as often, however, societies don't provide enough legitimate opportunities for everyone to succeed. As a result, some people experience strain. Most of them will force themselves to adhere to social norms despite the strain (Merton called this "conformity"). The rest adapt in one of four ways. They may drop out of conventional society ("retreatism"). They may reject the goals of conventional society but continue to follow its rules ("ritualism"). They may protest against convention and support alternative values ("rebellion"). Or they may find alternative and illegitimate means of achieving their society's goals ("innovation")—that is, they may become criminals. The American Dream of material success starkly contradicts the lack of opportunity available to poor youths, said Merton. As a result, poor youths sometimes engage in illegal means of attaining legitimate ends. Merton would say that "Monster" Scott Kody became an innovator at the age of 11 and a rebel at 21.

Subcultural Theory

A second type of motivational theory, known as **subcultural theory,** emphasizes that adolescents like Kody are not alone in deciding to join gangs. Many similarly situated adolescents make the same kind of decision, rendering the formation and growth of the Crips and other gangs a *collective* adaptation to social conditions. Moreover, this collective adaptation involves the formation of a subculture with distinct norms and values. Members of this subculture reject the legitimate world that, they feel, has rejected them (Cohen, 1955).

The literature emphasizes three features of criminal subcultures. First, depending on the availability of different subcultures in their neighborhoods, delinquent youths may turn to different types of crime. In some areas, delinquent youths are recruited by organized crime, such as the Mafia. In areas that lack organized crime networks, delinquent youths are more likely to create violent gangs. Thus, the relative availability of different subcultures influences the type of criminal activity to which one turns (Cloward and Ohlin, 1960).

A second important feature of criminal subcultures is that their members typically spin out a whole series of rationalizations for their criminal activities. These justifications make their illegal activities appear morally acceptable and normal, at least to the members of the subculture. Typically, criminals deny personal responsibility for their actions ("What I did harmed nobody"). They condemn those who pass judgment on them ("I'm no worse than anyone else"). They claim their victims get what they deserve ("She had it coming to her"). And they appeal to higher loyalties, particularly to friends and family ("I had to do it because he dissed my gang"). The creation of such justifications and rationalizations enables criminals to clear their consciences and get on with the job. Sociologists call such rationalizations **techniques of neutralization** (Sykes and Matza, 1957; see also the case of the professional fence in Chapter 2, "Research Methods").

Finally, although deviants depart from mainstream culture, they are strict conformists when it comes to the norms of their own subculture. They tend to share the same beliefs, dress alike, eat similar food, and adopt the same mannerisms and speech patterns.

Whether among professional thieves (Conwell, 1937) or young gang members (Short and Strodtbeck, 1965), deviance is strongly discouraged *within* the subculture. Paradoxically, deviant subcultures depend on internal conformity.

The main problem with strain and subcultural theories is that they exaggerate the connection between class and crime. Many self-report surveys find, at most, a weak tendency for criminals to come disproportionately from lower classes. Some self-report surveys report no such tendency at all, especially among young people and for less serious types of crime (Weis, 1987). A stronger correlation exists between *serious street crimes* and class. Armed robbery and assault, for instance, are more common among people from lower classes. A stronger correlation also exists between *white-collar* crime and class. Middle- and upper-class people are most likely to commit white-collar crimes. Thus, generalizations about the relationship between class and crime must be qualified by taking into account the severity and type of crime (Braithwaite, 1981). Note also that official statistics usually exaggerate class differences because they are concerned only with street crime. Moreover, there is generally more police surveillance in lower class neighborhoods.

Learning Theory

Apart from exaggerating the association between class and crime, strain and subcultural theories are problematic because they tell us nothing about which adaptation someone experiencing strain will choose. Even when criminal subcultures beckon ambitious adolescents who lack opportunities to succeed in life, only a minority joins up. Most adolescents who experience strain and have the opportunity to join a gang reject the life of crime and become conformists and ritualists, to use Merton's terms. Why?

Edwin Sutherland (1939) addressed both the class and choice problems more than 60 years ago by proposing a third motivational theory, which he called the theory of **differential association.** The theory of differential association is still one of the most influential ideas in the sociology of deviance and crime. In Sutherland's view, a person learns to favor one adaptation over another due to his or her life experiences or socialization. Specifically, everyone is exposed to both deviant and nondeviant values and behaviors as they grow up. If you happen to be exposed to more deviant than nondeviant experiences, chances are you will learn to become a deviant yourself. You will come to value a particular deviant lifestyle and consider it normal. Everything depends, then, on the exact mix of deviant and conformist influences a person faces. For example, a substantial body of participant-observation and survey research has failed to discover widespread cultural values prescribing crime and violence in the inner city (Sampson, 1997: 39). Most inner-city residents follow *conventional* norms, and that is one reason most inner-city adolescents do not learn to become gang members. Those who do become gang members tend to grow up in very specific situations and contexts that teach them the value of crime.

Significantly, the theory of differential association holds for people in all class positions. For instance, Sutherland applied the theory of differential association in his pathbreaking research on white-collar crime. He noted that white-collar criminals, like their counterparts on the street, learn their skills from associates and share a culture that rewards rule breaking and expresses contempt for the law (Sutherland, 1949).

Constraint Theories

Motivational theories ask how some people are driven to break norms and laws. Constraint theories, in contrast, pay less attention to people's motivations. "How are deviant and criminal 'labels' imposed on some people?" "How do various forms of social control fail to impose conformity on them?" "How does the distribution of power in society shape deviance and crime?" These are the kinds of questions posed by constraint theorists.

Labeling Theory

A few years ago in Boston, a white man used a cell phone to relay to the police the gruesome murder of his wife by a young African-American male. Many people found the husband's story utterly convincing because they believed young African-American men are

According to Edwin Sutherland's theory of differential association, people who are exposed to more deviant than nondeviant experiences as they grow up are likely to become deviants. As in *The Sopranos,* having family members and friends in the Mafia predisposes one to Mafia involvement.

prone to violent crime. As the story spread, young African-American men in Boston were subject to increased harassment, both by police and ordinary citizens. Thus, more African-American men were perceived to be criminals not because they suddenly engaged in more crime but because they were increasingly *labeled* as criminals. The Boston incident was a pure case of labeling. The story was a fabrication. It was the respectable, white, middle-class husband who brutally killed his wife. He made up the story because he thought people would believe it—and he was right.

As the Boston incident suggests, if the term "deviant" or "criminal" is attached to a person or a group of people, the label may stick, to some degree irrespective of the actual behavior involved. This is the chief insight of **labeling theory**—that deviance results not just from the actions of the deviant but also from the responses of others, who define some actions as deviant and other actions as normal.

If an adolescent misbehaves in high school a few times, teachers and the principal may punish him. However, his troubles really begin if the school authorities and the police label him a "delinquent." Surveillance of his actions will increase. Actions that authorities would normally not notice or would define as of little consequence are more likely to be interpreted as proof of his delinquency. He may be ostracized from nondeviant cliques in the school and eventually socialized into a deviant subculture. Over time, immersion in the deviant subculture may lead the adolescent to adopt "delinquent" as his **master status,** or overriding public identity. More easily than we may care to believe, what starts out as a few incidents of misbehavior can get amplified into a criminal career because of labeling (Matsueda, 1988, 1992).

That labeling plays an important part in who gets caught and who gets charged with crime was demonstrated more than 30 years ago by Aaron Cicourel (1968). Cicourel examined the tendency to label rule-breaking adolescents "juvenile delinquents" if they came from families in which the parents were divorced. He found that police officers tended to use their discretionary powers to arrest adolescents from divorced families more often than adolescents from intact families who committed similar delinquent acts. Judges, in turn, tended to give more severe sentences to adolescents from divorced families than to adolescents from intact families who were charged with similar delinquent acts. Sociologists and criminologists then collected data on the social characteristics of adolescents who were charged as juvenile delinquents, "proving" that children from divorced families were more likely to become juvenile delinquents. Their finding reinforced the beliefs of police officers and judges. Thus, the labeling process acted as a self-fulfilling prophecy.

Control Theory

All motivational theories assume people are good and require special circumstances to make them bad. A popular type of constraint theory assumes people are bad and require special circumstances to make them good. For, according to control theory, the rewards of deviance and crime are many. Proponents of this approach argue that nearly everyone wants fun, pleasure, excitement, and profit. Moreover, they say, if we could get away with it, most of us would commit deviant and criminal acts to get more of these valued things. For control theorists, the reason most of us don't engage in deviance and crime is that we are prevented from doing so. The reason deviants and criminals break norms and laws is that social controls are insufficient to ensure their conformity.

Travis Hirschi and Michael Gottfredson first developed the control theory of crime (Hirschi, 1969; Gottfredson and Hirschi, 1990). They argued that adolescents are more prone to deviance and crime than adults because they are incompletely socialized and therefore lack self-control. Adults and adolescents may both experience the impulse to break norms and laws, but adolescents are less likely to control that impulse. Gottfredson and Hirschi went on to show that adolescents who are most prone to delinquency are likely to lack four types of social control. They tend to have few social *attachments* to parents, teachers, and other respectable role models, few legitimate *opportunities* for education and a good job, few *involvements* in conventional institutions, and weak *beliefs* in traditional values and morality. Because of the lack of control stemming from these sources, they are relatively free to act on their deviant impulses.

Other sociologists have applied control theory to gender differences in crime. They have shown that girls are less likely to engage in delinquency than boys because families typically exert more control over girls, supervising them more closely and socializing them to avoid risk (Hagan, Simpson, and Gillis, 1987; Peters, 1994). Sociologists have also applied control theory to different stages of life. Just as weak controls exercised by family and school are important in explaining why some adolescents engage in deviant or criminal acts, job and marital instability make it more likely that some adults will be unable to resist the temptations of deviance and crime (Sampson and Laub, 1993).

Labeling and control theories have little to say about why people regard certain kinds of activities as deviant or criminal in the first place. For the answer to that question, we must turn to conflict theory, a third type of constraint theory.

Conflict Theory

The day after Christmas, 1996, JonBenét Ramsey was found strangled to death in the basement of her parents' $800,000 home in Boulder, Colorado. The police found no footprints in the snow surrounding the house and no sign of forced entry. The FBI concluded that nobody had entered the house during the night when, according to the coroner, the murder took place. The police did find a ransom note saying that the child had been kidnapped. A linguistics expert from Vassar later compared the note with writing samples of the child's mother. In a 100-page report, the expert concluded that the child's mother was the author of the ransom note. It was also determined that all of the materials used in the crime had been purchased by the mother. Finally, it was discovered that JonBenét had been sexually abused. Although by no means an open-and-shut case, enough evidence was available to cast a veil of suspicion over the parents. Yet, apparently due to the lofty position of the Ramsey family in their community, the police treated them in an extraordinary way. On the first day of the investigation, the Commander of the Boulder police detective division designated the Ramseys an "influential family" and ordered that they be treated as victims, not suspects (Oates, 1999: 32). The father was allowed to participate in the search for the child. In the process, he may have contaminated crucial evidence. The police also let him leave the house unescorted for about an hour. This led to speculation that he might have disposed of incriminating evidence. Because the Ramseys are millionaires, they were able to hire accomplished lawyers who prevented the Boulder police from interviewing them for 4 months and a public relations team that reinforced the idea that the Ramseys were victims. A grand jury decided on October 13, 1999, that nobody would be charged with the murder of JonBenét Ramsey.

Regardless of the innocence or guilt of the Ramseys, the way their case was treated adds to the view that the law applies differently to rich and poor. That is the perspective of **conflict theory.** In brief, conflict theorists maintain that the rich and the powerful impose deviant and criminal labels on the less powerful members of society, particularly those who challenge the existing social order. Meanwhile, they are usually able to use their money and influence to escape punishment for their own misdeeds.

Steven Spitzer (1980) conveniently summarizes this school of thought. He notes that capitalist societies are based on private ownership of property. Moreover, their smooth functioning depends on the availability of productive labor and respect for authority. When thieves steal, they challenge private property. Theft is therefore a crime. When so-called "bag ladies" and drug addicts drop out of conventional society, they are defined as deviant because their refusal to engage in productive labor undermines a pillar of capitalism. When young, politically volatile students or militant trade unionists strike or otherwise protest against authority, they, too, represent a threat to the social order and are defined as deviant or criminal.

Of course, says Spitzer, the rich and the powerful engage in deviant and criminal acts too. But, he adds, they tend to be dealt with more leniently. Industries can grievously harm people by damaging the environment, yet serious charges are rarely brought against the owners of industry. White-collar crimes are less severely punished than street crimes, regardless of the relative harm they cause. Compare burglary and fraud, for example. Fraud almost certainly costs society more than burglary. But burglary is a street crime committed mainly by lower-class people, while fraud is a white-collar crime committed mainly by middle- and upper-class people. Not surprisingly, therefore, in the United States in 1992, some 82% of people tried for burglary were sentenced to prison and served an average of approximately 26 months, while only about 46% of people tried for fraud went to prison and served an average of around 14 months (Reiman, 1995: 125). Laws and norms may change along with shifts in the distribution of power in society. However, according to conflict theorists, definitions of deviance and crime, and also punishments for misdeeds, are always influenced by who's on top.

And so we see that many theories contribute to our understanding of the social causes of deviance and crime. Some forms of deviance and crime are better explained by one theory than another. Different theories illuminate different aspects of the process by which people are motivated to break rules and get defined as rule breakers. Our overview should make it clear that no one theory is best. Instead, taking many theories into account allows us to develop a fully rounded appreciation of the complex processes surrounding the social construction of deviance and crime.

SOCIAL CONTROL AND PUNISHMENT

Trends in Social Control

No discussion of crime and deviance would be complete without considering in some depth the important issues of social control and punishment. For all societies seek to ensure that their members obey norms and laws. All societies impose sanctions on rule breakers. However, the *degree* of social control varies over time and from one society to the next. *Forms* of punishment also vary. Below we focus on how social control and punishment have changed historically.

Consider first the difference between preindustrial and industrial societies. Beginning in the late 19th century, many sociologists argued that preindustrial societies are characterized by strict social control and high conformity, while industrial societies are characterized by less stringent social control and low conformity (Tönnies, 1957 [1887]). Similar differences were said to characterize small communities versus cities. As the old German proverb says, "city air makes you free."

There is much truth in this point of view. Whether they are fans of opera or reggae, connoisseurs of fine wine or marijuana, city dwellers in industrial societies find it easier than people in small preindustrial communities to belong to a group or subculture of their choice (see, for example, the discussion of homosexual communities in Chapter 9, "Sexuality and Gender"). In general, the more complex a society, the less likely many norms will be widely shared. In fact, in a highly complex society such as the United States today, it is difficult to find an area of social life in which everyone is alike, or where one group can impose its norms on the rest of society without resistance. The existence of more than 2,100 different religious groups in the United States today speaks volumes about the extent of social diversity in our society (Melton, 1996 [1978]).

Nonetheless, some sociologists believe that social control has intensified over time, at least in some ways. They recognize that individuality and deviance have increased, but insist that this has happened only within quite strict limits, beyond which it is now *more* difficult to move. In their view, many crucial aspects of life have become more regimented, not less.

Much of the regimentation of modern life is tied to the growth of capitalism and the state. Factories require strict labor regimes, with workers arriving and leaving at a fixed time and, in the interim, performing fixed tasks at a fixed pace. Workers initially rebelled against this regimentation since they were used to enjoying many holidays and a flexible and vague work schedule regulated only approximately by the seasons and the rising and setting of the sun. But they had little alternative as wage labor in industry overtook feudal arrangements in agriculture (Thompson, 1967). Meanwhile, institutions linked to the growth of the modern state or regulated by it—armies, police forces, public schools, health

Web Interactive Exercises
The Mass Media and Gun Control

care systems, and various other bureaucracies—also demanded strict work regimes, curricula, and procedures. These institutions existed on a much smaller scale in preindustrial times or did not exist at all. Today they penetrate our lives and sustain strong norms of belief and conduct (Foucault, 1977 [1975]).

Electronic technology makes it possible for authorities to exercise more effective social control than ever before. With millions of cameras mounted in public places and workplaces, some sociologists say we now live in a "surveillance society" (Lyon and Zureik, 1996). Spy cameras enable observers to see deviance and crime that would otherwise go undetected and take quick action to apprehend rule breakers. Moreover, when people are aware of the presence of spy cameras, they tend to alter their behavior. For example, attentive shoplifters migrate to stores lacking electronic surveillance. On factory floors and in offices, workers display more conformity to management-imposed work norms. On college campuses, students are inhibited from engaging in organized protests (Boal, 1998).

Thanks to computers and satellites, intelligence services in the United States, Britain, Canada, Australia, and New Zealand now monitor all international telecommunications traffic, always on the lookout for threats. As easily as you can find the word "anomie" in your sociology essay using the "search" function of your word processor, the United States National Security Agency can scan digitized telephone and e-mail traffic in many languages for key words and word patterns that suggest unfriendly activity (Omega Foundation, 1998). However, the system, known as Echelon, is also used to target sensitive business and economic secrets from Western Europe, and some people, including Senator Frank Church, have expressed the fear that it could be used on the American people, robbing them of their privacy. Meanwhile, credit information on 95% of American consumers is available for purchase, the better to tempt you with credit cards, marketing ploys, and junk mail. When you browse the Web, information about your browsing patterns is collected in the background by many of the sites you visit, again largely for marketing purposes. Most large companies monitor and record their employees' phone conversations and e-mail messages. These are all efforts to regulate behavior, enforce conformity, and prevent deviance and crime more effectively using the latest technologies available (Garfinkel, 2000).

A major development in social control that accompanied industrialization was the rise of the prison. Today, prisons figure prominently in the control of criminals the world over. Americans, however, have a particular affinity for the institution, as we will now see.

The Prison

When he was 22, Robert Scully was sent to San Quentin for robbery and dealing heroin. Already highly disturbed, he became more violent in prison and attacked another inmate with a makeshift knife. As a result, Scully was shipped off to Corcoran, one of the new maximum-security facilities that the state of California began opening in the early 1980s. He was thrown into solitary confinement. In 1990, Scully was transferred to the new "supermax" prison at Pelican Bay. There, he occupied a cell the size of a bathroom. It had a perforated sheet metal door. He received food through a hatch. Even exercise was solitary. When he was released on parole in 1994, he had spent 9 years in isolation.

One night in 1995 Scully was loitering around a restaurant with a friend. The owner, fearing a robbery, called the police. Deputy Sheriff Frank Rejo, a middle-aged grandfather looking forward to retirement, soon arrived at the scene. He asked to see a driver's license, but as Scully's friend searched for it, Scully pulled out a sawed-off shotgun and shot Rejo in the forehead. Scully and his friend were apprehended by police the next day.

Robert Scully was already involved in serious crime before he got to prison, but it was in San Quentin, Corcoran, and Pelican Bay that he became a murderer. This follows a pattern known to sociologists for a long time. Prisons are agents of socialization, and new inmates often become more serious offenders as they adapt to the culture of the most hardened, long-term prisoners (Wheeler, 1961). In Scully's case, psychologists and psychiatrists called in by the defense team said that things had gone even farther. Years of

Web Research Projects
Does Prison Deter Criminals?

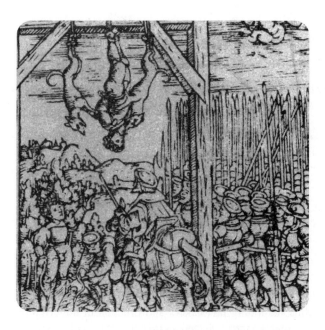

In preindustrial societies, criminals who committed serious crimes were put to death, often in ways that seem cruel by today's standards. One method involved hanging the criminal with starving dogs.

sensory deprivation and social isolation had so enraged and incapacitated Scully that it was impossible for him to think through the consequences of his actions. He had regressed to the point where his mental state was that of an animal able to act only on immediate impulse (Abramsky, 1999).

Because prison often turns criminals into worse criminals, it is worth pondering the institution's origins, development, and current dilemmas. As societies industrialized, imprisonment became one of the most important forms of punishment for criminal behavior (Garland, 1990; Morris and Rothman, 1995). In preindustrial societies, criminals were publicly humiliated, tortured, or put to death, depending on the severity of their transgression. In the industrial era, depriving criminals of their freedom by putting them in prison seemed less harsh, more "civilized" (Durkheim, 1973 [1899–1900]).

Some people still take a benign view of prisons, even seeing them as opportunities for *rehabilitation*. They believe that prisoners, while serving time, can be taught how to be productive citizens upon release. In the United States, this view predominated in the 1960s and early 1970s, when many prisons sought to reform criminals by offering them psychological counseling, drug therapy, skills training, college education, and other programs that would help at least the less violent offenders get reintegrated into society.

Today, however, most Americans scoff at the idea that prisons can rehabilitate criminals. We have adopted a much tougher line, as the case of Robert Scully shows. Some people see prison as a means of *deterrence*. In this view, people will be less inclined to commit crimes if they know they are likely to get caught and serve long and unpleasant prison terms. Others think of prisons as institutions of *revenge*. They think depriving criminals of their freedom and forcing them to live in poor conditions is fair retribution for their illegal acts. Still others see prisons as institutions of *incapacitation*. From this viewpoint, the chief function of the prison is simply to keep criminals out of society as long as possible to ensure they can do no more harm (Feeley and Simon, 1992; Simon, 1993; Zimring and Hawkins, 1995).

No matter which of these views predominates, one thing is clear. The American public has demanded that more criminals be arrested and imprisoned. And it has gotten what it wants (Gaubatz, 1995; Savelsberg, 1994). The nation's incarceration rate rose substantially in the 1970s, doubled in the 1980s, and doubled again in the 1990s.

Moral Panic

What happened between the early 1970s and the present to so radically change the United States prison system? In a phrase, the United States was gripped by **moral panic.** That is,

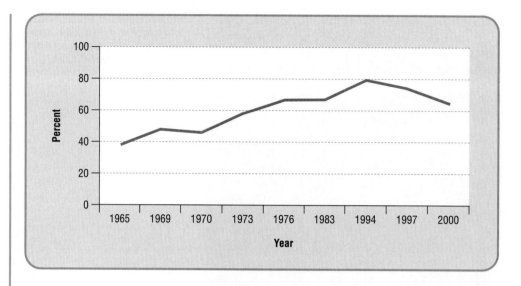

wide sections of the American public, including lawmakers and officials in the criminal justice system, were motivated by the fear that crime posed a grave threat to society's well-being (Cohen, 1972; Goode and Ben-Yehuda, 1994). The government declared a war on drugs. This resulted in the imprisonment of hundreds of thousands of nonviolent offenders. Sentencing got tougher, and many states passed a "three strikes and you're out" law. The law put three-time violent offenders in prison for life. The death penalty became increasingly popular. As Figure 6.6 shows, support for capital punishment more than doubled from 36% to 80% of the population between 1965 and 1994, although it has gradually fallen since then.[2]

Evidence of the moral panic was evident in crime *prevention,* too. For example, many well-to-do Americans had walls built around their neighborhoods, restricting access to residents and their guests. They hired private security police to patrol the perimeter and keep potential intruders at bay. Middle- and upper-class Americans installed security systems in their homes and steel bars in their basement windows. Many people purchased handguns in the belief they would enhance their personal security. The number of handguns in the United States is currently estimated at about 200 million. Some states even passed laws allowing people to conceal handguns on their person. In short, over the past 20 years or so, Americans have prepared themselves for an armed invasion and have decided to treat criminals much more toughly than in the past.

Are you part of the moral panic? Have you and your family taken special precautions to protect yourself from the "growing wave" of criminality in the United States? Even if you're not part of the moral panic, chances are you know someone who is. Therefore, to put things in perspective, you will need to recall an important fact from our discussion of recent trends in crime rates: According to FBI statistics, the moral panic of the 1980s and 1990s occurred during a period when all major crime indexes stabilized and then *fell* quite dramatically.

Why then the panic? Who benefits from it? We may mention several interested parties. First, the mass media benefit from moral panic because it allows them to rake in hefty profits. They publicize every major crime because crime draws big audiences, and big audiences mean more revenue from advertisers. Fictional crime programs draw tens of millions of additional viewers to their TVs, as the statistics cited at the beginning of this

---

[2]The fall in the percentage of Americans supporting the death penalty since 1994 is due to two main factors. First, an investigative series in *The Chicago Tribune* in 1999 examined all 285 Illinois capital trials since 1997 and found an astonishing number of disbarred defense councils, lying prosecutors, pseudoscientific evidence, and corrupt informants. This led the Republican governor of Illinois, George Ryan, long a death-penalty advocate, to declare a state moratorium on capital executions pending an investigation of the Illinois judicial system. In the first half of 2000, many other important Republicans began to publicly question the wisdom of the death penalty. Second, the Vatican expressed opposition to the death penalty in the 1997 edition of its catechism. This led many American Catholics to speak openly against the death penalty for the first time (Seeman, 2000). See page 165 for a discussion of capital punishment.

chapter show. Second, the crime prevention and punishment industry benefits from moral panic for much the same reason. Prison construction and maintenance firms, firearms manufacturers, and so forth are all big businesses that flourish in a climate of moral panic. Industries like these want Americans to own more guns and imprison more people, so they lobby hard in Washington and elsewhere for relaxed gun laws and invigorated prison construction programs. Third, some formerly depressed rural regions of the United States have become highly dependent on prison construction and maintenance for the economic well-being of their citizens. The Adirondack region of northern New York State is a case in point. Fourth, the criminal justice system is a huge bureaucracy with millions of employees. They benefit from moral panic because increased spending on crime prevention, control, and punishment secures their jobs and expands their turf. Finally, and perhaps most important, the moral panic is useful politically. Ever since the early 1970s, many politicians have based entire careers on get-tough policies. Party allegiance and ideological orientation matter less than you might think here; plenty of liberal Republicans (such as former Governor Nelson Rockefeller of New York) and Democrats (such as former Governor Mario Cuomo of New York) have done as much to build up the prison system as conservative Republicans (Schlosser, 1998).

Alternative Forms of Punishment

The two most contentious issues concerning the punishment of criminals are these: (a) Should the death penalty be used to punish the most violent criminals? (b) Should less serious offenders be incarcerated in the kinds of prisons we now have? In concluding this chapter, let us briefly consider each of these issues.

Capital Punishment

Although the United States has often been at the forefront of the struggle for human rights, it is one of the few industrial societies to retain capital punishment for the most serious criminal offenders.

Although the death penalty ranks high as a form of revenge, it is questionable whether it is much of a deterrent. There are two reasons for this. First, murder is often committed in a rage, when the perpetrator is not thinking entirely rationally. In such circumstances, the murderer is unlikely to coolly consider the costs and consequences of his or her

Sister Helean Prejean is a Catholic nun from Louisiana. Since 1981 she has been one of the country's most outspoken critics of the death penalty. Her book, *Dead Man Walking,* reflects on her experience with inmates on death row and raises important questions about the death penalty. *Dead Man Walking* was made into a critically acclaimed film starring Sean Penn and Susan Sarandon in 1995.

actions. Second, if rational calculation of consequences does enter into the picture, the perpetrator is likely to know that very few murders result in the death sentence. More than 20,000 murders take place in the United States every year. Only about 250 death sentences are handed out. Thus, there is only a 1.25% chance that a murderer will be sentenced to death. There is an even smaller chance that he or she will be executed.

Because the death penalty isn't likely to deter many people unless the probability of its use is high, some people take these figures as justification for sentencing more violent offenders to death. However, one must remember that capital punishment as it is actually practiced is hardly a matter of blind justice. This is particularly evident if we consider the racial distribution of people who are sentenced to death and executed. Murdering a white person is much more likely to result in a death sentence than murdering a black person. For example, in Florida in the 1970s, an African American who killed a white person was 40 times more likely to receive the death penalty than an African American who killed another African American. Moreover, a white person who murders a black person very rarely gets sentenced to death, but a black person who murders a white person is one of the most likely types of people to get the death penalty. Thus, of the 80 white people who murdered African Americans in Florida in the 1970s, not one was charged with a capital crime. In Texas, one was—out of 143 (Haines, 1996; Tonry, 1995; Black, 1989). Given this patent racial bias, we cannot view the death penalty as a justly administered punishment.

Sometimes people favor capital punishment because it saves money. They argue that killing someone outright costs less than keeping the person alive in prison for the rest of his or her life. However, after trials and appeals, a typical execution costs the taxpayer up to six times *more* than a 40-year stay in a maximum-security prison (Haines, 1996).

Finally, in assessing capital punishment, one must remember that mistakes are common. Nearly 40% of death sentences since 1977 have been overturned because of new evidence or mistrial (Haines, 1996).

Incarcerating Less Serious Offenders in Violent, "No Frills" Prisons

Most of the increase in the prison population over the past 20 years is due to the conviction of nonviolent criminals. Many of them were involved in drug trafficking, and many of them are first-time offenders. The main rationale for imprisoning such offenders is that incarceration presumably deters them from repeating their offence. Supposedly, it also deters others from engaging in crime. Arguably, therefore, the streets become safer by isolating criminals from society.

Unfortunately for the hypothesis that imprisoning more people lowers the crime rate, available data show a weak relationship between the two variables. True, between 1980 and 1986, the number of inmates in United States prisons increased 65% and the number of victims of violent crime decreased 16%. This is what one would expect to find if incarceration deterred crime. However, between 1986 and 1991, the prison population increased 51% and the number of victims of violent crime *increased* 15%—just the opposite of what one would expect to find if incarceration deterred crime. The same sort of inconsistency is evident if we examine the relationship between incarceration and crime across states. For example, in 1992 Oklahoma had a high incarceration rate and a low crime rate, while Mississippi had a low incarceration rate and a high crime rate. These cases fit the hypothesis that imprisonment lowers the crime rate. However, Louisiana had a high incarceration rate and a high crime rate while North Dakota had a low incarceration rate and a low crime rate. This is the opposite of what one would expect to find if incarceration deterred crime (Mauer, 1994). We can only conclude that, contrary to popular opinion, prison does not consistently deter criminals or lower the crime rate by keeping criminals off the streets.

Often, however, prison does teach inmates to behave more violently. The case of Robert Scully, who graduated from robbery to killing a police officer thanks to his experiences in the California prison system, is one example of this. Budgets for general education, job training, physical exercise, psychological counseling, and entertainment have been cut. Brutality in the form of solitary confinement, hard labor, and physical violence is on the increase. The result is a prison population that is increasingly enraged, incapacitated, lacking in job skills, and more dangerous upon release than upon entry into the system. Massachusetts Governor William F. Weld captured the spirit of the times when he said that prisons ought to be "a tour through the circles of hell" where inmates should learn only "the joys of busting rocks" (quoted in Abramsky, 1999). However, the new regime of United States prisons may have an effect just the opposite of that intended by Governor Weld. Between 1999 and 2010, an estimated 3.5 million first-time releases are expected from American prisons (Abramsky, 1999). We may therefore be on the verge of a real crime wave, one that will have been created by the very get-tough policies that were intended to deter crime. Ominously, the homicide rate spiked upward in 2001. That was mainly a result of the downturning economoy and the rising number of inmates being released from state and federal prisons, which increased from 474,300 in 1995 to 635,000 in 2001. According to Sgt. John Pasquarello of the Los Angeles Police Department, "[p]rison is basically a place to learn crime, so when these guys come out, we see many of them getting back into drug operations, and this leads to fights and killings" (quoted in Butterfield, 2001).

Is there a reasonable alternative to the kinds of prisons we now have? Although saying so may be unpopular, anecdotal evidence suggests that institutions designed to rehabilitate criminals and reintegrate them into society can work, especially for less serious offenders. They also cost less than the kind of prison system we have created.

Those are the conclusions some people have drawn from experience at McKean, a medium-security correctional facility opened in Bradford, Pennsylvania, in 1989. Dennis Luther, the warden at McKean, is a maverick who has bucked the trend in American corrections. Nearly half the inmates at McKean are enrolled in classes, many of them earning licenses in masonry, carpentry, horticulture, barbering, cooking, and catering that will help them get jobs when they leave. Recreation facilities are abundant, and annual surveys conducted in the prison show that inmates get into less trouble the greater their involvement in athletics. The inmates run self-help groups and teach Adult Continuing Education. Good behavior is rewarded. If a cellblock receives high scores for cleanliness and orderliness during weekly inspection, the inmates in the cellblock get special privileges, such as the use of TV and telephones in the evening. Inmates who consistently behave well are allowed to attend supervised picnics on Family Days. This helps them adjust to life on the outside. Inmates are treated with respect and are expected to take responsibility for their actions. For example, after a few minor incidents in 1992, Luther restricted inmates' evening activities. The restriction was meant to be permanent, but some inmates asked Luther if he would do away with the restriction provided the prison was incident free for 90 days. Luther agreed, and he has never had to reimpose the restrictions.

The effects of these policies are evident throughout McKean. The facility is clean and orderly. Inmates don't carry "shanks" (homemade knives). The per-inmate cost to taxpayers is below average for medium-security facilities and 28% lower than the average for all state prisons. That is partly because few guards are needed to maintain order. In McKean's first 6 years of operation, there were no escapes, no homicides, no sexual assaults, and no suicides. There were a few serious assaults on inmates and staff members, but the *annual* rate of assault at McKean is equal to the *weekly* rate of assault at other state prisons of about the same size. Senior staff members and a local parole officer claim that McKean inmates return to prison far less often than inmates of other institutions (Worth, 1995).

Thus, a cost-effective and workable alternative to the current prison regime may exist, at least for less serious offenders. It is also possible that some aspects of the McKean approach could have beneficial effects in the overcrowded, maximum-security prisons where violent offenders are housed and gangs proliferate. Dennis Luther thinks so, but we don't really know because it hasn't been tried. Nor is it likely to be tried anytime soon given the current climate of public opinion.

SUMMARY

1. Deviance involves breaking a norm. Crime involves breaking a law. Both crime and deviance evoke societal reactions that help define the seriousness of the rule-breaking incident.

2. The seriousness of deviant and criminal acts depends on the severity of the societal response to them, their perceived harmfulness, and the degree of public agreement about whether they should be considered deviant or criminal.

3. Acts that rank lowest on these three dimensions are called social diversions. Next come social deviations and then conflict crimes. Consensus crimes rank highest.

4. Definitions of deviance and crime are historically and culturally variable. These definitions are socially defined and constructed. They are not inherent in actions or the characteristics of individuals.

5. Power is a key element in defining deviance and crime. Powerful groups are generally able to create norms and laws that suit their interests. Less powerful groups are usually unable to do so.

6. The increasing power of women has led to greater recognition of crimes committed against them. However, no similar increase has occurred in the prosecution of white-collar criminals because the distribution of power between classes has not changed much in recent decades.

7. Crime statistics come from official sources, self-report surveys, and indirect measures. Each source has its strengths and weaknesses.

8. The crime wave of the 1960s and 1970s began to taper off in the 1980s and fell substantially in the 1990s. This was due to more policing, a smaller proportion of young men in the population, a booming economy, and perhaps a decline in the number of unwanted children resulting from the availability of abortion.

9. African Americans experience disproportionately high arrest, conviction, and incarceration rates. That is due to bias in the way crime statistics are collected, the low social standing of the African-American community, and racial discrimination in the criminal justice system.

10. Theories of deviance include motivational theories (strain theory, subcultural theory, and the theory of differential association) and constraint theories (labeling theory, control theory, and conflict theory). Different theories illuminate different aspects of the process by which people are motivated to break rules and get defined as rule breakers.

11. All societies seek to ensure that their members obey norms and laws by imposing sanctions on rule breakers. However, the degree and form of social control vary historically and culturally.

12. Although some sociologists say that social control is weaker and deviance is greater in industrial societies than in preindustrial societies, other sociologists note that in some respects social control is greater.

13. The prison has become an important form of punishment in modern industrial societies. Since the 1980s, the incarceration rate has shot up in the United States. Prisons now focus less on rehabilitation than on isolating and incapacitating inmates.

14. The mushrooming prison population is only one consequence of the moral panic that has engulfed the nation on the crime issue. In all aspects of crime prevention and punishment, most Americans have taken a "get-tough" stance.

15. A variety of commercial and political groups benefit from the moral panic over crime and therefore encourage it.

16. Although the death penalty ranks high as a form of revenge, it is doubtful whether it acts as a serious deterrent. Moreover, the death penalty is administered in a racially biased manner, does not save money, and sometimes results in tragic mistakes.

17. Rehabilitative correctional facilities are cost effective. They do work, especially for less serious offenders. However, they are unlikely to become widespread in the current political climate.

GLOSSARY

Conflict crimes are illegal acts that many people consider harmful to society. However, many people think they are not very harmful. They are punishable by the state.

Conflict theory holds that deviance and crime arise out of the conflict between the powerful and the powerless.

Consensus crimes are illegal acts that nearly all people agree are bad in themselves and harm society greatly. The state inflicts severe punishment for consensus crimes.

Constraint theories identify the social factors that impose deviance and crime (or conventional behavior) on people.

Control theory holds that the rewards of deviance and crime are ample. Therefore, nearly everyone would engage in deviance and crime if they could get away with it. The degree to which people are prevented from violating norms and laws accounts for variations in the level of deviance and crime.

Deviance occurs when someone departs from a norm.

Differential association theory holds that people learn to value deviant or nondeviant lifestyles depending on whether their social environment leads them to associate more with deviants or nondeviants.

Formal punishment takes place when the judicial system penalizes someone for breaking a law.

Informal punishment involves a mild sanction that is imposed during face-to-face interaction, not by the judicial system.

Labeling theory holds that deviance results not so much from the actions of the deviant as from the response of others, who label the rule breaker a deviant.

One's **master status** is one's overriding public identity.

A **moral panic** occurs when many people fervently believe that some form of deviance or crime poses a profound threat to society's well-being.

Motivational theories identify the social factors that drive people to commit deviant and criminal acts.

In **self-report surveys,** respondents are asked to report their involvement in criminal activities, either as perpetrators or victims.

Social constructionism is a school of sociological thought that emphasizes how some people are in a position to create norms and pass laws that define others as deviant or criminal.

Social control refers to methods of ensuring conformity.

Social deviations are noncriminal departures from norms that are nonetheless subject to official control. Some members of the public regard them as somewhat harmful while other members of the public don't.

A **social diversion** is a minor act of deviance that is generally perceived as relatively harmless and that evokes, at most, a mild societal reaction such as amusement or disdain.

People who are **stigmatized** are negatively evaluated because of a marker that distinguishes them from others.

Strain theory holds that people may turn to deviance when they experience strain. Strain results when a culture teaches people the value of material success and society fails to provide enough legitimate opportunities for everyone to succeed.

Subcultural theory argues that gangs are a collective adaptation to social conditions. Distinct norms and values that reject the legitimate world crystallize in gangs.

Street crimes include arson, burglary, assault, and other illegal acts disproportionately committed by people from lower classes.

Techniques of neutralization are the rationalizations that deviants and criminals use to justify their activities. Techniques of neutralization make deviance and crime seem normal, at least to the deviants and criminals themselves.

Victimless crimes involve violations of the law in which no victim has stepped forward and been identified.

White-collar crime refers to an illegal act committed by a respectable, high-status person in the course of work.

QUESTIONS TO CONSIDER

1. Has this chapter changed your view of criminals and the criminal justice system? If so, how? If not, why not?

2. Do you think different theories are useful in explaining different types of deviance and crime? Or do you think that one or two theories explain all types of deviance and crime while other theories are not very illuminating? Justify your answer using logic and evidence.

3. Do TV crime shows and crime movies give a different picture of crime in the United States than this chapter gives? What are the major differences? Which picture do you think is more accurate? Why?

WEB RESOURCES

Companion Web Site for This Book

http://sociology.wadsworth.com

Begin by clicking on the Student Resources section of the Web site. Choose "Introduction to Sociology" and finally the Brym and Lie book cover. Next, select the chapter you are currently studying from the pull-down menu. From the Student Resources page you will have easy access to InfoTrac College Edition®, MicroCase Online exercises, additional Web links, and many resources to aid you in your study of sociology, including practice tests for each chapter.

Infotrac Search Terms

These search terms are provided to assist you in beginning to conduct research on this topic by visiting http://www.infotraccollege.com/wadsworth.

Moral panic	Street crime
Prison	White collar crime
Stigma	

Recommended Web Sites

The FBI's Web site at http://www.fbi.gov is a rich resource on crime in the United States. For official statistics, click on "Uniform Crime Reports."

For a measure of how widespread corruption is in the governments of every country in the world, visit http://www.GWDG.DE/~uwvw/icr.htm.

What makes crime news? For interesting material on this subject, visit http://www.fsu.edu/crimdo/lecture2.html. See also http://www.fsu.edu/~crimdo/lecture4.html for material on the varieties of crime depicted by the mass media.

Cross-national data on rates of incarceration can be found at http://www.sentencingproject.org/pubs/tsppubs/9030data.html.

Michel Foucault. *Discipline and Punish* (New York: Vintage, 1977 [1975]). One of the most influential and stimulating analyses of the development of the prison.

John Hagan. *Crime and Disrepute* (Thousand Oaks, CA: Pine Forge Press, 1994). A clear and concise overview of crime, deviance, law, and punishment.

Mike Maguire, Rod Morgan and Robert Reiner, eds. *The Oxford Handbook of Criminology*. (Oxford: Clarendon Press, 1994). Everything you always wanted to know about crime and criminology. Although the focus is on Britain, the book is full of interesting ideas and covers current controversies.

Jerome G. Miller. *Search and Destroy: African-American Males in the Criminal Justice System*. (New York: Cambridge University Press, 1996). A modern classic on racial bias in the United States criminal justice system.

Jeffrey H. Reiman. *And the Poor Get Prison: Economic Bias in American Criminal Justice* (Boston: Allyn and Bacon, 1996). A modern classic on class bias in the United States criminal justice system.

INEQUALITY

IN THIS CHAPTER, YOU WILL LEARN THAT:

✦ Income inequality in the United States has been increasing since the mid-1970s.

✦ Income inequality is higher in the United States than in any other postindustrial society.

✦ As societies develop, inequality at first increases. Then, after passing the early stage of industrialization, inequality in society declines. In the postindustrial stage of development, inequality appears to increase again.

✦ Most theories of social inequality focus on its economic roots.

✦ Prestige and power are important noneconomic sources of inequality.

✦ While some sociologists used to think that talent and hard work alone determine one's position in the socioeconomic hierarchy, it is now clear that being a member of certain social categories limits one's opportunities for success. In this sense, social structure shapes the distribution of inequality.

7

STRATIFICATION: UNITED STATES AND GLOBAL PERSPECTIVES

PATTERNS OF STRATIFICATION

Shipwrecks and Inequality

Writers and filmmakers sometimes tell stories about shipwrecks and their survivors to make a point about social inequality. They use the shipwreck as a literary device. It allows them to sweep away all trace of privilege and social convention. What remains are human beings stripped to their essentials, guinea pigs in an imaginary laboratory for the study of wealth and poverty, power and powerlessness, esteem and disrespect.

The tradition began with Daniel Defoe's *Robinson Crusoe,* first published in 1719. Defoe tells the story of an Englishman marooned on a desert island. His strong will, hard work, and inventiveness turn the poor island into a thriving colony. Defoe was one of the first writers to portray capitalism favorably. He believed that people get rich if they possess the virtues of good businessmen—and stay poor if they don't.

The 1975 Italian movie, *Swept Away,* tells almost exactly the opposite story.[1] In the movie, a beautiful woman, one of the idle rich, boards her yacht for a cruise in the Mediterranean. She treats the hardworking deck hands in a condescending and abrupt way. The deck hands do their jobs but seethe with resentment. Then comes the storm. The yacht is shipwrecked. Only the beautiful woman and one handsome deck hand remain alive, swept up on a desert island. Now equals, the two survivors soon have passionate sex and fall in love.

All is well until the day of their rescue. As soon as they return to the mainland, the woman resumes her haughty ways. She turns her back on the deck hand, who is reduced again to the role of a common laborer. Thus, the movie sends the audience three harsh messages. First, it is possible to be rich without working hard because one can inherit wealth. Second, one can work hard without becoming rich. Third, something about the structure of society causes inequality, for it is only on the desert island, without society as we know it, that inequality disappears.

The most recent movie on the shipwreck-and-inequality theme is *Titanic* (see Box 7.1). At one level, the movie shows that class differences are important. For example, in first class, living conditions are luxurious while in third class they are cramped. Indeed, on the Titanic, class differences spell the difference between life and death. After the Titanic strikes the iceberg off the coast of Newfoundland, the ship's crew prevents second- and third-class passengers from entering the few available lifeboats. They give priority to rescuing first-class passengers.

As the tragedy of the Titanic unfolds, however, another contradictory theme emerges. Under some circumstances, we learn, class differences can be insignificant. In the movie, the sinking of the Titanic is the backdrop to a fictional love story about a wealthy young woman in first class and a working-class youth in the decks below. The sinking of the Titanic and the collapse of its elaborate class structure give the young lovers an opportunity to cross class lines and profess their devotion to one another. At one level, then, *Titanic* is an optimistic tale that holds out hope for a society where class differences matter little, a society much like that of the American Dream.

Robinson Crusoe, Swept Away, and *Titanic* raise many of the issues we address in this chapter. What are the sources of social inequality? Do determination, industry, and ingenuity shape the distribution of advantages and disadvantages in society, as *Robinson Crusoe* suggests? Or is *Swept Away* more accurate? Do certain patterns of social relations underlie and shape that distribution? Is *Titanic*'s first message still valid? Does social inequality still have big consequences for the way we live? What about *Titanic*'s second message? Can people act to decrease the level of inequality in society? If so, how?

---

[1] In July 2001, Madonna announced that she and her director husband, Guy Ritchie, are planning a remake of the movie.

BOX 7.1
SOCIOLOGY AT THE MOVIES

TITANIC (1997)

When the character played by Leonardo Di-Caprio goes to the dining hall for dinner in *Titanic* (1997), it turns out that a first-class ticket is not the only thing he is missing. He also lacks appropriate dinner attire and the knowledge and tastes expected of someone who can afford first-class travel. He is able to borrow a dinner jacket. However, the conversation at the dinner table is foreign to him. Being a working-class youth, he has little knowledge of the arts or other matters that interest the rich and the highly edu-cated. To make matters worse, his dinner companions look down on him as a member of a lower class. Only the character played by Kate Winslet sees something special in him. She is already falling in love.

The world of the luxury ship Titanic is highly stratified. Sumptuous ballrooms and suites are reserved for the rich. They are spacious, well appointed, and well serviced. Dingy and cramped living and sleeping quarters are reserved for the poor. Their cabins even lack windows. Still worse off are the workers in the boiler room. They must shovel coal for hours on end, suffering from heat, coal dust, and exhaustion.

In the movie, the characters played by Leonardo DiCaprio and Kate Winslet over-come class barriers and fall in love. However, their sentiments are clearly exceptional. Class differences matter even after the great ship strikes the iceberg. First-class passen-gers are given preference in access to the lifeboats. Lower class ticket holders and workers on the ship are left to drown.

What does *Titanic* teach us about social stratification? Should we conclude that so-cial stratification pervades all aspects of life? Or does love often overcome class dif-ferences? You can treat these questions em-pirically. Among the marriages and dating relationships in your circle of friends and relatives, what proportions involve partners from different classes? A high proportion suggests love often trumps class. A low pro-portion suggests class usually overcomes love, or doesn't even allow it to blossom.

To answer these questions, we first sketch the pattern of social inequality in the United States and globally. We pay special attention to change over time. We then critically review the major theories of **social stratification,** the way society is organized in layers or strata. We assess these theories in the light of logic and evidence. From time to time, we take a step back and identify issues that need to be resolved before we can achieve a more ade-quate understanding of social stratification, one of the fundamentally important aspects of social life.

Economic Inequality in the United States

How long would it take you to spend a million dollars? If you spent a thousand dollars a day, it would take you nearly 3 years. How long would it take you to spend a *billion* dol-lars? If you spent a thousand dollars a day, you couldn't spend the entire sum in a lifetime. It would take nearly 3,000 years to spend a billion dollars at the rate of $1,000 a day.[2] Thus, a billion dollars is an almost unimaginably large sum of money. Yet in 1995, George Soros, an American currency speculator, earned $1.5 billion. In contrast, the annual income of a full-time, minimum-wage worker was $8,840. The earnings of George Soros in 1995 were enough to hire nearly 170,000 minimum-wage workers for a year.

George Soros is not the richest person in America. In 1999, he ranked 57th, with $4 billion in accumulated wealth. We list the 20 richest Americans in Table 7.1. Their net worth ranges from $8 billion to $85 billion.

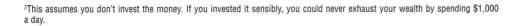

[2]This assumes you don't invest the money. If you invested it sensibly, you could never exhaust your wealth by spending $1,000 a day.

Web Interactive Exercises
Are the Rich Getting Richer and the Poor Getting Poorer?

✦ **TABLE 7.1** ✦

The Twenty Richest Americans, 1999

SOURCE: Forbes.com (1999a).

Name	Net Worth	Source
1. Bill Gates	$85 billion	Microsoft Corp.
2. Paul Allen	$40 billion	Microsoft Corp.
3. Warren Buffet	$31 billion	Berkshire Hathaway
4. Steven Ballmer	$23 billion	Microsoft Corp.
5. Michael Dell	$20 billion	Dell Computer Corp.
6. Helen Walton	$17 billion	Wal-Mart stores (inheritance)
7. John Walton	$17 billion	Wal-Mart stores (inheritance)
8. Alice Walton	$17 billion	Wal-Mart stores (inheritance)
9. S. Robson Walton	$17 billion	Wal-Mart stores
10. Jim Walton	$17 billion	Wal-Mart stores (inheritance)
11. Gordon Moore	$15 billion	Intel Corp.
12. Dupont Family	$13 billion	Dupont Corp. (inheritance)
13. Lawrence Ellison	$13 billion	Oracle Corp.
14. John Kluge	$11 billion	Metromedia Co.
15. Philip Anschutz	$11 billion	Oil, railroads, telecommunications
16. Mellon family	$10 billion	Various (inheritance)
17. Anthony & Barbara Cox	$9.7 billion	Cox Enterprises (inheritance)
18. Ann Cox Chambers	$9.7 billion	Cox Enterprises (inheritance)
19. Sumner Redstone	$9.4 billion	Viacom Inc.
20. Rockefeller family	$8 billion	Oil (inheritance)

Your wealth is what you own. For most adults, it includes a house (minus the mortgage); a car (minus the car loan); and some appliances, furniture, and savings (minus the credit card balance). Your income, on the other hand, is what you earn in a given period of time. In the United States and other societies, there is much inequality in the distribution of income and even more inequality in the distribution of wealth.

Unfortunately for sociologists, Americans are not required to report their wealth. As a result, wealth figures are sparse and based mainly on sample surveys. We know that about 40% of Americans owe more than they own. That is, their debt is greater than their assets. We also know that the richest 1% of American households own 37% of all national wealth, while the richest 10% own 72% (Albelda and Folbre, 1996; Brouwer, 1998; Folbre, 1995; Hacker, 1997; Levy, 1998; Wolff, 1996 [1995]). However, if we want more precise data on economic inequality, we must examine figures on annual income. Income has to be reported to the government, and income figures have been well mined by sociologists.

Students of social stratification often divide populations into categories of unequal size that differ in their lifestyle. These are often called "income classes." Table 7.2 shows how American households were divided into income classes in the second half of the 1990s. Alternatively, sociologists divide populations into a number of equal-sized statistical categories, usually called "income strata." Figure 7.1 adopts this approach. It divides the country's households into five income strata: the top 20% of income earners, the second 20%, and all the way down to the bottom 20%. It shows how total national income was divided among each of these fifths in 1974 and 1998. It illustrates two important facts. First, inequality has been rising in the United States for the past quarter of a century. In 1974, the top fifth of households earned 9.8 times more than the bottom fifth. By 1998, the top fifth of households earned 13.7 times more than the bottom fifth. If current trends continue, then by about 2005 the top 20% of households will earn more than the remaining 80%. Second, the middle 60% of income earners have been "squeezed" during the past quarter of a century, with their share of national income falling from 52.5% to 47.2% of the total. We conclude that, for about the past 25 years, the rich have been getting relatively richer while middle-income earners and the poor have been getting relatively poorer in the

Class	Percent of Households	Annual Household Income
Upper upper	0.8	$1 million +
Lower upper	9.7	$100,000-$999,999
Upper middle	21.1	$57,500-$99,999
Average middle	19.8	$37,500-$57,499
Lower middle or working	23.5	$20,000-$37,499
Lower	25.0	$0-$19,999
Total	99.9	

✦ **TABLE 7.2** ✦

Income Classes, Households, United States, 1998

Note: The total does not equal 100% due to rounding. The United States Census Bureau does not provide breakdowns of incomes over $100,000. Therefore, we estimated the breakpoint between the top two classes.

SOURCE: United States Bureau of the Census (2000h).

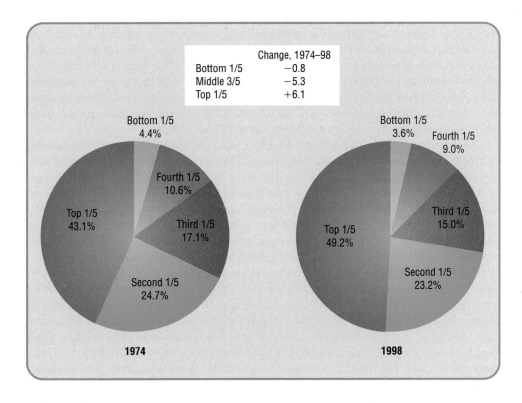

✦ **FIGURE 7.1** ✦

The Distribution of Total National Income Among Households, United States, 1974 and 1998

SOURCE: United States Bureau of the Census (1998a).

United States. (Note also that average earnings adjusted for inflation, or "average purchasing power," rose from the 1820s to the early 1970s and declined after that [Henwood, 1999]).

But what do these statistics mean? How do they reflect the everyday lives of real men and women? To flesh out the numbers, we now present brief sketches of ordinary people, most of them living in Silicon Valley (Santa Clara County), California. We focus on Silicon Valley not because it is typical but because it is the center of the United States computer industry, the heart of America's most recent economic boom, an extreme version of what has been happening to patterns of social stratification in the United States as a whole, and, according to some analysts, a harbinger of the way social stratification may develop in this country in coming decades. In particular, Silicon Valley illustrates well the growing gap between the "haves" and the "have-nots" in American society.

How Green Is the Valley?

Sociologists often divide society's upper class into two categories, the "upper-upper class" and the "lower-upper class" (see Table 7.2). The upper-upper class, comprising less than 1% of the United States population, used to be described as "old money." That is because people in this class inherited most of their wealth. Moreover, most of it was originally

Web Research Projects
The Super-Rich

earned in older industries such as banking, insurance, oil, real estate, and automobiles. "Old money" inhabited elite neighborhoods on the East Coast—places like Manhattan and Westchester Counties (New York); Fairfield, Somerset, and Bergen Counties (Connecticut); and Arlington (Virginia). It still does. Members of this class send their children to expensive private schools and high-prestige colleges. They belong to exclusive private clubs. They are overwhelmingly white and non-Hispanic. They live in a different world from most Americans (Baltzell, 1964).

In the past couple of decades, however, and especially in the 1990s, a substantial amount of "new money" entered the upper-upper class. New opportunities for entry were created mainly by booming high-tech industries. Among those with "new money," wealth is based less on inheritance than talent. "New money" is concentrated in high-tech meccas in the West—in places like San Francisco, San Mateo County, Santa Clara County (all in California), and King County (Washington) (Whitman, 2000). Larry Ellison, the flamboyant CEO of Oracle Corporation, the world's leading supplier of information management software, is perhaps the outstanding example of the new breed. The adopted son of a Chicago couple of modest means, Ellison started Oracle in 1977 with $1,200 after dropping out of college. In 1999, his personal net worth was $13 billion and *Forbes* magazine ranks him the 13th richest person in the United States. Such success stories notwithstanding, one thing remains constant: New members of the upper-upper class are still overwhelmingly white and non-Hispanic (Rothman and Black, 1998).

Let us take a moment to consider the history of new money in Santa Clara County, California, popularly known as Silicon Valley. John Doerr, an appropriately named venture capitalist in Menlo Park, California, has called Silicon Valley "the largest legal creation of wealth in the history of the planet" (quoted in Goodell, 1999: 65). He may be right. Fertilized by defense spending, Stanford University in Palo Alto spawned the region's electronics industry in the 1950s. By 1980, electronics had replaced prunes and apricots as the area's main product. Economic growth skyrocketed with the spread of the personal computer in the 1980s and the growing popularity of the Internet in the 1990s. High-tech industry in the Valley diversified and soon included not just microchip, computer, and software manufacturers, but also telecommunications and genetic engineering firms. By 1999, Santa Clara County was the home of 13 billionaires, several hundred people worth $25 million or more each, and 17,000 people worth more than $1 million each. And these figures exclude the value of people's homes—no trifle in Santa Clara county, where the median house price was $410,000 in 1999 and $760,000 in 2000, nearly four times the national average (Avery, 2000; Bernstein, 2000; Stacey, 1991: 20–6).

Fueled by high-tech industries, the 1990s witnessed the longest economic boom in American history. The many success stories of that remarkable decade were loudly trumpeted by the mass media. Often submerged beneath the good news, however, was a more

Bill Gates, the world's richest man, lives in a house with more than 66,000 square feet of floor space. It is valued at more than $53 million.

sobering reality: Although high-tech industries helped to change patterns of social stratification in the United States, the changes were not always positive. Specifically:

✦ *High-technology industry helped to create a new division in the lower-upper class—the "poor rich."* Composing nearly 10% of the United States population, members of the lower-upper class earn between $100,000 and $1 million a year (see Table 7.2). They rely mainly on earnings, not inheritance, for their wealth and income. They are typically highly educated and work as surgeons, dentists, corporate lawyers, engineers, successful businesspeople, and so forth. In the 1990s, many of these people grew even richer from their high-tech stock market investments. However, while high-technology industry helped to swell the ranks of those earning more than $100,000 a year during the 1990s, it also helped to give birth to a new division at the bottom of the lower-upper class—the "poor rich." These people earn more than $100,000 a year, yet are struggling to get by.

If you find it difficult to understand the predicament of the poor rich, consider Liliana and Peter Townshend (Goodell, 1999). They aspire to be among the millionaires of Silicon Valley. Yet they are far from their goal. Peter, 28, is a lawyer in one of the top law firms in the Valley. He earns $120,000 a year. Liliana, 27, is starting her own e-commerce business, which sells electronics and computer equipment to Spanish-speaking people all over the world. They bought their new 3,500-square-foot home in 1998 for $720,000. Yet, when you step inside the house, the first thing you notice is that it's bare. A few scraps of furniture—hand-me downs and items picked up at yard sales—dot the house. Their only luxuries are big new TVs on each floor. The Townshends regularly work 12 to 15 hours a day. Peter has been hospitalized twice for exhaustion. Liliana often wakes up at 3 A.M. in a panic that her business will fail. Peter is out of shape, he feels guilty about not making time to visit his parents, and Liliana complains about the marriage because she and Peter hardly ever see each other. They have two big mortgages on their house, and Liliana still owes $60,000 in student loans. Liliana buys food at Costco. When she visits her family in Los Angeles, she loads her car with groceries because food is so expensive in the Valley. "I feel like the poorest of the poor in Silicon Valley," says Liliana.

Like many other poor rich people, Liliana and Peter Townshend struggle despite their high family income. That is because they have assumed a mountain of debt and live in an area where the cost of living is very high. The Silicon Valley economic boom flooded the area with new residents who drove up demand for housing and just about everything else. Rent for a studio apartment in a bad neighborhood is $1,000 a month. Even gasoline costs almost 40% more than in the rest of the country.

✦ *High-technology industry squeezed the middle class and encouraged downward mobility.* The "middle class" consists of the nearly 65% of American households that earn more than $20,000 but less than $100,000 a year. Conventionally, the middle class is divided into roughly equal thirds: the "upper middle class," the "average middle class," and the "lower middle class" or "working class" (see Table 7.2).

Over a lifetime, an individual may experience considerable movement up or down the stratification system. Sociologists call this movement **vertical social mobility.** Movement up the stratification system ("upward mobility") is a constant theme in American literature and lore. Much sociological research has been conducted on this subject, and we review some of it below. Here, however, we note that about a quarter of Americans regularly report in surveys that their economic situation is deteriorating (Newman, 1988: 7, 21). In sociological terms, these people are "downwardly mobile." Downward mobility increases during periods of economic recession and especially during periods of economic restructuring, such as the United States experienced in the 1980s and early 1990s. In those years, layoffs and plant closings were common as computerized production became widespread and well-paying manufacturing jobs in steel, autos, and other industries were lost to low-wage countries such as Mexico and China. At the same time, computer technology and office reorganization allowed companies to fire many of their middle managers (see Chapter 5, "Interaction and Organization" and Chapter 10, "Work and the Economy"). Not even the computer industry was immune.

Take the case of David Patterson (Newman, 1988: 1–7). After growing up in the slums of Philadelphia, he managed to earn a business degree and land a good managerial job in California's thriving computer industry in the 1970s. His company transferred him to New York to take a more important executive job in the early 1980s. Two years later, he was fired in the midst of an industry-wide shakedown. After 9 months of failing to find work, he and his wife were forced to sell their house and move to a modest apartment in a nearby town. Their two teenage children grew furious with David. His wife began to express subtle doubts about his desire to find a new job. Gradually, the family's upper-middle-class friends stopped calling. When he listened to the news, David heard about all the plant closings and business restructuring taking place across the country. He knew there were good economic reasons for his plight. Nevertheless, as the months wore on, he grew depressed and started to ask: What's wrong with me? What have I done wrong? Like many people who lose their jobs, he forgot about the ups and downs of the nation's economy and blamed himself for his fate.

In Silicon Valley, even people who are employed in solid, middle-class jobs feel squeezed. Dan Hingle, 35, is a quality-assurance engineer at a start-up called Inter-Niche Technologies. He earns $50,000 a year but can only afford to live in a trailer park. "With $200,000, you can begin to approach a middle-class life," says Hingle. "How many people have jobs that pay $200,000? Not many. So people move out of the area, to where they can afford to buy a house, and commute an hour or two to work every day. That's fine, but then you're spending three or four hours on the road, and it's real easy to start hating life" (quoted in Goodell, 1999). In Los Altos, starting pay for police officers is around $40,000 a year. Only one of the town's 33 police officers can afford to live in Los Altos. In neighboring Los Gatos, 42 of 45 officers live out of town. In Palo Alto, the comparable figure is zero of 95. Officer Thomas Joy drives into Los Altos once a week, sleeps on his mother's couch for four nights, then drives home to his family. For the most part, teachers, nurses—all the people needed to run essential services in Silicon Valley—feel the same sort of squeeze.

✦ *High-technology industry lowered the value of unskilled work, swelling the ranks of the poor.* Thad Wingate earns almost $29,000 a year as a driver for a Silicon Valley courier service. By national standards, that makes him a member of the working or lower middle class. By Silicon Valley standards, however, that pushes him into the lower class. (Nationally, lower class households earn less than $20,000 a year.) In fact, Wingate lives in a shelter for homeless people in San Jose. In addition to the ex-convicts, recovering alcoholics, and mentally ill residents of the shelter, one finds a male nurse, a middle-aged trucker, a Puerto Rican woman in a McDonald's uniform, and other people who work full-time but are poor. San Jose's shelters and soup kitchens are booming. The soup kitchens served 83,000 people a month in 1999, up 27% from the year before.

Unskilled workers who are employed by big Silicon Valley corporations are victims of two increasingly popular corporate strategies, subcontracting and outsourcing (Bernstein, 2000). These strategies involve big corporations hiring smaller companies that, in turn, hire and supervise staff to do many of the manufacturing, janitorial, secretarial, and other routine jobs in the Valley. Big corporations favor this approach because it allows them to easily hire and fire workers as circumstances dictate. Some corporations also like this approach because it allows them to rely on subcontractors to keep wages down without getting involved in messy labor disputes. The 5,500 unionized janitors in Silicon Valley are mainly Mexican immigrants who earn at most about $18,000 a year, only about $1,000 above the official poverty line (Reed, 2000). To afford food and shelter, many of them must work second jobs and bunch into tiny apartments with friends and relatives. Twenty-two-year-old Alfredo Morales is a janitor who sleeps in an iron-casting shop in San Jose. He sends most of his wages back to his family in Mexico. He plans on moving back to Mexico as soon as he acquires some computer skills after hours. "I wouldn't bring my family here," he says. "It would be bringing them to greater misery" (Avery, 2000).

In sum, high-tech industry in Silicon Valley has encouraged much upward mobility. It has given a big boost to median income. However, it has also pushed up the cost of living and widened the gulf between the very rich and just about everyone else. In Silicon Valley, the middle class finds its standard of living deteriorating. The lower class is down and out. Silicon Valley is admittedly an extreme case. However, as our statistics on the distribution of national income show, growing income inequality is a countrywide story that has been developing for about 25 years. The stories sketched above give a human face to the statistics. Moreover, they suggest what may lie ahead as high-tech industries come to dominate the American economy.

International Differences

Global Inequality

Just as there are wide variations in income and wealth *within* countries, so there are wide variations in income and wealth *between* countries. For example, the United States is one of the richest countries in the world. Angola is a world apart. Angola is an African country of 11 million people. About 650,000 of its citizens have been killed in a civil war that has been raging since 1975, when the country gained independence from Portugal. Angola is one of the poorest nations on earth. Most Angolans live in houses made of cardboard, tin, and cement blocks. Most of these houses lack running water. Average income is about $1,000 a year. There is one telephone for every 140 people and one television for every 220 people. Inflation runs at about 90% per year. Adding to the misery of Angola's citizens are the millions of land mines that lay scattered throughout the countryside, regularly killing and maiming innocent passersby. Approximately 85% of the population survive on subsistence agriculture.

Angola itself is a highly stratified society. In fact, the gap between rich and poor is much wider than in the United States. That is because multinational companies such as Exxon and Chevron drill for oil in Angola. Oil exports account for nearly half the country's wealth. In the coastal capital of Luanda, an enclave of North Americans who work for Exxon and Chevron live in gated and heavily guarded communities containing luxury homes, tennis courts, swimming pools, maids, and SUVs. Here, side by side in the city of Luanda, people form a gulf that is as large as that separating Angola from the United States (see Figure 7.2).

✦ **FIGURE 7.2** ✦

Inequality on a Global Scale: Angola, 2000

Left: An amputated Angolan boy, victim of a land mine, hops home in Luanda. Right: United States and Canadian children relax at home in Luanda's North American enclave, built by Exxon for its supervisors and executives and their families.

◆ **TABLE 7.3** ◆

United Nations Indicators of Human Development, 1995, Top Five and Bottom Five Countries

Note: The United Nations publishes an annual index that combines three measures of "human development": life expectancy, adult literacy, and GDP per capita. This table lists the 5 countries that scored highest and lowest on the index in 1995 and gives the value of each measure for each of the 10 countries.

SOURCE: United Nations (1998b:130-2).

Country and Overall Rank	Life Expectancy (years)	Adult Literacy (percent)	GDP per Capita ($United States)
1. Canada	79.1	99.0	21,916
2. France	78.7	99.0	21,176
3. Norway	77.6	99.0	22,427
4. United States of America	76.4	99.0	26,977
5. Iceland	79.2	99.0	21,064
170. Burundi	44.5	35.3	637
171. Mali	47.0	31.0	565
172. Burkina Faso	46.3	19.2	784
173. Niger	47.5	13.6	765
174. Sierra Leone	34.7	31.4	625
Least Developed Countries	51.2	49.2	1,008
Industrial Countries	74.2	98.6	16,337
World	63.6	77.6	5,990

Some countries, like the United States, are rich. Others, like Angola, are poor. When sociologists study such differences *between* countries, they are studying **global inequality.** However, it is possible for country A and country B to be equally rich while, inside country A, the gap between rich and poor is greater than inside country B. When sociologists study such differences *within* countries, they are studying **cross-national variations in internal stratification.**

Consider global inequality for a moment. The United States, Canada, Japan, Australia, and a dozen or so Western European countries including Germany, France, and Britain are the world's richest postindustrial societies. The world's poorest countries cover much of Africa, South America, and Asia. Inequality between rich and poor countries is staggering. Nearly a fifth of the world's population lacks adequate shelter, and more than a fifth lacks safe water. About a third of the world's people are without electricity and more than two fifths lack adequate sanitation. In the United States, there are 626 phone lines for every 1,000 people, but in Cambodia, Congo, and Afghanistan there is only 1 line per 1,000 people. While annual health expenditure in the United States is $2,765 per person, the comparable figure for Tanzania and Sierra Leone is $4 and for Vietnam it is $3. The average educational expenditure for an American child is $11,329 per year. This compares with $57 in China, $46 in Mozambique, and $38 in Sri Lanka (United Nations, 1998b; see Table 7.3). People living in poor countries are also more likely than people in rich countries to experience extreme suffering on a mass scale. For example, because of political turmoil in many poor countries, an estimated 20–22 million people have been driven from their homes by force in recent years (Hampton, 1998). There are still about 27 million slaves in Mozambique, Sudan, and other African countries (Bales, 1999).

We devote much of Chapter 16 ("Population, Urbanization, and Development") to analyzing the causes, dimensions, and consequences of global inequality. You will learn that much of the wealth of the rich countries has been gained at the expense of the poor countries. Specifically, beginning centuries ago, the rich European countries turned large parts of the world into colonies that were used both as captive markets and as sources of cheap labor and raw materials. Even after the colonies gained political independence in the 20th century, most of them remained economically dependent on the rich countries. This helped to keep them poor. In Chapter 16, we also analyze the unique features of the few former colonies that have been able to escape poverty. Our analysis leads us to suggest possible ways of dealing with world poverty, one of the most vexing of social issues. For the moment, having merely described the problem of global inequality, we focus on a second type of international difference: cross-national variations in internal stratification.

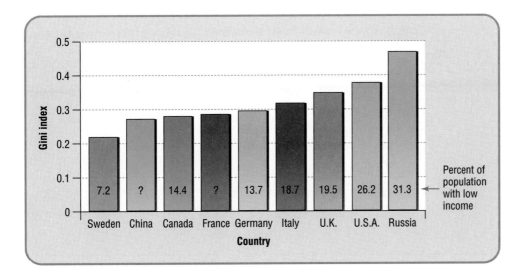

✦ **FIGURE 7.3** ✦

Household Income Inequality and Low Income, Selected Countries, 1992–1997

Note: If the Gini index = 1, all income is earned by one household. If the Gini index = 0, all income is shared equally by all households. The low income line is 50% of median income.

SOURCE: Luxembourg Income Study (1999a; 1999b).

Internal Stratification

Levels of global inequality aside, how does internal stratification differ from one country to the next? We can answer this question by first examining the **Gini index,** named after the Italian economist who invented it. The Gini index is a measure of income inequality. Its value ranges from zero to 1. A Gini index of zero indicates that every household in the country earns exactly the same amount of money. At the opposite pole, a Gini index of 1 indicates that a single household earns the entire national income. These are theoretical extremes. In the real world, most countries have Gini indexes between 0.2 and 0.5.

Figure 7.3 shows the Gini index for nine selected countries using the most recent cross-national income data available. Of the nine countries, Sweden has the lowest Gini index (.222) while the United States has the eighth highest (.375). Among these nine countries, only Russia's level of income inequality (Gini = .47) is higher than that of the United States. Figure 7.3 also shows the percentage of people in each country living below the low-income line (50% of median income). The data show that, among wealthy, postindustrial countries, the United States has the highest percentage of people living below the low-income line and the highest level of income inequality.

Development and Internal Stratification

What accounts for cross-national differences in internal stratification, such as those described above? Later in this chapter, you will learn that *political factors* explain some of the differences. For the moment, however, we focus on how *socioeconomic development* affects internal stratification.

Over the course of human history, as societies became richer and more complex, the level of social inequality first increased, then tapered off, then began to decline, and finally began to rise again (Lenski, 1966; Lenski, Nolan, and Lenski, 1995). Figure 7.4 illustrates the relationship between economic development and internal stratification. To account for this pattern, sociologists have analyzed how technology produces wealth, and how that wealth is controlled, in five types of societies:

Foraging Societies

For the first 90,000 years of human existence, people lived in nomadic bands of less than 100 people. To survive, they hunted wild animals and foraged for wild edible plants. Life was precarious. Some foragers and hunters were undoubtedly more skilled than others, but they did not hoard food. Instead, they shared food to ensure the survival of all band members. They produced little or nothing above what they required for subsistence. There were no rich and poor.

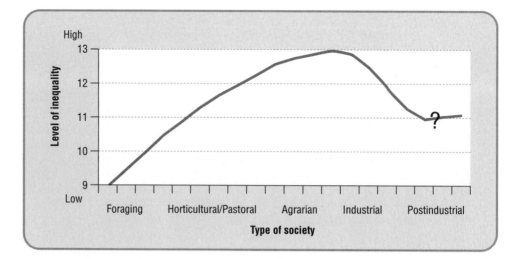

Horticultural and Pastoral Societies

About 12,000 years ago, people established the first agricultural settlements. They were based on horticulture (the use of small hand tools to cultivate plants) and pastoralism (the domestication of animals). These technological innovations enabled people to produce wealth, that is, a surplus above what they needed for subsistence. A small number of villagers controlled the surplus. Thus, significant social stratification emerged.

Agrarian Societies

About 5,000 years ago, people developed plow agriculture. By attaching oxen and other large animals to plows, farmers could increase the amount they produced. Again thanks to technological innovation, surpluses grew. With more wealth came still sharper social stratification.

Agrarian societies developed religious beliefs justifying steeper inequality. People came to believe that kings and queens ruled by "divine right." They viewed large landowners as "lords." Moreover, if you were born a peasant, you and your children were likely to

Vincent Van Gogh. *The Potato Eaters* (1889). Most people in agrarian societies were desperately poor. In the early 1800s in Ireland, for example, potatoes supplied about 80% of the peasant's diet. On average, each peasant consumed about 10 potatoes a day. The economic surplus was much larger than in horticultural and pastoral societies, but much of the surplus wound up in the hands of royalty, the aristocracy, and religious authorities.

remain peasants. If you were born a lord, you and your children were likely to remain lords. In the vocabulary of modern sociology, we say that stratification in agrarian societies was based more on **ascription** than **achievement.** That is, a person's position in the stratification system was determined more by the features he or she was born with ("ascribed characteristics") than his or her accomplishments ("achieved characteristics"). Another way of saying this is that there was little social mobility.

A nearly purely ascriptive society existed in agrarian India. Society was divided into **castes,** four main groups and many subgroups arranged in a rigid hierarchy. Being born into a particular caste meant you had to work in the distinctive occupations reserved for that caste and marry someone from the same or an adjoining caste. The Hindu religion strictly reinforced the system (Srinivas, 1952). For example, Hinduism explained people's place in the caste system by their deeds in a previous life. If you were good, you were presumably rewarded by being born into a higher caste in your next life. If you were bad, you were presumably punished by being born into a lower caste. Belief in the sanctity of caste regulated even the most mundane aspects of life. Thus, someone from the lowest caste could dig a well for a member of the highest caste, but once the well was dug, the well digger could not so much as cast his shadow on the well. If he did, the well was considered polluted and upper-caste people were forbidden to drink from it.

Caste systems have existed in industrial times. For example, the system of **apartheid** existed in South Africa from 1948 until 1992. While the white minority enjoyed the best jobs and other privileges, apartheid consigned the large black majority to menial jobs. It also prevented marriage between blacks and whites and erected separate public facilities for members of the two races. Asians and people of "mixed race" enjoyed privileges between these two extremes. However, apartheid was an exception. For the most part, industrialism causes a *decline* in inequality.

Industrial Societies

The Industrial Revolution began in Britain in the 1780s. A century later, it had spread to all of Western Europe, North America, Japan, and Russia.

The tendency of industrialism to lower the level of social stratification was not apparent in the first stages of industrial growth. If you've ever read a Dickens novel such as *Oliver Twist*, you know that hellish working conditions and deep social inequalities characterized early industrialism.

However, improvements in the technology and social organization of manufacturing soon made it possible to produce more goods at a lower cost per unit. This raised living standards for the entire population. Moreover, in industrial societies, birth was no longer destiny. Businesses required a literate, numerate, and highly trained work force. To raise profits, they were eager to identify and hire the most talented people. They encouraged everyone to develop their talents and rewarded them for doing so by paying higher salaries. Political pressure from below also played an important role in reducing inequality. Workers struggled for the right to form and join unions and expand the vote to all adult citizens. They used union power and their growing political influence to win improvements in the conditions of their existence. Although, as you will see later, barriers to mobility remained, social mobility became more widespread than ever before. Even traditional inequality between women and men began to break down due to the demand for talent and women's struggles to enter the paid work force on an equal footing with men. Why hire an incompetent man over a competent woman when you can profit more from the services of a capable employee? Put in this way, women's demands for equality made good business sense. For all these reasons, then, stratification declined as industrial societies developed.

Postindustrial Societies

It would be foolhardy to make definitive statements about long-term trends in social inequality in postindustrial societies. That is because the postindustrial era is only a few decades old. However, in the United States and other postindustrial countries, social inequality has been increasing for more than a quarter of a century. The concentration of wealth in the hands of the wealthiest 1% of Americans is higher today than anytime in the past 100 years. The gap between rich and poor is bigger today than it has been for at least 50 years.

Again, technological factors seem to be partly responsible for this trend. Many high-technology jobs have been created at the top of the stratification system over the past few decades. These jobs pay well. At the same time, new technologies have made many jobs routine. Routine jobs require little training. They pay poorly. And because the number of routine jobs is growing more quickly than the number of jobs at the top of the stratification system, the overall effect of technology today is to increase the level of inequality in society. We emphasize that this trend may be a function only of the early years of postindustrialism. However, when we examine these developments in detail in Chapter 10 ("Work and the Economy"), you will see that inequality seems likely to continue growing at least until the end of the first decade of the 21st century.

These, then, are the basic patterns and trends in the history of social stratification in the United States and globally. Bearing these descriptions in mind, we must now probe more deeply into the ways sociologists have explained social stratification. We begin with the theory of Karl Marx, who formulated a sweeping theory of social and historical development 150 years ago. The "engine" of Marx's theory—the driving force of history in his view—is the interaction of society's class structure with its technological base. Marx's work represents the first major sociological theory of social stratification, so it is worth considering in detail.

THEORIES OF STRATIFICATION

Classical Perspectives

Marx

In medieval Western Europe, peasants worked small plots of land owned by landlords. Peasants were legally obliged to give their landlords a set part of the harvest and to continue working for them under any circumstance. In turn, landlords were required to protect peasants from marauders. They were also obliged to open their storehouses and feed the peasants if crops failed. This arrangement was known as **feudalism** or serfdom. The peasants were called serfs.

According to Marx, by the late 1400s, several forces were beginning to undermine feudalism. Most important was the growth of exploration and trade, which increased the demand for many goods and services in commerce, navigation, and industry. By the 1600s and 1700s, some urban craftsmen and merchants had opened small manufacturing enterprises and saved enough capital to expand production. However, they faced a big problem. To increase profits, they needed more workers whom they could hire in periods of high demand and fire during slack times. Yet the biggest potential source of workers—the peasantry—was legally bound to the land. Thus, feudalism had to be destroyed so peasants could be turned into workers. In Scotland, for example, enterprising landowners recognized they could make more money raising sheep and selling wool than by having their peasants till the soil. So they turned their cropland into pastures, forcing peasants off the land and into the cities. The former peasants had no choice but to take jobs as urban workers.

In Marx's view, relations between workers and capitalists at first encouraged rapid technological change and economic growth. After all, capitalists wanted to adopt new tools, machines, and production methods so they could produce more efficiently and earn higher profits. But this had unforeseen consequences. In the first place, some capitalists were driven out of business by more efficient competitors and forced to become members of the working class. Together with former peasants pouring into the cities from the countryside, this caused the working class to grow. Second, the drive for profits motivated capitalists to concentrate workers in larger and larger factories, keep wages as low as possible, and invest as little as possible in improving working conditions. Thus, as the capitalist class grew richer and smaller, the working class grew larger and more impoverished, wrote Marx.

Marx thought that workers would ultimately become aware of belonging to the same exploited class. Their sense of **class consciousness** would, he wrote, encourage the growth of unions and workers' political parties. These organizations would eventually try to create a communist system in which there would be no private wealth. Instead, under communism, everyone would share wealth, said Marx (Marx, 1904 [1859]; Marx and Engels, 1972 [1848]).

We must note several points about Marx's theory. First, according to Marx, a person's **class** is determined by the *source* of his or her income, or, to use Marx's term, by one's "relationship to the means of production." For example, members of the capitalist class (or **bourgeoisie**) own means of production, including factories, tools, and land. However, they do not do any physical labor. They are thus in a position to earn profits. In contrast, members of the working class (or **proletariat**) do physical labor. However, they do not own means of production. They are thus in a position to earn wages. It is the source, not the amount, of income that distinguishes classes in Marx's view.

A second noteworthy point about Marx's theory is that it recognizes more than two classes in any society. For example, Marx discussed the **petty bourgeoisie.** This is a class of small-scale capitalists who own means of production but employ only a few workers or none at all. This situation forces them to do physical work themselves. In Marx's view, however, members of the petty bourgeoisie are bound to disappear as capitalism develops because they are economically inefficient. Just two great classes characterize every economic era, said Marx—landlords and serfs during feudalism, bourgeoisie and proletariat during capitalism.

Finally, it is important to note that some of Marx's predictions about the development of capitalism turned out to be wrong. Nevertheless, Marx's ideas about social stratification have stimulated thinking and research on social stratification until today, as you will see below.

A Critique of Marx

Marx's ideas strongly influenced the development of sociological conflict theory (see Chapter 1). Today, however, more than 120 years after Marx's death, it is generally agreed that Marx did not accurately foresee some specific aspects of capitalist development. In particular:

✦ Industrial societies did not polarize into two opposed classes engaged in bitter conflict. Instead, a large and heterogeneous middle class of "white-collar" workers emerged. Some of them are nonmanual employees. Others are professionals. Many of them enjoy higher income, wealth, and status than manual workers. With a bigger stake in capitalism than manual workers, nonmanual employees and professionals have generally acted as a stabilizing force in society.

✦ Marx correctly argued that investment in technology makes it possible for capitalists to earn high profits. However, he did not expect investment in technology also to make it possible for workers to earn higher wages and toil fewer hours under less oppressive conditions. Yet that is just what happened. Their improved living standard tended to pacify workers, as did the availability of various welfare-state benefits, such as unemployment insurance.

✦ Communism took root not where industry was most highly developed, as Marx predicted, but in semiindustrialized countries such as Russia in 1917 and China in 1948. Moreover, instead of evolving into classless societies, new forms of privilege emerged under communism. For example, in communist Russia, money income was more equal than in the West. However, membership in the Communist Party, and particularly membership in the so-called *nomenklatura,* a select group of professional state managers, brought special privileges. These included exclusive access to stores where they could purchase scarce Western goods at nominal prices, luxurious country homes, free trips abroad, and so forth. According to a Russian quip from the 1970s, "under capitalism, one class exploits the other, but under communism it's the other way around."

Weber

Writing in the early 1900s, Max Weber foretold most of these developments. For example, he did not think communism would create classlessness. He also understood the profound significance of the growth of the middle class. As a result, Weber developed an approach to social stratification much different from Marx's.

Weber, like Marx, saw classes as economic categories (Weber, 1946 [1922]: 180–95). However, he did not think a single criterion—ownership versus nonownership of property—determines class position. Class position, wrote Weber, is determined by one's "market situation," including the possession of goods, opportunities for income, level of education, and degree of technical skill. From this point of view, there are four main classes according to Weber: large property owners, small property owners, propertyless but relatively highly educated and well-paid employees, and propertyless manual workers. Thus, white-collar employees and professionals emerge as a large class in Weber's scheme of things.

If Weber broadened Marx's idea of class, he also recognized that two types of groups other than class have a bearing on the way a society is stratified: status groups and parties. **Status groups** differ from one another in the prestige or social honor they enjoy and in their style of life. Consider members of a particular minority ethnic community who have recently immigrated. They may earn relatively high income but endure relatively low prestige. The longer established members of the majority ethnic community may look down on them as the so-called vulgar "new rich." If their cultural practices differ from that of the majority ethnic group, their style of life may also become a subject of scorn. Thus, the position of the minority ethnic group in the social hierarchy does not derive just from its economic position but also from the esteem in which it is held.

In Weber's usage, **parties** are not just political groups but, more generally, organizations that seek to impose their will on others. Control over parties, especially large bureaucratic organizations, does not depend just on wealth or another class criterion. One can head a military, scientific, or other bureaucracy without being rich, just as one can be rich and still have to endure low prestige.

So we see why Weber argued that in order to draw an accurate picture of a society's stratification system, one must analyze classes, status groups, *and* parties as somewhat independent bases of social inequality (see Figure 7.5 and Table 7.4). But to what degree are they independent of one another? Weber said that the importance of status groups as a

✦ **FIGURE 7.5** ✦
Weber's Stratification Scheme

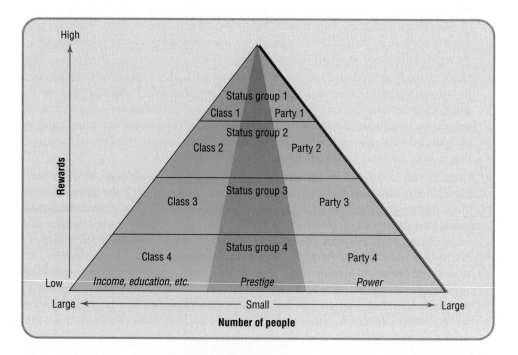

Occupation	Median Annual Income, 1998	Prestige Score, 1989
Airplane pilots and navigators	$71,916	61
Lawyers	$67,028	75
Physicians	$60,112	86
Aerospace engineers	$59,228	72
College faculty	$47,164	74
Computer programmers	$43,836	61
High school teachers	$38,272	66
Police and detectives	$33,592	60
Electricians	$33,436	51
Plumbers	$30,836	45
Median for All Occupations	$27,196	
Pre-kindergarten and kindergarten teachers	$20,644	55
Secretaries	$20,592	46
Taxi drivers	$19,708	28
Security guards	$19,500	42
Bank tellers	$17,160	43
Janitors	$17,004	22
Hairdressers and cosmetologists	$16,276	36
Cooks (except short-order)	$15,028	31
Textile sewing machine operator	$14,560	28

✦ **TABLE 7.4** ✦

Median Annual Income, Full-Time Workers, and Prestige Scores, Selected Occupations, United States

Note: Median annual income for the occupations listed above is taken from a 1998 labor force survey conducted by the federal government. Occupational prestige scores are based on sociological surveys in which respondents are asked to rank occupations in terms of the prestige attached to them. Prestige scores for all occupations range from 17 (miscellaneous food preparation occupations) to 86 (physicians). The correlation between median income and prestige scores for the occupations listed here is very high ($r = .84$).

SOURCE: Adapted from Interuniversity Consortium for Political and Social Research (2000); United States Department of Labor (1999c).

basis of stratification is greatest in precapitalist societies. Under capitalism, classes and parties (especially bureaucracies) become the main bases of stratification.

American Perspectives

Functionalism

Marx and Weber were Germans who wrote their major works between the 1840s and the 1910s. Inevitably, their theories bear the stamp of the age in which they wrote. The next major developments in the field occurred in mid-20th-century America. Just as inevitably, these innovations were colored by the optimism, dynamism, and prejudices of that time and place.

Consider first in this connection the **functional theory of stratification,** proposed by Kingsley Davis and Wilbert Moore at the end of World War II (Davis and Moore, 1945). Davis and Moore observed that some jobs are more important than others. A judge's work, for example, contributes more to society than the work of a janitor. This presents the society with a big problem: How can people be motivated to undergo the long training they need to serve as judges, physicians, and engineers and in other important jobs? After all, higher education is expensive. During training, one cannot earn much money. One must study long and hard rather than seek pleasure. Clearly, an incentive is needed to motivate the most talented people to train for the most important jobs. That "something," said Davis and Moore, is money. More precisely, social stratification is necessary (or "functional") because the prospect of high material rewards motivates people to undergo the sacrifices needed to get a higher education. Without substantial inequality, they conclude, the most talented people would have no incentive to become judges and so forth.

Although the functional theory of stratification may at first seem plausible, we can conduct what Max Weber called a "thought experiment" to uncover one of its chief flaws.

According to the functional theory of stratification, "important" jobs require more training than "less important" jobs. The promise of big salaries motivates people to undergo that training. Therefore, the functionalists conclude, social stratification is necessary. As the text makes clear, however, one of the problems with the functional theory of stratification is that it is difficult to establish which jobs are important, especially when one takes a historical perspective.

Imagine a society with just two classes of people—physicians and small family farmers. The farmers grow food. The physicians tend the ill. Then, one day, a rare and deadly virus strikes. The virus has the odd property of attacking only physicians. Within weeks, there are no more doctors in our imaginary society. As a result, the farmers are much worse off. Cures and treatments for their ailments are no longer available. Soon the average farmer lives fewer years than his or her predecessors. The society is less well off, although it survives.

Now imagine the reverse. Again we have a society composed of physicians and farmers. Again a rare and lethal virus strikes. This time, however, the virus has the odd property of attacking only farmers. Within weeks, the physicians' stores of food are depleted. A few more weeks, and the physicians start dying of starvation. The physicians who try to become farmers catch the virus and expire. Within months, there is no more society. Who then does the more important work, physicians or farmers? Our thought experiment suggests that farmers do, for without them society cannot exist.

From a historical point of view, we can say that *none* of the jobs regarded by Davis and Moore as "important" would exist without the physical labor done by people in "unimportant" jobs throughout the ages. To sustain the witch doctor in a tribal society, hunters and gatherers had to produce enough for their own subsistence plus a surplus to feed, clothe, and house the witch doctor. To sustain the royal court in an agrarian society, serfs had to produce enough for their own subsistence plus a surplus to support the lifestyle of members of the royal family. Government and religious authorities have taken surpluses from ordinary working people for thousands of years by means of taxes, tithes, and force. Among other things, these surpluses were used to establish the first institutions of higher learning in the 13th century. From these, modern universities developed.

So we see the question of which occupations are most important is not as clear-cut as Davis and Moore make it seem. To be sure, physicians today earn a lot more money than small family farmers and they also enjoy a lot more prestige. But that is not because their work is more important in any objective sense of the word. (On the question of why physicians and other professionals earn more than nonprofessionals, see Chapter 13, "Religion and Education").

Other problems with the functional theory of stratification were noted soon after Davis and Moore published their article (Tumin, 1953). We mention two of the most important criticisms here. First, the functional theory of stratification stresses how inequality helps society discover talent. However, it ignores the pool of talent lying undiscovered *because of* inequality. Bright and energetic adolescents may be forced to drop out of high school to help support themselves and their families. Capable and industrious high school graduates may be forced to forego a college education because they can't afford it. Inequality may encourage the discovery of talent, but only among those who can afford to take advantage of the opportunities available to them. For the rest, inequality prevents talent from being discovered.

A final problem with the functional theory of stratification is its failure to examine how advantages are passed from generation to generation. Like *Robinson Crusoe,* the functional theory correctly emphasizes that talent, hard work, and sacrifice often result in high occupational attainment and high material rewards. However, it is also the case that once people attain high class standing they use their power to maintain their position and promote the interests of their families regardless of how talented their children are. For example, inheritance allows parents to transfer wealth to children irrespective of their talent. Thus, glancing back at Table 7.1, we see that about half the 20 largest personal fortunes in the United States were inherited. The four Walton children are tied as the sixth richest people in America not because of their talent but because their father gave the Wal-Mart empire to them. Even rich people who do not inherit large fortunes often start near the top of the stratification system. Bill Gates, for example, is the richest person in the world. He did not inherit his fortune. However, his father was a partner in one of the most successful law firms in Seattle. Gates himself went to the most exclusive and expensive private schools in the city, followed by a stint at Harvard. In the late 1960s, his high school was one of the first in the nation to boast a computer terminal connected to a nearby university mainframe. Gates's early fascination with computers dates from this period. Gates is without doubt a highly talented man, but surely the advantages he was born with, and not just his talents, helped to elevate him to his present lofty status (Wallace and Erickson, 1992). An adequate theory of stratification must take inheritance into account, just as it must recognize how inequality prevents the discovery of talent and avoid making untenable assumptions about which jobs are important and which are not.

Social Mobility: Theory and Research

The cabin is built from rough logs cut from the wilderness. It is small and drafty. When the wind blows, the flame in the solitary lamp gutters. Outside, a cougar howls. Inside, the frontier family tends a small child lying on a simple quilt. Someday, the child will become President of the United States.

Is he Abraham Lincoln? Or Andrew Jackson perhaps? It hardly matters. What matters is the moral of the story. As every school child learns, in the United States anyone can become President, no matter how humble his—who knows? perhaps someday her—origins. The opportunity to rise to the top is the crux of the American Dream (Pessen, 1984). To many observers, it is the main difference between the United States and Europe.

Two hundred years ago, Britain and other European societies seemed to many observers to be based on class privilege. In this view, if you were born into a certain class, you were destined to go to certain schools, speak with a certain accent, and take a certain type of job. Some people managed to rise above their class origins, but not many. In contrast, the United States was widely viewed as the land of golden opportunity. With few people and abundant natural resources, it already enjoyed the highest standard of living in the world. Its western frontier beckoned adventurous migrants with gold rushes and vast tracts of fertile land. The United States came to be viewed as a land where hard work and talent could easily overcome humble origins. Almost anyone could strike it rich, it seemed. If Europe was based on class privilege, America was widely regarded as classless.

The rate of upward social mobility may have been higher in America than in Europe in the 19th century. As we will see in the following, however, there was little difference in mobility rates by the second half of the 20th century. It is revealing in this connection to contrast the biographies of Abraham Lincoln and Andrew Jackson with those of the front-running presidential hopefuls from both parties in 2000. All of the latter (George W. Bush, John McCain, Steve Forbes, Al Gore, and Bill Bradley) were born into millionaire families and attended elite colleges. Nonetheless, the *idea* that America is classless has persisted.[3]

In 1967, that idea was incorporated in Peter Blau and Otis Dudley Duncan's *The American Occupational Structure.* This book became one of the most influential works in

---

[3]Actually, even the rise of Lincoln and Jackson from poverty to the most powerful position in the land was not unaided. Lincoln's wife was the daughter of a banker. Jackson's wife was the daughter of a relatively well-to-do owner of a boarding house.

American sociology. Blau and Duncan set themselves the task of figuring out the relative importance of inheritance versus individual merit in determining one's place in the stratification system. To what degree is one's position based on ascription—that is, inheriting wealth and other advantages from one's family? To what degree is one's position based on achievement—that is, applying one's own talents to life's tasks? Blau and Duncan's answer was plain: Stratification in America is based mainly on individual achievement.

Blau and Duncan abandoned the European tradition of viewing the stratification system as a set of distinct groups. Marx, you will recall, distinguished two main classes by the source of their income. He was sure the bourgeoisie and the proletariat would become class conscious and take action to assert their class interests. Similarly, Weber distinguished four main classes by their market situation. He saw class consciousness and action as potentials that each of these classes might realize in some circumstances. In contrast, Blau and Duncan saw little if any potential for class consciousness and action in the United States. That is why they abandoned the entire vocabulary of class. For them, the stratification system is not a system of distinct classes at all, but a continuous hierarchy or ladder of occupations with hundreds of rungs. Each occupation—each rung on the ladder—requires different levels of education and generates different amounts of income.

To reflect these variations in education and earnings, Blau and Duncan created a **socioeconomic index of occupational status (SEI).** Using survey data, they found the average earnings and years of education of men employed full-time in various occupations. They combined these two averages to arrive at an SEI score for each occupation. (Similarly, other researchers combined income, education, and occupational prestige data to construct an index of **socioeconomic status** or **SES.**)

Next, Blau and Duncan used survey data to find the SEI of each respondent's current job, first job, and father's job, as well as the years of formal education completed by the respondent and the respondent's father. They showed how all five of these variables were related (see Figure 7.6). Their main finding was that the respondents' own achievements (years of education and SEI of first job) had much more influence on their current occupational status than did ascribed characteristics (father's occupation and years of education). Blau and Duncan concluded that the United States is a relatively open society in which individual merit counts for more than family background. This partly vindicated the functionalists and confirmed the core idea of the American Dream.

Blau and Duncan's study influenced a whole generation of stratification researchers in the United States and, to a lesser degree, other countries (Boyd et al., 1985; Erikson and Goldthorpe, 1992; Featherman and Hauser, 1978; Featherman, Jones, and Hauser, 1975; Grusky and Hauser, 1984). Their approach to studying social stratification, which focuses on the effects of family background and educational level on occupational achievement, became known as the "status attainment model."

Subsequent research confirmed that the rate of social mobility for men in the United States is high and that most mobility has been upward. However, since the early 1970s,

✦ FIGURE 7.6 ✦
Blau and Duncan's Model of Occupational Achievement

Note: Arrows indicate cause-and-effect relationships between variables, with the arrowheads pointing to effects. The association between father's education and occupation is not illustrated here for simplicity's sake. The thicker the line, the stronger the relationship between variables. The diagram shows that the respondents' own achievements (years of education and SEI of first job) had much more influence on their current occupational status than did ascribed characteristics (father's occupation and years of education).

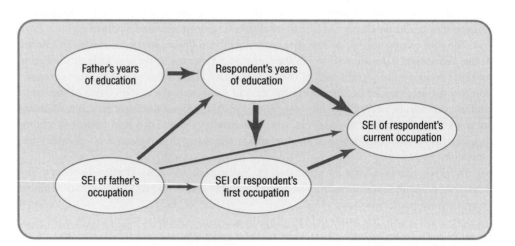

Horatio Alger Jr. wrote more than 100 books from the post-Civil War era until the end of the 19th century. He inspired Americans with tales of how courage, faith, honesty, hard work, and a little luck could help young people rise from rags to riches. Alger's novels lack appreciation of the social determinants of social stratification. Nonetheless, his ideas live on. Since 1947, they have been perpetuated by the Horatio Alger Association of Distinguished Americans. The Association honors the achievements of "outstanding individuals in our society who have succeeded in the face of adversity" according to its Web site (http://www.horatioalger.com/). On April 6, 2001, President George W. Bush welcomed Horatio Alger National Scholars and Association Members to the White House.

substantial downward mobility has occurred. This is apparent from our earlier discussion of trends in the distribution of income in the early stages of postindustrialism.

Researchers also found that mobility for men within a single generation (**intragenerational mobility**) is generally modest. Few people move "from rags to riches" or fall from the top to the bottom of the stratification system in a lifetime. On the other hand, mobility for men over more than one generation (**intergenerational mobility**) can be substantial.

In addition, research showed that most social mobility is due to change in the occupational structure. One of the most dramatic changes during the late 19th and early 20th centuries was the decline of agriculture and the rise of manufacturing. This caused a big drop in the number of farmers and a corresponding surge in the number of factory workers. A second dramatic change, especially apparent during the last third of the 20th century, was the decline of manufacturing and the rise of the service sector. This caused a big drop in the number of manual workers and a corresponding surge in the number of white-collar service workers. Mobility due to such changes in the occupational structure became known as **structural mobility.** Just as high tide raises all ships and low tide causes all ships to fall, structural mobility is a powerful force drawing individuals away from old occupations and into new ones.

Finally, research shows there are only small differences in rates of social mobility among the highly industrialized countries. The United States does not have an exceptionally high rate of upward social mobility (Lipset and Bendix, 1963). In fact, some countries, such as Australia and Canada, enjoy even higher upward mobility rates than the United States does (Tyree, Semyonov, and Hodge, 1979).

A Critique of Blau and Duncan

Blau and Duncan's approach to studying social stratification stimulated much research and led to important sociological insights, as we have seen. However, beginning in the 1970s, their stratification model was criticized from many angles. Most of the criticisms converged on a single point: Blau and Duncan's theory ignored much that was interesting and important in the study of social stratification.

To fully appreciate the significance of this criticism, one must know that Blau and Duncan sampled only men who were employed full-time. Women were excluded on the grounds that most of them were homemakers and not in the paid labor force. Whatever the merits of this exclusion in 1962, the year of Blau and Duncan's survey, it made little sense even 10 years later. It makes still less sense today, when the great majority of adult women are in the paid labor force.[4]

Similarly, Blau and Duncan did not sample part-time and unemployed workers. This may not have been a serious omission in the low-unemployment era of the early 1960s. However, it became serious in the 1970s and 1980s, when the unemployment rate rose. Compounding the problem is that the proportion of jobs that are part-time rose steadily from the 1970s on.

Finally, a disproportionately large number of unemployed and part-time workers are African American or Hispanic American and think of themselves as members of the working class. Therefore, these groups were underrepresented in Blau and Duncan's study. It would be a huge exaggeration to say that Blau and Duncan sampled only white middle-class men. But the percentage of women, working-class people, African Americans, and Hispanic Americans in their sample was smaller than in the United States population (Miller, 1998; Sørenson, 1992).

This raises a basic question about Blau and Duncan's study: How can one make valid claims about the relative importance of ascription versus achievement in American society when many of the most disadvantaged people in the society are excluded from one's analysis? The short answer is: "One cannot." It may be the case, as Blau and Duncan claimed, that merit is more important than inheritance in determining the social position of white middle-class men with full-time jobs. However, their study has little to say about the determinants of stratification among most adult Americans, who are not white middle-class men employed full-time.

Subsequent research yielded both good news and bad news for the Blau and Duncan status attainment model. The good news: Research confirmed that the process of status attainment is much the same for women and minorities as it is for white men. That is, years of schooling influence status attainment more than does father's occupation. This holds true for white men, women, African Americans, Hispanic Americans, and so forth. The bad news for the Blau and Duncan status attainment model is that if you compare people *with the same level of education and similar family backgrounds*, women and members of minority groups tend to attain lower status than white men (Featherman and Hauser, 1976; Hout, 1988; Hout and Morgan, 1975; McClendon, 1976; Stolzenberg, 1990; Tienda and Lii, 1987). These findings suggest that, to explain status attainment, it is insufficient to examine only the characteristics of *individuals*, such as their years of education, father's occupation, and so forth. It is also important to examine the characteristics of *groups*, such as whether some groups face barriers to mobility, regardless of the individual characteristics of their members (Horan, 1978). Such group barriers include racial and gender discrimination and being born in neighborhoods that make upward mobility unusually difficult due to poor living conditions. For instance, women and African Americans who are employed full-time earn less on average than white men with the same level of education (see Chapter 8, "Race and Ethnicity," and Chapter 9, "Sexuality and Gender"). The existence of such group disadvantages suggests that American society is not as open or "meritocratic" as Blau and Duncan make it out to be. There are group barriers to mobility, such as gender and race.

Could class in the Marxist or Weberian sense of the term act like race and gender, bestowing advantages and disadvantages on entire groups of people and perhaps even helping to shape their political views? Some sociologists think so. Their research shows that parents' wealth, education, and occupation are more important determinants of a person's occupation than Blau and Duncan's research suggest (Jencks et al., 1972; Rytina, 1992). Other sociologists, dissatisfied with the Blau and Duncan model, have revisited and

---

[4]The inclusion of married women in stratification studies raises the question of whether they should be located in terms of their own labor force characteristics or in terms of their own *and their husbands'* labor force characteristics. We share the opinion of Breen and Rottman (1995: 62-7) that the proper unit of analysis should be the household, not the individual. That is because each spouse is affected by the other's labor force characteristics and decision making is usually made at the household, not the individual, level. See also Clement and Myles (1994) and Crompton and Mann (1986).

updated the Marxist and Weberian concepts of class to make them more relevant to the late 20th century. We now briefly review their work.

THE REVIVAL OF CLASS ANALYSIS

An adequate theory of class stratification must do two things. First, it must specify criteria that distinguish a *small number* of distinct classes. Why small? Because the larger the number of classes specified by the theory, the more its picture of the stratification system will resemble Blau and Duncan's occupational ladder with its hundreds of rungs. Therefore, the less it will capture *class* differences in economic opportunities, political outlooks, and cultural styles. Second, an adequate theory of class stratification must spark research that demonstrates *substantial gaps* between classes in economic opportunities, political outlooks, and cultural styles. In other words, research must show that such differences are larger between classes than within them.

In the 1980s and early 1990s, sociologists on both sides of the Atlantic developed theories that meet the first requirement. In the United States, Erik Olin Wright updated Marx. In Britain, John Goldthorpe updated Weber. Both scholars create new "class maps" that specified criteria for distinguishing a small number of classes.

On the second requirement, the work of Goldthorpe and especially Wright has been less successful. That is, not enough research has been conducted to show that the classes distinguished by Wright and Goldthorpe differ substantially in terms of economic opportunities, political outlooks, and cultural styles. Especially sparse is research that would allow us to judge which of the two theories is superior in this regard. Nonetheless, as we will see, the research literature offers some tantalizing hints. Wright and Goldthorpe appear to be on the right track, and Goldthorpe's Weberian approach may be the more promising of the two.

Wright

Wright's update of Marx's class scheme is illustrated in Table 7.5 (Wright, 1985; 1997). Like Marx, Wright's basic distinction is between property owners and nonowners. The former earn profits, the latter wages and salaries.

Wright also distinguishes large, medium, and small owners. They differ from one another in terms of how much property they own and in terms of whether they have many employees, a few employees, or none at all.

If there are three propertied classes in Wright's theory, there are nine classes without property. These wage and salary earners differ from one another in two ways. First, they have different "skill and credential levels." That is, some wage and salary earners have more training and education than others. Second, wage and salary earners differ from one another in terms of their "organization assets." That is, some of them enjoy

✦ **TABLE 7.5** ✦
Wright's Typology of Classes, United States, 1980

SOURCE: Wright (1985: 88).

| Owners of Means of Production | Nonowners of Means of Production | | | Organizational Assets |
| | Skill Assets | | | |
	+	>0	−	
Bourgeoisie (hire, don't work) 2%	Expert Manager 4%	Semicredentialed Manager 6%	Uncredentialed Manager 2%	+
Small Employers (hire, work) 6%	Expert Supervisor 4%	Semicredentialed Supervisor 7%	Uncredentialed Supervisor 7%	>0
Petty Bourgeoisie (work, don't hire) 7%	Expert Nonmanager 3%	Semicredentialed Worker 12%	Proletarian 40%	−

more decision-making authority than others. The two extremes among the nonpropertied are "expert managers," who have high skill and credential levels combined with high organizational assets, and proletarians, who have low skill and credential levels combined with no organizational assets. The percentages in Table 7.5 show the proportion of the American labor force Wright found in each class in a survey he conducted in 1980.

Goldthorpe

For Goldthorpe, different classes are characterized by different "employment relations" (Goldthorpe, Llewellyn, and Payne, 1987 [1980]; Erikson and Goldthorpe, 1992). Goldthorpe's basic division in employment relations is among employers, self-employed people, and employees. He then makes finer distinctions within each of these broad groupings. For instance, he distinguishes between large and small employers, and between self-employed people in agriculture and those outside agriculture. Employees involved in service relationships, such as professionals, and those who have labor contracts, such as factory workers, are also viewed by Goldthorpe as different classes. He makes still finer distinctions based on educational and supervisory criteria. The result is an 11-class model of the stratification system (see Table 7.6).

Goldthorpe's class schema differs from Wright's in several important respects, two of which we mention here. First, the proletariat is a large and undifferentiated mass in Wright's model, amounting to 40% of the United States labor force. In contrast, Goldthorpe says there are three classes of workers, divided by skill and sector. Second, consistent with Marx's theory, Wright says that large employers form a separate class. Goldthorpe, however, groups large employers with senior managers, professionals, administrators, and officials. He believes that the common features of these occupational groups—level of income and authority, political interests, lifestyle, and so forth—transcend the Marxist divide between owners and nonowners.

Research should eventually help decide the merits of Wright's versus Goldthorpe's theories, but the jury is still out on this question (Crompton, 1993). To date, only a few attempts have been made to determine empirically how well the theories perform in comparison with each other. British sociologists have published the most comprehensive analysis of this issue to date (Marshall, Newby, Rose, and Vogler, 1988). They determined statistically whether Wright's or Goldthorpe's class models do a better job of explaining

✦ **TABLE 7.6** ✦
Goldthorpe's Typology of Classes

SOURCE: Goldthorpe (1987 [1980]).

Service Classes	Intermediate Classes	Working Classes
I Higher grade professionals, administrators, and officials; managers in large industrial enterprises; large proprietors	IIIa Routine nonmanual employees, higher grade (administration and commerce)	VI Skilled manual workers
II Lower grade professionals, administrators, and officials; higher grade technicians; managers in small industrial establishments; supervisors of nonmanual employees	IIIb Routine nonmanual employees, lower grade (sales and service)	VIIa Semiskilled and unskilled manual workers not in primary production
	IVa Small proprietors, artisans, etc., with employees	VIIb Agricultural and other workers in primary production
	IVb Small proprietors, artisans, etc., without employees	
	IVc Farmers and smallholders; other self-employed workers in primary production	
	V Lower grade technicians; supervisors of manual workers	

people's social mobility, voting intentions, class consciousness, and so forth. For all these variables, they found Goldthorpe's model superior.

THE NONECONOMIC DIMENSIONS OF CLASS

Prestige and Taste

The theories we have just reviewed focus on occupations, production relations, and employment relations. In short, they all emphasize the *economic* sources of inequality. However, as Weber correctly pointed out, inequality is not based on money alone. It is also based on prestige and power. These important dimensions of inequality have been somewhat neglected in recent writings on stratification. Therefore, in the following section, we discuss the political side of stratification. In this section, we focus on prestige or honor.

Weber, you will recall, said status groups differ from one another in terms of their lifestyles and the honor in which they are held. Here we may add that members of status groups signal their rank by means of material and symbolic culture. That is, they seek to distinguish themselves from others by displays of "taste" in fashion, food, music, literature, manners, travel, and so forth.

The difference between "good taste," "common taste," and "bad taste" is not inherent in cultural objects themselves. Rather, cultural objects that are considered to be in the best taste are generally those that are least accessible.

To explain the connection between taste and accessibility, let us compare Bach's *The Well-Tempered Clavier* with Gershwin's *Rhapsody in Blue*. A survey by French sociologist Pierre Bourdieu showed different social groups prefer these two musical works (Bourdieu, 1984: 17). Well-educated professionals, high school teachers, professors, and artists prefer *The Well-Tempered Clavier*. Less well-educated clerks, secretaries, and junior commercial and administrative executives favor *Rhapsody in Blue*. Why? The two works are certainly very different types of music. Gershwin evokes the jazzy dynamism of big-city America early in the 20th century, Bach evokes the almost mathematically ordered courtly life of early 18th-century Germany. But one would be hard-pressed to argue that *The Well-Tempered Clavier* is intrinsically *superior* music. Both are great art. Why then do more highly

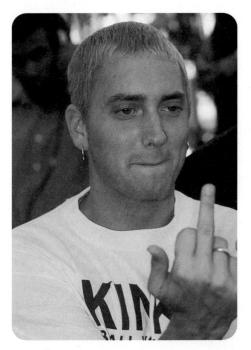

The differences among "good taste," "common taste," and "bad taste" are not inherent in cultural objects themselves. Rather, cultural objects that are considered to be in the best taste are generally those that are least accessible. Left: Eminem. Right: Johann Sebastian Bach.

educated people prefer *The Well-Tempered Clavier* to *Rhapsody in Blue?* Because, according to Bourdieu, during their education they acquire specific cultural tastes associated with their social position. These tastes help to distinguish them from people in other social positions. Many of them come to regard lovers of Gershwin condescendingly, just as many lovers of Gershwin come to think of Bach enthusiasts as snobs. These distancing attitudes help the two status groups remain separate.

Bach is known for such musical characteristics as counterpoint (playing two or more melodies simultaneously) and the fugue (in which instruments repeat the same melody with slight variations). His music is complex, and to really appreciate it one may require some formal instruction. Many other elements of "high culture," such as opera and abstract art, are similarly inaccessible to most people because fully understanding them requires special education.

However, it is not just education that makes some cultural objects less accessible than others. Purely financial considerations also enter the picture. A Mercedes costs four times more than a Ford, and a winter ski trip to Aspen can cost four times more than a week in a modest motel near the beach in Fort Lauderdale. Of course, one can get from point A to point B quite comfortably in a Ford and have a perfectly enjoyable winter vacation in south Florida. Still, most people would prefer the Mercedes and Aspen at least partly because they signal higher status. Access to tasteful cultural objects, then, is as much a matter of cost as education.

Often, rich people engage in conspicuous displays of consumption, waste, and leisure not because they are necessary, useful, or pleasurable but simply to impress their peers and inferiors (Veblen, 1899). This is evident if we consider how clothing acts as a sort of language that signals one's status to others (Lurie, 1981).

For thousands of years, certain clothing styles have indicated rank. In ancient Egypt, only people in high positions were allowed to wear sandals. The ancient Greeks and Romans passed laws controlling the type, number, and color of garments one could wear and the type of embroidery with which they could be trimmed. In medieval Europe, too, various aspects of dress were regulated to ensure that certain styles were specific to certain groups.

European laws governing the dress styles of different groups fell into disuse after about 1700. That is because a new method of control emerged as Europe became wealthier. From the 18th century on, the *cost* of clothing came to designate a person's rank. Expensive materials, styles that were difficult to care for, heavy jewelry, and superfluous trimmings became all the rage. It was not for comfort or utility that rich people wore elaborate powdered wigs, heavy damasked satins, the furs of rare animals, diamond tiaras, and patterned brocades and velvets. Such raiment was often hot, stiff, heavy, and itchy. One could scarcely move in many of these getups. And that was just their point—to prove not only that the wearer could afford enormous sums for handmade finery, but also that he or she didn't have to work to pay for them.

Today, we have different ways of using clothes to signal status. For instance, designer labels loudly proclaim the dollar value of garments. Another example: A great variety and quantity of clothing are required to maintain appearances. Thus, the well-to-do athletic type may have many different and expensive outfits that are "required" for jogging, hiking, cycling, aerobics, golf, tennis, and so forth. In fact, many people who really can't afford to obey the rules of conspicuous consumption, waste, and leisure feel compelled to do so anyway. As a result, they go into debt to maintain their wardrobes. Doing so helps them maintain prestige in the eyes of associates and strangers alike, even if their economic standing secretly falters.

The Role of Politics and the Plight of the Poor

Politics is a second noneconomic dimension of stratification that has received insufficient attention in recent work on inequality. Yet political life has a profound impact on the distribution of opportunities and rewards in society. Politics can reshape the class structure by changing laws governing people's right to own property. Less radically, politics can change the stratification system by entitling people to various welfare benefits and by re-distributing income through tax policies. We discuss the social bases of politics in detail in Chapter 11 ("Politics"). Here we illustrate the effect of politics on inequality by consider-ing how government policy has affected the plight of America's poor (see below for the standard definition of poverty).

John Lie recalls vividly the first time he saw real poverty. "It was when I was seven-teen," he says. "Of course, I must have seen poor people before. But I became conscious of poverty only after visiting New York City. Beneath the splendor of its famous skyscrapers—the Empire State Building, the Chrysler Building, Rockefeller Center—I noticed homeless people. I was shocked to see the most miserable-looking people I could imagine in what is probably the richest city in the world. I didn't know what to make of it, but I do re-member that Alan and Bill, the two friends I was traveling with, had completely different reactions."

"Alan," recalls John, "was outraged. He thought that something should be done to help the homeless right away. Alan didn't really have the vocabulary to express his thoughts clearly—after all, we had just graduated from high school. He blurted out something about how our economic and political systems are 'all screwed up.' In contrast, Bill blamed the poor people themselves for their condition. I remember him repeating that 'we live in a free society and we need to be responsible for ourselves.' The homeless just need to go find work, he said. The three of us weren't in the habit of talking about social issues, so I was surprised that two of my best friends disagreed so strongly."

Alan's and Bill's reactions are simplified versions of the two main currents of Amer-ican opinion on the subject of poverty. They correspond roughly to the Democratic and Re-publican positions. Broadly speaking, most Democrats want the government to play an important role in helping to solve the problem of poverty. Most Republicans want to re-duce government involvement with the poor so people can solve the problem themselves. At various times, each of these approaches to poverty has dominated public policy.

We can see the effect of government policy on poverty by examining fluctuations in the **poverty rate** over time. The poverty rate is the percentage of Americans who fall below the "poverty threshold." To establish the poverty threshold, the United States Department of Agriculture first determines the cost of an economy food budget. The poverty threshold is then set at three times that budget. It is adjusted for the number of peo-ple in the household, the annual inflation rate, and whether individual adult householders are under 65 years of age. In 1999, the poverty threshold for individual adult household-ers under the age of 65 was $8,667 per year. For a family of four with two children under the age of 18, the poverty threshold was $16,895. Figure 7.7 shows the percentage of Americans who lived below the poverty threshold from 1959–98. Between the late 1950s and the late 1970s, the poverty rate dropped dramatically from about 22% to around 11%. Then, between 1980 and 1984, it jumped to about 15%, fluctuating in the 13–15% range after that. In 1998, there were 34.5 million Americans living in poverty, 12.7% of the population.

Fluctuations in the poverty rate are related to political events. We can identify three policy initiatives that have been directed at the problem of poverty. The first dates from the mid-1930s. During the Great Depression (1929–39), a third of Americans were unem-ployed and many of the people lucky enough to have jobs were barely able to make ends meet. Remarkably, an estimated 68% of Americans were poor in 1940 (O'Hare, 1996: 13). In the middle of the Depression, Franklin Roosevelt was elected President. In response to the suffering of the American people and the large, violent strikes of the era, he introduced such programs as Social Security, Unemployment Insurance, and Aid to Families with De-pendent Children (AFDC) (see Chapter 17, "Collective Action and Social Movements"). For the first time, the federal government took responsibility for providing basic suste-nance to citizens who were unable to do so themselves. Due to Roosevelt's "New Deal" policies and rapidly increasing prosperity in the decades after World War II, the poverty rate fell dramatically, reaching 19% in 1964.

The second antipoverty initiative dates from the mid-1960s. In 1964, President Lyndon Johnson declared a "War on Poverty." In part, this initiative was a response to a

Dorothea Lange. "Migrant Mother, Nipomo California, 1936." During the Great Depression (1929–39), a third of Americans were unemployed and many of the people lucky enough to have jobs were barely able to make ends meet.

✦ **FIGURE 7.7** ✦
Poverty Rate, Individuals, United States, 1959-1998 (Percent)

SOURCE: United States Bureau of the Census (1999f).

[Figure 7.7: Line graph showing Poverty rate (y-axis, 10 to 25) vs. Year (x-axis, 59 to 99). The line starts near 22 in 1959, declines sharply to about 11 by the early 1970s, with annotations "President Johnson declares 'War on Poverty'" (arrow pointing near 1964 at ~20) and "Ronald Reagan elected President" (arrow pointing near 1980 at ~13), then rises to about 15 in the early 1980s and fluctuates between 13 and 15 through 1998.]

new wave of social protest. Millions of southern blacks who migrated to northern and western cities in the 1940s and 1950s were unable to find jobs. In some census tracts in Detroit, Chicago, Baltimore, and Los Angeles, black unemployment ranged from 26–41% in 1960. Suffering extreme hardship, many black Americans demanded at least enough money from the government to allow them to subsist. Some of them helped to organize the National Welfare Rights Organization to put pressure on the government to give more poor relief. Others took to the streets as race riots rocked the nation in the mid-1960s. President Johnson soon broadened access to AFDC and other welfare programs (Piven and Cloward, 1993 [1971]; 1977: 264-361). As a result, in 1973 the poverty rate dropped to 11.1%, the lowest it's ever been in this country.

Finally, the third initiative aimed at the poverty problem dates from 1980. The mood of the country shifted by the time President Ronald Reagan took power that year. Reagan assumed office after the social activism and rioting of the 1960s and 1970s had died down. He was elected in part by voters born in the 1950s and 1960s. These so-called "baby boomers" expected their standard of living to increase as quickly as that of their parents. Many of them were deeply disappointed when things didn't work out that way. Real household income (earnings minus inflation) remained flat in the 1970s and 1980s. Even that discouraging performance was achieved only thanks to the mass entry of women into the paid labor force. Reagan explained this state of affairs as the result of too much government. In his view, big government inhibits growth. In contrast, cutting government services and the taxes that fund those services stimulates economic growth. For example, Reagan argued that welfare causes long-term dependency on government handouts. This, he claimed, worsens the problem of poverty rather than solving it. As Reagan was fond of saying, "We fought the War on Poverty and poverty won" (quoted in Rank, 1994: 7). The appropriate solution, in Reagan's view, was to reduce poor relief. Cut welfare programs, he said, and welfare recipients will be forced to work. Cut taxes and taxpayers will spend more, thus creating jobs.

Reagan's message fell on receptive ears. The War on Poverty was turned into what one sociologist called a "war against the poor" (Gans, 1995). Because a large proportion of welfare recipients were African and Hispanic Americans, it has been argued that the war against the poor was fed by racist sentiment (Quadagno, 1994). The invidious stereotype of a young unmarried black woman having a baby in order to collect a bigger welfare check became common. The AFDC budget, expenditures for employee training, and many other government programs were cut sharply. Poverty rates rose. The number of homeless people jumped from 125,000 in 1980 to 402,000 by 1987–88 (Jencks, 1994). One of the main reasons for this increase was the erosion of government support for public housing (Rossi, 1989; Liebow, 1993).[5]

Research conducted in the 1990s showed that many of the beliefs underlying the war against the poor are inaccurate. In particular:

✦ *Myth 1: The overwhelming majority of poor people are African- or Hispanic-American single mothers with children.* While about a quarter of African and Hispanic Americans were poor in 1998, fully 46% of the poor were non-Hispanic whites (see Figure 7.8). In 1996, female-headed families represented just 38% of the poor. Another 34% lived in married-couple families, 22% lived alone or with

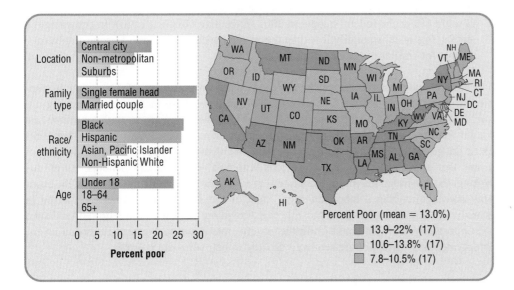

✦ **FIGURE 7.8** ✦
Poverty in the United States, 1997–98

SOURCE: United States Bureau of the Census (1999d).

---

[5]Just as the Reagan administration was cutting welfare, it lowered the top personal tax bracket. This substantially increased the amount of disposable income in the hands of the wealthiest Americans (Phillips, 1990). In this way, politics helped to increase the level of inequality in American society.

nonrelatives, and 6% lived in male-headed families with no wife present (O'Hare, 1996: 11).

◆ *Myth 2: People are poor because they don't want to work.* Over 40% of the poor *did* work in 1998, at least part time. Moreover, many of the poor were not of working age. Over half were either under 18 or over 65. Only 12% of the poor were between 18 and 65 and did not work at all (United States Bureau of the Census, 1999g: vi). Nearly all these people were unable to work for reasons of health or disability—or because they were single mothers who had to stay at home to care for their children due to the unavailability of affordable child care. The following comment of a 32-year-old never-married mother of two is typical of welfare recipients' attitude to work: "I feel better about myself when I'm working than when I'm not. Even if I had a job and every penny went to living from pay day to pay day, it doesn't bother me because I feel like a better person because I am going to work" (quoted in Rank, 1994: 111).

◆ *Myth 3: Poor people are trapped in poverty.* In fact, the poverty population is dynamic. People are always struggling to move out of poverty. They often succeed, at least for a time. Only about 12% of the poor remain poor 5 or more years in a row (O'Hare, 1996: 11).

◆ *Myth 4: Welfare encourages married women with children to divorce so they can collect welfare, and it encourages single women on welfare to have more children.* Because some welfare payments increase with the number of children in the family, some people believe that welfare mothers have more children to get more money from the government. In fact, women on welfare have a lower birthrate than women in the general population (Rank, 1994). Moreover, welfare payments are very low and recipients therefore suffer severe economic hardship. This is hardly an incentive to go on welfare. In the words of a 51-year-old divorced welfare mother: "I can't see anybody that would ever settle for something like this just for the mere fact of getting a free ride, because it's not worth it" (quoted in Rank, 1994: 168).[6] As for the argument that people with children often divorce to collect welfare, the most rigorous study of welfare recipients in the United States to date found that welfare programs "have little effect on the likelihood of marriage and divorce" (Rank, 1994: 169).

◆ *Myth 5: Welfare is a strain on the federal budget.* "Means-tested" welfare programs require that recipients meet an income test to qualify. Such programs accounted for a mere 6% of the federal budget in 2001 (Executive Office…, 2000). In percentage terms, this is substantially less than the amount spent by Western European governments for similar programs.

We conclude that sociological evidence does not support many of the arguments people use to justify current welfare reforms. Nor are current welfare reforms enjoying unqualified success in reducing poverty (see Box 7.2). Western Europe, with a poverty rate roughly half that of ours, seems to be having rather more success in this regard. That is because the governments of Western Europe have established job-training and child care programs that allow poor people to take jobs with livable wages and benefits. As of this writing, the American unemployment rate is at a 30-year low and many sectors of the economy are experiencing a labor shortage. Most of the 34.5 million Americans living in poverty are eager to work. It is possible that providing them with the required skills and the opportunity to work could bring the poverty rate down to Western European levels. Simply denying them welfare seems to do little to help alleviate poverty.[7]

---

[6]Incidentally, a study published by the Federal Reserve Bank of Cleveland found that eliminating the new-birth increment would save only 3% of the AFDC budget (Powers, 1994).

[7]It is unclear whether denying welfare even helps the taxpayer. That is because, to our knowledge, nobody has compared the direct tax saving gained from cutting welfare to the indirect cost of dealing with the consequences of widespread poverty, such as bigger Medicaid bills, larger budgets for police and prison services, and so forth.

BOX 7.2
IT'S YOUR CHOICE

REDESIGNING WELFARE

One of the most important experiments in American history is taking place today. Wisconsin is at the forefront of the experiment. In 1986, the state began testing pilot programs aimed at reducing AFDC dependency and increasing the economic self-sufficiency of families. Ten years later, President Clinton signed a federal welfare law replacing AFDC with a program of time limits and work requirements. Then, in 1997, Wisconsin introduced a tough new plan called W-2. It requires nearly all welfare recipients to work. It also imposes a strict 5-year limit on how long one can receive welfare assistance. Especially in Wisconsin, the age of welfare has passed. The age of "workfare" has dawned. Nationally, the welfare case-

load dropped 47% between 1994 and 1999. In Wisconsin, it dropped 91%.

Pointing to the drop in the welfare caseload, some observers consider the Wisconsin experiment a brilliant success. Closer inspection reveals mixed results (Newman, 1999; Massing, 1999). Roughly a third of former AFDC recipients in Wisconsin are working and earning $9 or $10 an hour, well above minimum wage. Another third are working at menial jobs close to minimum wage. The bottom third have few if any skills. Many of them suffer psychological, drug, or alcohol problems. For the most part, they have not been able to find work and many of them cannot pay rent. They live in Wisconsin's shelters and rely on handouts for food. The city's shelters have become so crowded that the Red Cross runs overflow sites in church basements during the winter. In 1999, the state of Wisconsin published the results of a survey of mothers who had left the welfare system in the first quarter of 1998. Sixty-nine percent said they were just barely making do. Thirty-three percent said they couldn't afford child care. Seventy-five percent of those who found jobs lost them within 9 months. Only 16% enjoyed earn-

ings above the poverty threshold. The proportion of poor children living with two married parents increased in Wisconsin and nationally between 1995 and 2000, due in part to welfare-to-work laws that encourage poor women to marry for the sake of greater financial stability. However, most of these arrangements appear to be high-conflict, unstable families that are no better for the children than single-parent families (Harden, 2001).

The mixed results of the Wisconsin experiment force us to reconsider welfare policy. In 1998, 59% of poor Americans did not work (United States Bureau of the Census, 1999d). Some of these people are too ill or disabled to take a job. Others are women with at least one young child and no male adult in the household. They typically lack the skills and child care services that would help them get a job. Without more child care and job training, it is questionable whether workfare can eliminate poverty in America. The challenge Americans face in the 21st century is to develop innovative programs that can prevent welfare dependency without punishing the powerless and the destitute. How we do so is your choice.

Politics and the Perception of Class Inequality

We expect you have had some strong reactions to our review of sociological theories and research on social stratification. You may therefore find it worthwhile to reflect more systematically on your own attitudes to social inequality. To start with, do you consider the family in which you grew up to have been lower class, working class, middle class, or upper class? Do you think the gaps between classes in American society are big, moderate, or small? How strongly do you agree or disagree with the view that big gaps between classes are needed to motivate people to work hard and maintain national prosperity? How strongly do you agree or disagree with the view that inequality persists because it benefits the rich and the powerful? How strongly do you agree or disagree with the view that inequality persists because ordinary people don't join together to get rid of it? Answering these questions will help you clarify the way you perceive and evaluate the American class structure and your place in it. If you take note of your answers, you can compare them with the responses of representative samples of Americans, which we review below.

Surveys show that few Americans have trouble placing themselves in the class structure when asked to do so. The General Social Survey has been asking Americans annually since 1972 whether they consider themselves "lower class," "working class," "middle class," or "upper class." Of the more than 36,000 respondents over the years, a mere 74 (0.2%) said either that they don't know which class they are in or that they are not members of any class. Just over 2% said they are upper class, and just over 5% said they are lower class. About 46% said they are working class, and about the same percentage said they are middle class. These percentages changed little from 1972 to 1998 (National Opinion Research Center, 1999; see also Jackman and Jackman, 1983; Vanneman and Cannon, 1987).

If Americans see the stratification system as divided into classes, they also know that the gaps between classes are relatively large. For instance, one study compared respondents in New Haven, Connecticut, and London, England. It found that Americans correctly perceive more inequality in their society than the British do in theirs (Bell and Robinson, 1980; Robinson and Bell, 1978).

Do we think these big gaps between classes are needed to motivate people to work hard, thus increasing their own wealth and the wealth of the nation? Some Americans think so, but most don't. A survey conducted in 18 countries asked more than 22,000 respondents (including nearly 1,200 Americans) if large differences in income are necessary for national prosperity. Americans were among the most likely to *disagree* with that view (Pammett, 1997: 77).

So we know we live in a class-divided society. We also tend to think deep class divisions are not necessary for national prosperity. Why then do we think inequality continues to exist? The 18-nation survey cited above sheds light on this issue. One of the survey questions asked respondents how strongly they agree or disagree with the view that "inequality continues because it benefits the rich and powerful." Most Americans agreed with that statement. Only 23% disagreed with it in any way. Another question asked respondents how strongly they agree or disagree with the view that "inequality continues because ordinary people don't join together to get rid of it." Again, most Americans agreed. Only 30% disagreed in any way (Pammett, 1997: 77–8).

Despite widespread awareness of inequality and considerable dissatisfaction with it, most Americans are opposed to the government playing an active role in reducing inequality. Most of us don't want government to provide citizens with a basic income. We tend to oppose government job-creation programs. We even resist the idea that government should reduce income differences through taxation (Pammett, 1997: 81). Most Americans remain individualistic and self-reliant. On the whole, we persist in the belief that opportunities for mobility are abundant and that it is up to the individual to make something of those opportunities by means of talent and effort (Kluegel and Smith, 1986).

Significantly, however, all the attitudes summarized above vary by class position. For example, discontent with the level of inequality in American society is stronger at the bottom of the stratification system than at the top. The belief that American society is full of opportunities for upward mobility is stronger at the top of the class hierarchy than at the bottom. One finds considerably less opposition to the idea that government should reduce inequality as one moves down the stratification system. This permits us to conclude that, if Americans allow inequality to persist, it is because the *balance* of attitudes—and of power—favors continuity over change. We take up this important theme again in Chapter 11, where we discuss the social roots of politics.

SUMMARY

1. For the past quarter of a century, income inequality has been increasing in the United States.

2. Of all the highly industrialized countries, income inequality is highest in the United States.

3. Inequality increases as societies develop from the foraging to the early industrial stage. With increased industrialization, inequality declines. Inequality then increases in the early stages of postindustrialism.

4. Marx's theory of stratification distinguishes between classes on the basis of their role in the productive process. It predicts inevitable conflict between bourgeoisie and proletariat and the birth of a communist system.

5. Weber distinguished between classes on the basis of their "market relations." His model of stratification included four main classes. He argued that class consciousness may develop under some circumstances but is by no means inevitable. Weber also emphasized prestige and power as important noneconomic sources of inequality.

6. Davis and Moore's functional theory of stratification argues that (a) some jobs are more important than others, (b) people have to make sacrifices to train for important jobs, and (c) inequality is required to motivate people to undergo these sacrifices. In this sense, stratification is "functional."

7. Blau and Duncan viewed the stratification system as a ladder with hundreds of occupational ranks. Rank is determined by the income and prestige associated with each occupation. On the basis of their studies, they concluded that the United States enjoys an achievement-based stratification system.

However, many sociologists subsequently concluded that being a member of certain social categories limits one's opportunities for success. In this sense, social structure shapes the distribution of inequality.

8. Wright updated Marx's class schema by distinguishing three classes of property owners (based on capitalization and number of employees) and nine classes of nonowners of property (based on skill levels and organizational assets).

9. Goldthorpe revised Weber's class schema by distinguishing classes on the basis of their "employment relations" and then making finer distinctions on the basis of economic sector, skill, and so forth.

10. People often engage in conspicuous consumption, waste, and leisure to signal their position in the social hierarchy.

11. Politics often influences the shape of stratification systems by changing the distribution of income, welfare entitlements, and property rights.

12. Most Americans are aware of the existence of the class system and their place in it. They believe that large inequalities are not necessary to achieve national prosperity. Most Americans also believe that inequality persists because it serves the interests of the most advantaged members of society and because the disadvantaged don't join together to change things. However, most Americans disapprove of government intervention to lower the level of inequality.

GLOSSARY

An **achievement**-based stratification system is one in which the allocation of rank depends on a person's accomplishments.

Apartheid was a caste system based on race that existed in South Africa from 1948 until 1992. It consigned the large black majority to menial jobs, prevented marriage between blacks and whites, and erected separate public facilities for members of the two races. Asians and people of "mixed race" enjoyed privileges between these two extremes.

An **ascription**-based stratification system is one in which the allocation of rank depends on the features a person is born with.

The **bourgeoisie** are owners of the means of production, including factories, tools, and land. They do not do any physical labor. Their income derives from profits.

A **caste** system is an almost pure ascription-based stratification system in which occupation and marriage partners are assigned on the basis of caste membership.

Class in Marx's sense of the term is determined by one's relationship to the means of production. In Weber's usage, class is determined by one's "market situation." Wright distinguishes classes on the basis of relationship to the means of production, amount of property owned, organizational assets, and skill. For Goldthorpe, classes are determined mainly by one's "employment relations."

Class consciousness refers to being aware of membership in a class.

Cross-national variations in internal stratification are differences between countries in their stratification systems.

Feudalism was a legal arrangement in preindustrial Europe that bound peasants to the land and obliged them to give their landlords a set part of the harvest. In exchange, landlords were required to protect peasants from marauders and open their storehouses and feed the peasants if crops failed.

The **functional theory of stratification** argues that (a) some jobs are more important than others, (b) people have to make sacrifices to train for important jobs, and (c) inequality is required to motivate people to undergo these sacrifices.

The **Gini index** is a measure of income inequality. Its value ranges from zero (which means that every household earns exactly the same amount of money) to one (which means that all income is earned by a single household).

Global inequality refers to differences in the economic ranking of countries.

Intergenerational mobility is social mobility that occurs between generations.

Intragenerational mobility is social mobility that occurs within a single generation.

Parties, in Weber's usage, are organizations that seek to impose their will on others.

The **petty bourgeoisie,** in Marx's usage, is the class of small-scale capitalists who own means of production but employ only a few workers or none at all, forcing them to do physical work themselves.

The **poverty rate** is the percentage of people living below the poverty threshold, which is three times the minimum food budget established by the United States Department of Agriculture.

The **proletariat,** in Marx's usage, is the working class. Members of the proletariat do physical labor but do not own means of production. They are thus in a position to earn wages.

Social stratification refers to the way society is organized in layers or strata.

Blau and Duncan's **socioeconomic index of occupational status (SEI)** combines, for each occupation, average earnings and years of education of men employed full-time in the occupation.

Socioeconomic status (SES) combines income, education, and occupational prestige data in a single index of one's position in the socioeconomic hierarchy.

Status groups differ from one another in terns of the prestige or social honor they enjoy, and also in terms of their style of life.

Structural mobility refers to the social mobility that results from changes in the distribution of occupations.

Vertical social mobility refers to movement up or down the stratification system.

QUESTIONS TO CONSIDER

1. How do you think the American and global stratification systems will change over the next 10 years? Over the next 25 years? Why do you think these changes will occur?

2. Why do you think most Americans oppose more government intervention to reduce the level of inequality in society? In answering your question, think about the advantages that in-

equality brings to many people and the resources at their disposal for maintaining inequality.

3. Compare the number and quality of public facilities such as playgrounds and libraries in various parts of your community. How is the distribution of public facilities related to the socioeconomic status of neighborhoods? Why does this relationship exist?

WEB RESOURCES

Companion Web Site for This Book

http://sociology.wadsworth.com

Begin by clicking on the Student Resources section of the Web site. Choose "Introduction to Sociology" and finally the Brym and Lie book cover. Next, select the chapter you are currently studying from the pull-down menu. From the Student Resources page you will have easy access to InfoTrac College Edition®, MicroCase Online exercises, additional Web links, and many other resources to aid you in your study of sociology, including practice tests for each chapter.

Infotrac Search Terms

These search terms are provided to assist you in beginning to conduct research on this topic by visiting http://www.infotraccollege.com/wadsworth.

Class	**Poverty**
Class consciousness	**Social mobility**
Global inequality	

Recommended Web Sites

For information on the 400 richest people in America, visit the annual compilation of *Forbes* business magazine at http://www.forbes.com/tool/toolbox/rich400.

An excellent analysis of Americans' real earnings from the 1870s to the 1990s can be found at http://www.panix.com/~dhenwood/Stats_earns.html.

For recent government data and analysis of poverty in the United States, see http://www.census.gov/prod/99pubs/p60-207.pdf.

For United Nations data on global inequality, visit http://www.undp.org /hdro/98hdi1.htm.

SUGGESTED READINGS

Richard Breen and David B. Rottman. *Class Stratification: A Comparative Perspective* (New York: Harvester Wehatsheaf, 1995). A concise and up-to-date overview of major issues in the field.

Steve Brouwer. *Sharing the Pie: A Citizen's Guide to Wealth and Power and America* (New York: Holt, 1998). An accessible guide to income and wealth inequality in the contemporary United States.

David B. Grusky, ed. *Social Stratification: Class, Race, and Gender in Sociological Perspective*. (Boulder, CO: Westview, 1994). An anthology of key articles in the field of social stratification.

IN THIS CHAPTER, YOU WILL LEARN THAT:

✦ Race and ethnicity are socially constructed ideas. We use them to distinguish people based on perceived physical or cultural differences, with profound consequences for their lives.

✦ Racial and ethnic labels and identities change over time and place. Relations between racial and ethnic groups help to shape these labels and identities.

✦ In the United States, racial and ethnic groups are blending over time. However, this tendency is weaker among members of highly disadvantaged groups, such as African Americans and Native Americans.

✦ Identifying with a racial or ethnic group can be economically, politically, and emotionally advantageous.

✦ High levels of racial and ethnic inequality are likely to persist in the United States for the foreseeable future.

DEFINING RACE AND ETHNICITY

The Great Brain Robbery

One hundred and fifty years ago, Dr. Samuel George Morton of Philadelphia was the most distinguished scientist in the United States. When he died in 1851, the *New York Tribune* wrote that "probably no scientific man in America enjoyed a higher reputation among scholars throughout the world, than Dr. Morton" (quoted in Gould, 1996 [1981]: 81).

Among other things, Morton collected and measured human skulls. The skulls came from various times and places. Their original occupants were members of different races. Morton believed he could show that the bigger your brain, the smarter you were. He packed BB-sized shot into a skull until it was full. Next he poured the shot from the skull into a graduated cylinder. He then recorded the volume of shot in the cylinder. Finally, he noted the race of the person from whom each skull came. This, he thought, allowed him to draw conclusions about the average brain size of different races.

As he expected, Morton found that the races ranking highest in the social hierarchy had the biggest brains, while those ranking lowest had the smallest brains. He claimed that the people with the biggest brains were whites of European origin. Next were Asians. Then came Native Americans. The people at the bottom of the social hierarchy—and those with the smallest brains—were African Americans.

Morton's research had profound sociological implications, for he claimed to show that the system of social inequality in the United States and throughout the world had natural, biological roots. If, on average, members of some racial groups are rich and others poor, some highly educated and others illiterate, some powerful and others powerless, that was, said Morton, due to differences in brain size and mental capacity. Moreover, because he used science to show that Native Americans and African Americans *naturally* rest at the bottom of the social hierarchy, his ideas were used to justify two of the most oppressive forms of domination and injustice: colonization and slavery.

Despite claims to scientific objectivity, not a shred of evidence supported Morton's ideas. For example, in one of his three main studies, Morton measured the capacity of skulls robbed from Egyptian tombs. He found the average volume of black people's skulls was 4 cubic inches smaller than the average volume of white people's skulls. This seemed to prove his case. Today, however, we know that three main issues compromise his findings:

◆ Morton claimed to be able to distinguish the skulls of white and black people by the shapes of the skulls. However, even today archeologists cannot precisely determine race by skull shape. As a result, it is unclear whether the skulls Morton identified as "Caucasian" belonged to white people and those identified as "Negroid" and "Negro" belonged to black people.

◆ Morton's skulls formed a small, unrepresentative sample. Morton based his conclusions on only 72 specimens. This is a very small number on which to base any generalization. Moreover, those 72 skulls are not representative of the skulls of white and black people in ancient Egypt or any other time and place. They are just the 72 skulls that Morton happened to have access to. For all we know, they are highly unusual.

◆ Even if we ignore these first two problems, 71% of the skulls Morton identified as "Negroid" or "Negro" were women's, compared to only 48% of the skulls he identified as "Caucasian." Yet women's bodies are on average smaller than men's bodies. To make a fair comparison, Morton would have had to make sure that the sex composition of the white and black skulls was identical. He did not. Instead, he biased his findings in favor of finding larger white skulls. When we compare Morton's black and white female skulls, the white skulls are only 2 cubic inches bigger. When we compare his black and white male skulls, the white skulls are 1 cubic inch *smaller* (Gould, 1996 [1981]: 84, 91, 92).

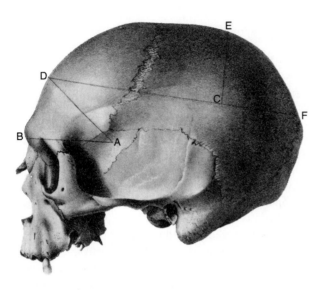

In the 19th century, brain size was falsely held to be one of the main indicators of intellectual capacity. Average brain size was incorrectly said to vary by race. Researchers who were eager to prove the existence of such correlations are now widely regarded as practitioners of a racist quasi-science.

Scientifically speaking, Morton's findings are meaningless. Yet they were influential for a long time. Some people still believe them. For example, less than 40 years ago, the author of an article about race in the *Encyclopedia Britannica,* the world's most authoritative general reference source, wrote that blacks have "a rather small brain in relation to their size" (Buxton, 1963: 864A). That claim was repeated in a controversial book published by a leading American press in the mid-1990s (Rushton, 1995). Yet there is no more evidence today than there was in 1850 that whites have bigger brains than blacks.

Race, Biology, and Society

Biological arguments about racial differences have grown more sophisticated over time. However, their scientific basis is just as shaky now as it always was.

In medieval Europe, some aristocrats saw blue veins underneath their pale skin. But they couldn't see blue veins underneath the peasants' suntanned skin. They concluded the two groups must be racially distinct. The aristocrats called themselves "blue bloods." They ignored the fact that the color of blood from an aristocrat's wound was just as red as the blood from a peasant's wound.

About 80 years ago, some Americans expressed the belief that racial differences in average IQ scores are based in biology. On average, Jews scored below non-Jews on IQ tests in the 1920s. This was used as an argument against Jewish immigration. "America must be kept American," proclaimed President Calvin Coolidge as he signed the 1924 Immigration Restriction Act (quoted in Gould, 1996 [1981]: 262). More recently, African Americans have on average scored below European Americans on IQ tests. Some people say this justifies cutting budgets for schools in the inner city, where many African Americans live. Why invest good money in inner-city schooling, such people ask, if low IQ scores are rooted in biology and therefore fixed (Herrnstein and Murray, 1994)? However, the people who argued against Jewish immigration and better education for inner-city African Americans ignored two facts. First, Jewish IQ scores rose as Jews moved up the class hierarchy and could afford better education. Second, enriched educational facilities have routinely boosted the intellectual development and academic achievement of inner-city African-American children (Campbell and Ramey, 1994; Frank Porter Graham Child Development Center, 1999; Gould, 1996 [1981]; Hancock, 1994; Steinberg, 1989 [1981]). Much evidence shows that the social environment in which one is raised and educated has a big impact on IQ and other standardized test scores. The evidence that racial differences in IQ scores are biologically based is about as strong as evidence showing that aristocrats have blue blood (Fischer, Hout,

Jankowski, Lucas, Swidler, and Voss, 1996; Maume, Cancio, and Evans, 1996; see Chapter 13, "Religion and Education").[1]

If one cannot reasonably maintain that racial differences in average IQ scores are based in biology, what about differences in singing ability or athletic prowess and crime rates? For example, some people insist that, for genetic reasons, African Americans are better than whites at singing and sports, and more prone to crime. Is there any evidence to support this belief?

At first glance, the supporting evidence might seem strong. Consider sports. Aren't 87% of NBA players and 75% of NFL players black? Don't West-African-descended blacks hold the 200 fastest 100-meter times, all under 10 seconds? Don't North and East Africans regularly win 40% of the top international distance-running honors yet represent only a fraction of 1% of the world's population (Entine, 2000)? While these facts are undeniable, the argument for the genetic basis of black athletic superiority begins to falter once we consider two additional points. First, no gene linked to general athletic superiority has yet been identified. Second, athletes of African descent do not perform unusually well in many sports, such as swimming, hockey, cycling, tennis, gymnastics, and soccer. The idea that people of African descent are in general superior athletes is simply untrue.[2]

Sociologists have identified certain *social* conditions leading to high levels of participation in sports (as well as entertainment and crime). These operate on all groups of people, whatever their race. Specifically, people who face widespread prejudice and discrimination often enter sports, entertainment, and crime in disproportionately large numbers for lack of other ways to improve their social and economic position. For such people, other avenues of upward mobility tend to be blocked. (**Prejudice** is an attitude that judges a person on his or her group's real or imagined characteristics. **Discrimination** is unfair treatment of people due to their group membership.) Thus, in the United States, Irish, Jews, Italians, Puerto Ricans, and African Americans formed successive waves of high-crime groups in the late 19th and 20th centuries. Their crime rates declined only to the degree that their economic and social standing improved (Bell, 1960). Similarly, it was not until the 1950s that prejudice and discrimination against American Jews began to decline appreciably. Until then, Jews played a prominent role in professional sports. For instance, when the New York Knicks played their first game on November 1, 1946, beating the Toronto Huskies 68–66, the starting lineup consisted of Ossie Schechtman, Stan Stutz, Jake Weber, Ralph Kaplowitz, and Leo "Ace" Gottlieb—a nearly all-Jewish squad (National Basketball Association, 2000). Koreans in Japan today are subject to much prejudice and discrimination. They often pursue careers in sports and entertainment. In contrast, Koreans in the United States face less prejudice and discrimination. Few of them become athletes and entertainers. Instead, they are often said to excel in engineering and science. As these examples show, then, social circumstances have a big impact on criminal, athletic, and other forms of behavior.

The idea that people of African descent are genetically superior to whites in athletic ability is the complement of the idea that they are genetically inferior to whites in intellectual ability.[3] Both ideas have the effect of reinforcing black–white inequality. For although there are probably fewer than 10,000 elite professional athletes in the United States, there are many millions of pharmacists, graphics designers, lawyers, systems analysts, police officers, nurses, and people in other interesting occupations that offer steady employment and good pay. By promoting only the Shaquille O'Neals of the world as suitable role models for African-American youth, the idea of "natural" black athletic superiority and intellectual inferiority in effect asks black Americans to bet on a high-risk proposition—that they will make it in professional sports. At the same time, it deflects attention from a

---

[1]Although sociologists commonly dispute a genetic basis of mean intelligence for *races*, evidence suggests that *individual* differences in intelligence are partly genetically transmitted (Bouchard, Lykken, McGue, Segal, and Tellegen, 1990; Lewontin, 1991: 19–37; Scarr and Weinberg, 1978; Schiff and Lewontin, 1986).

[2]It may be that genetic differences lead some groups to have physical characteristics that lend themselves to excellence in *particular* sports, but this is a different argument from the general argument we are criticizing here (Entine, 2000).

[3]The genetic argument also belittles the athletic activity itself by denying the role of training in developing athletic skill.

much safer bet—that they can achieve upward mobility through academic excellence (Doberman, 1997).

An additional problem with the argument that genes determine the specific behaviors of different racial groups is that one cannot neatly distinguish races based on genetic differences. A high level of genetic mixing has taken place between people of various races throughout the world. In the United States, for instance, white male slave owners often raped female African-American slaves. Similarly, many African Americans had children with Native Americans in the 18th and 19th centuries. Today, racial intermarriage is no longer a rarity, and it is becoming more common. About 50% of Asian Americans and 10% of African Americans now marry outside their racial group (Stanfield, 1997). A growing number of Americans are similar to Tiger Woods and Colin Powell. Woods claims he is of "Cablinasian" ancestry—part <u>Ca</u>ucasian, part <u>bl</u>ack, part <u>In</u>dian (Native American), and part <u>Asian</u>. Colin Powell's ancestry includes African, English, Irish, Scottish, Arawak Indian, and Jewish. As these examples illustrate, the difference between "black," "white," "Asian," and so forth is often anything but clear-cut.

Many scholars believe we all belong to one human race, which originated in Africa (Cavalli-Sforza, Menozzi, and Piazza, 1994). They argue that subsequent migration, geographical separation, and inbreeding led to the formation of more or less distinct races. However, particularly in modern times, humanity has experienced so much intermixing that race as a biological category has lost nearly all meaning. Some biologists and social scientists therefore suggest we drop the term "race" from the vocabulary of science (Angier, 2000).

Most sociologists, however, continue to use the term "race." That is because *perceptions* of race continue to affect the lives of most people profoundly. Everything from your wealth to your health is influenced by whether others see you as African American, white, Asian American, Native American, or something else. Race as a *sociological* concept is thus an invaluable analytical tool—if the user remembers that it refers to socially significant physical differences, such as skin color, rather than biological differences that determine behavioral traits.

Said differently, perceptions of racial difference are socially constructed and often arbitrary (Ferrante and Brown, 2001 [1998]). The Irish and the Jews in the United States were regarded as "blacks" by many people a hundred years ago, and today many northern Italians still think of southern Italians from Sicily and Calabria as "blacks" (Gilman, 1991; Ignatiev,1995; Roediger, 1991). During World War II in the United States, some people made arbitrary physical distinctions between Chinese allies and Japanese enemies that

Athletic heroes like Shaquille O'Neal are often held up as role models for African-American youth even though the chance of "making it" as a professional athlete is much less than the chance of getting a college education and succeeding as a professional. The cultural emphasis on African-American sports heroes also has the effect of reinforcing harmful and incorrect racial stereotypes about black athletic prowess and intellectual inferiority.

Colin Powell is of African, English, Irish, Scottish, Arawak Indian, and Jewish ancestry. Like an increasingly large number of Americans, it is difficult to say that he is a member of any particular race or ethnic group.

Chinese   *Japanese*

HOW TO TELL YOUR FRIENDS FROM THE JAPS

Of these four faces of young men (*above*) and middle-aged men (*below*) the two on the left are Chinese, the two on the right Japanese. There is no infallible way of telling them apart, because the same racial strains are mixed in both. Even an anthropologist, with calipers and plenty of time to measure heads, noses, shoulders, hips, is sometimes stumped. A few rules of thumb—not always reliable:

▶ Some Chinese are tall (average: 5 ft. 5 in.). Virtually all Japanese are short (average: 5 ft. 2½ in.).

▶ Japanese are likely to be stockier and broader-hipped than short Chinese.

▶ Japanese—except for wrestlers—are seldom fat; they often dry up and grow lean as they age. The Chinese often put on weight, particularly if they are prosperous (in China, with its frequent famines, being fat is esteemed as a sign of being a solid citizen).

▶ Chinese, not as hairy as Japanese, seldom grow an impressive mustache.

▶ Most Chinese avoid horn-rimmed spectacles.

▶ Although both have the typical epicanthic fold of the upper eyelid (which makes them look almond-eyed), Japanese eyes are usually set closer together.

▶ Those who know them best often rely on facial expression to tell them apart: the Chinese expression is likely to be more placid, kindly, open; the Japanese more positive, dogmatic, arrogant.

In Washington, last week, Correspondent Joseph Chiang made things much easier by pinning on his lapel a large badge reading "Chinese Reporter—NOT Japanese—Please."

▶ Some aristocratic Japanese have thin, aquiline noses, narrow faces and, except for their eyes, look like Caucasians.

▶ Japanese are hesitant, nervous in conversation, laugh loudly at the wrong time.

▶ Japanese walk stiffly erect, hard-heeled. Chinese, more relaxed, have an easy gait, sometimes shuffle.

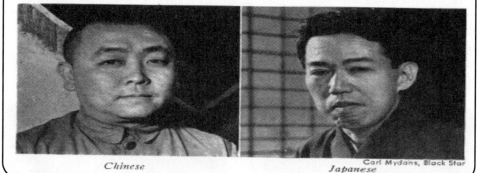

Chinese   *Japanese*   Carl Mydans, Black Star

helped to justify the American policy of placing Japanese Americans in internment camps (see Figure 8.1). These examples show that racial distinctions are social constructs, not biological "givens."

Finally, then, we can define **race** as a social construct used to distinguish people in terms of one or more physical markers, usually with profound effects on their lives. However, this definition raises an interesting question. If race is merely a social construct and not a useful biological term, why are perceptions of physical difference used to distinguish groups of people in the first place? Why, in other words, does race matter? Most sociologists believe race matters because it allows social inequality to be created and maintained. The English who colonized Ireland, the Americans who went to Africa looking for slaves, and the Germans who used the Jews as a scapegoat to explain their deep economic and political troubles after World War I, all set up systems of racial domination.

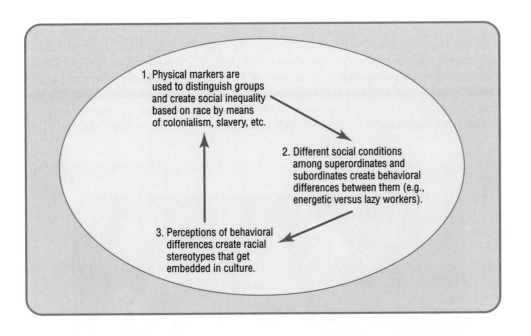

1. Physical markers are used to distinguish groups and create social inequality based on race by means of colonialism, slavery, etc.

2. Different social conditions among superordinates and subordinates create behavioral differences between them (e.g., energetic versus lazy workers).

3. Perceptions of behavioral differences create racial stereotypes that get embedded in culture.

(A **scapegoat** is a disadvantaged person or category of people whom others blame for their own problems.) Once colonialism, slavery, and concentration camps were established, behavioral differences developed between subordinates and superordinates. For example, African-American slaves and Jewish concentration camp inmates, with little motivating them to work hard except the ultimate threat of the master's whip, tended to do only the minimum work necessary to survive. Their masters noticed this and characterized their subordinates as inherently slow and unreliable workers (Collins, 1982: 66–9). In this way, racial stereotypes are born. The stereotypes then embed themselves in literature, popular lore, journalism, and political debate. This reinforces racial inequalities (see Figure 8.2). We thus see that race matters to the degree it helps to create and maintain systems of social inequality.

Ethnicity, Culture, and Social Structure

Race is to biology as ethnicity is to culture. A race is a category of people whose perceived *physical* markers are deemed socially significant. An **ethnic group** is composed of people whose perceived *cultural* markers are deemed socially significant. Ethnic groups differ from one another in terms of language, religion, customs, values, ancestors, and the like. However, just as physical distinctions don't *cause* differences in the behavior of various races, so cultural distinctions are often not by themselves the major source of differences in the behavior of various ethnic groups. In other words, ethnic values and other elements of ethnic culture have less of an effect on the way people behave than we commonly believe. That is because *social-structural* differences typically underlie cultural differences.

An example will help drive home the point. Many ethnic groups may be found in the African-American community (Waters, 2000). By far the largest is composed of descendants of former Southern slaves. Another is composed of Jamaican and other Caribbean or West Indian immigrants and their descendants. People often say that West Indians emphasize hard work, saving, investment, and education. Descendants of Southern black slaves, they claim, lack a culture emphasizing these values. As a result (the argument continues), descendants of Southern black slaves suffer from higher unemployment, lower income, and higher crime rates. As Nathan Glazer and Daniel Patrick Moynihan wrote, "the ethos of the West Indians, in contrast to that of the Southern Negro, emphasized saving, hard work, investment and education" (Glazer and Moynihan, 1963: 35).

To our knowledge, nobody ever conducted a scientific survey measuring the values of these two groups. The contrast sketched above may therefore be false or exaggerated.

However, for the sake of argument, let us accept the accuracy of the value contrast. Is it a sufficient explanation for behavioral differences between the two ethnic groups? It is not, according to sociologist Stephen Steinberg (1989 [1981]: 275–80; see also Kalmijn, 1996). Immigration records show that the first wave of West Indian immigrants, who came to the United States in the 1920s, was 89% literate. That is higher than the literacy rate for the United States population as a whole at the time. More than 40% of the immigrants were skilled workers. Many of them were highly educated professionals. Thus, the first West

BOX 8.1
SOCIOLOGY AT THE MOVIES

Spike Lee and Danny Aiello in *Do the Right Thing*.

DO THE RIGHT THING (1989)

"Dago, Wop, guinea, garlic breath, pizza-singin' spaghetti-bender" is what Mookie, the African-American pizza delivery man, calls his boss, Sal, the Italian-American pizzeria owner in Spike Lee's *Do the Right Thing*.

A European-American police officer shouts at a Puerto Rican American: "You Goya, bean-eating, fifteen-in-the-car, thirty-in-the-apartment, pointy red shoes-wearing, Puerto Rican, c___suckers."

A Puerto Rican American yells at a Korean-American grocer: "Little slanty-eyed, me-no-speak-American, own-every-fruit-and-vegetable-stand-in-New-York, bull-shit, Reverend Sun Myung Moon, Summer 88 Olympic kick-ass boxer, son of a bitch."

Clearly, the Bedford-Stuyvesant neighborhood in Brooklyn, New York, is not a scene of racial and ethnic harmony. But *Do the Right Thing* doesn't pull punches about ethnic and race relations in the United States.

The movie depicts a poor, multiethnic but largely African-American neighborhood populated by numerous memorable characters: the friendly and outgoing "mayor," the militant black activist, the disk jockey who offers a running commentary on the movie, and Mookie the pizza delivery man, played by Spike Lee himself. Although the residents are mainly African American, two businesses are owned by an Italian-American family (Sal's Famous Pizzeria) and a Korean-American family (the grocery store).

Do the Right Thing offers many interpretations of the situation. In fact, one of the great strengths of the movie is that it enables us to see the complexities of each character, understand what is likable and repugnant in each person, and develop an appreciation for just about everyone's point of view. For example, we come to understand why some African-American characters think "white" people (including Korean Americans) are oppressing blacks. We develop respect for Sal's wish to see people get along with one other. We sympathize with the Korean grocer's efforts to provide for his family despite an inhospitable racial and ethnic environment.

In addition to developing multiple points of view, a second strength of the movie is that it presents a fluid view of ethnicity, race, and opinions about these politically charged issues. Thus, we see how people cross boundaries, upsetting seemingly stable ethnic and racial distinctions. For instance, the Korean-American grocer at one point insists that *he* is oppressed and "black." Just about everyone is the object of racist insults and stereotypes. As in real life, what people say at one moment is logically inconsistent with what they say at other moments. Different social situations generate different views of race and ethnicity. Society, not logic, dictates opinion.

As should be apparent, this is not a movie that echoes the hopes and aspirations of the 1960s. Gone, it says, is the rhetoric of racial and ethnic harmony, assimilation, acculturation, and other words and phrases that describe the presumed peaceful coexistence of ethnic and racial groups in America. *Do the Right Thing* also takes a cynical view of the possibility of African-American progress.

Despite its strengths, one must ask whether the movie is entirely realistic. Is its characterization of ethnic and race relations accurate for the United States as a whole, just for poor neighborhoods in large urban areas, only for Bedford-Stuyvesant, or not at all? Are the ethnic and racial conflicts in the neighborhood aggravated by poverty and class divisions? If so, is class rather than race or ethnicity largely to blame for the conflict in the movie and in American society? The movie doesn't answer these questions, but it does raise them in a provocative manner. What are your answers?

Indian immigrants to the United States occupied a much higher class position than descendants of Southern black slaves. The latter were nearly all unskilled workers in the 1920s. If the West Indians exhibited greater interest in education, saving, and so forth, this was undoubtedly related to their class position. Skilled and literate people usually exhibit these traits more than unskilled and illiterate people. The latter often understand only too well that they have few opportunities to pursue higher education and a professional career.

That class was more important than ethnic values in determining economic success is also evident if we compare West Indian immigrants in New York and London in the 1950s and 1960s. The two groups came from the same place at the same time and presumably shared many values. However, the New York West Indians enjoyed more economic success than the London West Indians. Why? Largely because the percentage of white-collar and professional workers was nearly three times higher (27%) among the New Yorkers when they arrived in the United States. The New Yorkers started with a big advantage and capitalized on it.

Most sociologists stress how social-structural conditions rather than values determine the economic success or failure of racial and ethnic groups (Abelmann and Lie, 1995; Brym with Fox, 1989: 103–19; Lieberson, 1980; Portes and Rumbaut, 1990; Steinberg, 1989 [1981]: 82–105, 270–5). For example, people often praise Jews, Koreans, and other economically successful groups for emphasizing education, family, and hard work. People less commonly notice, however, that American immigration policy is highly selective. For the most part, the Jews and Koreans who arrived in the United States were literate, urbanized, and skilled. Some even came with financial assets. They confronted much prejudice and discrimination but far less than that reserved for descendants of Southern blacks. These *social-structural* conditions facilitated Jewish and Korean success. They gave members of these groups a firm basis on which to build and maintain a culture emphasizing education, family, and other middle-class virtues.

We conclude that it is misleading to claim that "[r]ace and ethnicity . . . are quite different, since one is biological and the other is cultural" (Macionis, 1997 [1987]: 321). As we have seen, both race and ethnicity are rooted in social structure, not biology and culture. The biological and cultural aspects of race and ethnicity are secondary to their sociological character. Moreover, the distinction between race and ethnicity is not as simple as the difference between biology and culture. As noted above for the Irish and the Jews, groups once socially defined as races may be later redefined as ethnic, even though they don't change biologically. Social definitions, not biology and not culture, determine whether a group is viewed as a race or an ethnic group. The interesting question from a sociological point of view is why social definitions of race and ethnicity change (see Box 8.1). We now consider that issue.

RACE AND ETHNIC RELATIONS

Labels and Identity

John Lie moved with his family from South Korea to Japan when he was a baby. He moved from Japan to Hawaii when he was 10 years old, and again from Hawaii to the American mainland when he started college. The move to Hawaii and the move to the mainland changed the way John thought of himself in ethnic terms.

In Japan, the Koreans form a minority group. (A **minority group** is a group of people who are socially disadvantaged, though they may be in the numerical majority, like the blacks in South Africa.) Before 1945, when Korea was a colony of Japan, some Koreans were brought to Japan to work as miners and unskilled laborers. The Japanese thought the Koreans who lived there were beneath and outside Japanese society (Lie, 2001). Not surprisingly, then, Korean children in Japan, including John, were often teased and occasionally beaten by their Japanese schoolmates. "The beatings hurt," says John, "but the

Personal Anecdote

The United States is becoming a more ethnically and racially diverse society due to a comparatively high immigration rate and a comparatively low birth rate among non-Hispanic whites. The United States Census Bureau estimates that around 2060 a minority of Americans will be non-Hispanic whites.

psychological trauma resulting from being socially excluded by my classmates hurt more. In fact, although I initially thought I was Japanese like my classmates, my Korean identity was literally beaten into me."

"When my family immigrated to Hawaii, I was sure things would get worse. I expected Americans to be even meaner than the Japanese. (By Americans, I thought only of white European Americans.) Was I surprised when I discovered that most of my schoolmates were not white European Americans, but people of Asian and mixed ancestry! Suddenly I was a member of a numerical majority. I was no longer teased or bullied. In fact, I found that students of Asian and non-European origin often singled out white European Americans (called *haole* in Hawaiian) for abuse. We even had a "beat up *haole* day" in school. Given my own experiences in Japan, I empathized somewhat with the white Americans. But I have to admit that I also felt a great sense of relief and an easing of the psychological trauma associated with being Korean in Japan.

"As the years passed, I finished public school in Hawaii. I then went to college in Massachusetts and got a job as a professor in Illinois. I associated with, and befriended, people from various racial and ethnic groups. My Korean origin became a less and less important factor in the way people treated me. There was simply less prejudice and discrimination against Koreans during my adulthood in the United States than in my early years in Japan. I now think of myself less as Japanese or Korean than as American. Sometimes I even think of myself as a Midwesterner. My ethnic identity has changed over time in response to the significance others have attached to my Korean origin. I now understand what the French philosopher Jean-Paul Sartre meant when he wrote that "the anti-Semite creates the Jew" (Sartre, 1965 [1948]: 43).

The details of John Lie's life are unique. But experiencing a shift in racial or ethnic identity is common. Social contexts, and in particular the nature of one's relations with members of other racial and ethnic groups, shape and continuously reshape one's racial and ethnic identity. Change your social context, and your racial and ethnic self-conception eventually changes too (Miles, 1989; Omi and Winant, 1986).

Consider Italian Americans. Around 1900, Italian immigrants thought of themselves as people who came from a particular town or perhaps a particular province, such as Sicily or Calabria. They did not usually think of themselves as Italians. Italy became a unified country only in 1861. A mere 40 years later, far from all of its citizens identified with their new nationality. In the United States, however, officials and other residents identified the newcomers as "Italians." The designation at first seemed odd to many of the new immigrants. However, over time it stuck. Immigrants from Italy started thinking of themselves as Italian-Americans because others defined them that way. A new ethnic identity was born (Yancey, Ericksen, and Leon, 1976).

As symbolic interactionists emphasize, the development of racial and ethnic labels, and ethnic and racial identities, is typically a process of negotiation. For example, members of a group may have a racial or ethnic identity, but outsiders may impose a new label on them. Group members then reject, accept, or modify the label. The negotiation between outsiders and insiders eventually results in the crystallization of a new, more or less stable ethnic identity. If the social context changes again, the negotiation process begins anew.

Hispanic Americans

You can witness the formation of new racial and ethnic labels and identities in the United States today. Consider, for instance, the terms "Hispanic American" and "Latino" (Darder and Torres, 1998).[4] People scarcely used these terms 30 or 40 years ago. Now they are common. According to the United States Bureau of the Census, more than 31 million Hispanic Americans live in the United States today. Due to continuing robust immigration and high fertility, the Bureau predicts they will number more than 96 million in 2050. They are the fastest-growing ethnic category in the country and will soon be the second biggest, next to non-Hispanic whites (see Figure 8.3).

Yet who is a Hispanic American? New York is home to about 10% of Hispanic Americans. In that state, a Hispanic American is most likely to be of Puerto Rican origin. More than 31% of Hispanic Americans live in California and more than 21% live in Texas. Hispanic Americans make up about a quarter of the population in those two states. In California and Texas, most Hispanic Americans are of Mexican descent and are known as "Chicanos." Florida is home to about 8% of Hispanic Americans. In that state, people of Cuban origin are by far the most numerous Hispanic Americans.

Besides varying degrees of knowledge of the Spanish language, what do members of these groups have in common? One survey shows that most of them do *not* want to be called "Hispanic American." Instead, they prefer being referred to by their national origin, that is, as Cuban Americans, Puerto Rican Americans and Mexican Americans. Many Hispanic Americans born in the United States want to be called simply "Americans" (de la Garza, DiSipio, Garcia, Garcia, and Falcon, 1992).

Members of these three groups also enjoy different cultural traditions, occupy different positions in the class hierarchy, and even vote differently. For example, there are many more middle-class and professional people among Americans of Cuban origin than among members of other major Hispanic groups in the United States. That is because a large wave of middle-class Cubans fled Castro's revolution and arrived in the Miami area in the late 1950s and early 1960s. There they formed an **ethnic enclave.** An ethnic enclave is a geographical concentration of ethnic group members who establish businesses that serve and employ mainly members of the ethnic group and reinvest profits in community businesses and organizations (Portes and Manning, 1991). Many Cubans who arrived in subsequent waves of immigration were poor, but because they could rely on a well-established and prosperous Cuban-American ethnic enclave for jobs and other forms of support, most of them soon achieved middle-class status themselves. The fact that the Cuban immigrants shared strong procapitalist and anticommunist values with the American public also helped their integration into the larger society. Not surprisingly given their background, Cuban Americans are more likely to vote Republican than Americans of Mexican and Puerto Rican origin.

Immigrants from Mexico and Puerto Rico tend to be members of the working class who have not completed high school. Their children typically achieve educational levels similar to that of non-Hispanic whites (Bean and Tienda, 1987). However, they have not

Web Research Projects
The Social Characteristics
of Hispanic Americans

---
[4]"Latino" (male) and "Latina" (female) seem to be the preferred usage in the West and "Hispanic American" in the East. "Latino/a" is the more inclusive term because it encompasses Brazilians of Portuguese origin but we use "Hispanic American" here because that is the term used by the United States Bureau of the Census.

✦ **FIGURE 8.3** ✦
Racial and Ethnic Composition, United States, 2000–2050 (projected)

SOURCE: United States Bureau of the Census (1999e).

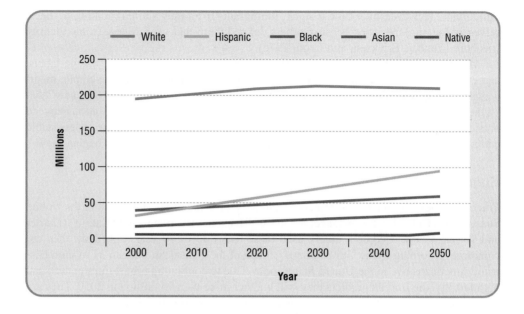

reached the average level of prosperity enjoyed by Cuban Americans due to the lower class origins of the community and its resulting weaker ethnic enclave formation. Chicanos and Puerto Rican Americans are more likely than Cuban Americans to vote Democrat.

So we see that the Hispanic-American community is highly diverse. In fact, not even knowledge of the Spanish language unifies Hispanic Americans. Some people who are commonly viewed as Hispanic American do not speak Spanish. Haitians, for example, speak French or a regional dialect of French. Still other people who are commonly viewed as Hispanic American reject the label. Chief among these are Mayans, indigenous Central Americans whom the Spanish colonized.[5]

Despite this internal diversity, however, the term "Hispanic American" is more widely used than it was 30 or 40 years ago. There are three main reasons for this:

✦ Many Hispanic Americans find it politically useful. Recognizing that power flows from group size and unity, Hispanic Americans have created national organizations to promote the welfare of their entire community. One of these is the National Council of La Raza, a nonprofit organization established in 1968 to "reduce poverty and discrimination, and improve life opportunities for Hispanic Americans" ("The National Council of La Raza," 2000). Thus, for certain purposes, Hispanic Americans from Cuba, Mexico, Puerto Rico, and other places find it convenient to play down what separates them and accentuate what they have in common.

✦ A second reason the term "Hispanic American" is becoming more common is that the government finds it useful for data collection and public policy purposes. Like "Asian American," "African American," and "Native American," "Hispanic American" is a convenient administrative term. For example, by collecting census data on the number of people who identify themselves as Hispanic American, the government is better able to allocate funding for Spanish language instruction in schools, ensure diversity in the workplace, and take other public policy actions that reflect the changing racial and ethnic composition of American society (see Box 8.2).

✦ Finally, "Hispanic American" is an increasingly popular label because non-Hispanic Americans find it convenient. There are about 25 countries in Central and South America and many ethnic divisions within those countries. It would be a tough job

---

[5] Similarly, particularly in the West, many indigenous Americans reject the label "Native American" as official, "white" terminology and proudly call themselves "Indians."

BOX 8.2
IT'S YOUR CHOICE

BILINGUAL EDUCATION

Ron Unz is an opponent of bilingual education. He argues that a "quarter of all the children in California public schools are classified as not knowing English . . . Of the ones who don't know English in any given year, only five or six percent learn English. Since the goal of the system, obviously, should be to make sure that these children learn English, we're talking about a system with an annual failure rate of 95 percent . . . Many of my friends are foreign immigrants. They came here when they were a variety of different ages. All of them agree that little children or even young teenagers can learn another language quickly, though only five percent of these children in California are learning English each year. And that's what I define as failure" (quoted in Public Broadcasting System, 1997).

In response, James Lyons of the National Association for Bilingual Education says: "It is not the case that bilingual education is failing children. There are poor bilingual education programs, just as there are poor programs of every type in our schools today. But bilingual education has made it possible for children to have continuous development in their native language, while they're in the process of learning English, something that doesn't happen overnight, and it's made it possible for children to learn math and science at a rate equal to English-speaking children while they're in the process of acquiring English" (quoted in Public Broadcasting System, 1997).

About 6% of all public school students were enrolled in bilingual education programs in the United States in 1997. That amounts to more than 3 million children. Bilingual education programs cost taxpayers $178 million (Public Broadcasting System, 1997). Is the expenditure worth it? Whatever the expense, is the ideal of bilingual education worth pursuing? The debate between Unz and Lyons touches on some important points concerning these issues. On the one hand, many bilingual education programs are not very effective. Furthermore, English is the main language spoken in the United States and should be taught to non-English speakers to ensure their economic progress; a less-than-fluent speaker of English is bound to do poorly not only in school but also in most workplaces. On the other hand, advocates argue that bilingual education helps nonnative speakers of English adjust to schools and keep up in other subjects. Bilingual education also recognizes the importance of learning a native language, such as Spanish, rather than simply assimilating to the dominant English-speaking culture. Some people argue that when students value their native language skills their self-esteem increases. Later, this improves their economic success.

What do you think? Should we make every public student learn only in English? What would we gain and lose by such a policy? Is it worthwhile keeping bilingual programs for students who wish to preserve their language skills? Should we encourage native language preservation just for large groups, such as Spanish-speaking students, or for small groups as well, such as students of Chinese or Russian origin? Perhaps in this age of globalization, everyone should participate in a bilingual education program. Would that make Americans less ethnocentric and better able to participate in global affairs?

for anyone to keep all those countries and ethnic groups straight in everyday speech. Lumping all Hispanic Americans together makes life easier for the majority group, though lack of sensitivity to a person's specific origins may sometimes be offensive to minority group members.

So we see that "Hispanic American" is a new ethnic label and identity. It did not spring fully formed one day from the culture of the group to which it refers. It was created out of social necessity and is still being socially constructed (Portes and Truelove, 1991). We can say the same about *all* ethnic labels and identities, even those that may seem most fixed and natural, such as "white" (Lieberson, 1991; Lieberson and Waters, 1986; Waters, 1990).

Ethnic and Racial Labels: Choice Versus Imposition

The idea that race and ethnicity are socially constructed does not mean that everyone can always choose their racial or ethnic identity freely. There are wide variations over time and from one society to the next in the degree to which people can exercise such freedom of choice. Moreover, in a given society at a given time, different categories of people are more or less free to choose. To illustrate these variations, we next discuss the way the government of the former Soviet Union imposed ethnicity on the citizens of that country. We then contrast imposed ethnicity in the former Soviet Union with the relative freedom of ethnic choice in the United States. In the United States case, we also underline the social forces that make it easier to choose one's ethnicity than one's race.

State Imposition of Ethnicity in the Soviet Union

Until it formally dissolved in 1991, the Soviet Union was the biggest and one of the most powerful countries in the world (see Chapter 10, "Work and the Economy," and Chapter 11, "Politics"). Stretching over two continents and 13 time zones, it was composed of 15 republics—Russia, Ukraine, Kazakhstan, and so forth—with a combined population larger than that of the United States. In each republic, the largest ethnic group was the so-called "titular" ethnic group of the republic: Russians in Russia, Ukrainians in Ukraine, Kazakhs in Kazakhstan, and so forth.[6] Over 100 minority ethnic groups also lived in the republics. As Figure 8.4 shows, in some republics the combined number of minority ethnic group members was greater than that of the titular ethnic group.

The vast size and ethnic heterogeneity of the Soviet Union required that its leaders develop strategies for preventing the country from falling apart at the seams. One such strategy involved weakening the boundaries between the republics so "a new historical community, the Soviet people" could come into existence (Bromley, 1982 [1977]: 270). The creation of a countrywide educational system and curriculum, the spread of the Russian language, and the establishment of propaganda campaigns trumpeting remarkable national achievements helped to create a sense of unity among many Soviet citizens.

A second strategy promoting national unity involved the creation of a system allowing power and privilege to be shared among ethnic groups. This was accomplished administratively through the "internal passport" system. Beginning in the 1930s, Soviet governments issued identity papers or internal passports to all citizens at the age of 16. The fifth entry in each passport noted the bearer's ethnicity. Adolescents were obliged to adopt the ethnicity of their parents. Only if the parents were of different ethnic backgrounds could a 16-year-old choose the ethnicity of the mother or the father.

The internal passport system enabled officials to apply strict ethnic quotas in recruiting people to institutions of higher education, professional and administrative positions, and political posts. Ethnic quotas were even used to determine where people could reside. Thus, ethnicity became critically important in determining some of the most fundamental aspects of one's life. To ensure the loyalty of the disparate and far-flung republics to the central government, officials granted advantages to members of titular ethnic groups living in their own republics. That is, you enjoyed the best opportunities for educational, occupational, and political advancement if you were a Russian living in Russia, a Ukrainian living in Ukraine, and so forth. If you happened to be a member of a titular ethnic group living outside your republic, you were at a disadvantage in this regard. And if you happened to be a member of a nontitular ethnic group, you were most disadvantaged (Brym with Ryvkina, 1994: 6–16; Karklins, 1986; Zaslavsky and Brym, 1983).

By thus organizing many basic social processes along ethnic lines, the government imposed ethnic labels on its citizens. Despite efforts to create a new "Soviet people," traditional ethnic identities remained strong. In 1991, when the Soviet Union ceased to exist, Russians knew they were Russians and Jews knew they were Jews mainly because the ethnicity entry in their internal passports had circumscribed their opportunities in life for more than 60 years. Only in 1997 did the Russian government finally introduce new internal passports without the notorious fifth entry, thus bringing an end to the era of administratively imposed ethnicity in that country.

Ethnic and Racial Choice in the United States

The situation in the United States is vastly different from that of the Soviet Union before 1991. Americans are freer to choose their ethnic identity than citizens of the Soviet Union were. The people with the most freedom to choose their ethnic or racial identity are white Americans whose ancestors came from Europe more than two generations ago (Waters, 1990). For example, identifying oneself as an Irish American no longer has negative implications, as it did in, say, 1900. Then, in a city like Boston, where a substantial number of Irish immigrants were concentrated, the English-Protestant majority typically regarded

---

[6]In Soviet terminology, these groups were called "nationalities."

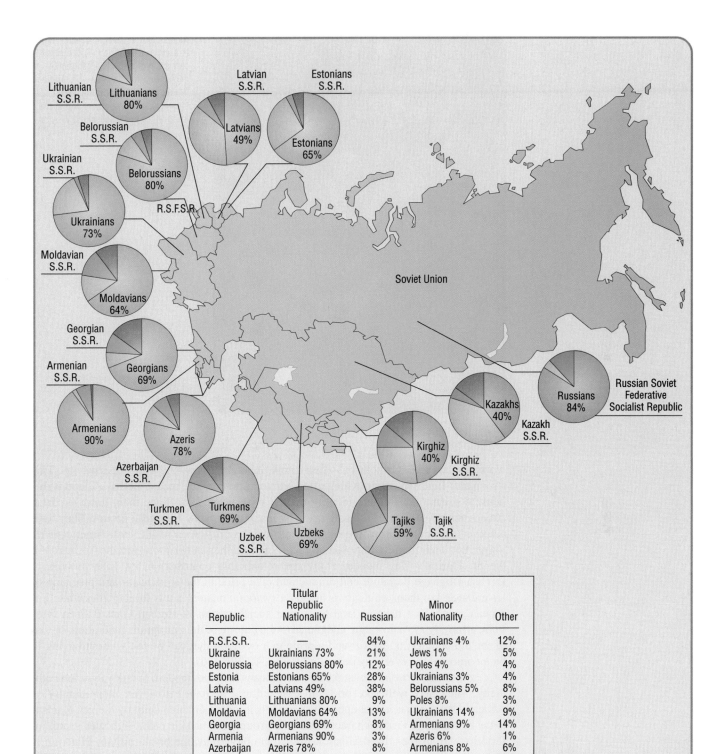

Republic	Titular Republic Nationality	Russian	Minor Nationality	Other
R.S.F.S.R.	—	84%	Ukrainians 4%	12%
Ukraine	Ukrainians 73%	21%	Jews 1%	5%
Belorussia	Belorussians 80%	12%	Poles 4%	4%
Estonia	Estonians 65%	28%	Ukrainians 3%	4%
Latvia	Latvians 49%	38%	Belorussians 5%	8%
Lithuania	Lithuanians 80%	9%	Poles 8%	3%
Moldavia	Moldavians 64%	13%	Ukrainians 14%	9%
Georgia	Georgians 69%	8%	Armenians 9%	14%
Armenia	Armenians 90%	3%	Azeris 6%	1%
Azerbaijan	Azeris 78%	8%	Armenians 8%	6%
Uzbek	Uzbeks 69%	11%	Tajiks 4%	16%
Kazakh	Kazakhs 40%	40%	Ukrainians 6%	14%
Tajik	Tajiks 59%	11%	Uzbeks 23%	7%
Turkmen	Turkmens 69%	13%	Uzbeks 9%	9%
Kirghiz	Kirghiz 48%	28%	Uzbeks 12%	14%

✦ **FIGURE 8.4** ✦
Ethnic Groups in the Soviet Union by Republic, 1979

SOURCE: Perry-Castañeda Library Map Collection (2000).

As Malcolm X noted, it doesn't matter to a racist whether an African American is a professor or a panhandler, a genius or a fool, a saint or a criminal. Where racism is common, racial identities are compulsory and at the forefront of one's self-identity.

working-class Irish Catholics as often drunk, inherently lazy, and born superstitious. This strong anti-Irish sentiment, which often erupted into conflict, meant the Irish found it difficult to escape their ethnic identity even if they wanted to. Since then, however, Irish Americans have followed the path taken by many other white European groups. They have achieved upward mobility and blended into the majority. As a result, Irish-Americans no longer find their identity imposed on them. Instead, they may *choose* whether to march in the St. Patrick's Day parade, enjoy the remarkable contributions of Irish authors to English-language literature and drama, and take pride in the athleticism and precision of Riverdance. For them, ethnicity is largely a *symbolic* matter, as it is for the other white European groups that have undergone similar social processes. Herbert Gans defines **symbolic ethnicity** as "a nostalgic allegiance to the culture of the immigrant generation, or that of the old country; a love for and a pride in a tradition that can be felt without having to be incorporated in everyday behavior" (Gans, 1979b: 436).

At the other extreme, most African Americans lack the freedom to enjoy symbolic ethnicity. They may well take pride in their cultural heritage. However, their identity as African Americans is not an option because it is imposed on them daily by racism. **Racism** is the belief that a visible characteristic of a group, such as skin color, indicates group inferiority and justifies discrimination. In his autobiography, the black militant Malcolm X poignantly noted how racial identity can be imposed on people. He described one of his black Ph.D. professors as "one of these ultra-proper-talking Negroes" who spoke and acted snobbishly. "Do you know what white racists call black Ph.D.s?" asked Malcolm X. "He said something like, 'I believe that I happen not to be aware of that . . . ' And I laid the word down on him, loud: 'N____!'" (Malcolm X, 1965: 284). Malcolm X's point is that it doesn't matter to a racist whether an African American is a professor or a panhandler, a genius or a fool, a saint or a criminal. Where racism is common, racial identities are compulsory and at the forefront of one's self-identity.

In sum, political and social processes structure the degree to which people are able to choose their ethnic and racial identities. Americans today are freer to choose their ethnic identity than citizens of the Soviet Union used to be. Members of *ethnic* minority groups in the United States today are freer to choose their identity than members of *racial* minority groups.

The contrast between Irish Americans and African Americans also suggests that relations between racial and ethnic groups can take different forms. For example, racial and ethnic groups can blend together as a result of residential integration, friendship, and intermarriage or they can remain separate due to hostility. We now discuss two theories—ecological theory and the theory of internal colonialism—that explain why forms of racial and ethnic relations vary over time and from place to place.

THEORIES OF RACE AND ETHNIC RELATIONS

Ecological Theory

Nearly a century ago, Robert Park proposed an influential theory of how race and ethnic relations change over time (Park, 1914; 1950). His **ecological theory** focuses on the struggle for territory. It distinguishes five stages in the process by which conflict between ethnic and racial groups emerges and is resolved:

1. *Invasion.* One racial or ethnic group tries to move into the territory of another. The territory may be as large as a country or as small as a neighborhood in a city.

2. *Resistance.* The established group tries to defend its territory and institutions against the intruding group. It may use legal means, violence, or both.

3. *Competition.* If the established group doesn't drive out the newcomers, the two groups begin to compete for scarce resources. These resources include housing, jobs, public park space, political positions, and so forth.

4. *Accommodation and Cooperation.* Over time, the two groups work out an understanding of what they should segregate, divide, and share. **Segregation** involves the spatial and institutional separation of racial or ethnic groups. For example, the two groups may segregate churches, divide political positions in proportion to the size of the groups, and share public parks equally.

5. *Assimilation.* **Assimilation** is the process by which a minority group blends into the majority population and eventually disappears as a distinct group. Park argued that assimilation is bound to occur as accommodation and cooperation allow trust and understanding to develop. Eventually, goodwill allows ethnic groups to fuse socially and culturally. Where there were formerly two or more groups, only one remains. Park agreed with the memorable image of America as "God's crucible, the great Melting Pot where all the races of Europe are melting and re-forming" (Zangwill, 1909: 37).

Park's theory stimulated important and insightful research (e.g., Suttles, 1968). However, it applies to some ethnic groups better than others. It applies best to whites of European origin. As Park predicted, many whites of European origin stopped thinking of themselves as Italian American or Irish American or German American after their families were in the United States for three or four generations. Today, they think of themselves just as "whites" (Lieberson, 1991). That is because, over time, they achieved rough equality with members of the majority group and, in the process, began to blend in with them. The story of the Irish is fairly typical. During the first half of the 20th century, Irish Americans experienced much upward mobility. By the middle of the century, they earned about as much as American Protestants of English origin. The tapering off of working-class Irish immigration prevented the average status of the group from falling. As their status rose, Irish Americans increasingly intermarried with members of other ethnic groups. Use of the Irish language, Gaelic, virtually disappeared. The Irish

Many white Americans lived in poor urban ghettos in the 19th and early 20th centuries. However, a larger proportion of them experienced upward mobility than was the case for African Americans in the second half of the 20th century. This photo shows the home of an Italian rag picker in a New York tenement in the late 19th century.

became less concentrated in particular cities and less segregated in certain neighborhoods. Finally, conflict with majority-group Americans declined (Greeley, 1974). With variations, a similar story may be told about Italian Americans, German Americans, and so forth.

However, the story does not apply to all Americans. Park's theory is too optimistic about the prospects for assimilation of African Americans, Native Americans, Hispanic Americans, and Asian Americans. It also fails to take into account the persistence of ethnicity among some middle-class whites of European origin into the third and fourth generation after immigration. For reasons we will now explore, racial minorities, and some European-origin ethnics, seem stuck between Park's third and fourth stages.

The Theory of Internal Colonialism

The main weakness of Park's theory is that it pays insufficient attention to the *social-structural* conditions that prevent some groups from assimilating. Robert Blauner examined one such condition, which he called **internal colonialism** (Blauner, 1972; Hechter, 1974). Blauner's work is important because it stimulated the development of a broad range of theories that emphasize the social-structural (and especially class) roots of race and ethnicity. These theories are much in evidence throughout this chapter.

Colonialism involves people from one country invading another. In the process, the invaders change or destroy the native culture. They gain virtually complete control over the native population. They develop the racist belief that the natives are inherently inferior. And they confine the natives to work that is considered demeaning. *Internal colonialism* involves the same processes but within the boundaries of a single country. Internal colonialism prevents assimilation by segregating the colonized in terms of jobs, housing, and social contacts ranging from friendship to marriage. To varying degrees, Russia, China, France, Great Britain, Canada, Australia, the United States, and other countries have engaged in internal colonialism. Let us consider the case of the United States.

Native Americans

The single word that best describes the treatment of Native Americans by European immigrants in the 19th century is expulsion. **Expulsion** is the forcible removal of a population from a territory claimed by another population.

Especially after the United States government passed the Indian Removal Act in 1830, white European Americans fought a series of wars against various Native-American groups. These wars were intended to evict native inhabitants from their lands. Relying on superior military technology and troop strength, the United States Army easily won. As General Thomas Jesup reported in 1838:

The villages of the Indians have all been destroyed; and their cattle, horses, and other stock, with nearly all their other property, taken or destroyed. The swamps and hammocks have been every where penetrated, and the whole country traversed from the Georgia line to the southern extremity of Florida; and the small bands who remain dispersed over the extensive region, have nothing of value left but their rifles (quoted in Wallace, 1993: 99).

The wars, as well as diseases brought to North America by European settlers, such as measles, killed more than half the Native-American population between 1800 and 1900. Gradually, the remaining "small bands" were placed on reservations under the rule of the Bureau of Indian Affairs. On the reservations, there were few good jobs and few opportunities for advancement.

Significantly, the Bureau adopted an assimilationist policy. It discouraged Native Americans from speaking their native languages and handing down their cultures and religions to their children. But this policy achieved just the opposite of what it was intended to accomplish. By oppressing Native Americans and segregating them on reservations, the American government caused Native-American resentment, anger, and solidarity to grow. Especially since the 1960s, Native Americans have begun to fight back culturally and politically. Today, the Native-American community enjoys renewed ethnic pride, a rebirth of cultural activities, and organized political action to protect and advance the rights of its members (Cornell, 1988; Nagel, 1996).

Chicanos

We can tell a similar story about Mexican Americans, or Chicanos. Motivated by the desire for land, the United States went to war with Mexico in 1848. The United States won. Arizona, California, New Mexico, Utah, and parts of Colorado and Texas became part of the United States. The war was justified in much the same way as the conquest of Native Americans. As an editorial in the *New York Evening Post* put it:

> The Mexicans are Indians—Aboriginal Indians . . . They do not possess the elements of an independent national existence. The Aborigines of this country have not attempted and cannot attempt to exist independently along side of us. Providence has so ordained it, and it is folly not to recognize the fact. The Mexicans are Aboriginal Indians, and they must share the destiny of their race (quoted in Steinberg, 1989 [1981]: 22).

For the past 150 years, millions of Chicanos have lived in the United States Southwest (Camarillo, 1979). Due mainly to discrimination, they are socially, occupationally, and residentially segregated from white European Americans. They were not forced onto reservations like Native Americans. However, until the 1970s, most Chicanos lived in ghettos or *barrios*, far from white European-American neighborhoods. Many of them still do. Many white European Americans continue to regard Chicanos as social inferiors. As a result, Chicanos interact mainly with other Chicanos. They still experience much job discrimination, and they still work mainly as agricultural and unskilled laborers with few prospects for upward mobility. (Ironically, some Mexicans who now work in California, Texas, and other lands from which their ancestors were expelled are called "illegal migrants.") Thus, as with Native Americans, high levels of occupational, social, and residential segregation prevent Chicanos from assimilating. Instead, especially since the 1960s, many Chicanos have taken part in a movement to renew their culture and protect and advance their rights (Gutiérrez, 1995).

African Americans

Most features of the internal colonialism model can also be used to explain the obstacles to African-American assimilation. For although the lands of Africa were not invaded and incorporated into the United States, many millions of Africans were brought here by force and enslaved. **Slavery** is the ownership and control of people. By about 1800, 24 million Africans had been captured and transported on slave ships to North, Central, and South

America. Due to violence, disease, and shipwreck, a mere 11 million survived the passage. Fewer than 10% of those 11 million arrived in the United States. However, because the birth rate of African slaves in the United States was higher than elsewhere in the Americas, nearly 30% of the black population in the New World was living in the United States by 1825. By the outbreak of the Civil War, there were 4.4 million black slaves in the United States. The cotton and tobacco economy of the American South depended completely on their labor (Patterson, 1982).

Slavery kept African Americans segregated from white society. Even after slavery was legally banned in 1863, they remained a race apart. So-called Jim Crow laws kept blacks from voting, attending white schools, and in general participating equally in a wide variety of social institutions. In 1896, the Supreme Court approved segregation when it ruled that separate facilities for blacks and whites are legal as long as they are of nominally equal quality (*Plessy v. Ferguson*). Most African Americans remained unskilled workers throughout this period.

In the late 19th and early 20th centuries, a historic opportunity to integrate the black population into the American mainstream presented itself. This was a period of rapid industrialization. The government could have encouraged African Americans to migrate northward and westward and get jobs in the new factories. There they could have enjoyed job training, steady employment, and better wages. But United States policy makers chose instead to encourage white European immigration. Between 1880 and 1930, 23 million Europeans came to the United States to work in the expanding industries. While white European immigrants made their first strides on the path to upward mobility, the opportunity to integrate the black population quickly and completely into the American mainstream was squandered (Steinberg, 1989 [1981]: 173–200).

Some jobs in northern and western industries did go to African Americans, who migrated from the South in substantial numbers from the 1910s on. They were able to compete against European immigrants in the labor market by accepting low wages. Sociologist Edna Bonacich developed a theory to explain why racial identities are reinforced in such "split labor markets." In brief, where low-wage workers of one race and high-wage workers of another race compete for the same jobs, high-wage workers are likely to resent the presence of low-wage competitors, and conflict is bound to result. Consequently, racist attitudes develop or get reinforced. This is certainly what happened during the early migration of African Americans northward. The split labor market fueled deep resentment, animosity, and even antiblack riots on the part of working-class whites, thus solidifying racial identities, both black and white (Bonacich, 1972).

Despite this conflict, black migration northward and westward continued. That is because social and economic conditions in the South were even worse. Already by the 1920s, the world center of jazz had shifted from New Orleans to Chicago. This as much as anything signaled the permanence and vitality of the new black communities. By the mid-1960s there were about 4 million African Americans living in the urban centers of the North and West.

The migrants from the South tended to congregate in low-income neighborhoods, where they sought inexpensive housing and low-skill jobs. Slowly—more slowly than was the case for white European immigrants—their situation improved. Many children of migrant blacks finished high school. Some finished college. Others established ethnic enterprises. Still others got jobs in the civil service. Residential segregation in poor neighborhoods decreased, and there was even some intermarriage with members of the white community (Lieberson, 1980).

In the 1960s, some sociologists observed these developments and expected African Americans to continue moving steadily up the social class hierarchy. As we will soon see, this optimism was only partly justified. Social-structural impediments, some new and some old, prevented many African Americans from achieving the level of prosperity and assimilation enjoyed by white Europeans. Thus, in the mid-1960s about a third of African Americans lived in poverty and the proportion is virtually unchanged today.

In concluding this chapter, we carry the story of African Americans and other racial minorities forward to the present. For the moment, however, we offer the following summary. Our thumbnail sketch of Native Americans, Chicanos, and African Americans

Jacob Lawrence. *The Migration of the Negro, Panel No. 57.* 1940–1941. Jacob Lawrence's "The Great Migration" series of paintings illustrates the mass exodus of African Americans from the South to the North in search of a better life. Lawrence's parents were among those who migrated in the first wave of the great migration (1916–1919).

shows that theories emphasizing the social-structural roots of race and ethnicity, such as the theory of internal colonialism and the theory of the split labor market, help to overcome the main weakness of Park's ecological theory. The groups that have had most trouble achieving upward mobility in the United States are those that were subjected to expulsion from their native lands and slavery. Expulsion and slavery left a legacy of racism that created social-structural impediments to assimilation—impediments such as forced segregation in low-status jobs and low-income neighborhoods. By focusing on factors like these, we arrive at a more realistic picture of the state of race and ethnic relations in the United States than is afforded by ecological theory alone.

Some Advantages of Ethnicity

The theory of internal colonialism emphasizes how social forces outside a racial or ethnic group force its members together, preventing their assimilation into the larger society. It focuses on the disadvantages of race and ethnicity. Moreover, it deals only with the most disadvantaged minorities. The theory has less to say about the internal conditions that promote group cohesion, and in particular about the value of group membership. Nor does it help us to understand why many European Americans continue to participate in the life of their ethnic communities, even if their families have been in the country more than two generations.

A review of the sociological literature suggests that three main factors enhance the value of ethnic group membership for some white European Americans who have lived in the country for many generations. These factors include the following:

1. *Ethnic group membership can have economic advantages.* The economic advantages of ethnicity are most apparent for immigrants, who comprise nearly 10% of the

◆ **FIGURE 8.5** ◆
Percent Foreign-Born, United States, 1900–1997

SOURCE: United States Bureau of the Census (1993: 2; 1999e)

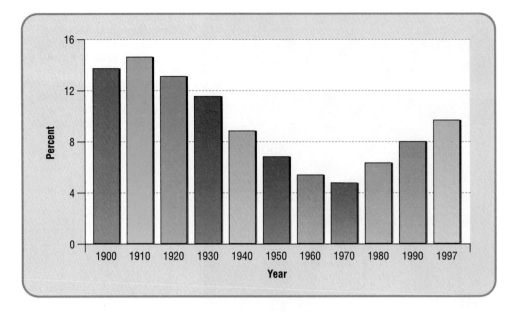

	1900			1960			1997	
Germany	2,663	(25.8)	Italy	1,257	(12.9)	Mexico	7,017	(27.2)
Ireland	1,615	(15.6)	Germany	990	(10.2)	Philippines	1,132	(4.4)
Canada	1,179	(11.4)	Canada	953	(9.8)	China & Hong Kong	1,107	(4.3)
Great Britain	1,167	(11.3)	Great Britain	765	(7.9)	Cuba	913	(3.5)
Sweden	582	(5.6)	Poland	748	(7.7)	Vietnam	778	(3.0)
Italy	484	(4.7)	Soviet Union	691	(7.1)	India	748	(2.9)
Soviet Union	423	(4.1)	Mexico	576	(5.9)	Dominican Republic	632	(2.5)
Poland	383	(3.7)	Ireland	338	(3.5)	El Salvador	607	(2.4)
Norway	336	(3.2)	Hungary	245	(2.5)	Great Britain	606	(2.4)
Austria	275	(2.7)	Czechoslovakia	228	(2.3)	Korea	591	(2.3)
Other	1,234	(11.9)	Other	2,947	(30.2)	Other	11,655	(45.2)
Total	10,341	(100.0)	Total	9,738	(100.0)	Total	25,779	(100.0)

◆ **TABLE 8.1** ◆

Top 10 Countries of Origin of Foreign-Born Americans, 1900, 1960, and 1997 (in thousands; percent of total foreign-born in parentheses)

SOURCE: Calculated from United States Bureau of the Census (1997; 1999e).

American population in 1997 (see Figure 8.5). However, some economic advantages extend into the third generation and beyond. Immigrants often lack extensive social contacts and fluency in English. Therefore, they commonly rely on members of their ethnic group to help them find jobs and housing. In this way, immigrant communities become tightly knit. Community solidarity is an important resource for "ethnic entrepreneurs" too. These are businesspeople who operate largely within their ethnic community. They draw on their community for customers, suppliers, employees, and credit. They can pass on their businesses to their children, who in turn can pass the businesses on to the next generation. In this way, strong economic incentives encourage some people to remain ethnic group members, even beyond the immigrant generation (Bonacich, 1973; Light, 1991; Portes and Manning, 1991).

2. *Ethnic group membership can be politically useful.* Consider, for instance, the way some European Americans reacted to the civil rights movement of the 1960s. Civil rights legislation opened new educational, housing, and job opportunities for African Americans. It also led to the liberalization of immigration laws. Until the passage of the 1965 Hart-Cellar Act, immigration from Asia, Africa, and Latin America was sharply restricted. Afterwards, most immigrants came from these regions (see Table 8.1). Some white European Americans felt threatened by the improved social

standing of African Americans and non-European immigrants. As a result, "racial minorities and white ethnics became polarized on a series of issues relating to schools, housing, local government, and control over federal programs" (Steinberg, 1989 [1981]: 50). Not coincidentally, many European Americans experienced renewed interest in their ethnic roots just at this time. Many sociologists believe the white ethnic revival of the 1960s and 1970s was a reaction to political conflicts with African, Asian, and Hispanic Americans. Such conflicts helped to strengthen ethnic group solidarity.

3. *Ethnic group membership tends to persist because of the emotional support it provides.* Like economic benefits, the emotional advantages of ethnicity are most apparent in immigrant communities but can endure for generations. Speaking the ethnic language and sharing other elements of one's native culture are valuable sources of comfort in an alien environment. Even beyond the second generation, however, ethnic group membership can perform significant emotional functions. For example, some ethnic groups experience unusually high levels of prejudice and discrimination involving expulsion or attempted genocide. (**Genocide** is the intentional extermination of an entire population defined as a "race" or a "people.") For people who belong to such groups, the resulting trauma is so severe it can be transmitted for several generations. In such cases, ethnic group membership offers security in a world still seen as hostile long after the threat of territorial loss or annihilation has disappeared (Bar-On, 1999). Another way in which ethnic group membership offers emotional support beyond the second generation is by providing a sense of rootedness. Especially in a highly mobile, urbanized, technological, and bureaucratic society such as ours, ties to an ethnic community can be an important source of stability and security (Isajiw, 1978).

The three factors listed above make ethnic group membership useful to some European Americans whose families have been in the country for many generations. It is also important to note in this connection that retaining ethnic ties beyond the second generation has never been easier. Inexpensive international communication and travel allow ethnic group members to maintain strong ties to their motherland in a way that was never possible in earlier times.

Immigration used to involve cutting all or most ties to one's country of origin. Travel by sea and air was expensive, long-distance telephone rates were prohibitive, and the occasional letter was about the only communication most immigrants had with their relatives in the old country. This lack of communication encouraged assimilation. Today, however, ties to the motherland are often maintained in ways that sustain ethnic culture. For example, nearly 400,000 Jews have emigrated from the former Soviet Union to the United States since the early 1970s. They frequently visit relatives in the former Soviet Union and Israel, speak with them on the phone, and use the Internet to exchange e-mail with them. They also receive Russian-language radio and TV broadcasts, act as conduits for foreign investment, and send money to relatives abroad (Brym with Ryvkina, 1994; Markowitz, 1993). This sort of intimate and ongoing connection with the motherland is typical of most recent immigrant communities in the United States. Thanks to inexpensive international travel and communication, some ethnic groups have become **transnational communities** whose boundaries extend between countries. Characteristically, the Ticuani Potable Water Committee in Brooklyn has been raising money for the farming community of Ticuani in the Mixteca region of Mexico for 25 years. Its seal reads *"Por el Progreso de Ticuani: Los Ausentes Siempre Presentes. Ticuani y New York."* Translation: "For the Progress of Ticuani: The Absent Ones Always Present. Ticuani and New York." The phrase "the absent ones always present" nicely captures the essence of transnational communities, whose growing number and vitality facilitate the retention of ethnic group membership beyond the second generation (Portes, 1996).

In sum, ethnicity remains a vibrant force in American society for a variety of reasons. Even some white Americans whose families arrived in this country generations ago have

In the late 19th and early 20th centuries, Ellis Island in New York harbor was the point of entry of more than 12 million Europeans to the United States. Today, more than a third of all Americans can trace their origins to a person who passed through Ellis Island. Here, a customs official attaches labels to the coats of a German immigrant family at the Registry Hall on Ellis Island in 1905.

Until the passage of the 1965 Hart-Cellar Act, immigration from Asia, Africa, and Latin America was sharply restricted. Today, most American immigrants come from Latin America and Asia.

reason to identify with their ethnic group. Bearing this in mind, what is the likely future of race and ethnic relations in the United States? We conclude by offering some tentative answers to that question.

THE FUTURE OF RACE AND ETHNICITY

The Declining Significance of Race?

Web Interactive Exercises
Are African Americans
Making Progress?

At 2:30 a.m. on June 7, 1998, James Byrd, Jr., was walking home along a country road near Jasper, Texas. Three men stopped and offered Byrd a ride. But instead of taking him home, they forced Byrd out of their pick-up truck. They beat him until he was unconscious. They then chained his ankles to the back of the truck and dragged him along the jagged road for nearly 3 miles. The ride tore Byrd's body into more than 75 pieces.

What was the reason for the murder? Byrd was a black man. His murderers are "white supremacists." They want the United States to be a white-only society, and they are prepared to use violence to reach their goal. At trial, the murderers expressed no remorse. Two of them wore tattoos suggesting membership in the racist Aryan Nation or the Ku Klux Klan. As such, they are part of a growing problem in the United States. More than 500 racist and Neo-Nazi organizations have sprung up in the country (Southern Poverty Law Center, 2000). In 1997, the FBI recorded 8,049 **hate crimes,** or criminal acts motivated by a person's race, religion, or ethnicity. By far the most frequent victims of hate crimes are African Americans, who comprise 39% of the total (United States Federal Bureau of Investigation, 1997).

White supremacists form only a tiny fraction of the American population. Many more Americans engage in subtle forms of racism. Thus, sociologists Joe Feagin and Melvin Sikes (1994) interviewed a sample of middle-class African Americans in 16 cities. They found their respondents often have trouble hailing a cab. If they arrive in a store before a white customer, a clerk commonly serves them afterwards. When they shop, store security often follows them around to make sure they don't shoplift. Police officers often stop middle-class African American men in their cars without apparent reason. Blacks are less likely than whites of similar means to receive mortgages and other loans (Oliver and Shapiro, 1995). In short, African Americans continue to suffer high levels of racial prejudice and discrimination, whether overt or covert (Hacker, 1992; Shipler, 1997).

Still, sociologist William Julius Wilson (1980 [1978]) and others believe race is declining in significance as a force shaping the lives of African Americans. Wilson argues that the civil rights movement helped to establish legal equality between blacks and whites. In 1954, the Supreme Court ruled against earlier decisions permitting school segregation (*Brown v. Board of Education*). The 1964 Civil Rights Act outlawed discrimination in public housing, employment, and the distribution of federal funds. It also supported school integration. The 1965 Voting Rights Act prohibited the systematic exclusion of blacks from the political process. The 1968 Civil Rights Act banned racial discrimination in housing. These reforms allowed a large black middle class to emerge, says Wilson. Today a third of the African-American population is middle class.

Many facts support Wilson's view that the social standing of blacks has improved. For example, in 1947, median family income among blacks was only 51.1% that of whites. Fifty years later, it stood at 61.2% (see Figure 8.6). The proportion of whites with 4 or more years of college tripled from 1960 to 1997. In the same period, the proportion of blacks with 4 or more years of college more than quadrupled. In 1997, 1.3% more whites but 5.1% fewer blacks lived below the poverty line than in 1970 (United States Bureau of the Census, 1998c: 167, 477). And public opinion polls suggest that whites are becoming more tolerant of blacks (see Figure 8.7) (Thernstrom and Thernstrom, 1997). Wilson, citing similar data, admits that the gap between blacks and whites remains substantial (see Figure 8.8). Still, he stresses, it is shrinking.

For Wilson, the third of African Americans who live below the poverty line are little different from other Americans in similar economic circumstances. He therefore calls for "color-blind" public policies that aim to improve the class position of the poor, such as job training and health care. He opposes race-specific policies, such as **affirmative action,** which apply racial quotas in education and at work. He feels that such policies disproportionately help middle-class African Americans while keeping poor African Americans poor (Wilson, 1996).

Despite the points noted above, some sociologists think Wilson exaggerates the declining significance of race. For Wilson's critics, racism remains a big barrier to black progress. Their case for the continuing impact of race is strengthened by statistical analyses of data on racial differences in wages and housing patterns. For example, they have shown that one's race had a *stronger* impact on wages in 1985 than in 1976, probably because the government retreated from antidiscrimination initiatives in the 1980s (Cancio,

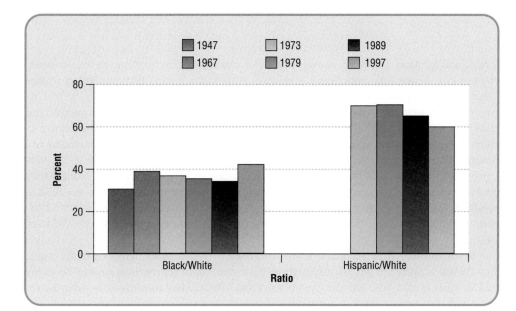

✦ **FIGURE 8.6** ✦
Median Family Income Ratios, Black/White and Hispanic/White, United States, 1947–1997

SOURCE: Mishel, Bernstein, and Schmitt (1999: 45).

◆ **FIGURE 8.7** ◆
White Prejudice and Discrimi-
nation Against Blacks, United
States, 1972–1996

SOURCE: National Opinion Research Center (1999).

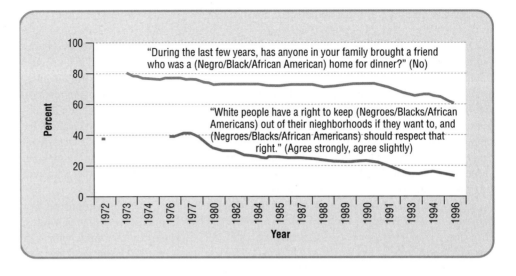

◆ **FIGURE 8.8** ◆
Social Standing by Race and
Hispanic Origin, United States,
1996

SOURCE: United States Bureau of the
Census (1998c: 167, 468).

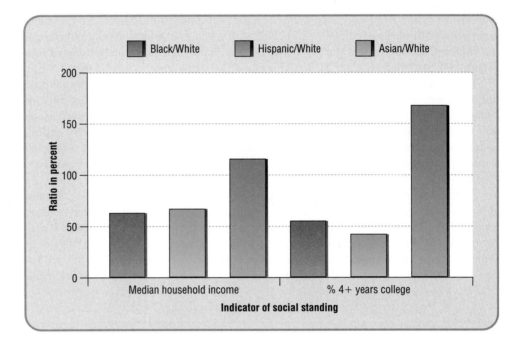

Evans, and Maume, 1996). Moreover, many African Americans continue to live in inner-
city ghettos. There they experience high rates of poverty, crime, divorce, teenage pregnancy, and unemployment.

Wilson says ghettos persist not so much because of racism but for three economic and
class reasons. First, since the 1970s, older manufacturing industries have closed down in
cities where the black working class was concentrated. This increased unemployment and
poverty. Second, many middle- and working-class African Americans with good jobs
moved out of the inner city. This deprived young people of successful role models they
could emulate. Third, the exodus of successful blacks eroded the inner-city's tax base at
precisely the same time that conservative federal and state governments were cutting budgets for public services. This added to the destitution of inner-city residents (Wilson,
1996).

While the economic and class factors discussed by Wilson undoubtedly explain much,
they do not explain the persistently high level of residential segregation among the many
middle-class blacks who left the ghettos since the 1960s. They moved to the suburbs, yet

City	Segregation Index
Detroit	87.6
Chicago	85.8
Cleveland	85.1
Milwaukee	82.8
New York	82.2
Philadelphia	77.2
St. Louis	77.0
Los Angeles-Long Beach	73.1
Birmingham	71.7
Baltimore	71.4

✦ **TABLE 8.2** ✦

The 10 Most Racially Segregated Major Metropolitan Areas in the United States, 1990

SOURCE: Massey and Denton (1993).

their neighborhoods are nearly as segregated as those in the inner city. That is why people sometimes call them "ghettos with grass."

Sociologists calculate a "segregation index" that measures the extent of the problem. The index has a value of zero if the percentage of nonwhites living on each city block is the same throughout the city. This is an unsegregated distribution. The index has a value of 100 if all nonwhites would have to move to produce an unsegregated distribution. In 1970, the segregation index in the 11 Northern cities with the biggest black populations stood at 84.5%. By 1990, it had fallen to 77.8%. This means that in 1990 the black population was only slightly less segregated than it was 20 years earlier. In the 9 Southern cities with the biggest black populations, the index of segregation fell from 75.5% to 66.5% over the same period (see Table 8.2). These figures suggest only modest improvement in housing segregation in recent decades.

As sociologists Douglas Massey and Nancy Denton (1993) show, black segregation in housing persists due to racism. Specifically, white homeowners are likely to move elsewhere if "too many" blacks move into a neighborhood. Meanwhile, real estate agents and mortgage lenders sometimes withhold information, refuse loans, and otherwise discourage blacks from moving into certain areas. They do this to protect real estate values.

The experience of Hispanic and especially Asian Americans is different. They are less segregated in housing than African Africans. Moreover, they are becoming desegregated faster. There is only one major exception to this pattern, and it supports the general argument about the continuing significance of race. Puerto Ricans are nearly as segregated in housing as blacks, and their situation is improving just as slowly. Massey and Denton (1987) say that is because the white majority typically views them as "black Hispanics." In short, race matters in housing, but being black appears to matter most due to widespread antiblack sentiment.

Some analysts explain the persistence of residential segregation and other aspects of racism as a reaction to black progress. Thus, survey research shows that European Americans often express resentment against the real and perceived advantages enjoyed by African Americans (Kinder and Sanders, 1996). In particular, arguments for affirmative action generate strong opposition (Bobo and Kluegel, 1993).

Supporters of affirmative action feel that African Americans should get preference if equally qualified people apply for a job or college entrance. In this way, the historical injustices of slavery, segregation, and discrimination can be corrected. Many whites object. They say affirmative action is a form of "reverse discrimination," or bias against European Americans. For them, America stands for equal opportunity, not racial preference. Most blacks favor affirmative action, but some oppose it. Opponents argue affirmative action demeans their accomplishments. It brands them as the "best black" rather than the best person for a job. Affirmative action, they say, contributes to belief in black inferiority (Carter, 1991).

✦ FIGURE 8.9 ✦
Opposition to Affirmative Action, United States, 1998

SOURCE: National Opinion Research Center (1999).

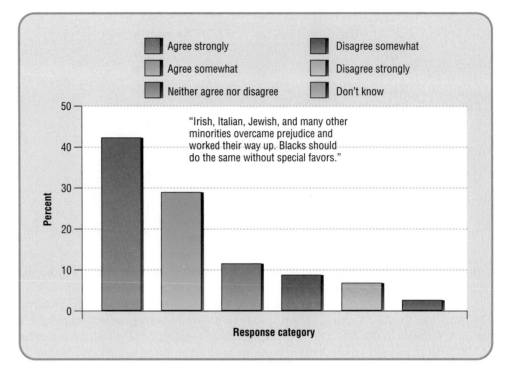

Today's political climate is unfavorable to affirmative action. According to the 1998 General Social Survey, fewer than 15% of Americans think blacks need "special favors" to work their way up (see Figure 8.9). Congress has ruled against giving preference to minority-owned firms in awarding small government contracts. California has eliminated affirmative action policies in state colleges.

Where do you fit in? Do you think the country needs affirmative action to correct historical injustices? Is it reasonable to expect African Americans to be satisfied with the rate of progress sketched above? What are the consequences of widespread black dissatisfaction? Do the benefits of affirmative action outweigh the costs?

A Vertical Mosaic

In 1800, the United States was a society based on slavery, expulsion, and segregation. Two centuries later, we are a society based on segregation, pluralism, and assimilation. (**Pluralism** is the retention of racial and ethnic culture combined with equal access to basic social resources.) Thus, on a scale of tolerance, the United States has come a long way in the past 200 years (see Figure 8.10).

In comparison with most other countries, too, the United States is a relatively tolerant land. In the 1990s, racial and ethnic tensions in some parts of the world erupted into wars of secession and attempted genocide. Conflict between Croats, Serbs, and other ethnic groups broke Yugoslavia apart. Russia is fighting a bloody war against its Chechen ethnic minority. In Rwanda in 1991, Hutu militia and soldiers massacred many thousands of Tutsi civilians. Three years later, Tutsi soldiers massacred many thousands of Hutu civilians. Comparing the United States to such poor countries may seem to stack the deck in favor of concluding that the United States is a relatively tolerant society. However, even when we compare the United States to other rich, stable, postindustrial countries, our society seems relatively tolerant by some measures (see Figure 8.11).

Due to such factors as intermarriage and immigration, the growth of tolerance in the United States is taking place in the context of increasing ethnic and racial diversity. By the time today's first-year college student is 75 or 80 years old, non-Hispanic whites will form a minority of the United States population for the first time in 350 years. The

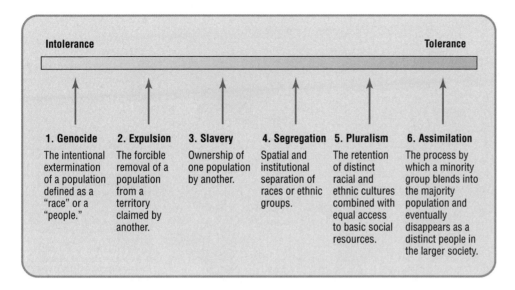

✦ FIGURE 8.10 ✦

Six Degrees of Separation: Types of Ethnic and Racial Group Relations

SOURCE: Adapted from Kornblum (1997 [1998]: 385).

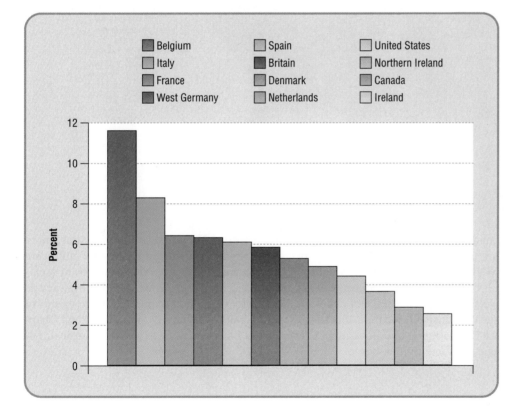

✦ FIGURE 8.11 ✦

Percent Opposed to Person of Another Race, Immigrants, or Foreign Workers Living Next Door, 12 Postindustrial Countries, 1990

SOURCE: Nevitte (1996: 231).

United States will be even more of a racial and ethnic mosaic than it is now (see Figure 8.12 and Figure 8.13).

If present trends continue, however, the racial and ethnic mosaic will be vertical, with some groups, such as African Americans, Native Americans, Puerto Rican Americans, and Chicanos, disproportionately clustered at the bottom. They will remain among the most disadvantaged groups in the country, enjoying less wealth, income, education, good housing, health care, and other social rewards than other ethnic and racial groups.

Policy initiatives could decrease the verticality of the American ethnic mosaic, thus speeding up the movement from segregation to pluralism and assimilation for the

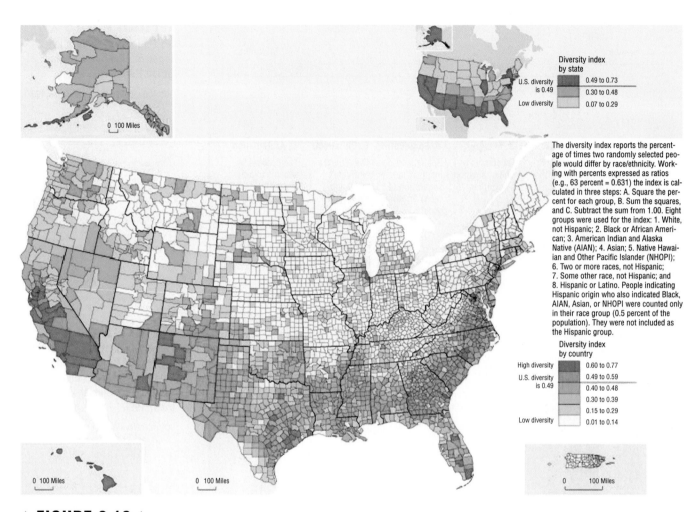

◆ FIGURE 8.12 ◆
Ethnic and Racial Diversity in the United States, 2000

SOURCE: United States Bureau of the Census (2001b).

country's most disadvantaged groups. Apart from affirmative action programs, more job training, improvements in public education, and subsidized health and child care would do much to promote equality. That is because these programs would be of greatest benefit to the most disadvantaged Americans. However, as we noted above, and as we elaborate in subsequent chapters, the country does not seem much in the mood for such expensive reforms at this time (see especially Chapter 12, "Families," and Chapter 13, "Religion and Education"). The United States is likely to remain a vertical mosaic for some time to come.

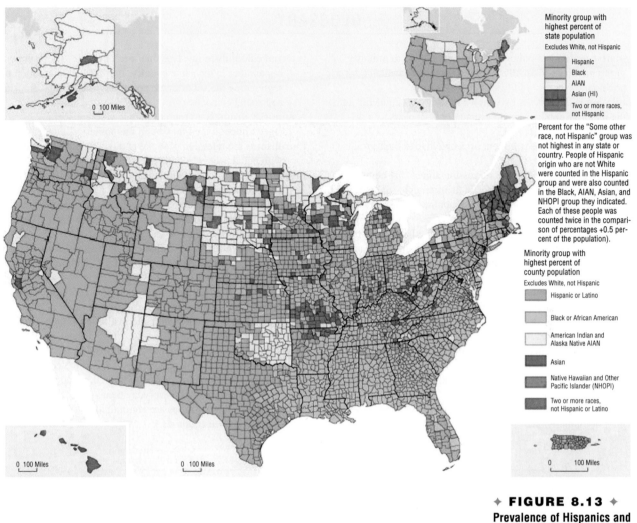

◆ FIGURE 8.13 ◆
Prevalence of Hispanics and Non-White Minorities, United States, 2000

SOURCE: United States Bureau of the Census (2001b).

SUMMARY

1. Race is not a purely biological category. Ethnicity is not a purely cultural category. Both are socially constructed ideas. We use them to distinguish people based on physical or cultural differences. These distinctions have profound consequences for people's lives.

2. Racial and ethnic labels and identities are variables. They change over time and place. Relations between racial and ethnic groups affect them. Cordial group relations hasten the blending of labels and identities.

3. Racial and ethnic groups are blending over time as members of society become more tolerant. However, this tendency is weak among members of highly disadvantaged groups. That is because such groups remain high segregated in jobs, housing, and social contacts. This is a historical legacy of slavery, expulsion, and legalized segregation. It is also the result of continuing racism.

4. Identifying with a racial or ethnic group can have economic, political, and emotional rewards. This accounts for the persistence of ethnic identity in many European-American families, even after they have been in the United States more than two generations.

5. Racial and ethnic inequalities are likely to persist in the foreseeable future. In addition to affirmative action programs, more job training, improvements in public education, and subsidized health care and child care would promote equality. However, the country does not seem to favor these expensive reforms at this time.

GLOSSARY

Affirmative action is a policy that gives preference to minority group members if equally qualified people are available for a position.

Assimilation is the process by which a minority group blends into the majority population and eventually disappears as a distinct group.

Discrimination is unfair treatment of people due to their group membership.

The **ecological theory** of ethnic succession argues that ethnic groups pass through five stages in their struggle for territory: invasion, resistance, competition, accommodation and cooperation, and assimilation.

An **ethnic enclave** is a spatial concentration of ethnic group members who establish businesses that serve and employ mainly members of the ethnic group and reinvest profits in community businesses and organizations.

An **ethnic group** is composed of people whose perceived cultural markers are deemed socially significant. Ethnic groups differ from one another in terms of language, religion, customs, values, ancestors, and the like.

Expulsion is the forcible removal of a population from a territory claimed by another population.

Genocide is the intentional extermination of an entire population defined as a "race" or a "people."

Hate crimes are criminal acts motivated by a person's race, religion or ethnicity.

Internal colonialism involves one race or ethnic group subjugating another in the same country. It prevents assimilation by segregating the subordinate group in terms of jobs, housing, and social contacts.

A **minority group** is a group of people who are socially disadvantaged although they may be in the numerical majority.

Pluralism is the retention of racial and ethnic culture combined with equal access to basic social resources.

Prejudice is an attitude that judges a person on his or her group's real or imagined characteristics.

Race is a social construct used to distinguish people in terms of one or more physical markers, usually with profound effects on their lives.

Racism is the belief that a visible characteristic of a group, such as skin color, indicates group inferiority and justifies discrimination.

A **scapegoat** is a disadvantaged person or category of people whom others blame for their own problems.

Segregation involves the spatial and institutional separation of racial or ethnic groups.

Slavery is the ownership and control of people.

Symbolic ethnicity is a nostalgic allegiance to the culture of the immigrant generation, or that of the old country, that is not usually incorporated in everyday behavior.

Transnational communities are communities whose boundaries extend between countries.

QUESTIONS TO CONSIDER

1. How do you identify yourself in terms of your race or ethnicity? Do conventional ethnic and racial categories, such as black, white, Hispanic, and Asian, "fit" your sense of who you are? If so, why? If not, why not?

2. Do you think racism is becoming more serious in the United States and worldwide? Why or why not? How do trends in racism compare to trends in other forms of prejudice, such as sexism? What accounts for similarities and differences in these trends?

3. What are the costs and benefits of ethnic diversity in your college? Do you think it would be useful to adopt a policy of affirmative action to make the student body and the faculty more ethnically and racially diverse? Why or why not?

WEB RESOURCES

Companion Web Site for This Book

http://sociology.wadsworth.com

Begin by clicking on the Student Resources section of the Web site. Choose "Introduction to Sociology" and finally the Brym and Lie book cover. Next, select the chapter you are currently studying from the pull-down menu. From the Student Resources page you will have easy access to InfoTrac College Edition®, MicroCase Online exercises, additional Web links, and many other resources to aid you in your study of sociology, including practice tests for each chapter.

Infotrac Search Terms

These search terms are provided to assist you in beginning to conduct research on this topic by visiting http://www.infotraccollege.com/wadsworth.

Affirmative action	**Ethnic enclave**
Assimilation	**Racism**
Discrimination	**Transnational community**

Recommended Web Sites

For a comprehensive guide to Web resources on race and ethnicity in the United States, go to http://www.georgetown.edu/crossroads/asw/race.html.

For recent FBI hate-crime statistics, go to http://www.fbi.gov/ucr/hc97all.pdf.

For a stimulating analysis of transnational ethnic communities by Alejandro Portes, former President of the American Sociological Association, go to http://www.prospect.org/archives/25/25port.html.

For United States Census Bureau projections of state populations by race and Hispanic origin from 1995 to 2005, visit http://www.census. gov/population/projections/state/stpjrace.txt.

SUGGESTED READINGS

Stephen Jay Gould. *The Mismeasure of Man,* rev. ed. (New York: Norton. 1996 [1981]). A Pulitzer-Prize-winning study of the abuse of science in the study of racial differences.

Alexander Saxton. *The Rise and Fall of the White Republic* (London: Verso, 1990). A provocative history of ethnic and racial divisions in the 19th-century United States.

Stephen Steinberg. *The Ethnic Myth: Race, Ethnicity, and Class in America,* updated ed. (Boston: Beacon Press, 1989 [1981]). A compelling analysis of the social-structural bases of race and ethnicity.

William Julius Wilson. *When Work Disappears: The World of the New Urban Poor* (New York: Knopf, 1996). Since 1970, global economic reorganization has substantially decreased the number of manufacturing jobs available to America's unskilled workers, who are disproportionately black. This book analyzes the negative consequences for the African-American community.

IN THIS CHAPTER, YOU WILL LEARN THAT:

✦ While biology determines sex, social structure and culture largely determine gender, or the expression of culturally appropriate masculine and feminine roles.

✦ The social construction of gender is evident in the way parents treat babies, teachers treat pupils, and the mass media portray ideal body images.

✦ The social forces pushing people to assume conventionally masculine or feminine roles are compelling.

✦ The social forces pushing people toward heterosexuality operate with even greater force.

✦ The social distinction between men and women serves as an important basis of inequality in the family and the workplace.

✦ Male aggression against women is rooted in gender inequality.

SEX VERSUS GENDER

Is It a Boy or a Girl?

On April 27, 1966, identical 8-month-old twin boys were brought to a hospital in the city of Winnipeg, Canada, to be circumcised. An electrical cauterizing needle—a device used to seal blood vessels as it cuts—was used for the procedure. However, due either to equipment malfunction or doctor error, the needle entirely burned off one baby's penis. The parents desperately sought medical advice. No matter whom they consulted, they were given the same prognosis. As one psychiatrist summed up baby John's future: "He will be unable to consummate marriage or have normal heterosexual relations; he will have to recognize that he is incomplete, physically defective, and that he must live apart . . ." (quoted in Colapinto, 1997: 58).

One evening, 7 months after the accident, the parents, now deeply depressed, were watching TV. They heard Dr. John Money, a psychologist from Johns Hopkins Medical School in Baltimore, say that he could *assign* babies a male or female identity. Money had successfully tested his idea on **hermaphrodites,** babies born with ambiguous genitals due to a hormone imbalance in the womb. For example, he argued that a boy born with a penis shorter than 1 inch should undergo surgery to remove the male genitals and construct a vagina. Immediately after surgery, the parents should treat the baby as a girl. They should never reveal her sexual status at birth. They should arrange for her to take regular doses of the female hormone, estrogen, beginning at puberty. The baby would then grow up to think of herself as a girl.

Until baby John, Money had never tested his idea on a child born unambiguously a boy or a girl. Therefore, when baby John's mother wrote to Dr. Money, he urged her to bring the baby to Baltimore. After several consultations and long deliberation, the parents gave the go-ahead. On July 3, 1967, baby John, now 22 months old, underwent surgical castration and reconstructive surgery. He became baby Joan. As the years passed, the child's parents tried to follow Dr. Money's instructions scrupulously. Joan wore lacy dresses and bonnets. She received dolls and skipping ropes for presents. She became a Girl Scout. At puberty, she took regular doses of estrogen.

In 1972, Dr. Money made the case of John/Joan public at a meeting of the American Association for the Advancement of Science in Washington, D.C. The experiment, he said, was an unqualified success. Joan was feminine in manner and appearance. According to *Time* magazine, this proved that "conventional patterns of masculine and feminine behavior can be altered." Furthermore, Money's work cast "doubt on the theory that major sexual differences, psychological as well as anatomical, are immutably set by the genes at conception" (quoted in Colapinto, 1997: 66). Dr. Money subsequently reported Joan's good progress in 1978 and 1985.

Then, in March 1997, a bombshell: Dr. Money, it emerged, had thoroughly doctored his reports. A biologist from the University of Hawaii and a psychiatrist from the Canadian Ministry of Health unleashed a big scientific scandal when they published an article in the *Archives of Adolescent and Pediatric Medicine*. It showed that John/Joan had in fact struggled against his/her imposed girlhood from the start. In December, a long and moving exposé of the case in *Rolling Stone* magazine gave further details.

The authors of these articles based their judgments on a review of medical records and extensive interviews with John/Joan's parents, twin brother, and John/Joan him/herself. They documented that John/Joan tried to tear off her first dress, wanted a toy razor like her brother's, refused to play with makeup and her toy sewing machine, insisted on playing with her brother's dump trucks and Tinker Toys, and strongly preferred to urinate standing up, even though it made a mess. John/Joan was a "tomboy." By the age of 7, she said she wanted to be a boy. She looked, walked, and talked like a boy and had stereotypical boys' interests. She detested Girl Scouts. At 6 or 7, she decided she wanted to be a garbage man when she grew up. She refused to play with girls in kindergarten and at school. At 12 she began taking estrogen, but only under protest.

Finally, in 1980, unable to suffer her imposed sexual identity any longer, Joan stopped taking estrogen and had her breasts surgically removed. She then had a penis surgically constructed. She was now John again. Subsequent surgeries allowed John to have sex with a woman at the age of 23. He married the woman 2 years later and adopted her three children from a previous marriage. John is now a devoted father. In 2000, he allowed his biography to be published; his real name is David Reimer, and he still lives in Winnipeg. After his long and painful journey, he enjoys a happy family life, although, understandably, he is still deeply troubled by his past.

The story of John/Joan introduces the first big question of this chapter. What makes us male or female? Of course, part of the answer is biological. Your **sex** depends on whether you were born with distinct male or female genitals and a genetic program that released either male or female hormones to stimulate the development of your reproductive system.

However, the case of John/Joan also shows that more is involved in becoming male or female than biological sex differences. Recalling his life as Joan, John said: "[E]veryone is telling you that you're a girl. But you say to yourself, 'I don't *feel* like a girl.' You think girls are supposed to be delicate and *like* girl things—tea parties, things like that. But I like to *do* guy stuff. It doesn't match" (quoted in Colapinto, 1997: 66; our emphasis). As this quotation suggests, being male or female involves not just biology but also certain "masculine" and "feminine" feelings, attitudes, and behaviors. Accordingly, sociologists distinguish biological sex from sociological **gender.** Your gender is composed of the feelings, attitudes, and behaviors typically associated with being male or female. **Gender identity** is your identification with, or sense of belonging to, a particular sex—biologically, psychologically, and socially. When you behave according to widely shared expectations about how males or females are supposed to act, you adopt a **gender role.**

Contrary to first impressions, the case of John/Joan suggests that, unlike sex, gender is not determined just by biology. Research shows that babies first develop a vague sense of being a boy or a girl at about the age of 1. They develop a full-blown sense of gender identity between the ages of 2 and 3 (Blum, 1997). We can therefore be confident that baby John already knew he was a boy when he was assigned a female gender identity at the age of 22 months. He had, after all, been raised as a boy by his parents and treated as a boy by his brother for almost 2 years. He had seen boys behaving differently from girls on TV and in storybooks. He had played only with stereotypical boys' toys. After his gender reassignment, the constant presence of his twin brother reinforced those early lessons on how boys ought to behave. In short, baby John's *social* learning of his gender identity was already far advanced by the time he had his sex-change operation. Dr. Money's experiment was thus bound to fail.

If gender reassignment occurs before the age of 18 months, it is usually successful (Green, 1974).[1] However, once the social learning of gender takes hold, as with baby John, it is apparently difficult to undo, even by means of reconstructive surgery, hormones, and parental and professional pressure. The main lesson we draw from this story is not that biology is destiny but that the social learning of gender begins very early in life.

Chapter Plan

The first half of this chapter helps you better understand what makes us male or female. We first outline two competing theories of gender differences. The first theory argues gender is inherent in our biological makeup and is merely reinforced by society. The second argues gender is constructed mainly by social influences. For reasons outlined below, we side with the second viewpoint.

After establishing our theoretical approach, we examine how people learn gender roles during socialization in the family and at school. Then we show how everyday social interactions and advertising reinforce gender roles.

---

[1]Still, a movement has emerged among some adults who had their sex assigned when they were infants to allow them to (a) choose their sex when they reach puberty or (b) continue living with ambiguous genitals if they wish to do so.

We next discuss how members of society enforce **heterosexuality**—the preference for members of the opposite sex as sexual partners. For reasons that are still poorly understood, some people resist and even reject the gender roles that are assigned to them because of their biological sex. When this occurs, negative sanctions are often applied to get them to conform or to punish them for their deviance. Members of society are often eager to use emotional and physical violence to enforce conventional gender roles.

The second half of the chapter examines one of the chief consequences of people learning conventional gender roles. Gender, as currently constructed, creates and maintains social inequality. We illustrate this in two ways. We first investigate why gender is associated with an earnings gap between women and men in the paid labor force. We then show how gender inequality encourages sexual harassment and rape. In concluding our discussion of sexuality and gender, we discuss some social policies that sociologists have recommended to decrease gender inequality and improve women's safety.

THEORIES OF GENDER

Essentialism[2]

As just noted, most arguments about the origins of gender differences in human behavior adopt one of two perspectives. Some analysts see gender differences as a reflection of naturally evolved dispositions. Sociologists call this perspective **essentialism** (Weeks, 1996: 15). That is because it views gender as part of the nature or "essence" of one's biological makeup. Other analysts see gender differences as a reflection of the different social positions occupied by women and men. Sociologists call this perspective **social constructionism** because it views gender as "constructed" by social structure and culture. We now summarize and criticize essentialism. We then turn to social constructionism.

Freud

Sigmund Freud (1977 [1905]) offered an early and influential essentialist explanation of male-female differences. He believed that differences in male and female anatomy account for the development of distinct masculine and feminine gender roles.

According to Freud, children around the age of 3 begin to pay attention to their genitals. As a young boy becomes preoccupied with his penis, he unconsciously develops a fantasy of sexually possessing the most conspicuous female in his life: his mother. Soon, he begins to resent his father because only his father is allowed to possess the mother sexually. Because he has seen his mother or another girl naked, the boy also develops anxiety that his father will castrate him for desiring his mother.[3] To resolve this fear, the boy represses his feelings for his mother. That is, he stores them in the unconscious part of his personality. In due course, this repression allows him to begin identifying with his father. This leads to the development of a strong, independent masculine personality.

In contrast, the young girl begins to develop a feminine personality when she realizes she lacks a penis. According to Freud:

> [Girls] notice the penis of a brother or playmate, strikingly visible and of large proportions, at once recognize it as the superior counterpart of their own small and inconspicuous organ, and from that time forward fall a victim to envy for the penis . . . She has seen it and knows that she is without it and wants to have it (quoted in Steinem, 1994: 50).

---

[2]We are grateful to Rhonda Lenton for her ideas on essentialism and its critique. See Lenton (2001).

[3]Freud called this set of emotions the "Oedipus complex" after the ancient Greek legend of Oedipus. Oedipus was abandoned as a child. When he became an adult he accidentally killed his father and unwittingly married his mother. Discovering his true relationship to his mother, he blinded himself and died in exile.

Due to her "penis envy," the young girl soon develops a sense of inferiority, according to Freud. She also grows angry with her mother, who, she naively thinks, is responsible for cutting off the penis she must have once had. She rejects her mother and develops an unconscious sexual desire for her father. Eventually, however, realizing she will never have a penis, the girl comes to identify with her mother. This is a way of vicariously acquiring her father's penis in Freud's view. In the "normal" development of a mature woman, the girl's wish to have a penis is transformed into a desire to have children. However, says Freud, since women are never able to resolve their penis envy completely, they are normally immature and dependent on men. This dependence is evident from the "fact" that women can be fully sexually satisfied only by vaginally-induced orgasm.[4] Thus, a host of gender differences in personality and behavior follows from the anatomical sex differences that children observe around the age of 3.

Sociobiology and Evolutionary Psychology

For the past 25 years, sociobiologists and evolutionary psychologists have offered a second essentialist theory. We introduced this theory in Chapter 3 ("Culture"). According to sociobiologists and evolutionary psychologists, all humans instinctively try to ensure their genes get passed on to future generations. However, men and women develop different strategies to achieve this goal. A woman has a bigger investment than a man in ensuring the survival of their offspring. That is because the woman produces only a small number of eggs during her reproductive life and, at most, can give birth to about 20 children. It is therefore in a woman's best interest to maintain primary responsibility for her genetic children and to look around for the best mate with whom to intermix her genes. He is the man who can best help support the children after birth. In contrast, most men can produce hundreds of millions of sperm every 24–48 hours. Thus, a man increases the chance his and only his genes will get passed on to future generations if he is promiscuous yet jealously possessive of his partners. Moreover, since men compete with other men for sexual access to women, men evolve competitive and aggressive dispositions that include physical violence. Women, says one evolutionary psychologist, are greedy for money, while men want casual sex with women, treat women's bodies as their property, and react violently to women who incite male sexual jealousy. These are "universal features of our evolved selves" that contribute to the survival of the human species (Buss, 1994: 211; 1998). Thus, from the point of view of sociobiology and evolutionary psychology, gender differences in behavior are based in biological differences between women and men.

A Critique of Essentialism

Sociologists have lodged four main criticisms against essentialist arguments such as those of Freud and the sociobiologists and evolutionary psychologists.

First, essentialists ignore the historical and cultural variability of gender and sexuality. There are wide variations from one society to the next in the level of gender inequality, the rate of male violence against women, criteria for mate selection, and all other gender differences that appear universal to the essentialists. This variability deflates the idea that biological constants account for innate behavioral differences between women and men. Three examples help illustrate this point.

1. Women's tendency to stress the "good provider" role in selecting male partners, and men's tendency to stress women's domestic skills, decrease in societies with low levels of gender inequality (Eagley and Wood, 1999). Thus, by changing the level of gender inequality in society you can change male and female criteria for mate selection.

---

[4]Freud called this set of emotions the "Electra complex" after the ancient Greek legend of Electra. Electra persuaded her brother to kill their mother and their mother's lover in order to avenge their father's murder. Incidentally, some sexologists call into question the existence of vaginal orgasm and stress the importance of clitoral stimulation (Masters and Johnson, 1966). This viewpoint emerged around the same time as the modern feminist movement and as more and more people came to view sexuality not just as a means of reproduction but also as a means of enjoyment.

Definitions of "male" and "female" traits vary across societies. For example, the ceremonial dress of male Wodaabe nomads in Niger may appear "feminine" by conventional North American standards.

2. Social situations involving competition and threat stimulate production of the hormone testosterone in women, causing them to act more aggressively. This happens when women become, say, corporate lawyers or police officers (Blum, 1997: 158–88). Thus, by allowing women to take jobs that stress competition and threat, you can change their level of aggressiveness.[5]

3. Gender differences are declining rapidly. Literally hundreds of studies, conducted mainly in the United States, show that women are developing traits that were traditionally considered masculine. Women have become more assertive, competitive, independent, and analytical in the last 3 decades (Twenge, 1997). They play more aggressive sports, choose more math and science courses, do better in standardized tests, take more nontraditional jobs, and earn more money than they used to. One recent study found that, if current trends continue, the difference between male and female math and science scores will disappear in 30–40 years (Nowell and Hedges, 1998: 210; Shea, 1994; see also Duffy, Gunther, and Walters, 1997; Tavris, 1992: 51–2). In what may be a first, standardized math tests administered in Ontario, Canada, in 1998 and 1999 to all Grade 3 and Grade 6 students found that girls outscored boys by 3% in Grade 3 and 2% in Grade 6 (Galt, 1999). It seems that by making school curricula and teaching methods less sexist, opening opportunities for women to study in college and get a wider variety of jobs, and so forth, a whole range of gender differences starts to disappear. As these examples show, then, gender differences are not inherent in men and women. They vary with social conditions.

The second problem with essentialism is that it tends to generalize from the average, ignoring variations within gender groups. On average, women and men do differ in some respects. For example, one of the best-documented gender differences is that men are on average more verbally and physically aggressive than women. However, when sociobiologists and evolutionary psychologists say men are *inherently* more aggressive than women, they make it seem as if this is true of all men and all women. As Figure 9.1 shows, however, it is not. When trained researchers measure verbal or physical aggressiveness, scores vary widely within gender groups. Aggressiveness is distributed so that there is considerable overlap between women and men. Thus, many women are more aggressive than the average man and many men are less aggressive than the average woman.

Third, no evidence directly supports the essentialists' major claims. Sociobiologists and evolutionary psychologists have not identified any of the genes that, they claim, cause male jealousy, female nurturance, the unequal division of labor between men and women,

[5]Women born with higher testosterone levels may gravitate to more stereotypically male jobs in the first place. But given what we know about how high-stress jobs increase testosterone levels, it also seems likely that biological tendencies are accentuated or dampened by occupational demands. This argument is reinforced by research on girls born with unusually high testosterone levels. They prefer rough, aggressive play, but that preference is almost always ratcheted down when they start playing with other girls, who direct them toward standard girls' games. Nature provides; society helps to decide (Blum, 1997: 158–88).

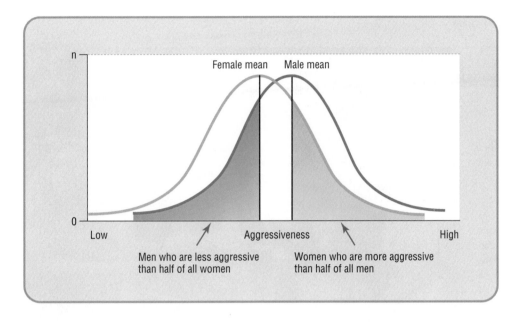

and so forth. Freudians have not collected any experimental or survey data that show boys are more independent than girls because of their emotional reactions to the discovery of their sex organs.

Finally, essentialists' explanations for gender differences ignore the role of power. Sociobiologists and evolutionary psychologists assume that existing behavior patterns help ensure the survival of the species. However, their assumption overlooks the fact that men are usually in a position of greater power and authority than women. Behavioral differences between women and men may therefore result not from any biological imperative but from men being in a position to establish their preferences over the interests of women. Indeed, from this point of view, sociobiology and evolutionary psychology may be seen as examples of the exercise of male power, that is, as a rationalization for male domination and sexual aggression. Much the same may be said of Freud's interpretation. *Must* young girls define themselves in relation to young boys by focusing on their lack of a penis? There is no reason young girls' sexual self-definitions cannot focus positively on their own reproductive organs, including their unique ability to bear children. Freud simply assumes men are superior to women and then invents a speculative theory that justifies gender differences.

Social Constructionism

Social constructionism is the main alternative to essentialism. We illustrate social constructionism by first considering how boys and girls learn masculine and feminine roles in the family and at school (see Box 9.1). We then show how gender roles are maintained in the course of everyday social interaction and through advertising in the mass media. We begin our discussion of social constructionism by examining the effect of an American icon—the Barbie doll—on girls' gender roles.

Gender Socialization

Barbie dolls have been around since 1959. Based on the creation of a German cartoonist, Barbie is the first modern doll modeled after an adult. (Lili, the German original, became a pornographic doll for men.) Some industry experts predicted mothers would never buy dolls with breasts for their little girls. Were *they* wrong! Mattel now sells about 10 million Barbies and 20 million accompanying outfits annually. The Barbie trademark is worth a billion dollars.

What do American girls learn when they play with Barbie? The author of a Web site devoted to Barbie undoubtedly speaks for millions when she writes: "Barbie was more

Web Interactive Exercises
Does Liberalism Cause Sex?

AMERICAN PIE (1999)

What could be more embarrassing than your father catching you masturbating? Perhaps your father bringing you pornography and condoms, and trying to give you a mini-course in sex education? *American Pie,* a popular movie released in the summer of 1999, portrayed these incidents while trying to make sense of sexual coming of age in the United States today.

American Pie focuses on the lives of four high school seniors in a middle-class Michigan suburb. They vow to lose their virginity by the time they graduate. The movie traces the sexual misadventures of the foursome. The jock with a golden heart first tries to seduce a college woman. His macho tactics fail miserably. He then joins a choir and begins to date a conservative young woman. Another one of the foursome studies a book of sexual "secrets" hoping to convince his girlfriend to have intercourse. The third character—the "nerd"—has no prospects. The central character—the boy with the pie—falters miserably with a willing foreign exchange student.

American Pie (1999).

Besides providing an amusing view of four recognizable types of boys in an urban American high school, the movie probes the place of sex in American life. It portrays the attempt to lose one's virginity as an important rite of passage full of missteps, embarrassment, humor, and hypocrisy. Losing one's virginity signifies coming of age—becoming an adult—for these four high school seniors.

We also learn through the longings and antics of these boys that sexual maturation means becoming "gendered." That is, as they develop their sexuality they learn what it means to be masculine in this time and place. Sexuality, we come to understand, helps us define ourselves.

Regardless of whether you've seen the movie, you undoubtedly remember all the talk about sex in high school. What exactly did you talk about when you talked about sex? Was it *just* sex? Or was it gender too? How did talk about sex help you define your masculinity or femininity? How do you think this process differs for girls and boys?

than a doll to me. She was a way of living: the Ideal Woman. When I played with her, I could make her do and be ANYTHING I wanted. Never before or since have I found such an ideal method of living vicariously through anyone or anything. And I don't believe I am alone. I am certain that most people have, in fact, lived their dreams with Barbie as the role player" (Elliott, 1995).

One dream that Barbie stimulates among many girls concerns body image. After all, Barbie is a scale model of a woman with a 40-18-32 figure. The scales that come with Workout Barbie always register a sprightly 110 lb. She can choose from many hundreds of outfits. And, judging from the Barbie sets available, she divides her days mainly between personal hygiene and physical fitness.[6] All this attention to appearance and physical perfection seems to be largely for the benefit of Ken. So when American girls play with Barbie, they learn to want to be slim, blond, and shapely and to exist mainly to please a pleasant man. The Scandinavian rock group Aqua put it well in their 1997 top-10 hit, "Barbie Girl": "Make me walk / Make me talk / I can act like a star / I can beg on my knees . . . You can touch / You can play / If you say / 'I'm always yours.'" Mattel tried to

---

[6]We say "mainly" because in recent years Mattel has released a few Barbie sets that portray Barbie as a professional.

A movement to market more gender-neutral toys emerged in the 1960s and 1970s. However, it has now been overtaken by the resumption of a strong tendency to market toys based on gender.

sue Aqua for its social commentary. A Los Angeles judge tossed the case out of court in May 1998.[7]

A comparable story, with competition and aggression its theme, could be told about how boys' toys, such as GI Joe, teach stereotypical male roles. True, a movement to market more gender-neutral toys arose in the 1960s and 1970s. However, it has now been overtaken by the resumption of a strong tendency to market toys based on gender. As *The Wall Street Journal* recently pointed out, "gender-neutral is out, as more kids' marketers push single-sex products" (Bannon, 2000: B1). For example, in 2000, Toys 'Я' Us took the wraps off a new store design that included a store directory featuring "Boy's World" and "Girl's World." The Boy's World section listed action figures, sports collectibles, radio remote-controlled cars, Tonka trucks, boys' role-playing games, and walkie-talkies. The Girl's World section listed Barbie dolls, baby dolls, collectible horses, play kitchens, housekeeping toys, girls' dress-up, jewelry, cosmetics, and bath and body products.

Yet toys are only part of the story of gender socialization, and hardly its first or final chapter. Research shows that, from birth, infant boys and girls who are matched in length, weight, and general health are treated differently by parents, and fathers in particular. Girls tend to be identified as delicate, weak, beautiful, and cute, boys as strong, alert, and well coordinated (Rubin, Provenzano, and Lurra, 1974). When viewing videotape of a 9-month-old infant, experimental subjects tend to label its startled reaction to a stimulus as "anger" if the child has earlier been identified by the experimenters as a boy, and as "fear" if it has earlier been identified as a girl, *whatever the infant's actual sex* (Condry and Condry, 1976). Parents, and especially fathers, are more likely to encourage their sons to engage in boisterous and competitive play and discourage their daughters from doing likewise. Parents tend to encourage girls to engage in cooperative, role-playing games (MacDonald and Parke, 1986). These different play patterns lead to the heightened development of verbal and emotional skills among girls. They lead to more concern with winning and the establishment of hierarchy among boys (Tannen, 1990). Boys are more likely than girls to be praised for assertiveness, and girls are more likely than boys to be rewarded for compliance (Kerig, Cowan, and Cowan, 1993). Given this early socialization, it seems perfectly "natural" that boys' toys stress aggression, competition, spatial manipulation, and outdoor activities, while girls' toys stress nurturing, physical attractiveness, and indoor activities (Hughes, 1995 [1991]). Still, what seems natural must be continuously socially reinforced. Presented with a choice between playing with a tool set and a dish set, preschool boys are about as likely to choose one as the other—unless the dish set is presented as a girl's toy and they think their fathers would view playing with it as "bad." Then, they tend to pick the tool set (Raag and Rackliff, 1998).

It would take someone who has spent very little time in the company of children to think they are passive objects of socialization. They are not. Parents, teachers, and other authority figures typically try to impose their ideas of appropriate gender behavior on

[7]Ironically, Lene Nystrom, the Norwegian lead singer of *Aqua*, had breast implants in 2000. She upset many women when she told Norway's leading daily newspaper, "I just want to be more feminine" (quoted in "Aqua Singer. . .," 2000).

children. But children creatively interpret, negotiate, resist, and self-impose these ideas all the time. Gender, we might say, is something that is done, not just given (West and Zimmerman, 1987). This is nowhere more evident than in the way children play.

Consider the fourth- and fifth-grade American classroom that sociologist Barrie Thorne (1993) observed. The teacher periodically asked the children to choose their own desks. With the exception of one girl, they always segregated *themselves* by gender. The teacher then drew upon this self-segregation in pitting the boys against the girls in spelling and math contests. These contests were marked by cross-gender antagonism and expression of within-gender solidarity. Similarly, when children played chasing games in the schoolyard, groups often *spontaneously* crystallized along gender lines. These games had special names, some of which, like "chase and kiss," had clear sexual meanings. Provocation, physical contact, and avoidance were all sexually charged parts of the game.

Although Thorne found that contests, chasing games, and other activities often involved self-segregation of boys and girls, she saw many cases of boys and girls playing together. She also noticed quite a lot of "boundary crossing." Boundary crossing involves boys playing stereotypically "girls'" games and girls playing stereotypically "boys'" games. The most common form of boundary crossing involved girls who were skilled at specific sports that were central to the boys' world—sports like soccer, baseball, and basketball. If girls demonstrated skill at these activities, boys often accepted them as participants. Finally, Thorne noticed occasions where boys and girls interacted without strain and without strong gender identities coming to the fore. For instance, activities requiring cooperation, such as a group radio show or art project, lessened attention to gender. Another situation that lessened strain between boys and girls, causing gender to recede in importance, was when adults organized mixed-gender encounters in the classroom and in physical education periods. On such occasions, adults legitimized cross-gender contact. Mixed-gender interaction was also more common in less public and crowded settings. Thus, boys and girls were more likely to play together and in a relaxed way in the relative privacy of their neighborhoods. By contrast, in the schoolyard, where they were under the close scrutiny of their peers, gender segregation and antagonism were more evident.

In sum, Thorne's research makes two important contributions to our understanding of gender socialization. First, children are actively engaged in the process of constructing gender roles. They are not merely passive recipients of adult demands. Second, while school children tend to segregate themselves by gender, boundaries between boys and girls are sometimes fluid and sometimes rigid, depending on social circumstances. In other words, the content of children's gendered activities is by no means fixed.

In her research on schoolchildren, sociologist Barrie Thorne noticed quite a lot of "boundary crossing" between boys and girls. Most commonly, boys accepted girls as participants in soccer, baseball, and basketball games if girls demonstrated skill at these sports.

This is not to suggest that adults have no gender demands and expectations. They do, and their demands and expectations contribute importantly to gender socialization. For instance, in most schools, teachers and guidance counselors still expect boys to do better in the sciences and math. They expect girls to achieve higher marks in English. Parents, for their part, tend to reinforce these stereotypes in their evaluation of different activities (Eccles, Jacobs, and Harold, 1990). Significantly, research comparing mixed- and single-sex schools shows that girls do much better in the latter. Sharlene Hesse-Biber and Gregg Lee Carter (2000: 99–100) summarize this research and are worth quoting at length:

> [In girls-only schools], female cognitive development is greater; female occupational aspirations and their ultimate attainment are increased; female self-confidence and self-esteem are magnified. Moreover, . . . females receive better treatment in the classroom; they are more likely to be encouraged to explore—and to have access to—wider curriculum opportunities; and teachers have greater respect for their work. Finally, females attending single-sex schools have . . . more egalitarian attitudes towards the role of women in society than do their counterparts in mixed-sex schools. . . . Single-sex schools accrue these benefits for girls for a variety of reasons . . . : (1) a diminished emphasis on "youth culture," which centers on athletics, social life, physical attractiveness, heterosexual popularity, and negative attitudes toward academics; (2) the provision of more successful same-sex role models (the top students in all subjects and all extracurricular activities [are] girls); (3) a reduction in sex bias in teacher-student interaction (there are [no] boys around [who] can be "favoured"); and (4) elimination of sex stereotypes in peer interaction (generally, cross-sex peer interaction in school involves male dominance, male leadership, and, often, sexual harassment).

Adolescents must usually start choosing courses in school by the age of 14 or 15. By then, their **gender ideologies** are well formed. Gender ideologies are sets of interrelated ideas about what constitutes appropriate masculine and feminine roles and behavior. One aspect of gender ideology becomes especially important around grades 9 and 10: adolescents' ideas about whether, as adults, they will focus mainly on the home, paid work, or a combination of the two. Adolescents usually make course choices with gender ideologies in mind. Boys are strongly inclined to consider only their careers in making course choices. Most girls are inclined to consider both home responsibilities and careers, although a minority considers only home responsibilities and another minority considers only careers. As a result, boys tend to choose career-oriented courses, particularly in math and science, more often than girls. In college, the pattern is accentuated. Young women tend to choose easier courses that lead to lower paying jobs because they expect to devote a large part of their lives to child rearing and housework (Hochschild with Machung, 1989: 15–18; Machung, 1989). The effect of these choices is to sharply restrict women's career opportunities and earnings in science and business (see Table 9.1). We examine this problem in depth in the second half this chapter.

The Mass Media and Body Image

The social construction of gender does not stop at the school steps. Outside school, children, adolescents, and adults continue to negotiate gender roles as they interact with the mass media. If you systematically observe the roles played by women and men in TV

Academic Major	Starting Salary	Percent Women in Occupation
1. Chemical Engineering	$42,758	18.8
2. Mechanical Engineering	39,852	6.2
3. Electrical Engineering	39,811	9.0
4. Industrial Engineering	37,732	15.7
5. Computer Science	36,964	29.2

✦ **TABLE 9.1** ✦
Academic Majors With the Highest Starting Salaries, United States, 1996–97

SOURCES: Calculated from ("Estimated Starting Salaries . . ." 1997; United States Department of Labor, 1998a).

programs and ads one evening, you will probably discover a pattern noted by sociologists since the 1970s. Women will more frequently be seen cleaning house, taking care of children, modeling clothes, and acting as objects of male desire. Men will more frequently be seen in aggressive, action-oriented, and authoritative roles. The effect of these messages on viewers is much the same as that of the Disney movies and Harlequin romances we discussed in Chapter 4 ("Socialization"). They reinforce the normality of traditional gender roles. As we will now see, many people even try to shape their bodies after the body images portrayed in the mass media.

The human body has always served as a sort of personal billboard that advertises gender. However, historian Joan Jacobs Brumberg (1997) makes a good case for the view that the importance of body image to our self-definition has grown over the past century. Just listen to the difference in emphasis on the body in the diary resolutions of two typical, white, middle-class American girls, separated by a mere 90 years. From 1892: "Resolved, not to talk about myself or feelings. To think before speaking. To work seriously. To be self restrained in conversation and actions. Not to let my thoughts wander. To be dignified. Interest myself more in others." From 1982: "I will try to make myself better in any way I possibly can with the help of my budget and baby-sitting money. I will lose weight, get new lenses, already got new haircut, good makeup, new clothes and accessories" (quoted in Brumberg, 1997: xxi).

As body image became more important for one's self-definition in the course of the 20th century, the ideal body image became thinner, especially for women. Thus, the first American "glamour girl" was Mrs. Charles Dana Gibson, who was famous in advertising and society cartoons in the 1890s and 1900s as the "Gibson Girl." According to the Metropolitan Museum of Art's Costume Institute, "[e]very man in America wanted to win her" and "every woman in America wanted to be her. Women stood straight as poplars and tightened their corset strings to show off tiny waists" (Metropolitan Museum of Art, 2000). As featured in the *Ladies Home Journal* in 1905, the Gibson Girl measured 38-27-45—certainly not slim by today's standards. During the 20th century, however, the ideal female body type thinned out. The "White Rock Girl," featured on the logo of the White Rock beverage company, was 5′4″ and weighed 140 lb. in 1894. In 1947, she had slimmed down to 125 lb. By 1970, she was 5′8″ and 118 lb. (Peacock, 2000).

Why did body image become more important to people's self-definition during the 20th century? Why was slimness stressed? Part of the answer to both questions is that more Americans grew overweight as their lifestyles became more sedentary. As they became

The "White Rock Girl," featured on the logo of the White Rock beverage company, dropped 15 pounds between 1894 (left) and 1947 (right).

The low-cal and diet food industry promotes an ideal of slimness that is often impossible to attain and that generates widespread body dissatisfaction.

better educated, they also grew increasingly aware of the health problems associated with being overweight. The desire to slim down was, then, partly a reaction to bulking up. But that is not the whole story. The rake-thin models who populate modern ads are not promoting good health. They are promoting an extreme body shape that is virtually unattainable for most people. They do so because it is good business. In 1990, the United States diet and lo-cal frozen entrée industry alone enjoyed revenues of nearly $700 million. Some 65 million Americans spent upwards of $30 billion in the diet and self-help industry in the pursuit of losing weight. The fitness industry generated $43 billion in revenue and the cosmetic surgery industry another $5 billion (Hesse-Biber, 1996: 35, 39, 51, 53). Bankrolled by these industries, advertising in the mass media blankets us with images of slim bodies and makes these body types appealing. Once people become convinced that they need to develop bodies like the ones they see in ads, many of them are really in trouble because these body images are very difficult for most people to attain.

Survey data show just how widespread dissatisfaction with our bodies is and how important a role the mass media play in generating our discomfort. For example, a 1997 survey of North American college graduates showed that 56% of women and 43% of men were dissatisfied with their overall appearance (Garner, 1997). Only 3% of the dissatisfied women, but 22% of the dissatisfied men, wanted to gain weight. This reflects the greater desire of men for muscular, stereotypically male physiques. Most of the dissatisfied men, and even more of the dissatisfied women (89%), wanted to lose weight. This reflects the general societal push toward slimness and its greater effect on women.

Figure 9.2 reveals gender differences in body ideals in a different way. It compares women's and men's attitudes towards their stomachs. It also compares women's attitudes toward their breasts with men's attitudes toward their chests. It shows, first, that women are more concerned about their stomachs than men are. Second, it shows that by 1997 men were more concerned about their chests than women were about their breasts. Clearly, then, people's body ideals are influenced by their gender. Note also that Figure 9.2 shows

✦ **FIGURE 9.2** ✦

**Body Dissatisfaction,
United States, 1972–1997
(in percent, n = 4,000)**

Note: The n of 4,000 refers to the 1997
survey only. The number of respondents
in the earlier surveys was not given.

SOURCE: Garner (1997).

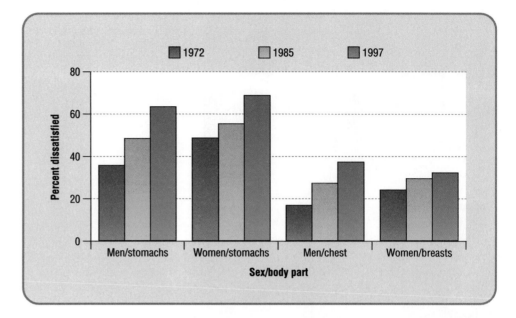

✦ **TABLE 9.2** ✦

**The Influence of Fashion
Models on Feelings About
Appearance, North America,
1997 (in percent; n = 4,000)**

SOURCE: Adapted from Garner (1997).

	Men	Women	Extremely Dissatisfied Women
I always or often:			
Compare myself to models in magazines	12	27	43
Carefully study the shape of models	19	28	47
Very thin or muscular models make me:			
Feel insecure about my weight	15	29	67
Want to lose weight	18	30	67

trends over time. North Americans' anxiety about their bodies increased substantially between 1972 and 1997.[8]

Table 9.2 suggests that advertising is highly influential in creating anxiety and insecurity about appearance, and particularly about body weight. Here we see that in 1997 nearly 30% of North American women compared themselves with the fashion models they saw in advertisements, felt insecure about their own appearance, and wanted to lose weight as a result. Among women who were dissatisfied with their appearance, the percentages were much larger, with about 45% making comparisons with fashion models and two thirds feeling insecure and wanting to lose weight. It seems safe to conclude that fashion models stimulate body dissatisfaction among many North American women.

Body dissatisfaction, in turn, motivates many women to diet. Because of anxiety about their weight, 84% of North American women said they had dieted in the 1997 survey. The comparable figure for men was 54%. Just how important is it for people to achieve their weight goals? According to the survey, it's a life or weight issue: 24% of women and 17% of men said they would trade more than 3 years of their lives to achieve their weight goals.

Body dissatisfaction prompts some people to take dangerous and even life-threatening measures to reduce. In the 1997 survey, 50% of female smokers and 30% of male smokers said they smoke to control their weight. Other surveys suggest that between 1% and 5% of American women suffer from anorexia, or refusal to eat enough to remain healthy.

---

[8]Slimness is somewhat less important for African-American women, who in general have healthier attitudes towards their bodies than white women (Molloy and Herzberger, 1998.)

About the same percentage of American female college students suffer from bulimia, or regular self-induced vomiting. For college men, the prevalence of bulimia is between 0.2% and 1.5% (Averett and Korenman, 1996: 305–6). In the United Kingdom, eating disorders are just as common, and the British Medical Association has warned that celebrities such as *Ally McBeal* star Calista Flockhart are contributing to a rise in anorexia and bulimia. However, British magazine editors are taking some responsibility for the problem. They recognize that waif-thin models are likely influencing young women to feel anxious about their weight and shape. As a result, the editors recently drew up a voluntary code of conduct that urges them to monitor the body images they portray, impose a minimum size for models, and use models of varying shapes and sizes ("British Magazines . . .," 2000). Whether similar measures are adopted in the United States remains to be seen.

Male-Female Interaction

The gender roles children learn in their families, at school, and through the mass media form the basis for their social interaction as adults. For instance, by playing team sports, boys tend to learn that social interaction is most often about competition, conflict, self-sufficiency, and hierarchical relationships (leaders vs. led). They understand the importance of taking center stage and boasting about their talents (Messner, 1995 [1989]). Since many of the most popular video games for boys exclude female characters, use women as sex objects, or involve violence against women, they reinforce some of the most unsavory lessons of traditional gender socialization (Dietz, 1998). On the other hand, by playing with dolls and baking sets, girls tend to learn that social interaction is most often about maintaining cordial relationships, avoiding conflict, and resolving differences of opinion through negotiation. They understand the importance of giving advice and not promoting themselves or being bossy.

Because of these early socialization patterns, misunderstandings between men and women are common. A stereotypical example: Harold is driving around lost. However, he refuses to ask for directions because doing so would amount to an admission of inadequacy and therefore a loss of status. Meanwhile, it seems perfectly "natural" to Sybil to want to share information, so she urges Harold to ask for directions. The result: conflict between Harold and Sybil (Tannen, 1990: 62).

Gender-specific interaction styles also have serious implications for who gets heard and who gets credit at work. Here are some examples uncovered by Deborah Tannen's research (1994a: 132–59):

✦   A female office manager doesn't want to seem bossy or arrogant. She is eager to preserve consensus among her coworkers. So she spends a good deal of time soliciting their opinions before making an important decision. She asks questions, listens attentively, and offers suggestions. She then decides. But her boss perceives her approach as indecisive and incompetent. He wants to recruit leaders for upper-management positions, so he overlooks the woman and selects an assertive man for a senior job that just opened up.

✦   A female technical director at a radio station wants to help a nervous new male soundboard operator do a good job. However, she is sensitive to the possibility that giving him direct orders may make him feel incompetent and cause him to do worse. So instead of instructing him, she starts a conversation about Macintosh computers, something he knows a lot about. This makes him feel capable and relaxed, and he sits back and puts his feet up. She then talks about some technical issues. She is careful to put everything in the context of an upcoming show (something the new soundboard operator couldn't possibly know about) rather than general technical knowledge (something he should have). Because of her sensitive management style, the show goes off without a hitch. Thankfully, the technical director's male supervisor did not come into the studio when she was making queries and the soundboard operator had his feet up. If the supervisor had arrived then, he could easily have concluded the technical director was so incompetent she had to get information from a subordinate who had just been hired.

✦ Male managers are inclined to say "I" in many situations where female managers are inclined to say "we"—as in "I'm hiring a new manager and I'm going to put him in charge of my marketing division" or "This is what I've come up with on the Lakehill deal." This sort of phrasing draws attention to one's personal accomplishments. In contrast, Tannen heard a female manager talking about what "we" had done when in fact she had done all the work alone. This sort of phrasing camouflages women's accomplishments.

The contrasting interaction styles illustrated above often result in female managers not getting credit for competent performance. That is why they sometimes complain about a **glass ceiling,** a social barrier that makes it difficult for them to rise to the top level of management. As we will soon see, factors other than interaction styles, such as outright discrimination and women's generally greater commitment to family responsibilities, also support the glass ceiling. Yet gender differences in interaction styles play an independent role in constraining women's career progress.

Homosexuality

The preceding discussion outlines some powerful social forces pushing us to define ourselves as conventionally masculine or feminine in behavior and appearance. For most people, gender socialization by the family, the school, and the mass media is compelling and it is sustained by daily interactions. A minority of people, however, resists conventional gender roles. For example, **transgendered** people are individuals who want to alter their gender by changing their appearance or resorting to medical intervention. About 1 in every 5,000 to 10,000 people in North America is transgendered. Some transgendered people are **transsexuals.** Transsexuals believe they were born with the "wrong" body. They identify with, and want to live fully as, members of the "opposite" sex. They often take the lengthy and painful path to a sex change operation. About 1 in every 30,000 people in North America is a transsexual (Nolen, 1999). **Homosexuals** are people who prefer sexual partners of the same sex, and **bisexuals** are people who prefer sexual partners of both sexes. People usually call homosexual men gay and homosexual women lesbians. The most comprehensive survey of sexuality in the United States shows that 2.8% of American men and 1.4% of American women think of themselves as homosexual or bisexual. However, 10.1% of men and 8.6% of women (a) think of themselves as homosexual or bisexual, (b) have had some same-sex experience, or (c) have had some same-sex desire (see Table 9.3) (Laumann, Gagnon, Michael, and Michaels, 1994: 299).

Homosexuality has existed in every society. Some societies, such as ancient Greece, have encouraged it. More frequently, however, homosexual acts have been forbidden. Until the late 18th and early 19th centuries in Western Europe and the United States, "unnatural acts" such as sodomy were punishable by death. However, homosexuals were not identified as a distinct category of people until the 1860s. That is when the term "homosexuality" was coined. The term "lesbian" is of even more recent vintage.

We do not yet understand well why some individuals develop homosexual orientations. Some scientists think the reasons are mainly genetic, others think they are chiefly hormonal, while still others point to life experiences during early childhood as the most

✦ **TABLE 9.3** ✦
Homosexuality in the United States, 1992 (in percent; n = 3,432)

SOURCE: Michael, Gagnon, Laumann, and Kolata (1994: 40).

	Men	Women
Identified themselves as homosexual or bisexual	2.8	1.4
Had sex with person of same sex in past 12 months	3.4	0.6
Had sex with person of same sex at least once since puberty	5.3	3.5
Felt desire for sex with person of same sex	7.7	7.5
Had some same-sex desire or experience or identified themselves as homosexual or bisexual	10.1	8.6

On April 1, 2001, the Netherlands recognized full and equal marriage rights for homosexual couples. Within hours, Dutch citizens were taking advantage of the new law. The Dutch law is part of a worldwide trend to legally recognize long-term same-sex unions.

important factor. We do know that sexual orientation does not appear to be a choice. According to the American Psychological Association, it "emerges for most people in early adolescence without any prior sexual experience . . . [it] is not changeable" ("American Psychological Association," 1998).

In any case, sociologists are less interested in the origins of homosexuality than in the way it is socially constructed, that is, in the wide variety of ways it is expressed and repressed (Foucault, 1990 [1978]; Weeks, 1986). It is important to note in this connection that homosexuality has become less of a stigma over the past century. Two factors are chiefly responsible for this, one scientific, the other political. In the 20th century, sexologists—psychologists and physicians who study sexual practices scientifically—first recognized and stressed the wide diversity of existing sexual practices. The American sexologist Alfred Kinsey was among the pioneers in this field. He and his colleagues interviewed thousands of men and women. In the 1940s, they concluded that homosexual practices were so widespread that homosexuality could hardly be considered an illness affecting a tiny minority (Kinsey, Pomeroy, and Martin, 1948).

Sexologists, then, provided a scientific rationale for belief in the normality of sexual diversity. However, it was sexual minorities themselves who provided the social and political energy needed to legitimize sexual diversity among an increasingly large section of the public. Especially since the middle of the 20th century, gays and lesbians have built large communities and subcultures, especially in major urban areas like New York and San Francisco. They have gone public with their lifestyles. They have organized demonstrations, parades, and political pressure groups to express their self-confidence and demand equal rights with the heterosexual majority. This has done much to legitimize homosexuality and sexual diversity in general.

Yet opposition to people who don't conform to conventional gender roles remains strong at all stages of the life cycle. When you were a child, did you ever laugh at a girl who, say, liked to climb trees and play with toy trucks? Did you ever call such a girl a "tomboy?" When you were a child, did you ever tease a boy who, say, liked to bake muffins while listening to Mozart? Did you ever call such a boy a "sissy?" If so, your behavior was not unusual. Children are typically strict about enforcing conventional gender roles. They often apply sanctions against playmates who deviate from convention.

Among adults, such opposition is just as strong. What is your attitude today toward transgendered people, transsexuals, and homosexuals? Do you, for example, think that sexual relations between adults of the same sex are always, or almost always, wrong? If

so, you are again not unusual. According to the 1998 General Social Survey, fully 64% of Americans believe that sexual relations between adults of the same sex are always, or almost always, wrong (National Opinion Research Center, 1999).

Antipathy to homosexuals is so strong among some people that they are prepared to back up their beliefs with force. A 1998 study of about 500 young adults in the San Francisco Bay area (probably the most sexually tolerant area in the United States) found that 1 in 10 admitted physically attacking or threatening people they believed were homosexuals. Twenty-four percent reported engaging in antigay name-calling. Among male respondents, 18% reported acting in a violent or threatening way and 32% reported name-calling. In addition, a third of those who had *not* engaged in antigay aggression said they would do so if a homosexual flirted with, or propositioned, them (Franklin, 1998).

Due to widespread animosity toward homosexuals, many people who have wanted sex with members of the same sex, or who have had sex with someone of the same sex, do not identify themselves as gay, lesbian, or bisexual. The operation of norms against homosexuality is evident in figures on the geographical distribution of people who identify themselves as homosexuals or bisexuals. More than 9% of the residents of America's 12 largest cities (excluding suburbs) identify themselves as homosexual or bisexual. That figure falls to just over 4% in the cities ranked 13–100 by size (again excluding suburbs), and it drops to a little over 1% in rural areas. Why? Because small population centers tend to be less tolerant of homosexuality and bisexuality. People with same-sex desires are therefore less likely to express and develop homosexual and bisexual identities in smaller communities. They are inclined to migrate to larger and more liberal cities where supportive, established gay communities exist (Michael, Gagnon, Laumann, and Kolata, 1994: 178, 182).

Recent research suggests some antigay crimes may result from repressed homosexual urges on the part of the aggressor (Adams, Wright, and Lohr, 1998). From this point of view, aggressors are "homophobic" or afraid of homosexuals because they cannot cope with their own, possibly subconscious, homosexual impulses. Their aggression is a way of acting out a denial of these impulses. However, while this psychological explanation may account for some antigay violence, it seems inadequate when set alongside the finding that fully half of all young male adults admitted to some form of antigay aggression in the San Francisco Bay area study cited above. An analysis of the motivations of these San Franciscans showed that some of them did commit assaults to prove their toughness and heterosexuality. Others committed assaults just to alleviate boredom and have fun. Still others believed they were defending themselves from aggressive sexual propositions. A fourth group acted violently because they wanted to punish homosexuals for what they perceived as moral transgressions (Franklin, 1998). It seems clear, then, that antigay violence is not just a question of abnormal psychology but a broad, cultural problem with several sources.

On the other hand, anecdotal evidence suggests that opposition to antigay violence is also growing in America. The 1998 murder of Matthew Shepard in Wyoming led to a public outcry. In the wake of his murder, some people called for a broadening of the definition of hate crime to include antigay violence (see Box 9.2). The 1999 movie *Boys Don't Cry* also raised awareness of the problem of antigay violence. The movie, for which Hilary Swank won the Best Actress Oscar, tells the true story of Teena Brandon, a young woman in Nebraska with a sexual identity crisis. She wants a sex-change operation but can't afford one. So she decides to change her name to Brandon Teena and "pass" as a man. She soon develops an intimate relationship with a woman by the name of Lana Tisdel. Tisdel knows Teena is anatomically a female. However, when two male members of Tisdel's family discover the truth about Teena, they beat, rape, and murder her. Teena's only transgression was that she wanted to be a man.

In sum, strong social and cultural forces lead us to distinguish men from women and heterosexuals from homosexuals. We learn these distinctions throughout the socialization process, and we continuously construct them anew in our daily interactions. Most people use positive and negative sanctions to ensure that others conform to conventional heterosexual gender roles. Some people resort to violence to enforce conformity and punish deviance.

Especially since the middle of the 20th century, gays and lesbians have built large communities and subcultures, especially in major urban areas such as New York and San Francisco. They have gone public with their lifestyles. They have organized demonstrations, parades, and political pressure groups to express their self-confidence and demand equal rights with the heterosexual majority. This has done much to legitimize homosexuality and sexual diversity in general.

BOX 9.2
IT'S YOUR CHOICE

HATE CRIME LAW AND HOMOPHOBIA

On October 7, 1998, Matthew Shepard, an undergraduate at the University of Wyoming, went to a campus bar in Laramie. From there, he was lured by Aaron James McKinney and Russell Henderson, both 21, to an area just outside town. McKinney and Henderson apparently wanted to rob Shepard. They wound up murdering him. They used the butt of a gun to beat Shepard's head repeatedly. According to the prosecutor at the trial of the two men, "[a]s

[Shepard] lay there bleeding and begging for his life, he was then bound to the buck fence" (quoted in CNN, 1998). The murderers left Shepard there in near-freezing temperatures. When a passerby saw Shepard several hours later, he thought the nearly dead man was a "scarecrow or a dummy set there for Halloween jokes" (quoted in CNN, 1998). Shepard died 5 days later.

One issue raised by Shepard's death concerns the definition of hate crime. Hate crimes are criminal acts motivated by a person's race, religion, or ethnicity. If hate motivates a crime, the law requires that the perpetrator be punished more severely than otherwise. For example, assaulting a person during an argument generally carries a lighter punishment than assaulting a person because he is an African American. Furthermore, the law (18 United States C. 245) permits federal prosecution of a hate crime only "if the crime was motivated by bias based on race, religion, national origin, or

color, and the assailant intended to prevent the victim from exercising a 'federally protected right' (e.g., voting, attending school, etc.)" (Human Rights Campaign, 1999). This definition excludes crimes motivated by the sexual orientation of the victim. According to the FBI, if crimes against gays, lesbians, and bisexuals were defined as hate crimes, they would have composed 14% of the total in 1998 (Human Rights Campaign, 1999). Matthew Shepard was 1 of 33 anti-gay murders in the United States in 1998, up from 14 the year before (Human Rights Campaign, 1999).

Do you think crimes motivated by the victim's sexual orientation are the same as crimes motivated by the victim's race, religion, or ethnicity? If so, why? If not, why not? Do you think crimes motivated by the sexual orientation of the victim should be included in the legal definition of hate crime? If so, why? If not, why not?

Our presentation also suggests the social construction of conventional gender roles helps to create and maintain social inequality between women and men. In the remainder of this chapter, we examine the historical origins and some of the present-day consequences of gender inequality.

GENDER INEQUALITY

The Origins of Gender Inequality

Contrary to what essentialists say, men have not always enjoyed much more power and authority than women. Substantial inequality between women and men has existed for only about 6,000 years. It was socially constructed. Three major sociohistorical processes account for the growth of gender inequality. Let us briefly consider each of them.

Long-Distance Warfare and Conquest

The anthropological record suggests that women and men were about equal in status in nomadic hunting-and-gathering societies, the dominant form of society for 90% of human history. Rough gender equality was based on the fact that women produced a substantial amount of the band's food, up to 80% in some cases (see Chapter 12, "Families"). The archeological record from "Old Europe" tells a similar story. Old Europe is a region stretching roughly from Poland in the north to the Mediterranean island of Crete in the south, and from Switzerland in the west to Bulgaria in the east (see Figure 9.3). Between 7,000 and 3,500 B.C.E., men and women enjoyed approximately equal status throughout the region. In fact, the religions of the region gave primacy to fertility and creator goddesses. Kinship was traced through the mother's side of the family. Then, sometime

✦ FIGURE 9.3 ✦
Old Europe

SOURCE: Gimbutas (1982:16).

Women's domestic role was idealized in the 19th century.

between 4,300 and 4,200 B.C.E., all this began to change. Old Europe was invaded by successive waves of warring peoples from the Asiatic and European northeast (the Kurgans) and the deserts to the south (the Semites). Both the Kurgan and Semitic civilizations were based on a steeply hierarchical social structure in which men were dominant. Their religions gave primacy to male warrior gods. They acquired property and slaves by conquering other peoples and imposed their religions on the vanquished. They eliminated, or at least downgraded, goddesses as divine powers. God became a male who willed that men should rule women. Laws reinforced women's sexual, economic, and political subjugation to men. Traditional Judaism, Christianity, and Islam all embody ideas of male dominance, and they all derive from the tribes who conquered Old Europe in the fifth millennium B.C.E. (Eisler, 1995 [1987]; see also Lerner, 1986).

Plow Agriculture

Long-distance warfare and conquest catered to men's strengths and so greatly enhanced male power and authority. Large-scale farming using plows harnessed to animals had much the same effect. Plow agriculture originated in the Middle East around 5,000 years ago. It required that strong adults remain in the fields all day for much of the year. It also reinforced the principle of private ownership of land. Since men were on average stronger than women, and since women were restricted in their activities by pregnancy, nursing, and childbirth, plow agriculture made men more powerful socially. Thus, land was owned by men and ownership was typically passed from father to eldest son (Coontz and Henderson, 1986).

The Separation of Public and Private Spheres

In the agricultural era, economic production was organized around the household. Men may have worked apart from women in the fields, but the fields were still part of the *family* farm. In contrast, during the early phase of industrialization, men's work moved out of the household and into the factory and the office. Most men became wage or salary workers. Some men assumed decision-making roles in economic and political institutions. Yet while men went public, most women remained in the domestic or private sphere. The idea soon developed that this was a natural division of labor. This idea persisted until the second half of the 20th century, when a variety of social circumstances, ranging from the introduction of the birth control pill to women's demands for entry into college, finally allowed women to enter the public sphere in large numbers.

So we see that, according to social constructionists, gender inequality derives not from any inherent biological features of men and women but from three main sociohistorical circumstances: the arrival of long-distance warfare and conquest, the development of plow agriculture, and the assignment of women to the domestic sphere and men to the public sphere during the early industrial era.

The Earnings Gap Today

After reading this brief historical overview, you might be inclined to dismiss gender inequality as a thing of the past. If so, your decision would be hasty. That is evident if we focus first on the earnings gap between men and women, one of the most important expressions of gender inequality today. In the first quarter of 2000, women over the age of 15 working full-time in the paid labor force earned only 78.3% of what men earned (United States Department of Labor, 2000b). Four main factors contribute to the gender gap in earnings. Let us consider each of them in turn (Bianchi and Spain, 1996; England, 1992).

Gender discrimination. In February 1985, when Microsoft, the software giant, employed about 1,000 people, it hired its first two female executives. According to a well-placed source who was involved in the hiring, both women got their jobs because Microsoft was trying to win a United States Air Force contract. Under the government's guidelines, it didn't have enough women in top management positions to qualify. The source quotes then 29-year-old Bill Gates, President of Microsoft, as saying: "Well, let's hire two women because we can pay them half as much as we will have to pay a man, and we can give them all this other 'crap' work to do because they are women" (quoted in Wallace and Erickson, 1992: 291).

This incident is a clear illustration of **gender discrimination,** rewarding women and men differently for the same work. Gender discrimination has been illegal in the United States since 1964. It has not disappeared, as the above anecdote confirms. However, anti-discrimination laws have helped to increase the **female–male earnings ratio,** that is, women's earnings as a percentage of men's earnings. The female–male earnings ratio increased 17.3% between 1960 and 2000. At that rate of improvement, women will be earning as much as men by 2050, around the time most first-year college students today retire (calculated from Feminist.com, 1999; United States Department of Labor, 2000b).

Heavy domestic responsibilities reduce women's earnings. Raising children can be one of the most emotionally satisfying experiences in life. However, it is so exhausting and time-consuming, and requires so many interruptions due to pregnancy and illness, it substantially decreases the time one can spend getting training and doing paid work. Since women are disproportionately involved in child rearing, they suffer the brunt of this economic reality. Women also do more housework and elderly care than men. Specifically, in most countries, including the United States, women do between two thirds and three quarters of all unpaid child care, housework, and care for the elderly (Boyd, 1997: 55). As a result, they devote fewer hours to paid work than men, experience more labor-force interruptions, and are more likely than men to take part-time jobs. Part-time jobs pay less per hour and offer fewer benefits than full-time work. Even when they work full-time in the paid labor force, women continue to shoulder a disproportionate share of domestic responsibilities, working, in effect, a "double shift" (Hochschild with Machung, 1989; see Chapter 12, "Families"). This affects how much time they can devote to their jobs and careers, with negative consequences for their earnings (Mahony, 1995; Waldfogel, 1997).

Women tend to be concentrated in low-wage occupations and industries. The third factor leading to lower earnings for women is that the courses they select in high school and college tend to limit them to jobs in low-wage occupations and industries. Thus,

Although women have entered many traditionally "male" occupations since the 1970s, they are still concentrated in lower paying clerical and service occupations and underrepresented in higher paying manual occupations.

✦ **TABLE 9.4** ✦
Women as a Percentage of Total Employed by Broad Occupational Division, United States, 1975 and 1995

SOURCE: Wooton (1997: 17).

Occupation	% Women, 1975	% Women, 1995
Managerial and professional		
Executive, administrative, and managerial	21.9	42.7
Professional	45.3	52.9
Technical, sales, and administrative support		
Technicians and related support	41.5	51.4
Sales	41.9	49.5
Administrative support, including clerical	77.2	79.5
Service		
Private household	97.5	95.5
Protective service	7.1	15.9
Other service	64.4	65.0
Precision production, craft, and repair	5.5	8.9
Operators, fabricators, and laborers		
Machine operators, assemblers, and inspectors	38.7	37.3
Transportation and material moving	4.8	9.5
Handlers, equipment cleaners, helpers, and laborers	16.9	19.1
Farming, forestry, and fishing	14.0	19.9
TOTAL	39.6	46.1

although women have made big strides since the 1970s, especially in managerial employment, they are still concentrated in lower paying clerical and service occupations and underrepresented in higher paying manual occupations (see Table 9.4). For example, over 95% of the people who provide private household services are women, compared to about 43% of the people with executive, administrative, and managerial jobs. Moreover, *within* the broad occupational divisions listed in Table 9.4, lower earnings are associated with occupations where women are concentrated (Hesse-Biber and Carter, 2000: 114–73).

Work done by women is commonly considered less valuable than work done by men because it is viewed as involving fewer skills. Finally, women tend to earn less than men because the skills involved in their work are often undervalued (Figart and Lapidus, 1996; Sorenson, 1994). Compare office machine repair technicians, 94.1% of whom were men in

1997, with prekindergarten and kindergarten teachers, 97.6% of whom were women. The man who repaired photocopiers earned an average of $548 a week while the woman who taught and played with 5-year-olds earned an average of $405 (United States Department of Labor, 1998a). It is, however, questionable whether it takes less training and skill to teach a young child the basics of counting and reading and cooperation and sharing than it takes to get a photocopier to collate paper properly. As this example suggests, we apply somewhat arbitrary standards to reward different occupational roles. In our society, these standards systematically undervalue the kind of skills needed for jobs where women are concentrated.

We thus see that the gender gap in earnings is based on several *social* circumstances rather than any inherent difference between women and men. This means that people can reduce the gender gap if they want to. Below, we discuss social policies that could create more equality between women and men. But first, to stress the urgency of such policies, we explain how the persistence of gender inequality encourages sexual harassment and rape.

Male Aggression Against Women

Serious acts of aggression between men and women are common. The great majority are committed by men against women. For example, in 1995, more than 340,000 rapes and sexual assaults were reported to the police in the United States. More than 90% of the victims were women, and nearly all the perpetrators were men (Maguire and Pastore, 1998: 198, 181). Among young singles, the rate of rape is higher than in the population as a whole. Thus, in a survey of acquaintance and date rape in American colleges, 7% of men admitted they attempted or committed rape in the past year. Eleven percent of women said they were victims of attempted or successful rape (Koss, Gidycz, and Wisniewski, 1987).

Why do men commit more frequent (and more harmful) acts of aggression against women than women commit against men? It is *not* because men on average are *physically* more powerful than women. Greater physical power is more likely to be used to commit acts of aggression when norms justify male domination and men have much more *social* power than women. When women and men are more equal socially, and norms justify gender equality, the rate of male aggression against women is lower. This is evident if we consider various types of aggressive interaction, including rape and sexual harassment (see also the discussion of wife abuse in Chapter 12, "Families").

Web Research Projects
Marital Rape

Rape

Some people think rapists are men who suffer a psychological disorder that compels them to achieve immediate sexual gratification even if violence is required. Others think rape occurs because of flawed communication. They believe some rape victims give mixed signals to their assailants by, for example, drinking too much and flirting with them.

Such explanations are not completely invalid. Interviews with victims and perpetrators show that some rapists do suffer from psychological disorders. Other offenders do misinterpret signals in what they regard as sexually ambiguous situations (Hannon, Hall, Kuntz, Van Laar, and Williams, 1995). But such cases account for only a small proportion of the total. Men who rape women are rarely mentally disturbed, and it is abundantly clear to most assailants that they are doing something their victims strongly oppose.

What then accounts for rape being as common as it is? A sociological answer is suggested by the fact that rape is sometimes not about sexual gratification at all. Some rapists cannot ejaculate. Some cannot even achieve an erection. Significantly, however, all rape involves domination and humiliation as principal motives. It is not surprising, therefore, that some rapists are men who were physically or sexually abused in their youth. They develop a deep need to feel powerful as psychological compensation for their early powerlessness. Other rapists are men who, as children, saw their mothers as potentially hostile figures who needed to be controlled, or as mere objects available for male gratification. They saw their fathers as emotionally cold and distant. Raised in such an atmosphere, rapists learn not to empathize with women. Instead, they learn to want to dominate them (Lisak, 1992).

Psychological factors aside, certain *social* situations also increase the rate of rape. One such situation is war. In war, conquering male soldiers often feel justified in wanting to humiliate the vanquished, who are powerless to stop them. Rape is often used for this

A San Francisco billboard suggests men still need reminding that no means no.

purpose, as was especially well documented in the ethnic wars that accompanied the breakup of Yugoslavia in the 1990s (Human Rights Watch, 1995).

Aggressiveness is also a necessary and important part of police work. Spousal abuse is therefore common among police officers. One United States study found that 37% of anonymously interviewed police wives reported spousal abuse. Several other surveys of police officers put the figure in the 40% range (Roslin, 2000). "It's a horrible, horrible problem," says Penny Harrington, former chief of police in Portland, Oregon, and now head of the Los Angeles–based National Center for Women and Policing. "Close to half of all 911 calls are due to family violence," says Harrington. "If the statistics are true, you've got a two-in-five chance of getting a batterer coming to answer your call" (quoted in Roslin, 2000: 46).

The relationship between male dominance and rape is also evident in research on college fraternities. Many college fraternities tend to emphasize male dominance and aggression as a central part of their culture. Thus, sociologists who have interviewed fraternity members have shown that most fraternities try to recruit members who can reinforce a macho image and avoid any suggestion of effeminacy and homosexuality. Research also shows that fraternity houses that are especially prone to rape tend to sponsor parties that treat women in a particularly degrading way. Thus, by emphasizing a very narrow and aggressive form of masculinity, some fraternities tend to facilitate rape on college campuses (Boswell and Spade, 1996; Martin and Hummer, 1989; Sanday, 1990).

Another social circumstance that increases the likelihood of rape is participation in athletics. Of course, the overwhelming majority of athletes are not rapists. However, there are proportionately more rapists among men who participate in athletics than among nonathletes (Welch, 1997). That is because many sports embody a particular vision of masculinity in North American culture: competitive, aggressive, and domineering. By recruiting men who display these characteristics and by encouraging the development of these characteristics in athletes, sports can contribute to "off-field" aggression, including sexual aggression. Furthermore, among male athletes, there is a distinct hierarchy of sexual aggression. Male athletes who engage in contact sports are more prone to be rapists than other athletes. There are proportionately even more rapists among athletes involved in collision and combative sports, notably football (Welch, 1997).

Rape, we conclude, involves using sex to establish dominance. The incidence of rape is highest in situations where early socialization experiences predispose men to want to control women, where norms justify the domination of women, and where a big power imbalance between men and women exists.

Sexual Harassment

There are two types of sexual harassment. **Quid pro quo harassment** takes place when sexual threats or bribery are made a condition of employment decisions. (The Latin phrase "quid pro quo" means "something for something.") **Hostile environment harassment** involves sexual jokes, comments, and touching that interferes with work or creates an unfriendly work setting. Research suggests that relatively powerless women are the most likely to be sexually harassed. Moreover, sexual harassment is most common in work settings that exhibit high levels of gender inequality and a culture justifying male domination of women. Specifically, women who are young, unmarried, and employed in nonprofessional jobs are most likely to become objects of sexual harassment, particularly if they are temporary workers, the ratio of women to men in the workplace is low, and the organizational culture of the workplace tolerates sexual harassment (Rogers and Henson, 1997; Welsh, 1999).

Ultimately, then, male aggression against women, including sexual harassment and rape, is encouraged by a lesson most of us still learn at home, in school, at work, through much of organized religion, and in the mass media—that it is natural and right for men to dominate women. To be sure, recent decades have witnessed important changes in the way women's and men's roles are defined. Nevertheless, in the world of paid work, in the household, in government, and in all other spheres of life, men still tend to command substantially more power and authority than women. Daily patterns of gender domination, viewed as legitimate by most people, get built into our courtship, sexual, family, and work norms. From this point of view, male aggression against women is simply an expression of male authority by other means.

This does not mean that all men endorse the principle of male dominance, much less that all men are inclined to rape or engage in other acts of aggression against women. Many men favor gender equality, and most men never rape or abuse a woman. However, the fact remains that many aspects of our culture legitimize male dominance, making it seem valid or proper. For example, pornography, jokes at the expense of women, and whistling and leering at women might seem mere examples of harmless play. At a subtler, sociological level, however, they are assertions of the appropriateness of women's submission to men. Such frequent and routine reinforcements of male authority increase the likelihood that some men will consider it their right to assault women physically or sexually if the opportunity to do so exists or can be created. "Just kidding" has a cost. For instance, researchers have found that college men who enjoy sexist jokes are most likely to report engaging in acts of sexual aggression against women (Ryan and Kanjorski, 1998).

We thus see that male aggression against women and gender inequality are not separate issues. Gender inequality is the foundation of aggression against women. In concluding this chapter, we consider how gender inequality can be decreased in the coming decades. As we proceed, you should bear in mind that gender equality is not just a matter of justice. It is also a question of safety.

Toward 2050

The 20th century witnessed growing equality between women and men in many countries. In the United States, the decline of the family farm made children less economically useful and more costly to raise. As a result, women started having fewer children. The industrialization of America, and then the growth of the economy's service sector, increased demand for women in the paid labor force (see Figure 9.4). This gave them substantially more economic power and also encouraged them to have fewer children. The legalization and availability of contraception made it possible for women to exercise unprecedented control over their own bodies. The women's movement fought for, and won, increased rights for women on a number of economic, political, and legal fronts. All these forces brought about a massive cultural shift, a fundamental reorientation of thinking on the part of many Americans about what women could and should do in society.

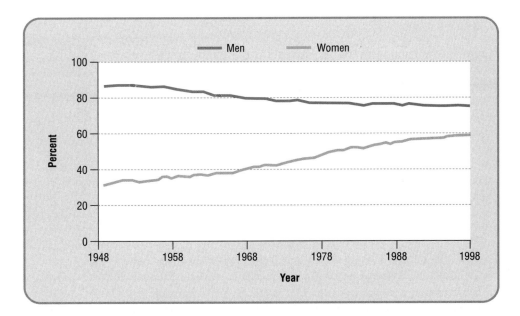

◆ FIGURE 9.4 ◆

Labor Force Participation Rate by Sex, United States, 1948–1999 (in percent)

Note: Figures are expressed as percent of men and women 16 years of age and over. The 1999 figure is for March only.

SOURCE: United States Department of Labor (1998b; 1999b).

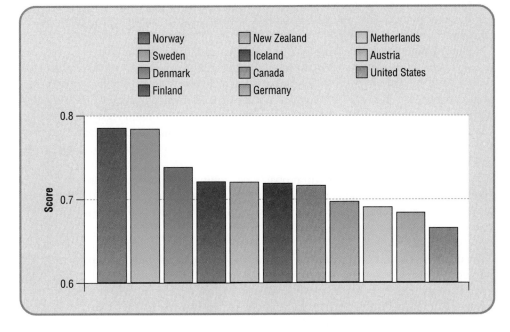

◆ FIGURE 9.5 ◆

Countries With Highest Scores on Gender Empowerment Measure, 1998

SOURCE: United Nations Development Program (1999).

One indicator of the progress of women is the "Gender Empowerment Measure" (GEM). The GEM is computed by the United Nations. It takes into account women's share of seats in parliament (the House of Representatives in the United States), women's share of administrative, managerial, professional, and technical jobs, and women's earning power. A score of 1.0 indicates equality with men on these three dimensions.

As Figure 9.5 shows, the Scandinavian countries (Norway, Sweden, Denmark, and Finland) were the most gender-egalitarian countries in the world in 1998. They had GEM scores ranging from 0.725 to 0.79. This means Scandinavian women are about three quarters of the way to equality with men on the three dimensions tapped by the GEM. The United States ranked 11th in the world with a GEM score of .675. This means American women are about two thirds of the way to equality with men.

In general, there is more gender equality in rich than in poor countries. Thus, the top 11 countries, shown in Figure 9.5, are all rich. In contrast, the lowest GEM scores are found in the poor countries of sub-Saharan Africa. This suggests gender equality is a function of economic development.

However, our analysis of the GEM data suggests that there are some exceptions to the general pattern. They show gender equality is also a function of government policy. Thus, in some of the former communist countries of Eastern Europe (such as Poland, Hungary, Slovakia, the Czech Republic, and Latvia) gender equality is *higher* than one would expect given their level of economic development. Meanwhile, in some of the Islamic countries (such the United Arab Emirates, Bahrain, Saudi Arabia, and Algeria), gender equality is *lower* than one would expect given their level of economic development. These anomalies exist because the former communist countries made gender equality a matter of public policy while many Islamic countries do just the opposite. To cite just one extreme case, in 1996 authorities in the Islamic country of Afghanistan made it illegal for girls to attend school and women to work in the paid labor force. (The situation has improved since the overthrow of the Taliban regime in 2001.)

The GEM figures suggest American women still have a long way to go before they achieve equality with men. We have seen, for example, that the gender gap in earnings is shrinking but will disappear only in 2050—and then only if it continues to diminish at the 1960–2000 rate. That is a big "if," because progress is never automatic.

In 1963, Congress passed the Equal Pay Act. It requires equal pay for the same work. Soon after, Congress passed Title VII of the Civil Rights Act. It prohibits employers from discriminating against women. These laws were important first steps in diminishing the gender gap in earnings. Since the mid-1960s, people in favor of closing the gender gap have recognized that we need additional laws and social programs to create gender equality.

Socializing children at home and in school to understand that women and men are equally adept at all jobs is important in motivating girls to excel in nontraditional fields. **Affirmative action,** which involves hiring more qualified women to diversify organizations, is important in helping to compensate for past discrimination in hiring.[9] However, without in any way belittling the need for such initiatives, we should recognize their impact will be muted if women continue to undertake disproportionate domestic responsibilities and if occupations containing a high concentration of women continue to be undervalued in money terms.

Two main policy initiatives will probably be required in coming decades to bridge the gender gap in earnings. One is the development of a better child care system. The other is the development of a policy of comparable worth. Let us consider both these issues in turn.

Child Care

High-quality, government-subsidized, affordable child care is widely available in most Western European countries, but not in the United States (see Chapter 12, "Families"). As a result, many American women with small children are either unable to work outside the home or able to work outside the home only on a part-time basis.

For women who have small children and work outside the home, child care options and the quality of child care vary by social class (Annie E. Casey Foundation, 1998; Gormley, 1995; Murdoch, 1995). Affluent Americans can afford to hire nannies and send their young children to expensive day-care facilities that enjoy a stable, relatively well-paid, well-trained staff and a high ratio of caregivers to children. These features yield high-quality child care. In contrast, the day-care centers, nursery schools, and preschools to which middle-class Americans typically send their children have higher staff turnover, relatively poorly paid, poorly trained staff, and a lower ratio of caregivers to children. Fewer than a third of American children in child care attend such facilities, however. More than two thirds—mainly from lower middle-class and poor families—use family child care homes or rely on the generosity of extended family members or neighbors. Overall, the quality of child care is lowest in family child care homes.

A third of all day-care facilities in the United States do not meet children's basic health and safety needs. This is true of about 12% of day-care centers, 13% of family child

[9]Affirmative action has also been applied to other groups that experience high levels of discrimination, such as African, Native, and Hispanic Americans.

care providers regulated by government, and 50% of family child care providers unregulated by government (calculated from Annie E. Casey Foundation, 1998; Gormley, 1995; Murdoch, 1995). In 1997, veterinary assistants earned a median wage of $7.34 an hour, parking lot attendants $6.38, and child care workers $6.12. One interpretation of these figures is that our society considers tending pets and cars more important than looking after young children.

The welfare-to-day-care initiative recently taken by 29 states adds weight to this interpretation. Governments throughout the country recognize the crisis in child care. They also want to get people off welfare and into the work force. So, in 1997, they began recruiting thousands of welfare mothers to start at-home, for-profit day-care facilities, thus hoping to kill two birds—welfare and child care—with one stone. But these welfare mothers were given few resources and little training. Most of them have little formal education and live in substandard housing. Many of them could undoubtedly make good child care providers. That, however, would require years of training, substantial subsidies, and ongoing support from outside sources (Dickerson, 1998).

Many companies, schools, and religious organizations in the United States provide high-quality day care. However, until the situation described above changes, women, particularly those in the middle and lower classes, will continue to suffer economically from the lack of accessible, affordable day care.

Comparable Worth

In the 1980s, researchers found women earn less than men partly because jobs in which women are concentrated are valued less than jobs in which men are concentrated. They therefore tried to establish gender-neutral standards by which they could judge the dollar value of work. These standards include such factors as the education and experience required to do a particular job and the level of responsibility, amount of stress, and working conditions associated with it. Researchers felt that, by using these criteria to compare jobs in which women and men are concentrated, they could identify pay inequities. The underpaid could then be compensated accordingly. In other words, women and men would receive equal pay for jobs of **comparable worth,** even if they did different jobs.

A number of United States states have adopted laws requiring equal pay for work of comparable worth. Minnesota leads the country in this regard. However, the laws do not apply to most employers ("Comparable Worth . . . ," 1990). Moreover, some comparable-worth assessments have been challenged in the courts. The courts have been reluctant to agree that the devaluation of jobs in which women are concentrated is a form of discrimination (England, 1992: 250). Only broad, new federal legislation is likely to change this state of affairs. However, no federal legislation on comparable worth is on the drawing boards. Most business leaders seem opposed to such laws since their implementation would cost them many billions of dollars.

THE WOMEN'S MOVEMENT

Improvements in the social standing of women do not depend just on the sympathy of government and business leaders. Progress on this front has always depended in part on the strength of the organized women's movement. This is likely to be true in the future too. In concluding this chapter, it is therefore fitting to consider the state of the women's movement and its prospects.

The "first wave" of the women's movement emerged in the 1840s. Drawing a parallel between the oppression of black slaves and the oppression of women, first-wave feminists made a number of demands, chief among them the right to vote. They finally achieved that goal in 1920, the result of much demonstrating, lobbying, organizing, and persistent educational work.

In the mid-1960s, the "second wave" of the women's movement started to grow. Second-wave feminists were inspired in part by the successes of the civil rights movement.

The "first wave" of the women's movement emerged in the 1840s. The movement achieved its main goal—the right to vote for women—in 1920 as a result of much demonstrating, lobbying, organizing, and persistent educational work.

They felt that women's concerns were largely ignored in American society despite persistent and pervasive gender inequality. Like their counterparts more than a century earlier, they held demonstrations, lobbied politicians, and formed women's organizations to further their cause. They advocated equal rights with men in education and employment, the elimination of sexual violence, and women's control over reproduction. One focus of their activities was mobilizing support for the Equal Rights Amendment (ERA) of the Constitution. The ERA stipulates equal rights for men and women under the law. The ERA was approved by the House of Representatives in 1971 and the Senate in 1972. However, it fell 3 states short of the 38 needed for ratification in 1982. Since then, no further attempt has been made to ratify the ERA.

Beyond the basic points of agreement noted above, there is considerable intellectual diversity in the modern feminist movement concerning ultimate goals. Three main streams may be distinguished (Tong, 1989).

Liberal feminism is the most popular current in the women's movement today. Its advocates believe the main sources of women's subordination are learned gender roles and the denial of opportunities to women. Liberal feminists advocate nonsexist methods of socialization and education, more sharing of domestic tasks between women and men, and extending to women all the educational, employment, and political rights and privileges men enjoy.

Socialist feminists regard women's relationship to the economy as the main source of women's disadvantages. They believe the traditional nuclear family emerged along with inequalities of wealth. In their opinion, once men possessed wealth, they wanted to ensure their property would be transmitted to their children, particularly their sons. They accomplished this in two ways. First, men exercised complete economic control over their property, thus ensuring it would not be squandered and would remain theirs and theirs alone. Second, they enforced female monogamy, thus ensuring their property would be transmitted only to *their* offspring. Thus, according to socialist feminists, the economic and sexual oppression of women has its roots in capitalism. Socialist feminists also assert that the reforms proposed by liberal feminists are inadequate. That is because they can do little to help working-class women, who are too poor to take advantage of equal educational and work opportunities. Socialist feminists conclude that only the elimination of private property and the creation of economic equality can bring about an end to the oppression of all women.

Radical feminists, in turn, find the reforms proposed by liberals and the revolution proposed by socialists inadequate. Patriarchy—male domination and norms justifying that domination—is more deeply rooted than capitalism, say the radical feminists. After all, patriarchy predates capitalism. Moreover, it is just as evident in self-proclaimed communist societies as it is in capitalist societies. Radical feminists conclude that the very idea of gender must be changed to bring an end to male domination. Some radical feminists argued that new reproductive technologies, such as *in vitro* fertilization, are bound to be helpful in this regard because they can break the link between women's bodies and childbearing (see Chapter 12, "Families"). But the revolution envisaged by radical feminists goes beyond the realm of reproduction to include all aspects of male sexual dominance. From their point of view, pornography, sexual harassment, restrictive contraception, rape, incest, sterilization, and physical assault must be eliminated in order for women to reconstruct their sexuality on their own terms.

This thumbnail sketch by no means exhausts the variety of streams of contemporary feminist thought. For example, since the mid-1980s, *antiracist* and *postmodernist* feminists have criticized liberal, socialist, and radical feminists for generalizing from the experience of white women and failing to see how women's lives are rooted in particular historical and racial experiences (hooks, 1984). These new currents have done much to extend the relevance of feminism to previously marginalized groups.

Partly due to the political and intellectual vigor of the women's movement, some feminist ideas have gained widespread acceptance in American society over the past three decades (see Table 9.5). For example, 1998 General Social Survey data show that 83% of Americans approve of married women working in the paid labor force. Some 78% think women are as well suited to politics as men. And 64% regard women's rights issues as important or very important. Support for these ideas has grown in recent decades. Still, many people, especially men, oppose the women's movement. In fact, in recent years several antifeminist men's groups have sprung up to defend traditional male privileges.[10] It is apparently difficult for some men to accept feminism because they feel that the social changes advocated by feminists threaten their traditional way of life and perhaps even their sexual identity.

The "second wave" of the women's movement started to grow in the mid-1960s. Members of the movement advocated equal rights with men in education and employment, the elimination of sexual violence, and women's control over reproduction.

[10]Profeminist men's groups, such as the National Organization for Men Against Sexism, also exist but seem to have a smaller membership (National Organization . . . , 2000).

✦ **TABLE 9.5** ✦
**Attitudes To Women's Issues,
United States, 1972–96
(in percent)**

SOURCE: National Opinion Research
Center (1999).

	1972–82	1983–87	1996
Approve of married women working in paid labor force	70	80	83
Women suited for politics	54	63	78
Women's rights issue one of the most important/important	—	58	64
Favor preferential hiring of women	—	—	27
Think of him/herself as a feminist	—	—	22
Women can best improve their position through women's rights groups	—	15	—

Personal Anecdote

Our own experience suggests that traditional patterns of gender socialization weigh heavily on many men. For example, John Lie grew up in a patriarchal household. His father worked outside the home, and his mother stayed home to do nearly all the housework and child care. "I remember my grandfather telling me that a man should never be seen in the kitchen," recalls John, "and it is a lesson I learned well. In fact, everything about my upbringing—the division of labor in my family, the games I played, the TV programs I watched—prepared me for the life of a patriarch. I vaguely remember seeing members of the 'women's liberation movement' staging demonstrations on the TV news in the early 1970s. Although I was only about 11 or 12 years old, I recall dismissing them as slightly crazed, bra-burning man haters. Because of the way I grew up and what I read, heard, and saw, I assumed the existing gender division of labor was natural. Doctors, pilots, and professors should be men, I thought, and people in the 'caring' professions, such as nurses and teachers, should be women.

"But socialization is not destiny," John insists. "Entirely by chance, when I got to college I took some courses taught by female professors. It is embarrassing to say so now, but I was surprised that they were so much brighter, more animated, and more enlightening than my male high school teachers had been. In fact, I soon realized that many of my best professors were women. I think this is one reason why I decided to take the first general course in women's studies offered at my university. It was an eye opener. I soon became convinced that gender inequalities are about as natural and inevitable as racial inequalities. I also came to believe that gender equality could be as enriching for men as for women. Sociological reflection overturned what my socialization had taught me. Sociology promised—and delivered. I think many college-educated men have similar experiences today, and I hope I now contribute to their enlightenment."

SUMMARY

1. The way culturally appropriate masculine and feminine roles are expressed depends on a variety of social conditions, especially the level of gender inequality.

2. Males and females are channeled into gender-appropriate roles by parents, teachers, and the mass media.

3. While society pushes people to assume conventionally masculine or feminine roles, it demands heterosexuality with even greater force.

4. The social distinction between men and women serves as a major basis of inequality in the family and the workplace.

5. The gender gap in earnings derives from outright discrimination against women, women's disproportionate domestic re-

sponsibilities, women's concentration in low-wage occupations and industries, and the undervaluation of work typically done by women.

6. Male aggression against women is rooted in gender inequality.

7. Among the major reforms that can help eliminate the gender gap in earnings and reduce the overall level and expression of gender inequality are (a) the development of an affordable, accessible system of high-quality day care; and (b) the remuneration of men and women on the basis of their work's actual worth.

GLOSSARY

Affirmative action involves hiring a woman if equally qualified men and women are available for a job, thus compensating for past discrimination.

Bisexuals are people who prefer sexual partners of both sexes.

Comparable worth refers to the equal dollar value of different jobs. It is established in gender-neutral terms by comparing jobs in terms of the education and experience needed to do them and the stress, responsibility, and working conditions associated with them.

Essentialism is a school of thought that sees gender differences as a reflection of biological differences between women and men.

The **female-male earnings ratio** is women's earnings expressed as a percent of men's earnings.

Your **gender** is your sense of being male or female and your playing masculine and feminine roles in ways defined as appropriate by your culture and society.

Gender discrimination involves rewarding men and women differently for the same work.

Gender identity is one's identification with, or sense of belonging to, a particular sex—biologically, psychologically, and socially

A **gender ideology** is a set of ideas about what constitutes appropriate masculine and feminine roles and behavior.

A **gender role** is the set of behaviors associated with widely shared expectations about how males or females are supposed to act.

The **glass ceiling** is a social barrier that makes it difficult for women to rise to the top level of management.

Hermaphrodites are people born with ambiguous genitals due to a hormone imbalance in their mother's womb.

Heterosexuals are people who prefer members of the opposite sex as sexual partners.

Homosexuals are people who prefer sexual partners of the same sex. People usually call homosexual men gay and homosexual women lesbians.

Hostile environment sexual harassment involves sexual jokes, comments, and touching which interferes with work or creates an unfriendly work setting.

Quid pro quo sexual harassment takes place when sexual threats or bribery are made a condition of employment decisions.

Your **sex** depends on whether you were born with distinct male or female genitals and a genetic program that released either male or female hormones to stimulate the development of your reproductive system.

Social constructionism is a school of thought that sees gender differences as a reflection of the different social positions occupied by women and men.

Transgendered people are individuals who want to alter their gender by changing their appearance or resorting to medical intervention.

Transsexuals believe they were born with the "wrong" body. They identify with, and want to live fully as, members of the "opposite" sex.

QUESTIONS TO CONSIDER

1. By interviewing your family members and using your own memory, compare the gender division of labor in (a) the households in which your parents grew up and (b) the household(s) in which you grew up. Then, imagine the gender division of labor you would like to see in the household you hope to live in about 10 years from now. What accounts for change over time in the gender division of labor in these households? Do you think your hopes are realistic? Why or why not?

2. In your own case, rank the relative importance of your family, your schools, and the mass media in your gender socialization. What criteria do you use to judge the importance of each socialization agent?

3. Systematically note the roles played by women and men on TV programs and ads one evening. Is there a gender division of labor on TV? If so, describe it.

4. Are you a feminist? If so, which of the types of feminism discussed in this chapter do you find most appealing? Why? If not, what do you find objectionable about feminism? In either case, what is the ideal form of gender relations in your opinion? Why do you think this form is ideal?

WEB RESOURCES

http://sociology.wadsworth.com

Begin by clicking on the Student Resources section of the Web site. Choose "Introduction to Sociology" and finally the Brym and Lie book cover. Next, select the chapter you are currently studying from the pull-down menu. From the Student Resources page you will have easy access to InfoTrac College Edition®, MicroCase Online exercises, additional Web links, and many other resources to aid you in your study of sociology, including practice tests for each chapter.

InfoTrac Search Terms

These search terms are provided to assist you in beginning to conduct research on this topic by visiting http://www.infotraccollege.com/wadsworth.

Gender	**Glass ceiling**
Gender discrimination	**Sexual harassment**
Gender role	

Recommended Web Sites

For a useful list of resources on gender and sexuality on the World Wide Web, go to http://www.georgetown.edu/crossroads/asw/gender.html.

The National Committee on Pay Equity is a coalition of more than 180 organizations working to eliminate sex- and race-based wage discrimination and to achieve pay equity. Visit their Web site at http://www.feminist.com/fairpay.

On the United Nations Gender Empowerment Measure, discussed in the text, see http://www.undp.org/hdro/98gem.htm.

In 1997, a conference was held in San Diego to analyze what the participants called the "National Sex Panic." For the proceedings of the conference on RealAudio, go to http://www.managingdesire.org/sexpanic/sexpanicindex.html.

SUGGESTED READINGS

Deborah Blum. *Sex on the Brain: The Biological Differences Between Men and Women* (New York: Penguin, 1997). The subtitle is a misnomer. This Pulitzer-Prize-winning science writer discusses not just the biological differences between men and women but the interaction between biology and environment.

Riane Eisler. *The Chalice and the Blade: Our History, Our Future* (New York: Harper Collins,1995 [1987]). The big picture on gender inequality. Eisler's brilliant examination of the archeological record uncovers the historical origins of gender in-equality and suggests that now, for the first time in 7,000 years, we are in a position to put an end to it.

Sharlene Hesse-Biber and Gregg Lee Carter. *Working Women in America: Split Dreams* (New York: Oxford University Press, 2000). A clear, brief, and comprehensive overview of how family, school, the mass media, and the economy intersect and structure women's work and aspirations in the paid labor force and the family.

IN THIS CHAPTER, YOU WILL LEARN THAT:

✦ Three work-related revolutions—one in agriculture, one in industry, and one in the provision of services—have profoundly altered the way people sustain themselves and the way people live.

✦ In the past few decades, the number of "good" jobs has grown but "bad" jobs have become even more numerous.

✦ With varying degrees of success, people seek to control work through unions, professional organizations, corporations, and markets.

✦ The growth of large corporations and global markets has shaped the transformation of work in recent decades and will shape the choices you face as a member of the labor force and a citizen.

THE PROMISE AND HISTORY OF WORK

Salvation or Curse?

The computerization of the office began in earnest about 20 years ago. Soon, the image of the new office was as familiar as a Dilbert cartoon. It was a checkerboard of $8' \times 8'$ cubicles. Three padded walls, $5'$ $6''$ high, framed each cubicle. Inside, a computer terminal sat on a desk. A worker quietly tapped away at a keyboard, seemingly entranced by the glow of a video screen.

Sociologist Shoshana Zuboff visited many such offices soon after they were computerized. She sometimes asked the office workers to draw pictures capturing their job experience before and after computerization. The pictures were strikingly similar. Smiles changed to frowns, mobility became immobility, sociability was transformed into isolation, freedom turned to regimentation. We reprint two of the workers' pictures in Figure 10.1. Work automation and standardization emerge from these drawings as profoundly degrading and inhuman processes (Zuboff, 1988).

The image conveyed by these drawings is only one view of the transformation of work in the Information Age. There is another, and it is vastly different. Bill Gates argues that computers reduce our work hours. They make goods and services less expensive by removing many distribution costs of capitalism. (Think of Amazon.com, which reduces the need for bookstores.) They allow us to enjoy our leisure time more (Gates with Myhrvold and Rinearson, 1996). This vision is well captured by the arresting December 1999 cover of *Wired* magazine, reprinted here as Figure 10.2. According to *Wired*, computers liberate us. They allow us to become more mobile and more creative. Computerized work allows our imaginations to leap and our spirits to soar.

These strikingly different images form the core questions of the sociology of work, and they will be our focus in this chapter. Is work a salvation or a curse? Or is it perhaps both at once? Is it more accurate to say that work has become more of a salvation or a curse over time? Or is work a salvation for some and a curse for others?

◆ FIGURE 10.1 ◆

One View of the Effects of Computers on Work

Shoshanna Zuboff asked office workers to draw pictures representing how they felt about their jobs before and after a new computer system was introduced. Here are "before" and "after" pictures drawn by two office workers. Notice how even the flower on one worker's desk wilted after the new computer system was introduced.

SOURCE: Zuboff (1988: 146–7).

Before

After

"Before I was able to get up and hand things to people without having someone say, what are you doing? Now, I feel like I am with my head down, doing my work."

Before

After

"My supervisor is frowning because we shouldn't be talking. I have on the stripes of a convict. It's all true. It feels like a prison in here."

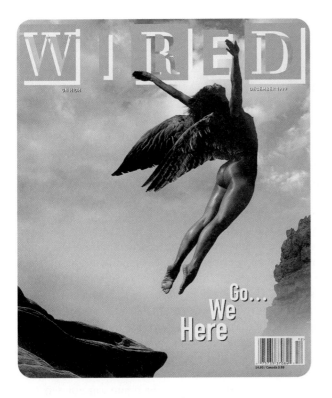

⁺ **FIGURE 10.2** ⁺
**Another View of the Effect
of Computers on Work**
Wired magazine is always on high
about the benefits of computer
technology.

SOURCE: *Wired* (1999).

To answer these questions, we first trace the evolution of work from preagricultural to postindustrial times. As you will see, we have experienced three work-related revolutions in the past 10,000 years. Each revolution has profoundly altered the way we sustain ourselves and the way we live. Next, we examine how job skills have changed over the past century. We also trace changes in the number and distribution of "good" and "bad" jobs over time and project these changes nearly a decade into the future. We then analyze how people have sought to control work through unions, professional organizations, corporations, and markets. Finally, we place our discussion in a broader context. The growth of large corporations and markets on a global scale has shaped the transformation of work over the past quarter century. Understanding these transformations will help you understand the work-related choices you face both as a member of the labor force and a citizen.

Three Revolutions

The **economy** is the institution that organizes the production, distribution, and exchange of goods and services. Conventionally, analysts divide the economy into three sectors. The *primary* sector includes farming, fishing, logging, and mining. In the *secondary* sector, raw materials are turned into finished goods; manufacturing takes place. Finally, in the *tertiary* sector, services are bought and sold. These services include the work of nurses, teachers, lawyers, hairdressers, computer programmers, and so forth. Often, the three sectors of the economy are called the "agricultural," "manufacturing," and "service" sectors. We follow that practice here.

Three truly revolutionary events have taken place in the history of human labor. In each revolution, a different sector of the economy rose to dominance. First came the agricultural revolution, then the industrial revolution, and finally the revolution in services (Gellner, 1988; Lenski, 1966).

The Development of Agriculture

Nearly all humans lived in nomadic tribes until about 10,000 years ago. Then, people in the fertile valleys of the Middle East, Southeast Asia, and South America began to herd cattle and grow plants using simple hand tools. Stable human settlements spread in these areas. About 5,000 years ago, farmers invented the plow. By attaching plows to large animals, they substantially increased the land under cultivation. **Productivity**—the amount produced for every hour worked—soared.

The Development of Modern Industry

International exploration, trade, and commerce helped to stimulate the growth of markets from the 15th century on. **Markets** are social relations that regulate the exchange of goods and services. In a market, prices are established by how plentiful goods and services are ("supply") and how much they are wanted ("demand"). About 225 years ago, the steam engine, railroads, and other technological innovations greatly increased the ability of producers to supply markets. This was the era of the Industrial Revolution. Beginning in England, the Industrial Revolution spread to Western Europe, North America, Russia, and Japan within a century.

The Development of the Service Sector

Even in preagricultural societies, a few individuals specialized in providing services rather than producing goods. For example, a person considered adept at tending to the ill, forecasting the weather, or predicting the movement of animals might be relieved of hunting responsibilities to focus on these services. However, such jobs were rare because productivity was low. Nearly everyone had to do physical work for the tribe to survive. Even in early agricultural societies, it took 80 to 100 farmers to support one nonfarmer (Hodson and Sullivan, 1995 [1990]: 10). Only at the beginning of the 19th century in Western Europe and North America did productivity increase to the point where a quarter of the labor force could be employed in services. By 1960, over half the labor force of the highly industrialized countries was providing services. Forty years later, the figure was close to 75% (see Figure 10.3). The rapid change in the composition of the labor force during the final decades of the 20th century was made possible by the computer. The computer automated many manufacturing and office procedures. It created jobs in the service sector as quickly as it eliminated them in manufacturing. Thus, the computer is to the service sector as the steam engine was to manufacturing and the plow was to agriculture.

Besides increasing productivity and causing shifts between sectors in employment, the agricultural, industrial, and service revolutions altered the way work was socially organized. For one thing, the **division of labor** increased. That is, work tasks became more specialized with each successive revolution. In preagricultural societies there were four main jobs: hunting wild animals, gathering wild edible plants, raising children, and tending to the tribe's spiritual needs. In contrast, a postindustrial society like the United States boasts tens of thousands of different kinds of jobs.

In some cases, increasing the division of labor involves creating new skills. Some new jobs even require long periods of study. Foremost among these are the professions. In other cases, increasing the division of labor involves breaking a complex range of skills into a series of simple routines. For example, a hundred years ago, a butcher's job involved knowing how to dissect an entire cow. In today's meat-packing plant there are large-stock scalpers, belly shavers, crotch busters, gut snatchers, gut sorters, snout pullers, ear cutters, eyelid removers, stomach washers (also known as belly bumpers), hind leg pullers, front leg toenail pullers, and oxtail washers. A different person performs each routine. The tasks are repetitive and require a narrow range of skills.

If the division of labor increased as one work revolution gave way to the next, then social relations among workers also changed. In particular, work relations became more hierarchical. While work used to be based on cooperation among equals, it now involves superordinates exercising authority and subordinates learning obedience. Owners oversee executives. Executives oversee middle managers. Middle managers oversee ordinary workers. Increasingly, work hierarchies are organized bureaucratically. That is, clearly defined positions and written goals, rules, and procedures govern the organization of work.

Clearly, the increasing division of labor changed the nature of work in fundamental ways. But did the *quality* of work improve or worsen as jobs became more specialized? That is the question we now address.

One of the great engineering and construction feats of the late 20th century was the "Chunnel" linking France and Britain under the English Channel. Here, construction workers carry an I-beam.

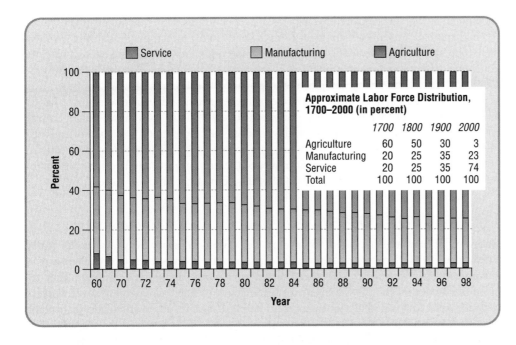

✦ FIGURE 10.3 ✦
Civilian Employment by Economic Sector, United States, 1960–1998 (in percent)

SOURCE: De Long (1998: 23); United States Department of Labor (1999a).

"GOOD" VERSUS "BAD" JOBS

John Lie once got a job as a factory worker in Honolulu. "The summer after my sophomore year in high school," John recalls, "I decided it was time to earn some money. I had expenses, after all, but only an occasionally successful means of earning money: begging my parents. Scouring the 'help wanted' ads in the local newspaper, I soon realized I wasn't really qualified to do anything in particular. Some friends at school suggested I apply for work at a pineapple-canning factory. So I did.

At the factory, an elderly man asked me a few questions and hired me. I was elated—but only for a moment. A tour of the factory floor ruined my mood. Row upon row of conveyor belts carried pineapples in various states of disintegration. Supervisors hastened the employees to work faster yet make fewer mistakes. The sickly sweet smell, the noise, and the heat were unbearable. After the tour, the interviewer announced I would get the graveyard shift (11 p.m. to 7 a.m.) at minimum wage.

"The tour and the prospect of working all night finished me off. Now dreading the prospect of working in the factory, I wandered over to a mall. I bumped into a friend there. He told me a bookstore was looking for an employee (nine to five, no pineapple smell, and air-conditioned, although still minimum wage). I jumped at the chance. Thus, my career as a factory worker ended before it ever began.

"A dozen years later, just after I got my Ph.D., I landed one of my best jobs ever. I was teaching for a year in South Korea. However, my salary hardly covered my rent. I needed more work desperately. Through a friend of a friend, I found a second job as a business consultant in a major corporation. I was given a big office with a panoramic view of Seoul and a personal secretary who was both charming and efficient. I wrote a handful of sociological reports that year on how bureaucracies work, how state policies affect workers, how the world economy had changed in the past two decades, and so on. I got to accompany the president of the company on trips to the United States. I spent most of my days reading books. I also went for long lunches with colleagues and took off several afternoons a week to teach."

What is the difference between a "good" job and a "bad" job, as these terms are usually understood? As this anecdote illustrates, bad jobs don't pay much and require the performance of routine tasks under close supervision. Working conditions are unpleasant and sometimes dangerous. Bad jobs require little formal education. In contrast, good jobs often require higher education. They pay well. They are not closely supervised, and they encourage the worker to be creative in pleasant surroundings. Other distinguishing features of good and bad jobs are not apparent from the anecdote. Good jobs offer secure employment, opportunities for promotion, health insurance, and other fringe benefits. In a bad job, you can easily be fired, you receive few if any fringe benefits, and the prospects for promotion are few. That is why bad jobs are often called "dead-end" jobs.

Most jobs fall between the two extremes sketched above. They have some mix of good and bad features. But what can we say about the overall mix of jobs in the United States? Are there more good than bad jobs? And what does the future hold? Are good or bad jobs likely to become more plentiful? What are your job prospects? These are tough questions, not least because some conditions that influence the mix of good and bad jobs are unpredictable. Nonetheless, sociological research sheds some light on these issues.

The Deskilling Thesis

One view of how jobs are likely to develop was proposed nearly 30 years ago by Harry Braverman (1974). Braverman argued that capitalists are always eager to organize work to maximize their profits. Therefore, they break complex tasks into simple routines. They replace labor with machines wherever possible. They exert increasing control over workers to make sure they do their jobs more efficiently. As a result, work tends to become **deskilled** over time. In the 1910s, for example, Henry Ford introduced the assembly line with just this aim in mind. The assembly line enabled Ford to produce affordable cars for a mass market. It also forced workers to do highly specialized, repetitive tasks requiring little skill at a pace set by their supervisors. Around the same time, Frederick W. Taylor developed the principles of **scientific management.** After analyzing the movements of workers as they did their jobs, Taylor trained them to eliminate unnecessary actions and greatly improve their efficiency. Workers became cogs in a giant machine known as the modern factory.

Many criticisms were lodged against Braverman's deskilling thesis in the 1970s and 1980s. Perhaps the most serious criticism was that he was not so much wrong as irrelevant. That is, even if his characterization of factory work was accurate (and we will see below that in some respects it was not), factory workers represent only a small proportion of the labor force. They represent a smaller proportion with every passing year. In 1974, the year Braverman's book was published, less than a third of the United States labor force was employed in manufacturing. By 1998, the figure had fallen to just over a fifth (see Figure 10.3). Science fiction writer Isaac Asimov's claim that the factory of the future will employ only a man and a dog is clearly an exaggeration. (The man will be there to feed the dog, said Asimov. The dog will be there to keep the man away from the machines.) However, the manufacturing sector is shrinking and the service sector is expanding. The vital question, according to some of Braverman's critics, is not whether jobs are becoming

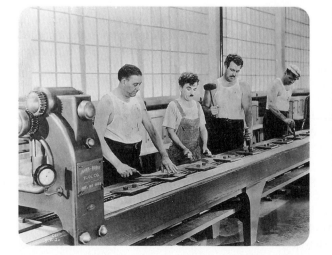

Charlie Chaplin's 1929 movie, "Modern Times," was a humorous critique of the factory of his day. In the movie, Chaplin gets a tick and moves like a machine on the assembly line. He then gets stuck on a conveyor belt and run through a machine. Finally, he is used as a test dummy for a feeding machine. The film thus suggests that workers were being used for the benefit of the machines rather than the machines being used for the benefit of the workers.

worse in manufacturing but whether good jobs or bad jobs are growing in services, the sector that accounts for three quarters of United States jobs today.

Shoshana Zuboff's analysis of office workers, mentioned at the beginning of this chapter, made it appear that Braverman's insights apply beyond the factory walls (Zuboff, 1988). She argued that the computerization of the office in the 1980s involved increased supervision of deskilled work. And she was right, at least in part. The computer did eliminate many jobs and routinize others. It allowed supervisors to monitor every keystroke, thus taking worker control to a new level. Today, employees who consistently fall behind a prescribed work pace or use their computers for personal e-mail, surfing the Web, and other pastimes can easily be identified and then retrained, disciplined, or fired. In the 1980s and 1990s, some analysts feared that good jobs in manufacturing were being replaced by bad jobs in services. From this point of view, the entire labor force was experiencing a downward slide (Bluestone and Harrison, 1982; Rifkin, 1995).

Part-Time Work

Adding to the fear of a downward slide was the growth of part-time work in America. The proportion of part-time workers in the United States labor force increased 46% between 1957 and 1996. In 1996, about a fifth of all people in the United States labor force were part-timers, working less than 35 hours a week (Tilly, 1996: 1–4).

For two reasons, the expansion of part-time work is not a serious problem in itself. First, some part-time jobs are good jobs in the sense we defined above. Second, some people want to work part-time and can afford to do so. For example, some people who want a job also want to devote a large part of their time to family responsibilities. Part-time work affords them that flexibility. Similarly, many high school and college students work part-time and are happy to do so. Perhaps you have joined the ranks of part-time retail clerks and fast-food servers to help pay for your college education or to earn money for a car or a vacation during March break.

Although the growth of part-time jobs is not problematic for voluntary part-time workers or people who have good part-time jobs, an increasingly large number of people depend on part-time work for the necessities of full-time living. And the plain fact is that most part-time jobs are bad jobs. Thus, part-time workers make up two thirds of the people working at or below minimum wage. Moreover, the fastest growing category of part-time workers is composed of *involuntary* part-timers. Today, according to official statistics, a quarter of part-time workers want to be working more hours. And official statistics underestimate the scope of the problem. Surveys show that about a third of women officially classified as voluntary part-time workers would work more hours if good child care or elder care were available (Henson, 1996; Tilly, 1996: 1–4).

The downside of part-time work is not only economic, however. Nor is it just a matter of coping with the dull routine, the routinization and standardization of procedures analyzed in our discussion of "McDonaldization" in Chapter 5 ("Interaction and Organization"). If you've ever had a part-time job, you know that one of its most difficult aspects involves maintaining your self-respect in the face of low pay, benefits, security, status, and creativity. In the words of Dennis, a McDonald's employee interviewed by sociologist Robin Leidner: "This isn't really a job. . . . It's about as low as you can get. Everybody knows it" (quoted in Leidner, 1993: 182). And, Dennis might have added, nearly everybody lets you know they know it.

Katherine Newman (1999: 89) studied fast-food workers in Harlem, many of whom are teenagers working part-time. These workers are trained, Newman noted, to keep smiling no matter how demanding or rude their customers may be. The trouble is you can only count backwards from a hundred so many times before feeling utterly humiliated. Anger almost inevitably boils over. According to Natasha, a young fast-food worker Newman interviewed:

> It's hard dealing with the public. There are good things, like old people. They sweet. But the younger people around my age are always snotty. They think they better than you because they not working [here]. . . . They told us that we just suppose to walk to the back

and ignore it, but when they in your face like that, you get so upset you have to say something. . . . I got threatened with a gun one time. 'Cause this customer had threw a piece of straw paper in the back and told me to pick it up like I'm a dog. I said, "No." And he cursed at me. I cursed at him back and he was like, "Yeah, next time you won't have nothing to say when I come back with my gun and shoot your ass." Oh, excuse me (quoted in Newman, 1999, 90-1).

In inner-city ghettos like Harlem, the difficulty of maintaining one's dignity as a fast-food worker is compounded by the high premium most young people place on independence, autonomy, and respect. Young people enjoy few opportunities for high-quality education and upward mobility in the inner city. Unable to derive dignity from such conventional achievements, they tend to define their self-worth by "macho" behavior codes (Anderson, 1990). The problem this creates for teenagers who take jobs in fast-food restaurants is that their constant deference to customers violates the norms of inner-city youth culture. Therefore, fast-food workers are typically stigmatized by their peers. They are frequently the brunt of insults and ridicule. Newman recounts the case of one fast-food worker in Harlem who tried valiantly to keep his job a secret from his friends and acquaintances. He wouldn't tell his friends where he was going when he left for work. He took a long, circuitous route to work so his friends wouldn't know where he was headed. He kept his uniform in a bag and put it on only at work. He lied to his friends when they asked him where he got his spending money. He even hid behind a large freezer when his friends came in for a burger (Newman, 1999: 97).

Fast-food workers in Harlem undoubtedly represent an extreme case of the indignity endured by part-timers. However, the problem exists in various guises in most part-time jobs. For instance, if you work as a "temp" in an office you are more likely than other office workers to be the victim of sexual harassment. You are especially vulnerable to unwanted advances because you lack power in the office and are considered "fair game" (Welsh, 1999). Thus, the form and depth of degradation may vary from one part-time job to another, but, as your own work experience may show, degradation seems to be a nearly universal feature of this type of deskilled work.

Web Research Projects
Sexual Harassment at Work

A Critique of the Deskilling Thesis

The deskilling thesis undoubtedly captures one important tendency in the development of work. However, it does not paint a complete picture. That is because analyses like Braverman's and Zuboff's are too narrowly focused. They analyze specific job categories near the bottom of the occupational hierarchy. As a result, they do not allow us to form an impression of what is happening to the occupational structure as a whole. For instance, Zuboff analyzed lower level service workers such as data entry personnel and routine claims processors in a health insurance company. It is unclear from her research what is happening at the top of the service sector—whether, for example, good professional and managerial jobs are becoming more numerous relative to clerical jobs.

Subsequent analyses comparing the *entire* manufacturing and service sectors cast doubt on whether Zuboff's argument can be generalized. For example, one report prepared for the United States Bureau of Labor Statistics showed a wide range in the distribution of earnings in the service sector. Thus, although there are many low-paying jobs in services, there are also many high-paying jobs. Moreover, the distribution of income in services is practically the same as that in manufacturing. On some measures of job quality, such as job security, service jobs are on average better than manufacturing jobs. Thus, the decline of the manufacturing sector and the rise of the service sector do not imply a downward slide of the entire labor force. To be sure, there are many bad jobs in the service sector. Yet there are also many good jobs, so one should avoid generalizing about the entire sector from case studies of just a few job categories (Meisenheimer, 1998).

Even if we examine the least skilled service workers—fast-food servers, video store clerks, parking lot attendants, and the like—we find reason to question one aspect of the deskilling thesis. For although these jobs are dull and pay poorly, they are not as "dead-end" as they are frequently made out to be. One analysis of United States census and survey data showed that the least skilled service workers tend to be under the age of 25 and hold their jobs only briefly. After working in these entry-level jobs for a short time, most men move on to blue-collar jobs and most women move on to clerical jobs. These are not great leaps up the socioeconomic hierarchy, but the fact they are common suggests that work is not all bleak and hopeless even at the bottom of the service sector (Jacobs, 1993; Myles and Turegun, 1994).

Braverman and Zuboff exaggerated the downward slide of the United States labor force because they underestimated the continuing importance of skilled labor in the economy. Assembly lines and computers may deskill many factory and office jobs. But if deskilling is to take place, then some members of the labor force must invent, design, advertise, market, install, repair, and maintain complex machines, including computerized and robotic systems. Most of these people have better jobs than the factory and office workers analyzed by Braverman and Zuboff. Moreover, although technological innovations kill off entire job categories, they also create entire new industries with many good jobs. Thirty years ago, Santa Clara County in California was best known for its excellent prunes. Today, it has been transformed into Silicon Valley, home to many tens of thousands of electronic engineers, computer programmers, graphics designers, venture capitalists, and so forth. Seattle, Austin, the Research Triangle in North Carolina, and Route 128 outside Boston are similar success stories. The introduction of new production techniques may even increase a country's competitive position in the world market. This can lead to employment gains at both the low and the high ends of the job hierarchy as consumers abroad rush to purchase relatively low-cost goods.

Rather than involving a downward shift in the entire labor force, it seems more accurate to think of recent changes in work as involving a declining middle or (to say the same thing differently) a polarization between good and bad jobs. Many good jobs are opening up at the top of the socioeconomic hierarchy. Even more mediocre and bad jobs are opening up at the bottom. There are fewer new jobs in the middle (Myles, 1988).

Job polarization is nowhere more evident than in Silicon Valley. Top executives in Silicon Valley earned up to $121 million each in 1999, up 70% from the year before. Even at less lofty levels, there are many thousands of high-paying, creative jobs in the Valley, over half of them dependent on the high-tech sector (Bjorhus, 2000). Amid all this wealth, however, the electronics assembly factories in Silicon Valley are little better than high-tech sweatshops. Most workers in the electronics factories earn less than 60% of the Valley's average wage. They are mainly immigrant women of Hispanic and Asian origin. In the factories they work long hours and are frequently exposed to toxic solvents, acids, and gases. Semiconductor workers therefore suffer industrial illnesses at three times the average rate for other manufacturing jobs. Three separate studies have found significantly higher miscarriage rates among women working in chemical handling jobs than in other manufacturing jobs. There are no unions in Silicon Valley's electronics assembly factories, and the industry has fought efforts to introduce union-scale wages and working conditions (Corporate Watch, 2000; see Figure 10.4). Although the opulent lifestyles of Silicon Valley's

✦ FIGURE 10.4 ✦
The Toshiba circuit board assembly line in Irvine, California.

✦ **TABLE 10.1** ✦
Expected Job Growth, Top 20 Occupations, United States, 1998–2008

*= college degree required

SOURCE: United States Department of Labor (2000c).

Occupation	Estimated New Jobs, 1998–2008	Education Required
1. Systems analyst	577,000	*Bachelor's degree
2. Retail salespersons	563,000	Short-term on-the-job training
3. All other sales and related	558,000	Moderate on-the-job training
4. Cashiers	556,000	Short-term on-the-job training
5. General managers, top executives	551,000	*Work experience plus degree
6. Truck drivers	493,000	Short-term on-the-job training
7. Office clerks	463,000	Short-term on-the-job training
8. Registered nurses	451,000	*Associate degree
9. Computer support specialists	439,000	*Associate degree
10. Personal care and home health aides	433,000	Short-term on-the-job training
11. Teacher assistants	375,000	Short-term on-the-job training
12. Janitors, cleaners, maids, house-keepers	365,000	Short-term on-the-job training
13. Nursing aides, orderlies, attendants	325,000	Short-term on-the-job training
14. Computer engineers	323,000	*Bachelor's degree
15. Teacher, secondary school	322,000	*Bachelor's degree
16. Office and administrative support supervisors and managers	313,000	Work experience in a related occupation
17. All other managers and administrators	305,000	*Work experience plus degree
18. Receptionists and information clerks	305,000	Short-term on-the-job training
19. Waiters and waitresses	303,000	Short-term on-the-job training
20. Security guards	294,000	Short-term on-the-job training

millionaires are often featured in the mass media, one must remember that a more accurate picture of the Valley—indeed, of work in the United States as a whole—is that of an increasingly polarized hierarchy.

Table 10.1 illustrates the trend toward job polarization in the United States. Based on projections by the United States Bureau of Labor Statistics, Table 10.1 estimates growth of the 20 fastest growing occupations for the period 1998-2008. Thirteen of the 20 jobs that are expected to grow fastest (65% of the total) require no higher education. Most of them demand only short-term on-the-job training. Twelve of them are low-level service jobs, including sales clerks, cashiers, home health workers, nursing aides, orderlies, janitors, waiters, and security guards. Five of the 20 fastest growing jobs (25%) require higher education and are at the high end in terms of earning power. They include several computer-related occupations (systems analysts, computer engineers, and computer support specialists), managers, and administrators. Two of the 20 jobs (10%) require higher education but are in the middle in terms of earning power: registered nurses and high school teachers. Thus, Table 10.1 suggests that, while the top of the service sector is growing, the bottom is growing more quickly and the middle is growing more slowly. This pattern is consistent with the pattern of growing income inequality discussed in Chapter 7 ("Stratification: United States and Global Perspectives"). There, you will recall, we noted growing inequality between the top 20% of income earners and the remaining 80%.

Labor Market Segmentation

The polarization of jobs noted above is taking place in the last of three stages of labor market development identified by David Gordon and his colleagues (Gordon, Edwards, and Reich, 1982). The period from about 1820 to 1890 was the period of *initial proletarianization* in the United States. During this period, craft workers in small workshops were replaced by a large industrial working class. Then, from the end of the 19th century until the start of World War II, the labor market entered the phase of *labor homogenization.* Extensive mechanization and deskilling took place during this stage. Finally, the third phase of labor market development is that of **labor market segmentation.** During this stage, which began after World War II and continues up to the present, the labor market has been divided into two distinct parts called the primary and secondary labor markets. In these different settings, workers have different characteristics. Specifically:

✦ The **primary labor market** is composed disproportionately of highly skilled or well-educated white males. They are employed in large corporations that enjoy high levels of capital investment. In the primary labor market, employment is relatively secure, earnings are high, and fringe benefits are generous.

✦ The **secondary labor market** contains a disproportionately large number of women and members of racial minorities, particularly African and Hispanic Americans. Employees in the secondary labor market tend to be unskilled and lack higher education. They work in small firms with low levels of capital investment. Employment is insecure, earnings are low, and fringe benefits are meager.

This characterization may seem to advance us only a little beyond our earlier distinction between good and bad jobs. However, proponents of labor market segmentation theory offer fresh insights into two important issues. First, they argue that work is found in different ways in the two labor markets. Second, they point out that social barriers make it difficult for individuals to move from one labor market to the other. To appreciate the significance of these points, it is vital to note that workers do more than just work. They also seek to control their work and prevent outsiders from gaining access to it. Some workers are more successful in this regard than others. Understanding the social roots of their success or failure permits us to see why the primary and secondary labor markets remain distinct. Therefore, we now turn to a discussion of forms of worker control.

Worker Resistance and Management Response

One criticism lodged against Braverman's analysis of factory work is that he inaccurately portrays workers as passive victims of management control. In reality, workers often resist the imposition of task specialization and mechanization by managers. They go on strike, change jobs, fail to show up for work, sabotage production lines, and so forth (Burawoy, 1979; Clawson, 1980; Gouldner, 1954).

Worker resistance has often caused management to modify its organizational plans. For example, Henry Ford was forced to double wages to induce his workers to accept the monotony, stress, and lack of autonomy associated with assembly line production. Even so, gaining the cooperation of workers proved difficult. Therefore, beginning in the 1920s, some employers started to treat their employees more like human beings than cogs in a giant machine. They hoped to improve the work environment and thus make their employees more loyal and productive.

In the 1930s, the **human relations school of management** emerged as a challenge to Frederick W. Taylor's scientific management approach. Originating in studies conducted at the Hawthorne plant of the Western Electric Company near Chicago, the human relations school of management advocated less authoritarian leadership on the shop floor, careful selection and training of personnel, and greater attention to human needs and employee job satisfaction (see Chapter 5, "Interaction and Organization").

Over the next 70 years, owners and managers of big companies in all the rich industrialized countries realized they had to make more concessions to labor if they wanted a loyal and productive work force. These concessions included not just higher wages, but more decision-making authority about product quality, promotion policies, job design, product innovation, company investments, and so forth. The biggest concessions to labor were made in countries with the most powerful trade union movements, such as Sweden

◆ **FIGURE 10.5** ◆
**Average Hours Worked per
Week, Selected Countries,
1997–1998**

SOURCE: "Mild Labor . . . " (1999).

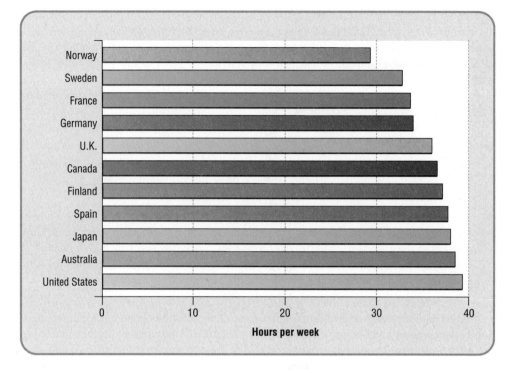

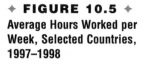

(see Chapter 11, "Politics," and Chapter 17, "Collective Action and Social Movements"). In these countries, a large proportion of eligible workers are members of unions (over 80% in Sweden). Moreover, unions are organized in nationwide umbrella organizations that negotiate directly with centralized business organizations and governments over wages and labor policy in general. At the other extreme among highly industrialized countries is the United States. Here, fewer than 14% of eligible workers are members of unions and there is no centralized, nationwide bargaining among unions, businesses, and governments. Two indicators of the relative inability of American workers to wrest concessions from their employers are given in Figures 10.5 and 10.6. On average, Americans work more hours per week than people in other rich industrialized countries. They also have fewer paid vacation days per year.

In the realm of industry-level decision making, too, American workers lag behind workers in Western Europe and Japan. We can see this if we briefly consider the two main types of decision-making innovations that have been introduced in the factories of the rich industrialized countries since the early 1970s:

1. *Reforms that give workers more authority on the shop floor* include those advanced by the **quality of work life** movement. "Quality circles" originated in Sweden and Japan. They involve small groups of a dozen or so workers and managers collaborating to improve both the quality of goods produced and communication between workers and managers. In some cases, this approach has evolved into a system that results in high productivity gains and worker satisfaction. For example, at Saab's main auto plant in Trolhattan, Sweden, the assembly line was eliminated (Krahn and Lowe, 1998: 239). Robots took over the arduous job of welding. In the welding area, groups of 12 workers program the computers, maintain the robots, ensure quality control, perform administrative tasks, and clean up. Elsewhere in the plant, autonomous teams of workers devote about 45 minutes to completing similar sets of integrated tasks. Workers build up inventory in "buffer zones" and decide themselves how to use it, thus introducing considerable flexibility into their schedule. All workers are encouraged to take advantage of many in-plant opportunities to upgrade their skills. Quality circles have been introduced in some American industries, including automotive and aerospace. However, they are less widespread in the United States than in Western Europe and Japan.

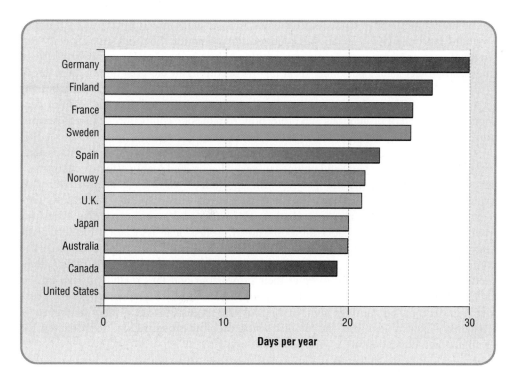

2. *Reforms that allow workers to help formulate overall business strategy* give workers more authority than quality circles. For example, in much of Western Europe workers are consulted not just on the shop floor but also in the boardroom. In Germany, this system is known as **codetermination.** German workers' councils review and influence management policies on a wide range of issues, including when and where new plants should be built and how capital should be invested in technological innovation. There are a few American examples of this sort of worker involvement in high-level decision making, mostly in the auto industry. Worker participation programs were widely credited with improving the quality of American cars and increasing the auto sector's productivity in the 1980s and 1990s, making it competitive again with Japanese carmakers. However, "[w]orker participation programs have had a difficult birth in North America. No broad policy agenda guides their development, and no systematic social theory lights their way" (Hodson and Sullivan, 1995 [1990]: 449).

Internationally, unions have clearly played a key role in increasing worker participation in industrial decision making since the 1920s and especially since the 1970s. To varying degrees, owners and managers of big corporations have conceded authority to workers in order to create a more stable, loyal, and productive work force. Understandably, workers who enjoy more authority in the workplace, whether unionized or not, have tried to protect the gains they have won. As we will now see, they have thereby contributed to the separation of primary from secondary labor markets.

Unions and Professional Organizations

Unions are organizations of workers that seek to defend and promote their members' interests. By bargaining with employers, unions have succeeded in winning improved working conditions, higher wages, and more worker participation in industrial decision making for their members.

With employers, unions have also helped to develop systems of labor recruitment, training, and promotion. These systems are sometimes called **internal labor markets** because they control pay rates, hiring, and promotions within corporations. At the same time, they reduce competition between a firm's workers and external labor supplies.

In an internal labor market, advancement through the ranks is governed by training programs that specify the credentials required for promotion. Seniority rules specify the length of time one must serve in a given position before being allowed to move up. These rules also protect senior personnel from layoffs according to the principle of "last hired, first fired." Finally, in internal labor markets, recruitment of new workers is usually limited to entry-level positions. In this way, the intake of new workers is controlled. Senior personnel are assured of promotions and protection from outside competition. For this reason, internal labor markets are sometimes called "labor market shelters."

Labor market shelters operate not only among unionized factory workers. Professionals, such as doctors, lawyers, and engineers, have also created highly effective labor market shelters. **Professionals** are people with specialized knowledge acquired through extensive higher education. They enjoy a high degree of work autonomy and usually regulate themselves and enforce standards through professional associations. The American Medical Association is probably the best-known professional association in the United States. Professionals exercise authority over clients and subordinates. They operate according to a code of ethics that emphasizes the altruistic nature of their work. Finally, they specify the credentials needed to enter their professions and thus maintain a cap on the supply of new professionals. This reduces competition, ensures high demand for their services, and keeps their earnings high. In this way, the professions act as labor market shelters, much like unions. (For further discussion of professionalization, see Chapter 13, "Religion and Education").

Barriers Between the Primary and Secondary Labor Markets

We saw above that workers in the secondary labor market do not enjoy the high pay, job security, and benefit packages shared by workers in the primary labor market, many of whom are members of unions and professional associations. We may now add that workers find it difficult to exit the "job ghettos" of the secondary labor market. That is because three social barriers make the primary labor market difficult to penetrate:

1. *Often, there are few entry-level positions in the primary labor market.* One set of circumstances that contributes to the lack of entry-level positions in the primary labor market is corporate "downsizing" and plant shutdowns. These took place on a wide scale in the United States throughout the 1980s and early 1990s (see below and Box 10.1). The lack of entry-level positions is especially acute during periods of economic recession. A recession is usually defined as a period of 6 months or more during which the economy shrinks and unemployment rises. The United States was in the grip of recession about one sixth of the time between November 1973 and October 2000. Big economic forces like plant shutdowns and recessions prevent upward mobility and often result in downward mobility. Consider the case of Jervis, a 19-year-old African American interviewed by Katherine Newman in her study of fast-food workers in Harlem. Like nearly all people in the secondary labor market, he wants steady work and upward mobility. Until 1991, he worked in a local fast-food restaurant and dreamt about moving up to a management position: "A decent job that gives you experience . . . that has a chain or something like that, that you could try to make some type of career where you don't always have to be locked into a position. It's a starting place. The bottom line was the overall benefit [of the job]: I moved ahead in life. That was good enough" (quoted in Newman, 1999: 252). Unfortunately, the recession of 1990–91 cut short Jervis's dream. He lost his job.

2. *A second barrier preventing people in the secondary labor market from penetrating the primary labor market is their lack of informal networks linking them to good job openings.* People often find out about job availability through informal networks of friends and acquaintances (Granovetter, 1995). These networks typically consist of people with the same ethnic and racial backgrounds. African and Hispanic Americans, who compose a disproportionately large share of workers in the secondary

BOX 10.1
SOCIOLOGY AT THE MOVIES

ROGER AND ME (1989)

Scene 1: An auto factory in Flint, Michigan. Scene 2: General Motors' chief executive officer, Roger Smith, announces he's closing the factory. Scene 3: Newspaper headlines proclaim that GM is opening plants in Mexico. Scene 4: Michael Moore, the director of the movie *Roger and Me* (1989), tries to interview Roger Smith so he can get him to face up to the consequences of his corporate decisions for the ordinary citizens of Flint. Moore is repeatedly rebuffed. Scene 5: During a gala Christmas party, Smith talks about generosity and "the total Christmas experience." The scene is interlaced with shots of the families of fired autoworkers being evicted from their homes. In one shot, a decorated Christmas tree is thrown on top of a family's belongings.

Roger and Me is an infuriating yet funny movie about deindustrialization and its impact on former GM workers in Flint. Beginning in the early 1980s, GM laid off tens of thousands of workers in the city and moved their jobs to Mexico. In 1980, Flint had 80,000 autoworkers. In 2000, it had 30,000 (Steinhart, 2000). Flint, once prosperous, was dubbed by *Money* magazine as the worst place to live in America.

Michael Moore directs *Roger and Me.*

In the movie, Michael Moore connects corporate decision making to everyday life in the United States and industrial policy abroad. We see how multinational corporations can close factories in the United States, thereby exporting jobs to low-wage countries like Mexico and overturning the lives of ordinary American workers. Moore's attempts to interview Smith and discuss these issues are consistently irreverent and hilarious. Typically, when he tries to see Smith at his office, he offers the security guards his Chuck E. Cheese discount card for identification.

Roger and Me was released when many scholars and politicians were expressing fears that the United States labor force was on a downward slide due to deindustrialization. In cities like Flint, some laid-off workers moved away. Others stayed but were unable to find work and so contributed to a rising poverty rate. Flint has never recovered from deindustrialization. In 2000, the city's unemployment rate was 7.7%, more than 2½ times Michigan's 2.9% rate. *Roger and Me* remains a testament to the suffering of laid-off employees when neither corporations nor governments assume any responsibility for compensating, retraining, and relocating them.

labor market, are less likely than whites to find out about job openings in the primary labor market, where the labor force is disproportionately white and non-Hispanic. Even within racial groups, the difference between getting good work and having none is sometimes a question of having the right connections. Jervis, for instance, certainly knew about the pockets of middle-class African Americans living on the tree-lined streets near his apartment. He saw them dressing in stylish suits and driving to and from work in nice cars. But he had no contact with these people and no sense of how he might form ties with them that could lead him to good, steady work. When Newman asked him, "How do people meet job contacts in the first place?" Jervis replied: "I'd say luck and destiny"—hardly a realistic sense of the kind of networking often required to get good work (Newman, 1999: 252).

3.   *Finally, mobility out of the secondary labor market is difficult because workers usually lack the required training and certification for jobs in the primary labor market.* What is more, due to their low wages and scarce leisure time, they cannot usually

◆ **FIGURE 10.7** ◆

From the Primary to the Secondary Labor Market

This figure shows 1995 median weekly earnings of full-time wage and salary workers in the United States by union status, sex, and race. Earnings decline as one moves from the core of the primary labor market (category 1) to the periphery of the secondary labor market (category 12). What happens to union status, sex, and race as one moves from core to periphery?

SOURCE: Hesse-Biber and Carter (2000: 125).

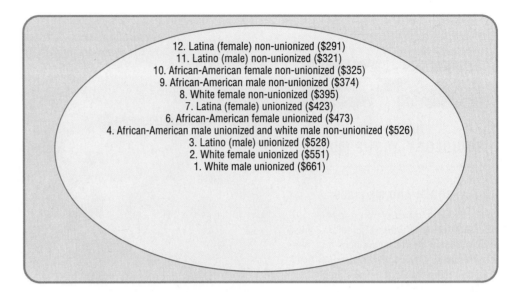

12. Latina (female) non-unionized ($291)
11. Latino (male) non-unionized ($321)
10. African-American female non-unionized ($325)
9. African-American male non-unionized ($374)
8. White female non-unionized ($395)
7. Latina (female) unionized ($423)
6. African-American female unionized ($473)
4. African-American male unionized and white male non-unionized ($526)
3. Latino (male) unionized ($528)
2. White female unionized ($551)
1. White male unionized ($661)

afford to upgrade either their skills or their credentials. When Newman interviewed Jervis, his self-confidence and his hope of rising to a management position were fading fast. He was even unable to find steady minimum-wage work in retail shops and fast-food restaurants, partly because older and more experienced workers were taking many of the jobs that had formerly been open to teenagers. With only a grade 10 education, his prospects were dim. He could not afford to go back to school. Due to impersonal economic forces, lack of network ties, and insufficient education he was stuck, probably permanently, on the margins of the secondary labor market.

The dollar effects of these barriers to entering the primary labor market are vividly illustrated by Figure 10.7. Figure 10.7 shows that the median weekly earnings of full-time wage and salary workers in the United States decline sharply as one moves from the core of the primary labor market (category #1) to the periphery of the secondary labor market (category #12). Notice also that men, non-Hispanic whites, and unionized workers tend to be concentrated in the primary labor market. Women, African and Hispanic Americans, and nonunion members tend to be concentrated in the secondary labor market. We conclude that the barriers to entering the primary labor market are neither color blind nor gender neutral.

The Time Crunch and Its Effects

Although the quality of working life is higher in the primary than in the secondary labor market, one must be careful not to exaggerate the differences. Overwork and lack of leisure have become central features of our culture, and this is true in both labor markets. We are experiencing a growing time crunch. All the adults in most American households work full-time in the paid labor force, and many adolescents work part-time. Some people work two jobs to make ends meet. Many office workers, managers, and professionals work 10, 12, or more hours a day due to tight deadlines, demands for high productivity, and a trimmed down work force (see Figure 10.8). According to the Director of the National Institute for Health and Safety, American full-time workers were putting in an average of 47 hours of work a week in the late 1990s, up nearly 10% from 20 years earlier (quoted in Solomon, 1999: 51). Add to this the heavy demands of family life and one can readily understand why stress, depression, aggression, and substance abuse are on the rise, both in the secondary and the primary labor markets. Three quarters of American workers believe there is greater stress on the job now than there was two decades ago (Solomon, 1999: 51). According to one Human Resources expert:

◆ FIGURE 10.8 ◆

Bedtime at Yahoo! Inc.
In high-tech industries, working on very little sleep is common. Here, David Filo, co-founder of Yahoo!, takes a nap under his desk.

SOURCE: Meri Simon, *San Jose Mercury News* (in Bliss, 2000).

People are on the edge, and acting it out in the workplace. They're yelling obscenities at each other, coming into work chronically late, throwing food in the cafeteria, and crying in the hallways. It's not just a matter of acting a little inappropriately any more. No, it's gotten far worse than that. Nice people are having trouble with alcohol, drugs, depression and acting aggressively at work.

And these aren't isolated instances—instead, they're a composite picture from chronic work distress, as well as difficulties trying to deal with personal life overload from marital problems, single parenthood, financial worries and the like. The stress is so great that people are snapping. And no one has to tell you that it's getting worse (Solomon, 1999: 48–9).

Stress is often defined as the feeling that one is unable to cope with life's demands given one's resources. Work is the leading source of stress throughout the world. In one study of office workers in 16 countries, including the United States, 54% of the respondents cited work as a current cause of stress in their lives and 29% cited money problems—which are also work related since one's job is the main source of income for most people ("Work-related Stress . . . ," 1995). The results of a nationwide survey released in 2001 show that nearly half of United States employees feel they are overworked, overwhelmed by how much work they have to do, or unable to find the time to step back and process or reflect on the work they are doing. Those who often or very often use cell phones, beepers, pagers, computers, e-mail, and fax machines are 9% more likely than other employees to experience high levels of feeling overworked. Women, baby boomers (people born between 1946 and 1964), and managers and professionals also have significantly above-average feelings of overwork (Galinsky, Kim, and Bond, 2001). The rate of severe depression is also on the rise. A survey of nine countries, including the United States, found that severe depression has increased in each succeeding generation since 1915. In some countries, people born after 1955 are three times more likely to experience serious depression than their grandparents ("The Changing Rate . . . ," 1992). According to a recent World Health Organization study, severe depression is the second leading contributor to "disease burden" (years lived with a disability) in the rich postindustrial countries. It is expected to rise to the number one position by 2020 (Vernarec, 2000).

There are three main reasons why leisure is on the decline and the pace of work is becoming more frantic for those who are employed in the paid labor force (Schor, 1992). First, big corporations are in a position to invest enormous and increasing resources in advertising. As we saw in Chapter 3 ("Culture"), advertising pushes Americans to consume goods and services at higher and higher levels all the time. Shopping has become an end in itself for many people, a form of entertainment, and, in some cases, a deeply felt "need." But consumerism requires money, so people have to work harder to shop more. Second, most corporate executives apparently think it is more profitable to have employees work more hours rather than hire more workers and pay expensive benefits for new employees.[1] Third, as we have just discussed, American workers are not in a position to demand reduced working hours and more vacation time because few of them are unionized. They lack clout and suffer the consequences in terms of stress, depression, and other work-related ailments.

THE PROBLEM OF MARKETS

One conclusion we can draw from the preceding discussion is that the secondary labor market is a relatively **free market.** That is, the supply of, and demand for, labor regulates wage levels and other benefits. If labor supply is high and demand is low, wages fall. If labor demand is high and supply is low, wages rise. People who work in the secondary labor market lack much power to interfere in the operation of the forces of supply and demand.

In contrast, the primary labor market is a more **regulated market.** Wage levels and other benefits are established not just by the forces of supply and demand but also by the power of workers and professionals. As we have seen, they are in a position to influence the operation of the primary labor market to their own advantage.

This suggests that the freer the labor market, the higher the resulting level of social inequality. In fact, in the freest markets, many of the least powerful people are unable to earn enough to subsist while a few of the most powerful people can amass unimaginably large fortunes. That is why the secondary labor market cannot be entirely free. The federal government had to establish a legal minimum wage to prevent the price of unskilled labor from dropping below the point at which people are literally able to make a living (see Box 10.2). Similarly, in the historical period that most closely approximates a completely free market for labor—late 18th-century England—starvation became so widespread and the threat of social instability so great the government was forced to establish a system of state-run "poor houses" that provided minimal food and shelter for people without means (Polanyi, 1957 [1944]).

The question of whether free or regulated markets are better for society lies at the center of much debate in economics and politics (Kuttner, 1997). For many economic sociologists, however, that question is too abstract. In the first place, regulation is not an either/or issue but a matter of degree. A market may be more or less regulated. Second, markets may be regulated by different groups of people with varying degrees of power and different norms and values. Therefore, the costs and benefits of regulation may be socially distributed in many different ways. Only sociological analysis can sort out the costs and benefits of different degrees of market regulation for various categories of the population. These, then, are the main insights of economic sociology as applied to the study of markets: (a) The structure of markets varies widely across cultures and historical periods, and (b) the degree and type of regulation depend on how power, norms, and values are distributed among various social groups (Lie, 1992).

The economic sociologist's approach to the study of markets is different from that of the dominant trend in contemporary economics, known as the "neoclassical" school (Becker, 1976; Mankiw, 1998). Instead of focusing on how power, norms, and values shape markets, neoclassical economists argue that free markets maximize economic growth. We may use the minimum wage to illustrate their point. According to neoclassical economists, if the minimum wage is eliminated entirely, everyone will be better off. For

---

[1]However, medical economists have recently shown that work-related illnesses are actually costing businesses far more than they realize in terms of absenteeism and low productivity (Vernarec, 2000). It is too early to tell whether this research will have an impact on hiring policies.

ACCOUNT OF THE

SALE of a WIFE, by J. NASH,

IN THOMAS-STREET MARKET,

On the 29th of May, 1823.

This day another of those disgraceful scenes which of late have so frequently annoyed the public markets in this country took place in St. Thomas's Market, in this city; a man (if he deserves the name) of the name of John Nash, a drover, residing in Rosemary-street, appeared there leading his wife in a halter, followed by a great concourse of spectators; when arrived opposite the Bell-yard, he publicly announced his intention of disposing of his better half by Public Auction, and stated that the biddings were then open; it was a long while before any one ventured to speak, at length a young man who thought it a pity to let her remain in the hands of her present owner, generously bid 6d.! In vain did the anxious seller look around for another bidding, no one could be found to advance one penny, and after extolling her qualities, and warranting her sound, and free from vice, he was obliged, rather than keep her, to let her go at that price. The lady appeared quite satisfied, but not so the purchaser, he soon repented of his bargain, and again offered her to sale, when being bid nine-pence, he readily accepted it, and handed the lady to her new purchaser, who, not liking the transfer, made off with her mother, but was soon taken by her purchaser, and claimed as his property, to this she would not consent but by order of a magistrate, who dismissed the case. Nash, the husband, was obliged to make a precipitate retreat from the enraged populace.

Copy of Verses written on the Occasion:

COME all you kind husbands who have scolding wives,
Who thro' living together are tired of your lives,
If you cannot persuade her nor good natur'd make her
Place a rope round her neck & to market pray take her

Should any one bid, when she's offer'd for sale,
Let her go for a trifle lest she should get stale,
If six-pence be offer'd, & that's all can be had,
Let her go for the same rather than keep a lot bad.

Come all jolly neighbours, come dance sing & play,
Away to the wedding where we intend to drink tea;
All the world assembles, the young and the old,
For to see this fair beauty, as we have been told.

Here's success to this couple to keep up the fun,
May bumpers go round at the birth of a son;
Long life to them both, and in peace & content
May their days and their nights for ever be spent.

Shepherd, Printer, No. 6, on the Broad Weir, Bristol.

An unregulated market creates gross inequalities. Here, an account of a wife sold by her husband for six-pence in Britain in 1823.

example, in a situation where labor is in low demand and high supply, wages will fall. This will increase profits. Higher profits will in turn allow employers to invest more in expanding their businesses. The new investment will create new jobs, and the rising demand for labor will drive wages up. Social inequality may increase, but eventually *everyone* will be better off thanks to the operation of the free market.[2]

---

[2]There is no necessary contradiction between one aspect of the sociologist's and the neoclassical economist's approaches to markets. As markets become less regulated, average wealth may grow and the gap between rich and poor may increase at the same time. As we saw in Chapter 7 ("Social Stratification: United States and Global Perspectives"), this is exactly what has been happening in the United States for nearly 30 years. Note also that a less influential school of economic thought, known as "institutionalism," is much closer to the sociological perspective on markets than the neoclassical school.

BOX 10.2
IT'S YOUR CHOICE

THE MINIMUM WAGE

"Flipping burgers at Mickey D's is no way to make a living," a young man once told John Lie. Having tried his hand at several minimum-wage jobs as a teenager, John knew the young man was right. At just over $5 an hour, a minimum-wage job may be fine for teenagers, many of whom are supported by their parents. However, it is difficult to live on one's own, much less to support a family, on a minimum-wage job, even if you work full-time. This is the problem with the minimum wage. It does not amount to a living wage for many people.

In 1999, nearly 12 million American workers held minimum-wage jobs. Of this group, 58% were women. One million of them were single mothers (Bernstein, Hartmann, and Schmitt, 1999). Taking inflation into account, the minimum wage rose from about $3 an hour to about $7 an hour between 1938 and 1968. Then it fell to about $5 an hour between 1968 and 1997 (see Figure 10.9). Today, a single mother working full-time at the minimum wage does not make enough to lift a family of three (herself and two children) above the poverty level. In 1998, her earnings would have been 18 percent below the poverty level (Bernstein, Hartmann, and Schmitt 1999). Because the minimum wage has fallen since the late 1960s, the percentage of workers earning poverty-level wages has increased. It now stands at nearly 29%. The percentage of workers earning poverty-level wages is much higher for women than men, and much higher for African Americans than others (see Table 10.2).

Given a million single mothers who cannot lift themselves and their children out of poverty even if they work full-time, many scholars and policy makers suggest raising the minimum wage. Others disagree. They fear that raising the minimum wage would decrease the number of available jobs. Others disagree in principle with government interference in the economy. Some scholars and policy makers even advocate the abolition of the minimum wage.

What do you think? Should the minimum wage be raised? Should someone working full-time be entitled to live above the poverty level? Or should businesses be entitled to hire workers at whatever price the market will bear? In thinking about this question, you should bear in mind what happened between 1996 and 1999. In late 1996 and 1997, the minimum wage increased. In the next couple of years, the employment rate of low-wage workers, and particularly single mothers, followed suit. That is, due to the booming economy, the percentage of low-wage workers rose dramatically (Bernstein, Hartmann, and Schmitt, 1999). What does this say about the relationship between the minimum wage and the employment rate of low-wage workers?

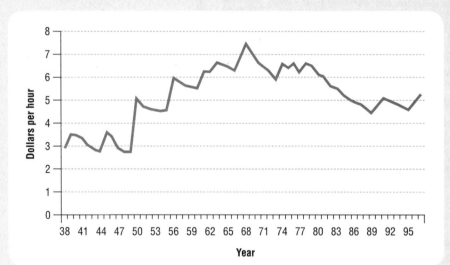

◆ **FIGURE 10.9** ◆

Value of the Federal Minimum Wage, United States, 1938–1997 (in 1998 dollars)

SOURCE: United States Department of Labor (1999f).

Year	All Americans			African Americans	
	Total	Men	Women	Men	Women
1973	23.5	12.8	39.1	28.4	50.7
1979	23.7	13.4	37.0	23.4	42.8
1989	28.5	21.2	36.8	33.2	43.6
1997	28.6	22.5	35.3	33.2	42.6

◆ **TABLE 10.2** ◆

Share of Workers Earning Poverty-Level Wages, by Gender and Race, United States, 1993–97 (in percent)

SOURCE: Mishel, Bernstein, and Schmitt (1999: 136, 141).

One difficulty with the neoclassical theory is that, in the real world, resistance to the operation of free markets increases as you move down the social hierarchy from the wealthiest and most powerful members of society to the poorest and least powerful. People at the low end of the social hierarchy usually fight against falling wages. The resulting social instability can disrupt production and investment. A social environment full of strikes, riots, and industrial sabotage is unfavorable to high productivity and new capital investment. That is why the economic system of a society at a given time is a more or less stable set of compromises between advocates and opponents of free markets. Markets are only as free as people are prepared to tolerate, and their degree of tolerance varies historically and between cultures (Berger and Dore, 1996; Doremus, Keller, Pauly, and Reich, 1998).

In the rest of this chapter, we offer several illustrations of how sociologists analyze markets. First, we compare capitalism and communism, the two main types of economic system in the 20th century. Second, we examine the ability of big corporations to shape markets in the United States today. Finally, we extend our analysis of corporate and free market growth to the global level. We identify advocates and opponents of these developments and sketch the main work-related decisions that face us in the early 21st century.

Capitalism and Communism

Capitalism

The world's dominant economic system today is **capitalism.** Capitalist economies have two distinctive features:

1. *Private ownership of property*. In capitalist economies, individuals and corporations own almost all the means of producing goods and services. Individuals and corporations are therefore free to buy and sell land, natural resources, buildings, manufactured goods, medical services, and just about everything else. Like individuals, **corporations** are legal entities. They can enter contracts and own property. However, corporate ownership has two advantages over individual ownership. First, corporations are taxed at a lower rate than individuals. Second, the corporation's owners are not normally liable if the corporation harms consumers or goes bankrupt. Instead, the corporation itself is legally responsible for damage and debt.

2. *Competition in the pursuit of profit*. The second hallmark of capitalism is that producers, motivated by the prospect of profits, compete to offer consumers desired goods and services at the lowest possible price. A purely capitalist economy is often called a laissez-faire system. *Laissez-faire* is French for "allow to do." In a laissez-faire system, the government does not interfere in the operation of the economy at all; it allows producers and consumers to do what they want. Adam Smith was an 18th-century Scottish economist who first outlined the operation of the ideal capitalist economy. According to Smith, everyone benefits from laissez-faire. The most efficient producers make profits while consumers can buy at low prices. If everyone pursues their narrow self-interest, unimpeded by government, the economy will achieve "the greatest good for the greatest number," said Smith (1981 [1776]).

In reality, no economy is purely laissez-faire. The state had to intervene heavily to create markets in the first place. For example, 500 years ago the idea that land is a commodity that can be bought, sold, and rented on the free market was utterly foreign to the native peoples who lived in the territory that is now North America. To turn the land into a marketable commodity, European armies had to force native peoples off the land and eventually onto reservations. Governments had to pass laws regulating the ownership, sale, and rent of land. Without the military and legal intervention of government, no market for land would exist.

Today, governments must also intervene in the economy to keep the market working effectively. For instance, governments create and maintain an economic infrastructure (roads, ports, etc.) to make commerce possible. They pass laws governing the minimum wage, occupational health and safety, child labor, and industrial pollution to protect workers and consumers from the excesses of corporations. If very large corporations get into financial trouble, they can expect the government to bail them out on the grounds that their bankruptcy would be devastating to the economy. For example, when Chrysler was facing financial ruin in the 1970s, the federal government stepped in with low-interest loans and other help to keep the corporation afloat. Similarly, in the 1980s, the government doled out $500 million of taxpayers' money to protect the nation's savings and loans institutions from collapse (Sherrill, 1990).

Governments also play an influential role in establishing and promoting many leading industries, especially those that require large outlays on research and development. Until the 1960s, the United States government covered two thirds of all research and development costs in the country, and it still covers nearly a third (Rosenberg, 1982; see Chapter 18, "Technology and the Global Environment," Figure 18.2). It developed the foundations of the Internet in the late 1960s. In the late 1980s, it gave $100 million a year to corporations working to improve the quality of the microchip (Reich, 1991). Occasionally, the government even resorts to armed force to ensure the smooth operation of the market, as, for example, when oil supplies are threatened by hostile foreign regimes (Lazonick, 1991).

As these examples of government intervention suggest, we should see Smith's ideal of a laissez-faire economy as just that: an ideal. In the real world, markets are free to varying degrees, but none is or can be entirely free. Which capitalist economies are the most free and which are the least free? The International Institute for Management Development in Switzerland publishes a widely respected annual index of competitiveness for 47 capitalist countries. The index is based on the amount of state ownership of industry and many other indicators of market freedom. Table 10.3 gives the overall scores for the 10 most competitive and 10 least competitive economies in 1999. The United States tops the list. It is the most competitive economy in the world, with a score of 100. Russia ranks last with a score of 38.

Communism

Like laissez-faire capitalism, communism is an ideal. **Communism** is the name Karl Marx gave to the classless society that, he said, is bound to develop out of capitalism. Socialism is the name he gave to the transitional phase between capitalism and communism. No country in the world is communist in the pure sense of the term. About two dozen countries in Asia, South America, and Africa consider themselves socialist. They include China, North Korea, Vietnam, and Cuba. As an ideal, communism is an economic system with two distinct features:

◆ **TABLE 10.3** ◆

The 10 Most Competitive and 10 Least Competitive Capitalist Economies in the World, 1999 (n = 47)

Note: Many countries, especially poor ones, are not on this list because comprehensive economic data are unavailable for them.

SOURCE: International Institute for Management Development (1999).

Country	Competitiveness Score	Country	Competitiveness Score
1. United States	100	38. South Korea	52
2. Singapore	86	39. India	50
3. Finland	83	40. Slovenia	50
4. Luxembourg	81	41. Czech Republic	49
5. The Netherlands	81	42. Republic of South Africa	48
6. Switzerland	80	43. Colombia	48
7. Hong Kong	80	44. Poland	48
8. Denmark	78	45. Venezuela	47
9. Germany	77	46. Indonesia	42
10. Canada	76	47. Russia	38
. . .			

1. *Public ownership of property*. Under communism, the state owns almost all the means of producing goods and services. Private corporations do not exist. Individuals are not free to buy and sell goods and services. The aim of public ownership is to ensure that all individuals have equal wealth and equal access to goods and services.

2. *Government planning*. Five-year state plans establish production quotas, prices, and most other aspects of economic activity. Political officials—not forces of supply and demand—design these state plans and determine what is produced, in what quantities, and at what prices. A high level of control of the population is required to implement these rigid state plans. As a result, democratic politics is not allowed to interfere with state activities. Only one political party exists—the Communist Party. Elections are held regularly, but only the Communist Party is allowed to run for office (Zaslavsky and Brym, 1978).

Several highly industrialized countries, including Sweden, Denmark, Norway, and, to lesser degree, France and Germany, are "democratic socialist" societies. Like the United States, they are prosperous and enjoy multiparty elections. However, their governments intervene in the economy much more than we are used to in the United States. At their core, however, and despite their name, the democratic socialist countries are capitalist. The great bulk of property is privately owned, and competition in the pursuit of profit is the main motive for business activity.

Until recently, the countries of Central and Eastern Europe were single-party, socialist societies. The most powerful of these countries was the Soviet Union, which was composed of Russia and 14 other socialist republics. In perhaps the most surprising and sudden change in modern history, the countries of the region started introducing capitalism and holding multiparty elections in the late 1980s and early 1990s.

The collapse of socialism in Central and Eastern Europe was due to several factors. For one thing, the citizens of the region enjoyed few civil rights. For another, their standard of living was only about half as high as that of people in the rich industrialized countries of the West. The gap between East and West grew as the arms race between the Soviet Union and the United States intensified in the 1980s. The standard of living fell as the Soviet Union mobilized its economic resources to try to match the quantity and quality of military goods produced by the United States. Dissatisfaction was widespread and expressed itself in many ways, including strikes and political demonstrations. It grew as television and radio signals beamed from the West made the gap between socialism and capitalism more apparent to the citizenry. Eventually, the communist parties of the region felt they could no longer govern effectively and so began to introduce reforms.

In the 1990s, some Central and East European countries were more successful than others in introducing elements of capitalism and raising their citizens' standard of living. The Czech Republic was most successful. Russia and most of the rest of the former Soviet Union were least successful. Many factors account for the different success rates, but perhaps the most important is the way different countries introduced reforms. The Czechs introduced both of the key elements of capitalism—private property and competition in the pursuit of profit. The Russians, however, introduced private property without much competition. Specifically, the Russian government first allowed prices to rise to market levels. This made many basic goods too expensive for a large part of the population, which was quickly impoverished. Next, the government sold off state-owned property to individuals and corporations. However, the only people who could afford to buy the factories, mines, oil refineries, airlines, and other economic enterprises were organized criminals and former officials of the Communist Party. They alone had access to sufficient capital and insider information about how to make the purchases (Handelman, 1995). The effect was to make Russia's level of socioeconomic inequality among the highest in the world. A crucial element lacking in the Russian reform was competition. A few giant corporations control nearly every part of the Russian economy. They tend not to compete against each other. Competition would drive prices down and efficiency up. However, these corporations are so big they can agree among themselves to set prices at levels that are most profitable for them. They thus have little incentive to innovate. Moreover, they are so big and rich they have enormous influence over government. When a few corporations are so big they can behave in this way, they are called **oligopolies.**[3] Russia is full of them. The lack of competition in Russia has prevented the country from experiencing much economic growth since reforms began more than a decade ago (Brym, 1996a; 1996b; 1996c).

The Corporation

Oligopolies can constrain innovation in all societies. They can force consumers to pay higher prices. They can exercise excessive influence on governments. However, in the United States and other Western countries, "antitrust" laws limit their growth. The 1890 Sherman Antitrust Act and the 1914 Clayton Act are the basic United States antitrust laws. They have prevented the largest corporations from gaining much more control of specific industries than they had in the 1930s. Thus, in 1935, the four largest firms in each manufacturing industry controlled, on average, 37% of sales in that industry. In 1992, the last year for which data are available as of this writing, the figure stood at 40% (Hodson and Sullivan, 1995 [1990]: 393; United States Bureau of the Census, 2000e).

However, the law has been only partly effective in stabilizing the growth of oligopolies. For instance, the government managed to break AT&T's stranglehold on the telecommunications market in the 1980s, but its efforts to break Microsoft into two smaller corporations seemed doomed to failure as of late 2001 (see Table 10.4). Moreover, when the four biggest corporations in an industry make 4 out of every 10 dollars in sales, it is hard to deny they are enormously powerful. The top 500 corporations in the United States control over two thirds of business resources and profit. This is a world apart from the early 19th century, when most business firms were family owned and served only local markets.

It is also important to note that an important effect of United States antitrust law is to encourage big companies to diversify. That is, rather than increasing their share of control

"The Monster Monopoly," an 1884 cartoon attacking John D. Rockefeller's Standard Oil Company. One of the most famous antitrust cases ever to reach the Supreme Court resulted in the breakup of Standard Oil in 1911. The Court broke new ground in deciding to dissolve the company into separate geographical units.

---

[3]A monopoly is a single producer that completely dominates a market.

Forum	Issue		
	Split the Company in Two	Stop Bundling Internet Explorer Browser, Microsoft Media, and MSN Internet Service With Windows	Stop Microsoft From Selling Windows to PC Makers Only If They Agree to Include Other Microsoft Software
Trial court, June 7, 2000	Yes	Yes	Yes
Appeals court, June 28, 2001	Reconsider	Reconsider	Yes
Department of Justice, September 6, 2001	No	No	Still considering

✦ **TABLE 10.4** ✦

The Rise and Decline of the Microsoft Antitrust Suit

in their own industry, corporations often move into new industries. Big companies that operate in several industries at the same time are called **conglomerates.** For example, in 2000, America Online (AOL), the world's biggest Internet service provider, took over Time-Warner, the entertainment giant, thereby forming a conglomerate. The takeover allowed AOL's business to grow, but because it grew outside the Internet industry, AOL avoided the charge of forming an oligopoly. Unlike oligopolies, conglomerates are growing rapidly in the United States. Big companies are swallowed up by still bigger ones in wave after wave of corporate mergers (Mizruchi, 1982; 1992).

Outright ownership of a company by a second company in another industry is only one way corporations may be linked. **Interlocking directorates** are another. Interlocking directorates are formed when an individual sits on the board of directors of two or more noncompeting companies. (As we saw, antitrust laws prevent an individual from sitting on the board of directors of a competitor.) For instance, in 1997 the Board of Directors of IBM included, among others, the Chair and CEO of Mobil Oil, the Chair and CEO of Ford, and the President of Mitsubishi ("IBM . . . ," 1997). In 2000, Ann McLaughlin, former Secretary of Labor in the Reagan administration, served on the boards of directors of Nordstrom, Kellogg, Marriott, Fannie Mae, and Microsoft, among other major corporations ("Ann McLaughlin . . . ," 2000). Such interlocks enable corporations to exchange valuable information and form alliances for their mutual benefit. They also create useful channels of communication to, and influence over, government (Mintz and Schwartz, 1985; Mintz, 1989; Useem, 1984).

Of course, small businesses continue to exist (Granovetter, 1984). In the United States, 85% of businesses have fewer than 20 employees. Fully 40% of the labor force works in firms with fewer than 100 employees. Small firms are particularly important in the service sector. However, compared with large firms, profits in small firms are typically low. Bankruptcies are common. Small firms usually use outdated production and marketing techniques. Jobs in small firms often offer low wages and meager benefits.

Most of the United States labor force now works in large corporations. Specifically, about a third of the labor force is employed in the 1,500 largest industrial, financial, and service firms. In the service sector, the biggest employer is Wal-Mart, with 867,500 employees in the United States in 1999. In manufacturing, the largest employer is General Motors, with 601,500 employees in the United States in 1999 ("Forbes 500 . . . ," 1999). As we will now see, however, even these figures underestimate the global reach and influence of the biggest corporations.

Globalization

In the 1980s and early 1990s, the United States was hit by a wave of corporate "downsizing" (Dudley, 1994; Gordon, 1996; Smith, 1990). Especially in the older manufacturing industries of the Northeast and Midwest—an area sometimes called the "rust belt"—hundreds of thousands of blue-collar workers and middle managers were fired. In places like Flint, Michigan, and Racine, Wisconsin, the consequences were devastating. Unemployment soared. Social problems such as alcoholism and wife abuse became acute.

Some people blamed government for the plant shutdowns. They said taxes were so high, big corporations could no longer make decent profits. Others blamed the unemployed themselves. They said powerful unions drove up the hourly wage to the point where companies like General Motors and Ford were losing money. Katherine Dudley studied the closure of an auto factory in Racine. She noticed that unemployed workers "who were once able to fulfill their obligations to family, community, and nation . . . have become culturally 'deviant' . . . [They] are no longer perceived as . . . hardworking [and] self-sacrificing . . . [but are] the target of national—and now even international—ridicule, censure, and shame" (Dudley, 1994: 161). Still others blamed the corporations. As soon as big corporations closed plants in places like Racine and Flint, they opened new ones in places like northern Mexico. Mexican workers were happy to earn only one sixth or one tenth as much as their American counterparts. The Mexican government was delighted to make tax concessions to attract the new jobs.

In the 1980s, workers, governments, and corporations got involved as unequal players in the globalization of the world economy. Japan and Germany had fully recovered from the devastation of World War II. With these large and robust industrial economies now firing on all cylinders, American-based multinationals were forced to cut costs and become more efficient to remain competitive. On a scale far larger than ever before, they began to build branch plants in many countries to take advantage of inexpensive labor and low taxes. Multinational corporations based in Japan and other highly industrialized countries did the same.

However, while multinational corporations could easily move investment capital from one country to the next, workers were rooted in their communities and governments were rooted in their nation-states. Multinationals thus had a big advantage over the other players in the globalization game. They could threaten to move plants unless governments and workers made concessions. They could play one government off another in the bidding war for new plants. And they could pick up and leave when it became clear that relocation would do wonders for their bottom line.

Today, more than 20 years after the globalization game began in earnest, it is easier to identify the winners than the losers. The clear winners are the stockholders of the multinational corporations, whose profits have soared. The losers, at least initially, were American blue-collar workers. To cite just one example, between 1980 and 1993, General

Some people in the rich, industrialized countries oppose the globalization of commerce. For example, the World Trade Organization (WTO) was set up by the governments of 134 countries in 1994 to encourage and referee global commerce. When the WTO met in Seattle in December 1999, 40,000 union activists, environmentalists, supporters of worker and peasant movements in developing countries, and other opponents of multinational corporations staged protests that caused property damage and threatened to disrupt the proceedings. Police and the National Guard replied with concussion grenades, tear gas, rubber bullets, and mass arrests. Similar protests have taken place at subsequent WTO meetings in other countries.

Motors cut its labor force by more than 30%. Between 1993 and 1999, the number of Americans employed by General Motors fell another 20%.

Even while these cuts were being made, however, some large American manufacturers were hiring. For instance, employment at Boeing grew more than 65% between 1993 and 1999. In the service sector, employment soared. For example, Wal-Mart employed twice as many people in 1999 as in 1993 ("Forbes 500 . . . ," 1999; Hodson and Sullivan, 1995 [1990]: 393). In 2000, unemployment in the United States hit a 38-year low. As a result, many analysts believe that the 1980s was a period of extremely difficult economic restructuring rather than the beginning of the decline of the American economy, as some people warned at the time.

Globalization in the Less Developed Countries

It is still too soon to tell whether the governments and citizens of the less developed countries will be losers or winners in the globalization game. On the one hand, it is hard to argue with the assessment of the rural Indonesian woman interviewed by Diane Wolf. She prefers the regime of the factory to the tedium of village life. In the village, the woman worked from dawn till dusk doing household chores, taking care of siblings, and feeding the family goat. In the factory, she earns less than a dollar a day sewing pockets on men's shirts. Yet because work in the factory is less arduous, pays something, and holds out the hope of even better work for future generations, the woman views it as nothing less than liberating (Wolf, 1992). Many workers in other regions of the world where branch plants of multinationals have sprung up in recent decades feel much the same way. A wage of $2 an hour is good pay in Mexico, and workers rush to fill jobs along Mexico's northern border with the United States.

Yet the picture is not all bright. The governments of developing countries attract branch plants by imposing few if any pollution controls on their operations. This has dangerous effects on the environment. Typically, fewer jobs are available than the number of workers who are drawn from the countryside to find work in the branch plants. This results in the growth of urban slums suffering from high unemployment and unsanitary conditions. High-value components are often imported. Therefore, the branch plants create few good jobs involving design and technical expertise. Finally, some branch plants—particularly clothing and shoe factories in Asia—exploit children and women, requiring them to work long workdays at paltry wages and in unsafe conditions.

Companies such as Nike and The Gap have been widely criticized for conditions in their overseas sweatshops. Nike is the market leader in sports footwear. It has been in the forefront of moving production jobs overseas to places like Vietnam and Indonesia. In Indonesia, Nike factory workers make about 10 cents an hour. That is why labor costs account for only about 4% of the price of a pair of Nike shoes. Workdays in the factories stretch as long as 16 hours. Substandard air quality and excessive exposure to toxic chemicals like toluene are normal. An international campaign aimed at curbing Nike's labor practices has had only a modest impact. For example, in 1999 wages in the Indonesian factories were raised about a penny an hour ("The Nike Campaign," 2000). In that same year, a Nike vice-president blasted human rights groups working to improve labor conditions in Nike's overseas factories. In a leaked letter to Vietnam's highest ranking labor official, he wrote that "United States human rights groups . . . are not friends of Vietnam." The letter also said their ultimate goal is to turn Vietnam into "a so-called democracy, modeled after the United States" (Press, 1999). In Vietnam, this amounts to a charge of subversion.

Meanwhile, The Gap has invested heavily in the Northern Mariana Islands near Guam. Strictly speaking, the Marianas are not a poor foreign country because they are a United States commonwealth territory with a status similar to that of Puerto Rico. But they might as well be. Garment manufacturing is the biggest source of income on the islands, and The Gap (which also owns Banana Republic and Old Navy) is the biggest employer. What attracts The Gap to the Marianas are below-minimum United States wage rates, duty-free access to United States markets, and the right to sew "Made in U.S.A." labels on clothes manufactured there (Bank of Hawaii, 1999). However, work conditions are horrific. Many workers live in guarded dormitories surrounded by barbed wire preventing their escape. They work 12 to 18 hours a day without overtime. California Congressman

George Miller is sponsoring legislation to shut down these sweatshops, but as of this writing he has met with little success ("Sweatshops.org," 2000). At the very least, then, cases like The Gap in the Marianas and Nike in Indonesia and Vietnam suggest that the benefits of foreign investment are unlikely to be uniformly beneficial for the residents of developing countries in the short term.

The Future of Work and the Economy

Although work and the economy have changed enormously over the years, one thing has remained constant for centuries. Businesses have always looked for ways to cut costs and boost profits. Two of the most effective means they have adopted for accomplishing these goals involve introducing new technologies and organizing the workplace in more efficient ways. Much is uncertain about the future of work and the economy. However, it is a pretty good bet that businesses will continue to follow these established practices.

Just how these practices will be implemented is less predictable. For example, it is possible to use technology and improved work organization to increase productivity by complementing the abilities of skilled workers. Worldwide, the automotive, aerospace, and computer industries have tended to adopt this approach. They have introduced automation and robots on a wide scale. They constantly upgrade the skills of their workers. And they have proven the benefits of small autonomous work groups for product quality, worker satisfaction, and therefore the bottom line. On the other hand, new technology and more efficient work organization can be used to replace workers, deskill jobs, and employ low-cost labor—mainly women and minority group members—on a large scale. Women are entering the labor force at a faster rate than men. Hispanic-, Asian-, and African-American workers are entering the labor force at a much faster rate than whites (see Figure 10.10). Competition from low-wage industries abroad remains intense. Therefore, the second option is especially tempting in some industries.

Our analysis suggests that each of the scenarios sketched above will tend to predominate in different industries. As a result, the polarization of the labor force between good jobs in the primary sector and bad jobs in the secondary sector is likely to continue in the foreseeable future. However, we have also suggested that workplace struggles have no small bearing on how technologies are implemented and work is organized. To a degree, therefore, the future of work and the economy is up for grabs.

Web Interactive Exercises
Social Capital

✦ **FIGURE 10.10** ✦
Percent Increase in United States Labor Force by Race and Hispanic Origin, 1998–2008 (projected; in percent)

SOURCE: United States Department of Labor (1999d).

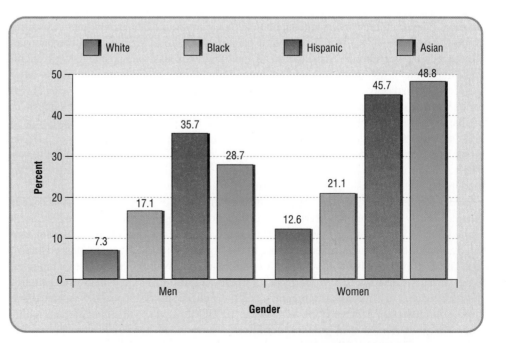

SUMMARY

1. The first work-related revolution began about 10,000 years ago when people established permanent settlements and started herding and farming. The second work-related revolution began 225 years ago when various mechanical devices such as the steam engine greatly increased the ability of producers to supply markets. The third revolution in work has been marked by growth in the provision of various services. It accelerated in the last decades of the 20th century with the widespread use of the computer.

2. Each revolution in work increased productivity and the division of labor, caused a sectoral shift in employment, and made work relations more hierarchical. However, for the past 30 years the degree of hierarchy has been lowered in some industries, resulting in productivity gains and more worker satisfaction.

3. Deskilling and the growth of part-time jobs are two of the main trends in the workplace in the 20th century. However, skilled labor has remained very important in the economy.

4. Good jobs are becoming more plentiful, but the number of bad jobs is growing even more rapidly. The result is polarization or segmentation of the labor force into primary and secondary labor markets. Various social barriers limit mobility from the secondary to the primary labor market.

5. Workers have resisted attempts to deskill and control jobs. As a result, business has had to make concessions by giving workers more authority on the shop floor and in formulating overall business strategy. Such concessions have been biggest in countries where workers are more organized and powerful.

6. Unions and professional organizations have established internal labor markets to control pay rates, hiring, and promotions in organizations and reduce competition with external labor supplies.

7. Markets are free or regulated to varying degrees. No market that is purely free or completely regulated could function for long. A purely free market would create unbearable inequalities, and a completely regulated market would stagnate.

8. Corporations are the dominant economic players in the world today. They exercise disproportionate economic and political influence by forming oligopolies, conglomerates, and interlocking directorates.

9. Growing competition between multinational corporations has led big corporations to cut costs by building more and more branch plants in low-wage, low-tax countries. Stockholders have profited from this strategy. However, the benefits for workers in both the industrialized and the less developed countries have been mixed.

GLOSSARY

Capitalism is the dominant economic system in the world today. Capitalist economies are characterized by private ownership of property and competition in the pursuit of profit.

Codetermination is a German system of worker participation that allows workers to help formulate overall business strategy. German workers' councils review and influence management policies on a wide range of issues, including when and where new plants should be built and how capital should be invested in technological innovation.

Communism is the name Karl Marx gave to the classless society that, he said, is bound to develop out of capitalism.

Conglomerates are large corporations that operate in several industries at the same time.

Corporations are legal entities that can enter into contracts and own property. They are taxed at a lower rate than individuals and their owners are normally not liable for the corporation's debt or any harm it may cause the public.

Deskilling refers to the process by which work tasks are broken into simple routines requiring little training to perform. Deskilling is usually accompanied by the use of machinery to replace labor wherever possible and increased management control over workers.

The **division of labor** refers to the specialization of work tasks. The more specialized the work tasks in a society, the greater the division of labor.

The **economy** is the institution that organizes the production, distribution, and exchange of goods and services.

In a **free market,** prices are determined only by supply and demand.

The **human relations school of management** emerged as a challenge to Taylor's scientific management approach in the 1930s. It advocated less authoritarian leadership on the shop floor, careful selection and training of personnel, and greater attention to human needs and employee job satisfaction.

Interlocking directorates are formed when an individual sits on the board of directors of two or more noncompeting companies.

Internal labor markets are social mechanisms for controlling pay rates, hiring, and promotions within corporations while reducing competition between a firm's workers and external labor supplies.

Labor market segmentation is the division of the market for labor into distinct settings. In these settings, work is found in different ways and workers have different characteristics. There is only a slim chance of moving from one setting to another.

Markets are social relations that regulate the exchange of goods and services. In a market, the prices of goods and services are established by how plentiful they are ("supply") and how much they are wanted ("demand").

Oligopolies are giant corporations that control part of an economy. They are few in number and they tend not to compete against one another. Instead, they can set prices at levels that are most profitable for them.

The primary labor market is composed mainly of highly skilled or well-educated white males. They are employed in large corporations that enjoy high levels of capital investment. In the primary labor market, employment is secure, earnings are high, and fringe benefits are generous.

Productivity refers to the amount of goods or services produced for every hour worked.

Professionals are people with specialized knowledge acquired through extensive higher education. They enjoy a high degree of work autonomy and usually regulate themselves and enforce standards through professional associations.

The **quality of work life** movement originated in Sweden and Japan. It involves small groups of a dozen or so workers and managers collaborating to improve both the quality of goods produced and communication between workers and managers.

In a **regulated market,** various social forces limit the capacity of supply and demand to determine prices.

Scientific management is a system of improving productivity developed in the 1910s by Frederick W. Taylor. After analyzing the movements of workers as they did their jobs, Taylor trained them to eliminate unnecessary actions and greatly improve their efficiency.

The secondary labor market contains a disproportionately large number of women and members of racial minorities, particularly African and Hispanic Americans. Employees in the secondary labor market tend to be unskilled and lack higher education. They work in small firms with low levels of capital investment. Employment is insecure, earnings are low, and fringe benefits are meager.

Unions are organizations of workers that seek to defend and promote their members' interests.

QUESTIONS TO CONSIDER

1. Women are entering the labor force at a faster rate than men. Hispanic, Asian, and African Americans are entering the labor force at a much faster rate than whites. What policies must companies adopt if they hope to see women and members of ethnic and racial minorities achieve workplace equality with white men?

2. The computer is widely regarded as a labor-saving device and has been adopted on a wide scale. Yet, on average, Americans work more hours per week now than they did 20 or 30 years ago. How do you explain this paradox?

3. Most of the less developed countries have been eager to see multinational corporations establish branch plants on their soil. What sorts of policies must less developed countries adopt to ensure maximum benefits for their populations from these branch plants? Would it be beneficial if the less developed countries worked out a common approach to this problem rather than competing against each other for branch plants?

WEB RESOURCES

Companion Web Site for This Book

http://sociology.wadsworth.com
Begin by clicking on the Student Resources section of the Web site. Choose "Introduction to Sociology" and finally the Brym and Lie book cover. Next, select the chapter you are currently studying from the pull-down menu. From the Student Resources page you will have easy access to InfoTrac College Edition®, MicroCase Online exercises, additional Web links, and many other resources to aid you in your study of sociology, including practice tests for each chapter.

InfoTrac Search Terms

These search terms are provided to assist you in beginning to conduct research on this topic by visiting http://www.infotraccollege. com/wadsworth.

Capitalism	**Deskilling**
Corporation	**Free market**
Communism	

Recommended Web Sites

A provocative analysis of the future of work can be found at http://www.panix.com/~dhenwood/Work.html.

The Bureau of Labor Statistics of the United States Department of Labor publishes estimates on the economy and labor market 10 years into the future, including projections of employment by industry and occupation. For these projections, go to http://www.bls.gov/text%5Fonly/emphome%5Ftxt.htm.

Robert Kuttner, founder and coeditor of *The American Prospect* magazine, presents a thought-provoking analysis of "The Limits of Markets" at http://www.prospect.org/archives/31/31kuttfs.html. See also the complementary analysis by George Soros, one of America's richest men, at http://www.theatlantic.com/issues/97feb/capital/capital.htm.

For discussion on worker burnout in the Internet economy, see http://www.disobey.com/netslaves. This popular site contains postings by former and current Internet employees, most in their 20s and 30s, who lament and analyze their predicament.

SUGGESTED READINGS

Randy Hodson and Teresa Sullivan. *The Social Organization of Work,* 2nd ed. (Belmont, CA: Wadsworth, 1995 [1990]). The latest edition of the definitive undergraduate textbook in the sociology of work.

Lawrence Mishel, Jared Bernstein, and John Schmitt. *The State of Working America, 1998-99.* (Ithaca, NY: Cornell University Press, 1999). An informative overview of work and the economy in the United States, full of useful statistics.

Karl Polanyi. *The Great Transformation: The Political and Economic Origins of Our Time.* (Boston: Beacon, 1957 [1944].) The classic account of the rise and decline of free-market society.

IN THIS CHAPTER, YOU WILL LEARN THAT:

✦ Political sociologists analyze the distribution of power in society and its consequences for political behavior and public policy.

✦ Sociological disputes about the distribution of power often focus on how social structures, and especially class structures, influence political life.

✦ Some political sociologists analyze how state institutions and laws affect political behavior and public policy.

✦ Three waves of democratization have swept the world in the last 175 years.

✦ Societies become highly democratic only when their citizens win legal protections of their rights and freedoms. This typically occurs when their middle and working classes become large, organized, and prosperous.

✦ Enduring social inequalities limit democracy even in the richest countries.

POLITICS

INTRODUCTION

The Tobacco War

In the spring of 1998, the tobacco war reached a decisive stage. Congress was ready to pass a bill that would cost the tobacco companies $516 billion in damages. The bill would also raise tobacco taxes by $1.10 a pack, limit cigarette advertising, and give Washington broad new powers to regulate the tobacco industry.

The public seemed eager to support the legislation. After all, 75% of the people would never have to pay the new tax because only a quarter of American adults smoked. And there was widespread alarm in the land. More than three decades of educational work by governments, schools, and health professionals made it common knowledge that one out of three smokers would die prematurely and probably wretchedly due to illnesses caused by smoking. Well-informed citizens knew that half a million Americans die *annually* from tobacco-related illnesses, more than *total* American casualties in World War II. They knew that about 90% of smokers started the habit by the age of 20. They knew that the percentage of grade 12 students who smoke rose from about 17% to nearly 25% between 1992 and 1997, mainly because of tobacco companies' marketing efforts.

Then, in 1998, there was the last straw. Documents released in a series of lawsuits against the tobacco industry revealed that tobacco companies were targeting teenagers in their ads, manipulating ammonia levels in tobacco to maximize nicotine addiction, and misrepresenting it all in public. (Some of these events were portrayed in the 1999 Oscar-winning film, *The Insider*.) Little wonder that polls showed strong public support for the antitobacco bill. The United States finally seemed ready to join the other rich industrialized countries in helping to stub out one of the world's leading health hazards.

Representatives of the tobacco industry did not, however, sit idly in the bleachers. They mobilized their allies, including retailers and smokers, to phone and write their members of Congress expressing outrage at the antitobacco bill. They tripled the budget for tobacco industry lobbyists. Legions of professional arm twisters wined, dined, and cajoled members of Congress to vote against the bill. And then the industry bankrolled a last minute $40 million national advertising blitz. The ad campaign gnawed away at traditional American sore points. According to the ads, the antitobacco bill was really a government

Chief executive officers of the major United States tobacco companies swear under oath at a 1994 congressional hearing that smoking is not addictive and does not cause any disease. This was a turning point in the battle against the tobacco industry.

tax grab. It would increase government regulation at the expense of individual freedom. It would allow antitobacco industry lawyers to earn exorbitant fees. And, just as Prohibition had encouraged liquor smuggling and the production of moonshine whiskey in the 1920s and early 1930s, the new law would encourage the import of contraband cigarettes. These arguments worked. The bill was defeated in June 1998. Just before the final vote, a *Wall Street Journal*/NBC poll found that 70% of Americans thought the bill's real aim was to raise new revenue. Only 20% said its purpose was to curb teen smoking (Centers for Disease Control, 2000; "European Tobacco Ban . . . ," 1998; Kluger, 1996; Leman, 1998; McKenna, 1998; "Monitoring the Future Study," 1998).

In separate deals, the 50 states eventually agreed to sign agreements with the tobacco companies worth a total of $246 billion, less than half the amount demanded in the federal bill. The money is intended to recover the cost of treating Medicaid-eligible smokers. However, the defeat of the federal bill raises important political questions. Does the outcome of the tobacco war illustrate the operation of "government of the people, by the people, for the people," as Abraham Lincoln defined democracy in his famous speech at Gettysburg? The war certainly allowed a diverse range of Americans to express conflicting views. It permitted them to influence their elected representatives. And, in the end, members of Congress did vote in line with the wishes of most American adults as expressed in public opinion polls. This suggests that Lincoln's characterization of American politics applied as well in 1998 as it did in 1863.

However, big business's access to a bulging war chest might lead one to doubt that Lincoln's definition applies. Few groups can put together $19 million for lobbyists, $3 million for political party contributions, and $40 million for public relations and advertising experts virtually overnight to sway the hearts and minds of the American people and their lawmakers. Should we therefore conclude that some people, especially big businessmen, are more equal than others?[1]

The tobacco war raises the question that lies at the heart of political sociology. What accounts for the degree to which a political system responds to the demands of all its citizens? As you will see, political sociologists have often answered this question by examining the effects of social structures, especially class structures, on politics. But while this approach contributes much to our understanding of political life, it is insufficient by itself. A fully adequate theory of democracy requires that we also examine how state institutions and laws affect political processes. By way of illustration, we show how voter registration laws bias American politics in favor of some groups at the expense of others.

From the mid-1970s till the early 1990s, a wave of competitive elections swept across many formerly nondemocratic countries. Most dramatically, elections were held in the former Soviet Union at the end of this period. Many Western analysts were ecstatic. However, by the mid-1990s, it became clear that their optimism was naïve. Often, the new regimes turned out to be feeble and limited democracies. As a result, political sociologists began to reconsider the social preconditions of democracy. We review some of their work in this chapter's third section. We conclude that genuine democracy is not based just on elections. In addition, large classes of people must win legal protection of their rights and freedoms for democracy to take root and grow. This has not yet happened in most of the world.

Some analysts believe that politics in the rich industrialized countries is less likely to be shaped by social inequality in the future. Others hold that the marriage of home computers and elections will allow citizens to get more involved in politics by voting often and directly on the Internet. Our reading of the evidence is different. In concluding this chapter, we argue that persistent social inequality is the major barrier to the progress of democracy in countries like the United States.

Before developing these themes, however, we define some key terms.

---

[1]We say business*men* advisedly. In 2000, only 46 women were on the list of America's 400 richest people. Of these, a mere six were self-made women. This suggests where the real power lies (DiCarlo, 2000).

What Is Politics? Core Concepts

Politics is a machine that determines "who gets what, when, and how" (Lasswell, 1936). **Power** fuels the machine. Power is the ability to control others, even against their will (Weber, [1947: 152]). Having more power than others gives you the ability to get more valued things sooner. Having less power than others means you get fewer valued things later. Political sociology's key task is figuring out how power drives different types of political machines.

The use of power sometimes involves force. For example, one way of operating a system for distributing jobs, money, education, and other valued things is by imprisoning people who don't agree with the system. In this case, people obey political rules because they are afraid to disobey. More often, however, people agree with the distribution system or at least accept it grudgingly. For instance, most people pay their taxes without much pressure from the IRS and their parking tickets without serving jail time. They recognize the right of their rulers to control the political machine. When most people basically agree with how the political machine is run, raw power becomes **authority.** Authority is legitimate, institutionalized power. Power is **legitimate** when people regard its use as valid or justified. Power is *institutionalized* when the norms and statuses of social organizations govern its use. These norms and statuses define how authority should be used, how individuals can achieve authority, and how much authority is attached to each status in the organization.

Max Weber (1947) wrote that authority can have one of three bases:

1. **Traditional authority.** Particularly in tribal and feudal societies, rulers inherit authority through family or clan ties. The right of a family or clan to monopolize leadership is widely believed to originate from the will of a god.

2. **Legal-rational authority.** In modern societies, authority is derived from respect for the law. Laws specify how one can achieve office. People generally believe these laws are rational. If someone achieves office by following these laws, their authority is respected.

3. **Charismatic authority.** Sometimes extraordinary, charismatic individuals challenge traditional or legal-rational authority. They claim to be inspired by a god or some higher principle that transcends other forms of authority. Most people believe this claim. One such principle is the idea that all people are created equal. Charismatic figures sometimes emerge during a **political revolution,** an attempt by many people to overthrow existing political institutions and establish new ones. Political revolutions take place when widespread and successful movements of opposition clash with crumbling traditional or legal-rational authority.

Politics takes place in all social settings. Such settings include intimate face-to-face relationships, families, and colleges. However, political sociology is mainly concerned with institutions that *specialize* in the exercise of power and authority. Taken together, these institutions form the **state.** The state consists of institutions that formulate and carry out a country's laws and public policies. In performing these functions, the state regulates citizens in **civil society,** the private sphere of social life (see Figure 11.1).

Citizens in civil society control the state to varying degrees. In an **authoritarian** state, citizen control is sharply restricted. In a **totalitarian** state it is virtually nonexistent. In a **democracy,** citizens exert a relatively high degree of control over the state. They do this partly by choosing representatives in regular, competitive elections.

In modern democracies, citizens do not control the state directly. They do so through several organizations. **Political parties** compete for control of government in regular elections. They put forward policy alternatives and rally adult citizens to vote. Special interest groups such as trade unions and business associations form **lobbies.** They advise politicians about their members' desires. They also remind politicians how much their members' votes, organizing skills, and campaign contributions matter. The **mass media** keep a watchful and

The three faces of authority according to Weber: traditional authority (King Louis XIV of France, circa 1670), charismatic authority (Vladimir Lenin, Bolshevik leader of the Russian Revolution of 1917), and legal-rational authority (Ronald Reagan, campaigning for the presidency of the United States).

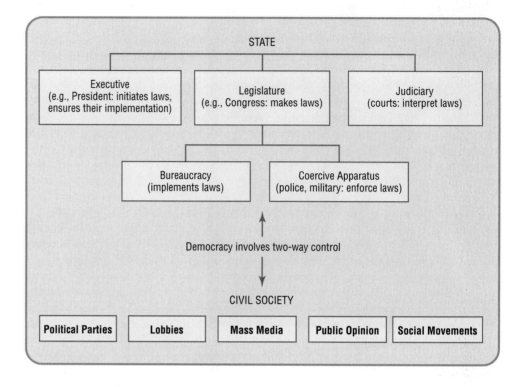

critical eye on the state. They keep the public informed about the quality of government. **Public opinion** refers to the values and attitudes of the adult population as a whole. It is expressed mainly in polls and letters to lawmakers. Public opinion gives politicians a reading of citizen preferences. Finally, when dissatisfaction with normal politics is widespread, protest sometimes takes the form of **social movements.** A social movement is a collective attempt to change all or part of the political or social order. As Thomas Jefferson wrote in a letter to James Madison in 1787, "a little rebellion now and then is a good thing" for democracy. It helps to keep government responsive to the wishes of the citizenry.

Bearing these definitions in mind, we now consider the merits and limitations of sociological theories of democracy.

THEORIES OF DEMOCRACY

Pluralist Theory

In the early 1950s, New Haven, Connecticut, was a city of about 150,000 people. It had seen better times. As in many other American cities, post–World War II prosperity and new roads had allowed much of the white middle class to resettle in the suburbs. This eroded the city's tax base. It also left much of the downtown to poor and minority-group residents. Some parts of New Haven became slums.

Beginning in 1954, Mayor Richard Lee decided to do something about the city's decline. He planned to attract new investment, eliminate downtown slums, and stem the outflow of the white middle class. Urban renewal was a potentially divisive issue. However, according to research conducted at the time, key decisions were made in a highly democratic manner. The city government listened closely to all major groups. It adopted policies that reflected the diverse wishes and interests of city residents.

The social scientists who studied New Haven politics in the 1950s are known as **pluralists** (Polsby, 1959; Dahl, 1961). They argued that the city was highly democratic because power was widely dispersed. They showed that few of the prestigious families in New Haven's *Social Register* were economic leaders in the community.[2] Moreover, neither economic leaders nor the social elite monopolized political decision making. Different groups of people decided various political issues. Some of these people had low status in the community. Moreover, power was more widely distributed than in earlier decades. The pluralists concluded that no single group exercised disproportionate power in New Haven.

The pluralists believed politics worked much the same way in the United States as a whole. America, they said, is a heterogeneous society with many competing interests and centers of power. None of these power centers can consistently dominate. The owners of United States Steel, for instance, may want tariffs on steel imports to protect the company's United States market. Meanwhile, the owners of General Motors may oppose tariffs on steel because they want to keep their company's production costs down. The idea that "industry" speaks with one voice is thus a myth. Competing interests exist even within one group. For instance, the automobile company with the lead in developing electric cars may favor clean air legislation now. An auto company lagging in its research effort may favor a go-slow approach to such laws. Because there is so much heterogeneity between and within groups, no single group can control political life. Sometimes one category of voters or one set of interest groups wins a political battle, sometimes another. Most often, however, politics involves negotiation and compromise between competing groups. Because no one group of people is always able to control the political agenda or the outcome of political conflicts, democracy is guaranteed.

Elite Theory

Elite theorists, C. Wright Mills (1956) chief among them, sharply disagreed. According to Mills, **elites** are small groups that occupy the command posts of America's most influential institutions. These include the country's two or three hundred biggest corporations, the executive branch of government, and the military. Mills wrote that the men who control these institutions make important decisions that profoundly affect all

Pluralist theory portrays politics as a neatly ordered game of negotiation and compromise in which all players are equal.

SOURCE: Brian Jones, "The Centre of the Universe," (1992).

---

[2]The *Social Register* is a listing of America's highest status families. First published in 1887, it now has about 40,000 entries.

In a nationally televised address on January 17, 1961, President Eisenhower sounded much like C. Wright Mills and other elite theorists when he warned of the "undue influence" of the "military-industrial complex" in American society. Maintaining a large, permanent military establishment is "new in the American experience," he said. An "engaged citizenry" offers the only effective defense against the "misplaced power" of the military-industrial lobby, according to Eisenhower.

members of society. Moreover, they do so without much regard for elections or public opinion.

Mills showed how the corporate, state, and military elites are connected. People move from one elite to another during their careers. Their children intermarry. They maintain close social contacts. They tend to be recruited from upper-middle and upper classes. Yet Mills denied these connections turn the three elites into a **ruling class.** A ruling class is a self-conscious and cohesive group of people, led by owners of big business, who act to advance their common interests. The three elites are relatively independent of one another, Mills insisted. They may see eye-to-eye on many issues, but each has its own sphere of influence. Conflict between elite groups is frequent (Mills, 1956: 277; Alford and Friedland, 1985: 199).

A Critique of Pluralism

Most political sociologists today question the pluralist account of American politics. That is because research has established the existence of large, wealth-based inequalities in political participation and political influence. As we will see, most political sociologists today are skeptical about some of C. Wright Mills's claims too. But on the whole they are more sympathetic to the elitist view.

Consider, for example, some results of the "Citizen Participation Study." In the early 1990s, a team of researchers surveyed a representative sample of more than 15,000 American adults. They asked respondents if they voted in the 1988 presidential campaign, how many contacts they had with public officials, how many hours they worked in the election campaign, and how many dollars they contributed to it. Then they calculated the percentage of each political activity that was undertaken by people in each income group. They found that people with higher incomes are more politically active, especially in those forms of political activity that are most influential.

Figure 11.2 compares the political participation of rich and poor Americans. The rich were defined as those who had family incomes of $125,000 per year or more. The poor were defined as those who had family incomes of less than $15,000 per year. So defined, the rich composed 3% of American citizens, the poor 18%. The ratio of rich to poor was .17:1 (because 3/18 = .17). Note how the ratio of rich to poor is higher for more influential political activities. For voting, the ratio of rich to poor is .29:1. For contacts with

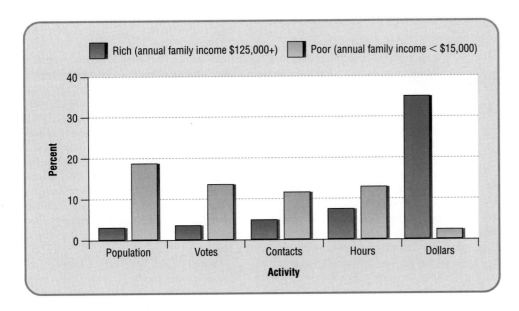

◆ **FIGURE 11.2** ◆

Percent of Political Activities Undertaken by Rich and Poor Americans, 1988

SOURCE: Verba, Schlozman, and Brady (1997).

public officials, the ratio is .50:1. For hours spent campaigning, the ratio is .62:1. And for dollars contributed to campaigns, the ratio is 17.5:1. In other words, the rich contribute 17.5 times more money to election campaigns than the poor although the poor are six times more numerous than the rich.

If money talks, does it speak with a single voice? In principle, well-to-do Americans may contribute the same amount of money to all candidates in an election campaign. If so, we would be wrong to assume that rich people share political interests and act in concert. A study of Political Action Committees (PACs) conducted by sociologist Dan Clawson and his associates shines light on this issue (Clawson, Neustadtl, and Scott, 1992).

In 1988, winning members of the House of Representatives spent an average of $388,000 on their election campaigns. Winning senators spent nearly 10 times as much—$3,745,000. Thus, *every week* of his or her term of office, a member of the House had to raise $3,700, and a senator $12,000, to finance an average winning campaign. PACs help them do that by collecting money from many contributors, pooling it, and then making donations to candidates (see Box 11.1). What do contributors to PACs expect in return for their money? Republican Senator and former presidential candidate Bob Dole answered the question delicately when he said "they expect something in return other than good government." One business donor put it more bluntly: "One question . . . raised in recent weeks had to do with whether my financial support in any way influenced several political figures to take up my cause. I want to say in the most forceful way I can, I certainly hope so" (quoted in Clawson, Neustadtl, and Scott, 1992: 9).

Figure 11.3 shows the contributions of the PACs that gave more than $100,000 in 1984 to 455 political races. It divides the races into three groups. In 1 case out of 15, large corporations were politically divided, giving only 1–2 times more money to one candidate than the other. In one case out of five, large corporations mainly supported one candidate, giving 2–9 times more money to one candidate than the other. And in three cases out of four, large corporations were politically unified, giving more than 9 times more money to one candidate than the other. These data suggest that big business is for the most part unified in its political views. It tends to favor one candidate over another.

Which candidate do corporate PACs tend to favor? The Republicans. When Clawson and his associates analyzed contributions by *all* large PACS, they found a sharp split between a unified business-Republican group on one side and a labor-women-environmentalist-Democratic group on the other.

This raises an interesting question. According to elite theorists, the distribution of power in America is heavily skewed toward the wealthy. The wealthy tend to support the

BOX 11.1
IT'S YOUR CHOICE

FINANCING POLITICAL CAMPAIGNS

"No matter what parliamentary tactics are used to prevent reform . . . no matter how fierce the opposition, no matter how personal, no matter how cynical this debate remains . . . I will persevere," proclaimed Senator John McCain. In contrast, Senator Mitch McConnell, who led the opposition, said: "I'd call [the reform movement] no progress whatsoever . . . I'd call it

. . . pretty dead" (quoted in Mitchell, 1999).

What was the issue that generated such heated rhetoric in the fall of 1999? It was the effort, led by Senators McCain and Feingold, to reform political campaign financing. They were especially upset about the role of "soft money." Campaign finance law caps each contribution to an *individual* candidate at $1,000. The reasoning behind the law is that politics should be about ideas, policies, and personalities, not money. After all, we wouldn't have much of a democracy if individuals could "buy" elections. Paradoxically, however, the law also allows "soft money" contributions. These are contributions to *political parties*, not individual candidates. Soft money is big money. By June 1999, the two major parties had raised nearly $60 million in soft money for the 2000 Presidential campaign, twice as

much as they had raised at the same point in the 1996 race (Dreyfus, 1999).

Why should we be concerned about the role of money in politics? The average winner in the 1998 Senate elections raised $5.2 million. The average loser raised $2.8 million. This suggests that money helps to win campaigns. Moreover, about 90% of the incumbents won in the 1996 and 1998 Senate elections. Thus, not only do candidates who raise more money tend to win. In addition, candidates with more money tend to be incumbents (Center for Responsive Politics, 2000).

What should be done? Campaign finance reform seems important. However, most elected representatives were helped by the existing system of campaign financing and so are unlikely to oppose it. Given that incumbents tend to win (in part by being able to raise more money), what can be done to reform campaign financing?

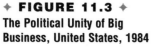

✦ FIGURE 11.3 ✦
The Political Unity of Big Business, United States, 1984

SOURCE: Clawson, Neustadtl, and Scott (1992: 160).

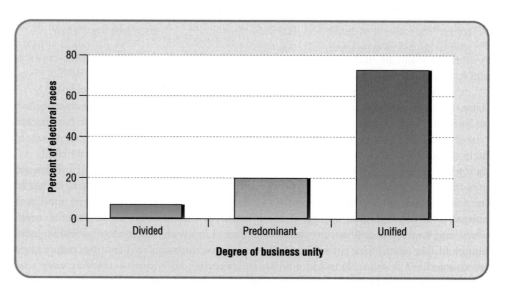

Republicans. Why then do Democrats often become President and get elected to Congress? As we will see, this question points to an important limitation of elite theory.

Power Resource Theory

In general, elite theorists believe it makes little difference whether Republicans or Democrats are in power. For them, elites always control society. Elections are little more than sideshows. Therefore, they believe, the victory of one party over another doesn't deserve much sociological attention.

We disagree. So do most political sociologists today. It matters a great deal to most citizens whether the party in office supports or opposes antitobacco laws, more military spending, weaker environmental standards, more publicly funded medical care, bigger government subsidies for child care, abortion on demand, gun control, and so forth. Elite

Web Interactive Exercises
How Electoral Laws Affect
Political Participation

theorists are correct to claim that most power is concentrated in the hands of the wealthy. But we still need a theory that accounts for the successes and failures of different parties and policies in different times and places.

This is where **power resource theory** is helpful. It focuses on how *variations* in the distribution of power affect the fortunes of parties and policies.

To understand power resource theory, first consider your own party preference. For many reasons, you may support one political party over another. For instance, your family may have a long tradition of voting for one party. You may have never really questioned that support. Maybe you support a party because you admire the energy, integrity, or track record of its leader. Or you might support a party because you agree with its policies on a range of issues. What factors lead *you* to prefer one party over another?

If a party's policies influence your vote, you're like many Americans. In fact, American voters cluster in two main policy groups. *Liberal* or left-wing voters promote extensive government involvement in the economy. Among other things, this means they favor a strong "social safety net" of health and welfare benefits to help the less fortunate members of society. As a result, liberal policies often lead to less economic inequality. In contrast, *conservative* or right-wing voters favor a reduced role for government in the economy. They favor a smaller welfare state and emphasize the importance of individual initiative in promoting economic growth. Economic issues aside, liberals and conservatives also tend to differ on social or moral issues. Liberals tend to support equal rights for women and racial and sexual minorities. Conservatives tend to support more traditional social and moral values.[3]

Table 11.1 shows how liberal and conservative sentiments were translated into support for Bill Clinton, a Democrat, and Bob Dole, a Republican, in 1996. Liberals tended to support Clinton. Conservatives tended to support Dole. A similar split was evident in 2000. For instance, fewer than 30% of homosexuals and supporters of reproductive choice supported Republican George W. Bush while 70% supported Democrat Al Gore (Frum, 2000). Do you think of yourself more as a Democrat or a Republican? Is your party choice related to the policies of the two parties?

The policies favored by different parties have different effects on different groups of people. Therefore, different parties tend to be supported by different classes, religious groups, races, and other groups. Do you think the policies of your preferred party favor the class, religious group, or race to which you belong? If so, how? If not, why not?

In most Western democracies, the main factor that distinguishes parties is differences in *class* support (Korpi, 1983: 35; Lipset and Rokkan, 1967; Manza, Hout, and Brooks, 1995). For example, as Table 11.2 shows, in the 1996 presidential race, low-income earners in the United States tended to support liberal, Democratic candidates. High-income earners tended to support conservative, Republican candidates. The 2000 presidential race revealed the same pattern ("Gore–Bush Race . . . ," 2000). This stands to reason because most Democrats favor policies that promote less inequality in society, such as universal health care.

The tendency for people in different classes to vote for different parties varies from one country to the next. The strength of this tendency depends on many factors. One of the

Candidate	Political Views		
	Liberal	Moderate	Conservative
Clinton (Democrat)	88	69	33
Dole (Republican)	52	31	67
Total	100	100	100
n	422	500	556

✦ TABLE 11.1 ✦
Support for Democratic and Republican Presidential Candidates by Political Views, 1996 (in percent; n = 1,478)

SOURCE: National Opinion Research Center (1999).

---

[3]Some people are liberal on economic issues and conservative on social issues or vice-versa. Many such people call themselves "moderates" rather than liberals or conservatives.

In the 2000 presidential election, Democrat Al Gore and Republican George W. Bush appealed to different segments of the American population. Low-income earners tended to support Gore, high-income earners, Bush. Gore won the vote of 90% of African Americans and 62% of Hispanic Americans. Male voters chose Bush by an 11% margin while female voters preferred Gore by the same margin. Fewer than 30% of homosexuals and supporters of reproductive choice supported Bush while 70% supported Gore.

✦ TABLE 11.2 ✦

Support for Democratic and Republican Presidential Candidates by Annual Family Income, 1996 (in percent; n = 1,365)

SOURCE: National Opinion Research Center (1999).

Candidate	Annual Family Income	
	$25,000/year	$25,000+/year
Clinton (Democrat)	75	55
Dole (Republican)	25	45
Total	100	100
n	421	944

most important is how socially organized or cohesive classes are (Brym with Fox, 1989: 57–91; Brym, Gillespie, and Lenton, 1989). For example, an upper class that can create PACs to support Republican candidates and lobbies to support conservative laws is more powerful than an upper class that cannot take such action. If an upper class makes such efforts while a working class fails to organize itself, right-wing candidates have a better chance of winning office. Conservative policies are more likely to become law. Similarly, a working class that can unionize many workers is more powerful than one with few unionized workers. That is because unions often collect money for the party that is more sympathetic to union interests. They also lobby on behalf of their members and try to convince members to vote for the pro-union party. If workers become more unionized while an upper class fails to organize itself, then left-wing candidates have an improved chance of winning office. Liberal policies are more likely to become law. This is the main insight of power resource theory: *Organization is a source of power. Change in the distribution of power between major classes partly accounts for the fortunes of different political parties and different laws and policies* (Korpi, 1983; Esping-Andersen, 1990; O'Connor and Olsen, 1998; Shalev, 1983).

We can see how power resource theory works by looking at Table 11.3. This table compares 18 industrialized democracies in the three decades after World War II. We divide the countries into three groups. In group one are countries like Sweden, where socialist parties usually control governments. (Socialist parties are more left wing than the Democrats in the United States.) In group two are countries like Australia, where socialist parties *sometimes* control, or share in the control of, governments. And in group three are

countries like the United States, where socialist parties rarely or never share control of governments. The group averages in column two show that socialist parties are generally more successful where workers are more unionized. The group averages in columns three and four show there is more economic inequality in countries that are weakly unionized and have no socialist governments. In other words, by means of taxes and social policies, socialist governments ensure that the rich earn a smaller percentage of national income and the poor form a smaller percentage of the population. Studies of pensions, medical care, and other state benefits in the rich industrialized democracies reach similar conclusions. In general, where working classes are more organized and powerful, disadvantaged people are economically better off (Myles, 1989 [1984]; O'Connor and Brym, 1988; Olsen and Brym, 1996).

Class is not the only factor that distinguishes parties. Historically, *religion* has also been an important basis of party differences. For example, in West European countries with large Catholic populations, such as Switzerland and Belgium, parties are distinguished partly by the religious affiliation of their supporters. In recent decades, *race* has become a cleavage factor of major and growing importance in some countries. In the United States in particular, African Americans have overwhelmingly supported the Democratic Party since the 1960s (Brooks and Manza, 1997b). In the 2000 presidential election, Al Gore won the vote of 90% of black voters and 62% of Hispanic voters ("Gore–Bush Race . . . ," 2000). Race is an increasingly important division in French politics too. This is due to heavy Arab immigration from Algeria, Morocco, and Tunisia since the 1950s and

✦ **TABLE 11.3** ✦

Some Consequences of Working Class Power in 18 Rich Industrialized Countries, 1946–76

Note: "Socialist share of government" is the proportion of seats in each cabinet held by socialist parties weighted by the socialist share of seats in parliament and the duration of the cabinet. "Percent poor" is the average percentage of the population living in relative poverty according to OECD standards with poverty line standardized according to household size.

SOURCE: Korpi (1983: 40, 196).

	Percent of Nonagricultural Workforce Unionized	Socialist Share of Government	Percent of Total National Income to Top 10% Earners	Percent Poor
Mainly Socialist Countries				
Sweden	71	High	21.3	3.5
Norway	46	High	22.2	5.0
Average	**68.5**		**21.8**	**4.3**
Partly Socialist Countries				
Austria	55	Medium	—	—
Australia	50	Medium	23.7	8.0
Denmark	49	Medium	—	—
Belgium	47	Medium	—	—
UK	44	Medium	23.5	7.5
New Zealand	42	Medium	—	—
Finland	39	Medium	—	—
Average	**46.6**		**23.6**	**7.8**
Mainly Nonsocialist Countries				
Ireland	36	Low	—	—
W. Germany	35	Low	30.3	3.0
Netherlands	30	Low	27.7	—
USA	27	Low	26.6	13.0
Japan	27	Low	27.2	—
Canada	26	Low	25.1	11.0
France	25	Low	30.4	16.0
Italy	23	Low	30.9	—
Switzerland	23	Low	—	—
Average	**28**		**28.3**	**10.8**

growing anti-immigration sentiment among a substantial minority of whites (Veugelers, 1997). Finally, a political *gender* gap is growing in some countries, especially the United States. Thus, male voters chose George W. Bush by an 11% margin in the 2000 presidential election, while female voters preferred Al Gore by the same margin ("Gore–Bush Race . . . ," 2000). Power resource theory focuses mainly on how the shifting distribution of power between working and upper classes affects electoral success. However, one can also use the theory to analyze the electoral fortunes of parties that attract different religious groups, races, gender groups, and so forth.

State-Centered Theory

Web Research Projects
Party Identification

Democratic politics is a contest among various classes, religious groups, races, and other collectivities to control the state for their own advantage. When power is substantially redistributed due to such factors as change in the cohesiveness of social groups, old ruling parties usually fall and new ones take office.

Note, however, that a winner-take-all strategy would be nothing short of foolish. If winning parties passed laws that benefit only their supporters, they might cause mass outrage and even violent opposition. Yet it would be bad politics to allow opponents to become angry, organized, and resolute. After all, winners want more than just a moment of glory. They want to be able to enjoy the spoils of office over the long haul. To achieve stability, they must give people who lose elections a say in government. That way, even determined opponents are likely to recognize the government's legitimacy. Pluralists thus make a good point when they say that democratic politics is about accommodation and compromise. They only lose sight of how accommodation and compromise typically give more advantages to some than others, as both elite theorists and power resource theorists stress.

There is, however, more to the story of politics than conflict between classes, religious groups, races, and so forth. Theda Skocpol and other **state-centered theorists** show how the state itself can structure political life, no matter how power is distributed at a given moment (Block, 1979; Skocpol, 1979; Evans, Rueschemeyer, and Skocpol, 1985). Their argument is a valuable supplement to power resource theory.

To illustrate how state structures shape politics, consider a common American political practice: nonvoting. Fewer than 49% of American citizens voted in the 1996 presidential election.[4] In the 2000 presidential election, voter turnout was only slightly better at about 51%—and this after a truly massive mobilization effort on the part of all political parties. With the exception of such minor reversals, voter turnout has been falling in the United States since World War II. Apart from Switzerland, the United States now has the lowest voter turnout of any democracy in the world (Piven and Cloward, 1989 [1988]: 5). How can we explain this troubling fact?

The high rate of nonvoting is partly a result of voter registration law, a feature of the American political structure, not of the current distribution of power. In every democracy, laws specify voter registration procedures. In some countries, citizens are registered to vote automatically when they receive state-issued identity cards at the age of 18. In other countries, state-employed canvassers go door to door before each election to register voters. Only in the United States do individual citizens have to take the initiative to go out and register themselves in voter registration centers. However, many American citizens are unable or unwilling to register. As a result, the United States has a proportionately smaller pool of eligible voters than the other democracies. Only about 70% of American citizens are registered to vote. True, since the National Registration Act took effect in 1995, it has been possible to register by mail, when renewing a driver's license, and when applying for welfare and disability services. However, the percentage of American adults registered to vote increased only about 2% between 1995 and 1998. The new "motor voter" law has had little impact on actual voter turnout (Ganz, 1996; Quinn, 2000).

---

[4] A mere 36% went to the polls for the 1998 congressional elections.

Apart from shrinking the pool of eligible voters, American voter registration law has a second important consequence. Because some *types* of people are less able and inclined to register than others, a strong bias is introduced into the political system. Specifically, the poor are less likely to register than the better off. People without much formal education are less likely to register than the better educated. Members of disadvantaged racial minority groups, especially African Americans, are less likely to register than whites. Thus, American voter registration law is a pathway to democracy for some, a barrier to democracy for others. Even the new motor voter law appears to have benefited mainly middle-class Americans rather than the disadvantaged (Brains, 1999; Grofman, 2000). Here we have "democracy's unresolved dilemma" (Lijphart, 1997). As Seymour Martin Lipset, America's leading political sociologist, explains:

> [W]hen the vote is low, this almost always means that the socially and economically disadvantaged groups are underrepresented in government. The combination of a low vote and a relative lack of organization among the lower-status groups means that they will suffer from neglect by the politicians who will be receptive to the wishes of the more privileged, participating, and organized strata (Lipset, 1981 [1960]: 226–7).

In short, the American political system is less responsive than other rich democracies to the needs of the disadvantaged for two main reasons. First, as we saw in our discussion of power resource theory, the working class is comparatively nonunionized and therefore weak. Second, as state-centered theory suggests, the law requires citizen-initiated voter registration, one result of which is that the vote is in effect taken away from many disadvantaged people.

In general, state structures resist change. *Constitutions* anchor their foundations. Only a large majority of federally elected representatives and state legislatures can change the constitution. *Laws* gird the upper stories of state structures. Some laws help to keep potentially disruptive social forces at bay. Voter registration law is a case in point. *Ideology* reinforces the whole edifice. All states create anthems, flags, ceremonies, celebrations,

During the Great Depression (1929–1939), widespread poverty, unemployment, bankruptcy, and strike violence led to the election of Franklin Delano Roosevelt as Democratic president. Today's unemployment insurance, old-age pension, and public assistance programs all originated in Roosevelt's "New Deal." The photo on the right shows FDR signing the Social Security Bill, August 14, 1935.

sporting events, and school curricula that stimulate patriotism and serve in part to justify existing political arrangements.

Despite these anchors, girders, and reinforcements, big shocks do sometimes reorient American public policy and cause a major shift in voting patterns. In the past 110 years, these shocks have occurred about every four decades—in the 1890s, the 1930s, and the 1970s:

1. In the 1890s, industrial unrest was widespread. Western and Southern farmers re-volted against the established parties. In 1896, these rebellious forces mounted a Democratic-Populist challenge to the Republicans of the North and the wealthy Democrats of the South. Their presidential candidate, William Jennings Bryan, won nearly 48% of the vote in the 1896 election. But America's elites learned an impor-tant lesson from Bryan's challenge. They instituted electoral reforms—including voter registration laws—that made possible the domination of the pro-business Republican party in the North and the pro-plantation-owner Democratic party in the South. Low voter turnout and the effective disenfranchisement of many poor and black voters date from this era (Piven and Cloward, 1989 [1988]: 26–95).

2. The next big shock to the American political system came during the Great Depres-sion of the 1930s. Widespread unemployment, bankruptcy, and strike violence led to the election of Franklin Delano Roosevelt as Democratic President. This time, America's business elite was too devastated and divided by the Depression to stave off the Democratic threat. Today's unemployment insurance, old-age pension, and public assistance programs all originated in Roosevelt's "New Deal" (Leman, 1977). The reform wave unleashed by the New Deal did not end until the 1960s. During that decade, the civil rights movement inspired sweeping constitutional change that extended many of the benefits of American citizenship to African Americans.

3. In the 1970s, America faced another economic crisis. Following their post–World War II reconstruction, Japan and West Germany emerged as major competitive threats to American manufacturers. Inflation became a serious issue after Middle East oil producers demanded much higher prices for oil. Industrial workers struck on a large scale for higher wages to compensate for inflation. Reacting to these threats, members of the American business elite became more politically organized and unified than ever before. Some sociologists argue that they formed a truly cohesive ruling class. These business leaders funded PACs, lobbies, and research

The Selma-to-Montgomery march for black voting rights was a pivotal mo-ment in the civil rights movements. It led to Congress's enactment of the Voting Rights Act in 1965.

institutes to promote conservative policies. Lowering taxes, cutting state funding of social programs, and creating a more favorable regulatory environment for business topped their list of priorities (Akard, 1992; Clawson, Neustadtl, and Scott, 1992; Domhoff, 1983; Schwartz, 1987; Useem, 1984; Vogel, 1996). The 1980 Presidential victory of Republican Ronald Reagan capped the resurgence of post–World War II conservatism in American politics.

In sum, political sociology has made good progress since the 1950s. Each of the field's major schools has made a useful contribution to our appreciation of political life (see Table 11.4). Pluralists teach us that democratic politics is about compromise and the accommodation of all group interests. Elite theorists teach us that, despite compromise and accommodation, power is concentrated in the hands of high-status groups, whose interests the political system serves best. Power resource theorists teach us that, despite the concentration of power in society, substantial shifts in the distribution of power do occur, and they have big effects on voting patterns and public policies. And state-centered theorists teach us that, despite the influence of the distribution of power on political life, state structures exert an important effect on politics too.

We now turn to an examination of the historical development of democracy, its sociological underpinnings, and its future.

	Pluralist	Elitist	Power Resource	State-Centered
How is power distributed?	Dispersed	Concentrated	Concentrated	Concentrated
Who are the main power holders?	Various groups	Elites	Upper class	State officials
On what is their power based?	Holding political office	Controlling major institutions	Owning substantial capital	Holding political office
What is the main basis of public policy?	Will of all citizens	Interests of major elites	Balance of power between classes, etc.	Influence of state structures
Do lower classes have much influence on politics?	Yes	No	Sometimes	Sometimes

✦ **TABLE 11.4** ✦
Four Sociological Theories of Democracy Compared

THE FUTURE OF DEMOCRACY

Two Cheers for Russian Democracy

In 1989, the Institute of Sociology of the Russian Academy of Science invited Robert Brym and nine other sociologists to attend one of a series of seminars in Moscow. The seminars were designed to acquaint some leading sociologists in the Soviet Union with Western sociology. The country was in the midst of a great thaw. Totalitarianism was melting, leaving democracy in its place. Soviet sociologists had never been free to read and research what they wanted. Now they were eager to learn from North American and European scholars (Brym, 1990).

Or at least so it seemed, according to Robert. One evening about a dozen of the sociologists were sitting around comparing the merits of Canadian whiskey and Russian vodka. Soon, conversation turned from Crown Royal versus Moskovskaya to Russian politics. "You must be so excited about what's happening here," Robert said to his Russian hosts. How long do you think it will be before Russia will have multiparty elections? Do you think Russia will become a liberal democracy like the United States or a socialist democracy like Sweden?"

One white-haired Russian sociologist slowly rose to his feet. His colleagues privately called him "the dinosaur." It soon became clear why. "*Nikogda*," he said calmly and deliberately—"never." "*Nikogda*," he repeated, his voice rising sharply in pitch, volume,

While Versace does brisk business in Moscow, the streets are filled with homeless people. That is because the richest 10% of Russians earned 15 times more than the poorest 10%, making Russia one of the most inegalitarian countries in the world.

and emphasis. Then, for a full minute he explained that capitalism and democracy were never part of Russia's history. Nor could they be expected to take root in Russian soil. "The Russian people," he proclaimed, "do not want a free capitalist society. We know 'freedom' means the powerful are free to compete unfairly against the powerless, exploit them, and create social inequality."

Everyone else in the room disagreed with the dinosaur's speech, in whole or in part. But not wanting to cause any more upset, we turned the conversation back to lighter topics. After 15 minutes, someone reminded the others that we had to rise early for tomorrow's seminars. The evening ended, its great questions unanswered.

Today, more than a decade later, the great questions of Russian politics remain unanswered. And it now seems there was some truth in the dinosaur's speech after all. Russia first held multiparty elections in 1991. Surveys found that most Russians favored democracy over other types of rule. However, support for democracy soon fell because the economy collapsed.

The government had formerly fixed prices. It now allowed prices to rise to levels set by the market. Consumer goods soon cost 10 or 12 times more than just a year earlier. Many enterprises shut down because they were too inefficient to stay in business under market conditions. This led to an unemployment rate of about 20%. Even when the state kept businesses alive by subsidizing them, they paid many workers irregularly. Sometimes workers went months without a paycheck. Many Russians were barely able to make ends meet. According to official estimates, 39% of the population lived below the poverty line in 1999.

At the other extreme, profitable businesses and valuable real estate formerly owned by the government were sold to private individuals and companies. The lion's share went to senior members of the Communist Party and organized crime syndicates. These were the only two groups with enough money and inside knowledge to take advantage of the sell-off. They became fantastically wealthy. Moscow is said to have more Mercedes-Benz automobiles per capita than any other city in the world. By 1994, the richest 10% of Russians earned 15 times more than the poorest 10%. The level of income inequality in Russia is one of the highest in the world (Brym, 1996a; 1996b; 1996c; Gerber and Hout, 1998; Handelman, 1995; Remnick, 1998).

Democratic sentiment weakened as economic conditions worsened (Whitefield and Evans, 1994). In elections held in 1995 and 1996, support for democratic parties plunged as support for communist and extreme right-wing nationalist parties surged (Brym, 1995; 1996d). Nationwide surveys conducted in 38 countries between 1995 and 1997 found that as many as 97% of the citizens of some countries viewed democracy as the ideal form of government. Russia ranked last, at a mere 51% (Klingemann, 1999). Democracy allowed a few people to enrich themselves at the expense of most Russians. Therefore, many citizens equated democracy not with freedom but with distress.

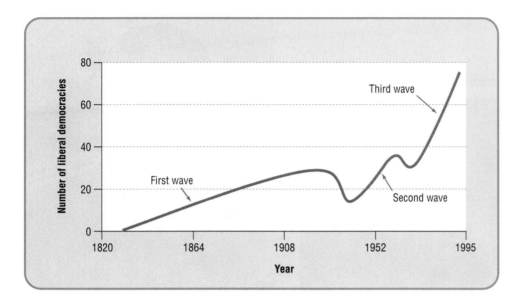

✦ **FIGURE 11.4** ✦
The Three Waves of Democratization, 1828–1995

SOURCE: Diamond (1996: 28); Huntington (1991: 26).

Russia's political institutions reflect the weakness of Russian democracy. Power is concentrated in the Presidency to a much greater degree than in the United States. The parliament and the judiciary do not act as checks on executive power. Only a small number of Russians belong to political parties. Voting levels are low. And minority ethnic groups are sometimes treated arbitrarily and cruelly. Clearly, Russian democracy has a long way to go before it can be considered on a par with democracy in the West.

The limited success of Russian democracy raises an important question. What social conditions must exist for a country to become fully democratic? That is the question to which we now turn. To gain some perspective, we first consider the three waves of democratization that have swept the world in the past 175 years (Huntington, 1991: 13–26; see Figure 11.4).

The Three Waves of Democracy

The first wave of democratization began when over half the white adult males in the United States became eligible to vote in the 1828 presidential election. By 1926, 33 countries enjoyed at least minimally democratic institutions. These countries included most of Western Europe, the British Dominions (Australia, Canada, and New Zealand), Japan, and four Latin American countries (Argentina, Colombia, Chile, and Uruguay). However, just as an undertow begins when an ocean wave recedes, a democratic reversal occurred between 1922 and 1942. During that period, fascist, communist, and militaristic movements caused two thirds of the world's democracies to fall under authoritarian or totalitarian rule.

The second wave of democratization took place between 1943 and 1962. Allied victory in World War II returned democracy to many of the defeated powers, including West Germany and Japan. The beginning of the end of colonial rule brought democracy to some states in Africa and elsewhere. Some Latin American countries formed limited and unstable democracies. However, even by the late 1950s, the second wave was beginning to exhaust itself. Soon, the world was in the midst of a second democratic reversal. Military dictatorships replaced many democracies in Latin America, Asia, and Africa. A third of the democracies in 1958 were authoritarian regimes by the mid-1970s.

The third and biggest wave of democratization began in 1974 with the overthrow of military dictatorships in Portugal and Greece. It crested in the early 1990s. In Southern and Eastern Europe, Latin America, Asia, and Africa, a whole series of authoritarian regimes fell. In 1991, Soviet communism collapsed. By 1995, 117 of the world's 191 countries were democratic in the sense that their citizens could choose representatives in regular,

Nigeria celebrated independence from Britain in 1960 (left). In 1993, General Sani Abacha (right) annulled the presidential election, became head of state, and began a reign of brutal civil rights violations. The world's third wave of democratization was drawing to a close.

competitive elections. That amounts to 61% of the world's countries containing nearly 55% of the world's population (Diamond, 1996: 26).

The third wave seems less dramatic, however, if we bear in mind that these figures refer to **formal democracies**—countries that hold regular, competitive elections. Many of these countries are not **liberal democracies.** That is, like Russia, they lack the freedoms and constitutional protections that make political participation and competition meaningful. In formal but nonliberal democracies, substantial political power may reside with a military that is largely unaffected by the party in office. Certain cultural, ethnic, religious, or regional groups may not be allowed to take part in elections. The legislative and judicial branches of government may not constrain the power of the executive branch. Citizens may not enjoy freedom of expression, assembly, and organization. Instead, they may suffer from unjustified detention, exile, terror, and torture. At the end of 1995, 40% of the world's countries were liberal democracies, 21% were nonliberal democracies, and 39% were nondemocracies (Diamond, 1996: 28). The number of liberal democracies in the world fell nearly 2% between 1991 and 1995. Some new democracies, including large and regionally influential countries like Russia, Nigeria, Turkey, Brazil, and Pakistan, experienced a decline in freedoms and protections. It seems this marks the end of the third wave (United States Information Agency, 1998–99).

The Social Preconditions of Democracy

Liberal democracies emerge and endure when countries enjoy considerable economic growth, industrialization, urbanization, the spread of literacy, and a gradual decrease in economic inequality (Huntington, 1991: 39–108; Lipset, 1981 [1960]: 27–63, 469–76; 1995; Moore, 1967; Rueschemeyer, Stephens, and Stephens, 1992; Zakaria, 1997). Economic development creates middle and working classes that are large, well organized, literate, and well off. When these classes become sufficiently powerful, their demands for civil liberties and the right to vote and run for office have to be recognized. If powerful middle and working classes are not guaranteed political rights, they sweep away kings, queens, landed aristocracies, generals, and authoritarian politicians in revolutionary upsurges. In contrast, democracies do not emerge where middle and working classes are too weak to wrest big political concessions from predemocratic authorities. In intermediate cases—where, say, a country's military is about as powerful a political force as its middle and working classes—democracy is precarious and often merely formal. The history of unstable democracies is largely a history of internal military takeovers (Germani and Silvert, 1961).

Apart from the socioeconomic conditions noted above, favorable external political and military circumstances help liberal democracy endure. Liberal democracies, even strong ones like France, collapse when they are defeated by fascist, communist, and military regimes and empires. They revive when democratic alliances win World Wars and authoritarian empires break up. Less coercive forms of outside political intervention are sometimes effective too. For example, in the 1970s and 1980s, the European Union helped liberal democracy in Spain, Portugal, and Greece by integrating these countries into the West European economy and giving them massive economic aid.

In sum, powerful, prodemocratic foreign states and strong, prosperous middle and working classes are liberal democracy's best guarantees. It follows that liberal democracy will spread in the less economically developed countries only if they prosper and enjoy support from a confident United States and European Union, the world centers of liberal democracy.

Recognizing the importance of the United States and the European Union in promoting democracy in many parts of the world should not obscure two important facts, however. First, the United States is not always a friend of democracy. For example, between the end of World War II and the collapse of the Soviet Union in 1991, democratic regimes that were sympathetic to the Soviet Union were often destabilized by the United States and replaced by antidemocratic governments. American leaders were willing to export arms and offer other forms of support to antidemocratic forces in Iran, Chile, Nicaragua, Guatemala, and other countries because they believed it was in the United States' political and economic interest to do so (see Chapter 16, "Population, Urbanization, and Development").

Second, just because the United States promotes democracy in many parts of the world, we should not assume that liberal democracy has reached its full potential in this country. We saw otherwise in our discussion of the limited participation and influence of disadvantaged groups in American politics. It seems fitting, therefore, to conclude this chapter by briefly assessing the future of liberal democracy in America.

Some analysts think the home computer will soon increase Americans' political involvement. They think it will help solve the problem of unequal political participation by bringing more disadvantaged Americans into the political process. Others think growing affluence means there are fewer disadvantaged Americans to begin with. This makes economic or material issues less relevant than they used to be. In the concluding section, we raise questions about both contentions. We argue that political participation is likely to remain unequal in the foreseeable future. Meanwhile, issues concerning economic inequality are likely to remain important for most people. Liberal democracy can realize its full potential only if both problems—political and economic inequality—are adequately addressed.

Electronic Democracy

On October 20, 1935, the *Washington Post* ran a full-page story featuring the results of the first nationwide poll. The story also explained how the new method of measuring public opinion worked. George Gallup was the man behind the poll. In the article, he said that polls allow the people to reclaim their voice: "After one hundred and fifty years we return to the town meeting. This time the whole nation is within the doors" (quoted in London, 1994: 1). Gallup was referring to the lively New England assemblies that used to give citizens a direct say in political affairs. He viewed the poll as a technology that can bring the town hall to the entire adult population of the country.

Gallup's idea seems naïve today. Social scientists have shown that polls often allow politicians to mold public opinion, not just reflect it. For example, they can word questions to increase the chance of eliciting preferred responses. They can then publicize the results to serve their own ends (Ginsberg, 1986; see also Chapter 2, "Research Methods"). From this point of view, polls are little different from other media events that are orchestrated by politicians to sway public opinion (see Box 11.2).

Recently, however, some people have greeted one new technology with the same enthusiasm that Gallup lavished on polls. Computers linked to the Internet could allow citizens to debate issues and vote on them directly. This could give politicians a clear signal of how public policy should be conducted. Some people think that, in an era of low and declining political participation, computers can revive American democracy. Public opinion would then become the law of the land (Westen, 1998).

It is a grand vision, but flawed. Social scientists have conducted more than a dozen experiments with electronic public meetings. They show that even if the technology needed for such meetings were available to everyone, interest is so limited that no more than a third of the population would participate (Arterton, 1987).

Subsequent experience supports this conclusion. The people most likely to take advantage of electronic democracy are those who have access to personal computers and the Internet. They form a privileged and politically involved group. They are not representative of the American adult population. This is apparent from Table 11.5. The table contains data from the most respected ongoing survey of World Wide Web users. Compared to the general population, American Web users are younger, better educated, wealthier, and contain a higher proportion of men, whites, and people in occupations requiring substantial computer use. It is also significant that nearly 83% of American Web users are registered to vote. That compares to about 70% in the voting age population as a whole. We conclude that if electronic democracy becomes widespread, it will probably reinforce the same inequalities in political participation that plague American democracy today. It is likely to help form a digital divide.

BOX 11.2
SOCIOLOGY AT THE MOVIES

WAG THE DOG (1998)

The President is caught having an affair with his aide. What should be done? Tell the truth? Deny having sex with the woman? Get people to change their definition of "having sex?" Hire a media consultant? Start a war to distract the public?

Wag the Dog is a film about a sex scandal that embroils a fictional President. To cover up his sexual misconduct, the fictional President's advisors hire a Hollywood movie producer. The advisors and the producer create a fake crisis in a small, poor, remote country to divert attention from the President. The producer films a newsreel, complete with computer-generated special effects, of a young girl fleeing a military skirmish. Television news programs air it. Seeing a young girl in distress, the public registers strong support for United States intervention in the war.

Anne Heche, Dustin Hoffman, and Robert DeNiro conspire to cover up the President's sexual misdeeds by staging a phony war in *Wag the Dog*.

Rather than reality informing political decisions, political convenience creates reality in the world of *Wag the Dog*. Although some people may feel that the movie is too cynical and even paranoid in its caricature of American politics, others would argue that it convincingly captures a key element of American political life.

In fact, the movie seemed to predict a real-life event. Several months after its release, President Clinton's affair with Monica Lewinsky became a major scandal. For about a year, few people could fail to bring up the Lewinsky affair when talk turned to politics. Politics seemed to focus on what "having sex" means. Moreover, in an eerie replication of the movie's plot, President Clinton ordered the bombing of Iraq soon after news of the affair broke. Regardless of the legitimacy of the military action, many concerned observers criticized Clinton's military tactic as a way to divert the media's attention from his personal problems. Many viewers of the movie found it unsettling to see reality follow a movie script.

What does *Wag the Dog* tell us about American politics? Does it merely caricature our media-obsessed, cynical view of politics? Or does it capture a slice of reality? Is it even useful to insist on the distinction between media and politics?

Under 41 Years Old	56.4
Male	64.2
Completed college or higher	58.5
Annual household income $50,000+	46.9
White	88.1
African American	2.3
Registered to vote	82.8

✦ TABLE 11.5 ✦

The Digital Divide: Social Characteristics of World Wide Web Users, United States, October–December 1998 (in percent; n = 4,254)

Note: These data come from the ongoing Georgia Tech survey of Web users. People learn about the survey on the Web and volunteer to participate in it. This produces sample bias. Respondents tend to be experienced and skilled Web users. However, comparing these results with survey results based on random samples suggests that the figures reported here are accurate (Hoffman and Novak, 1998: 390–1).

SOURCE: Georgia Tech (1999).

Postmaterialism

Believers in electronic democracy think the computer will solve the problem of unequal political participation. **Postmaterialists** believe that economic or material issues are becoming less important in American politics. They argue as follows. Liberal democracies are less stratified than both nonliberal democracies and nondemocracies. That is, in liberal democracies, the gap between rich and poor is less extreme and society as a whole is more prosperous. In fact, say the postmaterialists, prosperity and the moderation of stratification have reached a point where they have fundamentally changed political life in America. They claim that, as recently as 50 years ago, most people were politically motivated mainly by their economic or material concerns. As a result, parties were distinguished from one another chiefly by the way they attracted voters from different classes. Now, however, many if not most Americans have supposedly had their basic material wants satisfied. Particularly young people who grew up in prosperous times are less concerned with material issues, such as whether their next paycheck can feed and house their family. They are more concerned with postmaterialist issues, such as women's rights, civil rights, and the environment. The postmaterialists conclude that the old left–right political division, based on class differences and material issues, is being replaced. The new left–right political division, they say, is based on age differences and postmaterialist issues (Clark and Lipset, 1991; Clark, Lipset, and Rempel, 1993; Inglehart, 1997).

While America is certainly more prosperous and less stratified than the less developed countries of the world, the postmaterialists are wrong to think that affluence is universal in America. Nor is inequality decreasing. We saw in Table 11.3 that the United States has one of the highest poverty rates of the 18 rich industrialized countries (see also Chapter 7, "Stratification: United States and Global Perspectives"). Here we may add that poverty is particularly widespread among youth—just the people who, in the postmaterialist view, are the most affluent and least concerned with material issues. In 2000, the unemployment rate for people between the ages of 16 and 19 was about four times the rate for the whole labor force (United States Department of Labor, 2000a). Moreover, between the late 1960s and early 1990s, the percentage of children under the age of 18 living in poverty *after welfare payments* doubled, rising to about 22%. This makes the United States number one in child poverty among the 18 rich industrialized countries (Rainwater and Smeeding, 1995). These figures suggest that, today, more new voters are poor than at any time in the past 40 years.

In addition, inequality is not decreasing. People with a college degree have seen their real incomes rise substantially since the early 1970s. However, people without a college degree, who account for 70% of the American work force, have seen their real incomes rise only slightly or decline (Bluestone and Rose, 1997).

Under these circumstances, we should not be surprised that bread-and-butter issues are still important for most voters. In fact, research shows that class is just as important in influencing voting as it was 40 years ago. Class voting fluctuates from election to election. It varies from one rich industrialized country to the next. But disadvantaged people still tend to vote for parties on the left and advantaged people for parties on the right. To varying degrees, Americans, and voters in other liberal democracies, still tend to vote according to their material interests. There is no denying that people are inserting

✦ **FIGURE 11.5** ✦

The Class Cleavage in United States Presidential Elections, 1960–1992

SOURCE: Brooks and Manza (1997b).

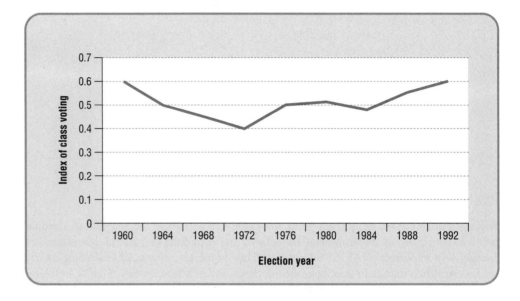

postmaterialist issues into today's political debates. It seems, however, that most people are layering these issues on top of old ones, not replacing them (Brooks and Manza, 1994; 1997a; 1997b; Hout, Brooks, and Manza, 1993; Manza, Hout, and Brooks, 1995; Weakliem, 1991). Figure 11.5 illustrates the American case. It shows change over time in the influence of class on voting. Although class voting dropped slightly from 1960 to 1972, it started to rise afterwards. In 1992, class voting stood at the same level as in 1960.

So we arrive at the big dilemma of American politics. Problems of economic inequality continue to loom large, but it is doubtful they will be addressed in a serious way unless disadvantaged Americans get more politically involved. Yet unequal political participation shows no sign of evaporating (Valelly, 1999). No mystery surrounds the identification of reforms needed to bring more disadvantaged people into the political process. For example, based on comparative research, researchers have determined that removing burdensome voter registration laws would increase participation rates in the United States by 8–15% (Lijphart, 1997). However, there is little political will to undertake such reforms now.

A solution may have to come from outside normal politics. We saw how big shocks reoriented American public policy in the 1890s, 1930s, and 1970s. Perhaps we need something similar to decrease political and economic inequality in the future. The full realization of Lincoln's "government of the people, by the people, for the people" may require one of those "little rebellions" called for by Thomas Jefferson. As you think about this possibility, ask yourself some related questions: Who is most likely to lead such rebellions? Under what circumstances might such rebellions come about? To what extent would you support or oppose them? Why?

SUMMARY

1. The level of democracy in a society depends on how power is distributed. When power is concentrated in the hands of few people, society is less democratic.

2. Pluralists correctly note that democratic politics is about negotiation and compromise. However, they fail to appreciate that economically advantaged groups have more power than disadvantaged groups.

3. Elite theorists correctly note that power is concentrated in the hands of advantaged groups. However, they fail to appreciate how variations in the distribution of power influence political behavior and public policy.

4. Power resource theorists usefully focus on changes in the distribution of power in society and their effects. However, they fail to appreciate what state-centered theorists emphasize—

that state institutions and laws also affect political behavior and public policy.

5. Many new democracies that emerged from the most recent wave of democratization are formal, not liberal democracies. Their citizens enjoy regular, competitive elections but lack legal protection of rights and freedoms.

6. Citizens win legal protection of rights and freedoms when their middle and working classes become large, organized, and prosperous; and when powerful, friendly, pro-democratic foreign states support them.

7. Believers in electronic democracy and postmaterialism think the United States has reached a new and higher stage of democratic development. However, enduring social inequalities prevent even the most advanced democracies from being fully democratic.

GLOSSARY

Authoritarian states sharply restrict citizen control of the state.

Authority is legitimate, institutionalized power.

Charismatic authority is based on belief in the claims of extraordinary individuals to be inspired by God or some higher principle.

Civil society is the private sphere of social life.

In a **democracy,** citizens exercise a high degree of control over the state. They do this mainly by choosing representatives in regular, competitive elections.

An **elite** is a group that controls the command posts of an institution.

Elite theory holds that small groups occupying the command posts of America's most influential institutions make the important decisions that profoundly affect all members of society. Moreover, they do so without much regard for elections or public opinion.

Formal democracy involves regular, competitive elections.

Legal-rational authority is typical of modern societies. It derives from respect for the law. Laws specify how one can achieve office. People generally believe these laws are rational. If someone achieves office by following these laws, people respect their authority.

Legitimate governments are those that enjoy a perceived right to rule.

A **liberal democracy** is a country whose citizens enjoy regular, competitive elections *and* the freedoms and constitutional protections that make political participation and competition meaningful.

Lobbies are organizations formed by special interest groups to advise and influence politicians.

The **mass media** in a democracy help to keep the public informed about the quality of government.

Pluralist theory holds that power is widely dispersed. As a result, no group enjoys disproportionate influence and decisions are usually reached through negotiation and compromise.

Political parties are organizations that compete for control of government in regular elections. In the process, they give voice to policy alternatives and rally adult citizens to vote.

A **political revolution** is the overthrow of political institutions by an opposition movement and its replacement by new institutions.

Postmaterialism is a theory which claims that growing equality and prosperity in the rich industrialized countries have resulted in a shift from class-based to value-based politics.

Power is the ability to control others, even against their will.

Power resource theory holds that the distribution of power between major classes partly accounts for the successes and failures of different political parties.

Public opinion is composed of the values and attitudes of the adult population as a whole. It is expressed mainly in polls and letters to lawmakers and gives politicians a reading of citizen preferences.

A **ruling class** is a self-conscious, cohesive group of people in elite positions. They act to advance their common interests and are lead by corporate executives.

Social movements are collective attempts to change all or part of the political or social order.

The **state** consists of the institutions responsible for formulating and carrying out a country's laws and public policies.

State-centered theory holds that the state itself can structure political life to some degree independently of the way power is distributed between classes and other groups at a given time.

In a **totalitarian** state, citizens lack almost any control of the state.

Traditional authority, the norm in tribal and feudal societies, involves rulers inheriting authority through family or clan ties. The right of a family or clan to monopolize leadership is widely believed to be derived from God's will.

QUESTIONS TO CONSIDER

1. Analyze any recent election. What issues distinguish the competing parties or candidates? What categories of the voting population are attracted by each party or candidate? Why?

2. Younger people are less likely to vote than older people. How would power resource theory explain this?

3. Do you think the United States will become a more democratic country in the next 25 years? Will a larger percentage of the population vote? Will class and racial inequalities in political participation decline? Will public policy more accurately reflect the interests of the entire population? Why or why not?

WEB RESOURCES

Companion Web Site for This Book

http://sociology.wadsworth.com

Begin by clicking on the Student Resources section of the Web site. Choose "Introduction to Sociology" and finally the Brym and Lie book cover. Next, select the chapter you are currently studying from the pull-down menu. From the Student Resources page you will have easy access to InfoTrac College Edition®, MicroCase Online exercises, additional Web links, and many other resources to aid you in your study of sociology, including practice tests for each chapter.

InfoTrac Search Terms

These search terms are provided to assist you in beginning to conduct research on this topic by visiting http://www.infotraccollege. com/wadsworth.

Civil society   **Ruling class**
Legitimacy   **Voting**
Political party

Recommended Web Sites

For everything you always wanted to know about campaign funding and the 2000 United States presidential race, go to http://www.opensecrets.org/2000elect/index/AllCands.htm.

For a good listing of Web resources on American politics, go to http://www.georgetown.edu/crossroads/asw/pol.html.

The most important organization of international governance is the United Nations. For the UN Web site, go to http://www.un.org.

There is a growing gender gap in American politics. Women are moving to the Democrats and men to the Republicans. For insightful analysis of this phenomenon, go to http://www.theatlantic.com/issues/96jul/gender/gender.htm.

SUGGESTED READINGS

Dan Clawson, Alan Neustadtl and Denise Scott. *Money Talks: Corporate PACS and Political Influence* (New York: Basic Books, 1992). Shows how PACS influence and dominate the political process in America.

Samuel Huntington. *The Third Wave: Democratization in the Late Twentieth Century* (Norman, OK: University of Oklahoma Press, 1991). An account of the sweep and limits of democracy around the world.

Frances Fox Piven and Richard A. Cloward. *Why Americans Don't Vote* (New York: Pantheon, 1989 [1988]). Shows how disadvantaged Americans have been disenfranchised from politics.

IN THIS CHAPTER, YOU WILL LEARN THAT:

✦ The traditional "nuclear" family is less common than it used to be. Several new family forms are becoming more popular. The frequency of one family form or another varies by class, race and ethnicity, sexual orientation, and culture.

✦ One of the most important forces underlying change from the traditional nuclear family is the entry of most women into the paid labor force. Doing paid work increases women's ability to leave unhappy marriages and control whether and when they will have children.

✦ Marital satisfaction increases as one moves up the class structure, where divorce laws are liberal, when teenage children leave the home, in families where housework is shared equally, and among spouses who enjoy satisfying sexual relations.

✦ The worst effects of divorce on children can be eliminated if there is no parental conflict and the children's standard of living does not fall after divorce.

✦ The decline of the traditional nuclear family is sometimes associated with a host of social problems, such as poverty, welfare dependency, and crime. However, policies have been adopted in some countries that reduce these problems.

Personal Anecdote

INTRODUCTION

One Saturday morning, the married couple who live next door to Robert Brym and his family asked Robert for advice on new speakers they wanted to buy for their sound system. So Robert volunteered to go shopping with them at a nearby mall. They told him they also wanted to buy two outdoor garbage cans at a hardware store. Robert told them he didn't mind waiting.

"After they made the purchases, we returned to their minivan in the mall's parking lot," says Robert. "The wife opened the trunk, cleared some space, and said to her husband, 'Let's put the garbage cans back here.'

"Meanwhile, the husband had opened the side door. He had already put the speakers on the back seat and was struggling to do the same with the second garbage can. 'It's okay,' he said, 'I've already got one of them part way in here.'

"'Oh,' laughed the wife, 'I can judge space better than you and you'll never get that in there. Bring it back here.'

"'You know,' answered the husband, 'we don't always have to do things your way. I'm a perfectly intelligent person. I think there's room up here and that's where I'm going to put this thing. You can put yours back there or stick it anywhere else you like.'

"'Why are you yelling at me?' snapped the wife.

"'I'm not yelling,' shouted the husband. 'I'm just saying that I know as well as you what fits where. There's more than one way—your way—to do things.'

"So, the wife put one garbage can in the trunk, the husband put one in the back seat (it was, by the way, a very tight squeeze) and we piled into the car for the drive home. The husband and the wife did not say a word to each other. When we got back to our neighborhood, I said I was feeling tired and asked whether I could perhaps hook up their speakers on Sunday. Actually, I wasn't tired. I just had no desire to referee round two. I went home, full of wonder at the occasional inability of presumably mature adults to talk rationally about something as simple as how to pack garbage cans into a minivan.

"However, trivializing the couple's argument in this way prevented me from thinking about it sociologically. If I had been thinking like a sociologist, I would have at least recognized that, for better or for worse, our most intense emotional experiences are bound up with our families. We love, hate, protect, hurt, express generosity toward, and envy nobody as much as our parents, siblings, children, and mates. Little wonder, then, that most people are passionately concerned with the rights and wrongs, the dos and don'ts, of family life. Little wonder that family issues lie close to the center of political debate in this country. Little wonder that words, gestures, and actions that seem trivial to an outsider can hold deep meaning and significance for family members."

Because families are emotional minefields, few subjects of sociological inquiry generate as much controversy. Much of the debate centers on a single question: Is the family in decline and, if so, what should be done about it? The question is hardly new. A contributor to the *Boston Quarterly Review* of October 1859 wrote: "The family, in its old sense, is disappearing from our land, and not only our free institutions are threatened but the very existence of our society is endangered" (quoted in Lantz, Schultz, and O'Hara, 1977: 413). The same sentiment was expressed in September 1998. While criticizing President Clinton's affair with Monica Lewinsky, Democratic Senator (later vice-presidential hopeful) Joseph Lieberman of Connecticut stressed that "the decline of the family is one of the most pressing problems we are facing" (quoted in McKenna, 1998). This alarm, or one much like it, is sounded whenever the family undergoes rapid change, and particularly when the divorce rate increases (see Box 12.1).

Today, when some people speak about "the decline of the family," they are referring to the **nuclear family.** The nuclear family is composed of a cohabiting man and woman who maintain a socially approved sexual relationship and have at least one child. Others are referring more narrowly to what might be called the **traditional nuclear family.** The traditional nuclear family is a nuclear family in which the wife works in the home without

BOX 12.1
SOCIOLOGY AT THE MOVIES

AMERICAN BEAUTY (1999)

His wife wants to kill him. So does his daughter. His daughter's boyfriend, who has been supplying him with marijuana, is willing to kill him on his girlfriend's behalf. The boyfriend's father, a retired Marine, is convinced his son is having a homosexual affair with him. So he wants to kill him too.

What does the character played by Kevin Spacey do to make so many people so angry in *American Beauty*? He gives up the pretenses of a middle-class, suburban husband. He returns in spirit (and in body as well by exercising furiously) to his teenage self. He quits his job as a magazine writer and gets a new one as a cook at a fast-food franchise. He trades in his "boring" late-model sedan for an old sports car. He is no longer willing to continue his loveless marriage or to suffer his daughter's taunts. He lusts after his teenage daughter's best friend. However, giving up such important family roles—dependable breadwinner, solid citizen, loving husband, sympathetic father—has big consequences. It enrages people enough to want to kill him.

American Beauty offers a depressing portrait of suburban American family life. The wife is a frustrated real estate broker.

Kevin Spacey and Mena Suvari in *American Beauty*.

When she has an affair with a successful real estate broker, she appears more interested in advancing her career than in seeking pleasure or love. The pleasure and love she does experience seem to derive mainly from her lover's high status. The retired Marine is an angry, violent, and obsessive-compulsive man. He has drained the life out of his wife. He is also homophobic, although it turns out that he harbors homosexual longings, as homophobes sometimes do (see Chapter 9, "Sexuality and Gender"). In fact, about the only people who seem genuinely happy are the homosexual couple next door, who deviate from the suburban norm of heterosexual marriage.

Clearly, suburban America is far from being a utopia. Well-manicured lawns and beautiful gardens sometimes mask deep frustrations and pathologies. What makes the Kevin Spacey character so subversive, however, is that he spurns the comforts of middle-class, suburban life and endangers the well-established norms of the nuclear family. But how should we understand the movie's challenge to conventional suburban family life? Does the movie merely illustrate a psychological problem—a mid-life crisis? Or does it illustrate a deeper, sociological problem—the collapse of conventional gender and family roles in a rapidly changing society?

pay while the husband works outside the home for money. This makes him the "primary provider and ultimate authority" (Popenoe, 1988: 1).

In the 1940s and 1950s, many sociologists and much of the American public considered the traditional nuclear family the most widespread and ideal family form. However, for reasons we will examine below, the percentage of married-couple families fell from about 78% to 53% of all households between 1950 and 1997 (see Figure 12.1). Between 1950 and 1999, the percentage of women over the age of 16 in the paid labor force increased from around 34% to 60%. As a result, only a minority of American adults live in traditional nuclear families today. Many new family forms have become popular in recent decades (see Table 12.1).

Some sociologists, many of them functionalists, view the decreasing prevalence of the married-couple family and the rise of the "working mother" as an unmitigated disaster (e.g., Popenoe, 1998; 1996). In their view, rising rates of crime, illegal drug use, poverty, and welfare dependency (among other social ills) can be traced to the fact that so many American children are not living in two-parent households with stay-at-home mothers.

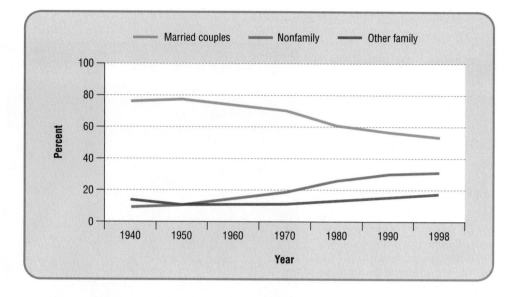

Traditional Nuclear Family	New Alternatives
Legally married	Never-married singlehood, nonmarital cohabitation
With children	Voluntary childlessness
Two-parent	Single-parent (never-married or previously married)
Permanent	Divorce, remarriage (including binuclear family involving joint custody, stepfamily)
Male primary provider, ultimate authority	Egalitarian marriage (including dual-career and commuter marriage)
Sexually exclusive	Extramarital relationships (including sexually open marriage, swinging, and intimate friendships)
Heterosexual	Same-sex intimate relationships or households
Two-adult household	Multi-adult households (including multiple spouses, communal living, affiliated families, and multigenerational families)

They call for various legal and cultural reforms to shore up the traditional nuclear family. For instance, they want to make it harder to get a divorce and they want people to place less emphasis on individual happiness at the expense of family responsibility.

Other sociologists, influenced by conflict and feminist theories, disagree with the functionalist assessment (e.g., Baca Zinn and Eitzen, 1993; Collins and Coltrane, 1991; Coontz, 1992; Skolnick, 1991). In the first place, they argue that it is inaccurate to talk about *the* family, as if this important social institution assumed or should assume only a single form. They emphasize that families have been structured in many ways and that the diversity of family forms is increasing as people accommodate to the demands of new social pressures. Second, they argue that changing family forms do not necessarily represent deterioration in the quality of people's lives. In fact, such changes often represent *improvement* in the way people live. They believe the decreasing prevalence of the traditional nuclear family and the proliferation of diverse family forms has benefited many men, women, and children and has not harmed other children as much as the functionalists think. They also believe that various economic and political reforms, such as the creation of an affordable nationwide day-care system, could eliminate most of the negative effects of single-parent households.

This chapter touches on divorce, reproductive choice, single-parent families, day care, and other topics in the sociology of families. However, we have structured this chapter around the debate about the so-called decline of the American family. We first outline the functional theory of the family because the issues raised by functionalism are still a focus of sociological controversy (Mann, Grimes, Kemp, and Jenkins, 1997). Borrowing from the work of conflict theorists and feminists, we next present a critique of functionalism. In particular, we show that the nuclear family became the dominant and ideal family form only under specific social and historical conditions. Once these conditions changed, the nuclear family became less prevalent and a variety of new family forms proliferated. You will learn how these new family forms are structured and how their frequency varies by class, race, and sexual orientation. You will also learn that, while postindustrial families solve some problems, they are hardly an unqualified blessing. The chapter's concluding section therefore considers the kinds of policies that might help alleviate some of the most serious concerns faced by families today. Let us, then, first review the functionalist theory of the family.

FUNCTIONALISM AND THE NUCLEAR IDEAL

Functional Theory

For any society to survive, its members must cooperate economically. They must have babies. And they must raise offspring in an emotionally supportive environment so the offspring can learn the ways of the group and eventually operate as productive adults. Since the 1940s, functionalists have argued that the nuclear family is ideally suited to meet these challenges. In their view, the nuclear family performs five main functions. It provides a basis for regulated sexual activity, economic cooperation, reproduction, socialization, and emotional support (Murdock, 1949: 1–22; Parsons, 1955).

Functionalists cite the pervasiveness of the nuclear family as evidence of its ability to perform the functions listed above. To be sure, other family forms exist. **Polygamy** expands the nuclear unit "horizontally" by adding one or more spouses (almost always wives) to the household. Polygamy is still legally permitted in many less industrialized countries of Africa and Asia. However, the overwhelming majority of families are monogamous because they cannot afford to support several wives and many children. The **extended family** expands the nuclear family "vertically" by adding another generation—one or more of the spouse's parents—to the household. Extended families used to be common throughout the world. They still are in some places. However, according to the functionalists, the basic building block of the extended family (and of the polygamous family) is the nuclear unit.

George Murdock was a functionalist who conducted a famous study of 250 mainly preliterate societies in the 1940s. Murdock wrote: "Either as the sole prevailing form of the family or as the basic unit from which more complex familial forms are compounded, [the nuclear family] exists as a distinct and strongly functional group in every known society" (Murdock, 1949: 2). Moreover, the nuclear family, Murdock continued, is everywhere based on **marriage.** He defined marriage as a socially approved, presumably long-term, sexual and economic union between a man and a woman. It involves rights and obligations between spouses and between spouses and their children.

Let us consider the five main functions of marriage and the nuclear family in more detail:

1. *Sexual regulation.* Imagine a world without an institution that defines the boundaries within which legitimate sexual activity is permitted. Such a world would be disrupted by many people having sex wherever, whenever, and with whomever they pleased. An orderly social life would be difficult. Because marriage provides a legitimate forum for expressing the intense human need for sexual activity, says Murdock, it makes social order possible.

   Sex is not, however, the primary motive for marrying, he continues. After all, sex is readily available outside marriage. Only 54 of Murdock's 250 societies forbade or disapproved of premarital sex between nonrelatives. In most of the 250 societies, a married man could legitimately have an extramarital affair with one or more female relatives (Murdock, 1949: 5–6). It is hardly news that premarital and extramarital sex is common in contemporary America and other postindustrial societies. As a president of the University of California once said in a *Time* magazine interview: "I find that the three major administrative problems on a campus are sex for the students, athletics for the alumni, and parking for the faculty" (quoted in Ember and Ember, 1973: 317).

2. *Economic cooperation.* Why then, apart from sex, do people marry? Murdock's answer is this: "By virtue of their primary sex difference, a man and a woman make an exceptionally efficient cooperating unit" (Murdock, 1949: 7). On average, women are physically weaker than men. Historically, pregnancy and nursing have restricted women in their activities. Therefore, writes Murdock, they can best perform lighter tasks close to home. These tasks include gathering and planting food, carrying water, cooking, making and repairing clothing, making pottery, and caring for children. Most men possess superior strength. They can therefore specialize in lumbering, mining, quarrying, land clearing, and house building. They can also range farther afield to hunt, fish, herd, and trade (Murdock, 1937). According to Murdock, this division of labor enables more goods and services to be produced than would otherwise be possible. People marry partly due to this economic fact. In Murdock's words: "Marriage exists only when the economic and the sexual are united into one relationship, and this combination occurs only in marriage" (Murdock, 1949: 8).

3. *Reproduction.* Before the invention of modern contraception, sex often resulted in the birth of a baby. According to Murdock, children are an investment in the future. Already by the age of 6 or 7, children in most societies do some chores. Their economic value to the family increases as they mature. When children become adults, they often help support their elderly parents. Thus, in most societies, there is a big economic incentive to having children.

4. *Socialization.* The investment in children can be realized only if adults rear the young to maturity. This involves not only caring for them physically but, as you saw in Chapter 4 ("Socialization"), teaching them language, values, beliefs, skills, religion, and much else. Talcott Parsons (1955: 16) regarded socialization as the "basic and irreducible" function of the family.

5. *Emotional support.* Parsons also noted that the nuclear family universally gives its members love, affection, and companionship. He stressed that, in the nuclear family, it is mainly the mother who is responsible for ensuring the family's emotional well-being. She develops what Parsons calls the primary "expressive" role because she is the one who bears children and nurses them. It falls on the husband to take on the more "instrumental" role of earning a living outside the family (Parsons, 1955: 23). The fact that he is the "primary provider" makes him the ultimate authority.

Foraging Societies

Does functionalism provide an accurate picture of family relations at any point in human history? To assess the adequacy of the theory, let us briefly consider family patterns in the two settings that were apparently foremost in the minds of the functionalists. We first discuss families in preliterate, foraging societies. In such societies, people subsist by hunting animals and gathering wild edible plants. Most of the cases in Murdock's sample are foraging societies. We then discuss families in urban and suburban middle-class America in the 1950s. The functionalists whose work we are reviewing lived in such families themselves.

Foraging societies are nomadic groups of 100 or fewer people. As we would expect from Murdock's and Parsons's analysis, a gender division of labor exists among foragers. Most men hunt and most women gather. Women also do most of the child care. However, research on foragers conducted since the 1950s, on which we base our analysis, shows that

There is rough gender equality among the !Kung-San, a foraging society in the Kalahari Desert in Botswana. That is partly because women play such a key economic role in providing food.

men often tend babies and children in such societies (Leacock, 1981; Lee, 1979; Turnbull, 1961). They often gather food after an unsuccessful hunt. In some foraging societies, women hunt. In short, the gender division of labor is less strict than Murdock and Parsons thought. What is more, the gender division of labor is not associated with large differences in power and authority. Overall, men have few if any privileges that women don't also enjoy. Relative gender equality is based on the fact that women produce up to 80% of the food.

Foragers travel in small camps or bands. The band decides by consensus when to send out groups of hunters. When they return from the hunt, they distribute game to all band members based on need. Each hunter does not decide to go hunting based on his or her own nuclear family's needs. Each hunter does not distribute game just to his or her own nuclear family. Contrary to what Murdock wrote, it is the band, not the nuclear family, that is the most efficient social organization for providing everyone with his or her most valuable source of protein.

In foraging societies, children are considered an investment in the future. However, it is not true that people always want more children for purposes of economic security. In fact, too many children are considered a liability. Subsistence is uncertain in foraging societies, and when band members deplete an area of game and edible plants, they move elsewhere. As a result, band members try to keep the ratio of children to productive adults low. In a few cases, such as the pre-20th century Inuit ("Eskimos"), newborns were occasionally allowed to die if the tribe felt its viability was threatened by having too many mouths to feed.

Life in foraging societies is highly cooperative. For example, women and men care for—and women even breast-feed—each other's children. Despite Parsons's claim that socialization is the "basic and irreducible" function of the nuclear family, it is the band, not the nuclear family, that assumes responsibility for child socialization in foraging societies. Socialization is more a public than a private matter. As a 17th-century Innu man from northern Quebec said to a French Jesuit priest who was trying to convince him to adopt European ways of raising children: "Thou hast no sense. You French people love only your own children; but we all love all the children of our tribe" (quoted in Leacock, 1981: 50).

In sum, recent research on foraging societies calls into question many of the functionalists' generalizations. In foraging societies, relations between the sexes are quite egalitarian. Children are not viewed just as an investment in the future. Each nuclear unit does not execute the important economic and socialization functions in private. On the contrary, cooperative band members execute most economic and socialization functions in public.

Let us now assess the functionalist theory of the family in the light of evidence concerning American middle-class families in the years just after World War II.

The American Middle Class in the 1950s

Functionalists recognized that the productive function of the family was less important after World War II than it had been in earlier times. In their view, the socialization and emotional functions of the family were now most important (Parsons, 1955). Thus, on the 19th-century family-owned farm or ranch, the wife had played an indispensable productive role while the husband was out in the field or on the range. She took responsibility for the garden, the dairy, the poultry, and the management of the household. The children also did crucial chores with considerable economic value. But in the typical urban or suburban nuclear family of the late 1940s and 1950s, noted the functionalists, only one person played the role of breadwinner. That was usually the husband. Children enjoyed more time to engage in the play and leisure-time activities that were now considered necessary for healthy development. For their part, most women got married, had babies, and stayed home to raise them. Strong normative pressures helped to keep women at home. Thus, in the 1950s, sociologist David Riesman called a woman's failure to obey the strict gender division of labor a "quasi-perversion." *Esquire* magazine called women's employment in the paid labor force a "menace." *Life* magazine called it a "disease" (quoted in Coontz, 1992: 32).

Web Interactive Exercises
Is the Woman's Place in the Home?

As a description of family patterns in the 15 years after World War II, functionalism has its merits. During the Great Depression (1929–39) and the war (1939–45), millions of Americans were forced to postpone marriage due to widespread poverty, government-imposed austerity, and physical separation. After this long and dreadful ordeal, many Americans just wanted to settle down, have children, and enjoy the peace, pleasure, and security that family life seemed to offer. Conditions could not have been better for doing just that. The immediate postwar era was one of unparalleled optimism and prosperity. Real per capita income rose 35% between 1945 and 1960. The percentage of Americans who owned their own homes jumped from 43% in 1940 to 62% in 1960. Government assistance in the form of the GI Bill and other laws helped to make the late 1940s and 1950s the heyday of the traditional nuclear family. This assistance took the form of guaranteed, tax-deductible mortgages, subsidized college education and health care for veterans, big income tax deductions for dependents, and massive road-building projects that opened the suburbs for commuters. People got married younger. They had more babies. They got divorced less. Increasingly, they lived in married-couple families (see Table 12.2). Middle-class women engaged in what has been called an "orgy of domesticity" in the postwar years, devoting increasing attention to child rearing and housework. They also became increasingly concerned with the emotional quality of family life as love and companionship became firmly established as the main motivation for marriage (Coontz, 1992: 23–41; Skolnick, 1991: 49–74).

However, as sociologist Andrew J. Cherlin meticulously shows, the immediate postwar period was in many respects a historical aberration (Cherlin, 1992 [1981]: 6–30). Trends in divorce, marriage, and childbearing show a gradual *weakening* of the nuclear family from the second half of the 19th century until the mid-1940s, and continued weakening after the 1950s. Specifically, throughout the 19th century, the **divorce rate** rose. The divorce rate is the number of divorces that occur in a year for every 1,000 people in the population. Meanwhile, the **marriage rate** fell. The marriage rate is the number of marriages that occur in a year for every 1,000 people in the population. The **total fertility rate** also fell. The total fertility rate is the average number of children born to women of the same age over their lifetime. In contrast, the divorce rate fell only between 1946 and 1958. The marriage rate took a big jump only in the 2 years following World War II. And the fertility rate rose only for women who reached childbearing age between 1930 and the mid-1950s. By the late 1950s or early 1960s, the earlier trends reasserted themselves. Only the

	1940s	1950s
Percent of women age 20–24 never-married	48.0	20.0
Divorce rate (per 1,000 population)	4.3	2.1
Total fertility rate for white women age 20	2.6	3.1
Total fertility rate for nonwhite women age 20	3.2	3.9
Married couples as percent of all families	84.4	87.8

◆ **TABLE 12.2** ◆

The Family in Numbers: The 1940s and 1950s Compared

Note: Most figures were read from graphs and are therefore approximate.

SOURCES: Adapted from Cherlin (1992 [1981]: 9, 19, 21); United States Bureau of the Census (1999a).

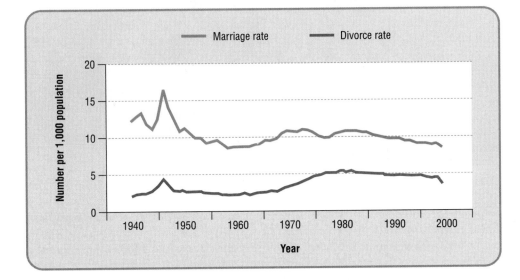

◆ **FIGURE 12.2** ◆

Marriages and Divorces, United States, 1940–1998 (per 1,000 population)

SOURCES: *Monthly Vital Statistics Report* (1995a; 1995b; 1998); *National Vital Statistics Reports* (1999b).

peculiar historical circumstances of the postwar years, noted above, temporarily reversed them (see Figure 12.2).

The functionalists, we may conclude, generalized too hastily from the era they knew best—the period of their own adulthood. Contrary to what they thought, the big picture from the 19th century till the present is that of a gradually weakening nuclear family. Let us now consider the conditions that made other family forms more prevalent.

CONFLICT AND FEMINIST THEORIES

I hadn't really wanted to marry at all. I wanted to make something of myself, not just give it away. But I knew if I didn't marry I would be sorry. Only freaks didn't. I knew I had to do it quickly, too, while there was still a decent selection of men to choose from . . . I was twenty . . .

Though I wanted to be a good wife, from the beginning I found it impossible to subdue my desires. I was in fierce competition with my husband, though Frank, completely absorbed in his own studies, was probably unaware of it. He believed he had married an impulsive girl, even a supergirl, but not a separate, feeling woman . . . Though we had agreed to study like fury till our money ran out and then take turns getting jobs, at bottom we knew it would be he who would get the degrees and I who would get the jobs (Shulman, 1997 [1969]: 163, 173).

This quotation is from *Memoirs of an Ex-Prom Queen*, an exposé of the plight of the "all-American girl" in the 1950s. It shows a side of family life entirely obscured by the functionalists. As the passage suggests, and as the novel establishes in biting and sometimes depressing detail, postwar families did not always operate like the smoothly functioning, happy, white, middle-class, mother-householder, father-breadwinner household

1950s TV classics such as *Father Knows Best* portrayed smoothly functioning, happy, white, middle-class, mother-householder, father-breadwinner families.

portrayed every week in 1950s TV classics such as *Leave it to Beaver, Father Knows Best,* and *Ozzie and Harriet.*

Many men and women felt coerced into getting married, trapped in their families, unable to achieve the harmony, security, and emotional satisfaction they had been promised. As a result, the nuclear family was often a site of frustration and conflict. Surveys show that only about a third of working-class couples and two thirds of middle-class couples were happily married (Barnard, 1972). Wives were less satisfied with marriage than husbands. They reported higher rates of depression, distress, and feelings of inadequacy. Dissatisfaction seems to have been especially high among the millions of women who, during World War II, had operated cranes in steel mills, greased locomotives, riveted the hulls of ships, worked the assembly lines in munitions factories, planted and harvested crops, and felled giant redwoods. They were universally praised for their dedication and industry during the war. Many of them did not want to leave these well-paying jobs that gave them gratification and independence. Management fired most of them anyway and downgraded others to lower paying, "women's" jobs to make room for returning soldiers. The tedium of domestic labor must have been especially difficult for many of these women to accept (Coontz, 1992: 23–41; Skolnick, 1991: 49–74).

Also, many families were simply too poor to participate in the functionalists' celebration of the traditional nuclear unit. For example, to support their families, some 40% of African-American women with small children had to work outside their homes in the 1950s, usually as domestics in upper-middle-class and upper-class white households. A quarter of these black women headed their own households. Thus, to a degree not recognized by the functionalists, the existence of the traditional nuclear family among well-to-do whites depended in part on many black families *not* assuming the traditional nuclear form.

Unlike the functionalists, Marxists had long seen the traditional nuclear family as a site of gender conflict and a basis for the perpetuation of social inequality. In the 19th century, Marx's close friend and co-author, Friedrich Engels, argued that the traditional nuclear family emerged along with inequalities of wealth. For once wealth was concentrated in the hands of a man, wrote Engels, he became concerned about how to transmit it to his children, particularly his sons. How could a man safely pass on an inheritance, asked

Engels? Only by controlling his wife sexually and economically. Economic control ensured that the man's property would not be squandered and would remain his and his alone. Sexual control, in the form of enforced female monogamy, ensured that his property would be transmitted only to his offspring. It follows from Engels's analysis that only the elimination of private property and the creation of economic equality—in a word, communism—can bring an end to the traditional nuclear family and the arrival of gender equality (Engels, 1970 [1884]: 138–9).

Engels was right to note the long history of male economic and sexual domination in the traditional nuclear family. After all, a hundred years ago in the United States, any money a wife might earn typically belonged to her husband. As recently as 40 years ago, an American wife could not rent a car, take a loan, or sign a contract without her husband's permission. It was only about 25 years ago that it became illegal for a husband to rape his wife.

However, Engels was wrong to think that communism would eliminate gender inequality in the family. Gender inequality is as common in societies that call themselves communist as in those that call themselves capitalist. For example, as one American researcher concluded, the Soviet Union left "intact the fundamental family structures, authority relations, and socialization patterns crucial to personality formation and sex-role differentiation. Only a genuine sexual revolution (or, as we prefer to put it, a "gender revolution") could have shattered these patterns and made possible the real emancipation of women" (Lapidus, 1978: 7).

Because gender inequality exists in noncapitalist (including precapitalist) societies, most feminists believe something other than, or in addition to, capitalism accounts for gender inequality. In their view, *patriarchy*—male dominance and norms justifying that dominance—is more deeply rooted in the economic, military, and cultural history of humankind than the classical Marxist account allows (see Chapter 9, "Sexuality and Gender"). For them, only a "genuine gender revolution" can alter this state of affairs.

Just such a revolution in family structures, authority relations, and socialization patterns picked up steam in the United States and other Western countries about 40 years ago, although its roots extend back to the 18th century. As you will see, the revolution is evident in the rise of romantic love and happiness as bases for marriage, women's increasing control over reproduction due to their use of contraceptives, and women's increasing participation in the system of higher education and the paid labor force, among other factors. We next consider some consequences of the gender revolution for the selection of mates, marital satisfaction, divorce, reproductive choice, housework, and child care. We begin by considering the sociology of mate selection.

In the 1950s, married women often hid their frustrations with family life.

Mimi Matte *Family Outing*

POWER AND FAMILIES

Love and Mate Selection

Most Americans take for granted that marriage ought to be based on love (see Figure 12.3). Our assumption is evident, for example, in the way most popular songs in the United States celebrate love as the sole basis of long-term intimacy and marriage ("Billboard Hot 100," 2000). In contrast, most of us view marriage devoid of love as tragic.

Yet in most societies throughout human history, love has had little to do with marriage. Some languages, such as the Chinese dialect spoken in Shanghai, even lack a word for love. Historically and across cultures, marriages were typically arranged by third parties, not by brides and grooms themselves. The selection of marriage partners was based mainly on calculations intended to maximize the prestige, economic benefits, and political advantages accruing to the families from which the bride and groom came. For a family of modest means, a small dowry might be the chief gain from allowing their son to marry a certain woman. For upper-class families, the benefits were typically bigger but no less strategic. In the early years of industrialization, for example, more than one old aristocratic family in economic decline scrambled to have its offspring marry into a family of the upstart bourgeoisie. Giuseppe di Lampedusa's *The Leopard,* probably the greatest Italian novel of the 20th century, deals in an especially moving way with just such a strategic marriage (di Lampedusa, 1991 [1958]).

The idea that love should be important in the choice of a marriage partner first gained currency in 18th-century England with the rise of liberalism and individualism, philosophies that stressed freedom of the individual over community welfare (Stone, 1977). However, the intimate linkage between love and marriage that we know today emerged only in the early 20th century, when Hollywood and the advertising industry began to promote self-gratification on a grand scale. For these new spinners of fantasy and desire, an important aspect of self-gratification was heterosexual romance leading to marriage (Rapp and Ross, 1986).

✦ **FIGURE 12.3** ✦

The Components of Love
According to psychologist Robert Sternberg, love can be built from three components: passion (erotic attraction), intimacy (confiding in others and shared feelings), and commitment (intention to remain in the relationship). In actual relationships, these components may be combined in various ways to produce different kinds of love. The fullest love requires all three components. Research shows that, in long-term relationships, passion peaks fairly quickly and then tapers off. Intimacy rises more gradually but remains at a higher plateau. Commitment develops most gradually but also plateaus at a high level.

SOURCE: Sternberg (1986).

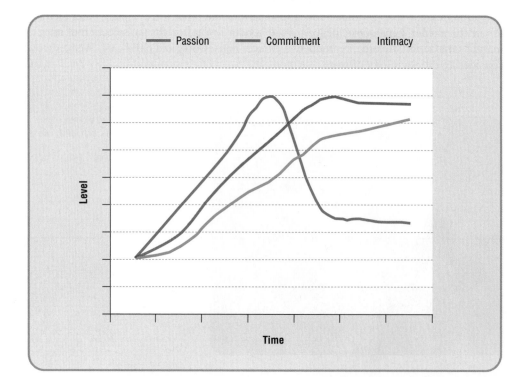

Hollywood glamorized heterosexual, romantic love and solidified the intimate linkage between love and marriage that we know today.

Clark Gable and Vivien Leigh in *Gone with the Wind* (1939).

Still, it would be a big mistake to think that love alone determines mate selection in our society—far from it. Three sets of social forces influence whom you are likely to fall in love with and marry (Kalmijn, 1998: 398–404):

1. *Marriage resources.* Potential spouses bring certain resources with them to the "marriage market." They use these resources to attract mates and compete against rivals. These resources include financial assets, status, values, tastes, and knowledge. Most people want to maximize the financial assets and status they gain from marriage, and they want a mate who has similar values, tastes, and knowledge. As a result, whom you fall in love with and choose to marry is determined partly by the assets you bring to the marriage market.

2. *Third parties.* A marriage between people from two different groups may threaten the internal cohesion of one or both groups. Therefore, to varying degrees, families, neighborhoods, communities, and religious institutions raise young people to identify with the groups they are members of and think of themselves as different from members of other groups. They may also apply sanctions to young people who threaten to marry outside the group. As a result, whom you fall in love with and choose to marry is determined partly by the influence of these third parties.

3. *Demographic and compositional factors.* The probability of marrying inside one's group increases with the group's size and geographical concentration. Conversely, if you are a member of a small group or a group that is dispersed geographically, you stand a greater chance of having to choose an appropriate mate from outside your group. There may simply be too few "prospects" in your group from which to choose (Brym, 1984; Brym, Gillespie, and Gillis, 1985). In addition, the ratio of men to women in a group influences the degree to which members of each sex marry inside or outside the group. For instance, war and incarceration may eliminate many male group members as potential marriage partners. This may encourage female group members to marry outside the group or forego marriage altogether. Finally, because people usually meet potential spouses in "local marriage markets"—schools, colleges, places of work, neighborhoods, bars, and clubs—the degree to which these settings are socially segregated also influences mate selection. You are more likely to marry outside your group if local marriage markets are socially heterogeneous. As a result, whom you fall in love with and choose to marry is

determined partly by the size, geographical dispersion, and sex ratio of the groups you belong to and the social composition of the local marriage markets you frequent.

As a result of the operation of these three sets of social forces, the process of falling in love and choosing a mate is far from random. The percentage of people who marry inside their group is more than 90% for African Americans, 75% for Asian Americans, 65% for Hispanic Americans, and 25% for European Americans. About 80% of Protestants and Jews, and 60% of Catholics, marry within their group. There is also a fairly strong correlation (about r = .55) between the educational attainment of husbands and wives (Kalmijn, 1998: 406–8). We are freer than ever before to fall in love with and marry anyone we want. As in all things, however, social forces still constrain our choices to varying degrees.

Marital Satisfaction

Just as mate selection came to depend more on romantic love over the years, so marital stability came to depend more on having a happy rather than a merely useful marriage. This change occurred because women in the United States and many other societies have become more autonomous, especially over the past 40 years or so. That is, one aspect of the gender revolution is that women are freer than ever to leave marriages in which they are unhappy.

One factor that contributed to women's autonomy was the introduction of the birth control pill in the 1960s. The birth control pill made it easier for women to delay childbirth and have fewer children. A second factor that contributed to their autonomy was the entry of millions of women into the system of higher education and the paid labor force (Cherlin, 1992 [1981]: 51–2, 56; Collins and Coltrane, 1991: xxv). For once women enjoyed a source of income independent of their husbands, they gained the means to decide the course of their own lives to a greater extent than ever before. A married woman with a job outside the home is less tied to her marriage by economic necessity than a woman who works only at home. If she is deeply dissatisfied with her marriage, she can more easily leave. Reflecting this new reality, laws were changed in the 1960s to make divorce easier and divide property between divorcing spouses more equitably. In 1979, the divorce rate reached a historic high and has declined only a little since then. Women initiate most divorces.

If marital stability now depends largely on marital satisfaction, what are the main factors underlying marital satisfaction? The sociological literature emphasizes five sets of forces (Collins and Coltrane, 1991: 394–406; 454–64):

1. *Economic forces.* Money issues are the most frequent subjects of family quarrels, and money issues loom larger when there isn't enough money to satisfy a family's needs and desires. Accordingly, marital satisfaction tends to fall and the divorce rate to rise as you move down the socioeconomic hierarchy. The lower the social class and the lower the educational level of the spouses, the more likely it is that financial pressures will make them unhappy and the marriage unstable. Marital dissatisfaction and divorce are also more common among groups with high poverty rates. Such groups include spouses who marry in their teens and African Americans. In contrast, the marital satisfaction of both husbands and wives generally *increases* when wives enter the paid labor force. This is mainly because of the beneficial financial effects. However, if *either* spouse spends so much time on the job that he or she neglects the family, marital satisfaction falls.

2. *Divorce laws.* Many surveys show that, on average, married people are happier than unmarried people. Moreover, when people are free to end unhappy marriages and remarry, the average level of happiness increases among married people. Thus, the level of marital happiness has increased in the United States over the past few decades, especially for wives, partly because it has become easier to get a divorce.

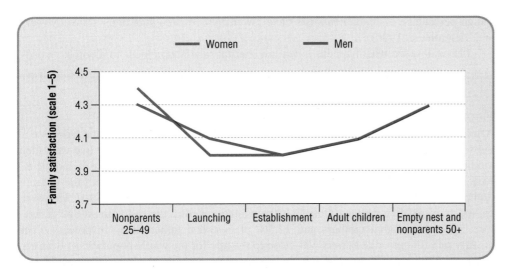

✦ **FIGURE 12.4** ✦
Family Satisfaction and the Family Life Cycle, United States, 1998

SOURCE: Keller (2000).

For the same reason, in countries where getting a divorce is more difficult (e.g., Italy and Spain), husbands and wives tend to be less happy than in countries where getting a divorce is easier (e.g., the United States and Canada) (Stack and Eshleman, 1998).

3. *The family life cycle.* About a quarter of divorces take place in the first 3 years of a first marriage, and half of all divorces take place by the end of the 7th year. However, for marriages that last longer, marital satisfaction reaches a low point after about 15 to 20 years. Marital satisfaction generally starts high, falls when children are born, reaches a low point when children are in their teenage years, and rises again when children reach adulthood (Rollins and Cannon, 1974). Figure 12.4 illustrates the effect of the family life cycle on marital satisfaction using recent survey data. Nonparents and parents whose children have left home (so-called "empty-nesters") enjoy the highest level of marital satisfaction. Parents who are just starting families or who have adult children living at home enjoy intermediate levels of marital satisfaction. Marital satisfaction is lowest during the "establishment" years, when children are attending school. Although most people get married at least partly to have children, it turns out that children, and especially teenagers, usually put big emotional and financial strains on families. This results in relatively low marital satisfaction.

4. *Housework and child care.* Marital happiness is higher among couples who share housework and child care. The farther couples are from an equitable sharing of domestic responsibilities, the more tension there is among all family members (Hochschild with Machung, 1989).

5. *Sex.* Having a good sex life is associated with marital satisfaction. Contrary to popular belief, surveys show that sex generally improves during a marriage. Sexual intercourse is also more enjoyable and frequent among happier couples. From these findings, some experts conclude that general marital happiness leads to sexual compatibility (Collins and Coltrane, 1991: 344). However, the reverse may also be true. Good sex may lead to a good marriage. After all, sexual preferences are deeply rooted in our psyches and our earliest experiences. They cannot easily be altered to suit the wishes of our partners. If spouses are sexually incompatible, they may find it hard to change, even if they communicate well, argue little, and are generally happy on other grounds. On the other hand, if a husband and wife are sexually compatible, they may work harder to resolve other problems in the marriage for the sake of preserving their good sex life. Thus, the relationship between marital satisfaction and sexual compatibility is probably reciprocal. Each factor influences the other.

Religion, we note, has little effect on level of marital satisfaction. But religion does influence the divorce rate. Thus, states with a high percentage of regular churchgoers and a high percentage of fundamentalists have lower divorce rates than other states (Sweezy and Tiefenthaler, 1996).

Let us now see what happens when low marital satisfaction leads to divorce.

Divorce

Economic Effects

After divorce, the most common pattern is a rise in the husband's income and a decline in the wife's. That is because husbands tend to earn more, children typically live with their mother, and support payments are often inadequate. Figure 12.5 shows some economic effects of divorce in the United States. Support payments were awarded to half the 9.9 million custodial mothers and a quarter of the 1.6 million custodial fathers in 1991. However, payments were often meager. Only half the parents received the full amount due. Therefore, 35% of custodial mothers and 12.5% of custodial fathers lived in poverty. (The poverty rate for custodial fathers was close to the rate for the whole population.) Between 1996 and 1998, the Clinton administration created a new felony offense for people who flee across state lines to avoid paying child support. It also developed a new computerized collection system to track parents across state lines, and new penalties and incentives to get states to cooperate in tracking so-called deadbeat parents (Clinton, 1998). These policies are likely to force more delinquent parents to pay child support.

Emotional Effects

While divorce enables spouses to leave unhappy marriages, serious questions have been raised about the emotional consequences of divorce for children, particularly in the long term. Some scholars claim that divorcing parents are simply trading the well-being of their children for their own happiness. What does research say about this issue?

Research shows that children of divorced parents tend to develop behavioral problems and do less well in school than children in intact families. They are more likely to engage in delinquent acts and to abuse drugs and alcohol. They often experience an emotional crisis, particularly in the first 2 years after divorce. What is more, when children of divorced parents become adults, they are less likely than children of nondivorced parents to be happy. They are more likely to suffer health problems, depend on welfare, earn low

✦ **FIGURE 12.5** ✦

The Economic Aftermath of Divorce, United States, by Sex, 1991 (in millions)

SOURCE: Scoon-Rogers and Lester (1995:7).

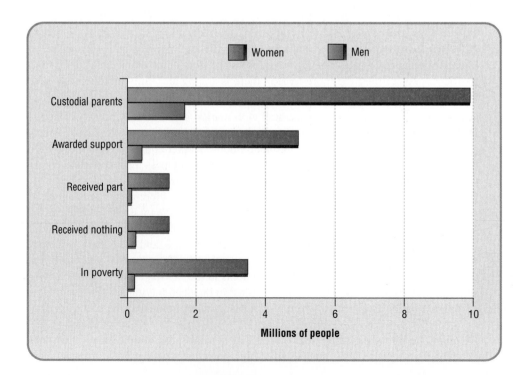

The United States divorce rate reached a historic high in 1979 and has declined since then.

Andrew Benyei *Pink Couch* (1993).

incomes, and experience divorce themselves. In one California study, almost half the children of divorced parents entered adulthood as worried, underachieving, self-deprecating, and sometimes angry young men and women (Wallerstein and Blakeslee, 1989; Wallerstein, Lewis, and Blakeslee, 2000). Clearly, divorce can have serious, long-term, negative consequences for children.

However, much of the research that seems to establish a link between divorce and long-term negative consequences for children is based on families who seek psychological counseling. Such families are a small and unrepresentative minority of the population. By definition, they have more serious emotional problems than the large majority, who do not need psychological counseling after divorce. One must be careful not to generalize from such studies. Another problem with much of this research is that some analysts fail to ask whether factors other than divorce might be responsible for the long-term distress experienced by many children of divorced parents.

Researchers who rely on representative samples and examine the separate effects of many factors on children's well-being provide the best evidence on the consequences of divorce for children. For example, Amato and Keith reanalyzed the results of 92 relevant studies (Amato and Keith, 1991). They showed that on average the overall effect of divorce on children's well-being is not strong and is declining over time. They also found three factors that account for much of the distress among children of divorce:

A high level of parental conflict creates a long-term distress among children. Divorce without parental conflict does children much less harm. Children in divorced families have a higher level of well-being on average than children in high-conflict, intact families.

1. *A high level of parental conflict.* A high level of parental conflict creates long-term distress among children. Divorce without parental conflict does children much less harm. In fact, children in divorced families have a higher level of well-being on average than children in high-conflict, *intact* families. The effect of parental conflict on the long-term well-being of children is substantially greater than the effect of the next two factors discussed by Amato and Keith.

2. *A decline in living standards.* By itself, the economic disadvantage experienced by most children in divorced families exerts a small impact on their well-being. Nonetheless, it is clear that children of divorce who do not experience a decline in living standards suffer less harm.

3. *The absence of a parent.* Children of divorce usually lose a parent as a role model, source of emotional support, practical help, and supervision. By itself, this factor

also has a small effect on children's well-being, even if the child has continued contact with the noncustodial parent.

Subsequent studies confirm these generalizations and add an important observation. Many of the behavioral and adjustment problems experienced by children of divorce existed before the divorce took place. We cannot therefore attribute them to the divorce itself (Cherlin, Furstenberg, Chase-Lansdale, Kiernan, Robins, Morrison, and Teitler, 1991; Entwisle and Alexander, 1995; Furstenberg and Cherlin, 1991; Stewart, Copeland, Chester, Malley, and Barenbaum, 1997).

In sum, claiming that divorcing parents trade the well-being of their children for their own happiness is an exaggeration. High levels of parental conflict have serious negative consequences for children, even when they enter adulthood. In such high-conflict situations, divorce can benefit children. Increased state intervention, such as the initiatives taken by the Clinton administration, can ensure that children of divorce do not experience the decline in living standards that often has long-term negative consequences for them. By itself, the absence of a parent has a small negative effect on children's well-being. But this effect is getting smaller over time, perhaps in part because divorce is so common it is no longer a stigma.

Reproductive Choice

We have seen that the power women gained from working in the paid labor force put them in a position to leave a marriage if it made them deeply unhappy. Another aspect of the gender revolution women are experiencing is that they are increasingly able to decide what happens in the marriage if they stay. For example, women now have more say over whether they will have children and, if so, when they will have them and how many they will have.

Children are increasingly expensive to raise. They no longer give the family economic benefits as they did, say, on the family farm. Most women want to work in the paid labor force, many of them to pursue a career. As a result, most women decide to have fewer children, to have them farther apart, and to have them at an older age. Indeed, 1 out of 20 couples does not have children at all, and among college graduates the figure is 3 out of 20.

Women's reproductive decisions are carried out by means of contraception and abortion. The United States Supreme Court struck down laws prohibiting birth control in 1965. Abortion first became legal in various states around 1970. Today, public opinion polls show that most Americans think women should be free to make their own reproductive choices. A substantial minority, however, is opposed to abortion.

Because Americans are sharply divided on the abortion issue, "right-to-life" versus "pro-choice" activists have been clashing since the 1970s. Right-to-life activists want to repeal laws legalizing abortion. Pro-choice activists want these laws preserved. Both groups have tried to influence public opinion and lawmakers to achieve their aims. For example, as a result of pressure from the right-to-life lobby, RU-486, the so-called morning after pill, was introduced in the United States years after it was available in Western Europe. A few extreme right-to-life activists (almost all men) have resorted to violence.

What are your views on abortion? Do you think your opinions are influenced by your social characteristics (income, education, occupation, religiosity, etc.)? In thinking about this issue, you will find it useful to know that right-to-life activists tend to be homemakers in religious, middle-income families. They argue that life begins at conception. Therefore, they say, abortion destroys human life and is morally indefensible. They advocate adoption instead of abortion. In their opinion, the pro-choice option is selfish, expressing greater concern for career advancement and sexual pleasure than moral responsibility.

In contrast, pro-choice activists tend to be women pursuing their own careers. They are more highly educated, less religious, and better off financially than right-to-life activists. They argue that every woman has the right to choose what happens to her own body and that bearing an unwanted child can harm not only a woman's career but the child too. For example, unwanted children are more likely to be neglected or abused. They are also more likely to get in trouble with the law due to inadequate adult supervision and discipline. Furthermore, according to pro-choice activists, religious doctrines claiming that life begins at conception are arbitrary. In any case, they point out, such ideas have no place in law because they violate the constitutionally guaranteed separation of church and state (Collins and Coltrane, 1991: 510–13). So what is your view? And to what degree is it influenced by your social characteristics?

As sociologists Randall Collins and Scott Coltrane note, it seems likely that a repeal of abortion laws would return us to the situation that existed in the 1960s. They claim that, on a per capita basis, roughly as many abortions took place then as now. But because they were illegal, abortions were expensive, hard to obtain, and posed more dangers to women's health. Clearly, if abortion laws were repealed, poor women and their unwanted children would suffer most. Taxpayers would wind up paying bigger bills for welfare and medical care.

Reproductive Technologies

For most women, exercising reproductive choice means being able to prevent pregnancy and birth by means of contraception and abortion. For some women, however, it means *facilitating* pregnancy and birth by means of reproductive technologies. As many as 15% of couples are infertile. With a declining number of desirable children available for adoption, and a persistent and strong desire by most people to have children, demand is strong for techniques to help infertile couples, some lesbian couples, and some single women have babies.

Fertilizing an egg *in vitro.*

There are four main reproductive technologies. In *artificial insemination,* a donor's sperm is inserted in a woman's vaginal canal or uterus during ovulation. In *surrogate motherhood,* a donor's sperm is used to artificially inseminate a woman who has signed a contract to surrender the child at birth in exchange for a fee. In *in vitro fertilization,* eggs are surgically removed from a woman and joined with sperm in a culture dish, and an embryo is then transferred back to the woman's uterus. Finally, various *screening techniques* are used on sperm and fetuses to increase the chance of giving birth to a baby of the desired sex and end pregnancies deemed medically problematic.

These procedures raise several sociological and ethical issues. We may mention two here (Achilles, 1993). The first problem is discrimination. Most reproductive technologies are expensive. Surrogate mothers charge $10,000 or more to carry a child. *In vitro* fertilization can cost $100,000 or more. Obviously, poor and middle-income earners who happen to be infertile cannot afford these procedures. In addition, there is a strong tendency for members of the medical profession to deny single women and lesbian couples access to reproductive technologies. In other words, the medical community discriminates not just against those of modest means but against nonnuclear families.

A second problem introduced by reproductive technologies is that they render the terms "mother" and "father" obsolete or at least vague. Is the mother the person who donates the egg, carries the child in her uterus, or raises the child? Is the father the person who donates the sperm or raises the child? As these questions suggest, a child conceived through a combination of reproductive technologies and raised by a heterosexual couple could have as many as three mothers and two fathers! This is not just a terminological problem. If it were, we could just introduce new distinctions like "egg mother," "uterine mother," and "social mother" to reflect the new reality. The real problem is social and legal. The question of who has what rights and obligations to the child, and what rights and obligations the child has vis-à-vis each parent, is unclear. This lack of clarity has already caused anguished court battles over child custody (Franklin and Ragone, 1999).

Public debate on a wide scale is needed to decide who will control reproductive technologies and to what ends. On the one hand, reproductive technologies may bring the greatest joy to infertile people. They may also prevent the birth of children with diseases such as muscular dystrophy and multiple sclerosis. On the other hand, reproductive technologies may continue to benefit mainly the well to do, reinforce traditional family forms that are no longer appropriate for many people, and cause endless legal wrangling and heartache.

Housework and Child Care

As we have seen, women's increased paid labor force participation, their increased participation in the system of higher education, and their increased control over reproduction transformed several areas of family life. Despite this far-ranging gender revolution, however, one domain remains largely resistant to change: housework, child care, and senior care. This fact was first documented in detail by sociologist Arlie Hochschild. She showed that even women who work full-time in the paid labor force usually begin a "second shift" when they return home. There, they prepare meals, help with homework, do laundry, clean the toilets, and so forth (Hochschild with Machung, 1989).

To be sure, there has been *some* change as men take a more active role in the day-to-day running of the household. But the change is modest. For example, one study conducted in the late 1980s compared full-time female homemakers in first marriages with wives in first marriages who worked 30 hours or more per week outside the home. The wives working full-time in the paid labor force did only 1 hour and 10 minutes less housework per day than the full-time homemakers. Husbands of women working full-time in the paid labor force did a mere 37 minutes more housework per day than husbands of full-time homemakers (calculated from Demo and Acock, 1993; see Figure 12.6). Studies estimate that, on average, American men now do 20–35% of the housework and child care (Shelton and John, 1996: 299).

Even these figures do not reveal the whole picture, however. Men tend to do low-stress chores that can often wait a day or a week. These jobs include mowing the lawn, repairing the car, and preparing income tax forms. They also play with their children more than they used to. In contrast, women tend to do higher stress chores that cannot wait. These jobs include getting kids dressed and out the door to school every day, preparing dinner by 6:00 p.m., washing clothes twice a week, and the like. In short, the picture is hardly that of a revolution (Harvey, Marshall, and Frederick, 1991).

Two main factors shrink the gender gap in housework, child care, and senior care. First, the smaller the difference between the husband's and the wife's earnings, the more equal the division of household labor. Apparently, women are routinely able to translate earning power into domestic influence. Put bluntly, their increased status enables them to get their husbands to do more around the house. In addition, women who earn relatively high incomes are also able to use some of their money to pay outsiders to do domestic work.

✦ **FIGURE 12.6** ✦

The Division of Domestic Labor by Woman's Work Status, United States, 1987–1988

SOURCE: Demo and Acock (1993:329).

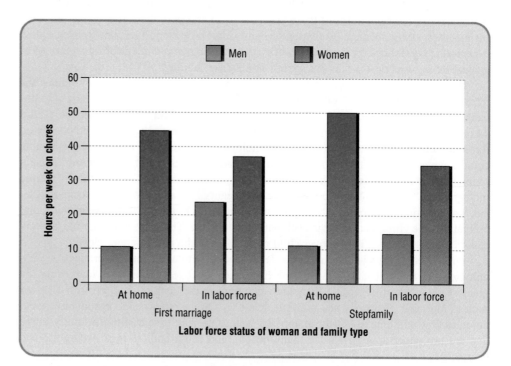

The double day.

Attitude is the second factor that shrinks the gender gap in domestic labor. The more husband and wife agree there *should* be equality in the household division of labor, the more equality there is. Seeing eye-to-eye on this issue is often linked to both spouses having a college education (Greenstein, 1996). Thus, if there is going to be greater equality between men and women in doing household chores, two things have to happen. There must be greater equality between men and women in the paid labor force and broader cultural acceptance of the need for gender equality.

Equality and Wife Abuse

Above we noted that more egalitarian couples are generally more happily married. Said differently, marital satisfaction increases as the statuses of husband and wife approach equality. Does it follow that higher levels of gender equality also result in lower rates of wife abuse? As we will now see, it does.

We must first note that wife abuse is widespread in American society. A 1997 Gallup poll found that 22% of women, compared to 8% of men, reported physical abuse by a spouse or companion at least once in the past (see Figure 12.7). Some of that abuse is severe. In about 2% of American families every year, husbands kick their wives, bite them, hit them with a fist, threaten to use a knife or gun, or use a knife or gun (Straus, 1995).

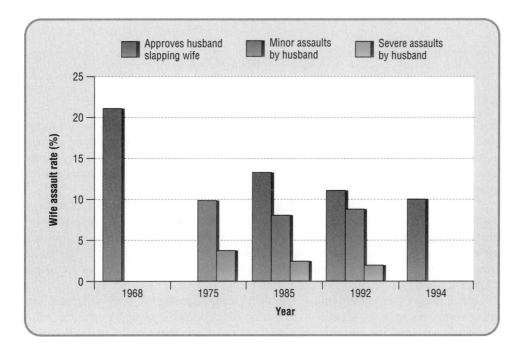

✦ **FIGURE 12.7** ✦

Spousal Violence Against Wives, United States, 1968–1994 (in percent)

Note: Figures include cohabiting but unmarried couples. Data on all indicators are not available for some years shown.

SOURCE: Straus (1995).

Although about as many wives commit such acts of violence against their husbands, the husbands are about seven times more likely to injure their wives physically than vice-versa.

Sociologist Murray A. Straus, the world's leading authority on spouse abuse, has established that severe forms of wife assault occur in all categories of the population. However, he and other researchers have also shown that severe wife assault is most common in lower class, less highly educated families, where men are more likely to believe that male domination is justified. Severe wife abuse is also more common among couples who witnessed their mothers being abused and who were themselves abused when they were children (Gelles, 1997 [1985]; Smith, 1990). Thus, male domination in both childhood socialization and current family organization increases the likelihood of wife abuse.

In addition, Straus (1994) showed that high levels of wife assault are associated with gender inequality in the larger society. He first constructed a measure of wife assault for each state using data from the 1985 National Family Violence Survey (n = 6,002). The measure shows the percentage of couples in each state in which the wife was physically assaulted by her partner during the 12 months preceding the survey. He then used government data to measure gender inequality in each state. His measure of gender inequality taps the economic, educational, political, and legal status of women. He found that wife assault and gender inequality vary proportionately. In other words, as gender equality increases—as women and men become more equal in the larger society—wife assault declines. The conclusion one must draw from this research is clear. The incidence of wife assault is highest where early socialization experiences predispose men to behave aggressively toward women, where norms justify the domination of women, and where a big power imbalance between men and women exists. Figure 12.7 shows a pattern of slowly declining of wife abuse in America between 1968 and 1994. This tendency, like many of the other trends we have discussed, may be attributed in large measure to the growing status and power of women in American society.

Summing up, we can say that conflict theorists and feminists have performed a valuable sociological service by emphasizing the importance of power relations in structuring family life. A substantial body of research shows that the gender revolution of the past 40 years has influenced the way we select mates, our reasons for being satisfied or dissatisfied with marriage, our propensity to divorce, the reproductive choices women make, the distribution of housework and child care, and the level of wife abuse—in short, all aspects of family life. As you will now learn, the gender revolution has also resulted in a much greater diversity of family forms.

FAMILY DIVERSITY

Sexual Orientation

In 1993, three couples showed up at the Hawaii Department of Health in Honolulu to apply for marriage licenses. They were turned away. The grounds? Improper sexual orientation. Two couples were lesbian, one was gay. According to Hawaiian officials, the couples were not legally entitled to form families. Therefore, marriage licenses could not be granted.

The couples took the case to court. In December 1996, Judge Kevin S. C. Chang of the Honolulu Circuit Court ordered the state to stop denying marriages to same-sex couples because doing so would violate the anti-sex-discrimination provisions of Hawaii's state constitution. The case was subsequently referred to the Supreme Court of Hawaii. In December 1999, the Supreme Court of Hawaii ruled that same-sex marriages are illegal (Supreme Court of the State of Hawai'i, 1999).

After Judge Chang's 1996 ruling, President Clinton signed a law denying federal benefits to same-sex spouses. Sixteen states hastily adopted laws denying recognition of homosexual marriages. Sixteen more states passed similar laws in the next 4 years. However,

One of the lesbian couples who showed up at the Hawaii Department of Health in Honolulu in 1993 to apply for a marriage license. The couple was turned away on the grounds of improper sexual orientation. Their case ultimately made its way to the Supreme Court of Hawaii, which ruled in December 1999 that same-sex marriages are illegal.

the battle continues over whether homosexual couples can form legally recognized unions and receive pension, medical, tax, and other benefits enjoyed by married couples. The debate grew in December 1997, when the state of New Jersey allowed homosexual and unmarried couples to adopt children. It intensified in 2000, when Vermont became the first state to pass a law allowing same-sex couples the legal rights and benefits of marriage.

Cambodia (Kampuchea) and the Netherlands are the only countries in the world that extend full and equal marriage rights to homosexuals. However, as of this writing, seven other countries allow homosexuals to register their partnerships under the law. They also recognize these partnerships as having some or all of the legal rights of marriage. These countries include Denmark (along with its dependency, Greenland), Hungary, Norway, Sweden, France, Iceland, Spain, and Germany. Canada and Slovenia seem close to passing similar legislation. (The Canadian province of British Columbia has already done so.) In the United States, there is more opposition to same-sex marriages than in these other countries. As of this writing, 32 states have passed laws opposing same-sex marriage and a nationwide Harris poll taken in February 2000 shows that about 56% of Americans oppose such unions (see Figure 12.8). Nevertheless, the direction of change is clear. Amid sharp controversy, the legal and social definition of "family" is being broadened to include cohabiting, same-sex partners in long-term relationships (Religious Tolerance.org, 2000).

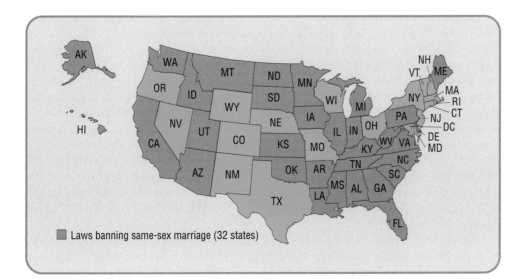

Laws banning same-sex marriage (32 states)

✦ **FIGURE 12.8** ✦
States With Laws Banning Same-Sex Marriages, 2000

SOURCE: National Gay and Lesbian Task Force (2000).

Research shows that most homosexuals, like most heterosexuals, want a long-term, intimate relationship with one other adult (Baca Zinn and Eitzen, 1993: 423). In fact, in Denmark, where homosexual couples can register partnerships under the law, the divorce rate for registered homosexual couples is lower than for heterosexual married couples (ReligiousTolerance.org, 2000). Not surprisingly, therefore, more than 2 million same-sex couples live together in the United States today. At least half of them are raising children. Most of these children are offspring of previous, heterosexual marriages. Some are adopted. Others result from artificial insemination.

Many people believe that children brought up in homosexual families will develop a confused sexual identity, exhibit a tendency to become homosexuals themselves, and suffer discrimination from children and adults in the "straight" community. Unfortunately, there is little research in this area. Much of the research is based on small, unrepresentative samples. Nevertheless, the research findings are consistent. They suggest that children who grow up in homosexual families are much like children who grow up in heterosexual families. For example, a 14-year study assessed 25 young adults who were the offspring of lesbian families and 21 young adults who were the offspring of heterosexual families (Tasker and Golombok, 1997). The researchers found the two groups were equally well adjusted and displayed little difference in sexual orientation. Two respondents from the lesbian families considered themselves lesbians, while all of the respondents from the heterosexual families considered themselves heterosexual.

Homosexual and heterosexual families do differ in some respects. Lesbian couples with children record higher satisfaction with their partnerships than lesbian couples without children. In contrast, among heterosexual couples, it is the childless who record higher marital satisfaction (Koepke, Hare, and Moran, 1992). On average, the partners of lesbian mothers spend more time caring for children than the husbands of heterosexual mothers. Because children usually benefit from adult attention, this must be considered a plus. Finally, homosexual couples tend to be more egalitarian than heterosexual couples, sharing most decision-making and household duties equally. That is because they tend to consciously reject traditional marriage patterns. The fact that they have the same gender socialization and earn about the same income also encourages equality (Baca Zinn and Eitzen, 1993: 424). In sum, available research suggests that raising children in lesbian families has no apparent negative consequences for the children. Indeed, there may be some benefits for all family members.

Race and Adaptations to Poverty

Single-Mother Families

We have seen how families differ from one another due to variations in the sexual orientation of adult family heads. Now let us examine how they vary across racial and ethnic groups in terms of the number of adults who head the family (Baca Zinn and Eitzen, 1993: 109–27; Cherlin 1981 [1992]: 91–123; Collins and Coltrane, 1991: 233–69). Figure 12.9 focuses on the country's two most common family types (two-parent and single-mother) and three largest racial and ethnic categories (white, African American, and Hispanic American). It shows that whites have the lowest incidence of single-mother families. African Americans have by far the highest. In all racial/ethnic groups, the proportion of single-mother families has been increasing in recent decades, but the increase has been most dramatic among African Americans. Thus, among African Americans in 1970, there were 1.9 two-parent families for every single-mother family. By 1997, there were more than 1.3 single-mother families for every two-parent family. The last few years of the 20th century witnessed a reversal in the trend toward more single-mother families in the African-American community. Nevertheless, single-mother families still outnumber two-parent families (Harden, 2001; see also "It's Your Choice" in Chapter 7).

Some single-parent families result from separation, divorce, or death. Others result from people not getting married in the first place. Marriage is an increasingly unpopular institution, particularly among African Americans. This is clear from statistics on births to

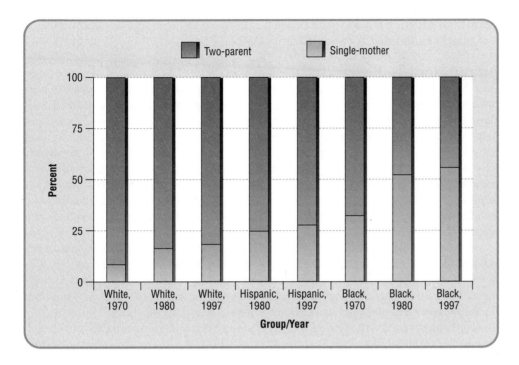

✦ **FIGURE 12.9** ✦
Families with Own Children Under 18 by Race and Hispanic Origin, United States, 1970, 1980, and 1997

Note: 1970 data on Hispanics and single-father families are not available.
SOURCES: Baca Zinn and Eitzen (1993: inside back cover); Bryson and Casper (1998: 4).

unmarried mothers. In 1996 among whites, more than 25% of births were to unmarried mothers. Among Hispanics, the figure was nearly 41%. Among African Americans, it was just under 70% (*Monthly Vital Statistics Report,* 1997).

What accounts for the decline of the two-parent family among African Americans (Cherlin, 1992 [1981])? Although some scholars trace the decline of the African-American two-parent family back to slavery (Jones, 1986), rapid decline began around 1925. By then, the mechanization of the cotton economy in the South had displaced many black agricultural laborers and sharecroppers. They were forced to migrate northward. In the North, they competed fiercely for industrial jobs. Due to discrimination, however, they suffered higher rates of unemployment than any other group in America. Thus, ever since about 1925, proportionately fewer black men have been able to help support a family. As a result, proportionately fewer stable two-person families have formed. Similarly, the decline of manufacturing industries in the Northeast and the movement of many blue-collar jobs to the suburbs in the 1970s and 1980s eliminated many secure, well-paying jobs for blacks and caused their unemployment rate to rise. It is precisely in this period that the rate of increase in African-American single-mother families skyrocketed.

Apart from unemployment, a second factor explaining the decline of the two-parent family among African Americans is the declining ratio of eligible black men to black women. This has three sources. First, largely because of the disadvantaged economic and social position of the African-American community, a disproportionately large number of black men are imprisoned, have been murdered, and suffer from drug addiction (see Chapter 6, "Deviance and Crime"). Second, because the armed forces represent one of the best avenues of upward mobility for African-American men, a disproportionately large number of them have enlisted and been killed in action. Third, a black man is nearly twice as likely as a black woman to marry a nonblack, and intermarriage has increased to nearly 10% of all marriages involving at least one black person. For all these reasons, there are relatively fewer black men available for black women to marry (Anderson, 1999; Cherlin, 1992 [1981]; Wilson, 1987).

The third main factor explaining the decline of the two-parent family in the black community concerns the relative earnings of women and men. In recent decades, the average income of African-American women has increased. Meanwhile, the earning power of African-American men has fallen. As a result, African-American women are more economically independent than ever. On average, they have less to gain in purely economic

By 1997, there were more than 1.3 single-mother families for every two-parent family in the African-American community.

terms from marrying a black man. Economically speaking, marriage has thus become a less attractive alternative for them (Cherlin, 1992 [1981]).

Adaptations to Poverty

Poor African-American women have adapted to harsh economic realities in creative ways that testify to their resilience. In particular, they have developed strong kinship and friendship networks that enable them to survive with few resources. Members of the network help each other with child care. They share money when they have it. They lend each other household items. They give each other hand-me-downs. These adaptations to poverty have a downside, however. If an individual is lucky enough to come into a windfall that could help remove her and her children from poverty—a job or a modest inheritance, for example—the money is quickly used up by the network and the individual remains poor. In other words, the network is a source of survival *and* a shackle that helps keep poor black women impoverished (Stack, 1974).

Just as poor black women can keep their families functioning thanks to the assistance of grandmothers, aunts, sisters, and friends, so Hispanics have developed a family system in which the godfather (*padrino*) and godmother (*madrina*) often act as coparents, providing child care and emotional and financial support as needed. Indeed, Hispanic-American families tend to rely on entire extended kin networks for social support.[1]

Note, however, that the tendency to rely on the extended kin network for social support declines with migration status and upward mobility. The best evidence for this comes from a recent study of extended kin among Mexican Americans (Glick, 1999). Using data from the census and a survey of income dynamics, the study found that reliance on extended kin is stronger among immigrant Hispanics than among those born in the United States. Moreover, reliance on extended kin declines as socioeconomic status increases. American-born, middle- and upper-class Hispanic Americans are less in need of social support from extended kin networks than immigrant, working-class, and poor Hispanic Americans. These findings suggest that the prominence of extended kin networks among Hispanic Americans is at least in part a function of class. Just as the incidence of single-mother families and reliance on extended kin networks drops off sharply among middle-class and upper-middle-class black families, so in the Hispanic community, class position, and not just culture, shapes family structure.

Web Research Projects
Family Values

FAMILY POLICY

Having discussed several aspects of the decline of the traditional nuclear family and the proliferation of diverse family forms, we can now return to the big question posed at the beginning of this chapter: Is the decline of the nuclear family a bad thing for society? Said differently, do two-parent families—particularly those with stay-at-home moms—provide the kind of discipline, role models, help, and middle-class lifestyle that children need to stay out of trouble with the law and grow up to become well-adjusted, productive members of society? Conversely, are family forms other than the traditional nuclear family the main source of teenage crime, poverty, welfare dependency, and other social ills (see Box 12.2)?

The answer suggested by research is clear: yes and no (Houseknecht and Sastry, 1996; Popenoe, 1988; 1991; 1992, 1993, 1996; Sandqvist and Andersson, 1992). Yes, the decline of the traditional nuclear family can be a source of many social problems. No, it doesn't have to be that way.

The United States is a good example of how social problems can emerge from nuclear family decline. Sweden is a good example of how such problems can be averted. Table 12.3 illustrates this. The top panel of Table 12.3 shows that *on most indicators of nuclear*

---

[1]Strong social support from the extended family is, incidentally, one of the main reasons (along with a diet high in vegetable content) why Hispanic Americans have a lower death rate from cancer and heart disease than non-Hispanic whites (Braus, 1994).

Indicators of Nuclear Family "Decline"	USA	Sweden	#1 "Decline"
Median age at first marriage			
Men	26.5	29.4	Sweden
Women	24.4	27.1	Sweden
Percentage of 45–49 population never married			
Men	5.7	15.4	Sweden
Women	5.1	9.1	Sweden
Nonmarital birth rate	25.7	50.9	Sweden
One-parent households with children <15 as % of all households with children <15	25.0	18.0	USA
% of mothers in labor force with children <3	51.0	84.0	Sweden
Total fertility rate	2.0	2.0	Tie
Average household size	2.7	2.2	Sweden
Indicators of Child Well-Being	**USA**	**Sweden**	**#1 Well Being**
Mean reading performance score at 14	5.14	5.29	Sweden
% of children in poverty			
Single-mother households	59.5	5.2	Sweden
Two parent-households	11.1	2.2	Sweden
Death rate of infants from abuse	9.8	0.9	Sweden
Suicide rate for children 15–19 (per 100,000)	11.1	6.2	Sweden
Juvenile delinquency rate (per 100,000)	11.6	12.0	USA
Juvenile drug offense rate (per 100,000)	558	241	Sweden

family decline, Sweden leads the United States. In Sweden, a smaller percentage of people get married. People usually get married at a later age than in the United States. The proportion of births outside of marriage is twice as high as in the United States. A much larger proportion of Swedish than American women with children under the age of 3 work in the paid labor force.

The bottom panel of Table 12.3 shows that *on most measures of children's well-being, Sweden also leads the United States.* Thus, in Sweden, children enjoy higher average reading test scores than in this country. The poverty rate in two-parent families is only one tenth the United States rate, while the poverty rate in single-parent families is only one twelfth as high. The rate of infant abuse is one eleventh the United States rate. The rate of juvenile drug offences is less than half as high. Sweden does have a higher rate of juvenile delinquency than the United States. However, the lead is slight and concerns only minor offences. Overall, then, the decline of the traditional nuclear family has gone farther in Sweden than in the United States, but children are much better off on average. How is this possible?

One possible explanation is that Sweden has something the United States lacks: a substantial family support policy. When a child is born in Sweden, a parent is entitled to 360 days of parental leave at 80% of his or her salary and an additional 90 days at a flat rate. Fathers can take an additional 10 days of leave with pay when the baby is born. Parents are entitled to free consultations at "well baby clinics." Like all citizens of Sweden, they receive free health care from the state-run system. Temporary parental benefits are available for parents with a sick child under the age of 12. One parent can take up to 60 days off per sick child per year at 80% of salary. All parents can send their children to heavily government-subsidized, high-quality day care. Finally, Sweden offers its citizens generous direct cash payments based on the number of children in each family.[2]

Among industrialized countries, the United States stands at the other extreme. Since the Family and Medical Leave Act was passed in 1993, a parent is entitled to 12 weeks of *unpaid* parental leave. About 44 million citizens have no health care coverage. Health care is at a low standard for many millions more. There is no system of state day care and no direct cash payments to families based on the number of children they have. The value of

♦ **TABLE 12.3** ♦

The "Decline" of the Nuclear Family and the Well-Being of Children: The United States and Sweden Compared

SOURCE: Adapted from Houseknecht and Sastry (1996).

Painting class in a state-subsidized day care facility in Stockholm, Sweden.

the dependent deduction on income tax has fallen by nearly 50% in current dollars since the 1940s. Thus, when an unwed Swedish woman has a baby, she knows she can rely on state institutions to maintain her standard of living and help give her child an enriching social and educational environment. When an unwed American woman has a baby, she is pretty much on her own. She stands a good chance of sinking into poverty, with all the negative consequences that has for her and her child.

In the United States, three criticisms are commonly raised against generous family support policies. First, some people say they encourage long-term dependence on welfare, illegitimate births, and the breakup of two-parent families. However, research shows that the divorce rate and the rate of births to unmarried mothers are not higher when welfare payments are more generous (Ruggles, 1997; Sweezy and Tiefenthaler, 1996).

Nor is welfare dependency widespread in America. African-American teen mothers are often thought to be the group most susceptible to chronic welfare dependence. Kathleen Mullan Harris (1997) studied 288 such women in Baltimore. She found that 29% were never on welfare. Twenty percent were on welfare only once and for a very brief time. Twenty-three percent cycled on and off welfare—off when they could find work, on when they couldn't. The remaining 28% were long-term welfare users. However, most of these teen mothers said they wanted a decent job that would allow them to escape life on welfare. That is why half of those on welfare in any given year were concurrently working.

A second criticism of generous family support policies focuses on child care. Some critics say nonfamily child care is bad for children under the age of 3. In their view, only parents can provide the love, interaction, and intellectual stimulation infants and toddlers need for proper social, cognitive, and moral development. The trouble with this argument is that, explicitly or implicitly, it compares the quality of child care in upper-middle-class families with the quality of child care in most existing day-care facilities in the United States. Yet, as we saw in Chapter 9 ("Sexuality and Gender"), existing child care facilities in the United States are often of poor quality. They are characterized by high turnover of poorly paid, poorly trained staff and a high ratio of caregivers to children. When studies compare family care with day care involving a strong curriculum, a stimulating environment, plenty of caregiver warmth, low turnover of well-trained staff, and a low ratio of caregivers to children, they find day care has no negative consequences for children over

---

[2]We are grateful to Gregg Olsen for this information.

BOX 12.2
IT'S YOUR CHOICE

THE PRO-FATHERHOOD CAMPAIGN

If you were watching David Letterman in September 1999, you might have seen a commercial that starts like this: "When young bull elephants from a national park in South Africa were moved to different locations without the presence of an adult male, they began to wantonly kill other animals. When an adult male was relocated with them, the delinquent behavior stopped." From a panoramic view of elephants, the commercial switched to a basketball court where an African-American man is hugging an African-American boy. The voice-over said: "Without the influence of their dads, kids are more likely to get into trouble, too. Just a reminder how important it is for fathers to spend time with their children" (quoted in Davidoff, 1999: 28).

The National Fatherhood Initiative sponsored the commercial. It is part of a nationwide campaign to emphasize the importance of fatherhood to family life in particular and society in general. The 1999 Responsible Fatherhood Act pledged more than $150 million to "allow states to implement programs that promote stable and married families and support responsible fatherhood" (quoted in Davidoff, 1999: 29).

Nobody can disagree with supporting fatherhood and stable family life. However, one problem with pro-fatherhood policies is that they are often intended to replace welfare programs, which are simultaneously being cut by both the federal and state governments. Moreover, critics of pro-fatherhood policies argue that what is essential for the healthy development of children is not just a father, or for that matter even a mother. As psychologists have shown, what is essential is a lasting and loving relationship with at least one adult (Silverstein and Auerbach, 1999). In other words, it is not necessarily the presence of a nuclear family that ensures healthy family life. Insofar as the fatherhood initiative supports only one kind of family, it devalues other forms of family life, including single-parent households, homosexual couples, and so on.

What do you think? Should we support the fatherhood initiative? Or does it lead us to ignore other social problems, such as poverty? Does the focus on fatherhood devalue other forms of family?

---

the age of 1 (Clarke-Stewart, Gruber, and Fitzgerald, 1994). A recent study of more than 6,000 American children found that a mother's employment outside the home does have a very small negative effect on the child's self-esteem, later academic achievement, language development, and compliance. However, this effect was apparent only if the mother returned to work within a few weeks or months of giving birth. Moreover, the negative effects usually disappeared by the time the child reached the age of 5 (Harvey, 1999). Research also shows that day care has some benefits, notably an enhanced ability to make friends. The benefits of high-quality day care are even more evident in low-income families, which often cannot provide the kind of stimulating environment offered by high-quality day care.

The third criticism lodged against generous family support policies is that they are expensive and have to be paid for by high taxes. This is true. Swedes, for example, are more highly taxed than the citizens of any other country. They have made the political decision to pay high taxes, partly to avoid the social problems and associated costs that sometimes emerge when the traditional nuclear family is replaced with other family forms and no institutions are available to help family members in need. The Swedish experience teaches us, then, that there is a clear tradeoff between expensive family support policies and low taxes. It is impossible to have both, and the degree to which any country favors one or the other is a political choice.

SUMMARY

1. The traditional nuclear family consists of a father-provider, mother-homemaker, and at least one child.

2. Today, only a minority of American adults live in traditional nuclear families. Many different family forms have proliferated in recent decades. The frequency of these forms varies by class, race, and sexual orientation.

3. In the 1950s, functionalist theory held that the traditional nuclear family is necessary because it performs essential functions in all societies. However, the theory is inaccurate.

4. Marxists stress how families operate to reproduce class inequality, while feminists stress how they operate to reproduce gender inequality.

5. The entry of women into the paid labor force increases their power to leave unhappy marriages and control whether and when they would have children. It does not, however, have a big effect on the sexual division of labor in families.

6. Marital satisfaction is lower at the bottom of the class structure, where divorce laws are strict, when children reach their teenage years, in families where housework is not shared equally, and among couples who do not have a good sexual relationship.

7. The effects of divorce on children are worst if there is a high level of parental conflict and the children's standard of living drops.

8. Growing up in a lesbian household has no known negative effects on children.

9. People sometimes blame the decline of the traditional nuclear family for increasing poverty, welfare dependence, and crime. However, some countries have adopted policies that largely prevent these problems.

GLOSSARY

The **divorce rate** is the number of divorces that occur in a year for every 1,000 people in the population.

The **extended family** expands the nuclear family "vertically" by adding another generation—one or more of the spouses' parents—to the household.

Marriage is a socially approved, presumably long-term, sexual and economic union between a man and a woman. It involves reciprocal rights and obligations between spouses and between parents and children.

The **marriage rate** is the number of marriages that occur in a year for every 1,000 people in the population.

A **nuclear family** consists of a cohabiting man and woman who maintain a socially approved sexual relationship and have at least one child.

Polygamy expands the nuclear family "horizontally" by adding one or more spouses (usually women) to the household.

The **total fertility rate** is the average number of children born to women of the same age over their lifetime.

A **traditional nuclear family** is a nuclear family in which the husband works outside the home for money and the wife works for free in the home.

QUESTIONS TO CONSIDER

1. Do you agree with the functionalist view that the traditional nuclear family is the ideal family form for the United States today? Why or why not?

2. Ask your grandparents and parents how many people lived in their household when they were your age. Ask them to identify the role of each household member (mother, brother, sister, grandfather, boarder, etc.) and to describe the work done by each member inside and outside the household. Compare the size, composition, and division of labor of your household with that of your grandparents and parents. How have the size, composition, and division of labor of your household changed over three generations? Why have these changes occurred?

WEB RESOURCES

Companion Web Site for This Book

http://sociology.wadsworth.com

Begin by clicking on the Student Resources section of the Web site. Choose "Introduction to Sociology" and finally the Brym and Lie book cover. Next, select the chapter you are currently studying from the pull-down menu. From the Student Resources page you will have easy access to InfoTrac College Edition®, MicroCase Online exercises, additional Web links, and many other resources to aid you in your study of sociology, including practice tests for each chapter.

InfoTrac Search Terms

These search terms are provided to assist you in beginning to conduct research on this topic by visiting http://www.infotraccollege.com/wadsworth.

Divorce	Marriage
Extended family	Nuclear family
Family values	

Recommended Web Sites

"Marriage and Family Processes" at http://www.trinity.edu/mkearl/family.html contains a wide range of valuable resources on family sociology.

You can find Online tests and quizzes concerning love and relationships at http://dir.yahoo.com/society_and_culture/relationships/quizzes_and_tests.

"Kinship and Social Organization" at http://www.umanitoba.ca/anthropology/kintitle.html is an Online interactive tutorial that teaches you about variations in patterns of descent, marriage, and residence using five case studies.

Visit these sites for statistics on families (http://www.cdc.gov/nchs/fastats/marriage.htm), interracial families (http://www. census.gov/population/www/socdemo/interrace.html), divorce (http://www.cdc.gov/nchs/fastats/divorce.htm), sexual behavior (http://purelove.org/statistics/index.html), and same-sex marriage (http://www.religioustolerance.org/hom_marr.htm).

SUGGESTED READINGS

Andrew J. Cherlin. *Marriage, Divorce, Remarriage,* revised and enlarged ed. (Cambridge, MA: Harvard University Press, 1992 [1981]). A concise and rock-solid presentation of some major issues in sociology of the family. Makes excellent use of demographic and survey data.

Stephanie Coontz. *The Way We Really Are: Coming to Terms with America's Changing Families* (New York: Basic Books, 1997). A well-written account from a feminist perspective of how history and sociology can help us come to grips with the problems facing American families today.

David Popenoe. *Life Without Father: Compelling New Evidence that Fatherhood and Marriage are Indispensable for the Good of Children and Society* (New York: Martin Kessler Books, 1996). The authoritative conservative view. This book relates today's major social problems to the breakdown of the traditional nuclear family.

Lillian B. Rubin. *Families on the Fault Line* (New York: Harper Collins, 1994). A compassionate, in-depth account of the strains experienced by working-class families in America due to the economic upheavals of recent decades.

IN THIS CHAPTER, YOU WILL LEARN THAT:

✦ The structure of society and one's place in it influence one's religious beliefs and practices.

✦ Under some circumstances, religion creates societal cohesion, while under other circumstances it promotes social conflict. When religion creates societal cohesion, it also reinforces social inequality.

✦ Religion governs fewer aspects of most people's lives than in the past. However, a religious revival has taken place in the United States and other parts of the world in recent decades and many people still adhere to religious beliefs and practices.

✦ Adults who were brought up in religious families attend religious services more frequently than adults who were brought up in nonreligious families. Attendance also increases with age and varies by race.

✦ Secular schools have largely replaced the church and religious schools as educational institutions. Today, the educational system is second in importance only to the family as an agent of socialization.

✦ The educational system often creates social cohesion. In the process, it also reinforces existing class, racial, and ethnic inequalities.

INTRODUCTION

Robert Brym started writing the first draft of this chapter just after coming home from a funeral. "Roy was a fitness nut," says Robert, "and cycling was his sport. One perfect summer day, he was out training with his team. I wouldn't be surprised if the sunshine and vigorous exercise turned his thoughts to his good fortune. At 41, he was a senior executive in a medium-size mutual fund firm. His boss, who treated him like a son, was grooming him for the presidency of the company. Roy had three vivacious children, ranging in age from 1 to 10, and a beautiful, generous, and highly intelligent wife. He was active in community volunteer work and everyone who knew him admired him. But on this particular summer day, he suddenly didn't feel well. He dropped back from the pack. He then suffered a massive heart attack. Within minutes, he was dead.

"During *shiva*, the ritual week of mourning following the death of a Jew, hundreds of people gathered in the family's home and on the front lawn. I never felt such anguish before. When we heard the steady, slow, clear voice of Roy's 10-year-old son solemnly intoning the mourner's prayer, we all wept. And we asked ourselves and each other the inevitable question: Why?"

In 1902, the great American psychologist William James observed that this question lies at the root of all religious belief. Religion is the common human response to the fact that we all stand at the edge of an abyss. It helps us cope with the terrifying fact that we must die (James, 1976 [1902]: 116). It offers us immortality, the promise of better times to come, and the security of benevolent spirits who look over us. It provides meaning and purpose in a world that might otherwise seem cruel and senseless.

The motivation for religion may be psychological, as James argued. However, the content and intensity of our religious beliefs, and the form and frequency of our religious practices, are influenced by the structure of society and our place in it. In other words, the religious impulse takes literally thousands of forms. It is the task of the sociologist of religion to account for this variation. Why does one religion predominate here, another there? Why is religious belief more fervent at one time than another? Under what circumstances does religion act as a source of social stability and under what circumstances does it act as a force for social change? Are we becoming more or less religious? These are all questions that have occupied the sociologist of religion, and we will touch on all of them here. Note that we will not have anything to say about the truth of religion in general or the value of any religious belief or practice in particular. These are questions of faith, not science. They lie outside the province of sociology.

The cover of *Time* magazine once proclaimed "God is dead." As a sociological observation, the assertion is preposterous. In 1998, 63% of respondents who answered a General Social Survey (GSS) question on the subject had no doubt God exists. Another 29% said they believe in God or some higher power at least some of the time. Only 5% said they don't know whether God exists. And a mere 3% said they don't believe in God (National Opinion Research Center, 1999). By this measure (and by other measures we will examine below), God is still very much alive in America. Nonetheless, as we will show, the scope of religious authority has declined in the United States and many other parts of the world. That is, religion governs fewer aspects of life than it used to. Some Americans still look to religion to deal with all of life's problems. But more and more Americans expect that religion can help them deal with only a restricted range of spiritual issues. Other institutions—medicine, psychiatry, criminal justice, education, and so forth—have grown in importance as the scope of religious authority has declined.

Foremost among these other institutions is the system of education. Organized religion used to be the main purveyor of formal knowledge and the most important agent of socialization apart from the family. Today, the education system is the main purveyor of formal knowledge and the most important agent of socialization apart from the family. It is this displacement of religion by the education system that justifies our analyzing religion and education side-by-side in a single chapter.

When a 1998 nationwide poll asked Americans to indicate the areas in which we are spending too little money, the nation's education system topped the list. Seventy percent of Americans with an opinion on the subject said we are not spending enough on education (National Opinion Research Center, 1999). What are our chief educational concerns? Low academic standards and lack of discipline. We will address both these issues below, paying particular attention to the way they are related to the larger problem of social inequality. By taking this approach, we follow tradition. Sociologists of education have long been interested in the relationship between education and inequality. Some say that education promotes upward mobility. Others argue that education faithfully reproduces inequality generation after generation. As you will see, the evidence offers stronger support for the second argument. Plenty of scope thus remains for educational reform. Through various parent-teacher initiatives, mentoring, and improved funding, individual citizens, politicians, and educational authorities can do much to improve academic standards and deal with the discipline problem, thus helping the school system perform more like a road to opportunity for all members of society. Before dealing with this issue, however, we examine the influence of society on religion.

CLASSICAL APPROACHES IN THE SOCIOLOGY OF RELIGION

Durkheim and the Problem of Order

Somebody once said Super Bowl Sunday is second only to Christmas as a religious holiday in the United States. Do you agree with that opinion? Before making up your mind, consider the following facts. The largest TV audience in history was recorded in 1996, when 138.5 million Americans watched Super Bowl XXX. Audience size fell to 135.2 million for Super Bowl XXXIV in 2000. Still, it is clear that few events attract the attention and enthusiasm of Americans as much as the annual football classic (Columbia Broadcasting System, 2000; "Super Bowl TV Ratings . . . ," 1999).

Apart from drawing a huge audience, the Super Bowl generates a sense of what Durkheim would have called "collective effervescence." That is, the Super Bowl excites us by making us feel part of something larger than us: the St. Louis Rams, the Tennessee Titans, the institution of American football, the competitive spirit of the United States itself. For several hours each year, Super Bowl enthusiasts transcend their everyday lives and experience intense enjoyment by sharing the sentiments and values of a larger collective. In their fervor, they banish thoughts of their own mortality. They gain a glimpse of eternity as they immerse themselves in institutions that will outlast them and athletic feats that people will remember for generations to come.

So do you think the Super Bowl is a religious event? There is no god of the Super Bowl (although some people wanted to elevate St. Louis's Cinderella quarterback Kurt Warner to that position in 2000 after his game-winning touchdown pass late in the game). Nonetheless, the Super Bowl meets Durkheim's definition of a religious experience. Durkheim said that when people live together, they come to share common sentiments and values. These common sentiments and values form a **collective conscience** that is larger than any individual. On occasion, we experience the collective conscience directly. This causes us to distinguish the secular, everyday world of the **profane** from the religious, transcendent world of the **sacred**. We designate certain objects as symbolizing the sacred. Durkheim called these objects **totems.** We invent certain public practices to connect us with the sacred. Durkheim referred to these practices as **rituals.** The effect (or function) of rituals and of religion as a whole is to reinforce social solidarity, said Durkheim. Durkheim would have found support for his theory in research showing that the suicide rate dips during the 2 days preceding Super Bowl Sunday and on Super Bowl Sunday itself, just as it does for the last day of the World Series, July 4th, Thanksgiving Day, and other collective celebrations (Curtis, Loy, and Karnilowicz, 1986). This pattern is consistent with

From a Durkheimian point of view, Super Bowl Sunday can be considered a religious holiday.

Durkheim's theory of suicide, which predicts a lower suicide rate when social solidarity increases (see Chapter 1, "A Sociological Compass").

Durkheim would consider the Super Bowl trophy and the team logos to be totems. The insignias represent groups we identify with. The trophy signifies the qualities that professional football stands for: competitiveness, sportsmanship, excellence, and the value of teamwork. The football game itself is a public ritual that is enacted according to strict rules and conventions. We suspend our everyday lives as we watch the ritual being enacted. The ritual heightens our experience of belonging to certain groups, increases our respect for certain institutions, and strengthens our belief in certain ideas. These groups, institutions, and ideas all transcend us. Thus, the game is a sacred event in Durkheim's terms. It cements society in the way Durkheim said all religions do (Durkheim, 1976 [1915]). Do you agree with this Durkheimian interpretation of the Super Bowl? Why or why not? Do you see any parallels between the Durkheimian analysis of the Super Bowl and sports in your community or college?

Religion, Conflict, and Inequality

Durkheim's theory of religion is a functionalist account. It clearly offers some useful insights into the role of religion in society. However, critics lodge two main criticisms against it. First, it overemphasizes religion's role in maintaining social cohesion. In reality, religion often incites social conflict. Second, when religion does increase social cohesion, it often reinforces social inequality. Durkheim ignores this issue too.

Consider first the role of religion in maintaining inequality. It was Marx who first stressed how religion often tranquilizes the underprivileged into accepting their lot in life. He called religion "the opium of the people" (Marx, 1970 [1843]: 131).

Evidence for Marx's interpretation may be drawn from many times and places. In medieval and early modern Europe, Christianity promoted the view that the Almighty ordains social inequality. In the words of an Anglican verse:

> The rich man at his castle,
> The poor man at his gate.
> God made them high or lowly
> And ordered their estate.

Nor were Western Christians alone in justifying social hierarchy on religious grounds. In Russian and other Slavic languages, the words for rich (*bogati*) and God (*bog*) have the same root. This suggests that wealth is God-given and perhaps even that it makes the wealthy godlike. The Hindu scriptures say that the highest caste sprang from the lips of the supreme creator, the next highest caste from his shoulders, the next highest from his thighs, and the lowest, "polluted" caste from his feet. And the Koran, the holy book of Islam, says that social inequality is due to the will of Allah (Ossowski, 1963: 19–20).

In the United States today, most people do not think of social hierarchy in such rigid terms—quite the opposite. Most people celebrate the alleged *absence* of social hierarchy. This is part of what sociologist Robert Bellah calls our **civil religion,** a set of quasi-religious beliefs and practices that binds the population together and justifies our way of life (Bellah, 1975). When we think of America as a land of golden opportunity, a country in which everyone can realize the American Dream (regardless of race, creed, or color), a place in which individualism and free enterprise ensure the maximum good for the maximum number, we are giving voice to America's civil religion. The National Anthem, the Stars and Stripes, and great public events like the Super Bowl help to make us feel at ease with our way of life, just as the Anglican verse helped the British feel comfortable with their stratification system hundreds of years ago. Paradoxically, however, our civil religion may also help to divert attention from the many inequalities that persist in American society. Strong belief in the existence of equal opportunity, for instance, may lead people to overlook the lack of opportunity that remains in our society (see Chapter 7, "Social Stratification: United States and Global Perspectives"). In this manner, America's civil religion functions much like the old Anglican verse cited above although its content is markedly different.

We can also find plenty of examples to illustrate religion's role in facilitating and promoting conflict. One case that springs immediately to mind is that of the African-American community. In the South in the 1940s, whites sometimes allowed African Americans to sit at the back of their churches. More often, African Americans had to worship in separate churches of their own. These separate black churches formed the breeding ground of the Civil Rights movement in the 1950s and 1960s (Morris, 1984). Their impact was both organizational and inspirational. Organizationally, black churches supplied the ministers who formed the leadership of the civil rights movement. They also supplied the congregations within which marches, boycotts, sit-ins, and other forms of protest were coordinated. In addition, ideas from Christian doctrine inspired the protesters. Among the most powerful of these was the notion that African Americans, like the Jews in Egypt, were slaves who would be freed. It was, after all, Michael—regarded by Christians as the patron saint of the Jews—who rowed the boat ashore. Some white segregationists reacted strongly against efforts at integration, often meeting the peaceful protesters with deadly violence. But the South was never the same again. Religion had helped to promote the conflict needed to make the South a more egalitarian and racially integrated place.

In sum, religion can maintain social order under some circumstances, as Durkheim said. When it does so, however, it often reinforces social inequality. Moreover, under other circumstances religion can promote social conflict.

Weber and the Problem of Social Change

If Durkheim highlighted the way religion contributes to social order, Max Weber stressed the way religion can contribute to social change. Weber captured the core of his argument in a memorable image: If history is like a train, pushed along its tracks by economic and political interests, then religious ideas are like railroad switches, determining exactly which tracks the train will follow (Weber, 1946: 280).

Weber's most famous illustration of his thesis is his short book, *The Protestant Ethic and Spirit of Capitalism*. Like Marx, Weber was interested in explaining the rise of modern capitalism. And, again like Marx, he was prepared to recognize the "fundamental importance of the economic factor" in his explanation (Weber, 1958 [1904–5]:

26). But Weber was also bent on proving the one-sidedness of any *exclusively* economic interpretation.

Weber made his case by first noting that the economic conditions Marx said were necessary for capitalist development existed in Catholic France during the reign of Louis XIV. Yet the wealth generated in France by international trade and commerce tended to be consumed by war and the luxurious lifestyle of the aristocracy rather than invested in the growth of capitalist enterprise. For Weber, what prompted vigorous capitalist development in non-Catholic Europe and North America was a combination of (a) favorable economic conditions such as those discussed by Marx and (b) the spread of certain moral values by the Protestant reformers of the 16th century and their followers.

For specifically religious reasons, wrote Weber, followers of the Protestant theologian John Calvin stressed the need to engage in intense worldly activity, to display industry, punctuality, and frugality in their everyday life. In the view of men like John Wesley and Benjamin Franklin, people could reduce their religious doubts and assure a state of grace by working diligently and living simply. Many Protestants took up this idea. Weber called it the Protestant ethic (Weber, 1958 [1904–5]: 183).

According to Weber, the Protestant ethic had wholly unexpected economic consequences. Where it took root, and where economic conditions were favorable, early capitalist enterprise grew most robustly. Weber made his case even more persuasive by comparing Protestant Western Europe and North America with India and China. In Weber's view, Protestantism was constructed on the foundation of two relatively rational religions: Judaism and Catholicism. These religions were rational in two senses. First, their followers abstained from magic. Second, they engaged in legalistic interpretation of the holy writ. In contrast, said Weber, Buddhism in India and Confucianism in China had strong magical and otherworldly components. This hindered worldly success in competition and capital accumulation. As a result, capitalism developed slowly in Asia, said Weber (Weber, 1963).

In application, two problems have confronted Weber's argument. First, the correlation between the Protestant work ethic and the strength of capitalist development is weaker than Weber thought. In some places, Catholicism has coexisted with vigorous capitalist growth and Protestantism with relative economic stagnation (Samuelsson, 1961 [1957]).

Second, Weber's followers have not always applied the Protestant ethic thesis as carefully as Weber did. For example, since the 1960s, the economies of Taiwan, South Korea, Hong Kong, and Singapore have grown quickly. Some scholars argue that Confucianism

The port of Singapore. Some scholars argue that Confucianism in East Asia acted much like Protestantism in 19th-century Europe, invigorating rapid economic growth by virtue of its strong work ethic. This not only ignores the fact that Weber himself regarded Confucianism as a brake on economic growth in Asia, it also plays down the economic and political forces that stimulated economic development in the region. This is the kind of one-sided explanation that Weber warned against.

in East Asia acted much like Protestantism in 19th-century Europe, invigorating rapid economic growth by virtue of its strong work ethic (Lie, 1998). This not only ignores the fact that Weber himself regarded Confucianism as a brake on economic growth in Asia. In addition, it plays down the economic and political forces that stimulated economic development in the region. This is just the kind of one-sided explanation that Weber warned against (see Chapter 16, "Population, Urbanization, and Development").

Despite these problems, Weber's treatment of the religious factor underlying social change is a useful corrective to Durkheim's emphasis on religion as a source of social stability. Along with Durkheim's work, Weber's contribution stands as one of the most influential insights into the influence of religion on society.

THE RISE, DECLINE, AND PARTIAL REVIVAL OF RELIGION

Secularization

In 1651, the British political philosopher Thomas Hobbes described life as "poore, nasty, brutish, and short" (Hobbes, 1968 [1651]: 150). His description fit the recent past. The standard of living in medieval and early modern Europe was abysmally low. On average, a person lived only about 35 years. The forces of nature and human affairs seemed entirely unpredictable. In this context, magic was popular. It offered easy answers to mysterious, painful, and capricious events.

As material conditions improved, popular belief in magic, astrology, and witchcraft gradually lost ground (Thomas, 1971). Christianity substantially replaced them. The better and more predictable times made Europeans more open to the teachings of organized religion. In addition, the Church campaigned vigorously to stamp out opposing belief systems and practices. The persecution of witches in this era was partly an effort to eliminate competition and establish a Christian monopoly over spiritual life.

The Church succeeded in its efforts. In medieval and early modern Europe, Christianity became a powerful presence in religious affairs, music, art, architecture, literature, and philosophy. Popes and saints were the rock musicians and movie stars of their day. The Church was the center of life in both its spiritual and worldly dimensions. Church authority was supreme in marriage, education, morality, economic affairs, politics, and so forth.

The persecution of witches in the early modern era was partly an effort to eliminate competition and establish a Christian monopoly over spiritual life.
Burning of Witches by Inquisition in a German Marketplace. After a drawing by H. Grobert.

European countries proclaimed official state religions. They persecuted members of religious minorities.

In contrast, a few hundred years later, Max Weber remarked on how the world had become thoroughly "disenchanted." By the turn of the 20th century, he said, scientific and other forms of rationalism were replacing religious authority. His observations formed the basis of what came to be known as the **secularization thesis,** undoubtedly the most widely accepted argument in the sociology of religion until the 1990s. According to the secularization thesis, religious institutions, actions, and consciousness are unlikely to disappear, but they are certainly on the decline worldwide (Tschannen, 1991).

Religious Revival

Despite the consensus about secularization that was still evident in the 1980s, many sociologists modified their judgments in the 1990s. There were two reasons for this. In the first place, accumulated survey evidence showed that religion was not in an advanced state of decay. Actually, in many places, such as the United States, it was in robust health (Greeley, 1989). Consider in this connection Table 13.1 and Figure 13.1, which contain data from

✦ **TABLE 13.1** ✦

The Importance of God and Church Attendance in 12 Post-industrial Societies, 1990 (in percent)

Note: All numbers are rounded. For column 1, the question read: "And how important is God in your life? (10 means very important and 1 means not at all important)". The figures indicate the percentage who indicated scores of 8, 9 or 10. For column 2, the question read: "Apart from weddings, funerals and christenings, about how often do you attend religious services these days?"

SOURCE: The World Values Survey, cited in Nevitte (1996: 210).

	God Is Important in My Life	I Attend Religious Services Once a Week or More
United States	70	44
Ireland	65	81
Northern Ireland	63	50
Canada	51	27
Italy	53	38
Spain	36	29
Belgium	30	27
Britain	28	14
West Germany	30	18
Netherlands	27	20
France	20	10
Denmark	13	3

✦ **FIGURE 13.1** ✦

The Social Condition of Religion, United States, 1972–1998 (in percent; n = 37,724)

SOURCE: National Opinion Research Center (1999).

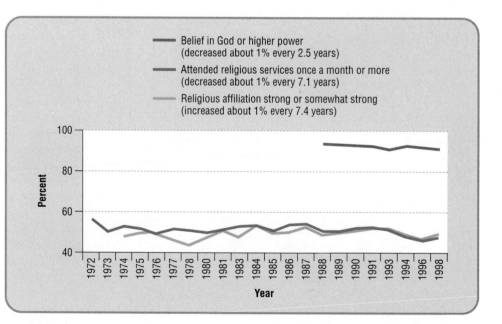

Belief in God or higher power (decreased about 1% every 2.5 years)

Attended religious services once a month or more (decreased about 1% every 7.1 years)

Religious affiliation strong or somewhat strong (increased about 1% every 7.4 years)

two of the most authoritative polls in the world. The GSS has been conducted out of the University of Chicago every year since 1972. The cross-national World Values Survey (WVS) was conducted out of the University of Michigan in two waves, one in 1981, the second in 1990. (The third wave of the WVS was conducted in 1999–2000 and its results had not been made public as this book went to press.)

Table 13.1 compares the United States with 11 other postindustrial countries. In 1990, the United States ranked first in the percentage of respondents who said God is important to them. It ranked third (next to Ireland and Northern Ireland) in the percentage of people who attend religious services once a week or more. By these measures, the United States is one of most religious postindustrial countries in the world.

Figure 13.1 shows the resilience of religion in America over time. Whether we focus on the percentage of Americans who believe in God or some higher power, the percentage who attend religious services once a month or more, or the percentage who claim their religious affiliation is "strong" or "somewhat strong," we see little change over the past quarter of a century. In 1998, about 90% of Americans believed in God or some higher power, and approximately 50% attended religious services frequently and felt at least somewhat strongly affiliated to their religion. True, the number of people who frequently attend religious services fell at the rate of 1% every 7.1 years between 1972 and 1998. The number of people who believe in God or some higher power fell by 1% every 2.5 years between 1988 and 1998. However, these rates of decline are gradual. And they must be balanced against the fact that the number of people who felt at least somewhat strongly affiliated to their religion *increased* by 1% every 7.4 years between 1972 and 1998.

This last statistic reflects the intensification of religious belief and practice among a large minority of Americans in recent decades. This is the second circumstance that caused many sociologists to revise their judgment about the secularization thesis. Since the 1960s, fundamentalist religious organizations have rapidly increased their membership, especially among Protestants (Finke and Starke, 1992). **Fundamentalists** interpret their scriptures literally, seek to establish a direct, personal relationship with the higher being(s) they worship, and are relatively intolerant of nonfundamentalists (Hunter, 1991). In the United States, Christian fundamentalists often support conservative social and political issues, which is why religion in American politics is resurgent (Bruce, 1988). For example, in 1998 only one of four Americans belonging to a fundamentalist denomination agreed with the statement that abortion is acceptable if "the woman wants it for any reason." In contrast, nearly two thirds of Americans belonging to a liberal denomination agreed

Since the 1960s, fundamentalist religious organizations have rapidly increased their membership, especially among Protestants.

with that statement (National Opinion Research Center, 1999). In 1980 and 1984, Jerry Falwell's conservative Moral Majority supported President Reagan's successful bids for the presidency. In 1988, conservative Christian Pat Robertson ran for the Republican presidential nomination, as did Pat Buchanan in 1992. The conservative Christian Coalition continues to lobby hard in Washington today (Smith, 2000).

In this same period, religious movements became dominant forces in many less developed countries. Hindu nationalists form the government in India. An Islamic revival swept Iran and other countries in the Middle East, Africa, and Asia. Religious fundamentalism became a worldwide phenomenon, sometimes leading to violence as a means of establishing fundamentalist ideas and institutions as the bases of nation-states (Juergensmeyer, 2000). Meanwhile, the Catholic Church played a critically important role in undermining communism in Poland, and Catholic "liberation theology" animated the successful fight against right-wing governments in Latin America (Kepel, 1994 [1991]; Segundo, 1976 [1975]; Smith, 1991). All these developments amount to a religious revival that was quite unexpected in, say, 1970.

The Revised Secularization Thesis

The developments reviewed above led some sociologists to revise the secularization thesis in the 1990s. The revisionists acknowledge that religion has become increasingly influential in the lives of some individuals and groups over the past 30 years. They insist, however, that the scope of religious authority has kept on declining in most people's lives. That is, for most people, religion has less and less to say about education, family issues, politics, and economic affairs even though it may continue to be an important source of spiritual belief and practice. In this sense, secularization continues (Chaves, 1994; Yamane, 1997).

According to the **revised secularization thesis,** in most countries, worldly institutions have broken off (or "differentiated") from the institution of religion over time. One such worldly institution is the education system. Religious bodies used to run schools and colleges that are now run almost exclusively by nonreligious authorities. Moreover, like other specialized institutions that separated from the institution of religion, the education system is generally concerned with worldly affairs rather than spiritual matters. In fact, the United States Constitution enshrines the separation of church and state in education (as well as other public spheres). That is why state-sanctioned prayer and the compulsory recitation of prayers are not allowed in public schools.

✦ **TABLE 13.2** ✦

The Perceived Adequacy of the Church in 12 Postindustrial Countries, 1990 (in percent)

Note: All numbers are rounded. The survey question read: "Generally speaking, do you think that your church is giving, in your country, adequate answers to: (a) the moral problems and needs of the individual? (b) the problems of family life?"

SOURCE: The World Values Survey, cited in Nevitte (1996: 215).

	Church Adequate: Moral Problems (change since 1981 in parentheses)	Church Adequate: Family Problems (change since 1981 in parentheses)
United States	67 (−5)	70 (−5)
Ireland	42 (−13)	36 (−15)
Northern Ireland	55 (1)	59 (4)
Canada	55 (−8)	55 (−8)
Italy	52 (3)	45 (−3)
Spain	43 (−3)	43 (3)
Belgium	42 (−6)	37 (−5)
Britain	36 (−2)	38 (−2)
West Germany	40 (−1)	34 (−4)
Netherlands	36 (−2)	33 (−4)
France	38 (−7)	28 (−8)
Denmark	20 (−2)	13 (−1)

The overall effect of the differentiation of secular institutions has been to make religion applicable only to the spiritual part of most people's lives. Because the scope of religious authority has been restricted, people look to religion for moral guidance in everyday life less often than they used to (see Table 13.2). Moreover, most people have turned religion into a personal and private matter rather than one imposed by a powerful, authoritative institution. Said differently, people feel increasingly free to combine beliefs and practices from various sources and traditions to suit their own tastes. As supermodel Cindy Crawford said in a *Redbook* interview in 1992: "I'm religious but in my own personal way. I always say that I have a Cindy Crawford religion—it's my own" (quoted in Yamane, 1997: 116). No statement could more adequately capture the decline of religion as an authoritative institution suffusing all aspects of life.

RELIGION IN THE UNITED STATES

Church, Sect, and Cult

The latest edition of the *Encyclopedia of American Religions* lists more than 2,100 religious groups that are active in this country (Melton, 1996 [1978]). While each of these organizations is undoubtedly unique in some respects, sociologists generally divide religious groups into just three types: churches, sects, and cults (Troeltsch, 1931 [1923]; Stark and Bainbridge, 1979; see Table 13.3).

In the sociological sense of the term, a **church** is any bureaucratic religious organization that has accommodated itself to mainstream society and culture. As a result, it may endure for many hundreds if not thousands of years. The bureaucratic nature of a church is evident in the formal training of its leaders, its strict hierarchy of roles, and its clearly drawn rules and regulations. Its integration into mainstream society is evident in its teachings, which are generally abstract and do not challenge worldly authority. In addition, churches integrate themselves into the mainstream by recruiting members from all classes of society.

Churches take two main forms. First are **ecclesia,** or state-supported churches. For example, Christianity became the state religion in the Roman Empire in 392 CE and Islam is the state religion in Pakistan, Afghanistan, and other countries today. State religions impose advantages on members and disadvantages on nonmembers. Tolerance of other religions is low in societies with ecclesia.

Alternatively, churches may be pluralistic, allowing diversity within the church and expressing tolerance of nonmembers. Pluralism allows churches to increase their appeal by allowing various streams of belief and practice to coexist under their overarching authority. These subgroups are called **denominations.** Baptists, Methodists, Lutherans, Presbyterians, and Episcopalians form the major Protestant denominations in the United States. The major Catholic denominations are Roman Catholic and Orthodox. The major Jewish denominations are Orthodox, Conservative, Reform, and Reconstructionist. The major Muslim denominations are Shiite and Sunni. Many of these denominations are

Web Research Projects
"Cults" and the Brainwashing Controversy

✦ **TABLE 13.3** ✦
Church, Sect, and Cult Compared

	Church	Sect	Cult
Integration into society	High	Medium	Low
Bureaucratization	High	Low	Low
Longevity	High	Low	Low
Leaders	Formally trained	Charismatic	Charismatic
Class base	Mixed	Low	Various but segregated

◆ **FIGURE 13.2** ◆
Religious Preference, United States, 1998 (in percent; n = 2,832)

SOURCE: National Opinion Research Center (1999).

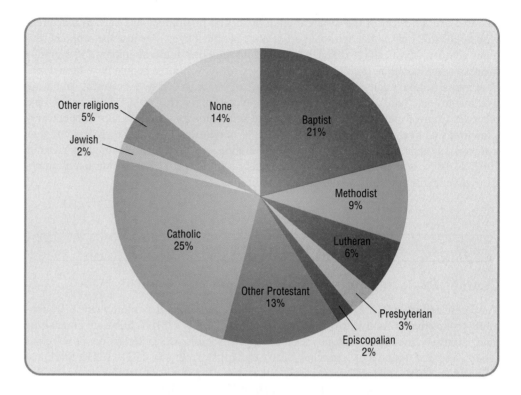

divided into even smaller groups. Figure 13.2 shows the percentage of Americans who belonged to the major religions and Protestant denominations in 1998.

Although, as noted, churches draw their members from all social classes, some churches are more broadly based than others. This is clear from Figure 13.3. According to the General Social Survey, about half the American population defines itself as lower or working class and half as middle or upper class (see Chapter 7, "Stratification: United States and Global Perspectives"). However, the Baptist denomination is more than 60% lower or working class. At the other extreme, about 75% of Presbyterians and Episcopalians, and close to 90% of Jews, are middle and upper class.

Sects often form by breaking away from churches due to disagreement about church doctrine. Sometimes, sect members choose to separate themselves geographically, as the Amish do in their small farming communities in Pennsylvania, Ohio, and Indiana. However, even in urban settings, strictly enforced rules concerning dress, diet, prayer, and intimate contact with outsiders can separate sect members from the larger society. Hasidic Jews in New York and other large American cities prove the viability of this isolation strategy. Sects are less integrated into society and less bureaucratized than churches. They are often led by **charismatic** leaders, men and women who claim to be inspired by supernatural powers and whose followers believe them to be so inspired. These leaders tend to be relatively intolerant of religious opinions other than their own. They tend to recruit like-minded members mainly from lower classes and marginal groups. Worship in sects tends to be highly emotional and based less on abstract principles than immediate personal experience (Stark, 1985: 314). Many sects are short-lived, but those that do persist tend to bureaucratize and turn into churches. If religious organizations are to enjoy a long life, they require rules, regulations, and a clearly defined hierarchy of roles.

Cults are small groups of people deeply committed to a religious vision that rejects mainstream culture and society. Cults are generally led by charismatic individuals. They tend to be class-segregated groups. That is, a cult tends to recruit members from only one segment of the stratification system, high, middle, or low. For example, many American cults today recruit nearly all their members from among the college educated. Some of these cults seek converts almost exclusively on college campuses (Kosmin, 1991). Because

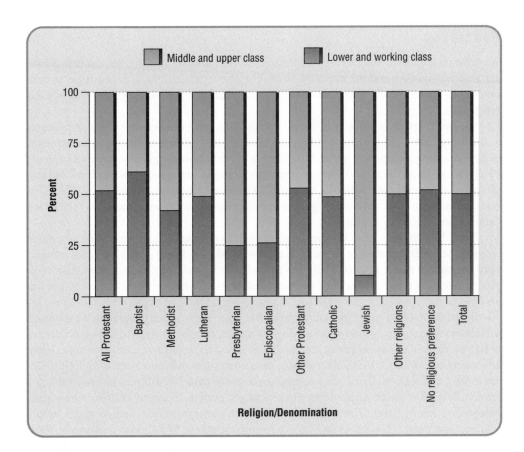

✦ **FIGURE 13.3** ✦
Religious Preference by Class, United States, 1998 (in percent; n = 2,788)

SOURCE: National Opinion Research Center (1999).

Even in urban settings, strictly enforced rules concerning dress, diet, prayer, and intimate contact with outsiders can separate sect members from the larger society.

they propose a radically new way of life, cults tend to recruit few members and soon disappear. There are, however, exceptions—and some extremely important ones at that. Jesus and Mohammed were both charismatic leaders of sects. They were so compelling, they and their teachings were able to inspire a large number of followers, including rulers of states. Their cults were thus transformed into churches.

Religiosity

We have reviewed the major classical theories of religion and society, the modern debate about secularization, and the major types of religious organizations. It is now time to consider some social factors that determine how important religion is to people, that is, their **religiosity.**

We can measure religiosity in various ways. Strength of belief, emotional attachment to a religion, knowledge about a religion, frequency of performing rituals, and frequency of applying religious principles in daily life all indicate how religious a person is (Glock, 1962). Ideally, one ought to examine many measures to get a fully rounded and reliable picture of the social distribution of religiosity. For simplicity's sake, however, we focus on just one measure here: how often people attend religious services. Again, we turn to the General Social Survey for insights.

Table 13.4 divides 1998 GSS respondents into two groups: those who said they attend religious services less than once a month and those who said they attend religious services once a month or more. It then subdivides respondents by their age, race, and whether their mother attended religious services frequently when they were children. (Data on this last variable come from 1989, the most recent year it was measured in the GSS.)

Some fascinating patterns emerge from the data. First, older people attend religious services more frequently than younger people. In fact, people 70 years of age and older are nearly twice as likely as people in the 18–29 age group to attend services frequently. There are two reasons for this. First, older people have more time and more need for religion. Because they are not usually in school, employed in the paid labor force, or busy raising a family, they have more opportunity than younger people to go to church, synagogue, mosque, or temple. And because elderly people are generally closer than younger people to illness and death, they are more likely to require the solace of religion. To a degree, then, attending religious services is a life-cycle phenomenon. That is, we can expect younger people to attend religious services more frequently as they age. But there is another issue at stake here too. Different age groups live through different times, and elderly people reached maturity when religion was a more authoritative force in society than it is today. A person's current religiosity depends partly on whether he or she grew up in more religious times. Thus, although young people are likely to attend services more often as they age, they are unlikely ever to attend services as frequently as elderly people do today.

Second, Table 13.4 shows that frequent church attendance is more common among African Americans than whites. This is undoubtedly because of the central political and

✦ **TABLE 13.4** ✦
Social Factors Influencing How Often Americans Attend Religious Services, 1998 (in percent)

Note: All numbers are rounded.

SOURCE: National Opinion Research Center (1999).

	Attends Religious Services . . .		
Age (n = 2,784)	Less Than Once a Month	Once a Month or More	Total
18–29	66	35	100
30–39	53	47	100
40–49	51	48	100
50–59	49	51	100
60–69	46	54	100
70+	36	64	100
Race (n = 2,599)			
White	54	46	100
Black	37	63	100
Mother's Attendance of Religious Services During Respondent's Youth (n = 966)			
Less than once a month	39	61	100
Once a month or more	18	82	100

cultural role played by the church historically in helping African Americans cope with, and combat, slavery, segregation, discrimination, and prejudice.

Third, respondents whose mothers attended religious services frequently are more likely to do so themselves. Religiosity is partly a *learned* behavior. Whether parents give a child a religious upbringing is likely to have a lasting impact on the child. The New Testament recognizes this: "Whosoever shall not receive the kingdom of God as a little child, he shall not enter therein" (Mark 10:15). Table 13.4 shows that children of frequent churchgoers are more than twice as likely as children of infrequent churchgoers to become frequent churchgoers themselves.

This is by no means an exhaustive list of the factors that determine frequency of attending religious services. However, this brief overview suggests that religiosity depends on opportunity, need, and learning. The people who attend religious services most often are those who were taught to be religious as children, who need organized religion for political reasons or due to their advanced age, and who have the most time to go to services.

Religiosity is partly a learned behavior. Whether parents give a child a religious upbringing is likely to have a lasting impact on the child.

The Future of Religion

A significant religious revival is taking place in the United States and other countries today. However, secularization seems to be the long-term trend, both in the United States and globally. In the 21st century, gradual secularization is likely to continue, although we can expect reversals at various times and in various places. A substantial minority of people will undoubtedly continue to want deep involvement with religious organizations, practices, and beliefs. Most others will probably want to maintain at least a minimal connection to organized religion. This will allow them to give added meaning to important events associated with celebration and mourning throughout the life cycle. The plain fact is, however, that the percentage of Americans expressing no religious preference increased from 3% in 1957 to 14% in 1998. The percentage who reported that religion is important in their lives fell from 75% in 1952 to around 50% in 1998. As Weber said, the world is gradually becoming "disenchanted," even in the postindustrial world's most religious countries, as institutions such as the educational system take over some of the functions formerly performed by religion. In this context, it is important to understand how the rising influence of education came about. That is our next task.

EDUCATION

The Rise of Mass Schooling

By the time you finished high school you had spent nearly 13,000 hours in a classroom. This fact alone suggests that the educational system has displaced organized religion as the main purveyor of formal knowledge. It also suggests that the educational system is second only to the family as an important agent of socialization (see Chapter 4, "Socialization").

Three hundred years ago only a small minority of people learned to read and write. A century ago most people in the world never attended school. Even as late as 1950 only about 10% of the world's countries boasted systems of compulsory mass education (Meyer, Ramirez, and Soysal, 1992). Today the situation is vastly different. Every country in the world has a system of mass schooling. While the average American completed just over 9 years of school in 1950, he or she completed nearly 16 years of schooling by 1995. In 1996, almost 26% of Americans in the 25–64 age group had graduated from college. By this measure, the United States is one of the most highly educated societies in the world (United States Department of Education, 1999: 128).

What accounts for the spread of mass schooling? Sociologists usually highlight three factors: the Protestant Reformation, the democratic forces unleashed by the French and American Revolutions, and industrialization:

✦ The Catholic Church relied on priests to convey dogmas to believers. However, in the early 16th century, Martin Luther, a German monk, began to criticize the Catholic Church. Protestantism grew out of his criticisms. The Protestants believed the Bible alone, and not Church doctrines, should guide Christians. They expected Christians to have more direct contact with the word of God than was allowed by the Catholic Church. Accordingly, Protestants needed to be able to read the scriptures for themselves. The rise of Protestantism was thus a spur to popular literacy.

✦ The populations of France, the United States, and other new democracies demanded access to centers of learning, which had previously been restricted to the wealthy. In the United States, educational opportunities expanded gradually in the 19th and 20th centuries (Katznelson and Weir, 1985; Brint and Karabel, 1989). Women began to enter colleges in the late 19th century and surged into the higher education system in the 1960s. Similarly, African Americans successfully fought for the right to enter schools previously restricted to white students (Kirp, 1982). The 1954 Supreme Court decision on *Brown v. Board of Education* was a watershed in the movement toward educational equality for African Americans. It struck down an 1896 Supreme Court decision that established the doctrine of "separate but equal" or segregation. According to the *Brown v. Board of Education* ruling, separate is unequal by definition (Orfield and Eaton, 1996). Figure 13.4 shows the gradual improvement in the **educational attainment** of ethnic minorities and women between 1960 and 1998. (Educational attainment refers to number of years of school completed, and should not be confused with **educational achievement,** which refers to how much students actually learn.)

✦ Before industrialization, people did not require formal education for many kinds of work (e.g., farming), while they could learn other occupations on the job (e.g., carpentry). In contrast, factories and offices called for literate and numerate workers. Furthermore, education became the key to a person's economic success. Figure 13.5 shows how each successive educational degree boosts earning power in the United States. In 1997, a professional's earning power was more than six times greater than a high school dropout's.

There is an interesting exception to the pattern in Figure 13.5. People who earn a doctorate (PhD) typically earn less than professionals such as doctors and lawyers. Yet PhDs usually have at least as many years of formal education as professionals. As you will learn

✦ **FIGURE 13.4** ✦

Percent With 4 or More Years of College, United States, by Ethnicity, Race, and Gender, 1960–1998

Note: The latest year for which data on Asian Americans are available is 1996.

SOURCE: United States Bureau of the Census (1999f: 169).

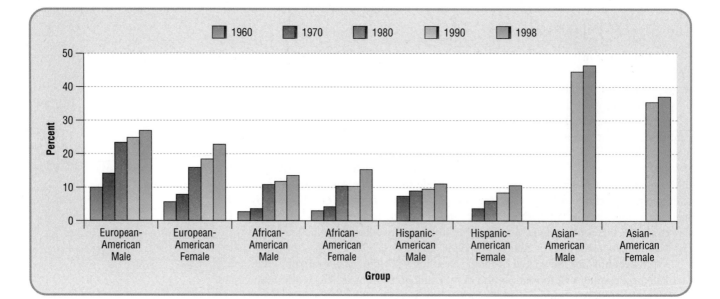

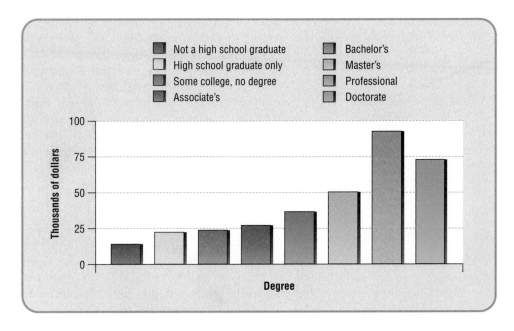

✦ **FIGURE 13.5** ✦
Average Yearly Earnings by Highest Degree Attained, United States, 1997 (in thousands of dollars)

SOURCE: United States Bureau of the Census (1998c).

in the next section, the reasons for this anomaly are connected to the phenomenon of "professionalization."

Credential Inflation and Professionalization

Over the years, the United States and other countries have experienced what Randall Collins calls **credential inflation** (Collins, 1979). That is, it takes even more certificates and diplomas to qualify for a given job. A century ago, for example, many professors, even at the most prestigious universities, did not hold a doctoral degree. Today, most professors are PhDs.

Part of the reason for credential inflation is the increasing technical requirements of many jobs. For example, because aircraft engines and avionics systems are more complex than they were, say, 75 years ago, working as an airplane mechanic today requires more technical expertise. Certification ensures that the airplane mechanic can meet the higher technical demands of the job. As Collins points out, however, in many jobs there is a poor fit between one's credentials and one's specific responsibilities. On-the-job training, not a diploma or a degree, often gives people the skills they need to get the job done. Yet, according to Collins, credential inflation takes place partly because employers find it a convenient sorting mechanism. For example, an employer may assume that an Ivy League graduate has certain manners, attitudes, and tastes that will be useful in a high-profile managerial position. Just as family background used to serve as a way of restricting high-status occupations to certain people, credentials serve that purpose today.

Credential inflation is also fueled by **professionalization.** Professionalization occurs when members of an occupation insist that people earn certain credentials in order to enter the occupation. Professionalization ensures that standards are maintained. It also keeps earnings high. After all, if "too many" people enter a given profession, the cost of services offered by that profession is bound to fall. This helps explain why, on average, physicians and lawyers earn more than college professors with a PhD. The American Medical Association and the American Bar Association are powerful organizations that regulate and effectively limit entry into the medical and legal professions. American professors have never been in a position to form such powerful organizations.

Because professionalization promotes high standards and high earnings, it has spread widely. Even some clowns now consider themselves professionals. Thus, there is a World Clown Association (WCA) that has turned rubber noses and big shoes into a serious

Since professionalization promotes high standards and high earnings, it has spread widely. Even some clowns now consider themselves professionals. Thus, there is a World Clown Association (WCA) that has turned rubber noses and big shoes into a serious business.

business. At its 18th annual conference in 2000, the WCA held seminars on a wide variety of subjects including "character development," "on-target marketing," "incredible bubbles," and "simple but impressive balloons" (Prittie, 2000). Sociologist David K. Brown tells the story of a friend who had been a successful plumber for more than 20 years but tired of the routine and decided to become a clown (Brown, 1995: xvii). His friend quickly found that becoming a clown is not just a matter of buying a costume and acting silly. First, he had to enter the "Intensive Summer Clown Training Institute" at a local college. The Institute awarded him a certificate signifying his competence as a clown. Based on his performance at the Institute, the prestigious Ringling Brothers Clown School in Florida invited him to enroll. Even a clown, it seems, needs credentials these days, and a really good clown can dream of going on to clown graduate school.

The Functions of Education

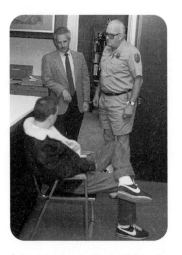

Schools encourage the development of a separate youth culture that often conflicts with parents' values.

Durkheim emphasized the role of schools in socializing the young and in promoting social integration. Human beings, he said, are torn between egoistic needs and moral impulses. Like religion in an earlier era, state-run educational institutions must ensure that the moral side predominates. By instilling a sense of authority, discipline, and morality in children, schools make society cohesive (Durkheim, 1956; 1961).

Contemporary sociologists have acknowledged Durkheim's point and broadened it. They point to a variety of manifest or intended functions performed by schools. Schools try to teach the young to view their nation with pride, respect the law of the land, think of democracy as the best form of government, and value capitalism (Callahan, 1962). Schools also transmit knowledge and culture from generation to generation, fostering a common cultural identity in the process. In recent decades, that common identity has been based on respect for the cultural diversity of American society. In addition, it is often said that schools identify talent and skills, making sure that the brightest and most industrious students are selected and trained for the most challenging jobs. Schools have played a particularly important role in assimilating the disadvantaged, minorities, and immigrants into American society (Fass, 1989).

Schooling performs important latent or unintended functions too. Schools encourage the development of a separate youth culture that often conflicts with parents' values (Coleman, 1961). At the college level, educational institutions bring potential mates together, thus serving as a "marriage market." Schools perform a useful custodial service by keeping children under close surveillance for much of the day and freeing parents to work in the paid labor force. Colleges, by keeping millions of young people temporarily out of the full-time paid labor force, restrict job competition and support wage levels (Bowles and Gintis, 1976). Finally, because they can encourage critical, independent thinking, educational institutions sometimes become "schools of dissent" that challenge authoritarian regimes and promote social change (Brower, 1975; Freire, 1972).

Reproducing Inequality

Web Interactive Exercises
Education and Child Poverty

Many Americans believe we enjoy equal access to basic schooling. They think schools identify and sort students based only on merit. They regard the educational system as an avenue of upward mobility. From this point of view, the best and the brightest are bound to succeed whatever their economic, ethnic, racial, or religious background. The school system is the American Dream in action.

While some sociologists agree with this functionalist assessment, they are in a minority (Bell, 1973). Most sociologists find the conflict perspective on education more credible. In their view, the benefits of education are unequally distributed and tend to **reproduce the existing stratification system** (Jencks et al., 1972). We have already seen how schools help to reproduce the existing system of gender inequality (Chapter 9, "Sexuality and Gender"). Here we may add that they function similarly with respect to class. In other words, children from families at the bottom of the stratification system tend to get tracked

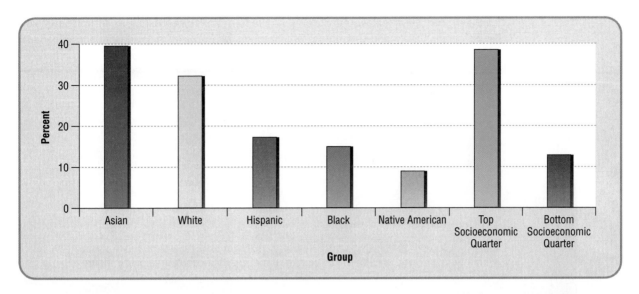

into low- and middle-ability classes. When they finish school, they tend to wind up with jobs that keep them at the bottom of the stratification system. Meanwhile, children from families at the top of the stratification system tend to get tracked into middle- and high-ability classes. When they finish school, they tend to wind up with jobs that keep them at the top of the stratification system. Exceptions to this rule abound, but the basic pattern is clear (see Figure 13.6).

The sociological literature emphasizes three social mechanisms that operate within the school system to reproduce inequality:

♦ *Unequal funding.* Jonathan Kozol (1991) compared average spending in Chicago city schools with spending in an upper-middle-class, suburban Chicago school. He found spending per pupil was 78% higher in the suburban school. The suburban school offered a wide range of college-level courses and boasted the latest audio-visual, computer, photographic, and sporting equipment. Meanwhile, many schools in inner-city Chicago neighborhoods lacked adequate furniture and books. On an average day in the city, even teachers were unavailable for 190 classrooms, affecting 57,000 students. This sort of disparity repeats itself throughout the country. Why? Because school funding is almost always based on local property taxes. In wealthy communities, where property is worth a lot, people can be taxed at a lower rate than in poor communities and still generate more school funding per pupil. (Do the arithmetic: Taxing $100,000 of property at 3% yields $3,000, while taxing $50,000 of property at 5% yields only $2,500.) Thus, wide variations in the wealth of communities and a system of school funding based mainly on local property taxes ensure that most children from poor families learn inadequately in ill-equipped schools while most children from well-to-do families learn well in better-equipped schools.

♦ *Testing and tracking.* Most schools in the United States are composed of children from various socioeconomic, racial, and ethnic backgrounds. Testing and tracking maintain social inequality in these schools. IQ tests sort students, who are then channeled into high-ability, middle-ability, and low-ability classrooms based on test results. As we have seen, this stratifies classrooms by socioeconomic status, race, and ethnicity, much like the larger society (see Figure 13.6).[1]

Nobody denies that students vary in their abilities and that high-ability students require special challenges to reach their full potential. Nor does anyone deny that the underprivileged tend to score low on IQ tests. For example, the difference

♦ **FIGURE 13.6** ♦
Percent of Eighth-Grade United States Public School Students in High-Ability Classes, by Ethnicity, Race, and Socioeconomic Status (n = 14,000)

SOURCE: Kornblum (1997 [1988]: 542).

---

[1]Recall also our discussion in Chapter 8 ("Race and Ethnicity") of how the channeling of many African-American students into sports rather than academic excellence limits their upward mobility (Doberman, 1997).

American schools are characterized by staggering inequalities in the resources available to students.

between the average white person's IQ and the average black person's IQ is 15 points. The controversial question is whether IQ is mainly genetic or social in origin. If IQ is genetic in origin, then it cannot be changed, so improving the quality of schooling for the underprivileged is arguably a waste of money (Herrnstein and Murray, 1994). If IQ is mainly social in origin, IQ tests and tracking only reinforce social differences that could otherwise be reduced by changing the social circumstances of students.

Most sociologists think IQ is mainly a reflection of social standing, not genetic endowment (Fischer et al., 1996). They believe that members of underprivileged groups tend to score low on IQ tests because they don't have the training and the cultural background needed to score high. To support their argument, they point to cases where changing social circumstances resulted in changes in IQ scores. For instance, in the first decades of the 20th century, most Jewish immigrants in America tested well below average on IQ tests. This was sometimes used as an argument against Jewish immigration (Gould, 1996 [1981]; Steinberg, 1989 [1981].) Today, most American Jews test above average in IQ. Even though the genetic makeup of Jews hasn't changed in the past century, why the change in IQ scores? Sociologists point to upward mobility. During the 20th century, most American Jews worked hard and moved up the stratification system. As their fortunes improved, they made sure their children had the skills and the cultural resources needed to do well in school. Average IQ scores rose as the social standing of Jews improved.

Here is an even more dramatic example of the effect of social circumstances on intelligence and school performance. The 300 black and Latino students at the Hostos-Lincoln Academy of Science in the South Bronx were all written off as probable dropouts by their eighth-grade counselors. Yet most seniors in the school now take honors and college-level classes. Eighty percent of them go to college, well above the national average. The reason? The City of New York designated Hostos-Lincoln a special school in 1987. It is small, well equipped, attentive to individual students, and demanding. It stresses team teaching and a safe, family-like environment. Is Hostos-Lincoln an exception? No. A study of 820 American high schools shows that where similar programs are introduced, students from 8th to 12th grades achieve 30% higher scores in math and 24% higher scores in reading compared with students in traditional schools (Hancock, 1994).

Investing more in the education of the underprivileged can produce results (Schiff and Lewontin, 1986).

◆ *The self-fulfilling prophecy of low minority group achievement.* When black and white children begin school, their achievement test scores are fairly close. However, the longer they stay in school, the more black students fall behind. By the sixth grade, blacks in many school districts are two full grades behind whites in achievement. Clearly, something happens in school to increase the gap between black and white students—and thereby help to reproduce the stratification system. It has been suggested that one mechanism responsible for the growing gap between black and white students is the self-fulfilling prophecy that minority students are likely to achieve little (on the self-fulfilling prophecy, see also Chapter 4, "Socialization"). Teachers often expect African Americans, Latinos, and Native Americans to do poorly in school, and they often do poorly as a result. Rather than being treated as a valued person with good prospects, a minority student is often under suspicion of intellectual inferiority and often feels rejected by teachers, white classmates, and the curriculum. Consequently, minority students often cluster together in resentment and defiance of authority. Many of them eventually reject academic achievement as a goal. Discipline problems, ranging from apathy to disruptive and illegal behavior, can result. In contrast, challenging minority students, giving them emotional support and encouragement, giving greater recognition in the curriculum to the accomplishments of their group, creating an environment in which they can relax and achieve— all these strategies explode the self-fulfilling prophecy and improve academic performance (Steele, 1992). Anecdotal evidence supporting this argument may be found in the 1988 movie *Stand and Deliver,* based on the true-life story of high school math teacher Jaime Escalante. Escalante refused to write off his East Los Angeles Chicano pupils as "losers" and inspired them to remarkable achievements in calculus (see Box 13.1).

In sum, schools reproduce the stratification system due to unequal funding, IQ testing and tracking, and the self-fulfilling prophecy that minority students are bound to do poorly. These social mechanisms increase the chance that minority students will in fact earn low grades and wind up with jobs closer to the bottom than the top of the occupational structure.

School Standards

When many students perform poorly in school, you can expect people to start complaining about low educational standards. And, in fact, many Americans believe that the public school system has turned soft if not rotten. They argue that the youth of Japan and South Korea spend long hours concentrating on the basics of math, science, and language. Meanwhile, American students spend fewer hours in school and study more nonbasic subjects that are of little practical value. From this point of view, art classes, drama programs, excessive attention to athletics, and sensitivity to cultural diversity in the American school curriculum are harmful distractions. If students don't spend more school time on subjects that "really" matter, the United States can expect to suffer declining economic competitiveness in the 21st century.

Worrisome evidence apparently proving the inferiority of the American school system was made public by the United States Department of Education in 1997 and 1998. The United States participated in the Third International Math and Science Study (TIMSS). This massive cross-national research effort tested more than half a million students on their math and science skills in grades 4 and 8 and the final year of school. The most arresting finding of the study is presented in Figure 13.7. Compared to grade 4 students in 25 other countries, American fourth graders did respectably well. They placed eighth in their combined math and science scores. However, the relative standing of American students dropped as they progressed through the school system. By the time they reached the last

BOX 13.1
SOCIOLOGY AT THE MOVIES

Edward James Olmos in *Stand and Deliver.*

STAND AND DELIVER (1988)

Garfield High School in East Los Angeles was on the verge of losing its accreditation in the early 1980s because so many of its students were failing. Because they were mostly Chicano and poor, the looming loss of accreditation was widely considered regrettable but hardly surprising.

Enter Jaime Escalante, a tough and engaging idealist who quit his promising job in the computer industry to work twice the hours and earn half the money teaching at Garfield. *Stand and Deliver* is the story of how Escalante, brilliantly played by Edward James Olmos, inspired 18 students to study math in school, after school, on Saturdays, and during the summer so they could take the Advanced Placement Calculus Exam. Only 2% of high school students nationwide even attempt the exam. All 18 of the Garfield students passed, many with high grades. In 2 years, students with poor and failing grades—gang members, students with after-school jobs, students with onerous responsibilities taking care of their younger siblings and elder family members—registered the best performance in the Advanced Placement Calculus Exam in the southern California school system.

Stand and Deliver shows how hard their struggle was. Parents discouraged them. "Boys don't like you if you're too smart," one mother told her daughter. Other stu-

dents scorned them. One gang member requested three copies of his math book—one for the classroom, one for his locker, one for home—so he could avoid being seen walking around with a book in his hand. Teachers doubted them. "Our kids can't handle calculus," one teacher told Escalante at a staff meeting. The testing service even distrusted the students' test results because they were so good, forcing the students to take the test a second time just to prove there was no cheating and that a group of poor Chicano kids living in a *barrio* really could excel in advanced math. Throughout, Escalante persisted. "Students will rise to the level of expectations," he told his fellow teachers. The principal asked: "What do you need, Mr. Escalante?" "*Ganas*," Escalante replied, "that's all we need is *ganas*." *Ganas* is Spanish for desire. There must have been plenty of it around Garfield. The number of Garfield students who passed the Advanced Placement Calculus Exam rose every year from 18 in 1982 to 87 in 1987, the year before the movie was made.

Stand and Deliver is an inspiring story of how students' school performances can be improved if they are encouraged to think highly of themselves, if much is expected of them, and if they are inspired by a dedicated teacher. It also raises the question of how more extraordinary teachers like Jaime Escalante can be attracted to the teaching profession. We hold teachers to high standards. We expect them to get a college education. We entrust them with the intellectual and moral development of impressionable children. We expect them to work with dedication and inspiration. We expect their behavior to be morally impeccable and a model to their students. We maintain these high standards because we think education is so important and we love our children. Yet we pay teachers relatively poorly. How can this be explained? Why is there such a big gap between the high standards to which we hold teachers and the amount we're willing to pay them? What are the consequences for students in the public school system?

year of school, Americans placed 18th out of 21 countries. This means that, on average, graduating American students are on a par with students in Italy and Russia in their math and science skills. They are far behind students in most of Western Europe, Japan, and Canada.

Despite the alarms sounded by the Department of Education and various business and political organizations in the United States, some analysts judge the TIMSS test results misleading (Bracey, 1998; Schrag, 1997). In the first place, most of the countries that participated in the study did not follow sampling guidelines. For political reasons, many of them excluded groups of students whom educational administrators thought would do poorly on the exam. These countries artificially inflated their TIMSS scores. Second, different countries have different kinds of secondary school systems. For instance, some keep students in school for 14 years, while others, like the United States, have 12-year systems.

Many Americans believe that the public school system has turned soft if not rotten. They argue that the youth of Japan and South Korea spend long hours concentrating on the basics of math, science, and language. Meanwhile, American students spend fewer hours in school and study more nonbasic subjects that are of little practical value. As the text shows, the reality is more complex than this simple characterization.

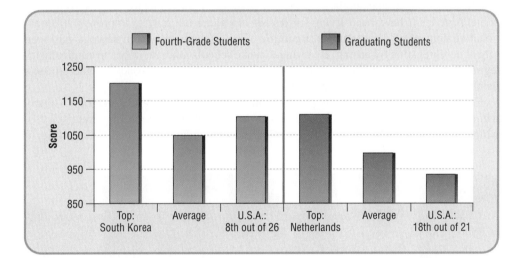

✦ **FIGURE 13.7** ✦

Combined Math and Science Scores, Fourth-Grade and Graduating Students, 1995: United States in Comparative Perspective

SOURCE: United States Department of Education (1997; 1998).

Moreover, some countries have higher dropout rates than the United States has and siphon off poor academic performers to trade schools and job-training programs before they graduate high school. This leaves only the top academic performers in the last year of high school. In contrast, the United States tries to make sure that as many students as possible graduate high school because this enhances the quality of democracy, increases social cohesion in a culturally diverse society, and may improve economic performance. If one compares 12th-grade students from the inclusive United States system with 14th-grade students from an elite school system, the United States students are bound to look inferior. In short, international comparisons, like those made possible by the TIMSS study, are often misleading.

Recognizing the limitations of international comparisons is not an excuse for ignoring the academic flabbiness of many United States schools. It is an opportunity to recognize where the real problem lies. As one educator notes, "[t]he top third of American schools are world-class . . . , the next third are okay, and the bottom third are in terrible shape" (Bracey, 1998). It would probably do students a lot of good if expectations and standards were raised in the entire school system. However, the real crisis in American education can be found not in upper-middle-class suburban schools but in the schools that contain many disadvantaged minority students, most of them in the inner cities. We need to keep this in mind when discussing the sensitive issue of school reform.

Some Solutions to the School Crisis

"At the clubhouse, I work with Lakesha. She is a mentor, which means she knows a lot about computers. When she is not at the clubhouse, she is an engineer. She shows me how to do lots of fun things with computers like controlling LEGO robots. I want to learn about engineering in college."

—LATOYA PERRY, AGE 13

Sharp differences in educational attainment distinguish racial and ethnic groups in the United States. For example, in 1998, a quarter of non-Hispanic whites over the age of 24 had a bachelor's degree. In contrast, only 15% of African Americans and 11% of Hispanic Americans over the age of 24 had a bachelor's degree. Some 84% of non-Hispanic whites over the age of 24 had a high school diploma. This compares to just 76% of African Americans and 56% of Hispanic Americans (United States Bureau of the Census, 1999f: 169; see also Figure 13.4). The high school dropout rate for the current generation of American adolescents also varies sharply by racial and ethnic group. For white non-Hispanics, the dropout rate is 8%. For black students it is 14%, while for Latinos it is 30% (United States Department of Education, 2000: Table 108).

Because educational attainment is the single most important factor that determines income, Americans have been trying for decades to figure out how to improve the educational attainment of disadvantaged minorities. For 40 years, the main strategy has been school desegregation by busing. By trying to make schools more racially and ethnically integrated, many people hoped that the educational attainment of disadvantaged students would rise to the level of the more advantaged students.

Things did not work out as hoped. Instead of accepting busing and integration, many white families moved to all-white suburbs or enrolled their children in private schools, which are more racially and ethnically homogeneous than public schools. Meanwhile, in

When a 1998 nationwide poll asked Americans to indicate the areas in which we are spending too little money, the nation's education system topped the list. Seventy percent of Americans with an opinion on the subject said we are not spending enough on education (see Chapter 16, "Population, Urbanization, and Development," Table 16.5). Here, Sumaya Jackson, 12, center, blows a whistle as she walks in front of her father, Robert, with schoolmates from Manhattan's PS187 as they demonstrate at the New York State Supreme Court building in New York, Tuesday, October 12, 1999. A coalition of public schools advocates argued in court for a change in funding formulas, saying New York City's public school children are being cheated out of money for education. Robert Jackson helped spearhead the effort.

integrated public schools, gains on the part of minority students were limited by racial tensions, competition among academically mismatched students from widely divergent family backgrounds, and related factors (Parillo, Stimson, and Stimson, 1999: 169). Research shows that desegregation closes only 10–20% of the academic gap between black and white students (Jencks et al., 1972).

The limited success of desegregation has convinced many people that, instead of pouring money into busing, a wiser course of action would be to improve the quality of traditionally underfinanced minority schools. Many educators fear that ignoring integration will deny American students from different races and ethnic groups the opportunity to learn to work and live together. Nevertheless, the movement to focus on improving school quality is gaining momentum.

Proposals to improve school quality fall into three main categories (see Box 13.2 for a fourth, recent proposal). First are local initiatives that can improve schooling without changing the distribution or current cost of educational resources. Second are government initiatives that can improve schooling by redistributing existing resources and increasing school budgets. Third are solutions that look outside the school system and stress the need for comprehensive preschools for children, and job training and job creation for disadvantaged parents, as the main ways to encourage upward mobility and end poverty. Let us consider each of these types of reform in turn.

Local Initiatives

We already noted one local initiative in our discussion of the self-fulfilling prophecy that disadvantaged students are bound to do poorly in school. By challenging minority students, giving them emotional support and encouragement, preparing a curriculum that gives more recognition to the accomplishments of their group, and creating an environment in which they can relax and achieve, they do better in school.

BOX 13.2
IT'S YOUR CHOICE

VOUCHERS VERSUS PUBLIC EDUCATION

Despair over the quality of public education in the United States is widespread. Not surprisingly, therefore, people have suggested many ways of improving the schools.

Caroline Minter Hoxby recently proposed an idea that is intriguing but controversial. She believes that many problems in the education system stem from the fact that bad schools are never punished. For instance, a school's budget is unaffected if it has a high dropout rate and many of its students do poorly on their SATs. Many people feel this is as it should be. They think that slashing the budgets of bad schools would only serve to punish their students. Hoxby

has no interest in punishing students for the poor quality of their schools. Instead, she argues that public funding of schools should be cut and parents should be given school vouchers valued at $2,500 to $5,000 a year per child. Parents would be free to enroll their children in any school they want by giving the vouchers to the schools of their choice. In this way, children would be moved out of bad schools and into good ones. Schools' budgets would depend mainly on how many students enroll in them. This system, which effectively promises to privatize public education, would help children go to better schools and it would give the bad schools an incentive to improve, says Hoxby. Eventually, if administrators don't improve the bad schools, they would be forced to close them for lack of funding. Hoxby notes: "It would be like restaurants . . . New, successful schools would take over from old failing schools, and the old ones would disappear" (quoted in Cassidy, 1999: 147).

Increasing choice and enhancing competition are popular ideas. However, some analysts and organizations are critical of

Hoxby's proposal (e.g., Anti-Defamation League, 1999). They argue that the voucher system would undermine public education in the United States and stimulate the growth of private schools, thus making education more unequal than it is already. Unlike public schools, private schools discriminate on a variety of grounds. They reject students due to low academic achievement or discipline problems. Many of them charge tuition fees well in excess of the $2,500–$5,000 vouchers. Moreover, excellent schools in a privatized voucher system are likely to be located where most of them are already, in middle-class and upper-middle-class suburbs, beyond commuting distance for most underprivileged students from the inner city. It seems likely, therefore, that a privatized voucher system would widen the chasm between rich and poor, majority and minority, white and black.

Should there be more choice and competition in the public school system? What should the goals of education be—choice, better performance, or equal opportunity? How would you rank these goals? Why would you rank them in this way?

A second local initiative is the mentoring movement. It involves community members volunteering to work with disadvantaged students in schools, church basements, community centers, and housing projects. There, the mentors tutor students, socialize with them, act as role models, and impart practical skills.

An outstanding example of the mentoring movement is the series of drop-in "computer clubhouses" that have been established since 1993 for underprivileged youth between the ages of 10 and 16 (Resnick and Rusk, 1996). These computer clubhouses are not places of traditional classroom instruction with an emphasis on computers. Nor are they places where computer resources are simply made available to be used in whatever way young people want. Instead of surfing the Web, students make waves. That is, the mentors build on students' existing interests in cartoons, dance, music, and so forth to help them create specific design projects that use computer technology. Emilio sees a laser-light show and wants to create one himself. To do this, he has to glue mirrors to robotic motors and then learn how to write computer code to precisely control the motors' movements. That way, when he reflects laser light off the mirrors, he achieves the effects he wants. He has to learn much mathematical thinking along the way. Paul arrives from Trinidad without ever having used a computer but he likes to draw cartoon characters. Two years later he's learned enough skills at the computer clubhouse to land a part-time job designing Web pages for a local company. He hopes to pursue a career in computer animation and graphic design. Sandi, a Native American girl, wants to learn more about her heritage so she creates a multimedia show using text, graphics, photographs, and sound. The project impresses her mentor, who hopes that her example will stimulate other pupils to create similar projects. Latoya Perry, the 13-year-old African-American girl whose eloquent words opened this section, models herself after her mentor, Latesha, who teaches Latoya how to program robots. Latoya now wants to become an engineer herself. All these success stories come from the first year of operation of the computer clubhouse in Boston. They show how much creating an environment of trust and respect can accomplish. In many cases it is possible to partly or completely erase the effects of material disadvantage by providing adequate resources for education, building on students' existing interests, and developing concrete design projects rather than expecting students either to passively absorb information or learn on their own.

Redistributing and Increasing School Budgets

Although we could tell more success stories, we need to stress that local initiatives can only go so far to improve the quality of education in the United States. Computer clubhouses require expensive equipment, software, and Internet connections. The teaching profession needs higher salaries to attract more inspiring teachers and reduce classroom size. And no amount of teaching cultural diversity is going to buy desks, repair a leaky roof, or get rid of rodents scouring the school grounds for scraps of food. According to a 1998 report by the United States General Accounting Office, the investigative arm of Congress, a third of the country's schools need major repairs or outright replacement (Tornquist, 1998; see Figure 13.8). This requires money. The second set of proposed educational reforms speaks to the need for increased investment in education and redistributing existing resources (Reich, 1991; Thurow, 1999).

Where could the money come from and how could existing funds be redistributed? The current federal budget surplus is one potential source of funds. Part of the surplus could be used to fix dilapidated schools, buy equipment, hire teachers, and raise teachers' salaries to attract even more dedicated people to the profession. It would also be helpful if the federal or state governments collected school taxes and tied them at least in part to people's ability to pay. As in some other postindustrial countries, well-to-do people could be obliged to pay a higher *rate* of school tax than the less well-to-do, and funds for schools could then be distributed more equitably to communities, whatever their wealth.

Economic Reform and Comprehensive Preschools

The problem with the first two proposed solutions for the school crisis is that attempts to implement them have met with limited success. Sociologists began to understand how little

✦ **FIGURE 13.8** ✦

**Decaying Schools in
New York City**

Many of the nation's schools are old
and in a state of disrepair. In New
York City, for example, more than half
the schools are over half a century
old. According to a 1996 report,
20% of New York schools are in
need of immediate major repairs.
Leaky ceilings, broken lights, and
crumbling walls are common. A girl
was killed in 1997 when debris fell
from a dilapidated school in Brooklyn
(Tornquist, 1998). Here, Angelo
Castucci, vice principal of Hawthorne
Avenue School in Newark, N.J.,
stands in a closed classroom on
March 8, 2001, next to a wooden
brace used to keep the ceiling from
falling down. Even though a wing
of the 104-year-old school had to
be closed and exterior scaffolding
installed, the Hawthorne Avenue
School is not on the top ten list of
school construction projects in
Newark.

SOURCE: Tornquist (1998).

schools could do on their own to encourage upward mobility and end poverty in the 1960s, when sociologist James Coleman conducted a monumental survey of American schools (Coleman et al., 1966). Coleman began his research convinced that the educational achievement of black children was due to the underfunding of their schools. What he found was that differences in the quality of schools—measured by assessments of such factors as school facilities and curriculum—accounted at most for about a third of the variation in students' academic performance. At least two thirds of the variation in academic performance was due to inequalities imposed on children by their homes, neighborhoods, and peers.

Three and a half decades later, little research contradicts Coleman's finding. Consider the results of Project Head Start, a federal government program that provides modest educational, medical, and social services to economically disadvantaged preschool children and their families. Head Start was often hailed as a major accomplishment of President Johnson's War on Poverty. Yet follow-up studies found that gains in cognitive and socioemotional functioning registered by Head Start graduates disappeared within about 2 years of leaving the program. Similarly, between 1964 and 2000, Congress dispensed over $100 billion to schools under Title I of the Elementary and Secondary Education Act. Until 1994, most of the money was used for additional instruction for students who were falling behind, especially in reading. Yet a 1997 study conducted for the Department of Education found no academic differences between students who received assistance and those who did not. Money alone, it seems, does not buy educational equality (Traub, 2000: 55).

A Head Start classroom. Research suggests that programs aimed at increasing school budgets and encouraging local school reform initiatives need to be augmented by policies that improve the social environment of young, disadvantaged children *before and outside* school. Head Start is insufficient in this regard.

On the basis of the research just reviewed, we conclude that programs aimed at increasing school budgets and encouraging local school reform initiatives need to be augmented by policies that improve the social environment of young, disadvantaged children *before and outside* school (Hertzman, 2000). Most children do not go to school until they are 5 years old. When they enroll, school takes up less than half their waking day. Academic work takes up only half the school day. What happens the rest of the time, at home and with peers, has a major impact on academic and, eventually, occupational success. Children from disadvantaged homes do better in school if their parents create a healthy, supportive, and academically enriching environment at home and if peers do not lead children to a life of drugs, crime, and disdain for academic achievement. Policies aimed at helping to create these conditions—job training and job creation for parents, comprehensive child and family assistance programs that start when a child is born—would go a long way toward improving the success rate of programs that increase school budgets and encourage local reform initiatives.

A few model comprehensive child and family assistance programs exist. One such program is the Abecedarian Project in North Carolina, funded by the federal government. Another is a state-funded preschool program in Vineland, New Jersey. Children are enrolled in these programs at birth. The programs last all day. The teachers are trained professionals. The programs offer individualized child care, primary health care, nutritional supplements, adult literacy courses, and education and counseling for pregnant teenagers. Classes are small and richly stocked with books, art supplies, and computers. In effect, these programs form alternative communities, replacements for the troubled neighborhoods in which the children and their mothers would otherwise spend their days. They yield impressive results. A 1999 study of the Abecedarian Project analyzed 21-year-old graduates of the program who were enrolled as infants between 1972 and 1977. Ninety-eight percent of them are African American. Two decades after they first entered the preschool program, they scored higher than a control group on math and reading achievement tests. Those graduates who had their own child were nearly 2 years older when the child was born than comparable members of the control group. They were far more likely than members of the control group to be attending an educational institution at the time of the study and far more likely to have ever attended a 4-year college (Campbell and Ramey, 1994; Frank Porter Graham Child Development Center, 1999; see Figure 13.9).

Offering such programs to all poverty-level children in the United States would cost three times more than current Head Start programs, which are far inferior in quality and comprehensiveness. Effective job training and job creation programs for poor adults would

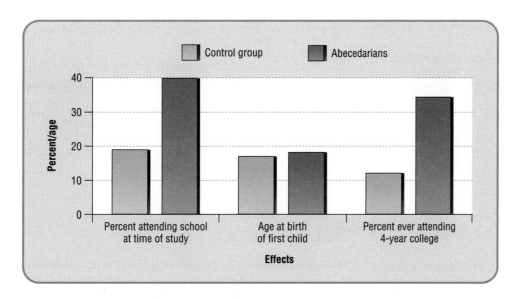

✦ **FIGURE 13.9** ✦

Some Effects of Comprehensive Preschool on 21-Year-Olds (98% African American; n = 111)

SOURCE: Campbell et al. (2002).

also be very costly. Realistically speaking, many people are likely to oppose such reforms at this time. After all, this is an era when school budgets in the inner cities are being cut. It is an era when many parents prefer to send their children to private schools or move to neighborhoods in the suburbs with excellent public schools if they can afford it (Orfield and Eaton, 1996). On the other hand, one must remember that 70% of Americans in the 1998 General Social Survey said too little money is being spent on the nation's schools. Education should be the country's number one priority according to the 1998 GSS (see Chapter 16, "Population, Urbanization, and Development," Table 16.5). This suggests that most Americans still want the education system to live up to its ideals and serve as a path to upward mobility.

SUMMARY

1. Durkheim argued that the main function of religion is to increase social cohesion by providing ritualized opportunities for people to experience the collective conscience.

2. Critics of Durkheim note that he ignored the ways religion can incite social conflict and reinforce social inequality.

3. Weber argued that religion acts like a railroad switch, determining the tracks along which history will be pushed by the force of political and economic interest. Protestantism, for example, invigorated capitalist development.

4. Critics of Weber note that the correlation between economic development and the predominance of Protestantism is not as strong as Weber thought. They also note that some of Weber's followers offer one-sided kind of explanations of the role of religion in economic development, which Weber warned against.

5. The secularization thesis holds that religious institutions, actions, and consciousness are on the decline worldwide.

6. Critics of the secularization thesis point out that there has been a religious revival in the United States and elsewhere over the past 30 years or so. They also note that survey evidence shows religion in the United States is not in an advanced state of decay.

7. The revised secularization thesis recognizes the religious revival and the resilience of religion but still maintains that the scope of religious authority has declined over time. The revisionists say that religion is more and more restricted to the realm of the spiritual; it governs fewer aspects of people's lives and is more a matter of personal choice than it used to be.

8. Frequency of attending religious services is determined by opportunity (how much time people have available for attending), need (whether people are in a social position that increases their desire for spiritual answers to life's problems), and learning (whether people were brought up in a religious household).

9. Secular schools have substantially replaced the church and religious schools as educational institutions. Today, the educational system is second in importance only to the family as an agent of socialization.

10. The rise of mass schooling was stimulated by the Protestant Reformation (which demanded that people have more direct contact with the word of God), the spread of democracy (which involved people demanding access to previously

restricted centers of learning), and industrialization (which required literate and numerate workers).

11. Credential inflation has taken place partly due to increased professionalization.

12. The educational system often creates social cohesion.

13. The educational system also reinforces existing class, racial, and ethnic inequalities. It reproduces the stratification system by means of unequal funding for schools, testing and tracking of students, and the operation of the self-fulfilling prophecy that poor and minority students will do poorly.

14. Educational standards are very low only in the bottom third of United States schools.

15. One type of solution to the school crisis involves local initiatives aimed at improving schools without changing the distribution or total cost of existing educational resources. A second type of solution to the school crisis involves redistributing existing resources and increasing school budgets. A third type of solution to the school crisis involves substantially improving the social environment of young, disadvantaged children before and outside school.

GLOSSARY

Charismatic leaders are men and women who claim to be inspired by supernatural powers and whose followers believe them to be so inspired.

A **church** is a bureaucratic religious organization that has accommodated itself to mainstream society and culture.

A **civil religion** is a set of quasi-religious beliefs and practices that bind a population together and justify its way of life.

The **collective conscience** is composed of the common sentiments and values that people share as a result of living together.

Credential inflation refers to the fact that it takes ever more certificates and diplomas to qualify for a given job.

Cults are small groups of people deeply committed to a religious vision that rejects mainstream culture and society.

Denominations are the various streams of belief and practice that some churches allow to coexist under their overarching authority.

Ecclesia are state-supported churches.

Educational achievement refers to how much students actually learn.

Educational attainment refers to number of years of school students complete.

Fundamentalists interpret their scriptures literally, seek to establish a direct, personal relationship with the higher being(s) they worship, and are relatively intolerant of nonfundamentalists.

The **profane** refers to the secular, everyday world.

Professionalization takes place when members of an occupation insist that people earn certain credentials in order to enter the occupation. Professionalization ensures standards and keeps professional earnings high.

Religiosity refers to how important religion is to people.

The **reproduction of stratification by the school system** involves schools stratifying students so that they tend to wind up in roughly the same position in the class structure as their parents.

The **revised secularization thesis** holds that worldly institutions break off from the institution of religion over time. As a result, religion governs an ever-smaller part of most people's lives and becomes largely a matter of personal choice.

Rituals are public practices designed to connect people to the sacred.

The **sacred** refers to the religious, transcendent world.

Sects usually form by breaking away from churches due to disagreement about church doctrine. Sects are less integrated into society and less bureaucratized than churches. They are often led by charismatic leaders, who tend to be relatively intolerant of religious opinions other than their own.

The **secularization thesis** says that religious institutions, actions, and consciousness are on the decline worldwide.

Totems are objects that symbolize the sacred.

QUESTIONS TO CONSIDER

1. Does the sociological study of religion undermine one's religious faith, make one's religious faith stronger, or have no necessary implications for one's religious faith? On what do you base your opinion? What does your opinion imply about the connection between religion and science in general?

2. What influence did religion have on politics in the 2000 United States Presidential election? Specifically, how influential was the religious right in influencing the platform of the Republican Party and attracting voters to that party? Did Joseph Lieberman's bid for the vice-presidency help or hurt the Democratic Party? (Lieberman is Jewish.) For information on this subject, use the CNN and *Time* search engines at http://www.cnn.com and http://www.time.com/time, respectively.

3. How would you try to solve the problem of unequal access to education? Do you favor any of the older approaches, such as busing children from poor districts to wealthier districts or greater federal control over education budgets? What do you think of the solutions to the education crisis discussed at the end of this chapter? Do you have some suggestions of your own?

WEB RESOURCES

Companion Web Site for This Book

http://sociology.wadsworth.com

Begin by clicking on the Student Resources section of the Web site. Choose "Introduction to Sociology" and finally the Brym and Lie book cover. Next, select the chapter you are currently studying from the pull-down menu. From the Student Resources page you will have easy access to InfoTrac College Edition®, MicroCase Online exercises, additional Web links, and many other resources to aid you in your study of sociology, including practice tests for each chapter.

InfoTrac Search Terms

These search terms are provided to assist you in beginning to conduct research on this topic by visiting http://www.infotraccollege.com/wadsworth.

Cult	**Fundamentalism**
Educational achievement	**Secularization**
Educational attainment	

Recommended Web Sites

Search for "religion" and "education" using the search engine of the General Social Survey at http://www.icpsr.umich.edu/GSS99/ search.htm. The search engine will retrieve dozens of research reports and tables based on this respected, ongoing survey of the American public.

Ontario Consultants on Religious Tolerance is an excellent Web site that provides basic, unbiased information on dozens of religions, religious tolerance and intolerance, religion and science, abortion and religion, and so forth. Visit the site at http://www.religioustolerance.org.

The United States Department of Education Web site at http://www.ed.gov:80/index.html contains a wealth of policy-related material and statistics on education, as does the Web site of the National Center for Education Statistics at http://www.nces. ed.gov.

For an inspiring account of a local initiative to improve the education of underprivileged children in the inner city, see Mitchel Resnick and Natalie Rusk "Access Is Not Enough: Computer Clubhouses in the Inner City," *The American Prospect* (27:July–August, 1996) pp. 60–68, on the World Wide Web at http://epn.org/prospect/27/27resn.html.

SUGGESTED READINGS

Gary Orfield and Susan E. Eaton. *Dismantling Desegregation: The Quiet Reversal of Brown v. Board of Education* (New York: New Press, 1996). Shows how recent government policies have effectively resegregated American schools.

James Traub. "What No School Can Do." *New York Times Magazine* 16 January, 2000: pp. 52–7, 68, 81, 90–1. A sobering account of the limited effects of schools on academic performance and a plea for high-quality preschools as a partial solution to the school crisis.

Max Weber. *The Protestant Ethic and the Spirit of Capitalism* (New York: Charles Scribner's Sons, 1958 [1904–5]). A classic; one of the most important works in the sociology of religion.

David Yamane. "Secularization on Trial: In Defense of a Neosecularization Paradigm." *Journal for the Scientific Study of Religion* (36: 1997) pp. 109–22. A lively analysis of critiques of the secularization thesis and a convincing reformulation of the thesis.

IN THIS CHAPTER, YOU WILL LEARN THAT:

✦ Movies, television, and other mass media sometimes blur the distinction between reality and fantasy.

✦ The mass media are products of the 19th and especially the 20th centuries.

✦ Historically, the growth of the mass media is rooted in the rise of Protestantism, democracy, and capitalism.

✦ The mass media make society more cohesive.

✦ The mass media also foster social inequality.

✦ Although the mass media are influential, audiences filter, interpret, resist, and even reject media messages if they are inconsistent with their beliefs and experiences.

✦ The interaction between producers and consumers of media messages is most evident on the new media frontier formed by the Internet, television, and other mass media.

THE MASS MEDIA

THE SIGNIFICANCE OF THE MASS MEDIA

Illusion Becomes Reality

The turn of the 21st century was thick with movies about the blurred line separating reality from fantasy. *The Truman Show* (1998) gave us Jim Carrey as an insurance sales agent who discovers that everyone in his life is an actor. He is the unwitting subject of a television program that airs 24 hours a day (see Box 14.1). In *The Matrix* (1999), Keanu Reaves finds that his identity and his life are illusions. Like everyone else in the world, Reaves is hardwired to a giant computer that uses humans as an energy source. The computer supplies people with nutrients to keep them alive and simulated realities to keep them happy. Similar blurring between reality and media-generated illusion is evident in *Pleasantville* (1998), *EdTV* (1999), and *Nurse Betty* (2000).

The most disturbing movie in this genre is *American Psycho* (2000). Based on a novel banned in some parts of North America when it was first published in 1991, the movie is the story of Patrick Bateman, Wall Street yuppie by day, cold and meticulous serial killer by night. Unfortunately, the public outcry over the horrifying murder scenes virtually drowned out the book's important sociological point. *American Psycho* is really about how people become victims of the mass media and consumerism. Bateman the serial murderer says he is "used to imagining everything happening the way it occurs in movies." When he kisses his lover, he experiences "the 70 mm image of her lips parting

BOX 14.1
SOCIOLOGY AT THE MOVIES

THE TRUMAN SHOW (1998)

Whenever Truman tries to leave his hometown, something happens to prevent his departure. When he is about to board a boat, he becomes afraid of the water. When he drives down the freeway, the police turn him back on some pretext or other. Although he is a middle-aged, married man, he has never stepped outside of his hometown.

The Truman Show presents the life of Truman, played by Jim Carrey. Truman turns out to be the main character in a television show. He doesn't know it, but since birth his life has been televised in the world's longest-running soap opera. Unlike most soap operas, however, Truman's life is on television 24 hours a day, 365 days a year. He lives in a town created by the television producers. Actresses and actors play all the people in his life—including his wife. A young woman, who once had a bit

The Truman Show, starring Jim Carrey.

part in the show, becomes an activist to free Truman. She even appears on the show to tell Truman the awful truth. Eventually, Truman manages to enter a new life by walking off the vast television set which had been his world until then.

The Truman Show suggests the mass media are so powerful they have collapsed the distinction between illusion and reality. Until he escapes, Truman's reality is the creation of the television producers. His life is synonymous with his television show. Does *The Truman Show* overstate the power and influence of the mass media? Is it really the case that some people have trouble distinguishing fact from fiction, media representations from real-life events? In answering this question, you might try researching the 1999 Columbine high school shootings on the Web to find out how well the teenagers who killed 13 of their fellow students were able to distinguish between reality and media-generated fantasy.

To the degree the mass media shape our reality, is it possible to walk off the set, like Truman did? If so, how? Do people even want to walk off the set? In 2000, there were over 11,000 cameras connected to the World Wide Web, many of them in college dorms, apartments, and frat houses. "Reality TV," spearheaded by programs like *Survivor* and *Big Brother,* was TV's hottest genre. Have the people who have connected cameras to their Web sites and starred on shows like *Survivor* and *Big Brother* become so many willing Trumans?

Many recent films—including *Nurse Betty,* starring Renée Zellweger and Morgan Freeman—suggest that the line between reality and media-generated illusion is becoming blurred.

and the subsequent murmur of 'I want you' in Dolby sound" (Ellis, 1991: 265). In Bateman's mind, his 14 murder victims are mere props in a movie in which he is the star. He feels no more empathy for them than an actor would for any other stage object. The mass media have so completely emptied him of genuine emotion, he even has trouble remembering his victims' names. At the same time, however, the mass media have so successfully infused him with consumer values he can describe his victims' apparel in great detail—styles, brand names, stores where they bought their clothes, even prices. Thus, in *American Psycho,* killer and killed are both victims of consumerism and the mass media.

In different ways, these movies suggest that the fantasy worlds created by the mass media are increasingly the only realities we know, every bit as pervasive and influential as religion was 500 or 600 years ago. Do you think this is an exaggeration dreamed up by filmmakers and novelists? If so, consider that of the 8,760 hours in a year, the average American spends 3,440 of them (39.3%) interacting with the mass media (calculated from United States Bureau of the Census, 1998c: 572). We spend more time watching TV, listening to the radio, going to the movies, reading newspapers and magazines, playing CDs, using the Internet, and so forth, than we do sleeping, working, or going to school. Figure 14.1 shows how we are expected to divide our time among the various mass media in 2001. Figure 14.2 shows how we are expected to divide our money. You might want to keep a tally of your activities for a couple of days to find out how you fit into this pattern of activity. Ask yourself, too, what you get out of your interactions with the mass media. Where do you get your ideas about how to dress, how to style your hair, and what music to listen to? Where do your hopes, aspirations, and dreams come from? If you're like most people, much of your reality is media generated. Media guru Marshall McLuhan, who coined the term "global village" in the early 1960s, said the media are extensions of the human body and mind (McLuhan, 1964). Forty years later, it is perhaps equally valid to claim that the human body and mind are extensions of the mass media (Baudrillard, 1983; 1988; Bourdieu, 1988).

The term **mass media** refers to print, radio, television, and other communication technologies. Often, "mass media" and "mass communication" are used interchangeably to refer to the transmission of information from one person or group to another. The word "mass" implies that the media reach many people. The word "media" signifies that communication does not take place directly through face-to-face interaction. Instead, technology intervenes or mediates in transmitting messages from senders to receivers.

✦ **FIGURE 14.1** ✦
Media Usage, United States, 2001 (hours per capita, projected)

SOURCE: United States Bureau of the Census (1998c: 572).

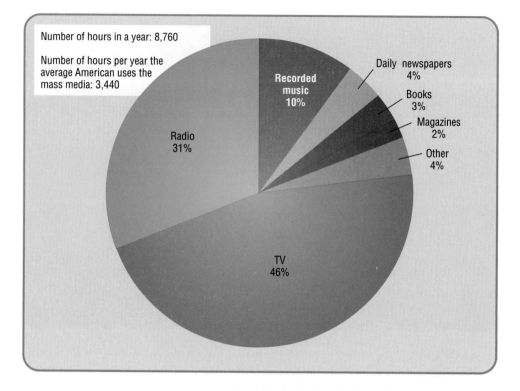

Number of hours in a year: 8,760

Number of hours per year the average American uses the mass media: 3,440

✦ **FIGURE 14.2** ✦
Media Usage, United States, 2001 (dollars per capita, projected)

SOURCE: United States Bureau of the Census (1998c: 572).

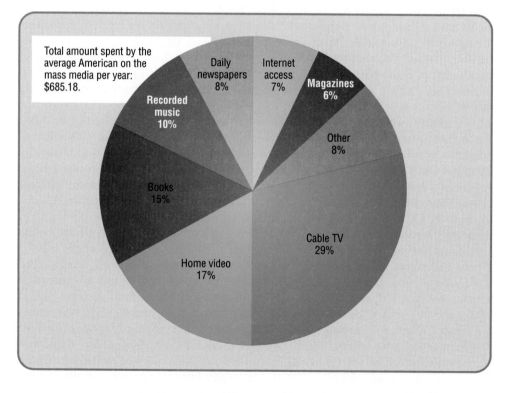

Total amount spent by the average American on the mass media per year: $685.18.

Furthermore, communication via the mass media is usually one-way, or at least one-sided. There are few senders (or producers) and many receivers (or audience members). Thus, most newspapers print a few readers' letters in each edition, but journalists and advertisers write virtually everything else. Ordinary people may appear on the Oprah Winfrey or David Letterman shows, enjoy play-along features on *Who Wants to Be a Millionaire?* and the *Today* show, and even delight in a slice of fame on *Survivor* or *Big Brother.* However, producers choose the guests and create the program content for these programs. Similarly, a handful of people may visit your personal Web site, but Yahoo.com boasted 235 million hits per day in March 1999 (Mosquera, 1999).

Usually, then, members of the audience cannot exert much influence on the mass media. They can only choose to tune in or tune out. And even tuning out is difficult because it excludes one from the styles, news, gossip, and entertainment most people depend on to grease the wheels of social interaction. Few people want to be cultural misfits. This does not mean that people are always passive consumers of the mass media. As noted below, we filter, interpret, and resist what we see and hear if it contradicts our experience and beliefs. Even so, in the interaction between audiences and media sources, the media sources usually dominate.

To fully appreciate the impact of the mass media on life today, we need to trace their historical development. That is the first task we set ourselves below. We then critically review theories of the mass media's effects on social life. As you will see, each of these theories contributes to our appreciation of media effects. Finally, we assess developments on the media frontier formed by the Internet, television, and other mass media. We show that, to a degree, the new media frontier blurs the distinction between producer and consumer and has the potential to make the mass media somewhat more democratic for those who can afford access.

The Rise of the Mass Media

John Lie once mentioned to his 11-year-old stepdaughter, Jessie, that when he grew up in Tokyo in the 1960s, not every household had a telephone. As someone who takes it for granted that she can pick up her brother's cell phone pretty much whenever she wants to call friends in California, Jessie was genuinely shocked. "How could you talk to your friends?" she asked.

"Well," John answered, "if a family were lucky enough to have a phone, it would be used not just by family members but by a whole network of relatives, friends, and neighbors. If you needed to call someone without a phone in his or her house, you had to call that person's closest neighbor who had a phone. The neighbor would then fetch the person you wanted to talk to. Some families even kept a little bowl next to their phone so nonfamily members could drop coins in to pay for their calls."

"Okay," said Jessie, "but that was in Japan. Here things were different, right?"

"They were different," replied John, "but in 1960 the United States was not all that far ahead of Japan in the use of telephones. One of my older colleagues told me that when he grew up in the United States in the 1940s, his parents had to make an appointment with the operator to phone long-distance. Long-distance calls were so expensive, they used to time their calls to make sure they didn't exceed the standard three-minute rate."

Jessie shook her head silently, seeming to wonder whether we were still getting around by horse and buggy when my colleague and I were children.

"What seems quaint is largely a matter of perspective," concludes John. "I didn't find the scarcity of telephones strange when I was growing up. I did, however, find it shocking that television broadcast signals could not reach my grandparents in rural South Korea."

Similarly, it may be difficult for you to imagine a world without the mass media. Yet, as Table 14.1 shows, most of the mass media are recent inventions. The first developed systems of writing appeared only about 5,500 years ago in Egypt and Mesopotamia (now southern Iraq). The print media became truly mass phenomena only in the 19th century. The inexpensive daily newspaper, costing a penny, first appeared in the United States in the 1830s. At that time, long-distance communication required physical transportation. To spread the news, you needed a horse or a railroad. In 1794, it took 44 days for news from New York to reach Cincinnati. In 1841, it took a week.

The newspaper was the dominant mass medium even as late as 1950 (Smith, 1980; Schudson, 1991). However, change was in the air as early as 1876, when a nationwide system of telegraphic communication was established (Pred, 1973). From that time on, long-distance communication no longer required physical transportation. The transformative power of the new medium was soon evident. For example, until 1883, hundreds of local

The newspaper was the dominant mass medium even as late as 1950.

◆ **TABLE 14.1** ◆

The Development of the Mass Media

SOURCES: Berners-Lee (1999); Croteau and Hoynes (1997: 9–10); "The Silent Boom" (1998).

Year (CE)	Media Development
Circa 100	Papermaking developed in China
Circa 1000	Movable clay type used in China
Circa 1400	Movable metal type developed in Korea
1450	Movable metal type used in Germany, leading to the Gutenberg Bible
1702	First daily newspaper, London's *Daily Courant*
1833	First mass-circulation newspaper, *New York Sun*
1837	Louis Daguerre invents a practical method of photography in France
1844	Samuel Morse sends the first telegraph message between Washington and Baltimore
1875	Alexander Graham Bell sends the first telephone message
1877	Thomas Edison develops the first phonograph
1895	Motion pictures are invented
1901	Italian inventor Guglielmo Marconi transmits the first transatlantic wireless message from England to Newfoundland
1906	First radio voice transmission
1920	First regularly scheduled radio broadcast, Pittsburgh
1921	First commercial TV broadcast
1922	Long-playing records (LPs) introduced
1939	Network TV begins in the United States
1952	VCR invented
1961	First cable television, San Diego
1969	First four nodes of the United States Defense Department's ARPANET (precursor of the Internet) set up at Stanford University, UCLA, UC Santa Barbara, and the University of Utah
1975	First microcomputer marketed
1983	Cell phone invented
1989	World Wide Web conceived by Tim Berners-Lee at the European Laboratory for Particle Physics in Switzerland

time zones existed in the United States—Michigan alone had 27. But virtually instant communication by telegraph made it possible to establish just four time zones (Carey, 1989). The telegraph thus gave new meaning to the old expression "times change."

Most of the electronic media are creatures of the 20th century. The first television signal was transmitted in 1928. Twenty-one years later, network TV began in the United States. The United States Department of Defense established ARPANET in 1969. It was designed as a system of communication between computers that would automatically find alternative transmission routes if one or more nodes in the network broke down due, say, to nuclear attack. The Internet grew out of ARPANET, which in turn begat the hyperlinked system of texts, images, and sounds known as the World Wide Web around 1991. By 2000, about 280 million people worldwide used the Web routinely. It was a quick trip. A mere 140 years separate Pony Express from the home videoconference.

Causes of Media Growth

The rise of the mass media can be explained by three main factors, one religious, one political, and one economic:

1. *The Protestant Reformation.* In the 16th-century Catholic Church, people relied on priests to tell them what was in the Bible. In 1517, however, Martin Luther protested certain practices of the Church. Among other things, he wanted people to develop a

The Gutenberg Bible.

more personal relationship with the Bible. Within 40 years, Luther's new form of Christianity, known as Protestantism, was established in half of Europe. Suddenly, millions of people were being encouraged to read. The Bible became the first mass media product in the West and by far the best-selling book.

The diffusion of the Bible and other books was made possible by technological improvements in papermaking and printing (Febvre and Martin, 1976 [1958]). The most significant landmark was Johannes Gutenberg's invention of the printing press. In the 50 years after Gutenberg produced his monumental Bible in 1455, more books were produced than in the previous 1,000 years. The printed book enabled the widespread diffusion and exchange of ideas. It contributed to the Renaissance (a scholarly and artistic revival that began in Italy around 1300 and spread to all of Europe by 1600) and the rise of modern science (Johns, 1998).

A remarkable feature of the book is its durability. Many electronic storage media became obsolete just a few years after being introduced. For instance, eight-track tapes are relics of the 1970s and $5^{1}/_{4}''$ floppy disks are icons of the early 1980s. They are barely remembered today. In contrast, books are still being published today, nearly 550 years after Gutenberg published his Bible. In fact, 68,175 new books and editions were published in the United States in 1998, and another 9,271 were imported (United States Bureau of the Census, 1998c: 581).

2. *Democratic movements.* A second force that promoted the growth of the mass media was political democracy. From the 18th century on, the ordinary citizens of France, the United States, and other countries demanded and achieved representation in government. At the same time, they wanted to become literate and gain access to previously restricted centers of learning. Democratic governments, in turn, depended on an informed citizenry and therefore encouraged popular literacy and the growth of a free press (Habermas, 1989).

Today, the mass media, and especially TV, mold our entire outlook on politics. Television was a source of presidential campaign news for only 51% of Americans in 1952, but by 1960 it was a source of campaign news for 87% of the public (Nie, Verba, and Petrocik, 1979 [1976]: 274). From a mass media point of view, the 1960 presidential election was significant in a second respect as well. It was the year of the first televised presidential debate—between John F. Kennedy and Richard Nixon. Sander Vanocur, one of the four reporters who asked questions during the debate, later recalled: "The people who watched the debate on their television sets apparently thought Kennedy came off better than Nixon. Those who heard the debate on radio thought Nixon was superior to Kennedy" (quoted in "The Candidates Debate," 1998). Kennedy smiled. Nixon perspired. Kennedy relaxed. Nixon fidgeted. The election was close, and most analysts believe that Kennedy got the edge simply because 70 million viewers thought he looked better on TV. Television was thus beginning to redefine the very nature of American politics. The 2000 presidential debates on TV were in some respects a rerun of 1960. Al Gore was widely perceived as wooden, nervous, pompous, and aggressive, George W. Bush as relaxed and

Americans now spend nearly 40% of their time interacting with the mass media.

friendly. Gore, the more experienced debater, was widely expected to win the debates, but many observers feel he lost them—and therefore the election—because of his problematic TV presence. Little wonder, then, that some analysts complain that television has oversimplified politics, reducing it to a series of catchy slogans and ever-shorter uninterrupted comments or "sound bites." (The average sound bite in nightly network news shrank from 42.3 seconds in 1968 to 8.4 seconds in 1992; see Thelen, 1996). From this point of view, candidates are marketed for high office like Kellogg's sells breakfast cereal, and a politician's stage presence is more important than his or her policies in determining success at the polls.

3. *Capitalist industrialization.* The third major force that stimulated the growth of the mass media was capitalist industrialization. Modern industries required a literate and numerate workforce. They also needed rapid means of communication to do business efficiently. Moreover, the mass media turned out to be a major source of profit in their own right. In 1997, box office receipts of the American movie industry topped $6 billion. CD sales amounted to more than $12 billion (Wright, 1998). As Table 14.2 shows, advertising revenues of popular magazines are phenomenal, and account for the bulk of most magazines' profits. Clearly, the mass media form a big business.

We conclude that the sources of the mass media lie deeply embedded in the religious, political, and economic needs of our society. Moreover, the mass media are among the most important institutions in our society today. How, then, do sociologists explain the effects of the mass media on society? To answer this question, we now summarize the relevant sociological theories.

✦ **TABLE 14.2** ✦
Paid Circulation and Advertising Revenue of Selected American Magazines, 1997

SOURCE: Wright (1998).

Magazine	Advertising Revenue ($ millions)	Paid Circulation
TV Guide	496.3	13,103,187
Better Homes & Gardens	377.5	7,605,187
Time	533.2	4,155,806
People	588.5	3,608,111
Sports Illustrated	548.6	3,223,810
Newsweek	408.5	3,177,407
United States News & World Report	239.2	2,224,003
PC Magazine	333.5	1,176,351
Business Week	329.7	901,891
Forbes	243.8	n.a.

THEORIES OF MEDIA EFFECTS

Functionalism

As societies develop, they become larger and more complex. The number of institutions and roles proliferate. Due to the sheer scale of the society, face-to-face interaction becomes less viable as a means of communication. As a result, the need increases for new means of coordinating the operation of the various parts of society. For example, people in Maine must have at least a general sense of what is happening in California and they need to share certain basic values with Californians if they are going to feel they are citizens of the same country. The mass media do an important job in this regard. The 19th-century German philosopher Georg Hegel once said that the daily ritual of reading the newspaper unites the secular world, just as the ritual of daily prayer once united the Christian world. Stated more generally, his point is valid. The nationwide distribution of newspapers, magazines, movies, and television shows cements the large, socially diverse, and geographically far-flung population of the United States. In a fundamental sense, the nation is an imagined community, and the mass media make it possible for us to imagine it (Anderson, 1990).

Thus, the mass media perform an important function by *coordinating* the operation of industrial and postindustrial societies. But, according to functionalist theorists, their significance does not stop there (Wright, 1975). In addition, the mass media are also important agents of *socialization*. Families have relinquished their former nearly exclusive right to transmit norms, values, and culture. The mass media have stepped into the breech. They reinforce shared ideals of democracy, competition, justice, and so forth (see Chapter 4, "Socialization").

A third function of the mass media involves *social control*. That is, the mass media help to ensure conformity. For example, news broadcasts, TV dramas, and "docutainment" programs such as *Cops* pay much attention to crime, and they regularly sing the praises of heroes who apprehend and convict criminals. By exposing deviants and showcasing law enforcement officials and model citizens, the mass media reinforce ideas about what kinds of people deserve punishment and what kinds of people deserve rewards. In this way, they reproduce the moral order. Some people think the Jenny Jones show and other similar programs are outlandish, and in a way they are. From a sociological point of view, however, they are also deeply conservative programs. For when television audiences get upset about marital infidelities and other outrages, they are reinforcing some of the most traditional norms of American society and thus serving as agents of social control. As Saul Bellow wrote in *Herzog,* "a scandal [is] after all a sort of service to the community" (Bellow, 1964: 18).

The mass media's fourth and final function is to provide *entertainment*. Television, movies, magazines, and so forth give us pleasure, relaxation, and momentary escape from the tension and tedium of everyday life. How often have you come home after a long and frustrating day at college or work, picked up the remote, channel surfed, concluded there's nothing really worth watching, but settled for a soap opera or some other form of easily digestible entertainment? It is precisely because some products of the mass media require little effort on the part of the audience that they are important. They relieve stress. Moreover, they do so in a way that doesn't threaten the social order. Without such escapes, who knows how our daily tensions and frustrations might express themselves?

Conflict Theory

Clearly, functionalism offers valuable insights into the operation of the mass media. However, the functional approach has been criticized by conflict theorists for paying insufficient attention to the social inequality fostered by the mass media. Specifically, conflict theorists say functionalism exaggerates the degree to which the mass media serve the interests of the entire society. They contend that some people benefit from the mass media more than others. In particular, the mass media favor the interests of dominant classes and political groups (Gitlin, 1983; Herman and Chomsky, 1988; Iyengar, 1991; Horkheimer and Adorno, 1986 [1944]).

Conflict theorists maintain there are two ways in which dominant classes and political groups benefit disproportionately from the mass media. First, the mass media broadcast beliefs, values, and ideas that create widespread acceptance of the basic structure of society, including its injustices and inequalities. Second, ownership of the mass media is highly concentrated in the hands of a small number of people and is highly profitable for them. Thus, the mass media are a source of economic inequality.

Media Ownership

Ownership of the mass media has certainly become more highly concentrated over time. Around 50 corporations controlled half of all media organizations in the United States in 1984. By 1993, about 20 corporations exercised this degree of media control (Bagdikian, 1997 [1983]). Between 1992 and 1996, the proportion of United States television stations owned by the 10 biggest owners nearly doubled. The proportion of radio stations owned by the 10 biggest owners more than quadrupled ("Media Mergers . . . ," 2000). United States book production, film production, newspaper publishing, and cable TV are each dominated by only six firms. A mere five firms dominate the United States music industry (McChesney, 1999).

It is not just the *degree* of media concentration that has changed. In addition, the *form* of media concentration began to shift in the 1990s (McChesney, 1999). Until the 1990s, media concentration involved mainly "horizontal integration." That is, a small number of firms tried to control as much production as possible in their particular fields (film production, newspapers, radio, television, etc.). In the 1990s, however, "vertical integration" became much more widespread. That is, media firms sought to control production and distribution in many fields. They became media "conglomerates." Today, a media conglomerate may own any combination of television networks, stations, and production facilities; magazines, newspapers, and book publishers; cable channels and cable systems; movie studios, theaters, and video store chains; sports teams and stadiums; Web portals and software companies. A media conglomerate can create content and deliver it in a variety of forms. For instance, it can make a movie, promote it on its TV and radio networks, and then spin off a TV series, CD, book, and merchandise—all delivered to the consumer in outlets owned by the conglomerate itself.

In the United States, the biggest media players include Disney, Viacom, and News Corporation. Other giant media conglomerates include Bertelsmann (Germany), Sony (Japan), and Vivendi (France and Canada). By listing the countries in which the principal owners of these media conglomerates reside, we do not wish to suggest that they operate solely within their national boundaries. On the contrary, today's media conglomerates are increasingly global in their operations. For example, BMG records and Universal Studios are well-known in the United States, but they are owned, respectively, by Bertelsmann and

Steve Case, Gerald Levin, and Ted Turner smile as they announce the AOL-Time Warner merger, creating the 600-pound gorilla among media conglomerates.

Vivendi. Moreover, citizenship is largely a matter of economic convenience for the owners of the media conglomerates. Thus, Rupert Murdoch, principal shareholder of News Corporation (which owns the Fox network, 20th Century Fox, *The New York Post,* etc.) and Edgar Bronfman, a principal shareholder in Vivendi (which owns Universal Studios, Polygram records, etc.), are both American citizens. However, Murdoch is an expatriate Australian, Bronfman an expatriate Canadian.

The media conglomerates listed above are big. However, the 600-pound gorilla among media conglomerates was born on January 10, 2000. On that day, the merger of two media behemoths into a single company, AOL/Time Warner, was announced. The market value of the new company: $350 billion, nearly as much as the Gross National Product of Argentina's 37 million people. The new company owns America Online, Warner Brothers, Time Warner Cable, CNN, HBO, *Time, People, Sports Illustrated, Life, Fortune,* DC Comics, Netscape, the Cartoon Network, the Atlanta Braves, the Atlanta Hawks, World Championship Wrestling, the Goodwill Games, Time Life Books, and the Book-of-the-Month Club, to name only a few of its best-known companies ("That's AOL Folks . . . ," 2000). Ted Turner, who owned 100 million shares of Time Warner, grew $3.5 billion richer in a single day when his stock soared on news of the merger.

Media Bias

Does the concentration of the mass media in fewer and fewer hands deprive the public of independent sources of information, limit the diversity of opinion, and encourage the public to accept their society as it is? Conflict theorists think so (see Box 14.2). They believe that when a few conglomerates dominate the production of news in particular, they squeeze out alternative points of view. Moreover, as Edward Hermann and Noam Chomsky (1988) argue, several mechanisms help to bias the news in a way that supports powerful corporate interests and political groups. These biasing mechanisms include advertising, sourcing, and flak:

✦ *Advertising.* Most of the revenue earned by television stations, radio stations, newspapers, and magazines comes from advertising by large corporations. According to Hermann and Chomsky, these corporations routinely seek to influence the news so it will reflect well on them. Thus, in one survey, 93% of newspaper editors said advertisers have tried to influence their news reports. Thirty-seven percent of newspaper editors admitted to actually being influenced by advertisers (Bagdikian, 1997 [1983]).

✦ *Sourcing.* Studies of news gathering show that most news agencies rely heavily for information on press releases, news conferences, and interviews organized by large corporations and government agencies. These sources routinely slant information to reflect favorably on their policies and preferences. Unofficial news sources are consulted less often. Moreover, unofficial sources tend to be used only to provide reactions and minority viewpoints that are secondary to the official story.

✦ *Flak.* Governments and big corporations routinely attack journalists who depart from official and corporate points of view. For example, tobacco companies have systematically tried to discredit media reports that cigarettes cause cancer. In a notorious case, the respected public affairs show, *60 Minutes,* refused to broadcast a damaging interview with a former Philip Morris executive because CBS was threatened with legal action by the tobacco company. (This incident is the subject of the 2000 Oscar-nominated movie *The Insider.*)

On the whole, the arguments of the conflict theorists are compelling. We do not, however, find them completely convincing (Gans, 1979a). After all, if 37% of newspaper editors have been influenced by advertisers, 63% have not. News agencies may rely heavily on government and corporate sources, but this doesn't stop them from routinely biting the hand that offers to feed them and evading flak shot their way. Examples of mainstream journalistic opposition to official viewpoints are plentiful. In the early 1970s, investigative reporters from *The Washington Post* unleashed the Watergate scandal. It eventually

BOX 14.2
IT'S YOUR CHOICE

MEDIA CONGLOMERATES

Who owns NBC? General Electric. Who owns ABC? Disney. Who owns CBS? Westinghouse. Who owns CNN? Turner Broadcasting. But who owns Turner Broadcasting? AOL/Time Warner, which also owns HBO, America Online, Warner Brothers, *Time, Sports Illustrated,* etc.

Hardly anyone doubts that the mass media, especially television, have a big impact on American society. However, people disagree whether concentrated ownership affects the content of broadcast news. Critics of concentrated ownership cite examples of corporations influencing the flow of information. For instance, Brian Ross, the leading investigative reporter for *20/20,* prepared a segment about Disney World in 1998. Ross claimed Disney was so lax in doing background checks on employees it had hired pedophiles. ABC killed the story before airtime. ABC, you will recall, is owned by Disney (McChesney, 1999). ABC News formally apologized to Philip Morris (a major TV advertiser through its subsidiary, Kraft Foods), for airing a report about the cigarette company's manipulation of ammonia levels in tobacco. Although the report was true, ABC apparently buckled to corporate pressure. One media critic has even gone so far as to suggest there are few stories critical of the military on the news because two of the owners of the four major TV networks (General Electric and Westinghouse) are large defense contractors (Miller, 1996).

Other analysts suggest that corporate ownership does not have a major impact on the televised flow of information. They argue that, despite some abuses, freedom of speech is protected. We hear plenty of stories critical of corporations and the government, and if we don't hear more it's mainly because the public isn't interested. Furthermore, they say, even if there were a problem with concentrated ownership, government intervention would likely create more problems than it solves.

What do you think? Does the high concentration of mass media ownership have an impact on the information we receive from the major television networks? Are different *types* of stories influenced to varying degrees and in different ways by the conglomeration of the mass media? If it could be determined that the problem is serious, what could be done about it?

---

resulted in the resignation of President Richard Nixon. In the late 1990s, *Time* published a cover story on "corporate welfare" in America. It showed that in the midst of the nation's biggest economic boom ever, and after nearly two decades of reducing welfare budgets to the poor, the federal government was giving $125 billion a year in subsidies to American businesses (Barlett and Steele, 1998). As these examples show, even mainstream news sources, although owned by media conglomerates, do not always act like the lap dogs of the powerful (Hall, 1980).

Still, conflict theorists make a valid point if they restrict their argument to how the mass media support core American values. In their defense of core values, the mass media *are* virtually unanimous. For example, the mass media enthusiastically support democracy and capitalism. We cannot think of a single instance of a major American news outlet advocating a fascist government or a socialist economy in the United States.

Similarly, the mass media virtually unanimously endorse consumerism as a way of life. As discussed in Chapter 3 ("Culture"), consumerism is the tendency to define oneself in terms of the goods and services one purchases. Endorsement of consumerism is evident in the fact that advertising fills the mass media and is its lifeblood. In the United States, expenditure on advertising is nearly a quarter the expenditure on all levels of education (United Nations, 1998). The average American is exposed to a staggering number of ads each day, some estimates placing the number in the thousands. Companies pay filmmakers to use their brand-name products conspicuously in their movies. The Orange Bowl is turned into the Federal Express Orange Bowl, and the Cotton Bowl becomes the Mobil Cotton Bowl. In some magazines, such as *Wired, Vogue,* and *Vanity Fair,* ads figure so prominently one must search for the articles. The American public responds to the ongoing advertising blitz by participating in the biggest buying spree in recorded history, even at the cost of ballooning credit card debt. Specifically, credit card debt is growing at the rate of about 10% per household per year. This growth rate is nearly twice as high among the bottom half of income earners as among the top half (Ewen, 1976; 1988; Twitchell, 1999; Yoo, 1998).

It is only when the mass media deal with news stories that touch on less central values that one may witness a diversity of media opinion. For example, as our Watergate and

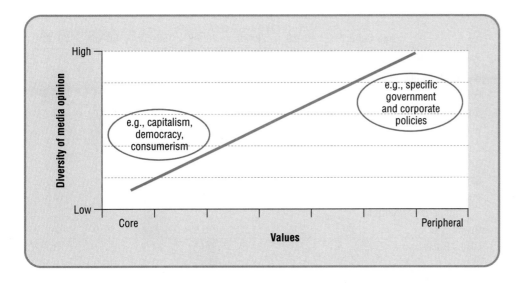

♦ **FIGURE 14.3** ♦
The Relationship Between Centrality of Values and Diversity of Media Opinion

corporate welfare examples show, specific government and corporate policies are often the subject of heated debate in the mass media.[1] Thus, despite the indisputable concentration of media ownership, the mass media are diverse and often contentious on specific issues that do not touch on core values (see Figure 14.3). To drive this point home, we must now say a few words about the sources of diversity in the mass media.

Diversity and the Mass Media

Both functionalists and conflict theorists stress how the mass media bridge social differences and reinforce society's core values. The two schools of thought differ in that functionalists regard core values as serving everyone's interests whereas conflict theorists regard them as favoring the interests of the rich and powerful. By focusing so tightly on core values, however, both approaches understate the diversity of the mass media.

Like all societies, the United States is differentiated by generation, class, gender, race, and ethnicity. To a degree, the mass media reflect this social diversity. Consider generational differences in musical taste (Bourdieu, 1984: 17). The 1993 General Social Survey asked Americans their feelings about 18 types of music. Table 14.3 shows the percentage of Americans in each age cohort who liked a particular type of music "very much." Big band and swing music are most popular among 70–89-year-olds. Gospel and bluegrass are most popular among 60–69-year-olds. Contemporary rock, reggae, heavy metal, rap, and new age music are most favored by 18–29-year-olds. Every American age cohort has distinct musical preferences. Tastes in books, magazines, television programs, and other mass media products are similarly differentiated by generation, class, gender, race, and ethnicity.[2]

Many social differences are widening in American society due to growing class inequality, increased immigration, the rise of a substantial middle class among racial minorities, and other social forces. The mass media must cater to this growing diversity or lose business. To entertain and sell products to a differentiated market, advertising and programming must appeal to specific market niches.

Technological advances, especially the computer, have made "niche marketing" possible. In the 1910s, when Henry Ford established the first assembly lines, his aim was to produce a standard automobile for a mass market. By standardizing production, Ford kept

---

[1]Conservatives often criticize the mass media for its allegedly liberal bias. We do not wish to enter into this highly charged political debate. It should be noted, however, that the conservative critique of the mass media focuses only on the media's analysis of specific government and corporate policies, not their analysis of stories relating to core values.

[2]Regional differences are often smaller. For example, market research shows that age and class affect color preferences in consumer products. Region of residence does not. Why not? According to Leatrice Eiseman, "[t]he reason is the [national] mass media, . . . fashion magazines and catalogs, home shopping shows, and big clothing chains all present the same options." Eiseman should know. She is a member of The Color Marketing Group (also known as "The Color Mafia"), a 1,300-member committee that meets regularly to help change the national palette of color preferences for consumer products in order to stimulate sales (Mundell, 1993).

Age Cohort / Type of Music	18–29	20–39	40–49	50–59	60–69	70–89
Big Band/Swing	5.0	7.9	13.0	22.8	32.5	37.7
Folk Music	3.7	5.5	10.0	12.8	12.0	14.6
Gospel	16.8	15.0	20.6	29.6	34.3	30.2
Bluegrass	6.7	4.7	9.4	13.3	16.3	10.4
Country and Western	19.0	22.4	25.8	30.5	27.7	23.6
Broadway Musical/Show Tunes	7.0	8.2	14.2	26.2	24.1	17.6
Mood/Easy Listening	10.0	12.6	18.2	23.2	21.1	14.7
Classical Music—Symphony & Chamber	13.7	16.3	19.4	22.2	18.7	17.0
Opera	3.0	1.3	3.9	8.4	7.9	7.1
Oldies Rock	27.3	31.6	35.8	24.6	12.7	6.1
Jazz	16.1	18.0	18.8	17.2	13.9	8.5
Latin/Mariachi/Salsa	3.7	3.7	9.4	6.9	4.2	3.3
Blues/Rhythm and Blues	16.0	13.5	15.5	13.8	12.0	10.9
Contemporary Rock	24.3	19.8	9.7	8.4	3.6	1.4
Reggae	14.0	5.3	4.5	2.0	.6	.9
Heavy Metal	11.0	2.1	.9	1.0	.6	.5
Rap	9.7	1.6	1.8	.5	0	.9
New Age/Space Music	4.4	4.2	3.9	1.5	1.2	.5

costs down and profits high. For about 60 years, this was the aim of all big businesses. Some people knew that Ford's production and marketing model was not ideal. They understood that consumer tastes were diverse. However, limits existed on what they could do to satisfy that diversity. That is because it was less expensive to produce standardized commodities for large masses of people than a great variety of products to suit the full range of consumer preferences. The computer changed all that. By allowing new forms of inventory control, short production runs, faster manufacturing, quick design changes, and so forth, computers allowed producers to cater to small market niches and increase profit margins. The mass media followed suit, diversifying its programming and advertising to reflect the needs of niche marketing.

Still, one must be careful not to exaggerate the degree to which the mass media reflect the diversity of American society. For despite improvements in recent decades, critics who are sympathetic to feminism and the plight of racial minorities have charged the mass media with persistent *numerical underrepresentation* and *biased portrayal* of minority groups. Let us consider these charges.

In 1998, the Screen Actors Guild, the professional organization of film and TV actors in the United States, sponsored a study of fictional television characters who appeared in prime time and daytime series, films, and animated cartoons in the periods 1991–92 and 1994–97 (Gerbner, 1998). Among other things, the study compared the percentage size of minority groups in the American population with their percentage representation in fictional TV roles. Figure 14.4 summarizes this part of the study. In Figure 14.4, a score of 100 indicates that the percentage of a group in the American population is the same as its percentage in fictional TV roles. A score of more than 100 suggests the degree of overrepresentation, a score of less than 100 the degree of underrepresentation. Figure 14.4 shows that in 1994–97 there were 29% *more* white men in fictional TV roles than white men in the United States population. At the same time, there were 28% *fewer* women in fictional TV roles than in the population, 46% fewer Native Americans, 61% fewer Asian–Pacific Americans, 66% fewer people 60 years of age or older, 76% fewer Hispanics, 88% fewer

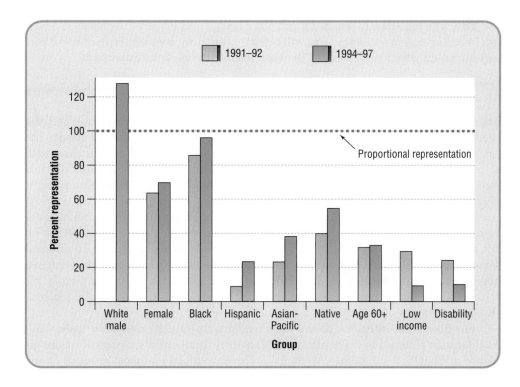

♦ **FIGURE 14.4** ♦

Representation of Minority Groups in Prime Time and Daytime Television, United States, 1991–1992 and 1994–1997

Note: Data for white males, 1991–1992 not reported.
SOURCE: Gerbner (1998).

people with disabilities, and 89% fewer poor people. These figures demonstrate that TV falls short of reflecting the diversity of American society in many crucial aspects.

In some ways, the situation is improving. For example, comparing 1991–92 with 1994–97, we see that racial minority representation in fictional TV roles is increasing. The representation of African Americans is now almost the same as their representation in the United States population. Nevertheless, the portrayal of women, racial minorities, the poor, and people with disabilities still tends to reinforce traditional, mainstream, negative stereotypes. For instance, fully 60% of fictional TV characters suffering from mental illness in the period 1994–97 were involved in crime or violence. Nonwhites tended to play comical or criminal characters rather than serious, heroic types.[3] Characters of different races often interacted professionally, sometimes interacted socially, but were rarely romantically involved. Asian Americans were typically portrayed as nerdy students, inscrutable martial arts masters, seductive Dragon Ladies, or clueless immigrants. Women were valued chiefly for their youth, sex appeal, and beauty, while older women were often associated with evil. Thus, in the 1990s, women playing fictional TV roles were on average younger than men and they became still younger relative to men in the course of the decade. Young women tended to play mainly romantic roles, but the proportion of villains among women increased with age. The tendency for villainy to increase among older men was much weaker (*Fall Colors II,* 2000; Gerbner, 1998).

In sum, the research findings reviewed here point to positive change in the way the mass media treat various minority groups. We have come a long way since the 1950s, when the only blacks on TV played butlers and buffoons. At the same time, however, research suggests that the mass media still have a long way to go before they cease reinforcing traditional class, race, and gender stereotypes in American society (Dines and Humez, 1995). To a degree, the mass media influence the way audiences think about the world. Therefore, the perpetuation of stereotypes acts as a brake on social change.

---

[3]Similarly, in a study of 4 years worth of stories in *Time, Newsweek,* and *United States News & World Report,* one researcher found that 53–66% of poor people were portrayed as African Americans while the actual rate was about 29%. This media bias reinforces the false idea that most poor people are African American (Fitzgerald, 1997).

That said, the *extent* to which TV and other mass media influence audiences is a subject of controversy in sociology. This will become clear as we consider the third major perspective on the sociological study of the mass media, the interpretive approach.

The Interpretive Approach

The final episode of *Buffy the Vampire Slayer* for the 1999 season included a scene of a shooting spree at a high school. However, the Fox TV network canceled the episode because it was scheduled to air shortly after two teenagers massacred 13 of their classmates at Columbine high school in Littleton, Colorado. In canceling the show, network officials were reacting to members of the public who shared the widespread belief that airing it might incite more school violence.

The view that the mass media powerfully influence a passive public is common, not least among functionalists and conflict theorists. Many people believe that violence on TV causes violence in real life, pornography on the magazine stands leads to immoral sexual behavior, and adolescents are more likely to start smoking cigarettes when they see popular movie stars lighting up. In a 1995 CBS/*New York Times* poll, for example, respondents said they thought television was the number one cause of teenage violence (Kolbert, 1995).

Just how much influence do the mass media actually exert over their audiences? The question is mired in controversy. For example, recall our discussion of media violence in Chapter 2 ("Research Methods"). There we found that most research on the subject is plagued by a big validity problem. Simply stated, experiments on media violence may not be measuring what they say they are. As a result, the degree to which TV violence encourages violent behavior is unclear. The sociological consensus seems to be that TV violence has a weak effect on a small percentage of viewers (Felson, 1996: 123).

There are other reasons for questioning the strength of media effects. For instance, researchers have known for half a century that people do not change their attitudes and behaviors just because the media tell them to do so. That is because the link between persuasive media messages and actual behavior is indirect. A **two-step flow of communication** takes place (Katz, 1957; Schiller, 1989; Schudson, 1995). In Step 1, respected people of high status evaluate media messages. They are the opinion leaders of a neighborhood or a community, people who are usually more highly educated, well-to-do, and/or politically powerful than others in their circle. Because of their high status, they exercise considerable independence of judgment. In Step 2, opinion leaders *may* influence the attitudes and behaviors of others. In this way, opinion leaders filter media messages. The two-step flow of communication limits media effects. If people are influenced to vote for certain candidates, buy certain products, or smoke cigarettes, it is less because the media tell them to and more because opinion leaders suggest they should.

Yet another persuasive argument that leads one to question the effects of the mass media comes from interpretive sociologists such as symbolic interactionists and interdisciplinary **cultural studies** experts. They use in-depth interviewing and participant observation to study how people actually interpret media messages.

British sociologist Stuart Hall, one of the foremost proponents of this approach, emphasizes that people are not empty vessels into which the mass media pour a defined assortment of beliefs, values, and ideas. Rather, audience members take an active role in consuming the products of the mass media. They filter and interpret mass media messages in the context of their own interests, experiences, and values. Thus, in Hall's view, any adequate analysis of the mass media needs to take into account both the production and the consumption of media products. First, he says, we need to study the meanings intended by the producers. Then we need to study how audiences consume or evaluate media products. Intended and received meanings may diverge; audience members often interpret media messages in ways other than those intended by the producers (Hall, 1980; Seiter, 1999).

Two sports examples from the summer of 2001: When Dale Earnhardt Jr. won a NASCAR race at the track where his famous father died in a crash a year earlier, some people cried "fix," just as they did when Cal Ripken Jr. hit a home run in the All-Star game. This suggests that people are often skeptical if not downright cynical about what they see on TV (Brady, 2001). A personal example: When John Lie's parents were preparing to emigrate to the United States in the late 1960s, his mother watched many American movies and television shows. One of her favorite TV programs was *My Three Sons,* a sitcom about three boys living with their father and grandfather. From the show she learned that boys wash dishes and vacuum the house in the United States. When the Lie family immigrated to Hawaii, John and his brother—but not his sister—had to wash dishes every night. When John complained, his mother reassured him that "in America, only boys wash dishes."

Even children's television viewing turns out to be complex when viewed through an interpretive lens. Research shows that young children clearly distinguish "make-believe" media violence from real-life violence (Hodge and Tripp, 1986). That is one reason why watching episode after episode of *South Park* has not produced a nation of *South Park* clones. Similarly, research shows differences in the way working-class and middle-class women relate to TV. Working-class women tend more than middle-class women to evaluate TV programs in terms of how realistic they are. This critical attitude reduces their ability to identify strongly with many characters, personalities, and story lines. For instance, working-class women know from their own experience that families often don't work the way they are shown on TV. They view the idealized, middle-class nuclear family depicted in many television shows with a mixture of nostalgia and skepticism (Press, 1991). Age also affects how one relates to television. Elderly viewers tend to be selective and focused in their television viewing. In contrast, people who grew up with cable TV and remote control often engage in channel surfing, conversation, eating, and housework, zoning in and out of programs in anything but an absorbed fashion (Press, 1991). The idea that such viewers are sponges, passively soaking up the values embedded in TV programs and then mechanically acting on them, is inaccurate.

We conclude that each of the theoretical approaches reviewed above contributes to our understanding of how the mass media influence us:

✦ *Functionalism* usefully identifies the main social effects of the mass media: coordination, socialization, social control, and entertainment. By performing these functions, the mass media help make social order possible.

✦ *Conflict theory* offers an important qualification. As vast moneymaking machines controlled by a small group of increasingly wealthy people, the mass media contribute to economic inequality and to maintaining the core values of a stratified social order.

✦ *Interpretive approaches* offer a second qualification. They remind us that audience members are people, not programmable robots. We filter, interpret, resist, and sometimes reject media messages according to our own interests and values. A full sociological appreciation of the mass media is obliged to recognize the interaction between producers and consumers of media messages.

In the final section of this chapter, we briefly explore the interaction between producers and consumers on the new media frontier. The new media frontier is formed by the Internet, television, and other mass media. As you will see, the new media frontier provides fresh opportunities for media conglomerates to restrict access to paying customers and accumulate vast wealth. Simultaneously, however, it gives consumers new creative capabilities, partially blurring the distinction between producer and consumer. The new media frontier, we conclude, has the potential to make the mass media somewhat more democratic—at least for those who can afford access.

THE NEW MEDIA FRONTIER

The contradictory tendencies of the new media frontier are evident on the Internet. Let us first consider the forces that restrict Internet access and augment the power of media conglomerates. We then discuss some countertrends.

Web Interactive Exercises
Is the Digital Divide Growing?

Access

The Internet requires an expensive infrastructure of personal computers, servers, and routers; an elaborate network of fiber-optic, copper twist, and coaxial cables; and many other components. This infrastructure has to be paid for, most of it by individual users. As a result, access is not open to everyone. Far from it. In the United States, for example, college-educated whites with above-average incomes are most likely to enjoy Internet access (see Chapter 11, "Politics," Table 11.5).

Nor is Internet access evenly distributed globally. As Figure 14.5 shows, in 2000 the United States was the overwhelming leader in getting its population connected. Moreover, the *rate* of Internet connectivity (per 1,000 people) in North America was more than double the rate in Western Europe, nearly 15 times higher than in Eastern Europe, and nearly 23 times higher than in South and Central America. It was more than 28 times higher than in the Asia-Pacific region and more than 68 times higher than in the Middle East and Africa. These figures show that international inequalities in Internet access mirror global inequalities overall (see Chapter 7, "Stratification: United States and Global Perspectives").

Web Research Projects
The Mass Media, Globalization, and Localization

Content

United States domination is even more striking when it comes to Internet content. Thus, roughly two thirds of the servers that provide content on the Internet are in the United States (Internet Software Consortium, 2000). Some analysts say that American domination of the Web is an example of **media imperialism.** Media imperialism is the control of a mass medium by a single national culture and the undermining of other national cultures. France and Canada are perhaps the countries that have spoken out most strongly against the perceived American threat to their national culture and identity. Some media analysts in those countries deeply resent the fact that the United States is the world's biggest exporter and smallest importer of mass media products, including Web content (see Figure 14.6).[4]

According to some media analysts, the Internet not only restricts access and promotes American content, it also increases the power of media conglomerates. This is most evident in the realm of **media convergence.** Media convergence is the blending of the World Wide Web, television, telephone, and other communications media into new, hybrid media forms. Many consumers find a PC too complex to operate and find TV limited in its functionality and entertainment value. Media convergence is intended to appeal mainly to such people (Dowling, Lechner, and Thielmann, 1998). The most visible form of media convergence today is interactive TV.

How will interactive TV operate in the near future? It will receive signals via cable, satellite dish, or fiber-optic telephone line. It will be connected to the Web through a built-in computer with a hard drive big enough to record 12 to 30 hours of programming. If you have interactive TV in your household, you will be able to program and record the exact blend of programs you want to watch from a long menu of specialty channels. You will be able to order feature-length movies, use e-mail, and hold videoconferences with people in

---

[4]The problem of media imperialism is felt particularly acutely in Canadian broadcasting because the country has a small population (slightly smaller than California's), is close to the United States, and is about 75% English speaking. Private broadcasters dominate the Canadian market and rely mainly on American entertainment programming. Moreover, the widespread use of cable and satellite dishes permits most Canadians to receive American programming directly from source (Knight, 1998 [1995]: 109–10).

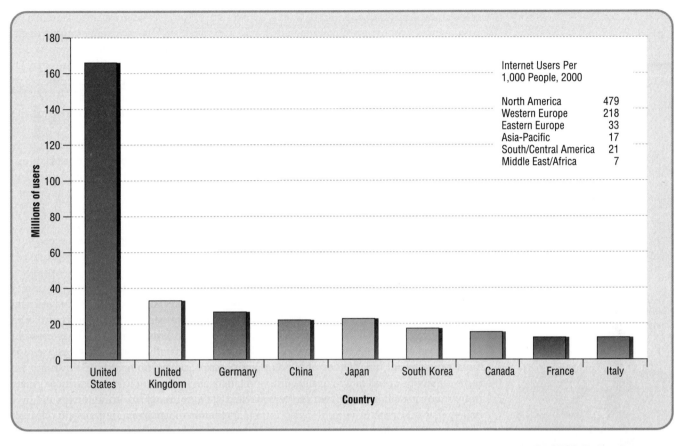

Internet Users Per
1,000 People, 2000

North America	479
Western Europe	218
Eastern Europe	33
Asia-Pacific	17
South/Central America	21
Middle East/Africa	7

♦ FIGURE 14.5 ♦

Internet Users Worldwide, 2001

SOURCES: Pastore (2001); "The World's Online Populations" (2001).

♦ FIGURE 14.6 ♦

Media Imperialism on Enaotai Island in West Papua New Guinea

SOURCE: Blake (2000) [Associated Press].

remote locations. You will be able to call up Web sites that offer additional information on the programs or movies you're watching, shop for a wide variety of goods, do your banking, and pay your utility bills. And you will be able to do all this from the convenience of your couch. Market researchers forecast that by 2004, 30 million United States households—about 27% of the total—will enjoy interactive TV. All the media giants are scrambling for market share. After all, the media conglomerates stand to earn many billions of dollars laying fiber-optic cable, building new media appliances, writing software, creating TV programs, and selling goods and services online (Davis, 2000).

The control of interactive TV by huge media conglomerates may seem like an old story. In some respects it is. Ownership of every mass medium has become more concentrated over time. Interactive TV seems poised to repeat the pattern more quickly than any other mass medium. Because entry costs are so high, only media giants can get involved.

However, this is a media story with a twist because consumers can *interact* with the new media. The big media conglomerates may be able to carve out a new and lucrative niche for themselves by merging the Internet and television. However, they can never fully dominate the Internet. That is because it is the first mass medium that makes it relatively easy for consumers to become producers.

Millions of people are not just passive users of the Internet. Instead, they help to create it. For example, the Web boasts millions of personal Web sites. As of June 2001, it contained nearly 1,800 role-playing communities (MUDs) with perhaps a million users ("The MUD Connector," 2001; see Chapter 4, "Socialization"). More than 11,000 cameras are connected to the Web. They open a window to the goings-on in people's offices, college dorms, apartments, and frat houses ("WebcamSearch.com," 2000). People have created nearly 158,000 formal discussion groups on the Internet ("Liszt's Usenet . . . ," 2000). Each group is composed of tens, hundreds, or thousands of individuals who discuss defined subjects by e-mail or in real time. Some discussion groups focus on particle physics. Others are devoted to banjoes, lawyer jokes, Russian politics, sadomasochism, and just about every other human activity imaginable. The groups are self-governing bodies with their own rules and norms of "netiquette" (McLaughlin, Osborne, and Smith, 1995; Sudweeks, McLaughlin, and Rafaeli, 1999). Entry costs are relatively low. All you need to join or create a Web site, a MUD, or a discussion group is your own PC, some free or inexpensive software, and an Internet connection. AOL/Time Warner and all the other media conglomerates have no control over the many thousands of communities that are proliferating online.

So we see that the image of the Internet as a medium that is subject to increasing domination by large conglomerates needs to be tempered by awareness of the contrary trend. Individual users are also making independent, creative contributions to Internet growth. Similarly, the view that the growth of the Web is an example of American media imperialism has not gone unchallenged. For where some media analysts see American media imperialism, others see globalization and postmodernization, social processes we introduced in Chapter 3 ("Culture"). From the latter point of view, all cultures, including that of the United States, are becoming less homogeneous and more fragmented as they borrow elements from each other. Inexpensive international travel and telecommunications make this cultural blending possible. Thus, if you look carefully at the Web, you will see that even United States sites adopt content liberally from Latin America, Asia, and elsewhere. Just as international influences can be seen in today's hairstyles, clothing fashions, foods, and popular music, so can the Internet be seen as a site of globalization (Hall, 1992).

Of course, nobody knows exactly how the social forces outlined above will play themselves out. A phenomenon like Napster emerges, enabling millions of people to freely share recorded music on the Web using a central server. Some analysts point to Napster as evidence of Internet democratization. Then the media conglomerates take Napster to court, forcing it to stop the giveaway on the grounds that it is effectively stealing royalties from musicians and profits from music companies. Some analysts see the court case as evidence of inevitable conglomeration on the Web. Then new Napster-like programs, such as Gnutella, emerge. They allow people to share recorded music on the Web *without* a central server, making them virtually impossible to shut down. (The name "Gnutella" comes

Shawn Fanning, creator of Napster. Napster allowed people to freely share recorded music on the Web but media conglomerates took Napster to court, forcing it to stop the giveaway on the grounds that it was effectively stealing royalties from musicians and profits from music companies. Since then, alternative music sharing programs, such as Gnutella and Morpheus, have become popular, suggesting there is no end in sight to the tug of war between the forces for and against media conglomerates on the Web.

from the popular nut-based paste, which spreads easily; see "Gnutella," 2000.) And so the tug of war between conglomeration and democratization continues, with no end in sight. One thing is clear, however. The speed of technological innovation and the many possibilities for individual creativity on the new media frontier make this an exciting era to be involved in the mass media and to study it sociologically.

SUMMARY

1. The mass media sometimes blur the line between reality and fantasy.

2. The mass media are means of transmitting information and entertainment from one person or group to another. The communication is typically from a few senders to many receivers.

3. The mass media became truly large-scale only when penny newspapers were published in the first half of the 19th century. The electronic media are products of the 20th century.

4. Three main historical forces stimulated the growth of the mass media. The Protestant Reformation of the 16th century encouraged people to read the Bible themselves. The democratic movements that began in the late 18th century encouraged people to demand literacy. Beginning in the late 19th century, capitalist industrialization required rapid means of communication and fostered the mass media as important sources of profit.

5. Functionalism stresses that the mass media act to coordinate society, exercise social control, and socialize and entertain people.

6. Conflict theory stresses that the mass media reinforce social inequality. They do this both by acting as sources of profit for the few people who control media conglomerates and by promoting core values that help legitimize the existing social order.

7. Although the mass media promote social cohesion, they also differentiate groups by generation, gender, class, and ethnicity.

8. Interpretive approaches to studying the mass media stress that audiences actively filter, interpret, and sometimes even resist and reject media messages according to their interests and values.

9. The interaction between the producers and consumers of media messages is most evident on the new media frontier formed by the Internet, television, and other mass media.

GLOSSARY

Cultural studies is an increasingly popular interdisciplinary area of media research. It focuses not just on the cultural meanings producers try to transmit but also on the way audiences filter and interpret mass media messages in the context of their own interests, experiences, and values.

The **mass media** are print, radio, television, and other communication technologies. The word "mass" implies that the media reach many people. The word "media" signifies that communication does not take place directly through face-to-face interaction. Instead, technology intervenes or mediates in transmitting messages from senders to receivers. Furthermore, communication via the mass media is usually one-way, or at least one-sided. There are few senders (or producers) and many receivers (or audience members).

Media convergence is the blending of the World Wide Web, television, and other communications media as new, hybrid media forms.

Media imperialism is the domination of a mass medium by a single national culture and the undermining of other national cultures.

The **two-step flow of communication** between mass media and audience members involves (a) respected people of high status and independent judgment evaluating media messages and (b) other members of the community being influenced to varying degrees by these opinion leaders. Due to the two-step flow of communication, opinion leaders filter media messages.

QUESTIONS TO CONSIDER

1. Locate Webcams in your region or community by searching http://www.webcamsearch.com. Who has set up these Webcams? For what purposes? How might they be used to strengthen ties among family members, friendship networks, special-interest groups, and so forth?

2. By phoning TV and radio stations in your community, find out which are locally controlled and which are controlled by large media companies. If there are no locally controlled stations, does this have implications for the kind of news coverage and public affairs programming you may be watching? If there are locally controlled stations in your community, do they differ in terms of programming content, audience size, and audience type from stations owned by large media companies? If you can observe such differences, why do they exist?

3. The 1999 movie, *Enemy of the State,* starring Will Smith, depicts one drawback of new media technologies: the possibility of intense government surveillance and violation of privacy rights. Do you think new media technologies are unqualified blessings or could they limit our freedom and privacy? If so, how? How could the capacity of new media technologies to limit freedom and privacy be constrained?

WEB RESOURCES

Companion Web Site for This Book

http://sociology.wadsworth.com

Begin by clicking on the Student Resources section of the Web site. Choose "Introduction to Sociology" and finally the Brym and Lie book cover. Next, select the chapter you are currently studying from the pull-down menu. From the Student Resources page you will have easy access to InfoTrac College Edition®, MicroCase Online exercises, additional Web links, and many other resources to aid you in your study of sociology, including practice tests for each chapter.

InfoTrac Search Terms

These search terms are provided to assist you in beginning to conduct research on this topic by visiting http://www.infotraccollege. com/wadsworth.

Cultural studies   **Media convergence**
Media bias   **Media imperialism**
Media concentration

Recommended Web Sites

"The Media and Communications Studies Site" at the University of Wales is one of the best sites on the Web devoted to the mass media. Among other interesting sections, it includes useful theoretical materials and resources on class, gender, and race in the mass media, mainly from a British perspective. Visit it at http://www.aber.ac.uk/media/Functions/mcs.html. Although the site contains some American materials, you might want to supplement it by visiting the University of Iowa's "Gender, Ethnicity and Race in Media" site at http://www.uiowa.edu/commstud/ resources/GenderMedia.

For useful resources on corporate ownership of the mass media and the social and political problems that result from increasingly concentrated media ownership, go to http://www.fair.org/ media-woes/corporate.html.

"Project Censored" is run out of Sonoma State University in California. It defines its main goal as follows: "to explore and publicize the extent of censorship in our society by locating stories about significant issues of which the public should be aware, but is not..." Visit the project site at http://www.projectcensored.org/ intro.htm.

"WebcamSearch.com" at http://www.webcamsearch.com is a directory with links to the more than 11,000 cameras broadcasting to the Web worldwide.

SUGGESTED READINGS

Manuel Castells. *The Information Age: Economy, Society and Culture,* 3 vols. (Oxford: Blackwell, 1996–1999). An ambitious effort to paint a comprehensive picture of the society of the "information age," in which the mass media play a key role.

Stuart Ewen. *PR!: A Social History of Spin* (New York: Basic, 1997). Shows how various mass media organizations manipulate images and messages.

Douglas Kellner. *Media Culture: Cultural Studies, Identity and Politics Between the Modern and the Postmodern* (New York: Routledge, 1995). Assesses the production, meaning, and reception of contemporary American mass culture. A balanced and comprehensive interpretation.

Jackson Lears. *Fables of Abundance: A Cultural History of Advertising in America* (New York: Basic, 1994). A fascinating history of American advertising, one of the main engines of media growth.

IN THIS CHAPTER, YOU WILL LEARN THAT:

✦ Health risks are unevenly distributed in human populations. Men and women, upper and lower classes, rich and poor countries, and privileged and disadvantaged racial and ethnic groups are exposed to health risks to varying degrees.

✦ The average health status of Americans is lower than the average health status of people in other rich postindustrial countries. That is partly because the level of social inequality is higher in the United States and partly because the health care system makes it difficult for many people to receive adequate care in this country.

✦ Medical successes created new problems. For instance, they allow people to live longer than they used to. This gives degenerative diseases such as cancer and heart disease more chance to develop. In turn, the increased incidence of such diseases raises new questions about when and how people should be allowed to die.

✦ The meaning people attach to aging and death varies historically and from one country to the next.

✦ The dominance of medical science is due to its successful treatments and the way doctors excluded competitors and established control over their profession and their clients.

✦ Patient activism, alternative medicine, and holistic medicine promise to improve the quality of health care in the United States and globally.

THE BLACK DEATH

In 1346, rumors reached Europe of a plague sweeping the East. Originating in Asia, the epidemic spread along trade routes to China and Russia. A year later, 12 galleys sailed from southern Russia to Italy. Diseased sailors were aboard. Their lymph nodes were terribly swollen and eventually burst, causing painful death. Anyone who came in contact with the sailors was soon infected. As a result, their ships were driven out of several Italian and French ports in succession. Yet the disease spread relentlessly, again moving along trade routes to Spain, Portugal, and England. Within two years, the Black Death, as it came to be known, killed a third of Europe's population. Six hundred and fifty years later, the plague still ranks as the most devastating catastrophe in human history (Herlihy, 1998; McNeill, 1976; Zinsser, 1935).

Today we know the cause of the plague was a bacillus that spread from lice to rats to people. It spread so efficiently because many people lived close together in unsanitary conditions. In the middle of the 14th century, however, nobody knew anything about germs. Therefore, Pope Clement VI sent a delegation to Europe's leading medical school in Paris to find out the cause of the plague. The learned professors studied the problem. They reported that a particularly unfortunate conjunction of Saturn, Jupiter, and Mars in the sign of Aquarius had occurred in 1345. The resulting hot, humid conditions caused the earth to emit poisonous vapors. To prevent the plague, they said, people should refrain from eating poultry, waterfowl, pork, beef, fish, and olive oil. They should not sleep during the daytime or engage in excessive exercise. Nothing should be cooked in rainwater. Bathing should be avoided at all costs.

We do not know if the Pope followed the professors' advice. We do know he made a practice of sitting between two large fires to breathe pure air. Because the plague bacillus is destroyed by heat, the practice may have saved his life. Other people were less fortunate. Some rang church bells and fired cannons to drive the plague away. Others burned incense, wore charms, and cast spells. But, apart from the Pope, the only people to have much luck in avoiding the plague were the well to do (who could afford to flee the densely populated cities for remote areas in the countryside) and the Jews (whose religion required that they wash their hands before meals, bathe once a week, and conduct burials soon after death).

Some of the main themes of the sociology of health, aging, and medicine are embedded in the story of the Black Death, or at least implied by it. First, recall that some groups were more likely to die of the plague than others. This is a common pattern. Health risks are always unevenly distributed. Women and men, upper and lower classes, rich and poor countries, and privileged and disadvantaged racial and ethnic groups are exposed to health risks to varying degrees. This suggests that health is not just a medical question but also a sociological issue. The first task we set ourselves below is to examine the sociological factors that account for the uneven distribution of health in society.

The story of the Black Death also suggests that health problems change over time. Epidemics of various types still break out, but there can be no Black Death where sanitation and hygiene prevent the spread of disease.[1] Today we are also able to treat many infectious diseases such as tuberculosis and pneumonia with antibiotics. These wonder drugs and many other lifesaving therapies were developed by 20th-century medical science.

However, our medical successes have created new problems. For instance, due to the overuse of antibiotics, resistant microbes have developed through mutation. We are now faced with the spread of forms of tuberculosis and other infectious diseases once considered eliminated. Similarly, medical successes allow people to live longer than they used to. **Life expectancy** is the average age at death of the members of a population. Life expectancy in the United States was 47 years in 1900. In 2000, it was 77 years. As a result

---

[1]One case that may approximate the Black Death in the 21st century is the spread of AIDS in sub–Saharan Africa. At the end of 1999, there were 16 African countries in which more than 10% of the adult population aged 15–49 was infected with HIV. In South Africa, nearly 20% of adults were infected, and in seven other countries in the southern cone of Africa, over 20% of adults were living with the virus. In Botswana, over 35% of adults were infected with HIV (United Nations, 2000).

	Deaths per 100,000 Population	Percent of Deaths	Percent Change Since 1979	Ratios Male:Female	Ratios Black:White
1900					
1. Pneumonia/influenza	202.2	11.8	—	—	—
2. Tuberculosis	194.4	11.3	—	—	—
3. Diarrhea/other intestinal	142.7	8.3	—	—	—
4. Heart disease	137.4	8.0	—	—	—
5. Stroke	106.9	6.2	—	—	—
6. Kidney disease	88.6	5.2	—	—	—
7. Accidents	72.3	4.2	—	—	—
8. Cancer	64.0	3.7	—	—	—
9. Senility	50.2	2.9	—	—	—
10. Bronchitis	40.3	2.3	—	—	—
All other causes	620.1	36.1	—	—	—
Total	1,719.1	100.0	—	—	—
1997					
1. Heart disease	271.6	31.4	−34.6	1.8	1.5
2. Cancer	201.6	23.3	−4.0	1.4	1.3
3. Stroke	59.7	6.9	−37.7	1.2	1.8
4. Lung disease	40.7	4.7	44.5	1.5	0.8
5. Accidents	35.7	4.1	−29.8	2.4	1.2
6. Pneumonia/influenza	32.3	3.7	15.2	1.5	1.4
7. Diabetes	23.4	2.7	37.8	1.2	2.4
8. Suicide	11.4	1.3	−9.4	4.2	0.6
9. Kidney disease	9.5	1.1	2.3	1.5	2.6
10. Liver disease	9.4	1.1	−38.3	2.3	1.2
11. Alzheimer's disease	8.4	1.0	1,250.0	0.9	0.7
12. Blood poisoning	8.4	1.0	82.6	1.2	2.8
13. Homicide/legal intervention	7.4	0.9	−21.6	3.8	6.0
14. AIDS	6.2	0.7	—	3.5	7.5
15. Hardening of arteries	6.0	0.7	−63.2	1.3	1.0
All other causes	132.9	15.4	—	—	—
Total	864.7	100.0	−17.0	1.6	1.5

✦ TABLE 15.1 ✦
Leading Causes of Death, United States, 1900 and 1997

SOURCES: Adapted from Centers for Disease Control and Prevention (1999c: 5); National Office of Vital Statistics (1947).

of increased life expectancy, degenerative conditions such as cancer and heart disease have an opportunity to develop in a way that was not possible a century ago (see Table 15.1, column 1). The ability to prolong life by technical means also raises new questions about when people should be allowed to die and whether under some circumstances medical personnel should be allowed to assist them in bringing life to a close. The sociology of aging examines how society copes with such consequences of a growing elderly population. It is the second major issue we examine in this chapter.

The story of the Black Death raises a third issue too. We cannot help being struck by the superstition and ignorance surrounding the treatment of the ill in medieval times. Remedies were often herbal but also included earthworms, urine, and animal excrement. People believed it was possible to maintain good health by keeping body fluids in balance. Therefore, cures that released body fluids were common. These included hot baths, laxatives, and diuretics, which increase the flow of urine. If these treatments didn't work, bloodletting was often prescribed. No special qualifications were required to administer medical treatment. Barbers doubled as doctors.

However, the backwardness of medieval medical practice, and the advantages of modern scientific medicine, can easily be exaggerated. For example, medieval doctors stressed the importance of prevention, exercise, a balanced diet, and a congenial environment in maintaining good health. We now know this is sound advice. On the other hand, one of the great shortcomings of modern medicine is its emphasis on high-tech cures rather than preventive and environmental measures. Therefore, in the third section of this chapter, we investigate not just the many wonderful cures and treatments brought to us by modern scientific medicine, but also its weaknesses. We also examine how the medical professions gained substantial control over health issues and promoted their own approach to well-being.

HEALTH AND INEQUALITY

Defining and Measuring Health

According to the World Health Organization (WHO), **health** is

> the ability of an individual to achieve his [or her] potential and to respond positively to the challenges of the environment . . . The basic resources for health are income, shelter and food. Improvement in health requires a secure foundation in these basics, but also information and life skills; a supportive environment, providing opportunities for making health choices among goods, services and facilities; and conditions in the economic, social and physical environments . . . that enhance health (World Health Organization, 2000).

The WHO definition lists in broad terms the main factors that promote good health. However, when it comes to *measuring* the health of a population, sociologists typically examine the negative: rates of illness and death. They reason that healthy populations experience less illness and longer life than unhealthy populations. This is the approach we follow here.

Assuming ideal conditions, how long can a person live? In the 21st century, the **maximum human life span** may well increase due to medical advances. So far, however, the record is held by Jeanne Louise Calment, a French woman who died in 1997 at the age of 122. (Other people claim to be older, but they lack authenticated birth certificates.)

Calment was an extraordinary individual. She took up fencing at 85, rode a bicycle until she was 100, gave up smoking at 120, and released a rap CD at 121 (Matalon, 1997). In contrast, only one in a hundred people in the world's rich countries now lives to be 100. Medical scientists tell us that the **maximum average human life span**—the average age of death for an entire population under *ideal* conditions—is likely to increase in this century. Now it is roughly 87 years (Olshansky, Carnes, and Cassel, 1990).

Unfortunately, conditions are nowhere ideal. Life expectancy throughout the world is less than 87 years. Figure 15.1 shows life expectancy in selected countries. Leading the list is Japan, where life expectancy was 81 years in 2000. Life expectancy was one to four years lower in the other rich postindustrial countries, including the United States. But in India, life expectancy was only 61 years. The poor African country of Niger suffered the world's lowest life expectancy at 41 years (Population Reference Bureau, 2000).

Accounting for the difference between the maximum average human life span and life expectancy is one of the main tasks of the sociologist of health. For example, while the maximum average human life span is 87 years, life expectancy in the United States is 77 years. This implies that, on average, Americans are being deprived of 10 years of life due to avoidable *social* causes (87 − 77 = 10). Avoidable social causes deprive the average citizen of Niger of more than 46 years of life (87 − 41 = 46). Clearly, social causes have a big—and variable—impact on illness and death. We must therefore discuss them in detail.

In the 21st century, the maximum human life span may increase due to medical advances. So far, the record is held by Jeanne Louise Calment, a French woman who died in 1997 at the age of 122.

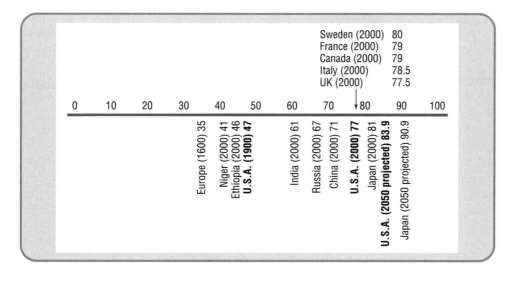

✦ FIGURE 15.1 ✦
Life Expectancy, Selected Countries (in years)

Note: We estimated some 2000 life expectancies by adding male and female life expectancies, dividing by 2, and adding 0.5 year to take account of the fact that the female to male sex ratio is greater than 1.

SOURCES: Population Reference Bureau (2000); Tuljapurkar, Li, and Boe (2000).

The Social Causes of Illness and Death

People get sick and die partly due to natural causes. One person may have a genetic predisposition to cancer. Another may come in contact with a deadly Ebola virus in the environment. However, over and above such natural causes of illness and death, we can single out three types of *social* causes:

1. *Human-environmental factors.* The environment constructed by humans poses major health risks. For example, more than 100 oil refineries and chemical plants are concentrated in a 75-mile strip between New Orleans and Baton Rouge. The area is commonly known as "cancer alley." That is because the petrochemical plants spew cancer-causing pollutants into the air and water. Local residents, overwhelmingly African American, are more likely than other Americans to get cancer because they breathe and drink high concentrations of these pollutants (Bullard, 1994 [1990]). Cancer alley is a striking illustration of how human-environmental conditions can cause illness and death (see also Chapter 18, "Technology and the Global Environment").

2. *Lifestyle factors.* Smoking cigarettes, excessive use of alcohol and drugs, poor diet, lack of exercise, and social isolation are among the chief lifestyle factors associated with poor health and premature death. For example, a third of the people who smoke are likely to die prematurely from smoking-related illnesses. This amounts to about half a million Americans annually. About 30% of all cancer deaths in the United States result from tobacco use. About 35% result from poor diet (Remennick, 1998: 17). Social isolation, too, affects one's chance of becoming ill and dying prematurely. Thus, unmarried people have a greater chance of dying prematurely than older people. At any age, the death of a spouse increases one's chance of dying while remarrying decreases one's chance of dying (Helsing, Szklo, and Comstock, 1981). Social isolation is a particularly big problem among elderly people who retire, lose a spouse and friends, and cannot rely on family members or state institutions for social support. Such people are prone to fall into a state of depression that contributes to ill health.

3. *Factors related to the public health and health care systems.* The state of a nation's health depends partly on public and private efforts to improve people's well-being and treat their illnesses. The **public health system** is composed of government-run programs that ensure access to clean drinking water, basic sewage and sanitation services, and inoculation against infectious diseases. The absence of a public health system is associated with high rates of disease and low life expectancy. The **health care system** is composed of a nation's clinics, hospitals, and other facilities for

ensuring health and treating illness. The absence of a system that ensures its citizens have access to a minimum standard of health care is also associated with high rates of disease and low life expectancy.

Exposure to all three sets of social causes of illness and death is strongly related to country of residence, class, race, and gender. We now consider the impact of these factors, beginning with country of residence.

Web Research Projects
The Changing Social
Distribution of AIDS/HIV

Country of Residence

AIDS is the leading cause of death in urban Haiti. Extreme poverty has forced many Haitians to become prostitutes. They cater mainly to tourists from North America and Europe. Some of those tourists carried HIV, the virus that leads to AIDS, and introduced it into Haiti. The absence of adequate health care and medical facilities makes the epidemic's impact all the more devastating (Farmer, 1992).

AIDS is also the leading cause of death in the poverty-stricken part of Africa south of the Sahara desert. Table 15.2 shows that in December 1999, over 8.5% of sub–Saharan Africans—24.5 million people—were living with AIDS/HIV (see also Figure 15.2). In

A health worker at Nazareth House in Cape Town, South Africa, lavishes care and attention on some of the 41 infected children in her care. Nearly a fifth of South Africa's adult population is infected with AIDS/HIV.

◆ **TABLE 15.2** ◆
A Global View of AIDS/HIV, December 1999

Note: The cumulative number of deaths due to AIDS on December 31, 1999, was about 14.5 million. More than 80% of these deaths had occurred in sub–Saharan Africa.

SOURCE: United Nations (2000).

	People Living With AIDS/HIV	Adult Rate (in percent)	AIDS Deaths, 1999
Began Late 70s– Early 80s			
Sub–Saharan Africa	24,500,000	8.57	2,200,000
Latin America	1,300,000	0.49	48,000
Caribbean	360,000	2.11	30,000
Western Europe	520,000	0.23	6,800
North America	900,000	0.58	20,000
Australia & New Zealand	15,000	0.13	120
Began Late 80s			
North Africa & Middle East	220,000	0.12	13,000
South & Southeast Asia	5,600,000	0.54	460,000
East Asia & Pacific	530,000	0.06	18,000
Began Early 90s			
Eastern Europe & Central Asia	420,000	0.21	8,500
Total	34,365,000	1.07	2,804,000

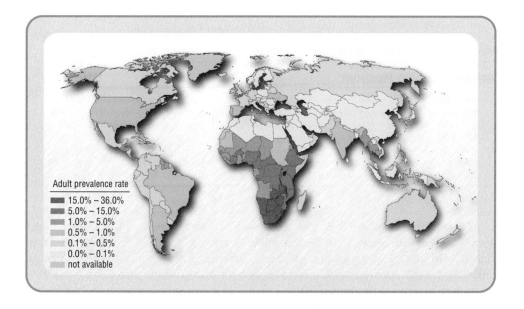

◆ **FIGURE 15.2** ◆
**Adult Prevalence Rate
of AIDS/HIV Infection**

SOURCE: UNAIDS/World Health
Organization.

contrast, 0.58% of North Americans and 0.23% of Western Europeans were living with AIDS/HIV. This means that AIDS/HIV is nearly 15 times more common in sub–Saharan Africa than in North America and 40 times more common than in Western Europe. AIDS was the 14th leading cause of death in the United States in 1997, accounting for 0.7% of all deaths that year (see Table 15.1). Despite the much greater prevalence of AIDS/HIV in sub–Saharan Africa, however, spending on research and treatment is concentrated overwhelmingly in the rich countries of North America and Western Europe. As the case of AIDS/HIV illustrates, global inequality influences the exposure of people to different health risks.

You might think that prosperity increases health due to biomedical advances, such as new medicines and diagnostic tools. If so, you are only partly correct. Biomedical advances do increase life expectancy. In particular, vaccines against infectious diseases have done much to improve health and ensure long life. However, the creation of a sound public health system was even more important in this regard. If a country can provide its citizens with clean water and a sewage system, epidemics decline in frequency and severity while life expectancy soars.

The industrialized countries started to develop their public health systems in the mid-19th century. Social reformers, concerned citizens, scientists, and doctors joined industrialists and politicians in urging governments to develop health policies that would help create a healthier labor force and citizenry (Goubert, 1989 [1986]; McNeill, 1976; Rosenberg, 1962). But what was possible in North America and Western Europe 150 years ago is not possible in many of the developing countries today. Most of us take clean water for granted. In contrast, more than 1 billion of the world's 6 billion people do not have access to a sanitary water supply (de Villiers, 1999).

Other indicators of health inequality for selected countries are given in Table 15.3. We see immediately that there is a positive association between national wealth and good health. Thus, the United States, Japan, and Canada are rich countries. They spend a substantial part of their wealth on health care. Many physicians and nurses service their populations. As a result, **infant mortality** (the annual number of deaths before the age of one for every 1,000 live births) is low. So is the rate of stunted growth due to malnutrition among children under the age of five. As noted above, rich countries also enjoy high life expectancy. But Mexico, which is poorer than the United States, Japan, and Canada, spends a smaller proportion of its wealth on health care. Accordingly, its population is less healthy in a number of respects. The sub–Saharan country of Niger is one of the poorest countries in the world. It spends little on health care, has few medical personnel, and suffers from high rates of malnutrition stunting and infant mortality.

	Health Expenditures as % of GDP	Physicians/ 100,000 Population	Nurses and Midwives/ 100,000 Population	Infant Mortality/ 1,000 Live Births	Children Immunized Against Measles (%)	Malnutrition Stunting, Children Under 5 (%)
USA	14.0	245	878	7	82	2
Japan	7.2	177	641	4	n.a.	0*
Canada	9.0	221	958	6	98*	1*
Mexico	4.2	107	40	31	97	23
Niger	2.0	3	17	115	42	40

◆ **TABLE 15.3** ◆

Health Indicators, Selected Countries, Mid-1990s

* = estimate

SOURCE: Adapted from World Health Organization (1999b).

Closer inspection of Table 15.3 reveals an anomaly, however. The United States spends nearly twice as much per person on health care as Japan and over 50% more than Canada. On average, Americans work nearly 2 months a year just to pay their medical bills. The United States has 11% more doctors per 100,000 people than Canada and 38% more than Japan. Yet the United States immunizes a smaller percentage of its children against measles, contains a higher percentage of malnourished children, and has a higher rate of infant mortality than these other countries. On one measure—immunization of children against measles—the United States is substantially behind Mexico. The American case shows that spending more money on health care does not always improve the health of a nation.

Class, Race, and Gender

What accounts for the American anomaly? Why do we spend far more on health care than any other country in the world yet wind up with a population that, on average, is less healthy than the population of other rich countries? Part of the answer is that the gap between rich and poor is greater in the United States than in Japan, Sweden, Canada, France, and other rich countries. In general, the higher the level of inequality in a country, the more unhealthy its population (Wilkinson, 1996). Because, as we saw in Chapter 7, the United States contains a higher percentage of poor people than other rich countries contain, its average level of health is lower. Moreover, because income inequality has widened in the United States since the early 1970s, health disparities between income groups have grown (Williams and Collins, 1995).

Health inequality manifests itself in many ways. For instance, the infant mortality rate in Harlem, New York's main African-American ghetto, is higher than in Bangladesh (Shapiro, 1992). Male life expectancy in poor, African-American areas of Washington, DC, is 15 years lower than in the nearby rich suburb of Fairfax, Virginia (Epstein, 1998).

One reason for this disparity is that the poor are more likely than others to be exposed to violence, high-risk behavior, and environmental hazards. As you know, poverty is more common among African Americans than white Americans. One would therefore expect African Americans to have higher mortality rates than whites. The last column of figures in Table 15.1 shows the ratio of black to white deaths for the 15 leading causes of death in 1997. Inspect it and you will see that for 11 of the 15 leading causes of death, African Americans have higher mortality rates than white Americans. Overall, African Americans have a 50% higher mortality rate than white Americans.

A second reason why the poor are less healthy than the well to do is that they cannot afford adequate, and in some cases even minimal, health care. Thus, in 1992, 36% of people who earned less than $14,000 a year had no health insurance, compared to only 4% of people making more than $50,000 (Folbre, 1995). In spite of Medicaid, most poor people are inadequately served. Only about half the poor receive Medicaid assistance. Furthermore, poor people typically live in areas where medical treatment facilities are inadequate. This is especially true in recent decades, when many public hospitals that served the poor were closed due to government budget cuts (Albelda and Folbre, 1996).

If poor people have less access than the well-to-do to doctors and hospitals, they also tend to have less knowledge about healthy lifestyles. For example, they are less likely to

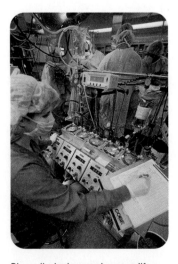

Biomedical advances increase life expectancy, but the creation of a sound public health system has even more dramatic effects.

know what constitutes a nutritious diet. This, too, contributes to their propensity to illness. Illness, in turn, makes it more difficult for poor people to escape poverty (Abraham, 1993).

Racial disparities in health status are largely, though not entirely, due to economic differences between racial groups. Thus, most studies show that blacks and whites *at the same income level* have similar health statuses. Nonetheless, the health status of African Americans is somewhat lower than the health status of European Americans even within the same income group. This suggests that racism affects health. It does so in three ways. First, income and other rewards do not have the same value across racial groups. For instance, due to discrimination, each year of education completed by an African American results in smaller income gains than it does for white Americans. Because, as we have seen, income is associated with good health, blacks tend to be worse off than whites at the same income level. Second, racism affects access to health services. That is because African Americans at all income levels tend to live in racially segregated neighborhoods with fewer health-related facilities. Third, the experience of racism induces psychological distress that has a negative effect on health status. For example, racism increases the likelihood of drug addiction and engaging in violence (Williams and Collins, 1995).

Increases in income have a bigger positive health impact on below-median income earners than on above-median income earners (see Figure 15.3). But inequality is not simply a matter of differential access to resources such as medical care and knowledge. Even among people who have the *same* access to medical resources, people of higher rank tend to live healthier and longer lives. Why? Researchers in the United States, Britain, and Canada have argued that people of high rank experience less stress because they are more in control of their lives. If you can decide when to work, how to work, and what to work on, if you can exercise autonomy and creativity at work, you are likely to be healthier than someone who lacks these freedoms. You not only have the resources to deal with stress. You also have the ability to turn it off. In contrast, subordinates in a hierarchy have little control over their work environment. They experience a continuous sense of vulnerability that results in low-level stress. Continuous low-level stress, in turn, results in reduced immune function, increased hardening of the arteries, increased chance of heart attack, and other ailments. In short, if access to medical resources is associated with improved health, so is lower stress—and both are associated with higher positions in the socioeconomic hierarchy (Epstein, 1998; Evans, 1999).

Finally, we must mention the health inequalities based on gender that some feminists emphasize. In a review of the relevant literature in the *New England Journal of Medicine,* one researcher concluded that such gender inequalities are substantial (Haas, 1998). Specifically:

◆ Gender bias exists in medical research. Thus, more research has focused on "men's diseases" (such as cardiac arrest) than on "women's diseases"(such as breast cancer). Similarly, medical research is only beginning to explore the fact that women

◆ **FIGURE 15.3** ◆
Mortality Rate by Group Income

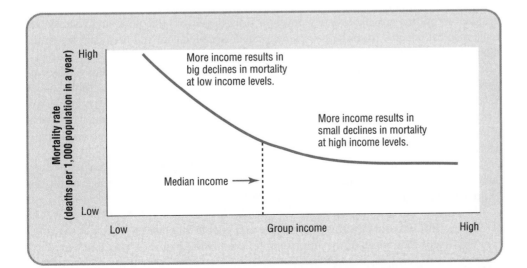

may react differently than men to some illnesses and may require different treatment regimes.

◆ Gender bias also exists in medical treatment. For example, women undergo fewer kidney transplants, various cardiac procedures, and other treatments than men.

◆ Because women live longer than men, they experience greater lifetime risk of functional disability and chronic illness, and greater need for long-term care. Yet more is spent on men's than women's health care in this country. (In contrast, Canadian health care spending for women and men, excluding expenditures related to childbirth, is about equal. This is probably due to the fact that Canada, unlike the United States, has a system of universal health insurance for a comprehensive range of health care services [Mustard, Kaufert, Kozyrskyj, and Mayer, 1998]. We analyze the American health care system below.)

◆ There are 40% more poor women than poor men in the United States (Casper, McLanahan, and Garfinkel, 1994: 597). Because, as we have seen, poverty contributes to ill health, we could expect improvements in women's economic standing to be reflected in improved health status for women.

In sum, although women live longer than men, gender inequalities have a negative impact on women's health. Women's health is negatively affected by differences between women and men in access to gender-appropriate medical research and treatment as well as the economic resources needed to secure adequate health care.

Health and Politics

Above we noted the existence of an "American anomaly." We spend more on health care than any other country, yet all the other rich postindustrial societies have healthier populations. One reason for this anomaly, as we have seen, is the relatively high level of social inequality in the United States. A second reason, which we will now examine, is the nature of the American health care system.

You will recall from our discussion in Chapter 1 ("A Sociological Compass") and elsewhere that conflict theory is concerned mainly with the question of how privileged groups seek to maintain their advantages and subordinate groups seek to increase theirs. As such, conflict theory is an illuminating approach to analyzing the American health care system. For the United States health care system is usefully seen as a system of privilege for some and disadvantage for others. It therefore contributes to the poor health of less well-to-do Americans.

Consider, for example, that the United States lacks a system of health insurance that covers the entire population. In other rich postindustrial countries, an average of about 80% of medical costs are paid by governments out of taxes. In the United States, the figure is about 40%. About 44 million Americans are not covered by health insurance at all. Another 40 million are inadequately covered. Only the elderly, the poor, and veterans receive medical benefits from the government under the Medicare and Medicaid programs (Starr, 1994 [1992]).

The vast majority of Americans are covered by private insurance programs run by employers and unions, although some people buy their own private coverage. Today, about 85% of employees buy their health coverage from health maintenance organization (HMOs) (Gorman, 1998). HMOs are private corporations. They collect regular payments from employers and employees. When an employee needs medical treatment, it is administered by the HMO.

Like all corporations, HMOs pursue profit. They employ four main strategies to keep their shareholders happy. Unfortunately, all four strategies lower the average quality of health care in the United States (Kuttner, 1998a; 1998b):

1. Some HMOs avoid covering sick people and people who are likely to get sick. This keeps their costs down. For example, if an HMO can show you had a medical

condition before you came under its care, the HMO won't cover you for that condition.

2. HMOs try to minimize the cost of treating sick people they can't avoid covering. Thus, HMOs have doctor-compensation formulas that reward doctors for withholding treatments that are unprofitable.

3. There have been allegations that some HMOs routinely inflate diagnoses to maximize reimbursements. For instance, at this writing, Columbia/HCA, the largest for-profit hospital chain in the nation, is under federal investigation for just that practice.

4. HMOs keep overhead charges high. In 1992, for example, administrative overhead composed 14% of total medical expenditures for private insurance companies. In contrast, administrative charges came to 4% for Medicaid and 2% percent for Medicare. In Canada, which has a government-run health care system that gives all citizens access to health care, administrative costs amount to 1% of expenditures (Folbre, 1995).

Despite these drawbacks, running health care institutions as for-profit organizations has one big advantage. Because functionalism tends to emphasize the contribution of social institutions to the smooth operation of society, it is an advantage that a functionalist analysis would undoubtedly highlight. Due to the fact that health organizations are so profitable, they can invest enormous sums in research, development, the latest diagnostic equipment, and high salaries to attract many of the best medical researchers and practitioners on the planet. As a result, the United States has the best health care system in the world—for those who can afford it.

The main supporters of the current United States health care system are the stockholders of the 1,500 private health insurance companies and the physicians and other health professionals who get to work with the latest medical equipment, conduct cutting-edge research, and earn high salaries. Thus, HMOs and the American Medical Association have been at the forefront of attempts to convince Americans that the largely private system of health care serves the public better than any state-run system could. Their efforts have been only partly successful. The 1998 General Social Survey asked a nationwide sample of Americans whether HMOs improve the quality of medical care. As Figure 15.4

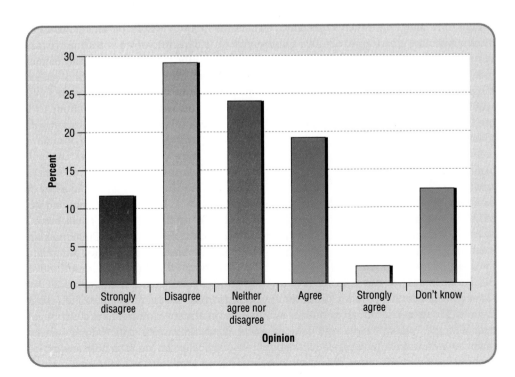

✦ FIGURE 15.4 ✦
"HMOs Improve the Quality of Care," United States, 1998 (in percent; n = 1,382)

SOURCE: National Opinion Research Center (1999).

shows, only 22% of Americans agree or strongly agree they do. In contrast, 41% disagree or strongly disagree. Nearly twice as many Americans disapprove of HMOs as approve of them (National Opinion Research Center, 1999).

Despite this overall negative evaluation of the private health care system, attempts to create a national system of health care in which everyone is covered regardless of his or her employment status or income level have failed (Hacker, 1997; Marmor, 1994; Quadagno, 1988). Most recently, Congress rejected President Clinton's 1993–94 Health Security proposal. Clinton was unable to unify political and public support for his proposal, partly because of the massive media campaign bankrolled by health insurance companies and Clinton's political opponents (Skocpol, 1996).

What is your opinion about American health care? Do you favor the existing, largely private system or would you like to see the United States move toward a state-run system like those of the other rich postindustrial countries? Would you personally benefit more from one type of system than the other? If so, why? Does the type of system from which you would personally benefit influence the type of system you think is best for the country? Do you think it is possible to create a health care system that encourages cutting-edge research and development, attracts the best medical personnel, *and* brings the average level of health care up to the level of the other rich postindustrial countries? If so, how would you structure such a health care system? If not, what compromises do you think different groups of Americans would have to make in order to achieve this ideal?

AGING AS A SOCIAL PROBLEM

If health care is an attempt to prolong life and improve its quality, aging is health care's relentless foe. Many people think of aging simply as a natural, biological process that inevitably thwarts our best attempts to delay death. Sociologists, however, see aging in a more complex light. For them, aging is also a deeply social phenomenon. Specifically, as we saw in Chapter 4 ("Socialization"), aging is a process of socialization, or learning new roles appropriate to different stages of life. The sociological nature of aging is also evident in the fact that its significance varies from one society to the next. That is, different societies attach different *meanings* to the progression of life through its various stages. Menopause, for example, occurs in all mature women. In the United States, it is often seen as a major life event. The old euphemism for menopause was the rather dramatic expression, "change of life." In contrast, menopause is a relatively minor matter in Japan. Moreover, while menopausal American women frequently suffer "hot flashes," menopausal Japanese women tend to complain mainly about "stiff shoulders" (Lock, 1993). In many Western countries, complaining about stiff shoulders is a classic symptom of having just given birth. As this example shows, the stages of life are not just natural processes but events deeply rooted in society and culture. As we will see, the same holds for death.

Aging and the Life Course

All individuals pass through distinct stages of life, which, taken together, sociologists call the **life course.** These stages are often marked by **rites of passage,** or rituals signifying the transition from one life stage to another (Fried and Fried, 1980). Circumcision, baptism, confirmation, the bar mitzvah and bat mitzvah, college convocation, the wedding ceremony, and the funeral are among the best-known rites of passage in the United States.

As we saw in Chapter 4 ("Socialization"), the duration of each stage of life differs from one society and historical period to the next. For example, there are no universal rules about when one becomes an adult. In preindustrial societies, adulthood arrived soon after puberty. In Japan, one becomes an adult at age 20. In the United States, adulthood arrives at age 18 (the legal voting age) or 21 (the legal drinking age).

Even the *number* of life stages varies historically and across societies. For instance, childhood was a brief and insignificant stage of development in medieval Europe (Ariès, 1962). In contrast, childhood is a prolonged stage of development in rich societies today, and adolescence is a new phase of development that was virtually unknown just a few hundred years ago (Kett, 1977). Increased life expectancy and the need for a highly educated labor force made childhood and adolescence possible and necessary.

Finally, although some life-course events are universal—birth, puberty, marriage, and death—not all cultures attach the same significance to them. Thus, ritual practices marking these events vary. For example, formal puberty rituals in many preindustrial societies are extremely important because they mark the transition to adult responsibilities. However, adult responsibilities do not immediately follow puberty in industrial and postindustrial societies because of the introduction of a prolonged period of childhood and adolescence. Therefore, formal puberty rituals are less important in such societies.

Sociological Aspects of Aging

As you pass through the life course, you learn new patterns of behavior that are common to people about the same age as you. Sociologically speaking, a category of people born in the same range of years is called an **age cohort.** For example, all Americans born between 1980 and 1989 form an age cohort. **Age roles** are patterns of behavior that are expected of people in different age cohorts. Age roles form an important part of our sense of self and others (Riley, Foner, and Waring, 1988). As we pass through the stages of the life course, we assume different age roles. To put it simply, a child is supposed to act like a child, an elderly person like an elderly person. We may find a 5-year-old dressed in a suit cute but look askance at a lone 50-year-old on a merry-go-round. "Act your own age" can be applied to people of all ages who do not conform to their age roles. Many age roles are informally known by character types, such as "rebellious teenager" or "wise old woman." We formalize some age roles by law. For instance, the establishment of minimum ages for drinking, driving, and voting formalizes certain aspects of the adolescent and adult age roles.

We find it natural that children in the same age cohort, such as preschoolers in a park, should play together, or that people of similar age cluster at parties. Differences across age cohorts are sufficiently large in the United States that some sociologists regard youth culture as a distinct subculture. Adolescents and teenagers—divided though they may be by gender, class, race, and ethnicity—frequently share common interests in music, movies, and so forth.

A **generation** is a special type of age cohort. Many people think of a generation as people born within a 15- to 30-year span. Sociologists, however, usually define a generation more narrowly. From a sociological point of view, a generation is composed of members of an age cohort who have unique and formative experiences during their youth. Age cohorts are statistically convenient categories, but most members of a generation are conscious of belonging to a distinct age group. For example, "baby boomers" are North Americans who were born in the prosperous years from 1946 to 1964. Most of them came of age between the mid-1960s and the early 1970s. Common experiences that bind them together include major historical events (e.g., the war in Vietnam, the Watergate break-in, and the subsequent resignation of President Nixon) and popular music (e.g., the songs of Bob Dylan, the Beatles, and the Rolling Stones). "Generation X" followed the baby boomers. Members of Generation X faced a period of slower economic growth and a job market glutted by the baby boomers. As a result, many of them resented having to take so-called "McJobs" when they entered the labor force. Thus, Douglas Coupland, the novelist who invented the term Generation X, cuttingly defined a McJob as a "low pay, low-prestige, low-dignity, low-benefit, no-future job in the service sector. Frequently considered a satisfying career choice by people who have never held one" (Coupland, 1991: 5).

Certain generationally defined moments sometimes help to crystallize the feeling of being a member of a particular generation. For instance, when you are elderly you will probably remember where you were when you heard about the terrorist attacks of

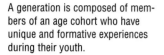

A generation is composed of members of an age cohort who have unique and formative experiences during their youth.

September 11, 2001. Such memories may someday help you to distinguish yourself from those who are too young to remember these tragic events. Finally, it should be noted that generations sometimes play a big role in history. Revolutionary movements, whether in politics or the arts, are sometimes led by members of a young generation who aggressively displace members of an older generation (Eisenstadt, 1956; Mannheim, 1952; Spitzer, 1973).

Age stratification refers to social inequality between age cohorts. It exists in all societies and may be observed in everyday social interaction. For example, there is a clear status hierarchy in most schools. On average, seniors enjoy higher status than sophomores while sophomores enjoy higher status than freshmen. However, the strength of age stratification differs across social contexts. For example, John Lie worked for a time in a South Korean corporation and noticed a practice alien to the United States. Whenever a new manager started work, everyone in the department who was older than the new manager either resigned or was reassigned. Given the importance of age seniority in South Korea, it was considered difficult for a manager to hold authority over older employees. Older employees in turn would have found it demeaning to be managed by a younger boss.

Some people think that ancient China and other societies were **gerontocracies,** that is, societies in which elderly men ruled, earned the highest incomes, and enjoyed the most prestige. While some societies did approximate this model, its extent has been exaggerated. Powerful, wealthy, and prestigious "elders" are often mature, but not necessarily the oldest, men. The United States today is typical of most societies, past and present, in this regard. For example, in the United States, median income gradually rises with age, reaching its peak in the 45–54 age cohort. Median income then declines for the oldest age cohorts (see Figure 15.5). Prestige and power follow the same course. This pattern reflects the fact that old age is usually not regarded as an unambiguous good. It denotes physical and mental decline and the nearness of death. Ambivalence about aging—especially as people reach the oldest age cohorts—is a cultural universal (Minois, 1989 [1987]). This ambivalence is accentuated by the common knowledge that the elderly face a host of social problems, the most serious of which we now examine.

Social Problems of the Elderly

Web Interactive Exercises
The Graying of America

If you've been to south Florida lately, you have a good idea of what the age composition of the United States will look like in 50 years. Figure 15.6 shows how the elderly have

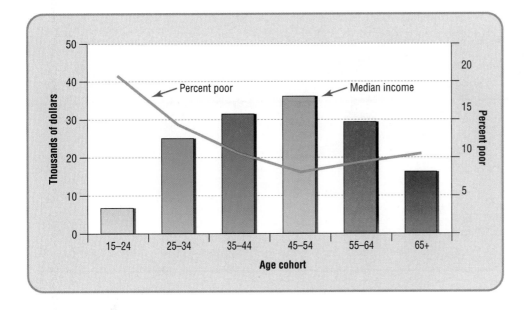

◆ **FIGURE 15.5** ◆
Median Income and Percent Poor by Age Cohort, United States, 1998

SOURCE: United States Bureau of the Census (1998:475, 478).

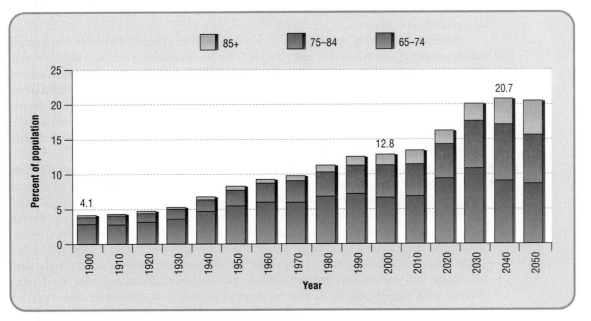

◆ **FIGURE 15.6** ◆
Elderly as Percent of United States Population, 1900–2050 (projected)

SOURCE: United States Administration on Aging (1999).

grown as a percentage of the United States population since 1900 and how this age cohort is expected to grow until 2050. In 1900, only about 4% of the United States population was 65 and over. Today, the figure is around 13%. By 2040, nearly 21% of Americans will be elderly. Thereafter, their weight in the United States population will start to decline. Figure 15.7 shows the geographical distribution of Americans 65 years and over.

Many sociologists of aging refer to elderly people who enjoy relatively good health—usually people between the ages of 65 and 74—as the "young old" (Neugarten, 1974; Laslett, 1991 [1989]). They refer to people 85 and over as the "old old." Figure 15.6 shows that the young old are expected to decline as a percentage of the American population after 2030. In contrast, the proportion of old old is expected to continue increasing. The rising number of old old concerns many people because they are most likely to suffer general physiological decline, life-threatening diseases, social isolation, and poverty.

Significantly, the sex ratio (the number of men compared to the number of women) falls with age. That is, because women live longer than men on average, there are many

♦ **FIGURE 15.7** ♦

Population Age 65+, United States, 1998

Do you see a pattern in the data? What kinds of states have a small percentage of elderly residents? What kinds of states have a large percentage of elderly residents?

SOURCE: American Association of Retired People (1999: 8).

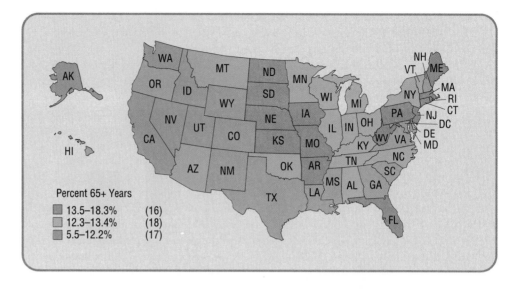

Percent 65+ Years

13.5–18.3%	(16)
12.3–13.4%	(18)
5.5–12.2%	(17)

more women than men among the elderly. This imbalance is most marked in the oldest age cohorts. In large part, therefore, poverty and related problems among the oldest Americans is a gender issue.

Economic inequality between elderly women and men is largely the result of women's lower earning power when they are younger. Women are entering the paid work force in increasing numbers, but there are still more women than men who are homemakers and do not work for a wage. Therefore, fewer women than men enjoy employer pensions when they retire. Moreover, as we saw in Chapter 9 ("Sexuality and Gender"), women who are in the paid labor force tend to earn less than men. When they retire, their employer pensions are generally inferior. As a result, the people most in need—the most elderly women—receive the fewest retirement benefits.

In addition to the old old and women, the categories of elderly people most likely to be poor include African Americans, people living alone, and people living in rural areas (Siegel, 1996). But declining income and poverty are not the only social problems faced by the elderly. In addition, the elderly are sometimes socially segregated in nursing homes, seniors' apartment buildings, and subdivisions with a high proportion of retired people. Because of the high value Americans place on participation in the paid labor force, the end of full-time work signifies the end of meaningful life for some people. And especially in a society that puts a premium on vitality and youth, such as the United States, being elderly is a social stigma. **Ageism** is prejudice about, and discrimination against, elderly people. Ageism is evident, for example, when elderly men are stereotyped as "grumpy." Ageism affects women more than men. Thus, the same person who considers some elderly men "distinguished-looking" may disparage elderly women as "haggard" (Banner, 1992).

Often, however, elderly people do not conform to the negative stereotypes applied to them. In the United States, 65 is usually taken to be the age when people become elderly. (Sixty-five used to be the age of mandatory retirement.) But just because someone is 65 or over does not mean he or she is decrepit and dependent. On the contrary, most people who retire from active working life are far from being a tangle of health problems and a burden on society. This is due to the medical advances of recent decades, the healthier lifestyles followed by many elderly people, and the improved financial status of the elderly.

Specifically, the housing arrangements of elderly people are not usually desolate and depressing (Hochschild, 1973; Myerhoff, 1978). Just over 4% of Americans 65 years and over lived in nursing homes in 1996, and many of these are of high quality (American Association for Retired People, 1999: 4).[2] In 2000, fully 70% of Americans 75 years and over lived in single-family detached homes. Moreover, surveys show that most Americans

About 4% of Americans 65 years and older live in nursing homes.

want to stay in their own home as long as possible. This desire is strongest among the oldest Americans and is increasing over time. Over four fifths of Americans 45 years and older say they would prefer to modify their homes and have help at home should such assistance become necessary (Bayer and Harper, 2000: 14, 24–5, 27–8).

Many elderly people own assets aside from their home, such as investments. Most elderly people receive private and public pensions. More than one out of eight Americans over the age of 65 work in the paid labor force, nearly half of them full-time (American Association for Retired People, 1999: 12). As a result, while the poverty rate does increase somewhat for people over the age of 54, the poverty rate among people 65 and over is only about half the poverty rate of people in the 15–24 age cohort (see Figure 15.5). Poverty among the elderly has fallen sharply since the 1960s. Finally, in comparison to other age cohorts, the elderly are better served by the social welfare system. Social Security and other programs geared to the elderly are relatively generous (see Box 15.1). The elderly are well covered by Medicare or health plans tied to the pension plan of their former union or employer.

One reason for the relative economic security of the elderly is that they are well organized politically. Their voter participation rate is above average, and they are over-represented among those who hold positions of political, economic, and religious power. Many groups seek to improve the status of the elderly, the Gray Panthers perhaps the best known among them. The American Association for Retired People is an effective lobby in Washington, DC (Morris, 1996). In part because of the activism of elderly people, discrimination based on age has become illegal in the United States. In fact, their activism may have led to a redistribution of resources away from young people. For example, educational funding has declined, but funding has increased for medical research related to diseases disproportionately affecting the elderly.

Death and Dying

It may seem odd to say so, but the ultimate social problem the elderly must face is their own demise. Why are death and dying *social* problems and not just religious, philosophical, and medical issues? A personal anecdote can help us begin to answer that question.

When John Lie was a teenager, he thought he would never live past the age of 30. For some reason, death was often on his mind. The feeling intensified after he visited his grandfather once in his native village in South Korea. "The old man took me to a mountain," John reminisces. "As we hiked up the slope, he showed me several mounds. 'Here your ancestors are buried,' the old man announced serenely. After a while, my grandfather showed me a plot that overlooked a wide vista of rice paddies. Smiling contentedly, he

---

[2]For people 85 years and over, the figure stood at nearly 20%.

BOX 15.1
IT'S YOUR CHOICE

THE SOCIAL SECURITY CRISIS

One of the greatest triumphs of public policy since the 19th century has been the development of the public health system. It led to a substantial increase in life expectancy. Another major public policy achievement is social welfare, especially as it applies to the elderly. The combination of Social Security, Medicare, and other government programs goes a long way toward ensuring that the elderly are not doomed to poverty and illness. It explains why, in spite of the abolition of a mandatory retirement age, the average age at retirement for American men fell from 69 in 1950 to 64 in 1994 (Urban Institute, 1998). Because more Americans feel finan-

cially secure in their old age, they can retire earlier.

A longer and more secure life span is a wonderful thing. However, some scholars and policy makers worry that a major crisis is looming. We may not be able to afford government programs for the elderly. That is because of the expected retirement of the "Baby Boom" generation, people born between 1946 and 1964. Many Americans born in this period contributed to social security and other measures to support the aged. As they begin to retire from the active labor force, however, fewer Americans will be contributing to government coffers. Currently, there are roughly three workers for every retiree. By 2030, the ratio is expected to be 2:1 (Urban Institute, 1998).

Some scholars argue that economic growth and higher immigration could offset the expected decline in the active labor force. However, others argue that we also need more concrete measures to deal with the expected crisis in government support for the elderly. Some of the possible policy proposals include the following:

◆ Encourage people to work longer, thereby having more workers support retirees.

◆ Increase national savings.

◆ Lower health care costs, especially the disproportionately higher burden of medical cost for the elderly.

◆ Provide government support only for the truly needy, thereby eliminating or lowering Social Security and other federal benefits for the well off.

What do you think? Should we worry about the expected crisis in Social Security and other government programs that support the elderly? If you expect the potential crisis to arise in your lifetime, what should you be doing now to avert the crisis? What are the advantages and disadvantages of each of the policy proposals listed above? What kind of lifestyle do you expect to lead when you are in your 60s and 70s? Do you think you will be working full-time? Or will you be fully or partly retired?

---

said: 'This is where I will be buried.' I was taken aback. Frankly, I had always seen my grandfather as a stereotypical grumpy old man. But there he was, contentedly thinking about his own demise. His composure puzzled me. I certainly felt nothing of the kind when he later suggested a location for my father's burial plot and even my own. In fact, the experience left me a little shaken. I wanted to think about death—especially my own—as little as possible.

"Years later, I realized that the contrast between my attitude toward death and that of my grandfather was not just a personal difference. It was also a historical and sociological difference. In most traditional societies, including the Korea my grandfather grew up in and Europe until early modern times, most people accepted death (Ariès, 1982). That was partly because the dying were not isolated from other people. They continued to interact with household members and neighbors, who offered them continuous emotional support. Moreover, because the dying had previous experience giving emotional support to other dying people, they could more easily accept death as part of everyday life.

"In contrast, in the United States today, dying and death are separated from everyday life. Most terminally ill patients want to die peacefully and with dignity at home, surrounded by their loved ones. Yet about 80% of Americans die in hospitals. Often, hospital deaths are hygienic, noiseless, and lonely (Nuland, 1994: 242–62). Dying used to be public. It is now private. The lack of social support makes dying a more frightening experience for many people (Elias, 1985 [1982]). In addition, our culture celebrates youth and denies death (Becker, 1973). We use diet, fashion, exercise, makeup, and surgery to prolong youth or at least the appearance of youth. This makes us less prepared for death than our ancestors were."

The reluctance of many Americans to accept death is clearly evident in the debate over euthanasia, also known as mercy killing or assisted suicide (Rothman, 1991). Various medical technologies, including machines able to replace the function of the heart and lungs,

can prolong life beyond the point that was possible in the past. This raises the question of how to deal with people who are near death. In brief, is it humane or immoral to hasten the death of terminally ill patients?

The American Medical Association's Council on Ethical and Judicial Affairs (AMA-CEJA) says it is the duty of doctors to withhold life-sustaining treatment if that is the wish of a mentally competent patient. The AMA-CEJA also endorses the use of effective pain treatment even if it hastens death. As a result, doctors and nurses make decisions every day about who will live and who will die (Zussman, 1992; 1997). Public opinion polls show that about three quarters of Americans favor this practice (Benson, 1999).

Euthanasia involves a doctor prescribing or administering medication or treatment that is *intended* to end a terminally ill patient's life. It is therefore a more active form of intervention than those noted in the preceding paragraph. Public opinion polls show that about two thirds of Americans favor physician-assisted euthanasia (Benson, 1999). Sixty percent of American doctors say they would be willing to perform euthanasia if it were legal. At least one survey shows that nearly 30% of American doctors have received a euthanasia request but only about 6% have ever complied with such a request (Meier, Emmons, Wallenstein, Quill, Morrison, and Cassel, 1998). The AMA-CEJA, the Catholic Church, some disabled people, and other groups oppose euthanasia.

Euthanasia is legal in the Netherlands and is likely to become legal in some other countries in the next few years. In Oregon, a physician-assisted suicide law took effect in October 1997. It allows doctors in that state to prescribe a lethal dose of drugs to terminally ill patients who choose not to prolong their suffering and have less than 6 months to live. In its first 2 years of operation, 43 patients took advantage of Oregon's physician-assisted suicide law (Sullivan, Hedberg, and Fleming, 2000). On the other hand, in a June 1997 ruling, the United States Supreme Court upheld as constitutional state laws that bar assisted suicide. The Court recognized, however, that "[t]hroughout the Nation, Americans are engaged in an earnest and profound debate about the morality, legality and practicality of physician assisted suicide. Our holding permits this debate to continue, as it should in a democratic society" (Longwood College Library, 2000). As the Supreme Court anticipates, euthanasia is bound to become a major political issue in coming decades as medical technologies for prolonging life improve, the number of elderly people increases, and the cost of medical care skyrockets. Extending the lives of terminally ill patients by all means possible will be upheld as an ethical imperative by some people. Others will regard it as immoral because it increases suffering and siphons scarce resources away from other pressing medical needs.

Summing up our discussion, we may say that the apparently *natural* processes of death and dying are in fact highly *social* phenomena. For example, social circumstances account for variations in the definition and duration of life stages and the rituals associated with the transition from one life stage to the next. Similarly, the condition of the elderly depends in part on how numerous and powerful the elderly are as a political force. Finally, our attitudes toward dying and death reflect both our cultural values and our social conventions.

In the last section of this chapter, we make a similar argument about medicine. **Medicine** is a social institution devoted to prolonging life by fighting disease and promoting health. It may seem to lie squarely in the realm of pure science. However, as you will now learn, society shapes medical practice every bit as much as it influences the processes of aging and dying. We can see this clearly by examining how the medical and psychiatric professions have increased their control over people in the past 150 years or so. In the following discussion, we begin this task by showing how forms of deviance that used to be considered the province of morality and the law have come increasingly under the sway of psychiatry. This, we argue, is only partly due to the scientifically proven benefits of psychiatric care. We then examine how medicine drove other competing professions out of the health care market. Again, this demonstrates that the type of health care we receive is a product not just of scientific considerations, but of social forces too.

Dr. Jack Kevorkian is the most prominent advocate of euthanasia in the United States. A study of 69 people who died with Kevorkian's assistance found that only 25% were terminally ill while a disproportionately large number of them were socially isolated women (divorced or never married) (Roscoe, Gragovic, and Cohen, 2000). These findings suggest that, in the absence of clinical safeguards, some groups are especially vulnerable to euthanasia and its misuse.

MEDICINE, POWER, AND CULTURE

The Medicalization of Deviance

You may recall from our discussion of deviance that one of the preoccupations of symbolic interactionism is the labeling process (see Chapter 6, "Deviance and Crime"). According to symbolic interactionists, deviance results not just from the actions of the deviant but also from the responses of others, who define some actions as deviant and other actions as normal.

Here we may add that the *type* of label applied to a deviant act may vary widely over time and from one society to another, depending on how that act is interpreted. Consider, for instance, the **medicalization of deviance.** The medicalization of deviance refers to the fact that, over time, "medical definitions of deviant behavior are becoming more prevalent in . . . societies like our own" (Conrad and Schneider, 1992 [1980]: 28–9). In an earlier era, much deviant behavior was labeled "evil." Deviants tended to be chastised, punished, and otherwise socially controlled by members of the clergy, neighbors, family members, and the criminal justice system. Today, however, a person prone to drinking sprees is more likely to be declared an alcoholic and treated in a detoxification center. A person predisposed to violent rages is more likely to be medicated. A person inclined to overeating is more likely to seek therapy and, in extreme cases, surgery. A heroin addict is more likely to seek the help of a methadone program. As these examples illustrate, what used to be regarded as willful deviance is now often regarded as involuntary deviance. More and more, what used to be defined as "badness" is defined as "sickness." As our definitions of deviance change, deviance is increasingly coming under the sway of the medical and psychiatric establishments (see Figure 15.8).

How did the medicalization of deviance come about? What are the major social forces responsible for the growing capacity of medical and psychiatric establishments to control our lives? To answer this question, we first examine the fascinating case of mental illness. As you will see, our changing definitions of mental illness show perhaps more clearly than any other aspect of medicine how thin a line separates science from politics in the field of health care.

The Political Sociology of Mental Illness

In 1974, a condition that had been considered a psychiatric disorder for more than a century ceased to be labeled as such by the American Psychiatric Association (APA). Did the condition disappear because it had become rare to the point of extinction? No. Did the discovery of a new wonder drug eradicate the condition virtually overnight? Again, no. In fact, in

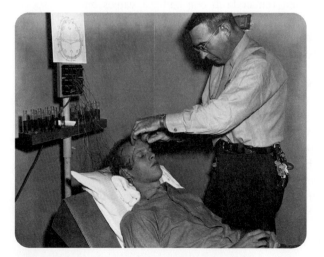

An example of the medicalization of deviance. A lobotomy is performed in Vacaville State Prison in California in 1961 to "cure" the inmate of criminality.

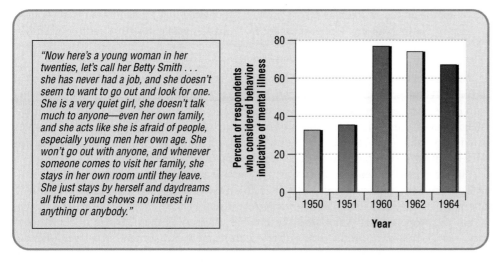

"Now here's a young woman in her twenties, let's call her Betty Smith . . . she has never had a job, and she doesn't seem to want to go out and look for one. She is a very quiet girl, she doesn't talk much to anyone—even her own family, and she acts like she is afraid of people, especially young men her own age. She won't go out with anyone, and whenever someone comes to visit her family, she stays in her own room until they leave. She just stays by herself and daydreams all the time and shows no interest in anything or anybody."

✦ **FIGURE 15.8** ✦

An Example of the Medicalization of Deviance

Five North American surveys conducted in the 1950s and 1960s presented respondents with the anecdote to the left. The graph shows the percentage of respondents who considered the behavior described in the anecdote evidence of mental illness. Notice the difference between the 1950s and the 1960s. (Nearly 100% of psychiatrists who evaluated the anecdote thought it illustrated "simple schizophrenia.")

1974 the condition was perhaps more widespread and certainly more public than ever before. But, paradoxically, just as the extent of the condition was becoming more widely appreciated, the "bible" of the APA, the *Diagnostic and Statistical Manual of Mental Disorders* (DSM), ceased to define it as a psychiatric disorder.

The "condition" we are referring to is homosexuality. In preparing the third edition of the DSM for publication, a squabble broke out among America's psychiatrists over whether homosexuality is in fact a psychiatric disorder. Gay and lesbian activists, who sought to destigmatize homosexuality, were partly responsible for a shift in the views of many psychiatrists on this subject. In the end, the APA decided that homosexuality is not a psychiatric disorder. They deleted the entry. The APA membership confirmed the decision in 1974.

The controversy over homosexuality was only one of several *political* debates that erupted among American psychiatrists in the 1970s and 1980s (Shorter, 1997: 288–327). Among others:

✦ The DSM task force initially decided to eliminate the term "neurosis" on the grounds that its role as a cause of mental disorder had never been proven experimentally. The decision outraged the psychoanalytic community. "Neurosis" was a keystone of their Freudian theories. So the psychoanalysts threatened to block publication of the third edition of DSM unless they were appeased. In 1979, the APA Board backed down, placing "neurosis" in parentheses after "disorder." This compromise had nothing to do with science.

✦ When veterans of the war in Vietnam started returning to the United States after 1971, they faced great difficulty reentering American society. The war was unpopular, so they were not universally greeted as heroes. The economy went into a tailspin in 1973, making jobs difficult to find. And the veterans had suffered high levels of stress during the war itself. Many of them believed their troubles were psychiatric in nature, and soon a nationwide campaign was underway, urging the APA to recognize "Post Traumatic Stress Disorder" (PTSD) in their manual. Many psychiatrists were reluctant to do so. Nonetheless, PTSD was listed in the third edition of the DSM. To be sure, the campaign succeeded partly on the strength of evidence that extreme trauma has psychological (and at times physiological) effects. But in addition, as one activist later explained, the PTSD campaign succeeded because "[we] were better organized, more politically active, and enjoyed more lucky breaks than [our] opposition" (Chaim Shatan, quoted in Scott, 1990: 308). Again, politics and not just science helped to shape the definition of a mental disorder.

✦ Feminists were unhappy that the 1987 edition of DSM contained listings such as "self-defeating personality disorder." The DSM said this disorder is twice as common among women as men. Feminists countered that the definition is an example of victim blaming. Under pressure from the feminists, the 1994 edition of DSM dropped the concept.

Some mental disorders have obvious organic causes, such as chemical imbalances in the brain. These organic causes can often be precisely identified. Often they can be treated with drugs or other therapies. Moreover, experiments can be conducted to verify their existence and establish the effectiveness of one treatment or another. However, the examples listed above show that the definition of a host of other mental disorders depends not just on scientific evidence but also on social values and political compromise.

In the mid-19th century there was just one mental disorder recognized by the federal government: idiocy/insanity. By 1975, the DSM recognized 106 mental disorders. The 1994 DSM lists 297 mental disorders. As the number of mental disorders has grown, so has the proportion of Americans presumably affected by them. In the mid-19th century, few people were defined as suffering from mental disorders. However, one respected survey conducted in 1992 found that fully 48% of Americans would suffer from a mental disorder—very broadly defined, of course—during their lifetimes. The most common mental disorder, supposedly affecting 17% of the population at some point in their lives: severe depression (Blazer, Kessler, McGonagle and Swartz, 1994; Shorter, 1997: 294).

Edward Shorter, one of the world's leading historians of psychiatry, notes that in psychiatric practice, definitions of mental disorders are often expanded to include ailments with dubious or unknown biological foundations (e.g., "minor depression," "borderline schizophrenia"; Shorter, 1997: 228). In addition, as we have seen, the sheer number of conditions labeled "mental disorder" has increased rapidly during the 20th century. We suggest four main reasons for expansion in the number and scope of such labels:

1. In the first place, as we saw in Chapter 10 ("Work and the Economy"), Americans are now experiencing more stress and depression than ever before, due mainly to the increased demands of work and the growing time crunch. Mental health problems are thus more widespread than they used to be. At the same time, traditional institutions for dealing with mental health problems are less able to cope with them. The weakening authority of the church and the weakening grip of the family over the individual leave the treatment of mental health problems more open to the medical and psychiatric establishments.

2. Second, there is inflation in the number of mental disorders because powerful organizations demand it. The United States Census Bureau first asked the American Medico-Psychological Association to classify mental disorders in 1908, and HMOs today demand precise diagnostic codes before paying for psychiatric care. Because public and private organizations find the classification of mental disorders useful, they have proliferated.

3. Third, the cultural context stimulates inflation in the number and scope of mental disorders. Probably more than any other people, Americans are inclined to turn their problems into medical and psychological issues, sometimes without inquiring deeply into the disadvantages of doing so. For example, in 1980 the term "attention deficit disorder" (ADD) was coined to label hyperactive and inattentive schoolchildren, mainly boys.[3] By the mid-1990s, doctors were writing 6 million prescriptions a year for Ritalin, an amphetamine-like compound that controls ADD. Evidence shows that some children diagnosed with ADD have problems absorbing glucose in the brain or suffer from imbalances in chemicals that help the brain regulate behavior (Optometrists Network, 2000). Yet the diagnosis of ADD is typically conducted clinically, i.e., by interviewing and observing children to see if they exhibit signs of

---

[3]Psychiatrists now recognize three types of ADD, but the distinguishing features of each type are not relevant to our discussion.

serious "inattention," "hyperactivity," and "impulsivity." This means that many children diagnosed with ADD may have no organic disorder at all. Some cases of ADD may be due to the school system failing to capture children's imagination. Some may involve children acting out because they are deprived of attention at home. Some may involve plain, old-fashioned boyish enthusiasm. A plausible case could be made that Tom Sawyer or Winnie the Pooh suffered from ADD (Shea, Gordon, Hawkins, Kawchuk, and Smith, 2000). However, once hyperactivity and inattentiveness in school are defined as a medical and psychiatric condition, officials routinely prescribe drugs to control the problem and tend to ignore possible social causes.

4. The fourth main reason for inflation in the number and scope of mental disorders is that various professional organizations have promoted it. Consider post-traumatic stress disorder. There is no doubt PTSD is a real condition, and that many veterans suffer from it. However, once the disorder was officially recognized in the 1970s, some therapists trivialized the term. By the mid-1990s some therapists were talking about PTSD "in children exposed to movies like *Batman*" (Shorter, 1997: 290). Some psychiatric social workers, psychologists, and psychiatrists may magnify the incidence of such mental disorders because doing so increases their stature and their patient load. Others may do so simply because the condition becomes "trendy." Whatever the motive, overdiagnosis is the result.

The Professionalization of Medicine

The preceding discussion shows that the diagnosis and treatment of some mental disorders is not a completely scientific enterprise. Social processes are at least as important as scientific principles in determining how we treat some mental disorders. Various mental health professions compete for patients, as do different schools of thought within professions. Practitioners offer a wide and sometimes confusing array of treatments and therapies. The American public spends billions of dollars a year on supposed cures; but in some cases their effectiveness is debatable, and some people remain skeptical about their ultimate worth.

In 1850, the practice of medicine was in an even more chaotic state. Herbalists, faith healers, midwives, druggists, and medical doctors vied to meet the health needs of the American public. A century later, the dust had settled. Medical science was victorious. Its first series of breakthroughs involved identifying the bacteria and viruses responsible for various diseases and then developing effective procedures and vaccines to combat them. These and subsequent triumphs in diagnosis and treatment convinced most people of the superiority of medical science over other approaches to health. Medical science worked, or at least it seemed to work more effectively and more often than other therapies.

It would be wrong, however, to think that scientific medicine came to dominate health care only because it produced results. A second, sociological reason for the rise and dominance of scientific medicine is that doctors were able to professionalize. As noted in Chapter 10, a profession is an occupation requiring extensive formal education. Professionals regulate their own training and practice. They restrict competition within the profession, mainly by limiting the recruitment of practitioners. They maximize competition with some other professions, partly by laying exclusive claim to a field of expertise. Professionals are usually self-employed. They exercise considerable authority over their clients. And they profess to be motivated mainly by the desire to serve their community even though they earn a lot of money in the process. Professionalization, then, is the process by which people gain control and authority over their occupation and their clients. It results in professionals enjoying high occupational prestige and income, and considerable social and political power (Johnson, 1972; Friedson, 1986; Starr, 1982).

The professional organization of American doctors is the American Medical Association (AMA). It was founded in 1847. It quickly set about broadcasting the successes of medical science and criticizing alternative approaches to health as quackery and charlatanism. By the early years of the 20th century, the AMA had convinced state licensing boards to certify only doctors who had been trained in programs recognized by the AMA. Soon, schools teaching other approaches to health care were closing down across the country. Doctors had never earned much. In the 18th century it was commonly said that "few lawyers die well, few physicians live well" (Illich, 1976: 58). But once it was possible to lay virtually exclusive claim to health care, big financial rewards followed. Today, American doctors in private practice earn on average about $200,000 a year, although income varies substantially by specialty.

The modern hospital is the institutional manifestation of the medical doctor's professional dominance. Until the 20th century, most doctors operated small clinics and visited patients in their homes. But the rise of the modern hospital was guaranteed by medicine's scientific turn in the mid-19th century. Expensive equipment for diagnosis and treatment had to be shared by many physicians. This required the centralization of medical facilities in large, bureaucratically run institutions that strongly resist deviations from professional conduct (see Box 15.2). Practically nonexistent until the Civil War, hospitals are now

BOX 15.2
SOCIOLOGY AT THE MOVIES

PATCH ADAMS (1998)

Patch Adams, played by Robin Williams, is suicidal. Checking into a mental hospital, he finds that the doctors, who are supposed to be helping him, are indifferent. In contrast, other patients help him overcome his suicidal urges. He resolves to become a doctor to help other patients.

Patch Adams, based on a real person of the same name, breathes humor and life into the dreary world of the modern hospital. As an intern, Adams finds that patients are identified by their ID number and disease. Doctors and nurses seem more concerned about medical charts than their patients. Finally, Adams startles a nurse by asking her about a patient: "What's her name?" The very idea that a patient may be something more than his or her medical records reveals the impersonal and bureaucratic nature of the modern doctor–patient relationship.

Patch Adams is intent on infusing personal care, humor, and humanity into the doctor–patient relationship. In dealing with children whose hair had fallen out because of chemotherapy, Patch plays a clown in order to bring smiles to their faces. He believes humor and laughter can be a great cure.

Patch Adams starring Robin Williams.

Not surprisingly, Adams faces resistance from medical school administrators. After all, he deviates significantly from the norm of impersonal professionalism. They attempt to expel him from the medical school. With support from his friends and patients, however, he manages to win a court battle to remain in medical school. In real life, Patch Adams goes on to become a medical doctor, who not only maintains a sense of humor, but also continues to live up to his ideals, including helping poor patients around the world.

Patch Adams is a sentimental movie, pitting the humorous individual against the grim organization. However, the critic Roger Ebert (1998) wrote: "To himself . . . [Patch Adams is] an irrepressible bundle of joy, a zany live wire who brings laughter into the lives of the sick and dying. To me, he's a pain in the wazoo. If this guy broke into my hospital room and started tap-dancing with bedpans on his feet, I'd call the cops." Here Ebert is saying that the norm of professionalism—grim and impersonal though it may be—may be preferable to the antics of Patch Adams. Do you agree with Ebert? Would you prefer your doctor to be a "human being," or simply to play his professional role efficiently and effectively? What are the health advantages and disadvantages of each approach to doctoring?

widespread, even though budget cutbacks have resulted in the closing of many hospitals in rural areas and inner cities over the past 20 years (Rosenberg, 1987).

Recent Challenges to Traditional Medical Science

Patient Activism

By the mid-20th century, the dominance of medical science in the United States was virtually complete. Any departure from the dictates of scientific medicine was considered deviant. Thus, when sociologist Talcott Parsons defined the **sick role** in 1951, he first pointed out that illness suspends routine responsibilities and is not deliberate. Then he stressed that people playing the sick role must want to be well and must seek competent help, cooperating with health care practitioners at all times (Parsons, 1951: 428 ff.). Must they? By Parsons's definition, a competent person suffering from a terminal illness cannot reasonably demand that doctors refrain from using heroic measures to prolong his or her life. And by his definition, a patient cannot reasonably question doctors' orders, no matter how well educated the patient and how debatable the effect of the prescribed treatment. Although Parsons's definition of the sick role may sound plausible to many people born before World War II, it probably sounds authoritarian and utterly foreign to most younger people.

That is because things have changed. The American public is more highly educated now than it was 50 years ago. Many people now possess the knowledge, the vocabulary, the self-confidence, and the political organization to participate in their own health care rather than passively accept whatever experts tell them. Increasingly, patients are taught to perform simple, routine medical procedures themselves. Many people now use the Internet to seek information about various illnesses and treatments.[4] Increasingly, they are uncomfortable with doctors acting like patriarchal fathers and patients like dutiful children. Doctors now routinely seek patients' informed consent for some procedures rather than deciding what to do on their own. Similarly, most hospitals have established ethics committees, which were unheard of only two decades ago (Rothman, 1991). These are responses to patients wanting a more active role in their own care.

Some recent challenges to the authority of medical science are organized and political. For example, when AIDS activists challenge the stereotype of AIDS as a "gay disease" and demand more research funding to help find a cure, they change research and treatment priorities in a way that could never have happened in, say, the 1950s or 1960s (Epstein, 1996). Similarly, when feminists support the reintroduction of midwifery and argue against medical intervention in routine childbirth, they are challenging the wisdom of established medical practice. The previously male-dominated profession of medicine considered the male body the norm and paid relatively little attention to women's diseases, such as breast cancer, and women's issues, such as reproduction. This, too, is now changing thanks to feminist intervention (Boston Women's Health Collective, 1998; Rothman, 1982; 1989; Schiebinger, 1993). And while doctors and the larger society traditionally treated people with disabilities like incompetent children, various movements now seek to empower them (Charlton, 1998; Zola, 1982). As a result, attitudes toward the disabled are changing.

Alternative Medicine

Other challenges to the authority of medical science are less organized and less political than those just mentioned. Consider, for example, alternative medicine. The most frequently used types of alternative medicine are chiropractic, acupuncture, massage therapy, and various relaxation techniques. Alternative medicine is used mostly to treat back problems, chronic headache, arthritis, chronic pain, insomnia, depression, and anxiety. Especially popular in the Western states, alternative medicine is most often used by highly

---

[4]There are at least two health-related dangers in using the Internet, however. First, some people may misinterpret information or assume that unreliable sources are reliable. Second, online relationships may lead to real-world meetings and so contribute to the spread of AIDS/HIV and other sexually transmitted diseases. Epidemiologists at the United States Centers for Disease Control and Prevention are now conducting a study on this subject (Roberts, 2000; SexQuiz.org, 2000).

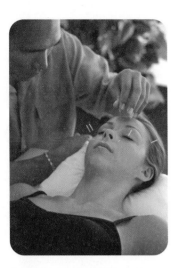

Acupuncture is one of the most widely accepted forms of alternative medicine.

educated, upper-income white Americans in the 25- to 49-year age group. A nationwide poll conducted in 1990 showed that 34% of Americans tried alternative medicine in the year prior to the survey. Most of them had *not* lost all faith in traditional medical science. Thus, 83% of them tried alternative medicine in conjunction with treatment from a medical doctor. Americans spent more money on alternative therapies than on hospitalization (Eisenberg, Kessler, Foster, Norlock, Calkins, and Delbanco, 1993). Two surveys conducted in 1998 show demand for alternative care is rising (American Chiropractic Association, 1999).

Despite its growing popularity, many medical doctors were hostile to alternative medicine until recently. They lumped all alternative therapies together and dismissed them as unscientific (Campion, 1993). By the late 1990s, however, a more tolerant attitude was evident in many quarters. For some kinds of ailments, physicians began to recognize the benefits of at least the most popular forms of alternative medicine. For example, a 1998 editorial in the respected *New England Journal of Medicine* admitted that the beneficial effect of chiropractic on low back pain is "no longer in dispute" (Shekelle, 1998). This change in attitude was due in part to new scientific evidence showing that spinal manipulation is a relatively effective and inexpensive treatment for low back pain (Manga, Angus, and Swan, 1993).

The medical profession's grudging acceptance of chiropractic in the treatment of low back pain indicates what we can expect in the uneasy relationship between scientific and alternative medicine in coming decades. Doctors will for the most part remain skeptical about alternative therapies unless properly conducted experiments demonstrate their beneficial effects. Most Americans probably agree with this cautious approach.

Holistic Medicine

Medical doctors understand that a positive frame of mind often helps in the treatment of disease. For example, research shows that strong belief in the effectiveness of a cure can by itself improve the condition of about a third of people suffering from chronic pain or fatigue (Campion, 1993). This is known as the **placebo effect.** Doctors also understand that conditions in the human environment affect people's health. There is no dispute, for example, about why so many people in "cancer alley" develop malignancies. However, despite their appreciation of the effect of mind and environment on the human body, traditional scientific medicine tends to respond to illness by treating disease symptoms as a largely physical and individual problem. Moreover, scientific medicine keeps subdividing into more specialized areas of practice that rely more and more heavily on drugs and high-tech machinery. Most doctors are less concerned with maintaining and improving health by understanding the larger mental and social context within which people become ill.

Traditional Indian and Chinese medicine takes a different approach. India's Ayurvedic medical tradition sees individuals in terms of the flow of vital fluids or "humors" and their health in the context of their environment. In this view, maintaining good health requires not only balancing fluids in the individual but also balancing the relationship between individuals and the world around them (Zimmermann, 1987 [1982]). In spite of significant differences, the fundamental outlook is similar in traditional Chinese medicine. Chinese medicine and its remedies, ranging from acupuncture to herbs, seek to restore individuals' internal balance, as well as their relationship to the outside world (Unschuld, 1985). Contemporary **holistic medicine,** the third and final challenge to traditional scientific medicine we will consider, takes a similar approach to these "ethnomedical" traditions. Practitioners of holistic medicine argue that good health requires maintaining a balance between mind and body, and between the individual and the environment.

Most holistic practitioners do not reject scientific medicine. However, they emphasize disease prevention. When they treat patients, they take into account the relationship between mind and body and between the individual and his or her social and physical environment. Holistic practitioners thus seek to establish close ties with their patients and treat them in their homes or other relaxed settings. Rather than expecting patients to react to illness by passively allowing a doctor to treat them, they expect patients to take an active role in maintaining their good health. And, recognizing that industrial pollution, work-related stress, poverty, racial and gender inequality, and other social factors contribute heavily to disease, holistic practitioners often become political activists (Hastings, Fadiman, and Gordon, 1980).

In sum, patient activism, alternative medicine, and holistic medicine represent the three biggest challenges to traditional scientific medicine today. Few people think of these challenges as potential replacements for scientific medicine. However, many people believe that, together with traditional scientific approaches, these challenges will help to improve the health status of people in the United States and throughout the world in the 21st century.

SUMMARY

1. The social causes of illness and death include human-environmental factors, lifestyle factors, and factors related to the public health and health care systems. All three factors are related to country of residence, class, race, and gender.

2. Health risks are lower among upper classes, rich countries, and privileged racial and ethnic groups than lower classes, poor countries, and disadvantaged racial and ethnic groups. In some respects related to health, men are in a more advantageous position than women.

3. While the United States has the world's most advanced health care system, the average health status of Americans is lower than the average health status of people in other rich postindustrial countries. That is partly because the level of social inequality is higher in the United States and partly because the health care system in this country makes it difficult for many people to receive adequate care.

4. Paradoxically, medical successes create new problems. For instance, they allow people to live longer than they used to. This gives degenerative diseases like cancer and heart disease the chance to spread. Medical successes also raise new questions about when and how people should be allowed to die.

5. People attach different meanings to aging and death in different societies and historical periods. Thus, the stages of life vary in number and significance across societies. So does anxiety about death.

6. The population of the United States is aging, and by 2040 over a fifth of Americans will be 65 or over. The fastest growing age cohort among the elderly is composed of people 85 years and older. The ratio of men to women falls with age.

7. Poverty is less widespread among people over 65 than among people younger than 45. Among the elderly, poverty is most widespread for those 85 and older, women, African Americans, people living alone, and people living in rural areas.

8. Over time, medical definitions of deviance have become more common. The recent history of psychiatry shows that social values and political compromise are at least as important as science in determining the classification of some mental disorders.

9. Medical science came to dominate the health care system partly because it proved to be so successful in treating the ill. In addition, dominance was assured by doctors excluding competitors and establishing control over their profession and their clients.

10. Several challenges to traditional scientific medicine promise to improve the quality of health care in the United States and globally. These include patient activism, alternative medicine, and holistic medicine.

GLOSSARY

An **age cohort** is a category of people born in the same range of years.

Age roles are norms and expectations about the behavior of people of different age cohorts.

Age stratification refers to social inequality between age cohorts.

Ageism is prejudice about and discrimination against old people.

Euthanasia (also known as mercy killing and assisted suicide) involves a doctor prescribing or administering medication or treatment that is intended to end a terminally ill patient's life.

A **generation** is an age group that has unique and formative historical experiences.

A **gerontocracy** is a society ruled by elderly people.

Health, according to the World Health Organization, is "the ability of an individual to achieve his [or her] potential and to respond positively to the challenges of the environment."

The **health care system** is composed of a nation's clinics, hospitals, and other facilities for ensuring health and treating illness.

Holistic medicine emphasizes disease prevention. Holistic practitioners treat disease by taking into account the relationship between mind and body and between the individual and his or her social and physical environment.

Infant mortality is the number of deaths before the age of 1 for every 1,000 lives births in a population in 1 year.

The **life course** refers to the distinct phases of life through which people pass. These stages vary from one society and historical period to another.

Life expectancy is the average age at death of the members of a population.

The **maximum average human life span** is the average age of death for a population under *ideal* conditions. It is currently about 85 years.

The **maximum human life span** is the longest an *individual* can live under current conditions. It is currently about 122 years.

The **medicalization of deviance** is the tendency for medical definitions of deviant behavior to become more prevalent over time.

Medicine is an institution devoted to fighting disease and promoting health.

The **placebo effect** is the positive influence on healing of strong belief in the effectiveness of a cure.

The **public health system** is composed of government-run programs that ensure access to clean drinking water, basic sewage and sanitation services, and inoculation against infectious diseases.

A **rite of passage** is a ritual that marks the transition from one stage of life to another.

Playing the **sick role,** according to Talcott Parsons, involves the nondeliberate suspension of routine responsibilities, wanting to be well, seeking competent help, and cooperating with health care practitioners at all times.

QUESTIONS TO CONSIDER

1. Because health resources are scarce, tough decisions have to be made about how they are allocated. For example, drug companies, physicians, hospitals, government research agencies, and other components of the health care system have to decide how much to invest in trying to prolong the life of the elderly versus how much to invest in improving the health of the poor. What do you think are the main factors that help different components of the health care system decide how to allocate resources between these two goals? Specifically, how important is the profit motive? Political pressure? Which components of the health system are most influenced by the profit motive? Which by political pressure? If you were in charge of a major hospital or government funding for health research, how would you divide your scarce resources between trying to prolong the life of the elderly and improving the health of the poor? Why? What pressures might be placed on you to act differently than you want?

2. What generation do you belong to? What is the age range of people in your generation? What are some of the defining historical events that have taken place in your generation? How strongly do you identify with your generation? Ask the same questions to someone of a different race, gender, or class. How do these variables seem to influence the experience of a generation?

3. Do you believe that patient activism and alternative medicine improve health care or detract from the efforts of scientifically trained physicians and researchers to do the best possible research and administer the best possible treatments? Because patient activists may not be scientifically trained, and because alternative therapies may not be experimentally proven, are there dangers inherent in these challenges to traditional medicine? On the other hand, do biases in traditional medicine detract from health care by ignoring the needs of patient activists and the possible benefits of alternative therapies?

WEB RESOURCES

Companion Web Site for This Book

http://sociology.wadsworth.com

Begin by clicking on the Student Resources section of the Web site. Choose "Introduction to Sociology" and finally the Brym and Lie book cover. Next, select the chapter you are currently studying from the pull-down menu. From the Student Resources page you will have easy access to InfoTrac College Edition®, MicroCase Online exercises, additional Web links, and many other resources to aid you in your study of sociology, including practice tests for each chapter.

InfoTrac Search Terms

These search terms are provided to assist you in beginning to conduct research on this topic by visiting http://www.infotraccollege. com/wadsworth.

Age discrimination   Life expectancy
Euthanasia   Public health
Health maintenance organizations

Recommended Web Sites

How long should you expect to live? Find out by answering questions on your health risks at http://www.msnbc.com/modules/ quizzes/lifex.asp.

The United States Centers for Disease Control and Prevention maintains a rich Web site at http://www.cdc.gov. It is full of up-to-date health information and statistics. The CDCP is conducting a study to see if the Internet is becoming a breeding ground for the spread of sexually transmitted diseases. To see the results of the study and participate in it, go to http://www.sexquiz.org.

Physicians for a National Health Program advocates a universal, comprehensive, national health care program. It has more than 9,000 members and chapters across the United States. Visit the Web site of these medical activists at http://www.pnhp.org.

Visit the Web site of the American Association for Retired People at http://www.aarp.org to find out about the many issues faced by retirees and the individual and collective actions they are taking to deal with these issues. Also of interest on this site are the list of research and reference resources at http://www.aarp.org/indexes/reference.html#center and the essay on "How to Write a Research Paper in Gerontology" by Harry R. Moody at http://research.aarp.org/ageline/modhome.html.

SUGGESTED READINGS

Peter Conrad and Joseph W. Schneider. *Deviance and Medicalization: From Badness to Sickness,* expanded ed. (Philadelphia : Temple University Press, 1992 [1980]). This classic won the 1981 Charles Horton Cooley Award of the Society for the Study of Social Interaction. In a series of case studies on mental illness, alcoholism, opiate addiction, and other forms of deviance, it shows how the growth of the medical and psychiatric professions led to the medicalization of deviance in the 20th century.

Peter Laslett. *A Fresh Map of Life: The Emergence of the Third Age* (Cambridge, MA: Harvard University Press, 1991

[1989]). A leading historical demographer reconsiders the role of the elderly.

Theda Skocpol. *Boomerang: Clinton's Health Security Effort and the Turn against Government in United States Politics* (New York: W. W. Norton, 1996). An analysis of the rise and fall of President Clinton's attempt to reform national health care.

Paul Starr. *The Social Transformation of American Medicine* (New York: Basic, 1982). In this Pulitzer-Prize-winning book, Starr traces the social, political, and economic rise of physicians, from their lowly status in the 18th century to the powerful positions they occupy today.

V

SOCIAL CHANGE

■

IN THIS CHAPTER, YOU WILL LEARN THAT:

✦ Many people think only natural conditions influence human population growth. However, social forces are important influences too.

✦ In particular, sociologists have focused on two major social determinants of population growth: industrialization and social inequality.

✦ Industrialization also plays a major role in causing the movement of people from countryside to city.

✦ Cities are not as anonymous and alienating as many sociologists once believed them to be.

✦ The spatial and cultural forms of cities depend on the level of development of the societies in which they are found.

✦ Global inequality results less from deficiencies in poor societies than from relations between rich and poor countries.

✦ Under certain circumstances, poor societies can move along the path toward prosperity.

C H A P T E R

16

POPULATION, URBANIZATION, AND DEVELOPMENT

INTRODUCTION

John Lie decided to study sociology because he was concerned about the poverty, pestilence, dictatorships, and natural disasters that he had read about in books and observed during his travels in Asia and Latin America. "Initially," says John, "I thought I would major in economics because it seemed to me that much of the misery of the poor countries was a result of economic factors. In my economics classes, I learned about the importance of birth-control programs to cap population growth, efforts to prevent the runaway growth of cities, and measures to spread Western knowledge, technology, and markets to people in less economically developed countries. My textbooks and professors assumed that if only the less developed countries would become more like the West, their populations, cities, economies, and societies would experience stable growth. Otherwise, the developing countries were doomed to suffer the triple catastrophe of overpopulation, too rapid urbanization, and economic underdevelopment.

"Equipped with this knowledge, I spent a summer in the Philippines working for an organization that offered farmers advice on how to promote economic growth. I assumed that, as in North America, farmers who owned large plots of land and used high technology would be more efficient and better off. However, I found the most productive villages were those in which most farmers owned *small* plots of land. In such villages, there was little economic inequality. The women in these villages enjoyed low birth rates, and the inhabitants were usually happier than the inhabitants of villages in which there was more inequality.

"What was going on? As I talked with the villagers, I came to realize that farmers who owned at least some of their own land had ample incentive to work hard. For the harder they worked, the more they earned. With a higher standard of living, they didn't need to have as many children to help them on the farm. In contrast, in villages with greater inequality, many farmers owned no land. Some of them leased land, but many worked as farm laborers for wealthy landlords. They didn't earn more for working harder, so their productivity and their standard of living were low. As a result, they wanted to have more children to help bring income into their households.

"Few Filipino farms could match the productivity of high-tech American farms because even large plots were small by American standards. Much high-tech agricultural equipment would have been useless there. Imagine trying to use a harvesting machine in a plot not much larger than some suburban backyards.

"Thus, my Western assumptions—that Filipino farmers needed big plots and high technology—turned out to be wrong. The Filipino farmers I met were knowledgeable and thoughtful about their needs and desires. When I started listening to them (and paying less attention to my economics textbooks and professors) I started understanding the real world of economic development. It was one of the most important sociological lessons I ever learned."

This chapter tackles the closely connected problems of population growth, urbanization, and economic development. It starts where John did in the Philippines, with the realization that not all is as it seems. Specifically, we first show that population growth is a process governed less by natural laws than by social forces. We argue that these social forces are not related exclusively to industrialization, as social scientists commonly believed just a few decades ago. Instead, social inequality also plays a major role in shaping population growth. We next turn to the problem of urbanization. Today, population growth is typically accompanied by the increasing concentration of the world's people in urban centers. As recently as 40 years ago, sociologists typically believed that cities are typically alienating and anomic (or "normless"). We argue that this view is an oversimplification. We also outline the social roots of the city's physical and cultural evolution from pre-industrial to postindustrial times. Finally, we turn to the distressing problem of global inequality among nations. We present and assess the major theoretical approaches to this problem in the light of available evidence and find some glimmers of hope in an otherwise troubling story.

POPULATION

The Population "Explosion"

Ten thousand years before the birth of Christ there were only about 6 million people in the world. By the time Christ was born, world population had risen to 250 million, and it increased to some 760 million by 1750. After that, world population skyrocketed. The number of humans reached 1 billion in 1804 and 5 billion in 1987 (see Figure 16.1). On July 1, 2001, there were an estimated 6.16 billion people in the world according to the United States Census Bureau (1998d). Where one person stood 12,000 years ago, there are now 1,025 people. Statistical projections suggest that, by 2100, there will be nearly 1,700 people. Of those 1,700, fewer than 250 will be standing in the rich countries of the world. More than 1,450 of them will be in the developing countries of South America, Asia, and Africa.

Many analysts project that, after passing the 10 billion mark around 2100, world population will level off. But given the numbers cited above, is it any wonder that some population analysts say we're now in the midst of a population "explosion"? Explosions are horrifying events. They cause widespread and severe damage. They are fast and unstoppable. And that is exactly the imagery some population analysts, or **demographers,** wish to convey (e.g., Ehrlich, 1968; Ehrlich and Ehrlich, 1990; see Figure 16.2). They have written many books, articles, and television programs dealing with the population explosion. You may have encountered some of these in your school or church. Images of an overflowing multitude in, say, Bangladesh, Nigeria, or Brazil remain fixed in our minds. Some people are frightened enough to refer to overpopulation as catastrophic. They link it to recurrent famine, brutal ethnic warfare, and other massive and seemingly intractable problems.

If this imagery makes you feel that the world's rich countries must do something about overpopulation, you're not alone. In fact, concern about the population "bomb" is as old as the social sciences. In 1798, Thomas Robert Malthus, a British clergyman of the Anglican faith, proposed a highly influential theory of human population (Malthus (1966 [1798]). As you will soon see, contemporary sociologists have criticized, qualified, and in part rejected his theory. But because much of the sociological study of population is, in effect, a debate with Malthus's ghost, we must confront the man's ideas squarely.

Web Interactive Exercises
World Population

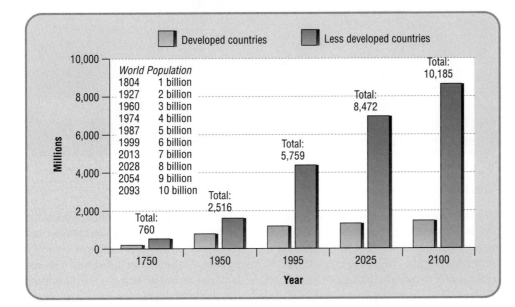

✦ FIGURE 16.1 ✦
World Population, 1750–2100 (in millions, projected)

SOURCES: Livi-Bacci (1992: 31); Merrick (1986: 12); United Nations (1993; 1998c).

Developed countries   Less developed countries

World Population	
1804	1 billion
1927	2 billion
1960	3 billion
1974	4 billion
1987	5 billion
1999	6 billion
2013	7 billion
2028	8 billion
2054	9 billion
2093	10 billion

Total: 760 (1750)
Total: 2,516 (1950)
Total: 5,759 (1995)
Total: 8,472 (2025)
Total: 10,185 (2100)

A "population explosion"? Hong Kong is one of the most densely populated places on earth.

The Malthusian Trap

Malthus's theory rests on two undeniable facts and a questionable assumption. The facts: people must eat, and they are driven by a strong sexual urge. The assumption: while food supply increases slowly and arithmetically (1, 2, 3, 4, etc.), population size grows quickly and geometrically (1, 2, 4, 8, etc.). Based on these ideas, Malthus concluded that "the superior power of population cannot be checked without producing misery or vice" (Malthus (1966 [1798]): 217–18.) Specifically, only two forces can hold population growth in check.

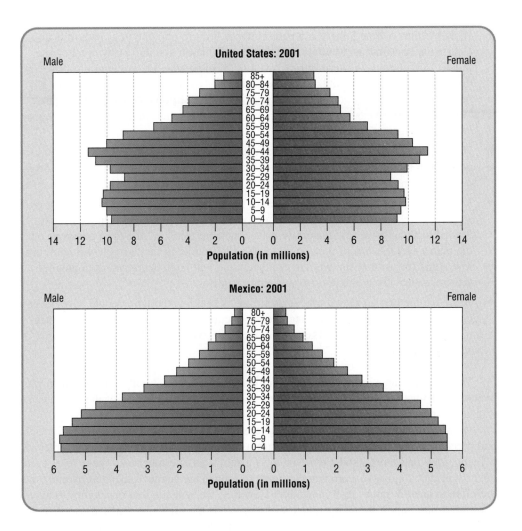

United States: 2001

Male / Female

Population (in millions)

Mexico: 2001

Male / Female

Population (in millions)

SOURCE: United States Bureau of the Census (2000c).

✦ FIGURE 16.2 ✦

How Demographers Analyze Population Changes and Composition

The main purpose of demography is to figure out why the size, geographical distribution, and social composition of human populations change over time. The basic equation of population change is $P2 = P1 + B − D + I − E$, where $P2$ is population size at a given time, $P1$ is population size at an earlier time, B is the number of births in the interval, D is the number of deaths in the interval, I is the number of immigrants arriving in the interval, and E is the number of emigrants leaving in the interval. One basic tool for analyzing the composition of a population is the "age–sex pyramid," which shows the number of males and females in each age cohort of the population at a given point in time. Age–sex pyramids for the United States and Mexico are shown here, projected by the United States Census Bureau for 2001. Why do you think they look so different? Compare your answer to that of the theory of the demographic transition, discussed in this section.

First are "preventive" measures, such as abortion, infanticide, and prostitution. Malthus called these "vices" because he morally opposed them and thought everyone else ought to also. Second are "positive checks," such as war, pestilence, and famine. Malthus recognized that positive checks create much suffering. Yet he felt they are the only forces that can be allowed to control population growth. Here, then, is the so-called **Malthusian trap:** a cycle of population growth followed by an outbreak of war, pestilence, or famine that keeps population growth in check. Population size might fluctuate, said Malthus, but it has a natural upper limit that Western Europe has reached.

Although many people supported Malthus's theory, others reviled him as a misguided prophet of doom and gloom (Winch, 1987). For example, people who wished to help the poor disagreed with Malthus. He felt such aid was counterproductive. Welfare, he said, would enable the poor to buy more food. With more food, they would have more children. And having more children would only make them poorer than they already were. Better leave them alone, said Malthus. That will reduce the sum of human suffering in the world.

Although in some respects compelling, events have cast doubt on several of Malthus's ideas. Specifically:

✦ Ever since Malthus proposed his theory, technological advances have allowed rapid growth in how much food is produced for each person on the planet. This is the opposite of the slow growth Malthus predicted. For instance, in the period 1991–93, India produced 23% more food per person than it did in 1979–81 and China produced 39% more. Moreover, except for Africa south of the Sahara, the largest increases in the food supply are taking place in the developing countries (Sen, 1994).

Albrecht Dürer, *The Four Horsemen of the Apocalypse* (woodcut, 1498). According to Malthus, only war, pestilence, and famine could keep population growth in check.

◆ If, as Malthus claimed, there is a natural upper limit to population growth, it is unclear what that limit is. Malthus thought the population couldn't grow much larger in late 18th-century Western Europe without "positive checks" coming into play. Yet the Western European population increased from 187 million people in 1801 to 321 million in 1900. It has now stabilized at about half a billion (McNeill, 1990). The Western European case suggests that population growth has an upper limit far higher than that envisaged by Malthus.

◆ Population growth does not always produce misery. For example, despite its rapid population increase over the past 200 years, Western Europe is one of the most prosperous regions in the world.

◆ Helping the poor does not generally result in the poor having more children. For example, in Western Europe, social welfare policies (unemployment insurance, state-funded medical care, paid maternity leave, pensions, etc.) are the most generous on the planet. Yet the size of the population is quite stable. In fact, as you will learn below, some forms of social welfare produce rapid and large *decreases* in population growth, especially in the poor, developing countries.

◆ Although the human sexual urge is as strong as Malthus thought, people have developed contraceptive devices and techniques to control the consequences of their sexual activity (Szreter, 1996). There is no necessary connection between sexual activity and childbirth.

The developments listed above all point to one conclusion. Malthus's pessimism was overstated. Human ingenuity seems to have enabled us to wriggle free of the Malthusian trap, at least for the time being.

We are not, however, home free. Today there are renewed fears that industrialization and population growth are putting severe strains on the planet's resources. As Chapter 18 ("Technology and the Global Environment") establishes, we must take these fears seriously. It is encouraging to learn that the limits to growth are as much social as natural, and therefore avoidable rather than inevitable. However, we will see that our ability to avoid the Malthusian trap in the 21st century will require all the ingenuity and self-sacrifice we can muster. For the time being, however, let us consider the second main theory of population growth, the theory of the demographic transition.

Demographic Transition Theory

According to **demographic transition theory,** the main factors underlying population dynamics are industrialization and the growth of modern cultural values (Notestein, 1945; Coale, 1974; Chesnais, 1992; see Figure 16.3). The theory is based on the observation that the European population developed in four distinct stages:

1. In the first, *preindustrial stage* of growth, a large proportion of the population died every year due to inadequate nutrition, poor hygiene, and uncontrollable disease. In other words, the **crude death rate** was high. The crude death rate is the annual number of deaths (or "mortality") per 1,000 people in a population. During this period, the **crude birth rate** was high too. The crude birth rate is the annual number of live births per 1,000 people in a population. In the preindustrial era, most people wanted to have as many children as possible. That was partly because relatively few children survived till adulthood. In addition, children were considered a valuable source of agricultural labor and a form of old age security in a society consisting largely of peasants and lacking anything resembling a modern welfare state.

2. The second stage of European population growth was the *early industrial or transition period*. At this stage, the crude death rate dropped. People's life expectancy, or average life span, increased because economic growth led to improved nutrition and hygiene. However, the crude birth rate remained high. With people living longer and

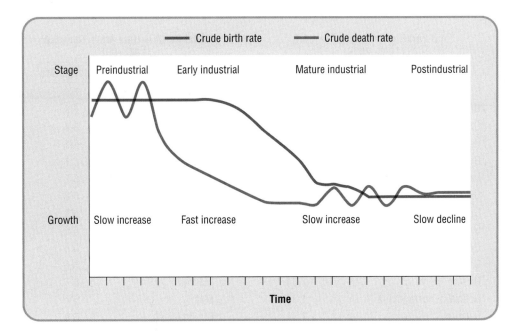

women having nearly as many babies as in the preindustrial era, the population grew rapidly. Malthus lived during this period of rapid population growth, and that accounts in part for his alarm.

3. The third stage of European population growth was the *mature industrial period*. At this stage, the crude death rate continued to fall. The crude birth rate fell even more dramatically. The crude birth rate fell because economic growth eventually changed people's traditional beliefs about the value of having many children. Having lots of children made sense in an agricultural society, where, as we have seen, children were a valuable economic resource. In contrast, children were more of an economic burden in an industrial society. That is because breadwinners worked outside the home for a wage or a salary and children contributed little if anything to the economic welfare of the family. Note, however, that the crude birth rate took longer to decline than the crude death rate did. That is because people's values often change more slowly than their technologies. People can put in a sewer system or a water purification plant to lower the crude death rate faster than they can change their minds about something as fundamental as how many children to have. Eventually, however, the technologies and outlooks that accompany modernity led people to postpone getting married and use contraceptives and other birth-control methods. As a result, population stabilized during the mature industrial period. This demonstrates the validity of one of the demographer's favorite sayings: "Economic development is the best contraceptive."

4. In the last decades of the 20th century, the **total fertility rate,** or the annual number of live births per 1,000 women in a population, continued to fall. In fact, it fell below the **replacement level** in some countries. The replacement level is the number of children each woman must have on average for population size to remain stable. Ignoring any inflow of settlers from other countries (**immigration** or **in-migration**) and any outflow to other countries (**emigration** or **out-migration**), the replacement level is 2.1. This means that, on average, each woman must give birth to slightly more than the two children needed to replace her and her mate. Slightly more than two children are required because some children die before they mature and reach reproductive age.

By the 1990s, some Europeans were worrying about declining fertility and its possible effects on population size. As you can see in Table 16.1, 21 countries, 18 of them in

Countries with the Lowest and Highest Total Fertility Rates

Note: Strict comparison between the two columns should be avoided because data collection is more sporadic in poor than in rich countries. As a result, the data in the right-hand column are on average 7 years older than the data in the left-hand column. Over that period, fertility rates in Africa declined somewhat.

SOURCES: United Nations (1997a; 1998a).

Total Fertility Rate Less Than 1.5 (1995–2000)		Total Fertility Rate More Than 6.0 (1987–96)	
Country	Fertility Rate	Country	Fertility Rate
Spain	1.15	Djibouti	6.0
Romania	1.17	Nigeria	6.0
Czech Republic	1.19	Senegal	6.0
Italy	1.20	Eritrea	6.1
Bulgaria	1.23	Madagascar	6.1
Latvia	1.25	Tanzania	6.2
Slovenia	1.26	Benin	6.3
Greece	1.28	Sierra Leone	6.3
Estonia	1.28	Zambia	6.5
Germany	1.30	Rwanda	6.6
China	1.32	Togo	6.6
Russia	1.34	Malawi	6.7
Bosnia & Herzegovina	1.35	Mali	6.7
Belarus	1.36	Burkina Faso	6.9
Portugal	1.37	Uganda	6.9
Hungary	1.37	Burundi	7.0
Ukraine	1.38	Niger	7.4
Slovakia	1.39	Ethiopia	7.7
Macau	1.40		
Austria	1.41		
Lithuania	1.42		
Japan	1.43	USA 1995–2000: 1.99	
Switzerland	1.47		

Europe, now have fertility rates below 1.5. In the period 1995–2000, 61 countries or areas of the world representing 44% of the world's population had a fertility rate below the replacement level. That is 10 more countries than in the period 1990–95. The United States was a member of this group with a fertility rate of 1.99 in the period 1995–2000 (United Nations, 1998a). Due to the proliferation of low-fertility societies, some scholars suggest that we have now entered a fourth, *postmodern stage* of population development. In this fourth stage of the demographic transition, the number of deaths per year exceeds the number of births (Van de Kaa, 1987).

As outlined above, the demographic transition theory provides a rough picture of how industrialization affects population growth. However, research has revealed a number of inconsistencies in the theory. Most of them are due to the theory placing too much emphasis on industrialization as the main force underlying population growth (Coale and Watkins, 1986). For example, demographers have found that reductions in fertility sometimes occur when standards of living stagnate or decline, not just when they improve due to industrialization. Thus, in Russia and some developing countries today, declining living standards have led to a deterioration in general health and a subsequent decline in fertility. Because of such findings, many scholars have concluded that an adequate theory of population growth must pay more attention to social factors other than industrialization, and in particular to the role of social inequality.

Population and Social Inequality

One of Malthus's staunchest intellectual opponents was Karl Marx. Marx argued that the problem of overpopulation is specific to capitalism (Meek, 1971). In his view, overpopulation is not a problem of too many people. Instead, it is a problem of too much poverty. Do away with the exploitation of workers by their employers, said Marx, and poverty will disappear. If a society is rich enough to eliminate poverty, then by definition its population is not too large. By eliminating poverty, one also solves the problem of overpopulation in Marx's view.

Marx's analysis makes it seem that capitalism can never generate enough prosperity to solve the overpopulation problem. He was evidently wrong. Overpopulation is not a serious problem in the United States or Japan or Germany today.[1] It *is* a problem in most of Africa, where capitalism is weakly developed and the level of social inequality is much higher than in the postindustrial societies. Still, a core idea in Marx's analysis of the overpopulation problem rings true. As some contemporary demographers argue, social inequality is a main cause of overpopulation. Below, we illustrate this argument by first considering how gender inequality influences population growth. Then, we discuss the effects of class inequality on population growth.

Gender Inequality and Overpopulation

The effect of gender inequality on population growth is well illustrated by the case of Kerala, a state in India with more than 30 million people. Kerala had a total fertility rate of 1.8 in 1991, half India's national rate and well below the replacement level of 2.1. How did Kerala achieve this remarkable feat? Is it a highly industrialized oasis in the midst of a semi-industrialized country, as one might expect given the arguments of demographic transition theory? To the contrary, Kerala is not highly industrialized. In fact, it is among the poorer Indian states, with a per capita income below the national average. Then has the government of Kerala strictly enforced a state childbirth policy similar to China's? The Chinese government strongly penalizes families that have more than one child and it allows abortion at 8½ months. As a result, China had a total fertility rate of just 2.0 in 1992 (Wordsworth, 2000). In Kerala, however, the government keeps out of its citizens' bedrooms. The decision to have children remains a strictly private affair.

The women of Kerala achieved a low total fertility rate because their government purposely and systematically raised their status over a period of decades (Franke and Chasin, 1992; Sen, 1994). The government helped to create a realistic alternative to a life of continuous childbearing and child rearing. It helped women understand that they could achieve that alternative if they wanted to. In particular, the government organized successful campaigns and programs to educate women, increase their participation in the paid labor force, and make family planning widely available. These government campaigns and programs resulted in Keralan women enjoying the highest literacy rate, the highest labor force participation rate, and the highest rate of political participation in India. Given their desire for education, work, and political involvement, most Keralan women want small families, so they use contraception to prevent unwanted births. Thus, by lowering the level of gender inequality, the government of Kerala solved its overpopulation problem. In general, where

women tend to have more power [the society has] low rather than high mortality and fertility. Education and employment, for example, often accord women wider power and influence, which enhance their status. But attending school and working often compete with childbearing and childrearing. Women may choose to have fewer children in order to hold a job or increase their education (Riley, 1997).

Class Inequality and Overpopulation

Unraveling the Keralan mystery is an instructive exercise. It establishes that population growth depends not just on a society's level of industrialization but also on its level of gender inequality. *Class* inequality influences population growth too. We turn to the South Korean case to illustrate this point.

---

[1]However, because Americans in particular consume so much energy and other resources, we have a substantial negative impact on the global environment. See Chapter 18, "Technology and the Global Environment."

In 1960, South Korea had a total fertility rate of 6.0. This prompted one American official to remark that "if these Koreans don't stop overbreeding, we may have the choice of supporting them forever, watching them starve to death, or washing our hands of the problem" (quoted in Lie, 1998: 21). Yet by 1989, South Korea's total fertility rate had dropped to a mere 1.6. Why? The first chapter in this story involves land reform, not industrialization. The government took land from big landowners and gave it to small farmers. Consequently, the standard of living of small farmers improved. This eliminated a major reason for high fertility. Once economic uncertainty decreased, so did the need for child labor and support of elderly parents by adult offspring. Soon, the total fertility rate began to fall. Subsequent declines in the South Korean total fertility rate were due to industrialization, urbanization, and the higher educational attainment of the population. But a decline in class inequality in the countryside first set the process in motion.

The reverse is also true. Increasing social inequality can lead to overpopulation, war, and famine. For example, in the 1960s the governments of El Salvador and Honduras encouraged the expansion of commercial agriculture and the acquisition of large farms by wealthy landowners. The landowners drove peasants off the land. The peasants migrated to the cities. There they hoped to find employment and a better life. Instead, they often found squalor, unemployment, and disease. Suddenly, two countries with a combined population of less than 5 million people had a big "overpopulation" problem. Competition for land increased and contributed to rising tensions. This eventually led to the outbreak of war between El Salvador and Honduras in 1969 (Durham, 1979).

Similarly, economic inequality helps to create famines. As Nobel prize winner Amartya Sen notes, "[f]amine is the characteristic of some people not *having* enough food to eat. It is not the characteristic of there not *being* enough food to eat" (Sen, 1981: 1; our emphasis). Sen's distinction is crucial, as his analysis of several famines shows. Sen found that, in some cases, while food supplies did decline, enough food was available to keep the stricken population fed. However, suppliers and speculators took advantage of the short supply. They hoarded grain and increased prices beyond the means of most people. In other cases, there was no decline in food supply at all. Food was simply withheld for political reasons, that is, to bring a population to its knees, or because many people were not considered entitled to receive it by the authorities. The source of famine, Sen concludes, is not underproduction or overpopulation but inequality of access to food (Drèze and Sen, 1989). In fact, even when starving people gain access to food, other forms of inequality may kill them. For instance, when food relief agencies delivered 3 million sacks of grain to prevent famine in Western Sudan in the mid-1980s, tens of thousands of people died anyway because they lacked clean water, decent sanitation, and vaccinations against various diseases (de Waal, 1989). Western aid workers failed to listen carefully to the Sudanese people about their basic medical and sanitary needs, with disastrous results.

Summing Up

A new generation of demographers has begun to explore how class inequality and gender inequality affect population growth (Levine, 1987; Seccombe, 1992; Szreter, 1996). Their studies drive home the point that population growth and its negative consequences do not stem from natural causes (as Malthus held). Nor are they only responses to industrialization and modernization (as demographic transition theory suggests). Instead, population growth is influenced by a variety of social causes, social inequality chief among them.

Some undoubtedly well-intentioned Western analysts continue to insist that people in the developing countries should be forced to stop multiplying at all costs. Some observers even suggest diverting scarce resources from education, health, and industrialization into various forms of birth control, including, if necessary, forced sterilization (Riedmann, 1993). They regard the presumed alternatives—poverty, famine, war, ethnic violence, and the growth of huge, filthy cities—as too horrible to contemplate. However, they fail to see how measures that lower social inequality help to control overpopulation and its consequences. Along with industrialization, lower levels of social inequality cause total fertility rates to fall.

URBANIZATION

We have seen that overpopulation remains a troubling problem due to lack of industrialization and too much gender and class inequality in much of the world. We may now add that overpopulation is in substantial measure an *urban* problem. Driven by lack of economic opportunity in the countryside, political unrest, and other factors, many millions of people flock to big cities in the world's poor countries every year. Thus, most of the fastest growing cities in the world today are in semi-industrialized countries where the factory system is not highly developed. As Table 16.2 shows, in 1900, 9 of the 10 biggest cities in the world were in industrialized Europe and the United States. By 2015, in contrast, 6 of the world's 10 biggest cities will be in Asia, 2 will be in Africa, and 2 will be in Latin America. Only 1 of the 10 biggest cities—Tokyo—will be in a highly industrialized country. Urbanization is, of course, taking place in the world's rich countries too. For instance, in North America, the urban population is expected to increase from 76% to 84% of the total population between 1996 and 2030. In Africa and Asia, however, the urban population is expected to increase much faster—from about 35% to 55% of the total population in the same time period (United Nations, 1997b).

1900		2015	
London	6.5	Tokyo	28.7
New York	4.2	Bombay	27.4
Paris	3.3	Lagos	24.4
Berlin	2.4	Shanghai	23.4
Chicago	1.7	Jakarta	21.2
Vienna	1.6	São Paulo	20.8
Tokyo	1.5	Karachi	20.6
Saint Petersburg	1.4	Beijing	19.4
Philadelphia	1.4	Dhaka	19.0
Manchester	1.3	Mexico City	18.8

✦ **TABLE 16.2** ✦
World's 10 Largest Cities, 1900 and 2015, Projected (in millions)

SOURCES: Department of Geography, Slippery Rock University (1997; 1998).

Mexico City during one of its frequent smog alerts. Of the world's 10 biggest cities in 2015, only one—Tokyo—will be in a highly industrialized country. All the others, including Mexico City, will be in developing countries.

Carcassone, France, a medieval walled city.

From the Preindustrial to the Industrial City

To a degree, urbanization results from industrialization. As you will learn below, many great cities of the world grew up along with the modern factory, which drew hundreds of millions of people out of the countryside and transformed them into urban, industrial workers. Industrialization is not, however, the whole story behind the growth of cities. As we have just seen, the connection between industrialization and urbanization is weak in the world's less developed countries today. Moreover, cities first emerged in Syria, Mesopotamia, and Egypt 5,000 or 6,000 years ago, long before the growth of the modern factory. These early cities served as centers of religious worship and political administration. Similarly, it was not industry but international trade in spices, gold, cloth, and other precious goods that stimulated the growth of cities in preindustrial Europe and the Middle East. Thus, the correlation between urbanization and industrialization is far from perfect (Bairoch, 1988 [1985]; Jacobs, 1969; Mumford, 1961; Sjöberg, 1960).

Preindustrial cities differed from those that developed in the industrial era in several ways. Preindustrial cities were typically smaller, less densely populated, built within protective walls, and organized around a central square and places of worship. The industrial cities that began to emerge at the end of the 18th century were more dynamic and complex social systems. A host of social problems, including poverty, pollution, and crime, accompanied their growth. The complexity, dynamism, and social problems of the industrial city were all evident in Chicago at the turn of the 20th century. Not surprisingly, therefore, it was at the University of Chicago that American urban sociology was born.

The Chicago School and the Industrial City

From the 1910s to the 1930s, the members of the **Chicago school** of sociology distinguished themselves by their vividly detailed descriptions and analyses of urban life, backed up by careful in-depth interviews, surveys, and maps showing the distribution of various features of the social landscape, all expressed in plain yet evocative language (Lindner, 1996 [1990]). Three of its leading members, Robert Park, Ernest Burgess, and Roderick McKenzie, proposed a theory of **human ecology** to illuminate the process of urbanization (Park, Burgess, and McKenzie, 1967 [1925]). Borrowing from biology and ecology, the theory highlights the links between the physical and social dimensions of cities and identifies the dynamics and patterns of urban growth.

The theory of human ecology, as applied to urban settings, holds that cities grow in ever-expanding concentric circles. It is sometimes called the "concentric zone model" of the city. Three social processes animate this growth (Hawley, 1950). **Differentiation** refers to the process by which urban populations and their activities become more complex and heterogeneous over time. For instance, a small town may have a diner, a pizza parlor, and a Chinese restaurant. But if that small town grows into a city, it will likely boast a variety of ethnic restaurants reflecting its more heterogeneous population.

Moreover, in a city, members of different ethnic and racial groups and socioeconomic classes may vie with one another for dominance in particular areas. Businesses may also try to push residents out of certain areas to establish commercial zones. When this happens, people are engaging in **competition,** an ongoing struggle by different groups to inhabit optimal locations. Finally, **ecological succession** takes place when a distinct group of people moves from one area to another, and another group moves into the old area to replace the first group. For example, a recurrent pattern of ecological succession involves members of the middle class moving to the suburbs, with working class and poor immigrants moving into the inner city.

In Chicago in the 1920s, differentiation, competition, and ecological succession resulted in the zonal pattern illustrated by Figure 16.4:

1. Zone 1 was the central business district (known in Chicago as "the Loop"). It contained retail shopping areas, office buildings, and entertainment centers. The land in this zone was the most valuable in the city.

2. Zone 2, the "zone of transition," was the area of most intense competition between residential and commercial interests. The businesses usually succeeded in driving out middle-class residents by bidding up the price of land and getting city governments to rezone the area for commercial use. Homes then declined in value, and cheap rental housing came to predominate. Eventually, the housing deteriorated into slums because speculators spent little on maintenance. They merely held onto the buildings until they could sell them for the commercial value of the land on which they stood. Where commercial development failed to materialize, the slums endured,

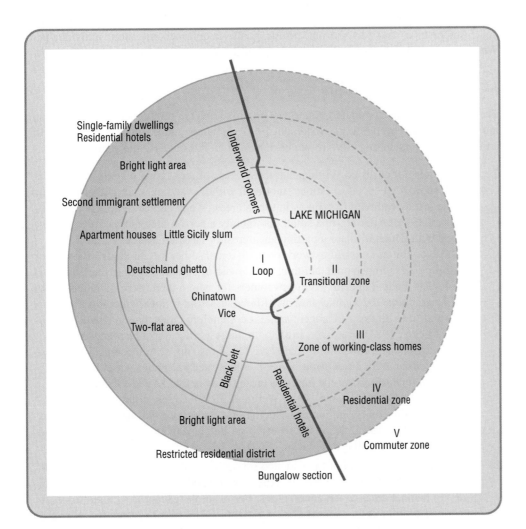

✦ **FIGURE 16.4** ✦
The Concentric Zone Model of Chicago, About 1920

SOURCE: Burgess (1961 [1925]).

attracting new immigrants, the poor, people with physical and mental disabilities, the unemployed, and criminals. The Chicago sociologists viewed this zone as "socially disorganized" because of its high level of deviance and crime.

3. When members of ethnic groups in Zone 2 could afford better housing, they moved to Zone 3, the "zone of working-class homes." These were mostly inexpensive, semidetached buildings.

4. The upwardly mobile offspring of families in Zone 3 usually moved to Zone 4, the "residential zone," containing small, middle-class, detached homes.

5. Zone 5, the "commuter zone," was where middle-, upper–middle-, and upper-class families lived in more expensive detached homes. These people also owned cars and commuted to work in the city.

For members of the Chicago school, the city was more than just a collection of socially segregated buildings, places, and people. It also involved a way of life they called **urbanism.** They defined urbanism as "a state of mind, a body of customs[,] . . . traditions, . . . attitudes and sentiments" specifically linked to city dwelling (Park, Burgess, and McKenzie, 1967 [1925]: 1). Louis Wirth (1938) developed this theme. According to Wirth, rural life involves frequent face-to-face interaction among a few people. Most of these people are familiar with each other, share common values and a collective identity, and strongly respect traditional ways of doing things. Urban life, in contrast, involves the absence of community and of close personal relationships. Extensive exposure to many socially different people leads city dwellers to become more tolerant than rural folk, said Wirth. However, urban dwellers also withdraw emotionally and reduce the intensity of their social interaction with others. In Wirth's view, interaction in cities is therefore superficial, impersonal, and focused on specific goals. People become more individualistic. Weak social control leads to a high incidence of deviance and crime.

After Chicago: A Critique

The Chicago school dominated American urban sociology for decades. It still inspires much interesting research (e.g., Anderson, 1990). However, three major criticisms of this approach to understanding city growth have gained credibility over the years.

1. One criticism focuses on Wirth's characterization of the "urban way of life." Research shows that social isolation, emotional withdrawal, stress, and other problems may be just as common in rural as in urban areas (Webb and Collette, 1977; 1979; Crothers, 1979). After all, in a small community a person may not be able to find anyone with which to share a particular interest or passion. Moreover, farm work can be every bit as stressful as work on an assembly line. Research also shows that urban life is less impersonal, anomic, and devoid of community than the Chicago sociologists made it appear. True, newcomers (of whom there were admittedly many in Chicago in the 1920s) may find city life bewildering if not frightening. Neighborliness and friendliness to strangers are less common in cities than in small communities (Fischer, 1981). However, even in the largest cities, most residents create social networks and subcultures that serve functions similar to those performed by the small community. Friendship, kinship, ethnic, and racial ties, as well as work and leisure relations, form the bases of these urban networks and subcultures (Fischer, 1984 [1976]; Jacobs, 1961; Wellman, 1979). Cities, it turns out, are clusters of many different communities. Sociologist Herbert Gans found such a rich assortment of close social ties in his research on Italian-Americans he was prompted to call them "urban villagers" (Gans, 1962).

2. A second major criticism of the Chicago school's approach to urban sociology focuses on the concentric zone model. True, specific activities and groups are concentrated in distinct areas of American cities. For example, except for a handful of integrated communities, such as Shaker Heights, Ohio, racial segregation in housing

remains a prominent feature of American urban life (Massey and Denton, 1993). However, the specific patterns discovered by the Chicago sociologists are most applicable to industrial cities in the first quarter of the 20th century. Thus, in pre-industrial cities, slums are more likely to be found on the outskirts and wealthy districts in the city core, while commercial and residential buildings are often not segregated (Sjöberg, 1960). After the automobile became a major means of transportation, some cities expanded not in concentric circles but in wedge-shaped sectors along natural boundaries and transportation routes (Hoyt, 1939). Others grew up around not one but many nuclei, each attracting similar kinds of activities and groups, as in Figure 16.5.

3. The third main criticism of the human ecology approach is that it presents urban growth as an almost natural process, slighting its historical, political, and economic foundations in capitalist industrialization. The Chicago sociologists' analysis of competition in the transitional zone came closest to avoiding this problem. However, their discussions of differentiation and ecological succession made the growth of cities seem almost like a force of nature rather than a process rooted in power relations and the urge to profit.

The so-called **new urban sociology,** heavily influenced by conflict theory, sought to correct this problem (Gottdiener and Hutchison, 2000 [1994]); Zukin, 1980). For new urban sociologists, urban space is not just an arena for the unfolding of social processes like differentiation, competition, and ecological succession. Instead, they see urban space as a set of *commodified* social relations. That is, urban space, like all commodities, can be

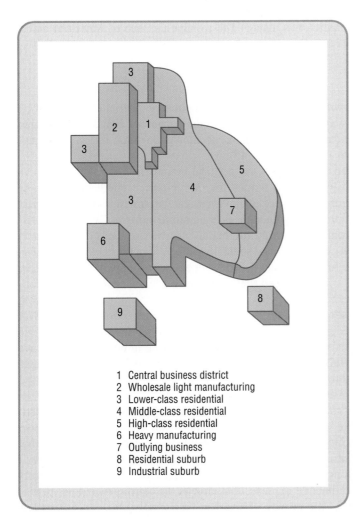

✦ FIGURE 16.5 ✦

The Multiple-Nuclei Model of a City

SOURCE: Harris and Ullman (1945).

1 Central business district
2 Wholesale light manufacturing
3 Lower-class residential
4 Middle-class residential
5 High-class residential
6 Heavy manufacturing
7 Outlying business
8 Residential suburb
9 Industrial suburb

bought and sold for profit. As a result, political interests and conflicts shape the growth pattern of cities. John Logan and Harvey Molotch (1987), for example, portray cities as machines fueled by a "growth coalition." This growth coalition is composed of investors, politicians, businesses, property owners, real estate developers, urban planners, the mass media, professional sports teams, cultural institutions, labor unions, and universities. All these partners try to get government subsidies and tax breaks to attract investment dollars. Reversing the pattern identified by the Chicago sociologists, this investment has been used to redevelop decaying downtown areas in many American cities since the 1950s. The Faneuil Hall Marketplace in Boston, Harborplace in Baltimore, South Street Seaport in New York, Grand Avenue in Milwaukee, Union Station in St. Louis, Bayside in Miami, and the Aloha Tower in Honolulu are all examples of such redevelopment projects from the 1970s.

According to Logan and Molotch, members of the growth coalition present redevelopment as a public good that benefits everyone. This tends to silence critics, prevent discussions of alternative ideas and plans, and veil the question of who benefits and who does not. In reality, the benefits of redevelopment are often unevenly distributed. Most redevelopments are "pockets of revitalization surrounded by areas of extreme poverty" (Hannigan, 1998a: 53). That is, local residents often enjoy few if any direct benefits from redevelopment. Indirectly, they may suffer when budgets for public schooling, public transportation, and other amenities are cut to help pay for development subsidies and tax breaks.

We can see the workings of the growth coalition in the histories of Los Angeles (Abelmann and Lie, 1995), Chicago (Cronon, 1991), Miami (Portes and Stepick, 1993), and other great American cities. This does not mean that the growth coalition is all-powerful. Community activism often targets local governments and corporations that seek unrestricted growth. Sometimes activists meet with success (Castells, 1983). Yet for the past 50 years, the growth coalition has managed to reshape the face of American cities, more or less in its own image (see Box 16.1).

The Corporate City

Due to the efforts of the growth coalition, the North American industrial city, typified by Chicago in the 1920s, gave way after World War II to the **corporate city.** John Hannigan defines the corporate city as "a vehicle for capital accumulation—that is, . . . a money-making machine" (Hannigan, 1998b [1995]: 345).

In the suburbs, urbanized areas outside the political boundaries of cities, developers built millions of single-family detached homes for the corporate middle class. These homes boasted large backyards and a car or two in every garage. A new way of life developed, which sociologists, appropriately enough, dubbed **suburbanism.** Every bit as distinctive as urbanism, suburbanism organized life mainly around the needs of children. It also involved higher levels of conformity and sociability than life in the central city (Fava, 1956). Suburbanism became fully entrenched as developers built shopping malls to serve the needs of the suburbanites. This reduced the need to travel to the central city for consumer goods.

The suburbs were at first restricted to the well to do. However, following World War II, brisk economic growth and government assistance to veterans put the suburban lifestyle within the reach of middle-class Americans. The lack of housing in city cores, extensive road-building programs, the falling price of automobiles, and the "baby boom" that began in 1946 also stimulated mushroom-like suburban growth. By 1970, more Americans lived in suburbs than in urban core areas. This remains the case today.

Due to the expansion of the suburbs, urban sociologists today often focus their attention not on cities but on entire "metropolitan areas." Metropolitan areas include downtown city cores and their surrounding suburbs. They also include two recent developments that indicate the continued decentralization of urban America: so-called exurbs, or rural residential areas within commuting distance of the city, and "edge cities," or exurban clusters of malls, offices, and entertainment complexes that arise at the convergence point of major highways (Garrau, 1991). The growth of exurban residential areas and edge cities

BOX 16.1
SOCIOLOGY AT THE MOVIES

BLADE RUNNER (1982)

A dark cloud of smog hangs menacingly over a city. Magnificent skyscrapers and brilliant, hundred-foot-high billboards pierce the clouds. Rich people live in the skyscrapers. Their lives are immaculate, orderly, and enhanced by the latest technologies. At ground level are the teeming poor. They live in shacks. Their lives are dirty and disorderly. Danger and crime are everywhere. Here, then, is a tale of two cities. It is Los Angeles in 2019. The movie is Ridley Scott's classic, *Blade Runner*.

Blade runners are police officers who hunt runaway "replicants," or human-like robots. Replicants live and work in space colonies as virtual slaves. They look and act like humans. They live only 4 years. Occasionally, however, some of them escape. It is the job of blade runners to hunt them down and terminate them. In the movie, Harrison Ford plays a blade runner searching for a group of runaway replicants. They want to find their designer before their 4 years expire so they can force him to extend their lives. In the course of his hunt, the character played by Ford falls in love with one of the replicants, played by Sean Young.

Blade Runner is science fiction, yet it mirrors aspects of urban life today. For instance, although the movie was made in the early 1980s, its depiction of Los Angeles as a tale of two cities—one rich, the other poor—is more realistic now than it was 20 years ago. Furthermore, the story of the blade runner and the replicants reminds the audience of an everyday drama that unfolds

Blade Runner, starring Harrison Ford.

in all large American cities. Every day, immigration control officials search out and terminate the residency of people trying desperately to extend their lives by working illegally in America. As you envisage the American city in the future, do you think it will look more or less like Los Angeles in *Blade Runner?*

since the 1970s has been stimulated by many factors. Among the most important are the mounting costs of operating businesses in city cores and the growth of new telecommunication technologies that allow businesses to operate in the exurbs. Home offices, mobile employees, and decentralized business locations are all made possible by these technologies.

City cores continued to decline as the middle class fled, pulled by the promise of suburban and exurban lifestyles and pushed by racial animosity and crime. Many middle-class people went farther afield, abandoning the Snowbelt cities in America's traditional industrial

The corporate city: New York.

✦ **TABLE 16.3** ✦
The 10 Largest Cities in the United States, 1998

Note: The figures in the first column are estimates for the cities proper, not their broader metropolitan regions, as in Table 16.2.

SOURCES: Adapted from United States Bureau of the Census (1998b; 1999c).

City	Estimated Population in Millions	Percent Growth Since 1990	Rank in 1950
1. New York	7.4	1.3	1
2. Los Angeles	3.6	3.2	4
3. Chicago	2.8	0.7	2
4. Houston	1.8	8.0	14
5. Philadelphia	1.4	−9.4	3
6. San Diego	1.2	9.9	31
7. Phoenix	1.2	21.3	99
8. San Antonio	1.1	14.1	25
9. Dallas	1.1	6.8	22
10. Detroit	1.0	−5.6	5

Web Research Projects
Welfare Reform

heartland and migrating to the burgeoning cities of the American Sunbelt in the South and the West (see Table 16.3 and Figure 16.6). As a result, particularly in Northeastern and Midwestern cities, tax revenues in the city core fell, even as more money was needed to sustain social welfare programs for the poor. In a spate of urban renewal in the 1950s and 1960s, many low-income and minority homes were torn down and replaced by high-rise apartment buildings and office towers in the city core. Nonetheless, large sections of downtown Detroit, Baltimore, Cleveland, and other cities remained in a state of decay.

The Postmodern City

Many of the conditions that plagued the industrial city—poverty, inadequate housing, structural employment—are still evident in cities today. However, since about 1970, a new urban phenomenon has emerged alongside the legacy of old urban forms. This is the **postmodern city** (Hannigan, 1995a). The postmodern city has three main features:

1.   The postmodern city is more *privatized* than the corporate city because access to formerly public spaces is increasingly limited to those who can afford to pay. Privatization is evident in the construction of closed-off "gated communities" in the suburbs. They have round-the-clock security guards stationed at controlled-access front gates and foot patrols on the lookout for intruders. Privatization is also apparent in downtown cores. There, gleaming office towers and shopping areas are built beside slums. Yet the two areas are separated by the organization of space and access. For instance, a series of billion-dollar, block-square structures have been built around Bunker Hill in Los Angeles. Nearly all pedestrian linkages to the surrounding poor

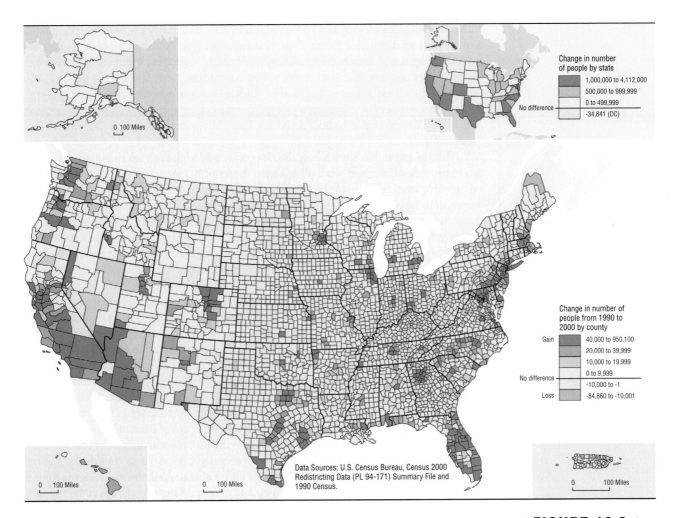

Change in number
of people by state

1,000,000 to 4,112,000
500,000 to 999,999
No difference ─ 0 to 499,999
-34,841 (DC)

Change in number of
people from 1990 to
2000 by county

Gain ─ 40,000 to 950,100
20,000 to 39,999
10,000 to 19,999
No difference ─ 0 to 9,999
-10,000 to -1
Loss ─ -84,860 to -10,001

Data Sources: U.S. Census Bureau, Census 2000
Redistricting Data (PL 94-171) Summary File and
1990 Census.

♦ FIGURE 16.6 ♦
**Difference in Number of
People, United States, 1990
to 2000**

United States Bureau of the Census
(2001).

immigrant neighborhoods have been removed. Barrel-shaped, "bum-proof" bus
benches prevent homeless people from sleeping on them. Trashcans are designed to
be "bag-lady-proof." Overhead sprinklers in Skid Row Park discourage overnight
sleeping. Public toilets and washrooms have been removed in areas frequented by
vagrants (Davis, 1990). Like the Renaissance Center in Detroit, the private areas of
downtown Los Angeles are increasingly intended for exclusive use by middle-class
visitors and professionals who work in the information sector, including financial
services, the computer industry, telecommunications, entertainment, and so forth.

2. The postmodern city is also more *fragmented* than the corporate city. That is, it
   lacks a single way of life, such as urbanism or suburbanism. Instead, a great variety
   of lifestyles and subcultures proliferate in the postmodern city. They are based on
   race, ethnicity, immigrant status, class, sexual orientation, and so forth.

3. The third characteristic of the postmodern city is that it is more *globalized* than the
   corporate city. According to Saskia Sassen (1991), New York, London, and Tokyo
   epitomize the global city. They are centers of economic and financial decision making. They are also sites of innovation, where new products and fashions originate.
   In short, they have become the command posts of the globalized economy and its
   culture.

The processes of privatization, fragmentation, and globalization are evident in the way
the postmodern city has come to reflect the priorities of the global entertainment industry.
Especially in the 1990s, the city and its outlying districts came to resemble so many Dis-
neyfied "Magic Kingdoms." These Magic Kingdoms are based on capital and technologies
from the United States, Japan, Canada, Britain, and elsewhere. Here we find the latest

entertainment technologies and spectacular thrills to suit nearly every taste. The postmodern city gets its distinctive flavor from its theme parks, restaurants and night clubs, waterfront developments, refurbished casinos, giant malls, megaplex cinemas, IMAX theaters, virtual reality arcades, ride simulators, sports complexes, book and CD megastores, aquariums, and hands-on science "museums." In the postmodern city, nearly everything becomes entertainment or, more accurately, combines entertainment with standard consumer activities. This produces hybrid activities like "shoppertainment," "eatertainment," and "edutainment."

John Hannigan has shown how the new venues of high-tech urban entertainment manage to provide excitement—but all within a thoroughly clean, controlled, predictable, and safe environment (Hannigan, 1998a). Not only are the new Magic Kingdoms kept spotless, in excellent repair, and fully temperature- and humidity-controlled, but they also provide a sense of security by touting familiar name brands and rigorously excluding anything and anybody that might disrupt the fun. For example, entertainment developments often enforce dress codes, teenager curfews, and rules banning striking workers and groups espousing social or political causes from their premises. The most effective barriers to potentially disruptive elements, however, are affordability and access. User surveys show that the new forms of urban entertainment tend to attract middle- and upper-middle-class patrons, especially whites. That is because they are pricey and many of them are in places that lack public transit and are too expensive for most people to reach by taxi.

Referring to the major role played by the Disney corporation in developing the new urban entertainment complexes, an architect once said that our downtowns would be "saved by a mouse" (quoted in Hannigan, 1998a: 193). But do the new forms of entertainment that dot the urban landscape increase the economic well-being of the communities in which they are established? Not much beyond creating some low-level, dead-end jobs (security guard, waiter, janitor). Do they provide ways of meeting new people, seeing old friends and neighbors, and in general improving urban sociability? Not really. You visit a theme park with family or friends, but you generally stick close to your group and rarely have chance encounters with other patrons or bump into acquaintances. Does the high-tech world of globalized urban entertainment enable cities and neighborhoods to retain and enhance their distinct traditions, architectural styles, and ambience? It would be hard to destroy the distinctiveness of New York, San Francisco, or Vancouver, but many large North American cities are becoming homogenized as they provide the same entertainment services—and the same global brands—as Tokyo, Paris, and Sydney. If the mouse is saving our cities, perhaps he is also gnawing away at something valuable in the process.

DEVELOPMENT

Global Inequality

The foregoing analysis shows that the *forms* assumed by urban social inequality in the United States have changed radically since the beginning of the 20th century. Nevertheless, American cities remain highly stratified places. If we now shift our attention from the national to the global level, we find an even more dramatic gap between rich and poor. In a Manhattan restaurant, pet owners can treat their cats to $100-a-plate birthday parties. In Cairo (Egypt) and Manila (The Philippines), the garbage dumps are home to entire families who sustain themselves by picking through the foul refuse. The startling difference between these two worlds forms the subject of this chapter's last section. How did it come about that "[a] fifth of the developing world's population goes hungry every night, a quarter lack access to even a basic necessity like safe drinking water, and a third live in a state of abject poverty—at such a margin of human existence that words simply fail to describe it" (United Nations, 1994: 2)? How did it transpire that most of the citizens of the 20 or so rich, industrialized countries spend more on cosmetics or alcohol or ice cream or pet food than it would take to provide basic education, or water and sanitation, or basic health and

A half-hour's drive from the center of Manila, the capital of the Philippines, an estimated 70,000 Filipinos live on a 55-acre mountain of rotting garbage, 150 feet high. It is infested with flies, rats, dogs, and disease. On a lucky day, residents can earn up to $5 retrieving scraps of metal and other valuables. On a rainy day, the mountain of garbage is especially treacherous. In July 2000, an avalanche buried 300 people alive. People who live on the mountain of garbage call it "The Promised Land."

✦ **TABLE 16.4** ✦

Global Priorities: Annual Cost of Various Goods and Services (in United States $ billion)

Note: Items in italics represent estimates of what they would cost to achieve. Other items represent estimated actual cost.

SOURCE: United Nations (1998b: 37)

Good or Service	Annual Cost (in United States $ billion)
Basic education for everyone in the world	6
Cosmetics in the United States	8
Water and sanitation for everyone in the world	9
Ice cream in Europe	11
Reproductive health for all women in the world	12
Perfumes in Europe and the United States	12
Basic health and nutrition for everyone in the world	13
Pet foods in Europe and the United States	17
Business entertainment in Japan	35
Cigarettes in Europe	50
Alcoholic drinks in Europe	105
Narcotic drugs in the world	400
Military spending in the world	780

nutrition for everyone in the world (see Table 16.4)? Let us see how sociologists explain this gaping inequality.

Theories of Development and Underdevelopment

Two main sociological theories claim to explain global inequality. According to **modernization theory,** global inequality results from various inadequacies of poor societies themselves. Specifically, modernization theorists say the citizens of poor societies lack sufficient *capital* to invest in Western-style agriculture and industry. They lack rational, Western-style *business techniques* of marketing, accounting, sales, and finance. As a result, their productivity and profitability remain low. They lack stable, Western-style *governments* that could provide a secure framework for investment. And, finally, they lack a

Western *mentality*—that is, values that stress the need for savings, investment, innovation, education, high achievement, and self-control in having children (Berger, 1963; Huntington, 1968; Inkeles and Smith, 1976; Rostow, 1960). It follows that people living in rich countries can best help their poor cousins by transferring Western culture and capital to them. Only then will the poor countries be able to cap population growth, stimulate democracy, and invigorate agricultural and industrial production. Government-to-government foreign aid can accomplish some of this. However, much also needs to be done to encourage Western businesses to invest directly in poor countries and to increase trade between rich and poor countries.

Critics have been quick to point out the chief flaw in modernization theory: For more than 200 years, the most powerful countries in the world deliberately impoverished the less powerful countries. It follows that an adequate theory of global inequality ought to focus on the relationship between rich and poor countries and not on the internal characteristics of poor countries themselves. Focusing on internal characteristics blames the victim rather than the perpetrator of the crime. This is the central argument of **dependency theory** (Baran, 1957; Frank; 1967; Wallerstein, 1974–89; Wolf, 1982).

According to dependency theorists, there was much less global inequality in 1750 than there is today. However, the industrial revolution enabled the Western European countries, Russia, Japan, and the United States to amass enormous wealth. They used their wealth to establish powerful armies and navies. Their armed forces subdued and then annexed or colonized most of the rest of the world between the middle of the 18th and the middle of the 20th centuries. The colonies were forced to become a source of raw materials, cheap labor, investment opportunities, and markets for the conquering nations. The colonizers thereby prevented industrialization and locked the colonies into poverty.

In the decades following World War II, nearly all of the colonies in the world became politically independent. However, say the dependency theorists, exploitation by direct political control was soon replaced by new means of achieving the same end: substantial foreign investment, support for authoritarian governments, and mounting debt. Let us consider each of these strategies in turn.

Substantial Foreign Investment

Multinational corporations invested heavily in the poor countries to siphon off wealth in the form of raw materials and profits. True, they created some low-paying jobs in the process. But they created many more high-paying jobs in the rich countries where the raw materials were used to produce manufactured goods. What is more, they sold part of the manufactured goods back to the poor, unindustrialized countries for additional profit.

Support for Authoritarian Governments

According to dependency theorists, multinational corporations and rich countries continued their exploitation of the poor countries in the postcolonial period by giving economic and military support to local authoritarian governments. These governments managed to keep their populations subdued most of the time. When this was not possible, Western governments sent in their own troops and military advisors, engaging in what became known as "gunboat diplomacy." The term itself was coined in colonial times. In 1839, the Chinese rebelled against the British importation of opium into China and the British responded by sending a gunboat up the Yangtze River, starting the Opium War. The war resulted in Britain winning control of Hong Kong and access to five Chinese ports; what started as gunboat diplomacy ended as a rich feast for British traders. In the postcolonial period, the United States has been particularly active in using gunboat diplomacy in Central America. A classic case occurred in Guatemala in the 1950s. In 1952, the democratic government of Guatemala began to redistribute land to impoverished peasants. Some of the land was owned by the United Fruit Company, a huge United States multinational corporation and the biggest landowner in Guatemala. Two years later, the CIA backed a right-wing coup in Guatemala, preventing land reform and allowing the United Fruit Company to continue its highly profitable business as usual.

In 1893, leaders of the British mission pose before taking over what became Rhodesia and is now Zimbabwe. To raise a volunteer army, every British trooper was offered about 9 square miles of native land and 20 gold claims. The Matabele and Mashona peoples were subdued in a 3-month war. Nine hundred farms and 10,000 gold claims were granted to the troopers and about 100,000 cattle were looted, leaving the native survivors without a livelihood. Forced labor was subsequently introduced by the British so that the natives could pay a £2 per year tax.

Mounting Debt

The governments of the poor countries struggled to create transportation infrastructures (airports, roads, harbors, etc.), build up their education systems, and deliver safe water and at least the most basic health care to their people. To accomplish these tasks, they had to borrow money from Western banks and governments. So it came about that debt—and the interest payments that inevitably accompany debt—grew every year. By 1997, the poor countries owed the rich countries more than 2.3 trillion dollars. That is 87% more than in 1980, allowing for inflation. At a 7% rate of interest, it would cost $162 billion a year to service that debt. These crushing interest payments leave governments of poor countries with far too little money for development tasks. Foreign aid helps, but because it amounted to less than $54 billion in 1997, it doesn't help much. At the end of the 20th century, the poor countries were paying roughly three times more interest to Western banks and governments than they were receiving as foreign aid (see Chapter 1, Figure 1.3)

Do you think the United States is spending too much, too little, or about the right amount on foreign aid? If you happen to think that we, like all members of the rich nations, have a responsibility to compensate for centuries of injustice and that we are therefore spending too little on foreign aid, you are in a minority. The General Social Survey periodically asks respondents whether the United States is spending too much, too little, or about the right amount on 15 items, including foreign aid. The results for 1998 are shown in Table 16.5. Foreign aid ranks last on Americans' list of priorities. The government seems

	Percent Answering "Too Little"
1. Improving the nation's education system	69.7
2. Improving and protecting the nation's health	66.9
3. Halting the rising crime rate	61.1
4. Improving and protecting the environment	60.0
5. Dealing with drug addiction	58.4
6. Social security	56.0
7. Solving problems of the big cities	45.6
8. Highways and bridges	38.4
9. Improving the conditions of Blacks	34.1
10. Parks and recreation	34.1
11. Mass transportation	31.9
12. The military, armaments, and defense	17.6
13. Welfare	16.0
14. Space exploration program	10.1
15. Foreign aid	6.7

✦ TABLE 16.5 ✦
National Priorities, United States, 1998 (in percent; n = 1,381)
"We are faced with many problems in this country, none of which can be solved easily or inexpensively. I'm going to name some of these problems, and for each one I'd like you to tell me whether you think we're spending too much money on it, too little money, or about the right amount. First, are we spending too much, too little, or about the right amount on . . . "

SOURCE: National Opinion Research Center (1999).

BOX 16.2
IT'S YOUR CHOICE

SHOULD THE UNITED STATES PROMOTE WORLD DEMOCRACY?

"In starting and waging a war, it is not right that matters, but victory," said Adolf Hitler (quoted in "A Survey . . . 1998: 10). The same mindset rationalizes state brutality today, from the Serbian attack on Bosnia and Kosovo to the Indonesian war against East Timor. Cherished ideals, such as political democracy and human rights, are trampled on daily by dictatorships and military governments.

There are different opinions as to what the United States government should do about this. One influential argument is that of Samuel Huntington. He argues that the United States and other Western nations should not be ethnocentric and impose Western values on people in other countries. If there are violations of democratic principles and human rights, they express in some way the indigenous values of those people. We should not intervene to stop nondemocratic forces and human rights abuse abroad.

Critics of Huntington argue that the ideals of democracy and human rights can be found in non-Western cultures too. If we explore Asian or African traditions, for instance, we soon find "respect for the sacredness of life and for human dignity, tolerance of differences, and a desire for liberty, order, fairness and stability" (quoted in "A Survey . . .," 1998: 10). Although there are despots who champion supposedly traditional values, there are people in Asia, Africa, and elsewhere who struggle for democracy and human rights. Beginning in the mid-1970s, the Indonesian government tried to smother democracy in East Timor, but the East Timorese fought valiantly for democracy.

What do you think the role of the United States should be? Should the United States government promote democracy and human rights? Or should we avoid what Huntington regards as ethnocentrism? Because foreign aid to authoritarian regimes may help nondemocratic forces and thereby stifle human rights, should we give foreign aid to such regimes?

responsive to this. As a proportion of the total federal budget, the United States spends a small and declining amount on foreign aid. Thus, between 1992 and 1997, the amount appropriated for this purpose decreased by 14% allowing for inflation. For the fiscal year 2000, the United States administration proposed a total foreign aid budget of just over $21 billion. That is barely 1% of the total federal budget and down slightly from the preceding year (United States Secretary of State, 1999; see Box 16.2).

Assessment

Few sociologists would deny that the dependency theorists are correct on one score. In both the colonial and postcolonial periods, Spain, Portugal, Holland, Britain, France, Italy, the United States, Japan, and Russia treated the world's poor with brutality to enrich themselves. Colonialism did have a devastating economic and human impact on the poor countries of the world. In the postcolonial era, the debt burden has crippled the development efforts of many poor countries.

That said, there remains a big question that research has not yet fully answered. Does foreign investment today have positive or negative effects on the developing countries? Much hinges on the answer to this question. Modernization theorists want more foreign investment in poor countries because they think it will promote economic growth and general well-being. They want trade and investment barriers to be dropped so free markets can bring prosperity to everyone. Dependency theorists diametrically oppose this strategy. They think foreign investment drains wealth out of poor countries. Therefore, they want the poor countries to revolt against the rich countries, throw up barriers to free trade and investment, and find their own paths to economic well-being.

Unfortunately, the results of research conducted to date on the effects of foreign investment lend strong support to neither side in this debate. Some analysts find foreign investment depresses economic growth in the poor countries in the medium to long term. This finding supports dependency theory (Bornschier and Chase-Dunn, 1985). Other analysts find foreign investment increases growth and the well-being of even the most impoverished citizens of the poor countries. This finding supports modernization theory (Firebaugh and Beck, 1994). Whether analysts reach one conclusion or the other depends on which variables they include in their analyses and which statistical techniques they use.

A 16th-century engraving by Theodore de Bry of Amerindians crocodile hunting. It was nearly a century after Europeans began subduing and conquering the Americas that de Bry provided the first high-quality European pictorial record of the New World.

There is, however, more to the story than that. Much research on the effects of foreign direct investment lumps together all poor countries. But there is good reason to believe that not all poor countries are alike. They have different histories and different social structures, and this may result in foreign investment having different effects in different times and places.

Immanuel Wallerstein proposes a variation on this theme. He argues that capitalist development has resulted in the creation of an integrated "world system" composed of three tiers. First are the **core** capitalist countries (the United States, Japan, Germany, etc.), which are major sources of capital and technology. Second are the **peripheral** countries (the former colonies), which are major sources of raw materials and cheap labor. Third are the **semiperipheral** countries (such as South Korea, Taiwan, and Israel), consisting of former colonies that are making considerable headway in their attempts to become prosperous (Wallerstein, 1974–89). To give just one dramatic illustration of this progress, South Korea and the African country of Ghana were among the poorest nations in the world in 1960. Ghana still is. But South Korea, the recipient of enormous foreign investment and aid, was nine times wealthier than Ghana in 1997 (as measured by per capita GNP; calculated from World Bank, 1999c: 193). Comparing the unsuccessful peripheral countries with the more successful semiperipheral countries presents us with a useful natural experiment. The comparison suggests circumstances that help some poor countries overcome the worst effects of colonialism.

The semiperipheral countries differ from the peripheral countries in four main ways (Hein, 1992; Kennedy, 1993: 193–227; Lie, 1998; Sanderson, 1995: 232–4).

Type of Colonialism

Around the turn of the 20th century, Taiwan and Korea became colonies of Japan. They remained so until 1945. However, in contrast to the European colonizers of Africa, Latin America, and other parts of Asia, the Japanese built up the economies of their colonies. They established transportation networks and communication systems. They built steel, chemical, and hydroelectric power plants. After Japanese colonialism ended, Taiwan and South Korea were thus at a big advantage compared to, say, Ghana at the time Britain gave up control of that country or Brazil when Portuguese rule ended there. South Korea and Taiwan could use the Japanese-built infrastructure as a springboard to development.

Seoul, capital of South Korea. South Korea is one of the semiperipheral countries that are making considerable headway in their attempts to become prosperous.

Geopolitical Position

By the end of World War II, the United States was the leading economic and military power in the world. However, it began to feel its supremacy threatened from the late 1940s on by the Soviet Union and China. Fearing that South Korea and Taiwan might fall to the communists, the United States poured unprecedented aid into both countries in the 1960s. It also gave them large low-interest loans and opened its domestic market to Taiwanese and South Korean products. Because Israel was seen by the United States as a crucially important ally in the Middle East, it, too, received special economic assistance. Other countries with less strategic importance to the United States received less help in their drive to industrialize.

State Policy

A third factor that accounts for the relative success of some countries in their efforts to industrialize and become prosperous concerns state policies. As a legacy of colonialism, the Taiwanese and South Korean states were developed on the Japanese model. They kept workers' wages low, restricted trade union growth, and maintained quasi-military discipline in the factories. Moreover, by placing high taxes on consumer goods, limiting the import of foreign goods, and preventing their citizens from investing abroad, they encouraged their citizens to put much of their money in the bank. This created a large pool of capital for industrial expansion. Finally, from the 1960s on, the South Korean and Taiwanese states gave subsidies, training grants, and tariff protection to export-based industries. These policies did much to stimulate industrial growth.

Social Structure

Taiwan and South Korea are socially cohesive countries. This makes it easy for them to generate consensus around development policies. It also allows them to get their citizens to work hard, save a lot, and devote their energies to scientific education.

Social solidarity in Taiwan and South Korea is based partly on the sweeping land reform they conducted in the late 1940s and early 1950s. By redistributing land to small farmers, both countries eliminated the class of large landowners, who usually oppose industrialization. A major source of social conflict was thus eliminated. In contrast, many countries in Latin America and Africa have not undergone land reform. The United States often intervened militarily to prevent land reform in Latin America. That is because United

States commercial interests profited handsomely from the existence of large plantations (LaFeber, 1993).

Another factor underlying social solidarity in Taiwan and South Korea is that both countries are more ethnically homogeneous than, say, countries in sub–Saharan Africa. British, French, and other West European colonizers often drew the borders of African countries to keep antagonistic tribes living side-by-side in the same jurisdiction. Keeping tribal tensions alive made it easier for imperial powers to rule. They could play one tribe off against another. However, this policy also caused much social and political conflict in postcolonial Africa. Today, the region suffers from frequent wars, coups, and uprisings. This high level of internal conflict acts as a barrier to economic development in Africa south of the Sahara desert.

On balance, then, it seems that certain conditions do permit foreign investment to have positive economic effects. Postcolonial countries that enjoy an industrial infrastructure, strategic geopolitical importance, strong states with strong development policies, and socially cohesive populations are in the best position to join the ranks of the rich countries in the coming decades. Countries that have *some* of these characteristics may also be expected to experience some economic growth and increase in the well-being of their populations in the near future. Such countries include Chile, Thailand, Indonesia, Mexico, and Brazil. In contrast, African countries south of the Sahara are in the worst position of all. They have inherited the most damaging consequences of colonialism, and they enjoy few of the conditions that could help them escape the history that has been imposed on them.

SUMMARY

1. Robert Malthus argued that while food supplies increase slowly, populations grow quickly. Because of these presumed natural laws, only war, pestilence, and famine can keep human population growth in check.

2. Rapid increases in food production, the existence of higher-than-expected upper limits to population size, the growth of large yet prosperous populations, the ability to provide generous social welfare and still maintain low population growth rates, and the widespread use of contraception have all cast doubt on Malthus's theory.

3. Demographic transition theory holds that the main factors underlying population dynamics are industrialization and the growth of modern cultural values. In the preindustrial era, both crude birth rates and crude death rates were high and population growth was therefore slow. In the first stages of industrialization, crude death rates fell, so population growth was rapid. As industrialization progressed and people's values about having children changed, the crude birth rate fell, resulting in slow growth again. Finally, in the postindustrial era, the crude death rate has risen above the crude birth rate in many societies. As a result, their populations slowly shrink unless in-migration augments their numbers.

4. Partially independent of the level of industrialization, the level of social inequality between women and men, and between classes, affects population dynamics, with lower levels of social inequality typically resulting in lower crude birth rates and therefore lower population growth rates.

5. Much urbanization is associated with the growth of factories. However, religious, political, and commercial need gave rise to cities in the preindustrial era. Moreover, the fastest growing cities in the world today are in semi-industrialized countries.

6. The members of the Chicago school famously described and explained the spatial and social dimensions of the industrial city. They developed a theory of human ecology that explained urban growth as the outcome of differentiation, competition, and ecological succession. They described the spatial arrangement of the industrial city as a series of expanding concentric circles. The main business, entertainment, and shopping area stood in the center, with the class position of residents increasing as one moved from inner to outer rings.

7. Subsequent research showed that the city was not as anomic as the Chicago sociologists made it appear and that the concentric zone pattern applied best to the industrial city in the first quarter of the 20th century.

8. The new urban sociology criticized the Chicago school for making city growth seem like an almost natural process, playing down the power conflicts and profit motives that prompted the evolution of cities.

9. The corporate city that emerged after World War II was a vehicle for capital accumulation that stimulated the growth of the suburbs and resulted in the decline of inner cities.

10. The postmodern city that took shape in the last decades of the 20th century is characterized by the increased globalization of culture, fragmentation of lifestyles, and privatization of space.

11. Modernization theory argues that global inequality is due to some countries lacking sufficient capital, Western values, rational business practices, and stable governments.

12. Dependency theory counters with the claim that global inequality results from the exploitative relationship between rich and poor countries.

13. An important test of the two theories concerns the effect of foreign investment on economic growth, but research on this subject is equivocal.

14. The poor countries best able to emerge from poverty have a colonial past that left them with industrial infrastructures. They also enjoy a favorable geopolitical position. They implement strong growth-oriented economic policies, and they have socially cohesive populations.

GLOSSARY

The **Chicago school** founded urban sociology in the United States in the first decades of the 20th century. Its members distinguished themselves by their vivid and detailed descriptions and analyses of urban life and their development of the theory of human ecology.

Competition in the theory of human ecology refers to the struggle by different groups for optimal locations in which to reside and set up their businesses.

The **core** capitalist countries are rich countries, such as the United States, Japan, and Germany, that are the major sources of capital and technology in the world.

The **corporate city** refers to the growing post–World War II perception and organization of the North American city as a vehicle for capital accumulation.

The **crude birth rate** is the annual number of live births per 1,000 women in a population.

The **crude death rate** is the annual number of deaths per 1,000 people in a population.

Demographers are social-scientific analysts of human population.

Demographic transition theory explains how changes in fertility and mortality affected population growth from preindustrial to postindustrial times.

Dependency theory views economic underdevelopment as the result of exploitative relations between rich and poor countries.

Differentiation in the theory of human ecology refers to the process by which urban populations and their activities become more complex and heterogeneous over time.

Ecological succession in the theory of human ecology refers to the process by which a distinct urban group moves from one area to another and a second group comes in to replace the group that has moved out.

Emigration or out-migration is the outflow of people from one country and their settlement in one or more other countries.

Human ecology is a theoretical approach to urban sociology that borrows ideas from biology and ecology to highlight the links between the physical and social dimensions of cities and identify the dynamics and patterns of urban growth.

Immigration or in-migration is the inflow of people into one country from one or more other countries and their settlement in the destination country.

In-migration (see immigration).

The **Malthusian trap** refers to a cycle of population growth followed by an outbreak of war, pestilence, or famine that keeps population growth in check.

Modernization theory holds that economic underdevelopment results from poor countries lacking Western attributes. These attributes include Western values, business practices, levels of investment capital, and stable governments.

The **new urban sociology** emerged in the 1970s and stressed that city growth is a process rooted in power relations and the urge to profit.

Out-migration (see emigration).

The **peripheral** countries are former colonies that are poor and are major sources of raw materials and cheap labor.

The **postmodern city** is a new urban form that is more privatized and socially and culturally fragmented and globalized than the corporate city.

The **replacement level** is the number of children that each woman must have on average in order for population size to remain stable. Ignoring any inflow of population from other countries and any outflow to other countries, the replacement level is 2.1.

The **semiperipheral** countries, such as South Korea, Taiwan, and Israel, consist of former colonies that are making considerable headway in their attempts to become prosperous.

Suburbanism is a way of life outside city centers that is organized mainly around the needs of children and involves higher levels of conformity and sociability than life in the central city.

The **total fertility rate** is the annual number of live births per 1,000 women in a population.

Urbanism is a way of life that, according to Louis Wirth, involves increased tolerance but also emotional withdrawal and specialized, impersonal, and self-interested interaction.

QUESTIONS TO CONSIDER

1. Do you think rapid global population growth is cause for alarm? If not, why not? If so, what aspects of global population growth are especially worrisome? What should be done about them?

2. Do you think of the city mainly as a place of innovation and tolerance or mainly as a site of crime, prejudice, and anomie?

Where does your image of the city come from? Your own experience? The mass media? Your sociological reading?

3. Should Americans do anything to help end global poverty? Why or why not? If you think Americans should be doing something to help end global poverty, then what should we do?

WEB RESOURCES

Companion Web Site for This Book

http://sociology.wadsworth.com
Begin by clicking on the Student Resources section of the Web site. Choose "Introduction to Sociology" and finally the Brym and Lie book cover. Next, select the chapter you are currently studying from the pull-down menu. From the Student Resources page you will have easy access to InfoTrac College Edition®, MicroCase Online exercises, additional Web links, and many other resources to aid you in your study of sociology, including practice tests for each chapter.

InfoTrac Search Terms

These search terms are provided to assist you in beginning to conduct research on this topic by visiting http://www.infotraccollege.com/wadsworth.

Dependency theory **Modernization theory**
Immigration **Urbanization**
Migration

Recommended Web Sites

The United States Census Bureau supports perhaps the richest demographic site on the World Wide Web. It contains mainly United States but also some international data, all easily accessible. Visit the site at http://www.census.gov.

Another useful, data-rich site is run by the Population Reference Bureau at http://www.prb.org.

The World Wide Web Library: Demography and Population Studies at http://demography.anu.edu.au/VirtualLibrary offers a comprehensive list of international Web sites devoted to the social scientific study of population.

United Nations statisticians have created indicators of human development for every country in the period 1990–99. The indicators are available at http://www.undp.org/hdro/indicators.html#developing.

The Urban Institute is a nonpartisan economic and social policy research organization located in Washington, DC. Visit its Web site at http://www.urban.org. Of particular interest is The Urban Institute's RealAudio program on "American Cities in the 21st Century."

SUGGESTED READINGS

John Hannigan. *Fantasy City: Pleasure and Profit in the Postmodern Metropolis* (New York: Routledge, 1998). An entertaining and perceptive analysis of the postindustrial city as an entertainment hub.

John Lie. *Han Unbound: The Political Economy of South Korea* (Stanford, CA: Stanford University Press, 1998). A case study of the successes and limitations of development in one of the world's leading semiperipheral countries.

Massimo Livi-Bacci. *A Concise History of World Population* (Cambridge, MA: Blackwell, 1992). A definitive introduction by one of the world's leading demographers.

Eric R. Wolf. *Europe and the People without History* (Berkeley, CA: University of California Press, 1982). A broad and rich analysis of colonization and development from a dependency theory perspective.

IN THIS CHAPTER, YOU WILL LEARN THAT:

✦ People sometimes lynch, riot, and engage in other forms of nonroutine group action to correct perceived injustices. Such events are rare, short-lived, spontaneous, and often violent. They subvert established institutions and practices. Nevertheless, most nonroutine collective action requires social organization, and people who take part in collective action often act in a calculated way.

✦ Collective action can result in the creation of one or more formal organizations or bureaucracies to direct and further the aims of its members. The institutionalization of protest signifies the establishment of a social movement.

✦ People are more inclined to rebel against existing conditions when strong social ties bind them to many other people who feel similarly wronged, when they have the time, money, and other resources needed to protest, and when political structures and processes give them opportunities to express discontent.

✦ For social movements to grow, members must make the activities, goals, and ideology of the movement consistent with the interests, beliefs, and values of potential recruits.

✦ The history of social movements is a struggle for the acquisition of constantly broadening citizenship rights—and opposition to those struggles.

CHAPTER

17

COLLECTIVE ACTION AND SOCIAL MOVEMENTS

INTRODUCTION

Robert Brym almost sparked a small riot once. "It happened in grade eleven," says Robert, "shortly after I learned that water combined with sulfur dioxide produces sulfurous acid. The news shocked me. To understand why, you have to know that I lived 60 miles east of the state of Maine and about 100 yards downwind of one of the largest pulp and paper mills in Canada. Waves of sulfur dioxide billowed day and night from the mill's smokestacks. The town's pervasive rotten-egg smell was a long-standing complaint in the area. But, for me, disgust turned to upset when I realized the fumes were toxic. Suddenly it was clear why many people I knew—especially people living near the mill—woke up in the morning with a kind of 'smoker's cough.' By the simple act of breathing we were causing the gas to mix with the moisture in our bodies and form an acid that our lungs tried to expunge, with only partial success.

"Twenty years later, I read the results of a medical research report showing that area residents suffered from rates of lung disease, including emphysema and lung cancer, significantly above the North American average. But even in 1968 it was evident my hometown had a serious problem. I therefore hatched a plan. Our high school was about to hold its annual model parliament. The event was notoriously boring, partly because, year in, year out, virtually everyone voted for the same party, the Conservatives. But here was an issue, I thought, that could turn things around. A local man, K. C. Irving, owned the pulp and paper mill. *Forbes* business magazine ranked him as one of the richest men in the world. I figured that when I told my fellow students what I had discovered, they would quickly demand the closure of the mill until Irving guaranteed a clean operation.

"Was *I* naive. As head of the tiny Liberal party, I had to address the entire student body during assembly on election day to outline the party platform and rally votes. When I got to the part of my speech explaining why Irving was our enemy, the murmuring in the audience, which had been growing like the sound of a hungry animal about to pounce on its prey, erupted into loud "boos." A couple of students rushed the stage. The principal suddenly appeared from the wings and commanded the student body to settle down. He then took me by the arm and informed me that, for my own safety, my speech was finished. So, I discovered on election day, was our high school's Liberal party. And so, it emerged, was my high school political career.

"This incident troubled me for many years, partly due to the embarrassment it caused, partly due to the puzzles it presented. Why did I almost spark a small riot? Why didn't my fellow students rebel in the way I thought they would? Why did they continue to support an arrangement that was enriching one man at the cost of a community's health? Couldn't they see the injustice? Other people did. Nineteen sixty-eight was not just the year of my political failure in high school. It was also the year that student riots in France nearly toppled that country's government. In Mexico, the suppression of student strikes by the government left dozens of students dead. In the United States of America, students at Berkeley, Michigan, and other colleges demonstrated and staged sit-ins with unprecedented vigor. They supported free speech on their campuses, an end to American involvement in the war in Vietnam, increased civil rights for American blacks, and an expanded role for women in public affairs. It was, after all, the 60s."

Robert didn't know it at the time, but by asking why students in Paris, Mexico City, and Berkeley rebelled while his fellow high school students did not, he was raising the main question that animates the study of collective action and social movements. Under what social conditions do people act in unison to change, or resist change to, society? That is the main issue we address in this chapter.

We have divided the chapter into three sections:

1. We first discuss the social conditions leading to the formation of lynch mobs, riots, and other types of nonroutine **collective action.** When people engage in collective action, they act in unison to bring about or resist social, political, and economic change (Schweingruber and McPhail, 1999: 453). Some collective actions are

"routine" and others are "nonroutine" (Useem, 1998: 219). Routine collective actions are usually nonviolent and follow established patterns of behavior in bureaucratic social structures. For instance, when Mothers Against Drunk Driving (MADD) lobbies for tougher laws against driving under the influence of alcohol, when members of a community organize a campaign against abortion or for freedom of reproductive choice, or when workers decide to form a union, they are engaging in routine collective action. Sometimes, however, "usual conventions cease to guide social action and people transcend, bypass, or subvert established institutional patterns and structures" (Turner and Killian, 1987 [1957]: 3). On such occasions, people engage in nonroutine collective action, which is often short-lived and sometimes violent. They may, for example, form lynch mobs and engage in riots. Until the early 1970s, it was widely believed that people who engage in nonroutine collective action lose their individuality and capacity for reason. Lynch mobs and riots were often seen as wild and uncoordinated affairs, more like stampedes of frightened cattle than structured social processes. As you will see, however, sociologists later showed that this portrayal is an exaggeration. It deflects attention from the social organization and inner logic of extraordinary sociological events.[1]

2. We next outline the conditions underlying the formation of **social movements.** Social movements are enduring and usually bureaucratically organized collective attempts to change (or resist change to) part or all of the social order. This is achieved by petitioning, striking, demonstrating, and establishing lobbies, unions, and political parties. We will see that an adequate explanation of institutionalized protest also requires the introduction of a set of distinctively sociological issues. These concern the distribution of power in society and the framing of political issues in ways that appeal to many people.

3. Finally, we make some observations about the changing character of social movements. We argue that the history of social movements is the history of attempts by underprivileged groups to broaden their members' citizenship rights and increase the scope of protest from the local to the national to the global level.

We begin by considering the lynch mob, a well-studied form of nonroutine collective action.

NONROUTINE COLLECTIVE ACTION: THE LYNCH MOB

The Lynching of Claude Neal

On October 27, 1934, a black man was lynched near Greenwood, a town in Jackson County, Florida. Claude Neal, 23, was accused of raping and murdering 19-year-old Lola Cannidy, a pretty white woman and the daughter of his employer. The evidence against Neal was not totally convincing. Some people thought he confessed under duress. But Neal's reputation in the white community as a "mean n_____," "uppity," "insolent," and "overbearing," helped to seal his fate (McGovern, 1982: 51). He was apprehended and jailed. Then, for his own safety, he was removed to the jailhouse in Brewton, Alabama, about 120 miles northwest of the crime scene.

When the white residents in and around Greenwood found that Neal had been taken from the local jail, they quickly formed a lynch mob to find him. Once word mysteriously leaked out that Neal was in Brewton, 15 men in three cars immediately headed west. They figured out a way to get the sheriff out of the Brewton jail by sending him on a wild goose

Web Interactive Exercises
Is the Klan History?

---

[1]Reflecting new research and theoretical perspectives, the older term, *collective behavior,* fell into disfavor in the 1990s. That is because *behavior* suggests a relatively low level of consciousness of self and therefore conduct that is not entirely rational. Following Weber (1947), *action* denotes greater consciousness of self and therefore more rationality.

Following the lynching, Claude Neal's body was strung up on a tree on the lawn of the Jackson County courthouse. Later, his murderers sold a photograph of Neal's hanging body as a postcard, from which this photo is taken.

chase. Then, entering the jail holding guns and dynamite, they threatened to blow up the place if the lone jailer did not hand over Neal. He complied. They then tied Neal's hands with a rope. Finally, they dumped him in the back seat of a car for the ride back to Jackson County. There, a mob of two or three thousand people soon gathered near the Cannidy house.

The mob was in a state of violent agitation. Drinking moonshine whiskey and shouting "We want the n_____," many of them "wanted to get their hands on [Neal] so bad they could hardly stand it," according to one bystander. However, the jail raiders feared the mob was uncontrollable and its members might injure each other in the frenzy to get at Neal. So they led their prisoner into the woods. An investigator interviewed one member of the company 10 days later. He described what next happened in these horrifying words:

> After taking the n_____ to the woods about four miles from Greenwood, they cut off his penis. He was made to eat it. Then they cut off his testicles and made him eat them and say he liked it. Then they sliced his sides and stomach with knives and every now and then somebody would cut off a finger or a toe. Red hot irons were used on the n_____ to burn him from top to bottom.

"From time to time during the torture," continued the investigator, "a rope would be tied around Neal's neck and he was pulled up over a limb and held there until he almost choked to death when he would be let down and the torture [would] begin all over again" (McGovern, 1982: 80).

Having thus disposed of Neal, the jail raiders tied a rope around his body. They attached the rope to a car and dragged the body several miles to the mob in front of the Cannidy house. There, several people drove knives into the corpse, "tearing the body almost to shreds" according to one report (McGovern, 1982: 81). Lola Cannidy's grandfather took his .45 and pumped three bullets into the corpse's forehead. Some people started kicking the body. Others drove cars over it. Children were encouraged to take sharpened sticks and drive them deep into the flesh of the dead man. Then some members of the crowd rushed to a row of nearby shacks inhabited by blacks and burned the dwellings to the ground. Others took the nude and mutilated body of Claude Neal to the lawn of the Jackson County courthouse, where they strung it up on a tree. Justice, they apparently felt, had now been served. Later, they sold a photograph of Neal's hanging body as a postcard.

Breakdown Theory

Until about 1970, most sociologists believed at least one of three conditions must be met for nonroutine collective action, such as Claude Neal's lynching, to emerge. First, a group of people must be economically deprived or socially rootless. Second, their norms must be strained or disrupted. Third, they must lose their capacity to act rationally by getting caught up in the supposedly inherent madness of crowds. Following Charles Tilly and his

Research on riots, crowds, and demonstrations shows that nonroutine collective action may be wild, but it is usually socially structured. Here, following an election campaign rally in Jakarta, Indonesia, a crowd attacks a supporter of an Islamic opposition party.

associates, we may group these three factors together as the **breakdown theory** of collective action. That is because all three factors assume that collective action results from the disruption or breakdown of traditional norms, expectations, and patterns of behavior (Tilly, Tilly, and Tilly, 1975: 4–6). At a more abstract level, breakdown theory may be seen as a variant of functionalism, for it regards collective action as a form of social imbalance that results from various institutions functioning improperly (see Chapter 1, "A Sociological Compass"). Specifically, most pre-1970 sociologists would have said that Neal's lynching was caused by one or more of the following factors:

1. *A background of economic deprivation experienced by impoverished and marginal members of the community.* The very year of Neal's lynching signals deprivation: 1934, the midpoint of the Great Depression of 1929–39. As fast as farm income fell in the collapsed cotton economy of Jackson County, bankruptcies and rural unemployment rose. Severe economic deprivation may have resulted in a general rise in tensions in the area, requiring only a spark for ignition. Moreover, blacks may have become the collective target of white frustration because fully a quarter of all black farmers in Jackson County owned their own land and received government aid from the Farm Credit Administration. In contrast, many whites were landless migrants from other states or dispossessed sharecroppers who may have resented blacks receiving federal funds (McGovern, 1982: 39–41).

   Often, say proponents of breakdown theory, it is not grinding poverty, or **absolute deprivation,** that generates collective action so much as **relative deprivation.** Relative deprivation refers to the growth of an intolerable gap between the social rewards people expect to receive and those they actually receive. Social rewards are widely valued goods including money, education, security, prestige, and so forth. Accordingly, people are most likely to rebel when rising expectations (brought on by, say, rapid economic growth and migration) are met by a sudden decline in received social rewards (due to, say, economic recession or war) (Davies, 1969; Gurr, 1970). From this point of view, the rapid economic growth of the "roaring 20s," followed by the economic collapse of 1929, would likely have caused widespread relative deprivation in Jackson County.

2. *The inherent irrationality of crowd behavior* is a second factor likely to be stressed in any pre-1970 explanation of the Neal lynching. Gustave Le Bon, an early interpreter of crowd behavior, wrote that an isolated person may be a cultivated individual. But in a crowd, the individual is transformed into a "barbarian," a "creature acting by instinct" possessing the "spontaneity, violence," and "ferocity" of "primitive beings" (Le Bon, 1969 [1895]): 28). Le Bon argued that this transformation occurs because people lose their individuality and will power when they join a crowd. Simultaneously, they gain a sense of invincible group power that derives from the crowd's sheer size. Their feeling of invincibility allows them to yield to instincts they would normally hold in check. Moreover, if people remain in a crowd long enough, they enter something like a hypnotic state. This makes them particularly open to the suggestions of manipulative leaders and ensures that extreme passions spread through the crowd like a contagious disease. (Sociologists call Le Bon's argument the **contagion** theory of crowd behavior.) For all these reasons, Le Bon held, people in crowds are often able to perform extraordinary and sometimes outrageous acts. "Extraordinary" and "outrageous" are certainly appropriate terms for describing the actions of the citizens of Jackson County in 1934.

3. *The serious violation of norms* is the third factor that pre-1970s sociologists would likely have stressed in trying to account for the Neal lynching. In the 1930s, intimate contact between blacks and whites in the South was strictly forbidden. In that context, black-on-white rape and murder were not just the most serious of crimes but the deepest possible violation of the region's norms. Neal's alleged crimes were therefore bound to evoke a strong reaction on the part of the dominant race. In general, before about 1970, sociologists highlighted the breakdowns in traditional

norms that preceded group unrest, sometimes referring to them as indicators of **strain** (Smelser, 1963: 47–8, 75).

Assessing Breakdown Theory

Can deprivation, contagion, and strain really explain what happened in the backwoods of Jackson County in the early hours of October 27, 1934? Can breakdown theory adequately account for collective action in general? The short answer is "no." Increasingly since 1970, sociologists have uncovered flaws in all three elements of breakdown theory. They have proposed alternative frameworks for understanding collective action. To help you appreciate the need for these alternative frameworks, let us reconsider the three elements of breakdown theory in the context of the Neal lynching.

DEPRIVATION.   Research shows no clear association between fluctuations in economic well-being (as measured by, say, the price of cotton in the South) and the number of lynchings that took place each year between the 1880s and the 1930s (Mintz, 1946). Moreover, in the case of the Neal lynching, the main instigators were not especially economically deprived. The men who seized Neal from the Brewton jail were middle- to lower-middle-class farmers, merchants, salesmen, and the like. They were economically solvent, even at the height of the Great Depression, with enough money, cars, and free time to take 2 days off work to organize the lynching. Nor were they socially marginal "outside agitators" or rootless, recent migrants to the region. They enjoyed good reputations in their communities as solid citizens, churchgoing men with a well-developed sense of civic responsibility. Truly socially marginal individuals did not take part in the lynching at all (McGovern, 1982: 67–8, 85). This fits a general pattern. In most cases of collective action, leaders and early joiners are well-integrated members of their communities, not outsiders (Brym, 1980; Brym and Economakis, 1994; Economakis and Brym, 1995; Lipset, 1971 [1951]). Levels of deprivation, whether absolute or relative, are not commonly associated with the frequency or intensity of outbursts of collective action (McPhail, 1994).

CONTAGION.   Despite its barbarity, the Neal lynching was not a spontaneous and unorganized affair. Sophisticated planning went into the Brewton jail raid. For example, decoying the sheriff took much cunning. Even the horrific events in front of the Cannidy home did not just erupt suddenly because of the crowd's "madness." For example, hours before Neal's body was brought to the Cannidy home, some adults got the idea of sharpening some long sticks, stacking them, and instructing children to use them to pierce the body. As this example shows, and as research on riots, crowds, and demonstrations has consistently confirmed, nonroutine collective action may be wild but it is usually structured. In the first place, nonroutine collective action is structured by ideas and norms that emerge in the crowd itself, such as the idea of preparing sharp sticks in the Neal lynching (Turner and Killian, 1987 [1957]). Second, nonroutine collective action is structured by the predispositions that unite crowd members and predate their collective action. The participants in the Neal lynching, for instance, were all predisposed to take part in it by racist attitudes. If they had not been similarly predisposed, they would never have assembled for the lynching in the first place (Berk, 1974; Couch, 1968; McPhail, 1991). Third, nonroutine collective action is structured by the *degree* to which different types of participants adhere to emergent and preexisting norms. Leaders, rank-and-file participants, and bystanders adhere to such norms to varying degrees (Zurcher and Snow, 1981). Fourth, preexisting social relationships among participants structure nonroutine collective action. For instance, relatives, friends, and acquaintances are more likely than strangers to cluster together and interact in crowds, riots, demonstrations, and lynchings (McPhail, 1991; McPhail and Wohlstein, 1983; Weller and Quarantelli, 1973)

STRAIN.   The alleged rape and murder of Lola Cannidy by Claude Neal did violate the deepest norms of the Old South in a pattern that was often repeated. Thus, data exist on 4,752 lynchings that took place in the United States between 1882 and 1964, when the last lynching was recorded. Three quarters of them were white lynchings of blacks. Nearly two

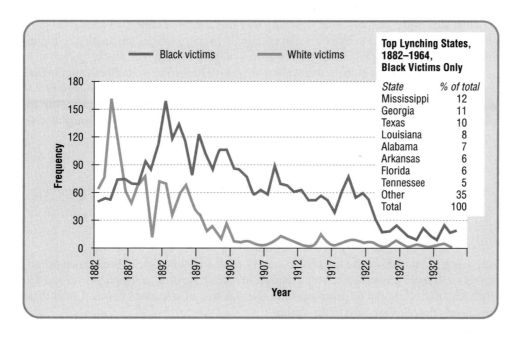

✦ **FIGURE 17.1** ✦
Frequency of Lynching, United States, 1882–1935

SOURCE: Williams (1970: 8–11).

thirds were motivated by alleged rapes or murders (calculated from Williams, 1970: 12–15).

However, lynching had deeper roots than the mere violation of norms governing black-white relations. Significantly, it was a chief means by which black farm workers were disciplined and kept tied to the southern cotton industry after the abolition of slavery threatened to disrupt the industry's traditional, captive labor supply.

Figure 17.1, which contains data on the annual frequency and geographical distribution of lynching, supports this interpretation. Note first that there were more white than black lynching victims in the first half of the 1880s. That is because, originally, lynching was not just an expression of a racist system of labor control. It was also a means by which people sought quick and brutal justice in areas with little government or police, whatever the alleged criminal's race. Only in 1886 did the number of black victims exceed the number of white victims for the first time. After that, as state control over criminal justice became more widespread, the number of white lynchings continued to decline. The practice was soon used almost exclusively to control blacks.

Second, notice that nearly two thirds of all lynchings took place in just eight contiguous southern states that formed the center of the cotton industry. Other data show the great majority of lynchings took place in rural areas, where, of course, cotton is farmed (McAdam, 1982: 89–90). Where cotton was king, lynching was his handmaiden.

Third, observe how the annual number of lynchings rose when the cotton industry's labor supply was most threatened. Thus, the peak in black lynchings occurred between 1891 and 1901: an annual average of 112. These were the years when the Populist Party, a coalition of black and white farmers, threatened to radically restructure southern agriculture and eliminate many of the white plantation owners' privileges. More lynching was one reaction to this danger to the traditional organization of agricultural labor. Finally, note that lynching disappeared as a form of collective action when the cotton industry lost its economic significance and its utter dependence on dirt-cheap black labor. Specifically, the organization of the southern cotton industry began to change after 1915 due to mechanization, the mass migration of black workers to jobs in northern industry, and other factors. By 1935, when the cotton industry's economic significance had substantially declined, "only" 18 lynchings of blacks took place. After that, the figure never again reached double digits, finally dropping to zero in 1965.

We conclude that lynching was a two-sided phenomenon. Breakdown theory alerts us to one side. Lynching was partly a reaction to the violation of norms that threatened to *disorganize* traditional social life in the South. But breakdown theory deflects attention from the other side of the phenomenon. Lynching was also a form of collective action that grew out of, and was intended to maintain, the traditional *organization* of the South's cotton industry. Without that organization, there was no lynching (Tolnay and Beck, 1995; Soule, 1992). And so we arrive at the starting point of post-1970 theories of collective action and social movements. For the past 30 years, most students of the subject have recognized that collective action is often not a short-term reaction to disorganization and deprivation. Instead, it is a long-term attempt to correct perceived injustice that requires a sound social-organizational basis.

SOCIAL MOVEMENTS

Solidarity Theory

According to breakdown theory, people usually rebel soon after social breakdown occurs. In this view, rapid urbanization, industrialization, mass migration, unemployment, and war often lead to the buildup of deprivations or the violation of important norms. Under these conditions, people soon take to the streets.

In reality, however, people often find it difficult to turn their discontent into an enduring social movement. Social movements emerge from collective action only when the discontented succeed in building up a more or less stable membership and organizational base. Once this is accomplished, they typically move from an exclusive focus on short-lived actions such as demonstrations to more enduring and routine activities. Such activities include establishing a publicity bureau, founding a newspaper, and running for public office. These and similar endeavors require hiring personnel to work full-time on various movement activities. Thus, the creation of a movement bureaucracy takes time, energy, and money. On these grounds alone, one should not expect social breakdown to quickly result in the formation of a social movement.

Research conducted since 1970 shows that, in fact, social breakdown often does not have the expected short-term effect. That is because several social-structural factors modify the effects of social breakdown on collective action. For example, Charles Tilly and his associates studied collective action in France, Italy, and Germany in the 19th and 20th centuries (Lodhi and Tilly, 1973; Snyder and Tilly, 1972; Tilly, 1979a; Tilly, Tilly, and Tilly, 1975). They systematically read newspapers, government reports, and other sources so they could analyze a representative sample of strikes, demonstrations, and acts of collective violence. (They defined acts of collective violence as events in which groups of people seized or damaged persons or property.) They measured social breakdown by collecting data on rates of urban growth, suicide, major crime, prices, wages, and the value of industrial production. Breakdown theory would be supported if they found that levels of social breakdown rose and fell with rates of collective action. But they did not. For example, as the top panel of Table 17.1 shows for France, nearly all the correlations between collective violence and indicators of breakdown are close to zero. This means that acts of collective violence did not increase in the wake of mounting social breakdown, nor did they decrease in periods marked by less breakdown.

Significantly, however, Tilly and his associates found stronger correlations between collective violence and some other variables. You will find them in the bottom panel of Table 17.1. These correlations hint at the three fundamental lessons of the **solidarity theory** of social movements, a variant of conflict theory (see Chapter 1, "A Sociological Compass") and the most influential approach to the subject since the 1970s:

1. Inspecting Table 17.1, we first observe that collective violence in France increased when the number of union members rose and decreased when the number of union members fell. Why? Because union organization gave workers more power, and that power increased their capacity to pursue their aims—if necessary, by demonstrating, striking, and engaging in collective violence. We can generalize from the French

Variable	Correlation With Frequency of Collective Violence
Breakdown variables	
Number of suicides	.00
Number of major crimes	−.16
Deprivation variables	
Manufactured goods prices	.05
Food prices	.08
Value of industrial production	.10
Real wages	.03
Organizational variables	
Number of union members	.40
Political process variable	
National elections	.17
State repression variable	
Days in jail	−.22

✦ **TABLE 17.1** ✦
Correlates of Collective Violence, France, 1830–1960

Notes: (a) Correlations can range from −1.0 (indicating a perfect, inversely proportional relationship) to 1.0 (indicating a perfect, directly proportional relationship). A correlation of 0 indicates no relationship. (b) The correlation between major crimes and the rate of collective violence is negative, but it should be positive according to breakdown theory. (c) The exact years covered by each correlation vary.

SOURCE: Adapted from Tilly, Tilly, and Tilly (1975: 81–2)

case as follows. Most collective action is part of a power struggle. The struggle usually intensifies as disadvantaged groups become more powerful relative to privileged groups. How do disadvantaged groups become more powerful? By gaining new members, getting better organized, and increasing their access to scarce resources, such as money, jobs, and means of communication (Bierstedt, 1974). French unionization is thus only one example of **resource mobilization,** a process by which groups engage in more collective action as their power increases due to their growing size and increasing organizational, material, and other resources (Gamson, 1975; Jenkins, 1983; McCarthy and Zald, 1977; Oberschall, 1973; Tilly, 1978; Zald and McCarthy, 1979).

2. Table 17.1 also shows that there was somewhat more collective violence in France when national elections were held. Again, why? Because elections gave people new political opportunities to protest. In fact, by providing a focus for discontent and a chance to put new representatives with new policies in positions of authority, election campaigns often serve as invitations to engage in collective action. When else do new political opportunities open up for the discontented? Chances for protest also emerge when influential allies offer support, when ruling political alignments become unstable, and when elite groups get divided and come into conflict with one another (Tarrow, 1994: 86–9; Useem, 1998). Said differently, collective action takes place and social movements crystallize not just when disadvantaged groups become more powerful but when privileged groups and the institutions they control get divided and therefore become weaker. As Harvard economist John Kenneth Galbraith once said about the weakness of the Russian ruling class at the time of the 1917 revolution, if someone manages to kick in a rotting door, some credit has to be given to the door. In short, this second important insight of solidarity theory links the timing of collective action and social movement formation to the emergence of new **political opportunities** (McAdam, 1982; Piven and Cloward, 1977; Tarrow, 1994).

3. The third main lesson of solidarity theory is that government reactions to protest influence subsequent protest (see Box 17.1). Specifically, governments can try to lower the frequency and intensity of protest by taking various **social control** measures (Oberschall, 1973: 242–83). These measures include making concessions to protesters, co-opting the most troublesome leaders (for example, by appointing them advisors), and violently repressing collective action. The last point explains the modest correlation in Table 17.1 between frequency of collective violence and

BOX 17.1
IT'S YOUR CHOICE

GOVERNMENT SURVEILLANCE OF SOCIAL MOVEMENTS

In 2001, a public policy debate emerged over whether the American government should have a free hand to spy on the activities of social movements. The debate was provoked by Osama bin Laden's terrorist network, *al-Qaeda*. *Al-Qaeda* originated in Afghanistan in the 1980s, where it helped radical anti-Western Muslim fundamentalists wage a successful war against the Soviet Union ("Hunting bin Laden," 1999). By 2000, bin Laden had operatives in 60 countries. He often used a satellite phone to communicate with them—until United States law enforcement officials revealed they were tapping his calls. Once he learned of these taps, bin Laden increased his use of another, more effective means of communication: sending messages that are easily encrypted but difficult to decode via the Internet (Kelley, 2001; McCullagh, 2000a). Such messages may have been used to help plan and coordinate the complex, almost simultaneous jet hijackings that resulted in the crash of an airliner in Pennsylvania and the destruction of the World Trade Center and part of the Pentagon on September 11, 2001.

Hiding messages in innocent looking packages is an old practice made easier by computers (Johnson and Jajodia, 1998). For example, using "steganography" programs freely available on the Web, one can hide messages inside photographs or MP3 files and then place the files on publicly accessible Web site, where they can be downloaded by operatives ("Welcome . . . ," 2001; "MP3stego," 2001).* A digitized photograph or an MP3 file is made up of many zeros and ones. Steganography programs insert a small number of additional zeros and ones into such files. The added zeros and ones are invisible when the photograph

The remains of the twin towers of the World Trade Center, destroyed by terrorists on September 11, 2001. Increased government surveillance of some social movements may be necessary to prevent disasters such as this. However, giving the government a free hand to increase surveillance of social movements could endanger democracy. Americans must therefore make politically difficult choices about which movements are dangerous and which could be beneficial.

is seen and inaudible when the MP3 file is heard. When decoded, however, they form a message. Such messages have been found in files posted in sports chat rooms and pornography sites, for example.

From the point of view of law enforcement officials, the problem is figuring out how to decode the messages. Supercomputers can be used, but they can take months to decode a single message. That is why some people think we need more government regulation, such as laws requiring that all steganography and other encryption programs be built with a "backdoor" or an encryption key that would allow officials to read coded messages quickly and easily.

Right-wing free-market organizations oppose this idea. They are wary of giving the government more power to invade people's privacy (McCullagh, 2000b). Opposition may be found on the left, too, because many liberals know that the government has a history of surveillance of popular social movements. For example, the civil rights movement, led by Dr. Martin Luther King Jr. and others, was under constant surveillance by the FBI and other law-enforcement

agencies. Many people now decry the government's attempt to control and even suppress the civil rights movement. They worry that government access to backdoors and encryption keys will only enhance its ability to spy on popular American movements and suppress them.

Here, then, we face one of democracy's great dilemmas: Democracy empowers both its citizens and its enemies. So should the government be allowed to increase its surveillance of social movements? Increasing government surveillance in general is likely to harm both the enemies of democracy and its champions. But failing to increase government surveillance is likely to help democracy's enemies. It may be that the only way out of this dilemma is to allow increased surveillance only of those movements that most Americans consider to be dangerous to society. Identifying those movements will require vigorous political debate and tough political decisions. Ultimately, it's your choice.

---

* *Steganography* comes from the Greek for "hidden writing."

governments throwing more people into jail for longer periods of time. For in France, more violent protest often resulted in more state repression. However, the correlation is modest because social control measures do not always have the desired effect. For instance, if grievances are very deeply felt, and yielding to protesters' demands greatly increases their hopes, resources, and political opportunities, government concessions may encourage protesters to press their claims further. And while the firm and decisive use of force usually stops protest, using force moderately or inconsistently often backfires. That is because unrest typically intensifies when protesters are led to believe that the government is weak or indecisive (Piven and Cloward, 1977: 27–36; Tilly, Tilly, and Tilly, 1975: 244).

Discussions of strain, deprivation, and contagion dominated analyses of collective action and social movements before 1970. Afterwards, analyses of resource mobilization, political opportunities, and social control dominated the field. Let us now make the new ideas more concrete. We do so by analyzing the ups and downs of one of the most important social movements in 20th-century America, the union movement, and its major weapon, the strike.

Strikes and the Union Movement in America

Workers have traditionally drawn three weapons from the arsenal of collective action to advance their interests: unions, political parties, and strikes. Unions enable groups of workers to speak with one voice and thus to bargain more effectively with their employers for better wages, working conditions, and benefits. The union movement brought us many things we take for granted today, such as the 8-hour day, 2-day weekends, health insurance, and pensions. In most of the advanced industrial democracies (although, as we saw in Chapter 11, not in the United States), workers have also created and supported labor or socialist parties. Their hope has been that, by gaining political influence, they can get laws passed that favor their interests. Finally, when negotiation and political influence get them nowhere, workers have tried to extract concessions from employers by withholding their labor. That is, they have gone on strike.

22 May, Minneapolis:

A crowd of 20,000 striking workers clashes with police, who kill 2 workers and wound 50. The workers virtually control the city. They are supporting truck drivers in the coal yards, whose employers have rejected their attempts to form a union. In their next confrontation, the police kill 2 workers and wound 67. The governor declares martial law. The employers finally accept a federal plan that leads to collective bargaining agreements with 500 Minneapolis employers.

23 May, Toledo:

Ten thousand militant workers assemble outside the Electric Auto-Lite factory, imprisoning 1,500 strikebreakers inside the plant. The sheriff orders his deputies to attack. The crowd fights back and several people are seriously injured. The authorities call in the Ohio National Guard, armed with machine guns and bayoneted rifles. The Guard fires into the crowd. They kill two people and wound many more, but the crowd refuses to disperse. Four more companies of Guardsmen are called up. With workers now threatening to shut down the entire city, Auto-Lite finally agrees to get rid of the strikebreakers and engage in federal mediation. The strikers win a 22 percent wage increase and limited union recognition.

2 July, San Francisco:

Seven hundred police officers storm dockworkers who have been on strike for 45 days. Twenty-five people are hospitalized. Two days later, the police charge again, hospitalizing 155 people and killing 2. The National Guard is called in to restore order. An eyewitness describes the funeral procession for the slain strikers as follows:

North Carolina mill workers on strike, 1934.

> In solid ranks, eight to ten abreast, thousands of strike sympathizers . . . Tramp-tramp-tramp. No noise except that. The band with its muffled drums and somber music . . . On the marchers came—hour after hour—ten, twenty, thirty thousand of them . . . A solid river of men and women who believed they had a grievance and who were expressing their resentments in this gigantic demonstration (quoted in Piven and Cloward, 1977: 125).

1 September, nationwide:

More than 375,000 textile workers are on strike. Employers hire armed guards who, with the National Guard, keep the mills open in Alabama, Mississippi, Georgia, and the Carolinas. The governor of Georgia declares martial law and sets up a detention camp for 2,000 strikers. Six strikers are killed in clashes with police in South Carolina. Another nine are killed elsewhere in the country. This brings the annual total of slain strikers to at least 40. Riots break out in Rhode Island, Connecticut, and Massachusetts, and National Guardsmen are on duty throughout New England.

It was 1934, one the bloodiest years of collective violence in American history. What spurred the mass insurgency? As we might suspect from our knowledge of resource mobilization theory, an important underlying cause was the rapid growth of the industrial working class over the preceding half century. By 1920, industrial workers made up 40 percent of the American labor force and were central to the operation of the economy. As a result, strikes were never more threatening to political and industrial leaders. On the other hand, workers were not well organized. Until 1933, they did not have the right to bargain collectively with their employers, so fewer than 12 percent of America's non-farm workers were union members. And they were anything but well-to-do. The economic collapse that began in 1929 brought unemployment to a full third of the work force and severely depressed the wages of those lucky enough to have jobs.

More than their ability to mobilize organizational and material resources, it was a new law that galvanized industrial workers by opening vast political and economic opportunities for them. In 1932, a nation in despair swept Franklin Delano Roosevelt into the

White House. He forged his "New Deal" legislation aimed at ending the Great Depression. One of his early laws was the 1933 National Industrial Recovery Act (NIRA). Section 7(a) of the NIRA specified workers' minimum wages and maximum hours of work. It also gave them the right to form unions and bargain collectively with their employers. Not surprisingly, industrial workers hailed the NIRA as a historic breakthrough. But employers challenged the law in the courts. And so the seesaw was set in motion. First the Supreme Court invalidated the NIRA. Then Congress reinstated the basic terms of the NIRA by passing the Wagner Act in 1935. Then the Wagner Act was largely ignored in practice. Finally, in 1937, the Supreme Court ruled the Wagner Act constitutional. In the interim, from 1933 to 1937, the promise of the new pro-union laws gave industrial workers new hope and determination. The workers thus armed, open class war rocked America.

From 1933 until the end of World War II, many millions of American workers joined unions. In 1945, unionization reached its historic peak (see Figure 17.2). In that year, 35.5% of non-farm employees were union members. Then the figure began to drop. **Union density** (union members as a percentage of non-farm workers) remained above 30% until the early 1960s. By 2000, it stood at a mere 13.5%. Today, the United States has the lowest union density of any rich industrialized country. What accounts for the post-1945 drop? Focusing on resource mobilization and political opportunities takes us a long way toward answering these questions:

1. *Resource mobilization.* The post-1945 drop in union density is partly a result of changes in America's occupational structure. The industrial working class has shrunk and therefore become weaker (Troy, 1986). In 1900, there was roughly one blue-collar (goods-producing) job in America for every white-collar (service-producing) job. By 1970, the blue-collar/white-collar ratio dropped to about 1:2. By 2000, it fell to about 1:3 (see Figure 10.3, Chapter 10, "Work and the Economy"). These figures show that blue-collar workers are an increasingly rare species. Yet it is precisely among blue-collar workers that unionism is strongest. True, unionization has increased among government workers. Since the early 1960s, they have enjoyed limited union rights. But this gain has not offset losses due to decline in the size of the industrial working class. Meanwhile, unions have scarcely penetrated the rapidly growing ranks of white-collar workers in the private sector. Usually better educated and higher paid, and with more prestige attached to their occupations than blue-collar workers, American private sector white-collar workers have traditionally resisted unionization.

The industrial working class has also been weakened by globalization and employer hostility to unions. As we saw in Chapter 10 ("Work and the Economy"), the globalization of production that began in the 1970s put American blue-collar

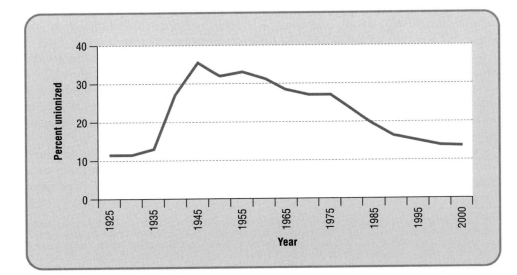

◆ **FIGURE 17.2** ◆
Unionization as Percent of Non-farm Workers, United States, 1925–2000

SOURCE: United States Department of Labor (2000d; 2000e).

An abandoned factory in East St. Louis. The globalization of production that began in the 1970s put American blue-collar workers in direct competition for jobs with overseas workers. Employers could now close American factories and relocate them in Mexico, China, and other countries unless American workers were willing to work for lower wages, fewer benefits, and less job security. American plant closings became increasingly common in the 1970s and 1980s, and unions were often forced to make concessions on wages and benefits.

workers in direct competition for jobs with overseas workers. Employers could now close American factories and relocate them in Mexico, China, and other countries unless American workers were willing to work for lower wages, fewer benefits, and less job security. American plant closings became increasingly common in the 1970s and 1980s, and unions were often forced to make concessions on wages and benefits. Growing ineffectiveness weakened unions and made them less popular among some workers. In addition, beginning in the 1970s, many American employers began to contest unionization elections legally. They also hired consulting firms in anti-union "information" campaigns aimed at keeping their workplaces union free. In some cases, they used outright intimidation to prevent workers from unionizing. Thus, a decline in organizational resources available to industrial workers was matched by an increase in anti-union resources mobilized by employers (Clawson and Clawson, 1999: 97–103).

2. *Political opportunities.* Apart from the erosion of the union movement's mass base, government action has limited opportunities for union growth since the end of World War II. This was evident as early as 1947, when Congress passed the Taft-Hartley Act in reaction to a massive post–World War II strike wave. Unions were no longer allowed to force employees to become members or to require union membership as a condition of being hired. The Taft-Hartley Act also allowed employers to replace striking workers. Unions thus became less effective—and therefore less popular—as vehicles for achieving workers' aims. Taft-Hartley remains the basic framework for industrial relations in America.

Resource mobilization theory teaches us that social organization usually facilitates collective action. The opposite also holds. Less social organization typically means less protest. We can see this by examining the frequency of strikes over time.

Comparing historical periods, we see that unusually low union density has helped to virtually extinguish the strike as a form of collective action in the United States.[2] This is apparent from Figure 17.3, which shows the annual number of strikes involving 1,000 or more workers from 1947 to 2000. Between 1947 and 1983, an annual average of 277 big strikes took place. In contrast, between 1984 and 2000, there was an annual average of only 45 big strikes. This indicates a major historical shift.

Over the short term, strikes have usually been more frequent during economic "booms" and less frequent during economic "busts" (Kaufman, 1982). That is the main reason we see year-to-year fluctuations in strike frequency in Figure 17.3. With more

---

[2]Note, however, that strike activity also tends to be low in countries with *high* levels of unionization. Sweden, for example, has the world's highest union density. It strike rate is low, however, because workers and their representatives have been involved in government policy making since World War II. Decisions about wages and benefits tend to be made in negotiations between unions, employer associations, and governments rather than on the picket line. We conclude that strike activity is highest in countries with intermediate levels of unionization.

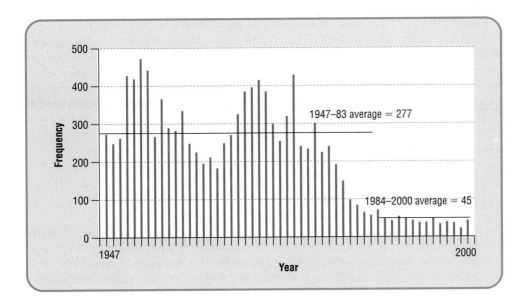

◆ **FIGURE 17.3** ◆
Frequency of Strikes With 1,000+ Workers, United States, 1947–2000

SOURCE: United States Department of Labor (2000f).

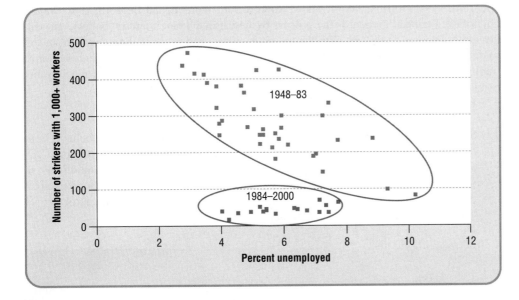

◆ **FIGURE 17.4** ◆
Unemployment and the Frequency of Big Strikes, United States, 1948–2000

SOURCE: United States Department of Labor (2000a; 2000f).

money, more job opportunities, and bigger strike funds in good times, workers can better afford to go out on strike to press their claims than during periods of high unemployment.

However, Figure 17.4 shows that the relationship between unemployment and strike frequency changed after 1983. Thus, the dots representing the years 1948–83 slope downward. This means that, whenever unemployment increased, big strikes were less common. In contrast, for the period 1984–99, a flat line replaces the downward slope. This means that, even in good times, workers avoided striking. It seems that, due partly to weak unions, many workers are now unable to use the strike weapon as a means of improving their wages and benefits.[3]

Since the mid-1990s, the American Federation of Labor-Congress of Industrial Organizations (AFL-CIO), the largest union umbrella organization in the United States,

---

[3]Other important factors making strikes less sensitive to the business cycle since the early 1980s include: (1) The willingness of employers to fire strikers and replace them with other workers. This practice renders strikes riskier from the worker's viewpoint. Using replacement workers—a practice that is outlawed in much of Western Europe and some Canadian provinces—was legalized in 1938 by the Supreme Court and became widespread after President Reagan fired the nation's striking air traffic controllers in 1980. (2) In addition, workers have come to regard strikes as riskier since the 1980s because income-replacing social welfare benefits have been cut. See Cramton and Tracy (1998); Schor and Bowles (1987).

has sought to reverse the trends in unionization described above. Specifically, it has tried to organize immigrants, introduce more feminist issues into its program in a bid to attract more women, and develop new forms of employee organization and representation that are more appropriate to a postindustrial society. The latter include coalitions with other social movements and councils that bring together all the unions in a city or other geographical area (Clawson and Clawson, 1999: 112–15). Whether these strategies will succeed in revitalizing the American union movement is unclear. Figures on union density up to 2000 suggest they have not yet reversed the downward slide of the union movement (see Figure 17.2).

Framing Discontent

As we have seen, solidarity theory helps to explain the emergence of many social movements. Still, the rise of a social movement sometimes takes strict solidarity theorists by surprise. So does the failure of an aggrieved group to press its claims by means of collective action. It seems, therefore, that something lies between (a) the capacity of disadvantaged people to mobilize resources for collective action and (b) the recruitment of a substantial number of movement members. That "something" is **frame alignment** (Benford, 1997; Goffman, 1974; Snow, Rochford Jr., Worden, and Benford, 1986; Valocchi, 1996). Frame alignment is the process by which individual interests, beliefs, and values either become congruent with the activities, ideas, and goals of the movement or fail to do so. Thanks to the efforts of scholars operating mainly in the symbolic interactionist tradition (see Chapter 1, "A Sociological Compass"), frame alignment has recently become the subject of sustained sociological investigation.

Frame alignment can be encouraged in several ways. For example:

1. Social movement leaders can reach out to other organizations that, they believe, contain people who may be sympathetic to their movement's cause. Thus, leaders of an antinuclear movement may use the mass media, telephone campaigns, and direct mail to appeal to feminist, antiracist, and environmental organizations. In doing so, they assume these organizations are likely to have members who would agree at least in general terms with the antinuclear platform.

The "Human Rights Now" tour, Los Angeles, 1988. LR: Peter Gabriel, Tracy Chapman, Youssou N'Dour, Sting, Joan Baez, and Bruce Springsteen. When bands play at protest rallies or festivals, it is not just for entertainment and not just because the music is relevant to a social movement's goals. The bands also attract nonmembers to the movement. This is one way of framing a social movement's goals to make them appealing to nonmembers.

2. Movement activists can stress popular values that have so far not featured prominently in the thinking of potential recruits. They can also elevate the importance of positive beliefs about the movement and what it stands for. For instance, in trying to win new recruits, movement members might emphasize the seriousness of the social movement's purpose. They might analyze the causes of the problem the movement is trying solve in a clear and convincing way. Or they might stress the likelihood of the movement's success. By doing so, they can increase the movement's appeal to potential recruits and perhaps win them over to the cause.

3. Social movements can stretch their objectives and activities to win recruits who are not initially sympathetic to the movement's original aims. This may involve a "watering down" of the movement's ideals. Alternatively, movement leaders may decide to take action calculated to appeal to nonsympathizers on grounds that have little or nothing to do with the movement's purpose. When rock, punk, or reggae bands play at nuclear disarmament rallies or gay liberation festivals, it is not necessarily because the music is relevant to the movement's goals. Nor do bands play just because movement members want to be entertained. The purpose is also to attract nonmembers. Once attracted by the music, however, nonmembers may make friends and acquaintances in the movement and then be encouraged to attend a more serious-minded meeting.

In short, there are many ways social movements can make their ideas more appealing to more people. All of them involve increasing the alignment between the way movement members and potential recruits frame issues (see Box 17.2).

Refrain: Back to 1968

Frame alignment theory stresses the face-to-face interaction strategies employed by movement members to recruit nonmembers who are like-minded, apathetic, or even initially opposed to the movement's goals. Resource mobilization theory focuses on the broad social-structural conditions that facilitate the emergence of social movements. One theory usefully supplements the other.

The two theories certainly help clarify the 1968 high school incident described at the beginning of this chapter. In light of our discussion, it seems evident that two main factors prevented Robert Brym from influencing his classmates when he spoke to them about the dangers of industrial pollution from the local pulp and paper mill.

First, he lived in a poor and relatively unindustrialized region of Canada where people had few resources they could mobilize on their own behalf. Per capita income and the level of unionization were among the lowest of any state or province in North America. The unemployment rate was among the highest. In contrast, K. C. Irving, who owned the pulp and paper mill, was so powerful that most people in the region could not even conceive the need to rebel against the conditions of life that he created for them. He owned most of the industrial establishments in the province. Every daily newspaper, most of the weeklies, all of the TV stations, and most of the radio stations were his too. Little wonder one rarely heard a critical word about his operations. Many people believed that Irving could make or break local governments single-handed. Should one therefore be surprised that mere high school students refused to take him on? In their reluctance, Robert's fellow students were only mimicking their parents, who, on the whole, were as powerless as Irving was mighty (Brym, 1979).

Second, many of Robert's classmates did not share his sense of injustice. Most of them regarded Irving as the great provider. They thought his pulp and paper mill, as well as his myriad other industrial establishments, gave many people jobs. They regarded that fact as more important for their lives and the lives of their families than the pollution problem Robert raised. Frame alignment theory suggests Robert needed to figure out ways of building bridges between their understanding and his. He did not. Therefore, he received an unsympathetic hearing.

BOX 17.2
SOCIOLOGY AT THE MOVIES

LAWRENCE OF ARABIA (1962)

One of the greatest movies of all time is a story about frame alignment. It is the Oscar-winning account of how British Colonel T. E. Lawrence, brilliantly played by Peter O'Toole, helped to turn divided Arab tribes into a united movement for national independence from Turkey.

The Arabs had fallen under Turkish rule in the 1500s and subsequently endured a deep political and cultural decline. However, in World War I (1914–8), Turkey fought against Britain, and Britain recognized in the Arabs a potential ally against the Turks. In the movie, the British use Lawrence to unite the Arabs against their Turkish overlords.

At first, the British military dismisses the squabbling Arab tribes as "a nation of sheep stealers." Enter Lawrence. He sees in them a real people and a potentially valuable ally against the Turks. As a result, he sets out to align the beliefs of the Arabs with his own thinking. He accomplishes this task by word and example. By force of personality he convinces tribal leaders that disunity will only ensure Arab status as a petty people, unable to gain its freedom and recapture the scientific, architectural, and literary glories it had achieved centuries earlier. Conversely, he argues, unity will ensure political freedom and cultural flowering.

Words, however, are not enough to galvanize any more than a few tribes. Lawrence understands he can effectively align the beliefs of the Arabs with his own ambitions only by showing *in practice* how unity creates power. And so, in the movie, he proposes a land attack on the Turkish-controlled port of Aqaba. It is an outlandish idea because a land attack requires the nearly impossible crossing of a long stretch of barren desert known as "The Sun's Anvil." It is, however, an idea that is strategically dazzling, for the Turkish guns at Aqaba are stationary and they point out to the Red Sea, not inland. Fighting thirst,

Arabs led by Colonel T. E. Lawrence attack a Turkish supply train in *Lawrence of Arabia*.

hunger, and fatigue, Lawrence leads his supporters across The Sun's Anvil. In Aqaba, the Turks, defenseless against the land attack, lose hundreds in battle and quickly capitulate. It is a turning point. The British military, now convinced that Lawrence can unite the Arabs, gives him guns and artillery to continue his campaign against the Turks. Arabs throughout the Middle East take pride in their military accomplishments. They emerge at the end of the war by no means a fully united national independence movement, but at least able to see the possibility of Arab unity.

Lawrence of Arabia is a great movie, but it is flawed because it gives too much credence to Lawrence's own, self-promoting account of events and not enough to other credible historical sources that emphasize the native origins of Arab nationalism. Thus, the roots of the Arab national movement lay deeper than the movie allows. The movement first began to stir more than 60 years before Lawrence arrived in the Middle East, having sprung up among semi-Westernized Arab intellectuals in urban centers like Beirut and Damascus in the late 1840s (Antonius, 1939: 35–60). Similarly, Arabs alone conceived and executed the all-important raid on Aqaba. Lawrence participated merely as "a trusted friend and companion-

in-arms" of Faisal, a tribal leader and later King of Iraq (Antonius, 1939: 323). Lawrence's exercise in frame alignment certainly helped stimulate the Arab national movement, but, as these examples illustrate, by overstating and romanticizing Lawrence's role, *Lawrence of Arabia* understates the Arabs' part in fashioning their own destiny.

The movie does, however, accurately portray the duplicity of the British, and it shows how they helped to arouse a more militant Arab nationalism. The British promised the Arabs independence after World War I in exchange for their support against the Turks. In 1916, however, they made a secret deal with France to divide up much of the region. After the war, Britain ruled part of the Middle East, France another. Increasingly, the United States, too, exercised substantial influence over parts of the region. This fueled anti-Western resentment on the part of the Arabs. Western support for the creation of the State of Israel in 1948 further inflamed Arab nationalism and anti-Westernism. So did subsequent Western political and military intervention to ensure access to the region's enormous oil reserves. Thus, what began as an exercise in frame alignment turned out to be the world's most intractable political problem in the early 21st century.

Try applying solidarity and frame alignment theories to times when *you* felt a deep sense of injustice against an institution such as a school, an organization, a company, or a government. Did you do anything about your upset? If not, why not? If so, what did you do? Why were you able to act in the way you did? Did you try to get other people to join you in your action? If not, why not? If so, how did you manage to recruit them? Did you reach the goal you set out to achieve? If not, why not? If so, what enabled you to succeed? If you've never been involved in a collective action to correct a perceived injustice, try analyzing a movie about collective action using insights gleaned from solidarity and frame alignment theories. A classic is *Norma Rae* (1979), starring Sally Field. Field won the Best Actress Oscar for her performance in this film as a Southern textile worker who joins with a labor organizer to unionize her mill.

THE HISTORY AND FUTURE OF SOCIAL MOVEMENTS

Attempts to synthesize theories of collective action and social movement formation are in their infancy (Diani, 1996; McAdam, McCarthy, and Zald, 1996; Tarrow, 1994). Still, we can briefly summarize what we have learned about the causes of collective action and social movement formation with the aid of Figure 17.5. Breakdown theory partly answers the question of *why* discontent is sometimes expressed collectively and in nonroutine ways. Industrialization, urbanization, mass migration, economic slowdown, and other social changes often cause dislocations that engender feelings of strain, deprivation, and injustice. Solidarity theory focuses on *how* these social changes may eventually facilitate the emergence of social movements. They cause a reorganization of social relations, shifting the balance of power between disadvantaged and privileged groups. Solidarity theory also speaks to the question of *when* collective action erupts and social movements emerge. The opening and closing of political opportunities, as well as the exercise of social control by authorities, helps to shape the timing of collective action. Finally, by analyzing the day-to-day strategies employed to recruit nonmembers, frame alignment theory directs our attention to the question of *who* is recruited to social movements. Altogether, then, the theories

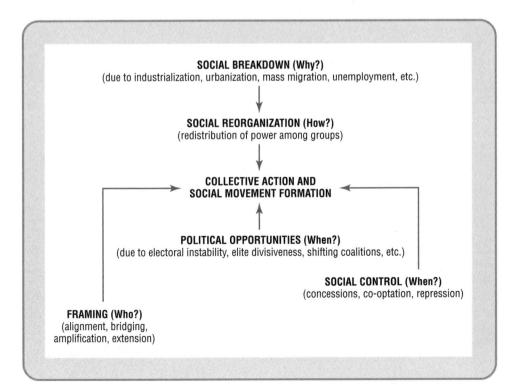

✦ FIGURE 17.5 ✦
Determinants of Collective Action and Social Movement Formation

we have considered provide a comprehensive picture of the why, how, when, and who of collective action and social movements.

Bearing this summary in mind, we can now turn to this chapter's final goal: sketching the historical development and future prospects of social movements in broad, rapid strokes. We begin three centuries ago.

The History of Social Movements

In 1700, social movements were typically small, localized, and violent. In Europe, poor residents of a city might riot against public officials in reaction to a rise in bread prices or taxes. Peasants on an estate might burn their landowner's barns (or their landowner) in response to his demand for a larger share of the crop. However, as the state grew, the form of protest changed. The state started taxing nearly all its citizens at higher and higher rates as government services expanded. It imposed a uniform language and often a common curriculum in a compulsory education system. It drafted most young men for army service. It instilled in its citizens all the ideological trappings of modern nationalism, from anthems to flags to historical myths.

As the state came to encompass most aspects of life, social movements changed in three ways. First, they became national in scope. That is, they typically directed themselves against central governments rather than local targets. Second, their membership grew. That was partly because potential recruits were now literate and could communicate using the printed word. In addition, big new social settings—factories, offices, densely populated urban neighborhoods—could serve as recruitment bases. Third, social movements became less violent. That is, their size and organization often allowed them to bureaucratize, stabilize, and become sufficiently powerful to get their way without frequent resort to extreme measures (Tilly, 1978; 1979a; 1979b; Tilly, Tilly, and Tilly, 1975).

Social movements often used their power to expand the rights of citizens. We may identify four stages in this process, focusing on Britain and the United States. In Britain, rich property owners fought against the king in the 18th century for **civil citizenship.** Civil

In medieval Europe, social movements were small, localized, and violent. For example, a medieval French historian reported that in 1358, "there were very strange and terrible happenings in several parts of the kingdom. . . . They began when some of the men from the country towns came together in the Beauvais region. They had no leaders and at first they numbered scarcely a hundred. One of them got up and said that the nobility of France . . . were disgracing and betraying the realm, and that it would be a good thing if they were all destroyed. At this they all shouted: 'He's right! He's right! Shame on any man who saves the gentry from being wiped out!' They banded together and went off, without further deliberation and unarmed except for pikes and knives, to the house of a knight who lived near by. They broke in and killed the knight, with his lady and children, big and small, and set fire to the house" (Froissart, 1968 [c. 1365]: 151).

©Bibliotheque nationale de F

The 15th Amendment to the Constitution gave African Americans the right to vote in 1870. However, most of them were unable to exercise that right, at least in the South, from the late 19th century until the 1960s because of poll taxes and literacy tests. The civil rights movement of the 1960s was in part a struggle over that issue. Some African Americans claim they were effectively denied the right to vote in the 2000 presidential election.

citizenship is the right to free speech, freedom of religion, and justice before the law. The male middle class and the more prosperous strata of the working class fought against rich property owners in the 19th century for **political citizenship.** Political citizenship is the right to vote and run for office. In early 20th-century Britain, women and poorer workers succeeded in achieving these same rights despite the opposition of well-to-do men in particular. During the remainder of the century, blue- and white-collar workers fought against the well to do for **social citizenship.** Social citizenship is the right to a certain level of economic security and full participation in social life with the help of the modern welfare state (Marshall, 1965).

The timing of the struggle for citizenship rights was different in the United States. In particular, universal suffrage for white males was won earlier in the 19th century than in Europe. This accounts in part for the greater radicalism of the European working class. It had to engage in a long and bitter struggle for the right to vote while its United States counterpart was already incorporated into the political system (Lipset, 1977). Another important distinguishing feature of the United States concerns African Americans. The 15th Amendment to the Constitution gave them the right to vote in 1870. However, most of them were unable to exercise that right, at least in the South, from the late 19th century until the 1960s. That was because of various restrictions on voter registration, including poll taxes and literacy tests. The civil rights movement of the 1960s was in part a struggle over this issue. It helped to create a community that is more politically radical than its white counterpart.

The Future of Social Movements

Partly because the success of the civil rights movement inspired them, so-called **new social movements** emerged in the 1970s (Melucci, 1980; 1995). What is new about new social movements is the breadth of their goals, the kinds of people they attract, and their potential for globalization. Let us consider each of these issues in turn.

Goals

Some new social movements promote the rights not of specific groups but of humanity as a whole to peace, security, and a clean environment. Such movements include the peace movement, the environmental movement, and the human rights movement. Other new social movements, such as the women's movement and the gay rights movement, promote the rights of particular groups that have been excluded from full social participation.

Web Research Projects
Ballots or Bricks? The Role of Violence in African-American Politics

Accordingly, the more than 1,200 gay rights groups in the United States have fought for laws that eliminate all forms of discrimination based on sexual orientation. They have also fought for the repeal of laws that discriminate on the basis of sexual orientation, such as antisodomy laws and laws that negatively affect parental custody of children. They have succeeded mainly at the county and local government levels. The women's movement has been more successful in getting laws passed.[4] In 1982, it even came close to having the Equal Rights Amendment (ERA) recognized as the 27th Amendment to the Constitution. The ERA is intended to eliminate discrimination based on sex. Approved by the House of Representatives in 1971 and by the Senate in 1972, the ERA fell just 3 states short of the 38 needed for ratification in 1982. For the past 30 years, the women's movement has been most successful in getting admission practices altered in professional schools, winning more freedom of reproductive choice for women, and opening up opportunities for women in the political, religious, military, educational, medical, and business systems (Whittier, 1995). The emergence of the peace, environmental, human rights, gay rights, and women's movements marked the beginning of a fourth stage in the history of social movements. This fourth stage involves the promotion of **universal citizenship,** or the extension of citizenship rights to all adult members of society and to society as a whole (Roche, 1995; Turner, 1986: 85–105).

Membership

New social movements are also novel in that they attract a disproportionately large number of highly educated, relatively well-to-do people from the social, educational, and cultural fields. Such people include teachers, college professors, journalists, social workers, artists, actors, writers, and student apprentices to these occupations. For several reasons, people in these occupations are more likely to participate in new social movements than are people in other occupations. Their higher education exposes them to radical ideas and makes those ideas appealing. They tend to hold jobs outside the business community, which often opposes their values. And they often get personally involved in the problems of their clients and audiences, sometimes even becoming their advocates (Brint, 1984; Rootes, 1995).

Globalization Potential

Finally, new social movements are new in that they have more potential for globalization than did old social movements.

Up until the 1960s, social movements were typically *national* in scope. That is why, for example, the intensity and frequency of urban race riots in the United States in the 1960s did not depend on such local conditions as the degree of black–white inequality in a given city (Spilerman, 1970; 1976). Instead, African Americans came to believe that racial problems are nationwide and capable of solution only by the federal government. Congressional and Presidential action (and lack of action) on civil rights issues, national TV coverage of race issues, and growing black consciousness and solidarity helped to create this belief.[5]

Many new social movements that gained force in the 1970s increased the scope of protest beyond the national level. For example, members of the peace movement viewed federal laws banning nuclear weapons as necessary. Environmentalists felt the same way about federal laws protecting the environment. However, environmentalists also recognized that the condition of the Brazilian rain forest affects climatic conditions worldwide. Similarly, peace activists understood that the spread of weapons of mass

[4]The "first wave" of the women's movement dates back to the 19th-century suffragettes, who struggled to win the vote for women. In the text, we refer to the "second wave" that emerged in the 1960s and whose goals are much broader.

[5]In the 1990s, improved data and statistical techniques confirmed that local levels of strain and deprivation did not influence the incidence of rioting. However, later analyses did find two local effects. First, riots were more frequent where job competition between blacks and whites was more intense. Second, riots tended to occur in close geographical proximity, suggesting that favorable outcomes of collective action in one city provided a model that was learned by people in neighboring cities (Myers, 1997; Olzak and Shanahan, 1996; Olzak, Shanahan, and McEneaney, 1996).

destruction can destroy all of humanity. Therefore, members of the peace and environmental movements pressed for *international* agreements binding all countries to protect the environment and stop the spread of nuclear weapons. Social movements went global.

The globalization of social movements was facilitated by inexpensive international travel and communication. New technologies made it easier for people in various national movements to work with like-minded activists in other countries. In the age of CNN, inexpensive jet transportation, fax machines, Web sites, and e-mail, it was possible not only to see the connection between apparently local problems and their global sources. It was also possible and increasingly desirable to act both locally and globally.

Consider the case of Greenpeace. Greenpeace is a highly successful environmental movement that originated in Vancouver in the mid-1970s. It now has offices in 41 countries, with its international office in Amsterdam (Greenpeace, 2000). Among many other initiatives, it has mounted a campaign to eliminate the international transportation and dumping of toxic wastes. Its representatives visited local environmental groups in African and other developing countries. They supplied the Africans with organizing kits to help them tie their local concerns to global political efforts. They also published a newsletter to keep activists up-to-date about legal issues. Thus, Greenpeace coordinated a global campaign that enabled weak environmental organizations in developing countries to act more effectively. Their campaign also raised the costs of continuing the international trade in toxic waste.

Greenpeace is hardly alone in its efforts to go global. In 1953, 110 international social movement organizations spanned the globe. By 1993, there were 631. About a quarter were human rights organizations, and about a seventh were environmental organizations. The latter are by far the fastest-growing organizational type (Smith, 1998: 97).

Even "old" social movements can go global due to changes in the technology of mobilizing supporters. In 1994, for example, the peasants of Chiapas, a southern Mexican province, started an uprising against the Mexican government. Oppressed by Europeans and their descendants for nearly 500 years, the poor, indigenous people of southern Mexico were now facing a government edict preventing them from gaining access to formerly communal farmland. They wanted the land for subsistence agriculture. But the government wanted to make sure the land stayed in the hands of large, Hispanic ranchers and farmers, who could earn foreign revenue by exporting goods to the United States and Canada under the terms of the new North American Free Trade Agreement. The peasants seized a large number of ranches and farms. A mysterious masked man known simply as "Subcomandante Marcos" was their leader (see Figure 17.6). Effectively using the Internet and the international mass media as his secret

Greenpeace is a highly successful global environmental movement that originated in Vancouver in the mid-1970s and now has offices in 41 countries. Here, Greenpeace activists try to stop a whaling ship.

✦ FIGURE 17.6 ✦
Subcomandante Marcos, Leader of the Zapatista National Liberation Army in Southern Mexico
Effectively using the Internet and the international mass media as his secret weapon against the Mexican government, Marcos led what the *New York Times* called "the first postmodern revolution," combining a peasant uprising with the World Wide Web, short-wave radio, and photo spreads in *Marie Claire*. By keeping the movement in the international public eye using modern technologies of communication, Marcos mobilized support abroad and limited the retaliatory actions of the Mexican government.

SOURCE: "Ejército Zapatista de Liberación Nacional" (2000).

weapon against the Mexican government, Marcos led what the *New York Times* called "the first postmodern revolution," combining a peasant uprising with the World Wide Web, short-wave radio, and photo spreads in *Marie Claire.* By ingeniously keeping the movement in the international public eye using modern technologies of communication, Marcos mobilized support abroad and limited the retaliatory actions of the Mexican government (*A Place . . .* , 1998; Jones, 1999).

The globalization of social movements can be further illustrated by coming full circle and returning to the anecdote with which we began this chapter. In 1991, Robert Brym visited his hometown. He hadn't been back in years. As he entered the city he vaguely sensed that something was different. "I wasn't able to identify the change until I reached the pulp and paper mill," says Robert. "Suddenly, it was obvious. The rotten-egg smell was virtually gone. I discovered that in the 1970s a local woman whose son developed a serious case of asthma took legal action against the mill and eventually won. The mill owner was forced by law to install a 'scrubber' in the main smokestack to remove most of the sulfur dioxide emissions. Soon, the federal government was putting pressure on the mill owner to purify the polluted water that poured out of the plant and into the local river system." Apparently, local citizens and the environmental movement had caused a deep change in the climate of opinion. This influenced the government to force the mill owner to spend millions of dollars to clean up his operation. It took decades, but what was political heresy in 1968 became established practice by 1991. That is because environmental concerns had been amplified by the voice of a movement that had grown to global proportions. In general, as this case illustrates, globalization helps to ensure that many new social movements transcend local and national boundaries and promote universalistic goals.

SUMMARY

1. In the short term, deprivation and strain due to rapid social change are generally *not* associated with increased collective action and social movement formation.

2. Mobs, riots, and other forms of crowd behavior may be wild and violent. However, social organization and rationality underlie much crowd behavior.

3. People are more inclined to rebel against the status quo when social ties bind them to many other people who feel similarly wronged and when they have the time, money, organization, and other resources needed to protest.

4. Collective action and social movement formation are more likely to occur when political opportunities allow them. Political opportunities emerge due to elections, increased support by influential allies, the instability of ruling political alignments, and divisions among elite groups.

5. Authorities' attempts to control unrest also influence the timing of collective action. They may offer concessions to insurgents, co-opt leaders, and employ coercion.

6. For social movements to grow, members must make the activities, goals, and ideology of the movement congruent with the interests, beliefs, and values of potential new recruits.

7. In 1700, social movements were typically small, localized, and violent. By mid-20th century, social movements were typically large, national, and less violent. In the late 20th century, new social movements developed broader goals, recruited more highly educated people, and developed global potential for growth.

8. The history of social movements is a struggle for the acquisition of constantly broadening citizenship rights. These rights include (a) the right to free speech, religion, and justice before the law, (b) the right to vote and run for office, (c) the right to a certain level of economic security and full participation in the life of society, and (d) the right of marginal groups to full citizenship and the right of humanity as a whole to peace and security.

GLOSSARY

Absolute deprivation is a condition of extreme poverty.

Breakdown theory suggests that social movements emerge when traditional norms and patterns of social organization are disrupted.

Civil citizenship recognizes the right to free speech, freedom of religion, and justice before the law.

Collective action occurs when people act in unison to bring about or resist social, political, and economic change. Some collective actions are routine. Others are nonroutine. Routine collective actions are typically nonviolent and follow established patterns of behavior in existing types of social structures. Nonroutine collective actions take place when usual conventions cease to guide social action and people transcend, bypass, or subvert established institutional patterns and structures.

Contagion is the process by which extreme passions supposedly spread rapidly through a crowd like a contagious disease.

Frame alignment is the process by which individual interests, beliefs, and values become congruent and complementary with the activities, goals, and ideology of a social movement.

New social movements became prominent in the 1970s. They attract a disproportionately large number of highly educated people in the social, educational, and cultural fields and universalize the struggle for citizenship.

Political citizenship recognizes the right to run for office and vote.

Political opportunities for collective action and social movement growth occur during election campaigns, when influential allies offer insurgents support, when ruling political alignments become unstable, and when elite groups become divided and conflict with one another.

Relative deprivation is an intolerable gap between the social rewards people feel they deserve and the social rewards they expect to receive.

Resource mobilization refers to the process by which social movements crystallize due to increasing organizational, material, and other resources of movement members.

Social citizenship recognizes the right to a certain level of economic welfare security and full participation in the social life of the country.

Social control refers to the means by which authorities seek to contain collective action, including co-optation, concessions, and coercion.

Social movements are enduring collective attempts to change part or all of the social order by means of rioting, petitioning, striking, demonstrating, and establishing lobbies, unions, and political parties.

Solidarity theory suggests that social movements are social organizations that emerge when potential members can mobilize resources, take advantage of new political opportunities, and avoid high levels of social control by authorities.

Strain refers to breakdowns in traditional norms that precede collective action.

Union density is the number of union members in a given location and time as a percentage of non-farm workers. It measures the organizational power of unions.

Universal citizenship recognizes the right of marginal groups to full citizenship and the rights of humanity as a whole.

QUESTIONS TO CONSIDER

1. How would you achieve a political goal? Map out a detailed strategy for reaching a clearly defined aim, such as a reduction in income tax or an increase in government funding of colleges. Who would you try to recruit to help you achieve your goal? Why? What collective actions do you think would be most successful? Why? To whose attention would these actions be directed? Why? Write a manifesto that frames your argument in a way that is culturally appealing to potential recruits.

2. Do you think that social movements will be more or less widespread in the 21st century than they have been in the 20th century? Why or why not? What kinds of social movements are likely to predominate?

WEB RESOURCES

Companion Web Site for This Book

http://sociology.wadsworth.com

Begin by clicking on the Student Resources section of the Web site. Choose "Introduction to Sociology" and finally the Brym and Lie book cover. Next, select the chapter you are currently studying from the pull-down menu. From the Student Resources page you will have easy access to InfoTrac College Edition®, MicroCase Online exercises, additional Web links, and many other resources to aid you in your study of sociology, including practice tests for each chapter.

InfoTrac Search Terms

These search terms are provided to assist you in beginning to conduct research on this topic by visiting http://www.infotraccollege.com/wadsworth.

Collective action	Resource mobilization
Frame alignment	Unions
Relative deprivation	

Recommended Web Sites

The use of the Internet to mobilize social movement support worldwide is well demonstrated by Mexico's Zapatista National Liberation Army in the southern province of Chiapas. Visit their Web site (mostly in Spanish, but with sections in English, French, and Portuguese) at http://www.ezln.org and read about their information warfare in *Wired* magazine at http://www.wired.com/news/print/0,1294,17633,00.html.

The environmental movement Greenpeace is one of the most successful cases of globalized protest. Greenpeace now has offices in 41 countries. Its Web site is at http://adam.greenpeace.org/information.shtml.

The National Organization for Women (NOW) is the largest organization of feminist activists in the United States, with half a million members and 550 chapters in all 50 states and the District of Columbia. Founded in 1966, NOW's goal has been to take action to bring about equality for all women. NOW's Web site is at http://www.now.org.

Formed in 1955, the American Federation of Labor-Congress of Industrial Organizations (AFL-CIO) is the largest union umbrella organization in the United States, with 13 million members in 68 unions. Its stated goal is "to bring social and economic justice to our nation by enabling working people to have a voice on the job, in government, in a changing global economy and in their communities." To learn more about the AFL-CIO, visit http://www.aflcio.org/home.htm on the Web.

SUGGESTED READINGS

Benjamin R. Barber. "Jihad vs. McWorld," *The Atlantic Monthly* (March 1992) pp. 53–63. A compelling analysis of the fragmenting and globalizing forces affecting social movements today.

Hanspeter Kreisi, et al. *New Social Movements in Western Europe* (Minneapolis, MN: University of Minnesota Press, 1995). One of the best empirical analyses of new social movements.

Frances Fox Piven and Richard A. Cloward. *Poor People's Movements: Why They Succeed, How They Fail* (New York: Vintage, 1977). An influential study of the unemployed workers' and industrial workers' movements of the 1930s and the civil rights and welfare rights movements of the 1960s. Argues that the poor have fared best by engaging in mass defiance and disruption.

Sidney Tarrow. *Power in Movement: Social Movements, Collective Action and Politics* (Cambridge, UK: Cambridge University Press, 1994). An excellent synthesis of resource mobilization and framing theories. Also underlines the importance of political structures in shaping discontent.

IN THIS CHAPTER, YOU WILL LEARN THAT:

✦ Some people think of technology as a useful magic that drives history forward.

✦ Others think of technology as a monster that has escaped human control and causes more harm than good.

✦ In reality, technology does transform society and history. But it is under human control because human need shapes technological growth.

✦ Increasingly, technological development has come under the sway of large multinational corporations and the military establishments of the major world powers.

✦ Widespread environmental degradation is the main negative consequence of technological development.

✦ Policy-oriented scientists, the environmental movement, the mass media, and respected organizations have to discover and promote environmental issues if they are to be turned into social problems. In addition, the public must connect the information learned from these groups to real-life events.

✦ Economically disadvantaged groups experience more environmental risks than economically advantaged groups.

✦ Most Americans are not prepared to pay the price of creating a safe environment, but repeated environmental catastrophes could easily change their minds.

✦ By helping to make the public aware of the environmental and other choices we face in the 21st century, sociology can play an important role in the evolution of human affairs.

18

TECHNOLOGY AND THE GLOBAL ENVIRONMENT

Technology: Savior or Frankenstein?

Technology *and* People Make History

How High Tech Became Big Tech

Environmental Degradation

The Social Construction of Environmental Problems

The Social Distribution of Risk

What Is to Be Done?

Evolution and Sociology

J. Robert Oppenheimer, the "father" of the atom bomb.

Web Interactive Exercises
The Surveillance Society

TECHNOLOGY: SAVIOR OR FRANKENSTEIN?

On August 6, 1945, the United States Air Force dropped an atomic bomb on Hiroshima. The bomb killed about 200,000 Japanese, almost all civilians. It hastened the end of World War II, thus making it unnecessary for American troops to suffer heavy losses in a land invasion of Japan.

Scholars interested in the relationship between technology and society also recognize that Hiroshima divided the 20th century into two distinct periods. We may call the period before Hiroshima the era of naive optimism. During that time, technology could do no wrong, or so, at least, it seemed to nearly all observers. **Technology** was widely defined as the application of scientific principles to the *improvement* of human life. It seemed to be driving humanity down a one-way street named progress, picking up speed with every passing year thanks to successively more powerful engines: steam, turbine, internal combustion, electric, jet, rocket, and nuclear. Technology produced tangible benefits. Its detailed workings rested on scientific principles that were mysterious to all but those with advanced science degrees. Therefore, most people regarded technologists with reverence and awe. They were viewed as a sort of priesthood whose objectivity allowed them to stand outside the everyday world and perform near-magical acts.

With Hiroshima, the blush was off the rose. Growing pessimism was in fact evident 3 weeks earlier, when the world's first nuclear bomb exploded at the Alamagordo Bombing Range in New Mexico. The bomb was the child of J. Robert Oppenheimer, who had been appointed head of the top-secret Manhattan Project just 28 months earlier. After recruiting what General Leslie Groves called "the greatest collection of eggheads ever," including three past and seven future Nobel prize winners, Oppenheimer organized the largest and most sophisticated technological project in human history up to that time. As an undergraduate at Harvard, Oppenheimer had studied Indian philosophy, among other subjects. On the morning of July 16, 1945, as the flash of intense white light faded, and the purplish fireball rose, sucking desert sand and debris into a mushroom cloud more than $7\frac{1}{2}$ miles high, Oppenheimer quoted from Hindu scripture: "I am become Death, the shatterer of worlds" (quoted in Parshall, 1998).

Oppenheimer's misgivings continued after the war. Having witnessed the destructive power he helped unleash, Oppenheimer wanted the United States to set an example to the only other nuclear power at the time, the Soviet Union. He wanted both countries to halt thermonuclear research and refuse to develop the hydrogen bomb. But the governments of the United States and the Soviet Union had other plans. When Secretary of State Dean

◆ **FIGURE 18.1** ◆
Nobel Prizes in Natural Science by Country, 1901–1996 (in percent)

SOURCES: Kidron and Segal (1995: 92–93); United States Bureau of the Census (1998c).

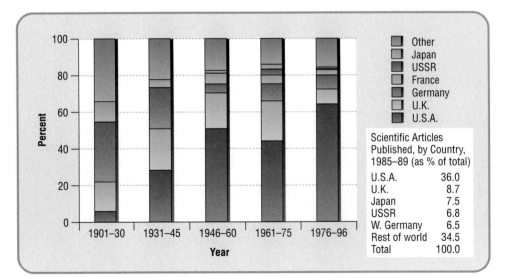

BOX 18.1
SOCIOLOGY AT THE MOVIES

The Matrix, starring Keanu Reaves.

THE MATRIX (1999)

"Have you ever felt that there's something not right in the world?" With these words, *The Matrix* introduces Thomas Anderson, played by Keanu Reeves. Respectable software programmer by day, notorious hacker by night, Anderson, who goes by the handle "Neo," has been plagued by the thought that there is something wrong with the world. "You don't know what it is, but it's there, like a splinter in your mind."

Neo knows somehow that something is wrong, but he cannot point to anything in particular. His moment of awareness comes when he encounters two legendary hackers, Morpheus and Trinity. They introduce him to the secret of his day world and the reality of the Matrix. It turns out that "reality" lived by Neo and others is a form of collective imagination made possible by a gigantic computer, the Matrix. In fact, Neo and other people are nothing more than power supplies housed in liquid-filled containers. They supply energy for the Matrix. The Matrix, in turn, supplies these "batteries" with images—making them feel they are living, not merely dreaming.

The Matrix is an exciting action-adventure film with extraordinary special effects. Beyond the glitz, however, there is much to ponder in the movie. It depicts a world where information technology and the *representation* of reality have taken over the seemingly stable reality of the physical world. Is this what *our* world is becoming? *The Matrix* also poses a classic sociological question about technology. Is technology always a means of improving human life? Or is it sometimes antagonistic to human values?

Acheson brought Oppenheimer to meet President Truman in 1946, Oppenheimer said, "Mr. President, I have blood on my hands." Truman later told Acheson, "don't bring that fellow around again" (quoted in Parshall, 1998).

Overall, Americans value science and technology highly. By a wide margin, the United States is the world leader in scientific research, publications, and elite achievements (see Figure 18.1). In 1998, 59% of Americans agreed that science and technology do more good than harm. Only 18% thought they do more harm than good. The remaining 26% were neutral on the subject (National Opinion Research Center, 1999). However, in the postwar years, a growing number of people, including Nobel prize winners who worked on the bomb, have come to share Oppenheimer's doubts (Feynman, 1999: 9–10). Indeed, they have extended those doubts not just to the peaceful use of nuclear energy but also to technology in general. Increasingly, ordinary citizens—and a growing chorus of leading scientists—are beginning to think of technology as a monster run amok, a Frankenstein rather than a savior (see Box 18.1; Joy, 2000; Kurzweil, 1999: 137–42).

It was only in the 1970s that a series of horrific disasters woke many people (including some sociologists) up to the fact that technological advance is not always beneficial, not even always benign. The most infamous technological disasters of the 1970s and 1980s include the following:

◆ An outbreak of "Legionnaires Disease" in a Philadelphia hotel in 1976 killed 34 people. It alerted the public to the possibility that the very buildings they live and work in can harbor toxic chemicals, lethal molds, and dangerous germs.

◆ In 1977, dangerously high levels of toxic chemicals were discovered leaking into the basements and drinking water of the residents of Love Canal, near Niagara Falls.

The Three-Mile Island nuclear facility.

This led to the immediate shutdown of an elementary school and the evacuation of residents from their homes.

◆ The partial meltdown of the reactor core at the Three Mile Island nuclear facility in Pennsylvania in 1979 caused lethal radioactive water and gas to pour into the environment. (A 1974 report by the Atomic Energy Commission said such an accident would likely occur only once in 17,000 years.)

◆ A gas leak at a poorly maintained Union Carbide pesticide plant in Bhopal, India, killed about 4,000 people in 1984 and injured 30,000, a third of whom died excruciating deaths in the following years.

◆ In 1986, the No. 4 reactor at Chernobyl, Ukraine, exploded, releasing 30 to 40 times the radioactivity of the blast at Hiroshima. It resulted in mass evacuations, more than 10,000 deaths, countless human and animal mutations, and hundreds of square miles of unusable cropland.

◆ In 1989, the Exxon Valdez ran aground in Prince William Sound, Alaska, spilling 11 million gallons of crude oil, producing a dangerous slick more than 1,000 miles long, causing billions of dollars of damage, and killing hundreds of thousands of animals.

By the mid-1980s, sociologist Charles Perrow was referring to events such as those listed above as **normal accidents.** The term "normal accident" recognizes that the very complexity of modern technologies ensures they will *inevitably* fail, though in unpredictable ways (Perrow, 1984). For example, a large computer program contains many thousands of conditional statements. They take the form: if x = y, do z; if a = b, do c. When in use, the program activates many billions of *combinations* of conditional statements. As a result, complex programs cannot be tested for all possible eventualities. Therefore, when rare combinations of conditions occur, they have unforeseen consequences that are usually minor, occasionally amusing, sometimes expensive, and too often dangerous. You experience normal accidents when your home computer "crashes" or "hangs." A few years ago, the avionics software for the F-16 jet fighter caused the jet to flip upside down whenever it crossed the equator. In January 1990, AT&T's entire long distance network was crippled for 9 hours due to a bug in the software for its routing switches. In Perrow's sense of the term, these are all normal accidents, although not as dangerous as the chemical and nuclear mishaps mentioned above.

A sea otter covered in oil spilled by the Exxon Valdez in 1989.

German sociologist Ulrich Beck also coined a term that stuck when he said we live in a **risk society.** A risk society is a society in which technology distributes danger among all categories of the population. Some categories, however, are more exposed to technological danger than others. Moreover, in a risk society, danger does not result from technological accidents alone. In addition, increased risk is due to mounting *environmental* threats. Environmental threats are more widespread, chronic, and ambiguous than technological accidents. They are therefore more stressful (Beck, 1992 [1986]; Freudenburg, 1997). New and frightening terms—"greenhouse effect," "global warming," "acid rain," "ozone depletion," "endangered species"—have entered our vocabulary. To many people, technology seems to be spinning out of control. From their point of view, it enables the production of ever more goods and services, but at the cost of breathable air, drinkable water, safe sunlight, plant and animal diversity, and normal weather patterns. In the same vein, Neil Postman (1992) refers to the United States as a **technopoly.** He argues that the United States is the first country in which technology has taken control of culture. Technology, he says, compels people to try to solve all problems using technical rather than moral criteria, although technology is often the source of the problems.

The latest concern of technological skeptics is biotechnology. Molecular biologists have mapped the entire human gene structure and are also mapping the gene structure of selected animals and plants. They can splice genes together, creating plants and animals with entirely new characteristics. As we will see, the ability to create new forms of life holds out incredible potential for advances in medicine, food production, and other fields. That is why the many advocates of this technology speak breathlessly of a "second genesis" and "the perfection of the human species." Detractors claim that, without moral and political decisions based on a firm sociological understanding of who benefits and who suffers from these new techniques, the application of biotechnology may be a greater threat to our well-being than any other technology ever developed.

These considerations suggest five tough questions. We tackle each of them below. First, is technology *the* great driving force of historical and social change? This is the opinion of cheerleaders and naysayers, those who view technology as our savior and those who fear it as a Frankenstein. In contrast, we argue that technology is able to transform society only when it is coupled with a powerful social need. People control technology as much as technology transforms people. Second, if some people do control technology, then exactly who are they? We argue against the view that scientific and engineering wizards are in control. The military and big corporations now decide the direction of most technological research and its application. Third, what are the most dangerous spin-offs of technology and how is risk distributed among various social groups? We focus on global warming, industrial pollution, the decline of biodiversity, and genetic pollution. We show that while these dangers put all of humanity at risk, the degree of danger varies by class, race, and country. In brief, the socially and economically disadvantaged are most at risk. Fourth, how can we overcome the dangers of environmental degradation? We argue that market and technological solutions are insufficient by themselves. In addition, much self-sacrifice and cooperation will be required. The fifth and final question underlies all the others. It is the question with which we began this book (p. 25): Why sociology?

Technology *and* People Make History

Russian economist Nikolai Kondratiev was the first social scientist to notice that technologies are invented in clusters. As Table 18.1 shows, a new group of major inventions has cropped up every 40–60 years since the Industrial Revolution. Kondratiev argued that these flurries of creativity cause major economic growth spurts beginning 10–20 years later and lasting 25–35 years each. Thus, Kondratiev subscribed to a form of **technological determinism,** the belief that technology is the major force shaping human society and history (Ellul, 1964 [1954]).

Is it true that technology helps shape society and history? Of course it is. James Watt developed the steam engine in Britain in the 1760s. It was the main driving force in the

◆ **TABLE 18.1** ◆
"Kondratiev Waves" of Modern Technological Innovation and Economic Growth

SOURCE: Adapted from Pacey (1983: 32).

Wave	Invention Dates	New Technologies	Base	Economic Growth Spurt
1	1760s–70s	Steam engine, textile manufacturing, chemistry, civil engineering	Britain	1780–1815
2	1820s	Railways, mechanical engineering	Britain, Continental Western Europe	1840–70
3	1870s–80s	Chemistry, electricity, internal combustion engine	Germany, United States	1890–1914
4	1930s–40s	Electronics, aerospace, chemistry	United States	1945–70
5	1970s	Microelectronics, biotechnology	United States, Japan	1985–?

mines, mills, factories, and railways of the Industrial Revolution. Gottlieb Daimler invented the internal combustion engine in Germany in 1883. It was the foundation stone of two of the world's biggest industries, automobiles and petroleum. John Atanasoff was among the first people to invent the computer in 1939 at Iowa State College (now University). It utterly transformed the way we work, study, and entertain ourselves. It also put the spurs to one of the most sustained economic booms ever. We could easily cite many more examples of how technology shapes history and transforms society.

However, if we probe a little deeper into the development of any of the technologies mentioned above, we notice a pattern: They did not become engines of economic growth until *social* conditions allowed them to do so. The original steam engine, for instance, was invented by Hero of Alexandria in the first century CE. He used it as an amusing way of opening a door. People then promptly forgot the steam engine. Some 1,700 years later, when the Industrial Revolution began, factories were first set up near rivers and streams, where waterpower was available. That was several years before Watt patented his steam engine. Watt's invention was all the rage once its potential became evident. But it did not cause the Industrial Revolution, and it was adopted on a wide scale only after the social need for it emerged (Pool, 1997: 126–7).

Similarly, Daimler's internal combustion engine became the basis of the automobile and petroleum industries thanks to changes in the social organization of work wrought by Henry Ford, the self-defeating business practice of Ford's main competitors, the Stanley brothers, and, oddly enough, an epidemic of hoof-and-mouth disease. When Ford incorporated his company in 1903, a steam-driven automobile, the Stanley Steamer, was his main competition. Many engineers then believed the Stanley Steamer was the superior vehicle on purely technical grounds. Many engineers still think so today. (For one thing, the Stanley Steamer didn't require a transmission system.) But while the Stanley brothers built a finely tooled automobile for the well to do, Ford tried to figure out a way of producing a cheap car for the masses. His inspiration was the meat-packing plants of Cincinnati and Chicago. In 1913, he modeled the first car assembly line after those plants. Only then did he open a decisive lead in sales over the Stanleys. The Stanleys were finally done in a few years later. An outbreak of hoof-and-mouth disease led officials to close down the public watering troughs for horses that were widely used in American cities. Owners of the Stanley Steamer used the troughs to replenish its water supply. So we see it would be wrong to say, along with strict technological determinists, that Daimler's internal combustion engine *caused* the growth of the car industry and then the petroleum industry. The car and petroleum industries grew out of the internal combustion engine only because an ingenious entrepreneur efficiently organized work in a new way and because a chance event undermined access to a key element required by his competitor's product (Pool, 1997: 153–5).

ORDVAC, an early computer developed at the University of Illinois, was delivered to the Ballistic Research Laboratory at the Aberdeen Proving Ground of the United States Army. Technology typically advances when it is coupled to an urgent social need.

Regarding the computer, Atanasoff stopped work on it soon after the outbreak of World War II. However, once the military potential of the computer became evident, its development resumed. The British computer Colossus helped to decipher secret German codes in the last 2 years of the war and played an important role in the Allied victory. The University of Illinois delivered one of the earliest computers, the ORDVAC, to the Ballistic Research Laboratory at the Aberdeen Proving Ground of the United States Army. Again we see how a new technology becomes a major force in society and history only after it is coupled with an urgent social need. We conclude that technology and society influence each other. Scientific discoveries, once adopted on a wide scale, often transform societies. But scientific discoveries are turned into useful technologies only when social need demands it.

How High Tech Became Big Tech

Enjoying a technological advantage usually translates into big profits for businesses and military superiority for countries. In the 19th century, gaining technological advantage was still inexpensive. It took only modest capital investment, a little knowledge about the best way to organize work, and a handful of highly trained workers to build a shop to manufacture stirrups or even steam engines. In contrast, mass-producing cars, sending a man to the moon, and other feats of 20th- and 21st-century technology require enormous capital investment, detailed attention to the way work is organized, and legions of technical experts. Add to this the intensely competitive business and geopolitical environment of the 20th and 21st centuries, and one can readily understand why ever larger sums have been invested in research and development over the past hundred years.

It was in fact already clear in the last quarter of the 19th century that turning scientific principles into technological innovations was going to require not just genius but substantial resources, especially money and organization. Thus, Thomas Edison established the first "invention factory" at Menlo Park, New Jersey, in the late 1870s. Historian of science Robert Pool notes:

> [T]he most important factor in Edison's success—outside of his genius for invention—was the organization he had set up to assist him. By 1878, Edison had assembled at Menlo Park a staff of thirty scientists, metalworkers, glassblowers, draftsmen, and others working under his close direction and supervision. With such support, Edison boasted that he could turn out "a minor invention every ten days and a big thing every six months or so" (Pool, 1997: 22).

The phonograph and the electric light bulb were two such "big things." Edison inspired both. Both, however, were also expensive team efforts, motivated by vast

✦ **FIGURE 18.2** ✦
Research and Development, United States, 1960 and 1997, by Source (in percent)

SOURCE: United States Bureau of the Census (1998c); Woodrow Federal Reserve Bank of Minneapolis (2000).

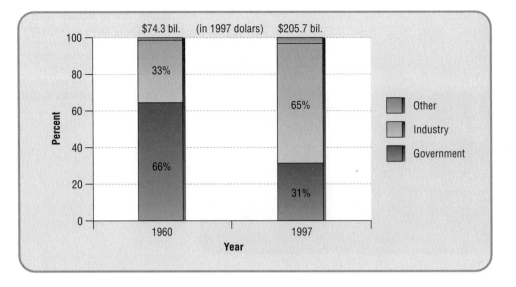

commercial possibilities. (Edison founded General Electric, the most profitable company in the world in 1999 and the second most valuable based on market capitalization; see "Global 1000," 1999.)

By the beginning of the 20th century, the scientific or engineering genius operating in isolation was only rarely able to contribute much to technological innovation. By mid-century, most technological innovation was organized along industrial lines. Entire armies of experts and vast sums of capital were required to run the new invention factories. The prototype of today's invention factory was the Manhattan Project, which built the nuclear bomb in the last years of World War II. By the time of Hiroshima, the manufacturing complex of the United States nuclear industry was about the same size as that of the United States automobile industry. The era of big science and big technology had arrived. Only governments and, increasingly, giant multinational corporations could afford to sustain the research effort of the second half of the 20th century.

As the 20th century ended, there seemed to be no upper limit to the amount that could be spent on research and development. The United States had fewer than 10,000 research scientists before World War I. Today, it has more than a million (Hobsbawm, 1994: 523). In 1997, American research and development spending reached $205.7 billion, up from $74.3 billion in 1960 (calculated in 1997 dollars to take account of inflation). During that same period, industry's share of spending rose from 33% to 65% of the total, while government's dropped from 66% to 31% (see Figure 18.2).

Because large multinational corporations now routinely invest astronomical sums in research and development to increase their chance of being the first to bring innovations to market, the time lag between new scientific discoveries and their technological application is continuously shrinking. That is clear from Figure 18.3, which shows how long it took five of the most popular new consumer products of the 1980s and 1990s to penetrate the United States market. It was fully 38 years after the VCR was invented in 1952 before the device achieved 25% market penetration. It took 18 years before the personal computer, invented in 1975, was owned by 25% of Americans. The World Wide Web, invented in 1991, took only 7 years to reach that level of market penetration.

Because of these developments, it should come as no surprise that military and profit-making considerations now govern the direction of most research and development. A reporter once asked a bank robber why he robs banks. The robber answered: "Because that's where the money is." This is hardly the only motivation prompting scientists and engineers to research particular topics. Personal interests, individual creativity, and the state of a field's intellectual development still influence the direction of inquiry. This is especially true for theoretical work done in colleges, as opposed to applied research funded by governments and private industry. It would, however, be naive to

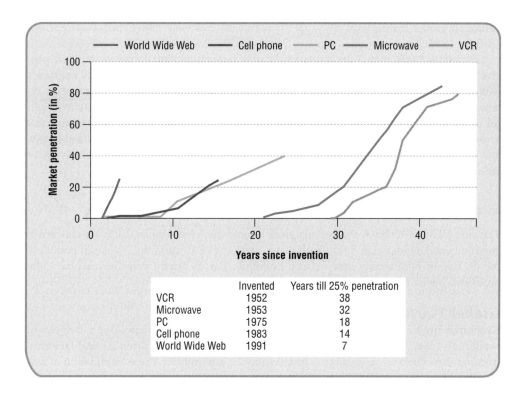

✦ FIGURE 18.3 ✦
Market Penetration by Years Since Invention, United States

SOURCE: "The Silent Boom" (1998).

think that practicality doesn't also enter the scientist's calculation of what he or she ought to study. Even in a more innocent era, Sir Isaac Newton studied astronomy partly because the explorers and mariners of his day needed better navigational cues. Similarly, Michael Faraday was motivated to discover the relationship between electricity and magnetism partly by his society's search for new forms of power (Bronowski, 1965 [1956]: 7–8). The connection between practicality and research is even more evident today. Many researchers—even many of those who do theoretically driven research in colleges—are pulled in particular directions by large research grants, well-paying jobs, access to expensive state-of-the-art equipment, and the possibility of winning patents and achieving commercial success. For example, many leading molecular biologists in the United States today have established genetic engineering companies, serve on their boards of directors, or receive research funding from them. In not a few cases, major pharmaceutical and agrochemical corporations have bought out these companies because they see their vast profit potential (Rural Advancement Foundation International, 1999). Even in the late 1980s, nearly 40% of the biotechnology scientists who

Research in biotechnology is big business. Even in the late 1980s, nearly 40% of the biotechnology scientists who belonged to the prestigious National Academy of Sciences had industry affiliations.

Due to global warming, glaciers are melting, the sea level is rising, and extreme weather events are becoming more frequent.

belonged to the prestigious National Academy of Sciences had industry affiliations (Rifkin, 1998: 56).

Economic lures, increasingly provided by the military and big corporations, have generated moral and political qualms among some researchers. Some scientists and engineers wonder whether work on particular topics achieves optimum benefits for humanity. Certain researchers are troubled by the possibility that some scientific inquiries may be harmful to humankind. However, a growing number of scientists and engineers recognize that to do cutting-edge research they must still any residual misgivings, hop on the bandwagon, and adhere to military and industrial requirements and priorities. That, after all, is where the money is.

Environmental Degradation

The side effect of technology that has given people the most serious cause for concern is environmental degradation. It has four main aspects: global warming, industrial pollution, the decline of biodiversity, and genetic pollution. Let us briefly consider each of these problems, beginning with global warming.

Global Warming

Ever since the Industrial Revolution, humans have been burning increasing quantities of fossil fuels (coal, oil, gasoline, natural gas, etc.) to drive their cars, furnaces, and factories. Burning these fuels releases carbon dioxide into the atmosphere. The accumulation of carbon dioxide allows more solar radiation to enter the atmosphere and less heat to escape. This is the so-called **greenhouse effect.** Most scientists believe that the greenhouse effect contributes to **global warming,** a gradual increase in the world's average surface temperature. Using data from NASA's Goddard Institute for Space Studies, Figure 18.4 graphs the world's annual average surface air temperature from 1866 to 2000 and the concentration of carbon dioxide in the atmosphere from 1866 to 1998. The graph shows a warming trend that mirrors the increased concentration of carbon dioxide in the atmosphere. It also shows that the warming trend intensified sharply in the last third of the 20th century. Between 1866 and 1965, average surface air temperature rose at a rate of 0.25 degree Celsius per century. From 1966 to 2000, average surface air temperature rose at a rate of 1.29 degrees Celsius per century.

Many scientists believe global warming is already producing serious climatic change. For as temperatures rise, more water evaporates. This causes more rainfall and bigger storms, which leads to more flooding and soil erosion, which in turn leads to less cultivable

✦ **FIGURE 18.4** ✦
Annual Mean Global Surface Air Temperature and Carbon Dioxide Concentration, 1866–2000

SOURCES: Goddard Institute for Space Studies (2001); Karl and Trenberth (1999: 102).

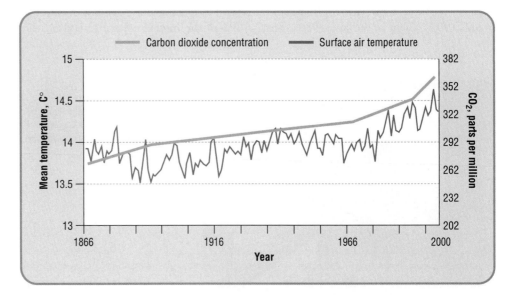

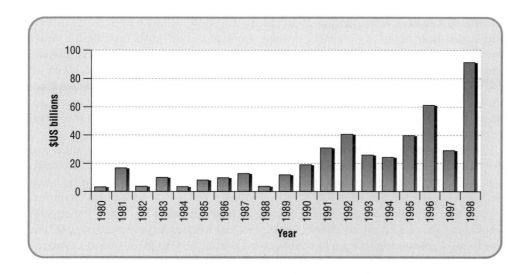

✦ **FIGURE 18.5** ✦
Worldwide Damage Due to "Natural" Disasters, 1980–1998 (in 1998 U.S. Dollars)

SOURCES: Abu-Nasr (1998); Vidal (1999).

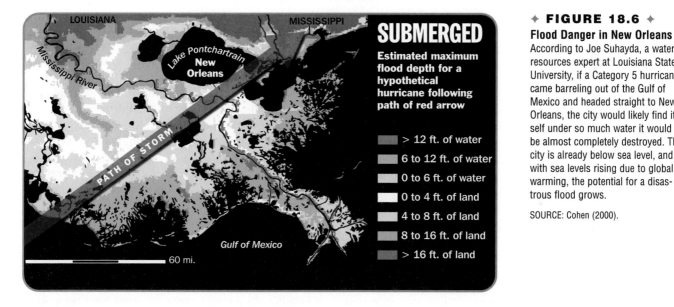

✦ **FIGURE 18.6** ✦
Flood Danger in New Orleans
According to Joe Suhayda, a water resources expert at Louisiana State University, if a Category 5 hurricane came barreling out of the Gulf of Mexico and headed straight to New Orleans, the city would likely find itself under so much water it would be almost completely destroyed. The city is already below sea level, and with sea levels rising due to global warming, the potential for a disastrous flood grows.

SOURCE: Cohen (2000).

land. People suffer and die all along the causal chain. This was tragically evident in 1998, when Hurricane Mitch caused entire mountainsides to collapse on poor villages in Guatemala and Honduras, killing thousands of inhabitants and ruining the fertile banana plantations of those countries.

Figure 18.5 graphs the worldwide dollar cost of damage due to "natural" disasters from 1980 to 1998. ("Natural" is in quotation marks because, as we have just seen, an increasingly large number of meteorological events are rendered extreme by human action.) Clearly, the damage caused by extreme meteorological events was on the upswing throughout the 1990s. This, however, may be only the beginning. It seems that global warming is causing the oceans to rise. That is partly because warmer water expands and partly because the partial melting of the polar ice caps puts more water in the oceans. In the 21st century, this may result in the flooding of some heavily populated coastal regions throughout the world. Just a 1-yard rise in the sea level would flood about 12% of the surface area of Egypt and Bangladesh and 0.5% of the surface area of the United States (Kennedy, 1993: 110; see Figure 18.6).

Industrial Pollution

Industrial pollution is the emission of various impurities into the air, water, and soil due to industrial processes. It is a second major form of environmental degradation. Every day, we release a witch's brew into the environment. The more common ingredients include household trash, scrap automobiles, residue from processed ores, agricultural runoff containing dangerous chemicals, lead, carbon monoxide, carbon dioxide, sulfur dioxide, ozone, nitrogen oxide, various volatile organic compounds, chlorofluorocarbons (CFCs), and various solids mixed with liquid droplets floating in the air. Most pollutants are especially highly concentrated in the United States Northeast and around the Great Lakes. Old, heavy, dirty industries are centered in these densely populated areas (United States Environmental Protection Agency, 2000).

Pollutants may affect us directly. For example, they seep into our drinking water and the air we breathe, causing a variety of ailments, particularly among the young, the elderly, and the ill. A dramatic natural experiment demonstrating the direct effect of air pollution on health occurred during the 1996 Atlanta Olympics. For the 17 days of the Olympics, asthma attacks among children in the Atlanta area plummeted 42%. When the athletes went home, the rate of asthma attacks among children immediately bounced back to normal levels. Epidemiologists soon figured out why. During the Olympics, Atlanta closed the downtown to cars and operated public transit around the clock. Vehicle exhaust fell, with an immediate benefit to children's health. Children's health deteriorated as soon as normal traffic resumed (Mittelstaedt, 2001).

Pollutants may also affect us indirectly. For instance, sulfur dioxide and other gases are emitted by coal-burning power plants, pulp and paper mills, and motor-vehicle exhaust. They form **acid rain.** This is a form of precipitation whose acidity eats away at, and eventually destroys, forests and the ecosystems of lakes. Another example: CFCs are widely used in industry and by consumers, notably in refrigeration equipment. They contain chlorine, which is responsible for the depletion of the **ozone layer** 5–25 miles above the earth's surface. Ozone is a form of oxygen that blocks ultraviolet radiation from the sun. Let more ultraviolet radiation reach ground level and, as we are now witnessing, rates of skin cancer and crop damage increase.

Radioactive waste deserves special attention. About 100 nuclear reactors are now generating commercial electricity in the United States. They run on enriched uranium or plutonium fuel rods. Once these fuel rods decay beyond the point where they are useful in the reactor, they become waste material. This waste is highly radioactive. It must decay about 10,000 years before humans can be safely exposed to it without special protective equipment. The spent fuel rods need to be placed in sturdy, watertight copper canisters and buried deep in granite bedrock where the chance of seismic disturbance and water seepage is small. The trouble is, most Americans are petrified at the prospect of having a nuclear waste facility anywhere near their families. As a result, spent fuel rods have been accumulating since the 1950s in "temporary" facilities. These are mainly pools of water near nuclear reactors. These facilities are a safety threat the American public has not really begun to deal with yet (Pool, 1997).

The Destruction of Biodiversity

The third main form of environmental degradation is the decline in **biodiversity,** the enormous variety of plant and animal species inhabiting the earth. Biodiversity changes as new species emerge and old species die off because they cannot adapt to their environment. This is all part of the normal evolutionary process. However, in recent decades the environment has become so inhospitable to so many species that the rate of extinction has greatly accelerated. Examination of fossil records suggests that, for millions of years, an average of one to three species became extinct annually. Today, about 1,000 species are becoming extinct annually (Tuxill and Bright, 1998: 41). In 11 countries, 10% or more of bird species are threatened with extinction. In 29 countries, 10% or more of mammal species are similarly threatened (Kidron and Segal, 1995: 14–15).

The extinction of species is impoverishing in itself, but it also has practical consequences for humans. For example, each species of animal and plant has unique properties. When scientists discover that a certain property has a medically useful effect, they get busy trying to synthesize the property in the laboratory. Treatments for everything from headaches to cancer have been found in this way. Indeed, about a quarter of all drugs prescribed in the United States today (including 9 of the top 10 in sales) include compounds first found in wild organisms. The single richest source of genetic material with pharmaceutical value is found in the world's rain forests, particularly in Brazil, where more than 30 million species of life exist. However, the rain forests are being rapidly destroyed by strip mining, the construction of huge pulp and paper mills and hydroelectric projects, and the deforestation of land by farmers and cattle grazers.

Similarly, fleets of trawlers belonging to the highly industrialized countries are now equipped with sonar to help them find large concentrations of fish. Some of these ships use fine mesh nets to increase their catch. They have been enormously "successful." Trawlers have depleted fish stocks in some areas of the world. In North America, for example, the depletion of cod, salmon, blue-fin tuna, and shark stocks has devastated fishing communities and endangered one of the world's most important sources of protein. All told, 11 of the world's 15 main fishing grounds and 69% of the world's main fish species are in decline (McGinn, 1998: 60).

Genetic Pollution

Genetic pollution is the fourth main form of environmental degradation. It refers to the health and ecological dangers that may result from artificially splicing genes together (Rifkin, 1998).

The genetic information of all living things is coded in a chemical called DNA. When members of a species reproduce, the characteristics of the mates are naturally transmitted to their offspring through DNA. **Recombinant DNA,** in contrast, is a technique developed by molecular biologists in the last few decades. It involves artificially joining bits of DNA from a donor to the DNA of a host. Donor and host may be of the same or different species. The donor DNA grows along with the host DNA, in effect creating a new form of life. For example, scientists inserted the gene that makes fireflies sparkle at night into a tobacco plant. The offspring of the plant had leaves that glowed in the dark. Researchers inserted human growth hormone into a mouse embryo. This created mice that grew twice as big and twice as fast as ordinary mice. Biologists combined embryo cells from a sheep and a goat and placed them in the womb of a surrogate animal. The surrogate animal then gave birth to an entirely new species, half sheep, half goat.

These wonders of molecular biology were performed in the mid-1980s and helped to dramatize and publicize the potential of recombinant DNA. Since 1990, governments and corporations have been engaged in a multibillion-dollar international effort to create a

Artificially splicing genes together may yield benefits as well as dangers. Woody Allen in *Sleeper* (1973).

complete genetic map of humans and various plants, microorganisms, and animal species. With human and other genetic maps in hand, and using recombinant DNA and related techniques, it is possible to design what some people regard as more useful animals and plants and superior humans. By 2000, scientists had identified the location and chemical structure of every one of the approximately 40,000 human genes. This will presumably enable them to understand the function of each gene. They can then detect and eliminate hereditary propensities to a wide range of diseases. Recombinant DNA will also enable farmers to grow disease- and frost-resistant crops with higher yields. It will allow miners to pour ore-eating microbes into mines, pump the microbes aboveground after they have had their fill, and then separate out the ore. This will greatly reduce the cost and danger of mining. Recombinant DNA will allow companies to grow plants that produce cheap biodegradable plastic and microorganisms that consume oil spills and absorb radioactivity. The potential health and economic benefits to humankind of these and many other applications of recombinant DNA are truly startling.

So are the dangers genetic pollution poses to human health and the stability of ecosystems (Rifkin, 1998: 67–115; Tokar, 2001). Consider, for example, the work of scientists at the National Institute of Allergy and Infectious Diseases. In the late 1980s, they introduced the genetic instructions for the human AIDS virus into mouse embryos. Subsequent generations of mice were born with AIDS and were used for research to find a cure for the disease. But what would happen if some of those mice got loose and bred with ordinary mice? In 1990, Dr. Robert Gallo, codiscoverer of the AIDS virus, and a team of other scientists reported in the respected journal *Science* that the AIDS virus carried by the mice could combine with other mouse viruses. This could result in a new form of AIDS capable of reproducing more rapidly and being transmitted to humans through the air. Recognizing this danger, scientists housed the AIDS mice in stainless steel glove boxes surrounded by a moat of bleach. They enclosed the entire apparatus in the highest level biosafety facility that exists. No mice have escaped so far, but the risk is still there.

Meanwhile, humans are already the recipients of transplanted bone marrow and hearts from baboons and pigs. While the animals are screened for known problems, critics point out such transplants could enable dangerous unknown viruses and retroviruses to jump between species and cause an epidemic among humans. If this seems farfetched, remember that the AIDS virus is widely believed to have jumped between a chimpanzee and a human in the late 1930s. By the end of 1999, the AIDS virus had killed about 14.5 million people worldwide and infected more than 34 million others (United Nations, 2000). Ominously, in 1997 scientists discovered a previously unknown pig virus that can infect humans. And in 2000, scientists reported that at least three known pig retroviruses could infect human cells (Van der Laan, Lockey, Griffeth, Frasier, Wilson, Onions, Hering, Long, Otto, Torbett, and Salomon, 2000).

Genetic pollution may also affect the stability of ecosystems. When a nonnative organism enters a new environment, it usually adapts without a problem. Sometimes, however, it unexpectedly wreaks havoc. Kudzu vine, Dutch elm disease, the gypsy moth, chestnut blight, starlings, Mediterranean fruit flies, zebra mussels, rabbits, and mongooses have all done just that. Now, however, the potential for ecological catastrophe has multiplied. That is because scientists are regularly testing genetically altered plants (effectively, nonnative organisms) in the field. Some have gone commercial, and many more will soon be grown on a wide scale. These plants are resistant to insects, disease, and frost. However, once their pollen and seeds escape into the environment, weeds, insects, and microorganisms will eventually build up resistance to the genes that resist herbicides, pests, and viruses. Thus, superbugs, superweeds, and superviruses will be born. We cannot predict the exact environmental consequences of these developments. However, the insurance industry refuses to insure genetically engineered crops against the possibility of their causing catastrophic ecological damage.

Global warming, industrial pollution, the decline of biodiversity, and genetic pollution threaten everyone. However, as you will now see, the degree to which they are perceived as threatening depends on certain social conditions being met. Moreover, the threats are not evenly distributed in society.

THE SOCIAL CONSTRUCTION OF ENVIRONMENTAL PROBLEMS

Environmental problems do not become social issues spontaneously. Before they can enter the public consciousness, policy-oriented scientists, the environmental movement, the mass media, and respected organizations must discover and promote them. People have to connect real-life events to the information learned from these groups. Because some scientists, industrial interests, and politicians dispute the existence of environmental threats, the public can begin to question whether environmental issues are in fact social problems that require human intervention. We must not, then, think of environmental issues as inherently problematic. Rather, they are contested phenomena. They can be socially constructed by proponents. They can be socially demolished by opponents. This is the key insight of the school of thought known as **social constructionism** (Hannigan, 1995b).

The controversy over global warming is a good example of how people create and contest definitions of environmental problems (Gelbspan, 1997; 1999; Hart and Victor, 1993; Mazur, 1998; Ungar, 1992; 1995; 1998; 1999). The theory of global warming was first proposed about a century ago. However, an elite group of scientists began serious research on the subject only in the late 1950s. They attracted no public attention until the 1970s. That is when the environmental movement emerged. The environmental movement gave new legitimacy and momentum to the scientific research and helped to secure public funds for it. Respected and influential scientists now began to promote the issue of global warming. The mass media, always thirsting for sensational stories, were highly receptive to these efforts. Newspaper and television reports about the problem began to appear in the late 1970s. They proliferated in the mid- to late 1980s. Between 1988 and 1991, the public's interest in global warming reached an all-time high. That was because frightening events helped to make the media reports more believable. For example, the summer of 1988 brought the worst drought in half a century. As crops failed, New York sweltered, and huge fires burned in Yellowstone National Park, *Time* magazine ran a cover story entitled "The Big Dry." It drew the connection between global warming and extreme weather. Many people got worried. Soon, respected organizations outside the scientific community, the mass media, and environmental movement—such as the insurance industry and the United Nations—expressed concern about the effects of global warming. By 1994, 59% of Americans with an opinion on the subject thought that using coal, oil, and gas contributes to the greenhouse effect (calculated from National Opinion Research Center, 1999).

By 1994, however, public concern with global warming had already passed its peak. The eruption of Mount Pinatubo in the Philippines pumped so much volcanic ash into the atmosphere, clouding the sunshine, that global surface air temperatures fell in 1992–93. Media reports about global warming sharply declined. The media, always thirsting for new scares to capture larger audiences, thought the story had grown stale. Some scientists, industrialists, and politicians began to question whether global warming was in fact taking place. They cited satellite data showing the earth's lower atmosphere had cooled in recent decades. They published articles and took out ads to express their opinion, thus increasing public skepticism.

With surface temperatures showing warming and lower atmospheric temperatures showing cooling, different groups lined up on different sides of the global warming debate. Those who had most to lose from carbon emission cuts emphasized the lower atmospheric data. This group included Western coal and oil companies, the member states of the Organization of Petroleum Exporting Countries (OPEC), and other coal- and oil-exporting nations. Those who had most to lose from the consequences of global warming or least to lose from carbon emission cuts emphasized the surface data. This group included insurance companies, an alliance of small island states, the European Union, and the United Nations. In the United States, the division was sufficient to prevent the government from acting. The Clinton–Gore administration pushed for a modest 7% cut in carbon emissions between 1990 and 2012. But the Republican-controlled Congress blocked the proposal. As a result, the United States is now the only industrialized country that has failed to legislate cuts in carbon emissions.

This could change in the near future. In August 1998, the global warming skeptics were dealt a serious blow when their satellite data were shown to be misleading. Until then, no one had taken into account that the satellites were gradually slipping from their orbits due to atmospheric friction, thus causing imprecise temperature readings. Allowing for the slippage, scientists from NASA and private industry now calculate that temperatures in the lower atmosphere are rising, just like temperatures on the earth's surface (Wentz and Schabel, 1998; Hansen, Sato, Ruedy, Lacis, and Glascoe, 1998). These new findings may finally help lay to rest the claims of the global warming skeptics. However, one thing is certain. As the social constructionists suggest, the power of competing interests to get their definition of reality accepted as the truth will continue to influence public perceptions of the seriousness of global warming.

In addition to being socially defined, environmental problems are socially distributed. That is, environmental risks are greater for some groups than others. Let us now examine this issue.

The Social Distribution of Risk

You may have noticed that after a minor twister touches down on some unlucky community in Texas or Kansas, TV reporters often rush to interview the surviving residents of trailer parks. The survivors stand amid the rubble that was their lives. They heroically remark on the generosity of their neighbors, their good fortune in still having their family intact, and our inability to fight nature's destructive forces. Why trailer parks? Small twisters aren't particularly attracted to them, but reporters are. That is because trailers are pretty flimsy in the face of a small tornado. They often suffer a lot of damage from twisters. They therefore make a more sensational story than the minor damage typically inflicted on upper-middle-class homes with firmly shingled roofs and solid foundations. This is a general pattern. Whenever disaster strikes—from the sinking of the Titanic to the fury of Hurricane Mitch—economically and politically disadvantaged people almost always suffer most. That is because their circumstances render them most vulnerable.

In fact, the advantaged often consciously put the disadvantaged in harm's way to avoid risk themselves. For example, oil refineries, chemical plants, toxic dumps, garbage incinerators, and other environmentally dangerous installations are more likely to be built in poor communities with a high percentage of African Americans or Hispanic Americans than in more affluent, mainly white communities. That is because disadvantaged people are often too politically weak to oppose such facilities and some may even value the jobs they create. Thus, in a study conducted in the mid-1980s, the number and size of hazardous waste facilities were recorded for every ZIP code area in the United States. At a time when about 20% of Americans were of African or Hispanic origin, ZIP code areas lacking any such facilities had, on average, a 12% minority population. ZIP code areas with one such facility had about a 24% minority population on average. And ZIP code areas with more than one such facility or with one of the five largest landfills in the United States had on average a 38% minority population. The study concluded that three out of five African Americans and Hispanic Americans live in communities with uncontrolled toxic waste sites (Szasz and Meuser, 1997: 100; Stretesky and Hogan, 1998). Similarly, the 75-mile strip along the lower Mississippi River between New Orleans and Baton Rouge has been nicknamed "cancer alley" because the largely black population of the region suffers from unusually high rates of lung, stomach, pancreatic, and other cancers. The main reason? This small area is the source of fully one quarter of the petrochemicals produced in the country, containing more than 100 oil refineries and chemical plants (Bullard, 1994 [1990]). A final example: Some poor Native American reservations have been targeted as possible interim nuclear waste sites. That is partly because states have little jurisdiction over reservations, so the usual state protests against such projects are less likely to prove effective. In addition, the Goshute tribe in Utah and the Mescalero Apaches in New Mexico have expressed interest in the project because of the money it promises to bring into their reservations (Pool, 1997: 247–8). Here again we see the recurrent pattern of what

Petrochemical plants between New Orleans and Baton Rouge form what local residents call "cancer alley." Here, the Union Carbide plant in Taft, Louisiana.

some analysts call **environmental racism** (Bullard, 1994 [1990]). This is the tendency to heap environmental dangers on the disadvantaged, and especially on disadvantaged racial minorities.

What is true for disadvantaged classes and racial groups in the United States also holds for the world's less developed countries. The underprivileged face more environmental dangers than the privileged (Kennedy, 1993: 95–121). In North America, Western Europe, and Japan, population growth is low and falling. Industry and government are eliminating some of the worst excesses of industrialization. In contrast, world population will grow from about 6 to 7 billion between 2000 and 2010, and nearly all of that growth will be in the less developed countries of the Southern Hemisphere. Moreover, Mexico, Brazil, China, India, and many other southern countries are industrializing rapidly. This is putting tremendous strain on their natural resources. Rising demand for water, electricity, fossil fuels, and consumer products is creating more polluted rivers, dead lakes, and industrial waste sites. At a quickening pace, rain forests, grazing land, cropland, and wetlands are giving way to factories, roads, airports, and housing complexes. Smog-blanketed megacities continue to sprawl. Eighteen of the world's 21 biggest cities are in less developed countries.

Given the picture sketched above, it should come as no surprise that, on average, people in less developed countries are more concerned about the environment than people in rich countries (Brechin and Kempton, 1994). However, the developing countries cannot afford much in the way of pollution control, so antipollution regulations are lax by North American, Western European, and Japanese standards. This is an incentive for some multinational corporations to site some of their most environmentally unfriendly operations in the Southern Hemisphere (Clapp, 1998). It is also the reason the industrialization of the less developed countries is proving so punishing to the environment. When car ownership grows from less than 1% to 10% of the population in China, and when 50 or 75 million Indians with motor scooters upgrade to cars, environmental damage may well be catastrophic. That is because the Chinese and the Indians simply cannot afford catalytic converters and electric cars. They have no regulations phasing in the use of these and other devices that save energy and pollute less.

For the time being, however, the rich countries do most of the world's environmental damage. That is because their inhabitants earn and consume more than the inhabitants of less developed countries. How much more? The richest fifth of humanity earns about 80 times more than the poorest fifth (up from 30 times more in 1950). In the past half century, the richest fifth doubled its per capita consumption of energy, meat, timber, steel, and copper and quadrupled its car ownership. In that same period, the per capita consumption of the poorest fifth hardly changed. The United States has only 4.5% of the world's population, but it uses about 25% of the earth's resources. It also produces more than 20% of global emissions of carbon dioxide, the pollutant responsible for about half of global warming (Ehrlich, Daily, Daily, Myers, and Salzman, 1997). Thus, the inhabitants of the Northern Hemisphere cause a disproportionately large share of the world's environmental problems, enjoy a disproportionate share of the benefits of technology, and live with fewer environmental risks than people in the Southern Hemisphere.

Social inequalities are also apparent in the field of biotechnology. For instance, the large multinational companies that dominate the pharmaceutical, seed, and agrochemical industries now routinely send anthropologists, biologists, and agronomists to all corners of the world. There they take samples of wild plants, the crops people grow, and human blood. They hope to find genetic material with commercial value in agriculture and medicine. If they discover genes with commercial value, the company they work for patents the discovery. This gives them the exclusive legal right to manufacture and sell the genetic material without compensating the donors. For example, Indian farmers and then scientists worked for a hundred generations discovering, skillfully selecting, cultivating, and developing techniques for processing the neem tree, which has powerful antibacterial and pesticidal properties. However, a giant corporation based in a rich country is now the sole commercial beneficiary of their labor. Monsanto (United States), Novartis (Switzerland), Glaxo Wellcome (United Kingdom), and other prominent companies in the life sciences call this "protection of intellectual property." Indigenous peoples and their advocates call it "biopiracy" (Rifkin, 1998: 37–66).

Finally, consider the possible consequences of people having their babies genetically engineered. This should be possible on a wide scale in 10 or 20 years. Free of inherited diseases and physical abnormalities, and perhaps genetically programmed to enjoy superior intellectual and athletic potential, these children would, in effect, speed up and improve the slow and imperfect process of natural evolution. That, at least, is the rosy picture sketched by proponents of the technology. In practice, because only the well to do are likely to be able to afford fully genetically engineered babies, the new technology could introduce an era of increased social inequality and low social mobility. Only the economically underprivileged would bear a substantial risk of genetic inferiority. This future was foreseen in the 1997 movie *Gattica*. The plot revolves around the tension between a society that genetically engineers all space pilots to perfection and a young man played by Ethan Hawke, who was born without the benefit of genetic engineering yet aspires to become a space pilot. Hawke's character manages to overcome his genetic handicap. It is clear from the movie, however, that his success is both illegal and extremely rare. The norm is rigid genetic stratification, and it is strongly sanctioned by state and society.

BOX 18.2
IT'S YOUR CHOICE

WEB-BASED LEARNING AND HIGHER EDUCATION

"I love it," says Carol Thibeault, a student at Central Connecticut State University. "With online classes, there's no set time that I have to show up. Sometimes I lug a heavy laptop onto the commuter bus and work on course files I've downloaded while I ride along. I even take my computer to the beach" (quoted in Maloney 1999: 19). Carol is not alone in expressing her enthusiasm for the new information technology in higher education. E-mail and the World Wide Web are now about as exotic as the telephone in the United States and other rich countries. Some scholars see "online education" as the future of higher education.

The advantages of Web-based education are many. Parents with children, or students with jobs, can learn at their own speed, on their own schedule, and in their own style. This will make learning easier and more enjoyable. Potentially, many students can be taught efficiently and effectively. This will lower the cost of higher education.

Although few would argue for its elimination, many people think too much dependence on the "virtual classroom" has drawbacks. Some professors argue it is difficult to control the quality of online educational materials and instruction. That is why dropout rates for distance education courses tend to be significantly higher than rates in conventional classrooms (Merisotis, 1999). Others suggest that distance education will spread primarily among low-cost, low-status institutions. At elite institutions, they say, classroom contact and discussion will become even more important. So while one group of students will enjoy a great deal of personal attention from faculty members, another group will receive only cursory and impersonal attention.

What do you think the role of Web-based learning should be in higher education? Do you think distance learning is superior or inferior to traditional classroom learning? Is it better to learn from a professor and other students in a "real" classroom as opposed to a "virtual" classroom? Do you think online education will lead to an increase in social inequality?

(For other examples of how new technologies can contribute to social inequality, see the discussion of job polarization in Chapter 10, electronic democracy in Chapter 11, and Box 18.2).

What Is to Be Done?

The Market and High-Tech Solutions

Some people believe the environmental crisis will resolve itself. More precisely, they think we already have two weapons that will work together to end the crisis: the market and high technology. The case of oil illustrates how these weapons can combine forces. If oil reserves drop or oil is withheld from the market for political reasons, the price of oil goes up. This makes it worthwhile for oil exploration companies to develop new technologies to recover more oil. When they discover more oil and bring it to market, prices fall back to where they were. This is what happened following the oil crises of 1973 (when prices tripled) and 1978–9 (when prices tripled again). Reserves are higher now than they were in the 1970s and 1980s, and, at the time of this writing, oil is relatively inexpensive again. Similarly, if too little rice and wheat are grown to meet world demand, the price of these grains goes up. This prompts agrochemical companies to invent higher yield grains. Farmers use the new grain seed to grow more wheat and rice, and prices eventually fall. This is what happened during the so-called "green revolution" of the 1960s. Projecting these experiences into the future, optimists believe global warming, industrial pollution, and other forms of environmental degradation will be dealt with similarly. In their view, human inventiveness and the profit motive will combine to create the new technologies we need to survive and prosper in the 21st century.

Some evidence supports this optimistic scenario. In recent years, we have adopted new technologies to combat some of the worst excesses of environmental degradation. For example, we have replaced brain-damaging leaded gas with unleaded gas. We have developed environmentally friendly refrigerants, allowing the production of ozone-destroying CFCs to plummet. In a model of international cooperation, rich countries have even subsidized the cost of replacing CFCs in the developing countries. Efficient windmills and solar panels are now common. More factories are equipped with high-tech pollution control devices, preventing dangerous chemicals from seeping into the air and water. We have introduced cost-effective ways to recycle metal, plastic, paper, and glass. New methods are being developed for eliminating carbon dioxide emissions from the burning of fossil fuels (Parson and Keith, 1998). In November 1999, Ford and General Motors took the wraps off their diesel–electric hybrid cars, five-passenger sedans that get as much as 108 miles to the gallon (see Figure 18.7). The widespread use of electric cars is perhaps only a decade

✦ **FIGURE 18.7** ✦
The Precept, General Motors' New Diesel–Electric Hybrid, Gets 108 Miles to the Gallon
General Motors Vice-Chairman Harry Pearce presents the Precept, a fuel-cell powered vehicle that gets 108 mpg and has a 500-mile range. High-tech inventions are one important part of the solution to the environmental crisis, but they are by no means sufficient.

SOURCE: General Motors (2000).

♦ **FIGURE 18.8** ♦

Air Pollutant Emission Projections, United States, 1990–2010 (in mil. short tons, projected)

SOURCE: Office of Air Quality Planning and Standards (1998: 5.4–5.8).

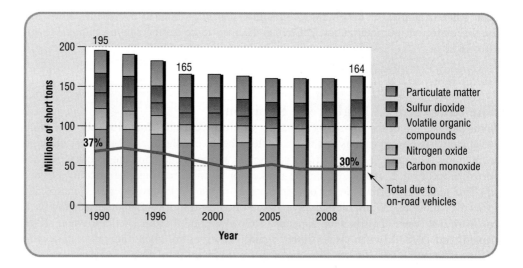

away. Figure 18.8 uses data from the United States Environmental Protection Agency to illustrate one consequence of these and related efforts. It shows actual production of five of the most common air pollutants in the United States between 1990 and 1998 and expected pollutant production between 1998 and 2010. Production of the five pollutants fell more than 15% between 1990 and 1999.

Clearly, market forces are helping to bring environmentally friendly technologies online. However, three factors suggest market forces cannot solve environmental problems on their own. First, price signals often operate imperfectly. Second, political pressure is often required to stimulate policy innovation. Third, markets and new technologies are not working quickly enough to deal adequately with the environmental crisis. Let us consider each of these issues in turn.

♦ *Imperfect price signals.* The price of many commodities does not reflect their actual cost to society. Gasoline in the United States costs about $1.40 a gallon on average at the time of this writing. But the *social* cost, including the cost of repairing the environmental damage caused by burning the gas, is $4 or more. In order to avoid popular unrest, the government of Mexico City charges consumers only about 10 cents a cubic meter for water. The actual cost to society is about 10 times that amount (Ehrlich, Daily, Daily, Myers, and Salzman, 1997). Due to these and many other price distortions, the market often fails to send signals that might result in the speedy adoption of technological and policy fixes.

♦ *Importance of political pressure.* Political pressure exerted by environmental social movement activists, community groups, and public opinion is often necessary to motivate corporate and government action on environmental issues. For instance, organizations like Greenpeace have successfully challenged the practices of logging companies, whalers, the nuclear industry, and other groups engaged in environmentally dangerous practices. Many less famous community associations have also played an important role in this regard (Brown, 1997). The antinuclear movement is an outstanding example of a movement that forced a substantial turnaround in government and corporate policy. For instance, in Germany, which obtains a third of its electricity from nuclear power, the antinuclear movement has had a major effect on public opinion, and in June 2000 the government decided to phase out all of the country's nuclear power plants within about 20 years. In the United States, no more nuclear power plants are planned. Again, the antinuclear movement must be credited with helping to change the public mood and bring about the halt in construction of new nuclear facilities.[1] All told, about 8% of Americans belong to groups committed

Web Research Projects
Who Are the Environmentalists?

---

[1] Recent statements by the Bush administration suggest this could change, however.

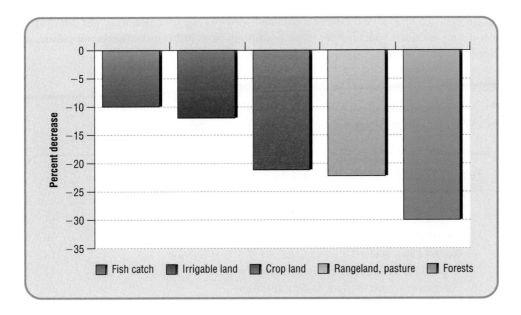

✦ FIGURE 18.9 ✦
Renewable Resources, World, Percent Change, 1990–2010 (projected)

SOURCE: Postel (1994:11).

to protecting the environment. About 10% have contributed money to such organizations (National Opinion Research Center, 1999). Without the political efforts of pro-environment individuals, organizations, and social movements, it is doubtful many environmental issues would be defined as social problems by corporations and governments.

✦ *Slow pace of change.* We saw above how price signals and new technologies have created pockets of environmental improvement, especially in the rich countries. However, it is unclear whether they can deal with the moral and political issues raised by biotechnology. Moreover, our efforts so far to clean up the planet are just not good enough. Returning to Figure 18.8, we observe that, after improving somewhat in the 1990s, United States air pollution is not expected to get any better between 1999 and 2010. Glancing back at Figure 18.4, we note that global warming continues to accelerate. Examining Figure 18.9, we see we can expect a substantial decrease in all of the world's renewable resources over the next decade. In 1993, 1,680 of the world's leading scientists, including 104 Nobel prize winners, signed the "World Scientists' Warning to Humanity." It stated: "A great change in our stewardship of the earth and the life on it is required, if vast human misery is to be avoided and our global home on this planet is not to be irretrievably mutilated . . . Human beings and the natural world are on a collision course" (Union of Concerned Scientists, 1993). Evidence suggests we still are.

The Cooperative Alternative

The alternative to the market and high-tech approach involves people cooperating to greatly reduce their overconsumption of just about everything. This strategy includes investing heavily in energy-saving technologies, environmental cleanup, and subsidized, environmentally friendly industrialization in the developing countries. It would require renewed commitment to voluntary efforts, new laws and enforcement bodies to ensure compliance, increased environmentally related research and development by industry and government, more environmentally directed foreign aid, and hefty new taxes to pay for everything (Livernash and Rodenburg, 1998). In addition, a cooperative strategy entails careful assessment of all the risks associated with biotechnology projects and consultation with the public before such projects are allowed to go forward. Profits from genetic engineering would also have to be shared equitably with donors of genetic material.

Is the solution realistic? Certainly not, at least not in the short term. In fact, it would probably be political suicide for anyone in the rich countries to propose the drastic measures listed above. Few drivers would be happy paying $4 a gallon for gas, for example. To be politically acceptable, three conditions have to be met. The broad public in North America, Western Europe, and Japan would have to be:

◆ Aware of the gravity of the environmental problem;

◆ Confident in the capacity of people and their governments to solve the problem; and

◆ Willing to make substantial economic sacrifices to get the job done.

◆ **TABLE 18.2** ◆

Public Opinion on Environmental Issues, United States, 1994 (in percent)

SOURCE: National Opinion Research Center (1999).

Proportion of Americans who think the following environmental problems are extremely/very/somewhat dangerous:	
1. Air pollution caused by cars	91
2. Air pollution caused by industry	94
3. Nuclear power stations	84
4. A rise in the world's temperature caused by the 'greenhouse effect'	82
5. Pollution of America's rivers, lakes, and streams	95
6. Pesticides and chemicals used in farming	84
7. *"It is just too difficult for someone like me to do much about the environment."*	
Strongly agree/agree	27
Neither agree nor disagree	17
Disagree/strongly disagree	56
8. *"Government should let businesses decide for themselves how to protect the environment, even if it means they don't always do the right thing, or government should pass laws to make businesses protect the environment, even if it interferes with business' right to make their own decisions."*	
Government should let businesses decide	11
Government should pass laws	89
9. *"We are faced with many problems in this country, none of which can be solved easily or inexpensively. I'm going to name some of these problems, and for each one I'd like you to tell me whether you think we're spending too much money on it, too little money, or about the right amount. Are we spending too much money, too little money, or about the right amount on improving and protecting the environment?"*	
Too little	61
About right	30
Too much	9
10. *"How willing would you be to pay much higher prices in order to protect the environment?"*	
Very/fairly willing	47
Neither willing nor unwilling	25
Not very/not at all willing	28
11. *"And how willing would you be to accept cuts in your standard of living in order to protect the environment?"*	
Very/fairly willing	32
Neither willing nor unwilling	23
Not very/not at all willing	45
12. *"And how willing would you be to pay much higher taxes in order to protect the environment?"*	
Very/fairly willing	34
Neither willing nor unwilling	21
Not very/not at all willing	45
13. *"How often do you make a special effort to sort glass or cans or plastic or papers and so on for recycling?"*	
Always/often/sometimes	87
Never	13
14. *"And how often do you cut back on driving a car for environmental reasons?"*	
Always/often/sometimes	36
Never	67

Data from the 1994 General Social Survey allow us to see whether these three conditions are being met in the United States. They paint a good news/bad news scenario. Nearly all Americans are aware of the environmental problem. As Table 18.2 shows, between 82% and 95% consider pollution and other environmental problems to be dangerous. Moreover, a solid majority (56%) think they can do something about environmental issues themselves, while a huge majority (89%) believe the government should pass more laws to protect the environment. Most Americans (61%) even say too little is being spent on environmental cleanup. All this is encouraging.

However, expressing environmental awareness and agreeing on the need for action is one thing. Biting the bullet is another. Fewer than half of Americans (47%) are willing to pay much higher prices to protect the environment. Fewer than a third (32%) are willing to accept cuts in their standard of living. Barely a third (34%) are willing to pay much higher taxes. Most Americans are prepared to protect the environment if it does not inconvenience them too much. Thus, 87% say they sort glass, cans, plastic, or paper for recycling. But when it is inconvenient, the numbers drop sharply. Only 36% say they have ever cut back on driving for environmental reasons.

Other surveys conducted in the United States and elsewhere reveal much the same pattern. Most people know about the environmental crisis. They want it dealt with. But they are unwilling to pay much of the cost themselves. The situation is reminiscent of American attitudes toward involvement in World War II. In 1939, when Britain and France went to war with Germany, most Americans considered the Nazi threat remote and abstract. They did not want to go to war. As German and Japanese aggression expanded, however, more Americans were willing to help their allies. Eventually, when it seemed Germany and Japan posed a real threat to the United States, the United States began providing supplies on favorable terms to Britain, Russia, and China. But the United States did not go to war until the Japanese attacked Pearl Harbor, crippling United States naval power in the Pacific and making it clear America had to fight to survive. This episode of American history teaches us that people are not usually prepared to make big personal sacrifices for seemingly remote and abstract goals. They are, however, prepared to sacrifice a great deal if the goals become much less remote and abstract. By extension, more and bigger environmental catastrophes may have to occur before more people are willing to take remedial action. Realistically speaking, it may well take one, two, or many environmental Pearl Harbors to get most Americans to make the necessary commitment to help save the planet. The good news is that there may still be time to act.

Evolution and Sociology

For many thousands of years, humans have done well on this planet. That is because we have created cultural practices, including technologies, that allowed us to adapt to and thrive in our environment. Nonetheless, there have been some failures along the way. Many tribes and civilizations are extinct. And our success to date as a species is no warrant for the future. If we persist in using technologies that create an inhospitable environment, Nature will deal with us in the same way it always deals with species that cannot adapt.

Broadly speaking, we have two survival strategies to cope with the challenges that lie ahead: competition and cooperation. Charles Darwin wrote famously about competition in *The Origin of Species* (1859). He observed that members of each species struggle against each other and against other species in their struggle to survive. Most of the quickest, the strongest, the best camouflaged, and the smartest live long enough to bear offspring. Most of the rest are killed off. Thus, the traits passed on to offspring are those most valuable for survival. Ruthless competition, it turns out, is a key survival strategy of all species, including humans.

In *The Descent of Man,* Darwin mentioned our second important survival strategy: cooperation. In some species mutual assistance is common. The species members that flourish are those that best learn to help each other (Darwin, 1871: 163). The Russian geographer and naturalist Petr Kropotkin (1908 [1902]) elaborated this idea. After spending 5 years studying animal life in Siberia, he concluded that "mutual aid" is at least as

important a survival strategy as competition. Competition takes place when members of the same species compete for limited resources, said Kropotkin. Cooperation occurs when species members struggle against adverse environmental circumstances. According to Kropotkin, survival in the face of environmental threat is best assured if species members help each other. Kropotkin also showed that the most advanced species in any group—ants among insects, mammals among vertebrates, humans among mammals—are the most cooperative. Many evolutionary biologists now accept Kropotkin's ideas (Gould, 1988). Recently, based on computer simulations involving competitive and cooperative strategies, mathematicians concluded that "cooperation [is] as essential for evolution as . . . competition" (Nowak, May, and Sigmund, 1995: 81).

As we have seen, a strictly competitive approach to dealing with the environmental crisis—relying on the market alone to solve our problems—now seems inadequate. Instead, it appears we require more cooperation and self-sacrifice. This involves substantially reducing consumption, paying higher taxes for environmental cleanup and energy-efficient industrial processes, subsidizing the developing countries to industrialize in an environmentally friendly way, and so forth. Previously, we outlined some grave consequences of relying too little on a cooperative survival strategy at this historical juncture. But which strategy you emphasize in your own life is, of course, your choice.

Similarly, throughout this book—when we discussed families, gender inequality, crime, race, population, and many other topics—we raised social issues lying at the intersection point of history and biography—yours and ours. We set out alternative courses of action and outlined their consequences. We thus followed our disciplinary mandate: helping people make informed choices based on sound sociological knowledge (Wilensky, 1997; see Figure 18.10). In the context of the present chapter, however, we can make an even bolder claim for the discipline. Conceived at its broadest, sociology promises to help in the rational and equitable evolution of humankind.

SUMMARY

1. Technology is not beyond human control. For while technologies routinely transform societies, they are adopted only when there is a social need for them.

2. Since the last third of the 19th century, technological development has increasingly come under the control of multinational corporations and the military establishments of the major world powers.

3. Research scientists and engineers who work for these organizations must normally adhere to their research priorities.

4. A substantial and growing minority of Americans is skeptical about the benefits of technology.

5. Four important negative consequences of technology are global warming, industrial pollution, the decline of biodiversity, and genetic pollution.

6. Disadvantaged classes, racial minorities, and developing countries are exposed to a disproportionately large share of the risks associated with environmental degradation.

7. Most Americans are unwilling to undergo the personal sacrifices required to deal with environmental degradation. However, that could easily change in the face of repeated environmental catastrophes.

8. Some analysts think the market and high technology will solve the environmental problem. However, three issues suggest these are insufficient solutions: imperfect price signals, the importance of political pressure, and the slow pace of change.

9. Sociology can play an important role sensitizing the public to the social issues and choices humanity faces in the 21st century. For example, sociology poses the choice between more competition and more cooperation as ways of solving the environmental crisis.

GLOSSARY

Acid rain is precipitation whose acidity destroys forests and the ecosystems of lakes. It is formed by sulfur dioxide and other gases emitted by coal-burning power plants, pulp and paper mills, and motor-vehicle exhaust.

Biodiversity refers to the enormous variety of plant and animal species inhabiting the earth.

Environmental racism is the tendency to heap environmental dangers on the disadvantaged, especially on disadvantaged racial minorities.

Genetic pollution refers to the potential dangers of mixing the genes of one species with those of another.

Global warming is the gradual worldwide increase in average surface temperature.

The **greenhouse effect** is the accumulation of carbon dioxide in the atmosphere that allows more solar radiation to enter the atmosphere and less solar radiation to escape.

Normal accidents are accidents that occur inevitably though unpredictably due to the very complexity of modern technologies.

The **ozone layer** is 5–25 miles above the earth's surface. It is depleted by CFCs. The depletion of the ozone layer allows more ultraviolet light to enter the earth's atmosphere. This increases the rate of skin cancer and crop damage.

Recombinant DNA involves taking a piece of DNA from one living species and inserting it into the DNA of another living species, where it grows along with the host DNA.

A **risk society** is a postmodern society defined by the way risk is distributed as a side effect of technology.

Social constructionism is a sociological approach to studying social problems such as environmental degradation. It emphasizes that social problems do not emerge spontaneously. Instead, they are contested phenomena whose prominence depends on the ability of supporters and detractors to make the public aware of them.

Technological determinism is the belief that technology is the main factor shaping human history.

Technology is the practical application of scientific principles.

Technopoly is a form of social organization in which technology compels people to try to solve all problems using technical rather than moral criteria, even though technology is often the source of the problems.

QUESTIONS TO CONSIDER

1. What are the main environmental problems in your community? How are they connected to global environmental issues? (See "Recommended Web Sites," below, for useful leads.)

2. Take an inventory of your environmentally friendly and environmentally dangerous habits. In what ways can you act in a more environmentally friendly way?

WEB RESOURCES

Companion Web Site for This Book

http://sociology.wadsworth.com
Begin by clicking on the Student Resources section of the Web site. Choose "Introduction to Sociology" and finally the Brym and Lie book cover. Next, select the chapter you are currently studying from the pull-down menu. From the Student Resources page you will have easy access to InfoTrac College Edition®, MicroCase Online exercises, additional Web links, and many other resources to aid you in your study of sociology, including practice tests for each chapter.

InfoTrac Search Terms

These search terms are provided to assist you in beginning to conduct research on this topic by visiting http://www.infotraccollege.com/wadsworth.

Environmental problems	Human Genome Project
Environmental racism	Technology
Global warming	

Recommended Web Sites

The Web site of the United States Environmental Protection Agency at http://www.epa.gov is a valuable educational tool. Of particular interest is the search engine at http://www.epa.gov/epahome/comm.htm. It allows you to discover the environmental issues in your community.

The Web sites of 54 environmental movements are listed at http://www.uccs.edu/socges/env-cl12.html.

Against All Reason is a provocative electronic journal devoted to "the radical nature of science as a route to knowledge and the radical critique of the social, political, and economic roles of science and technology." Go to http://www.human-nature.com/reason/index.html.

Good lists of Sociology of Science and Technology Web links can be found http://WWW.Trinity.Edu/mkearl/science.html and http://www.ualberta.ca/slis/guides/scitech/kmc.htm.

SUGGESTED READINGS

Lester R. Brown, Janet N. Abramovitz, Linda Starke et al. *State of the World 2001* (New York: Norton, 2001). This popular annual contains a rich compendium of facts and interpretations about the environmental condition of the planet. It also proposes workable solutions.

Robert Pool. *Beyond Engineering: How Society Shapes Technology* (New York: Oxford University Press, 1997). A lucid analysis of how social factors influence technological development, with particular emphasis on nuclear power.

Jeremy Rifkin. *The Biotech Century: Harnessing the Gene and Remaking the World* (New York: Jeremy P. Tarcher/Putnam, 1998). An alarming account of the potential and problems of the technology that promises to change humanity more than any other.

REFERENCES

Abelmann, Nancy, and John Lie. 1995. *Blue Dreams: Korean Americans and the Los Angeles Riots*. Cambridge, MA: Harvard University Press.

Abraham, Laurie Kaye. 1993. *Mama Might Be Better Off Dead: The Failure of Health Care in Urban America*. Chicago: University of Chicago Press.

Abramsky, Sasha. 1999. "When They Get Out." *The Atlantic Monthly* June. On the World Wide Web http://www.theatlantic.com/issues/99jun/9906prisoners.htm (29 April 2000).

Abu-Nasr, Donna. 1998. "Natural Disaster Costs Soar to World Record." *Globe and Mail* 28 November: A25.

Achilles, Rhona. 1993. "Desperately Seeking Babies: New Technologies of Hope and Despair." Pp. 214–29 in Bonnie J. Fox, ed. *Family Patterns, Gender Relations*. Toronto: Oxford University Press.

Adams, Henry E., Lester W. Wright, Jr., and Bethany A. Lohr. 1998. "Is Homophobia Associated with Homosexual Arousal?" *Journal of Abnormal Psychology* 105: 440–45.

Ad Critic.com. 2000. "Ad Critic: All Ads, All the Time." On the World Wide Web at http://www.adcritic.com (16 May).

Adler, Patricia A., and Peter Adler. 1998. *Peer Power: Preadolescent Culture and Identity*. New Brunswick, NJ: Rutgers University Press.

Akard, Patrick J. 1992. "Corporate Mobilization and Political Power: The Transformation of US Economic Policy in the 1970s." *American Sociological Review* 57: 587–615.

Albas, Daniel and Cheryl Albas. 1989. "Modern Magic: The Case of Examinations." *The Sociological Quarterly* 30: 603–13.

Albelda, Randy, and Nancy Folbre. 1996. *The War on the Poor: A Defense Manual*. New York: New Press.

Aldrich, Howard E. 1979. *Organizations and Environments*. Englewood Cliffs, NJ: Prentice-Hall.

Alford, Robert R., and Roger Friedland. 1985. *Powers of Theory: Capitalism, the State, and Democracy*. Cambridge, UK: Cambridge University Press.

Amato, Paul R., and Bruce Keith. 1991. "Parental Divorce and the Well-Being of Children: A Meta-Analysis." *Psychological Bulletin* 110: 26–46.

American Association of Retired People. 1999. *A Profile of Older Americans*. Washington, DC. On the World Wide Web at http://research.aarp.org/general/profile99.pdf (13 August 2000).

American Chiropractic Association. 1999. "Two More Surveys Show Demand for Alternative Care is Rising." On the World Wide Web at http://www.amerchiro.org/research/new_research.html (2 May 2000).

American Psychological Association. "Answers to Your Questions About Sexual Orientation and Homosexuality." 1998. On the World Wide Web at http://www.apa.org/pubinfo/orient.html (14 June 2000).

Anderson, Ben. 1999. "GOP Combats Census Sampling with Money, Logistics." *Conservative News Service*. On the World Wide Web at http://www.conservativenews.net/InDepth/archive/199903/IND19990316b.html (6 May 2000).

Anderson, Benedict O. 1990. *The Imagined Community*, rev. ed. London: Verso.

Anderson, Elijah. 1990. *Streetwise: Race, Class, and Change in an Urban Community*. Chicago: University of Chicago Press.

Anderson, Margo, and Stephen E. Feinberg. 2000. "Race and Ethnicity and the Controversy over the US Census." *Current Sociology* 48, 3: 87–110.

Angier, Natalie. 2000. "Do Races Differ? Not Really, DNA Shows." *The New York Times on the Web* 22 August. On the World Wide Web at http://www.nytimes.com/library/national/science/082200sci-genetics-race.html (24 August).

"Ann McLaughlin Named to Microsoft Board of Directors." 2000. *Microsoft PressPass* 27 January. On the World Wide Web at http://www.microsoft.com/PressPass/press/2000/Jan00/annMcLaughlinPR.asp (30 April).

Annie E. Casey Foundation. 1998. *Child Care You Can Count On: Model Programs and Policies*. Baltimore. On the World Wide Web at http://www.kidscount.org/publications_child/afford.htm (30 April 2000).

Anti-Defamation League. 1999. "School Vouchers: The Wrong Choice for Public Education." On the World Wide Web at http://www.adl.org/frames/front_vouchers.html (10 August 2000).

Antonius, George. 1939. *The Arab Awakening: The Story of the Arab National Movement*. Philadelphia: J. B. Lipincott.

"Aqua Singer Defends Breast Implants." *National Post* 6 April: F11.

Arendt, Hannah. 1977 [1963]. *Eichmann in Jerusalem: A Report on the Banality of Evil*, rev. ed. Harmondsworth, UK: Penguin.

Ariès, Phillipe. 1962 [1960]. *Centuries of Childhood: A Social History of Family Life*, Robert Baldick, trans. New York: Knopf.

_____. 1982. *The Hour of Our Death*. New York: Knopf.

Arnett, Jeffrey Jensen. 1995. "Adolescents' Uses of Media for Self-Socialization." *Journal of Youth and Adolescence* 24: 519–33.

Arterton, F. Christopher. 1987. *Teledemocracy: Can Technology Protect Democracy?* Newbury Park, CA: Sage Publications.

Averett, Susan, and Sanders Korenman. 1996. "The Economic Reality of *The Beauty Myth*." *Journal of Human Resources* 31: 304–30.

Avery, Simon. 2000. "The Digital Divide." *National Post* 3 June. On the World Wide Web at http://www.nationalpost.com (6 June).

Babbie, Earl. 2000 [1973]. *The Practice of Social Research*, rev. ed. of 9th ed. Belmont CA: Wadsworth.

Baca Zinn, M., and D. Stanley Eitzen. 1993 [1988]. *Diversity in American Families*, 3rd ed. New York: Harper Collins.

Bagdikian, Ben H. 1997 [1983]. *The Media Monopoly*, 5th ed. Boston: Beacon.

Bairoch, Paul. 1988 [1985]. *Cities and Economic Development: From the Dawn of History to the Present*. Christopher Braider, trans. Chicago: University of Chicago Press.

Bales, Kevin. 1999. *Disposable People: New Slavery in the Global Economy*. Berkeley, CA: University of California Press.

Ball, Howard, S. D. Berkowitz, and Mbulelo Mzamane, eds. 1998. *Multicultural Education in Colleges and Universities: A Transdisciplinary Approach*. Mahwah, NJ: Lawrence Erlbaum Associates.

Baltzell, E. Digby. 1964. *The Protestant Establishment: Aristocracy and Caste in America*. New York: Vintage.

Bank of Hawaii. 1999. *Commonwealth of the Northern Mariana Islands: Economic Report, October 1999*. On the World Wide Web at http://www.boh.com/econ/pacific/cnmi/1999/cnmi1999.pdf (23 June 2000).

Banner, Lois W. 1992. *In Full Flower: Aging Women, Power, and Sexuality*. New York: Knopf.

Bannon, Lisa. 2000. "Why Girls and Boys Get Different Toys." *The Wall Street Journal* 14 February: B1, B4.

Baran, Paul A. 1957. *The Political Economy of Growth*. New York: Monthly Review Press.

Barash, David. 1981. *The Whispering Within*. New York: Penguin.

Barber, Bernard. 1992. "Jihad vs. McWorld," *The Atlantic Monthly* March. On the World Wide Web at http://www.theatlantic.com/politics/foreign/barberf.htm (28 April 2000).

_____. 1996. *Jihad vs. McWorld: How Globalism and Tribalism are Reshaping the World*. New York: Ballantine Books.

Barlett, Donald L., and James B. Steele. 1998. "Corporate Welfare." *Time* 152, 19: 9 November. On the World Wide Web at http://www.time.com/time/magazine/1998/dom/981109/cover.html (11 January 2000).

Barnard, Chester I. 1938. *The Functions of the Executive*. Cambridge, MA: Harvard University Press.

Barnard, Jessie. 1972. *The Future of Marriage*. New York: World.

Bar-On, Dan. 1999. *The Indescribable and the Undiscussable: Reconstructing Human Discourse After Trauma*. Ithaca, NY: Cornell University Press.

Baudrillard, Jean. 1983. *Simulations*. New York: Semiotext(e).

_____. 1988 [1986]. *America*. Chris Turner, trans. London: Verso.

_____. 1988. *Selected Writings*, Mark Poster, ed. Stanford, CA: Stanford University Press.

Bauman, Zygmunt. 1991 [1989]. *Modernity and the Holocaust*. Ithaca, NY: Cornell University Press.

Bayer, Ada-Helen, and Leon Harper. 2000. *Fixing to Stay: A National Survey of Housing and Home Modification Issues*. Washington, DC: American Association for Retired People.

On the World Wide Web at http://research.aarp.org/il/home_mod.pdf (13 August 2000).

Bayles, Martha. 1994. *Hole in Our Soul: The Loss of Beauty and Meaning in American Popular Music*. Chicago: University of Chicago Press.

Bean, Frank D., and Marta Tienda. 1987. *The Hispanic Population of the United States*. New York: Russell Sage Foundation.

Beck, Ulrich. 1992 [1986]. *Risk Society: Towards a New Modernity*. Mark Ritter, trans. London: Sage.

Becker, Ernest. 1973. *The Denial of Death*. New York: Free Press.

Becker, Gary. 1976. *The Economic Approach to Human Behavior*. Chicago: University of Chicago Press.

Becker, Howard S. 1963. *Outsiders: Studies in the Sociology of Deviance*. New York: Free Press.

Bell, Daniel. 1961. *The End of Ideology*. New York: Collier.

_____. 1973. *The Coming of Post-Industrial Society: A Venture in Social Forecasting*. New York: Basic Books.

Bell, Wendell, and Robert V. Robinson.1980. "Cognitive Maps of Class and Racial Inequalities in England and the United States." *American Journal of Sociology* 86: 320–49.

Bellah, Robert A. 1975. *The Broken Covenant: American Civil Religion in a Time of Trial*. New York: Seabury Press.

Bellow, Saul. 1964. *Herzog*. New York: Fawcett World Library.

Benford, Robert D. 1997. "An Insider's Critique of the Social Movement Framing Perspective." *Sociological Inquiry* 67: 409–39.

Benson, John M. 1999. "End-of-Life Issues." *Public Opinion Quarterly* 63: 263–77.

Berger, Peter L. 1963. *Invitation to Sociology: A Humanistic Approach*. New York: Doubleday.

_____. 1967. *The Sacred Canopy: Elements of a Sociological Theory of Religion*. Garden City, NY: Doubleday.

_____. 1986. *The Capitalist Revolution: Fifty Propositions About Prosperity, Equality, and Liberty*. New York: Basic Books.

_____ and Thomas Luckmann. 1966. *The Social Construction of Reality: A Treatise in the Sociology of Knowledge*. Garden City, NY: Doubleday.

Berger, Suzanne, and Ronald Dore, eds. 1996. *National Diversity and Global Capitalism*. Ithaca, NY: Cornell University Press.

Berk, Richard A. 1974. *Collective Behavior*. Dubuque, IA: Wm. C. Brown.

Berners-Lee, Tim. "Tim Berners-Lee." 1999. On the World Wide Web at http://www.w3.org/People/Berners-Lee/Overview.html (2 May 2000).

Bernstein, Aaron. 2000. "Down and Out in Silicon Valley." *Business Week* 27 March: 76–92.

Bernstein, Jared, Heidi Hartmann, and John Schmitt. 1999. "The Minimum Wage Increase: A Working Woman's Issue." On the World Wide Web at http://www.epinet.org/Issuebriefs/Ib133.html (30 April 2000).

Besserer, Sandra. 1998. "Criminal Victimization: An International Perspective." *Juristat* 18: 6.

Best, Steven, and Douglas Kellner. 1999. "Rap, Black Rage, and Racial Difference." *Enculturation* 2: 2. On the World Wide Web at http://www.uta.edu/huma/enculturation/2_2/best-kellner.html (10 May 2000).

Bianchi, Suzanne M., and Daphne Spain. 1996. "Women, Work, and Family in America." *Population Bulletin* 51, 3: 2–48.

Bierstedt, Robert. 1963. *The Social Order*. New York: McGraw-Hill.

_____. 1974. "An Analysis of Social Power." Pp. 220–41 in *Power and Progress: Essays in Sociological Theory*. New York: McGraw-Hill.

"Billboard Top 100." 2000. On the World Wide Web at http://music.lycos.com/charts/showchart.asp?provider=billboard&chart=singles (19 August).

Birdwhistell, Ray L. 1970. *Kinesics and Context*. Philadelphia: Pennsylvania State University.

Bjorhus, Jennifer. 2000. "Gap Between Execs, Rank-and-file Grows Wider." *San Jose Mercury News* 18 June. On the World Wide Web at http://www.mercurycenter.com/premium/business/docs/disparity18.htm (20 June).

Black, Donald. 1989. *Sociological Justice*. New York: Oxford University Press.

Blake, Michael. 2000. "Rights for People, Not Cultures." *National Post* 18 August: A16.

Blau, Peter M. 1963 [1955]. *The Dynamics of Bureaucracy: A Study of Interpersonal Relationships in Two Government Agencies*, rev. ed. Chicago: University of Chicago Press.

_____. 1964. *Exchange and Power in Social Life*. New York: Wiley.

_____ and Otis Dudley Duncan. 1967. *The American Occupational Structure*. New York: Wiley.

Blauner, Robert. 1972. *Racial Oppression in America*. New York: Harper and Row.

Blazer, Dan G., Ronald C. Kessler, Katherine A. McGonagle, and Marvin S Swartz. 1994. "The Prevalence and Distribution of Major Depression in a National Community Sample: The National Comorbidity Survey." *American Journal of Psychiatry* 151: 979–86.

Bliss, Jeff. 2000. "Getting a Life Offline." *Financial Post* 29 June: C3.

Block, Fred. 1979. "The Ruling Class Does Not Rule." Pp. 128–40 in R. Quinney, ed. *Capitalist Society*. Homewood, IL: Dorsey Press.

Bluestone, Barry, and Bennett Harrison. 1982. *The Deindustrialization of America*. New York: Basic Books.

_____ and Stephen Rose. 1997. "Overworked and Underemployed: Unraveling an Economic Enigma." *The American Prospect* 31: 58–69. On the World Wide Web at http://www. prospect.org/archives/31/31bluefs.html (1 May 2000).

Blum, Deborah. 1997. *Sex on the Brain: The Biological Differences Between Men and Women*. New York: Penguin.

Blumberg, Paul. 1989. *The Predatory Society: Deception in the American Marketplace*. New York: Oxford University Press.

Blumer, Herbert. 1969. *Symbolic Interactionism: Perspective and Method*. Englewood Cliffs, NJ: Prentice-Hall.

Boal, Mark. 1998. "Spycam City." *The Village Voice*. 30 September–6 October. On the World Wide Web at http://www.villagevoice.com/issues/9840/boal.shtml (29 April 2000).

Bobo, Lawrence, and James R. Kluegel. 1993. "Opposition to Race Targeting: Self-Interest, Stratification Ideology, or Racial Attitudes." *American Sociological Review* 58: 443–64.

Bonacich, Edna. 1972. "A Theory of Ethnic Antagonism: The Split Labor Market." *American Sociological Review* 37: 547–59.

_____. 1973. "A Theory of Middleman Minorities." *American Sociological Review* 38: 583–94.

Bornschier, Volker, and Christopher Chase-Dunn. 1985. *Transnational Corporations and Underdevelopment*. New York: Praeger.

Boston Women's Health Book Collective, ed. 1998. *Our Bodies, Our Selves for the New Century: A Book by and for Women*. New York: Simon & Schuster.

Boswell, A. Ayres, and Joan Z. Spade. 1996. "Fraternities and Collegiate Rape Culture: Why Are Some Fraternities More Dangerous Places for Women?" *Gender and Society* 10: 133–47.

Bouchard, Thomas J., Jr., David T. Lykken, Matthew McGue, Nancy L. Segal, and Auke Tellegen. 1990. "Sources of Human Psychological Differences: The Minnesota Study of Twins Reared Apart." *Science* 250, 4978: 223–6.

Bourdieu, Pierre. 1977 [1972]. *Outline of a Theory of Practice*, Richard Nice, trans. Cambridge, UK: Cambridge University Press.

_____. 1984 [1979]. *Distinction: A Social Critique of the Judgment of Taste*, Richard Nice, trans. Cambridge, MA: Harvard University Press.

_____. 1998 [1996]. *On Television*. New York: New Press.

Bowles, Samuel, and Herbert Gintis. 1976. *Schooling in Capitalist America: Educational Reform and the Contradictions of Economic Life*. New York: Basic Books.

Boyd, Monica. 1997. "Feminizing Paid Work." *Current Sociology* 45: 49–73.

_____, John Goyder, Frank Jones, Hugh A. McRoberts, Peter Pineo, and John Porter. 1985. *Ascription and Achievement: Studies in Mobility and Status Attainment in Canada*. Ottawa: Carleton University Press.

Bracey, Gerald W. 1998. "Are U.S. Students Behind?" *The American Prospect* 37, March–April: 54–70. On the World Wide Web at http://www.prospect.org/archives/37/37bracfs.html (1 May 2000).

Brady, Erik. 2001. "Too Good To Be True?" *USA Today* 12 July On the World Wide Web at wysiwyg://14/http://www.usatoday.com/sports/stories/2001-07-12-cover.htm (21 July).

Brains, Craig Leonard. 1999. "When Registration Barriers Fall, Who Votes?" *Public Choice* 35: 161–76.

Braithwaite, John. 1981. "The Myth of Social Class and Criminality Revisited." *American Sociological Review* 46: 36–57.

_____. 1989. *Crime, Shame and Reintegration*. New York: Cambridge University Press.

Braus, Patricia. 1994. "Why Do Hispanics Have Lower Death Rates?" *American Demographics* 16, 5: 18–19.

Braverman, Harry. 1974. *Labor and Monopoly Capital: The Degradation of Work in the Twentieth Century*. New York: Monthly Review Press.

Breault, K. D. 1986. "Suicide in America: A Test of Durkheim's Theory of Religious and Family Integration, 1933–1980." *American Journal of Sociology* 92: 628–56.

Brechin, Steven R., and Willett Kempton. 1994. "Global Environmentalism: A Challenge to the Postmaterialism Thesis." *Social Science Quarterly* 75: 245–69.

Breen, Richard, and David B. Rottman. 1995. *Class Stratification: A Comparative Perspective*. New York: Harvester Wheatsheaf.

Brint, Stephen. 1984. "New Class and Cumulative Trend Explanations of the Liberal Political Attitudes of Professionals." *American Journal of Sociology* 90: 30–71.

_____ and Jerome Karabel. 1989. *The Diverted Dream: Community Colleges and the Promise of Educational*

Opportunity in America, 1900–1985. New York: Oxford University Press.

"British Magazines Agree to Ban Ultra-thin Models." 2000. *National Post* 23 June: A2.

Brokaw, Tom. 2001. "Into an Unknowable Future." *The New York Times on the Web*. On the World Wide Web at http://www.nytimes.com/2001/09/28/opinion/28BROK.html?todaysheadlines (28 September)

Bromley, Julian V. 1982 [1977]. *Present-Day Ethnic Processes in the USSR*. Moscow: Progress Publishers.

Bronowski, J. 1965 [1956]. *Science and Human Values*, revised ed. New York: Harper and Row.

Brooks, Clem, and Jeff Manza. 1994. "Do Changing Values Explain the New Politics? A Critical Assessment of the Postmaterialist Thesis." *Sociological Quarterly* 35: 541–70.

_____ and _____. 1997a. "Class Politics and Political Change in the United States, 1952–1992." *Social Forces* 76: 379–408.

_____ and _____. 1997b. "Social Cleavages and Political Alignments: U.S. Presidential Elections, 1960 to 1992." *American Sociological Review* 62: 937–46.

Brouwer, Steve. 1998. *Sharing the Pie: A Citizen's Guide to Wealth and Power in America*. New York: Holt.

Brower, David. 1975. *Training the Nihilists: Education and Radicalism in Tsarist Russia*. Ithaca, NY: Cornell University Press.

Brown, Andrew. 2001. "S-Tools v4." On the World Wide Web at http://members.tripod.com/steganography/stego/s-tools4.html (13 September).

Brown, David K. 1995. *Degrees of Control: A Sociology of Educational Expansion and Occupational Credentialism*. New York: Teachers College Press.

Brown, Lester R., Christopher Flavin, and Hilary French, et al. 2000. *State of the World 2000*. New York: W. W. Norton.

_____, Janet N. Abramovitz, and Linda Starke, et al. *State of the World 2001*. New York: W. W. Norton.

Brown, Lyn Mikel, and Carol Gilligan. 1992. *Meeting at the Crossroads: Women's Psychology and Girls' Development*. Cambridge, MA: Harvard University Press.

Brown, Phil. 1997. "Popular Epidemiology Revisited." *Current Sociology* 45, 3: 137–56.

Browne, Malcolm W. 1998. "From Science Fiction to Science: The Whole Body Transplant." *New York Times* 5 May: Sec. B, 16.

Bruce, Steve. 1988. *The Rise and Fall of the New Christian Right: Conservative Protestant Politics in America 1978–1988*. Oxford, UK: Clarendon Press.

Brumberg, Joan Jacobs. 1997. *The Body Project: An Intimate History of American Girls*. New York: Random House.

Brym, Robert J. 1979. "Political Conservatism in Atlantic Canada." Pp. 59–79 in Robert J. Brym and R. James Sacouman, eds. *Underdevelopment and Social Movements in Atlantic Canada*. Toronto: New Hogtown Press.

_____. 1980. *Intellectuals and Politics*. London: George Allen and Unwin.

_____. 1984. "Cultural versus Structural Explanations of Ethnic Intermarriage in the USSR: A Statistical Re-analysis." *Soviet Studies* 36: 594–601.

_____. 1990. "Sociology, *Perestroika*, and Soviet Society." *Canadian Journal of Sociology* 15: 207–15.

_____. 1995. "Voters Quietly Reveal Greater Communist Leanings." *Transition: Events and Issues in the Former Soviet Union and East-Central and Southeastern Europe* 1, 16: 32–5.

_____. 1996a. "The Ethic of Self-reliance and the Spirit of Capitalism in Russia." *International Sociology* 11: 409–26.

_____. 1996b. "Reevaluating Mass Support for Political and Economic Change in Russia." *Europe–Asia Studies* 48: 751–66.

_____. 1996c. "'The Third Rome' and 'The End of History:' Notes on Russia's Second Communist Revolution." *Canadian Review of Sociology and Anthropology* 33: 391–406.

_____. 1996d. "The Turning Point in the Presidential Campaign." Pp. 44–9 in *The 1996 Presidential Election and Public Opinion*. Moscow: VTsIOM. [In Russian.]

_____ and Evel Economakis. 1994. "Peasant or proletarian? Blacklisted Pskov Workers in St. Petersburg, 1913." *Slavic Review* 53: 120–39.

_____ with Bonnie J. Fox. 1989. *From Culture to Power: The Sociology of English Canada*. Toronto: Oxford University Press.

_____, Michael Gillespie, and A. Ron Gillis. 1985. "Anomie, Opportunity, and the Density of Ethnic Ties: Another View of Jewish Outmarriage in Canada." *Canadian Review of Sociology and Anthropology* 22: 102–12.

_____, Michael Gillespie, and Rhonda L. Lenton. 1989. "Class Power, Class Mobilization, and Class Voting: The Canadian Case." *Canadian Journal of Sociology* 14: 25–44.

_____ and Rhonda Lenton. 2001. *Love Online: A Report on Digital Dating in Canada*. Toronto: MSN.CA. On the World Wide Web at http://www.nelson.com/nelson/harcourt/sociology/newsociety3e/loveonline.pdf (20 December 2001).

_____ with the assistance of Rozalina Ryvkina. 1994. *The Jews of Moscow, Kiev and Minsk: Identity, Antisemitism, Emigration*. New York: New York University Press.

Bryson, Ken, and Lynne M. Casper. 1998. *Household and Family Characteristics: March 1997*. Washington, DC: United States Department of Commerce, Economics and Statistics Administration. On the World Wide Web at http://www.census.gov/prod/3/98pubs/p20-509.pdf (1 May 2000).

Bullard, Robert D. 1994 [1990]. *Dumping in Dixie: Race, Class and Environmental Quality*, 2nd ed. Boulder, CO: Westview Press.

Burawoy, Michael. 1979. *Manufacturing Consent: Changes in the Labor Process under Monopoly Capitalism*. Chicago: University of Chicago Press.

Burgess, Ernest. W. 1967 [1925]. "The Growth of the City: An Introduction to a Research Project." Pp. 47–62 in Robert E. Park, Ernest W. Burgess, and Roderick D. McKenzie. *The City*. Chicago: University of Chicago Press.

Burns, Tom, and G.M. Stalker. 1961. *The Management of Innovation*. London: Tavistock.

Buss, D. M. 1994. *The Evolution of Desire*. New York: Basic Books.

_____. 1998. "The Psychology of Human Mate Selection: Exploring the Complexity of the Strategic Repertoire." Pp. 405–29 in C. Crawford and D. L. Krebs, eds. *Handbook of Evolutionary Psychology: Ideas, Issues, and Applications*. Mahwah, NJ: Erlbaum.

Butterfield, Fox. 2001. "Killings Increase in Many Big Cities." *New York Times on the Web*. On the World Wide Web at http://www.nytimes.com/2001/12/21/national/21CRIM.html?todaysheadlines (21 December 2001).

Buxton, L. H. Dudley. 1963. "Races of Mankind," *Encyclopaedia Britannica* 18: 864–6. Chicago: Encyclopaedia Britannica, Inc.

Callahan, Raymond E. 1962. *Education and the Cult of Efficiency: A Study of the Social Forces That Have Shaped the Administration of the Public Schools*. Chicago: University of Chicago Press.

Camarillo, Albert. 1979. *Chicanos in a Changing Society: From Mexican Pueblos to American Barrios in Santa Barbara and Southern California, 1848–1930*. Cambridge, MA: Harvard University Press.

Campbell, Donald, and Julian Stanley. 1963. *Experimental and Quasi-Experimental Designs for Research*. Chicago: Rand McNally.

Campbell, Frances A., and Craig T. Ramey. 1994. "Effects of Early Intervention on Intellectual and Academic Achievement: A Follow-up Study of Children from Low-income Families." *Child Development* 65: 684–98.

Campion, Edward W. 1993. "Why Unconventional Medicine?" *New England Journal of Medicine* 328: 282.

Cancio, A. Silvia, T. David Evans, and Daivd J. Maume, Jr. 1996. "Reconsidering the Declining Significance of Race: Racial Differences in Early Career Wages." *American Sociological Review* 61: 541–56.

"The Candidates Debate." 1998. *MSNBC News*. On the World Wide Web at http://msnbc.com/onair/msnbc/TimeAndAgain/archive/ken-nix/Default.asp?cp1=1 (2 May 2000).

Carey, James. 1989. *Culture as Communication*. Boston: Unwin Hyman.

Carter, Stephen L. 1991. *Reflections of an Affirmative Action Baby*. New York: Basic Books.

Casper, Lynne M., Sara S. McLanahan, and Irwin Garfinkel. 1994. "The Gender-Poverty Gap: What Can We Learn From Other Countries?" *American Sociological Review* 59: 594–605.

Cassidy, John. 1999. "Schools Are Her Business." *New Yorker* 18–25 October: 144–60.

Castells, Manuel. 1983. *The City and the Grassroots: A Cross-Cultural Theory of Urban Social Movements*. Berkeley: University of California Press.

_____. 1996 *The Information Age: Economy, Society and Culture: The Rise of the Network Society*, vol 1. Oxford, UK: Blackwell.

Cavalli-Sforza, L. Luca, Paolo Menozzi, and Alberto Piazza. 1994. *The History and Geography of Human Genes*. Princeton, NJ: Princeton University Press.

Center for Responsive Politics. 2000. "Election Statistics at a Glance." On the World Wide Web at http://www.opensecrets.org/pubs/bigpicture2000/overview/stats.ihtml (1 May 2000).

Centers for Disease Control and Prevention, National Center for Health Statistics. 1947. "Deaths and Death Rates for Leading Causes of Death: Death Registration States, 1900–1940." Special tabulation prepared for the authors.

_____. 1995a. *Monthly Vital Statistics Report* 43, 9(S): 22 March.

_____. 1995b. *Monthly Vital Statistics Report* 43, 12(S): 14 July.

_____. 1997. *Monthly Vital Statistics Report* 46, 1(S)2: 11 September.

_____. 1998. *Monthly Vital Statistics Report* 46, 12: 28 July.

_____. 1999a. *National Vital Statistics Reports* 47, 19: 30 June.

_____. 1999b. *National Vital Statistics Reports* 47, 25: 5 October. On the World Wide Web at http://www.cdc.gov/nchs/data/nvs47_25.pdf (27 April 2000).

_____. 1999c. "Trend B: Table 291: Death Rates for 72 Selected Causes, by 5-Year Age Groups, Race, and Sex: United States, 1979–97 (Rates per 100,000 Population)." On the World Wide Web at http://www.cdc.gov/nchs/data/gm291_1.pdf and http://www.cdc.gov/nchs/data/gm291_2.pdf (27 April 2000).

_____. 2000. "Cumulative Age of Initiation of Cigarette Smoking — United States, 1991." On the World Wide Web at http://www.cdc.gov/tobacco/init.htm (1 May 2000).

Chambliss, Daniel F. 1996. *Beyond Caring: Hospitals, Nurses, and the Social Organization of Ethics*. Chicago: University of Chicago Press.

Chambliss, William J. 1989. "State-Organized Crime." *Criminology* 27: 183–208.

"The Changing Rate of Major Depression." 1992. *Journal of the American Medical Association* 268, 21: 3098–4005.

Charlton, James I. 1998. *Nothing About Us Without Us: Disability Oppression and Empowerment*. Berkeley, CA: University of California Press.

Chaves, Mark. 1994. "Secularization as Declining Religious Authority." *Social Forces* 72: 749–74.

Cherlin, Andrew J. 1992 [1981]. *Marriage, Divorce, Remarriage*, revised and enlarged ed. Cambridge, MA: Harvard University Press.

_____, Frank F. Furstenberg, Jr., P. Lindsay Chase-Lansdale, Kathleen E. Kiernan, Philip K. Robins, Donna Ruane Morrison, and Julien O. Teitler. 1991. "Longitudinal Studies of Effects of Divorce on Children in Great Britain and the United States." *Science* 252: 1386–9.

Chesnais, Jean-Claude. 1992 [1986]. *The Demographic Transition: Stages, Patterns, and Economic Implications*. Elizabeth Kreager and Philip Kreager, trans. Oxford, UK: Clarendon Press.

Choldin, Harvey M. 1994. *Looking for the Last Percent: The Controversy over Census Undercounts*. New Brunswick, NJ: Rutgers University Press.

Cicourel, Aaron. 1968. *The Social Organization of Juvenile Justice*. New York: Wiley.

Clapp, Jennifer. 1998. "Foreign Direct Investment in Hazardous Industries in Developing Countries: Rethinking the Debate." *Environmental Politics* 7, 4: 92–113.

Clark, Terry Nichols, and Seymour Martin Lipset. 1991. "Are Social Classes Dying?" *International Sociology* 6: 397–410.

_____, _____, and Michael Rempel. 1993. "The Declining Political Significance of Class." *International Sociology* 8: 293–316.

Clarke-Stewart, K. Alison, Christian P. Gruber, and Linda May Fitzgerald. 1994. *Children at Home and in Day Care*. Hillsdale, NJ: Lawrence Erlbaum.

Clawson, Dan. 1980. *Bureaucracy and the Labor Process: The Transformation of U.S. Industry, 1860–1920*. New York: Monthly Review Press.

_____ and Mary Ann Clawson. 1999. "What Has Happened to the US Labor Movement? Union Decline and Renewal" *Annual Review of Sociology* 25: 95–119.

_____, Alan Neustadtl, and Denise Scott. 1992. *Money Talks: Corporate PACS and Political Influence.* New York: Basic Books.

Clinton, William J. 1998. "Statement on Signing the Child Support Performance and Incentive Act of 1998." *Weekly Compilation of Presidential Documents* 34, 29: 1396.

Corporate Watch. 2000. "High Tech Sweatshops." On the World Wide Web at http://www.corpwatch.org/trac/gallery/sweat/27a.html (20 June 2000).

Couch, Carl J. 1968. "Collective Behavior: An Examination of Some Stereotypes." *Social Problems* 15: 310–22.

Croteau, David, and William Hoynes. 1997. *Media/Society: Industries, Images, and Audiences.* Thousand Oaks, CA: Pine Forge Press.

Clement, Wallace, and John Myles. 1994. *Relations of Ruling: Class and Gender in Postindustrial Societies.* Montreal: McGill-Queen's University Press.

Clinard, Marshall B., and Peter C. Yeager. 1980. *Corporate Crime.* New York: Free Press.

Cloward, Richard A., and Lloyd E. Ohlin. 1960. *Delinquency and Opportunity: A Theory of Delinquent Gangs.* New York: Free Press.

CNN.COM. 1998. "Prosecutor: Attackers Planned to Rob Gay Student." On the World Wide Web at http://www.htt.com/cnn.com/US/9811/19/shepard.01/index.html (1 November 1999).

_____. 1999. "From Little League to Madness: Portraits of the Littleton Shooters." 30 April. On the World Wide Web at www.cnn.com/SPECIALS/1998/schools/they.hid.it.well/index.html (29 April 2000).

Coale, Ansley J. 1974. "The History of Human Population." *Scientific American* 23, 3: 41–51.

_____ and Susan C. Watkins, eds. 1986. *The Decline of Fertility in Europe.* Princeton, NJ: Princeton University Press.

Cohen, Adam. 2000. "The Big Easy on the Brink." *Time* (Canadian edition) 10 July: 43.

Cohen, Albert. 1955. *Delinquent Boys: The Subculture of a Gang.* New York: Free Press.

Cohen, Stanley. 1972. *Folk Devils and Moral Panics: The Creation of the Mods and Rockers.* London: MacGibbon and Kee.

Colapinto, John. 1997. "The True Story of John/Joan." *Rolling Stone* 11 December: 54–73, 92–7.

Cole, Michael. 1995. *Cultural Psychology.* Cambridge, MA: Harvard University Press.

Coleman, James S. 1961. *The Adolescent Society.* New York: Free Press.

_____. 1990. *Foundations of Social Theory.* Cambridge, MA: Harvard University Press.

_____ et al. 1966. *Equality of Educational Opportunity.* Washington, DC: United States Department of Health, Education, and Welfare, Office of Education.

Collins, Randall. 1975. *Conflict Sociology: Toward an Explanatory Science.* New York: Academic Press.

_____. 1979. *The Credential Society: An Historical Sociology of Education and Stratification.* New York: Academic Press.

_____. 1982. *Sociological Insight: An Introduction to Nonobvious Sociology.* New York: Oxford University Press.

_____ and Scott Coltrane. 1991 [1985]. *Sociology of Marriage and the Family: Gender, Love, and Property*, 3rd ed. Chicago: Nelson-Hall.

Columbia Broadcasting System. 2000. "Super Bowl Ratings Up From Last Year." On the World Wide Web at http://cbs.sportsline.com/u/ce/multi/0,1329,1959564_59,00.html (9 August).

Combs, Sean "Puffy". 1999. *Forever.* New York: Bad Boy Entertainment, Inc. (Compact Disc.)

_____ and The Lox. 1997. "I Got the Power." On the World Wide Web at http://www.ewsonline.com/badboy/lyrpow.html (28 April 2000).

Commins, Patricia. 1997. "Foreign Sales Prop Up McDonald's." *Globe and Mail* 26 August: B8.

"Comparable Worth." 1990. *Issues in Ethics* 3, 2. On the World Wide Web at http://www.scu.edu/SCU/Centers/Ethics/publications/iie/v3n2/comparable.shtml (30 April 2000).

Comte, Auguste. 1975. *Auguste Comte: The Foundation of Sociology*, Kenneth Thompson, ed. New York: Wiley.

Condry, J. and S. Condry. 1976. "Sex Differences: The Eye of the Beholder." *Child Development* 47: 812–19.

Conrad, Peter, and Joseph W. Schneider. 1992 [1980]. *Deviance and Medicalization: From Badness to Sickness,* expanded ed. Philadelphia: Temple University Press.

Converse, Jean M., and Stanley Presser. 1986. *Survey Questions: Handrcrafting the Standardized Questionnaire.* Newbury Park, CA: Sage.

Conwell, Chic. 1937. *The Professional Thief: By A Professional Thief,* annotated and interpreted by Edwin H. Sutherland. Chicago: University of Chicago Press.

Cooley, Charles Horton. 1902. *Human Nature and the Social Order.* New York: Scribner's.

Coontz, Stephanie. 1992. *The Way We Never Were: American Families and the Nostalgia Trip.* New York: Basic Books.

_____. 1997. *The Way We Really Are: Coming to Terms With America's Changing Families.* New York: Basic Books.

_____ and Peta Henderson, eds. 1986. *Women's Work, Men's Property: The Origins of Gender and Class.* London: Verso.

Cornell, Stephen. 1988. *The Return of the Native: American Indian Political Resurgence.* New York: Oxford University Press.

Coupland, Douglas. 1991. *Generation X: Tales for An Accelerated Culture.* New York: St. Martin's Press.

Cramton, Peter, and Joseph Tracy. 1998. "The Use of Replacement Workers in Union Contract Negotiations: The U.S. Experience, 1980–89." *Journal of Labor Economics* 16: 667–701.

Creedon, Jeremiah. 1998. "God With a Million Faces." *Utne Reader* July–August: 42–8.

Crompton, Rosemary. 1993. *Class and Stratification: An Introduction to Current Debates.* Cambridge, UK: Polity Press.

_____ and Michael Mann, eds. 1986. *Gender and Stratification.* Cambridge, UK: Polity Press.

Cronon, William. 1991. *Nature's Metropolis: Chicago and the Great West.* New York: W. W. Norton.

Croteau, David, and William Hoynes *Media/Society: Industries, Images, and Audiences* (Thousand Oaks CA: Pine Forge Press, 1997).

Crothers, Charles. 1979. "On the Myth of Rural Tranquility: Comment on Webb and Collette." *American Journal of Sociology* 84: 429–37.

Crozier, Michel. 1964 [1963]. *The Bureaucratic Phenomenon.* Chicago: University of Chicago Press.

Curtis, James, John Loy and Wally Karnilowicz. 1986. "A Comparison of Suicide-Dip Effects of Major Sport Events and Civil Holidays." *Sociology of Sport Journal* 3: 1–14.

Dahl, Robert A. 1961. *Who Governs?* New Haven, CT: Yale University Press.

Darder, Antonia, and Rodolfo D. Torres, eds. 1998. *The Latino Studies Reader: Culture, Economy and Society.* Malden, MA: Blackwell.

Darwin, Charles. 1859. *On the Origin of Species by Means of Natural Selection.* London: John Murray.

_____. 1871. *The Descent of Man.* London: John Murray.

Davidoff, Judith. 1999. "The Fatherhood Industry: Welfare Reformers Set Their Sights on Wayward Dads." *The Progressive.* November: 28–31.

Davies, James C. 1969. "Toward a Theory of Revolution." Pp. 85–108 in Barry McLaughlin, ed. *Studies in Social Movements: A Social Psychological Perspective.* New York: Free Press.

Davies, Mark, and Denise B. Kandel. 1981. "Parental and Peer Influences on Adolescents' Educational Plans: Some Further Evidence." *American Journal of Sociology* 87: 363–87.

Davis, Jim. 2000. "AOL Previews TV Plans." *CNET News* 6 January. On the World Wide Web at http://news.cnet.com/category/0-1006-200-1516271.html (2 May 2000).

Davis, Kingsley, and Wilbert E. Moore. 1945. "Some Principles of Stratification." *American Sociological Review* 10: 242–9.

Davis, Mike. 1990. *City of Quartz: Excavating the Future in Los Angeles.* New York: Verso.

Dawidowicz, Lucy S. 1975. *The War Against the Jews, 1933–1945.* New York: Holt, Rinehart and Winston.

Dean, John. 2000. "Why Americans Don't Vote—And How that Might Change." On the World Wide Web at http://www.cnn.com/2000/LAW/11/columns/fl.dean.voters.02.11.07 (8 November 2000).

de la Garza, Rodolpho O., Luis DiSipio, F. Chris Garcia, John Garcia, and Angelo Falcon. 1992. *Latino Voices: Mexican, Puerto Rican, and Cuban Perspectives on American Politics.* Boulder, CO: Westview.

De Long, J. Bradford. 1998. *Global Trends: 1980–2015 and Beyond.* Ottawa: Industry Canada.

Demo, David H., and Alan C. Acock. 1993. "Family Diversity and the Division of Domestic Labor: How Much Have Things Really Changed? *Family Relations* 42: 323–31.

Democratic National Committee. 1998. "Democrats Fight to Make Sure that Every American is Counted in the 2000 Census." On the World Wide Web at http://www.democrats.org/archive/news/rel1998/rel060298.html (6 May 2000).

Denzin, Norman K. 1992. *Symbolic Interactionism and Cultural Studies: The Politics of Interpretation.* Oxford, UK: Blackwell.

Department of Geography, Slippery Rock University. 1997. "World's Largest Cities, 1900." On the World Wide Web at http://www.sru.edu/depts/artsci/ges/discover/d-6-8.htm (2 May 2000).

_____. 1998. "World's Largest Urban Agglomerations, 2015." On the World Wide Web at http://www.sru.edu/depts/artsci/ges/discover/d-6-9b.htm (2 May 2000).

Department of Justice, Canada. 1995. "A Review of Firearm Statistics and Regulations in Selected Countries." On the World Wide Web at http://www.cfc-ccaf.gc.ca/research/publications/reports/1990%2D95/reports/siter_rpt_en.html (29 April 2000).

Derber, Charles. 1979. *The Pursuit of Attention: Power and Individualism in Everyday Life.* New York: Oxford University Press.

de Villiers, Marq. 1999. *Water.* Toronto: Stoddart Publishing.

de Waal, Alexander. 1989. *Famine That Kills: Darfur, Sudan, 1984–1985.* Oxford, UK: Clarendon Press.

Diamond, Larry. 1996. "Is the Third Wave Over?" *Journal of Democracy* 7, 3: 20–37. On the World Wide Web at http://muse.jhu.edu/demo/jod/7.3diamond.html (1 May 2000).

Diani, Mario. 1996. "Linking Mobilization Frames and Political Opportunities: Insights from Regional Populism in Italy." *American Sociological Review* 61: 1053–69.

Dibbell, Julian. 1993. "A Rape in Cyberspace." *The Village Voice* 21 December: 36–42. On the World Wide Web at http://www.levity.com/julian/bungle.html (29 April 2000).

DiCarlo, Lisa. 2000. "32 Debut on the List." *Forbes.com.* On the World Wide Web at http://www.forbes.com/tool/toolbox/rich400 (22 September 2000).

Dickerson, Marla. 1998. "A New Army of Child-Care Workers." *Los Angeles Times* (12 August) Part A: 1.

Dietz, Tracy L. 1998. "An Examination of Violence and Gender Role Portrayals in Video Games: Implications for Gender Socialization and Aggressive Behavior." *Sex Roles* 38: 425–42.

di Lampedusa, Giuseppe Tomasi. 1991 [1958]. *The Leopard.* New York: Pantheon.

Dines, Gail, and Jean McMahon Humez. 1995. *Gender, Race, and Class in Media: A Text-Reader.* Thousand Oaks, CA: Sage.

Doberman, John. 1997. *Darwin's Athletes: How Sport Has Damaged Black America and Preserved the Myth of Race.* Boston: Houghton Mifflin.

Domhoff, G. William. 1983. *Who Rules America Now? A View for the 1980s.* New York: Touchstone.

Donahue III, John J., and Steven D. Levitt. 2001. "The Impact of Legalized Abortion on Crime." *Quarterly Journal of Economics* 116: 379–420.

Dore, Ronald. 1983. "Goodwill and the Spirit of Market Capitalism." *British Journal of Sociology* 34: 459–82.

Doremus, Paul N., William W. Keller, Louis W. Pauly, and Simon Reich. 1998. *The Myth of the Global Corporation.* Princeton, NJ: Princeton University Press.

Douglas, Jack D. 1967. *The Social Meanings of Suicide.* Princeton, NJ: Princeton University Press.

Douglas, Susan J. 1994. *Where the Girls Are: Growing Up Female with the Mass Media.* New York: Random House.

Dowling, Michael, Christian Lechner, and Bodo Thielmann. 1998. "Convergence—Innovation and Change of Market Structures Between Television and Online Services." *Electronic Marketing* 8, 4. On the World Wide Web at http://www.electronicmarkets.com/netacademy/publications.nsf/all_pk/1124 (2 May 2000).

Dreyfus, Robert. 1999. "Money 2000." *The Nation* 18 October: 11–17.

Drèze, Jean, and Amartya Sen. 1989. *Hunger and Public Action.* Oxford, UK: Clarendon Press.

DuBois, W. E. B. 1967 [1899]. *The Philadelphia Negro: A Social Study.* New York: Schocken.

Dudley, Kathryn Marie. 1994. *The End of the Line: Lost Jobs, New Lives in Postindustrial America*. Chicago: University of Chicago Press.

Duffy, Jim, Georg Gunther, and Lloyd Walters. 1997. "Gender and Mathematical Problem Solving." *Sex Roles* 37: 477–94.

Durham, William H. 1979. *Scarcity and Survival in Central America: Ecological Origins of the Soccer War*. Stanford, CA: Stanford University Press.

Durkheim, Émile. 1951 [1897]. *Suicide: A Study in Sociology*, G. Simpson, ed., J. Spaulding and G. Simpson, trans. New York: Free Press.

_____. 1956. *Education and Sociology*, Sherwood D. Fox, trans. New York: Free Press.

_____. 1961 [1925]. *Moral Education: A Study in the Theory and Application of the Sociology of Education*, Everett K. Wilson and Herman Schnurer, trans. New York: Free Press.

_____. 1973 [1899–1900]. "Two Laws of Penal Evolution." *Economy and Society* 2: 285–308.

_____. 1976 [1915]. *The Elementary Forms of the Religious Life*, Joseph Ward Swain, trans. New York: Free Press.

Dutton, Judy. 2000. "Detect His Lies Every Time." *Cosmopolitan* April: 126.

Eagley, Alice H., and Wendy Wood.1999. "The Origins of Sex Differences in Human Behaviour: Evolved Dispositions Versus Social Roles." *American Psychologist* 54: 408–23.

Ebert, Roger. 1998. "Patch Adams." *Chicago Sun-Times*. On the World Wide Web at http://www.suntimes.com/ebert/ebert_reviews/1998/12/122504.html (2 May 2000).

Eccles, J. S., J. E. Jacobs, and R. D. Harold. 1990. "Gender Role Stereotypes, Expectancy Effects and Parents' Socialization of Gender Differences." *Journal of Social Issues* 46: 183–201.

Economakis, Evel, and Robert J. Brym. 1995. "Marriage and Militance in a Working Class District of St. Petersburg, 1896–1913." *Journal of Family History* 20: 23–43.

Edel, Abraham. 1965. "Social Science and Value: A Study in Interrelations." Pp. 218–38 in Irving Louis Horowitz, ed. *The New Sociology: Essays in Social Science and Social Theory in Honor of C. Wright Mills*. New York: Oxford University Press.

e.Harlequin.com. 2000. "About eHarlequin.com." On the World Wide Web at http://eharlequin.women.com/harl/globals/about/00bkrd11.htm (17 May 2000).

Ehrlich, Paul R. 1968. *The Population Bomb*. New York: Ballantine.

_____ and Anne H. Ehrlich. 1990. *The Population Explosion*. New York: Simon & Schuster.

_____, Gretchen C. Daily, Scott C. Daily, Norman Myers, and James Salzman. 1997. "No Middle Way on the Environment." *Atlantic Monthly* 280, 6: 98–104. On the World Wide Web at http://www.theatlantic.com/issues/97dec/enviro.htm (2 May 2000).

Eichler, Margrit. 1988. *Nonsexist Research Methods: A Practical Guide*. Boston: Unwin Hyman.

Eisenberg, David M., Ronald C. Kessler, Cindy Foster, Frances E. Norlock, David R. Calkins, and Thomas L. Delbanco. 1993. "Unconventional Medicine in the United States—Prevalence, Costs, and Patterns of Use." *New England Journal of Medicine* 328: 246.

Eisenstadt, S. N. 1956. *From Generation to Generation*. New York: Free Press.

Eisler, Riane. 1995 [1987]. *The Chalice and the Blade: Our History, Our Future*. New York: Harper Collins.

"Ejército Zapatista de Liberación Nacional." 2000. On the World Wide Web at http://www.ezln.org (30 July 2000).

Ekman, Paul. 1978. *Facial Action Coding System*. New York: Consulting Psychologists Press.

Elias, Norbert. 1985 [1982]. *The Loneliness of the Dying*, Edmund Jephcott, trans. Oxford, UK: Blackwell.

Elliott, H. L. 1995. "Living Vicariously Through Barbie." On the World Wide Web at http://ziris.syr.edu/path/public_html/barbie/main.html (19 November 1998).

Ellis, Brett Easton. 1991. *American Psycho*. New York: Vintage.

Ellul, Jacques. 1964 [1954]. *The Technological Society*, trans. John Wilkinson. New York: Vintage.

Ember, Carol, and Melvin Ember. 1973. *Anthropology*. New York: Appleton-Century-Crofts.

Engels, Frederick. 1970 [1884]. *The Origins of the Family, Private Property and the State*, Eleanor Burke Leacock, ed., Alec West, trans. New York: International Publishers.

England, Paula. 1992. *Comparable Worth: Theories and Evidence*. Hawthorne, NY: Aldine de Gruyter.

Entine, John. 2000. *Taboo: Why Black Athletes Dominate Sports and Why We Are Afraid to Talk About It*. New York: Public Affairs.

Entwisle, Doris R., and Karl L. Alexander. 1995. "A Parent's Economic Shadow: Family Structure versus Family Resources as Influences on Early School Achievement." *Journal of Marriage and the Family* 57: 399–410.

Epstein, Helen. 1998. "Life and Death on the Social Ladder." *New York Review of Books* 45, 12: 16 July: 26–30.

Epstein, Steven. 1996. *Impure Science: AIDS, Activism, and the Politics of Knowledge*. Berkeley, CA: University of California Press.

Erikson, Robert, and John H. Goldthorpe. 1992. *The Constant Flux: A Study of Class Mobility in Industrial Societies*. Oxford, UK: Clarendon Press.

Esping-Andersen, Gøsta. 1990. *The Three Worlds of Welfare Capitalism*. Princeton, NJ: Princeton University Press.

"Estimated Starting Salaries: New College Graduates, 1996–97." 1997. On the World Wide Web at http://www.excite.com/eduframes/frame.dcg? (30 April 2000).

Estrich, Susan. 1987. *Real Rape*. Cambridge, MA: Harvard University Press.

Etzioni, Amitai. 1975. *A Comparative Analysis of Complex Organizations*, 2nd ed. New York: Free Press.

"European Tobacco Ban Receives Fresh Approval." 1998. *The Lancet* 351: 1568.

Evans, Peter B., Dietrich Rueschemeyer, and Theda Skocpol. 1985. *Bringing the State Back In*. Cambridge, UK: Cambridge University Press.

Evans, Robert G. 1999. "Social Inequalities in Health." *Horizons* (Policy Research Secretariat, Government of Canada) 2, 3: 6–7.

Everett-Green, Robert. 1999. "Puff Daddy: The Martha Stewart of Hip-Hop." *Globe and Mail* 4 September: C7.

Ewen, Stuart. 1976. *Captains of Consciousness: Advertising and the Social Roots of Consumer Culture*. New York: McGraw-Hill.

_____. 1988. *All Consuming Images: The Politics of Style in Contemporary Culture*. New York: Basic Books.

_____. 1997. *PR! A Social History of Spin*. New York: Basic Books.

Executive Office of the President of the United States. 2000. "A Citizen's Guide to the Federal Budget." On the World Wide Web at http://usgovinfo.about.com/newsissues/usgovinfo/gi/dynamic/offsite.htm?site=http://w3.access.gpo.gov/usbudget (8 June 2000).

"Face of the Web Study Pegs Global Internet Population at More than 300 Million." 2000. On the World Wide Web at http://www.angusreid.com/media/content/displaypr.cfm?id_to_view=1001 (2 October 2000).

Fall Colors II: Exploring the Quality of Diverse Portrayals on Prime Time Television. 2000. Oakland, CA: Children Now. On the World Wide Web at http://www.childrennow.org/media/fall-colors-2k/fc2-2k.pdf (5 August 2000).

Farmer, Paul. 1992. *AIDS and Accusation: Haiti and the Geography of Blame*. Berkeley, CA: University of California Press.

Fass, Paula S. 1989. *Outside In: Minorities and the Transformation of American Education*. New York: Oxford University Press.

Fava, Sylvia Fleis. 1956. "Suburbanism as a Way of Life." *American Sociological Review* 21: 34–7.

Feagin, Joe R., and Melvin P. Sikes. 1994. *Living with Racism: The Black Middle-Class Experience*. Boston: Beacon Press.

Fearon, E. R. 1997. "Human Cancer Syndrome: Clues to the Origin and Nature of Cancer." *Science* 278: 1043–50.

Featherman, David L., and Robert M. Hauser. 1976. "Sexual Inequalities and Socioeconomic Achievement in the U.S., 1962–1973." *American Sociological Review* 41: 462–83.

_____ and _____. 1978. *Opportunity and Change*. New York: Academic Press.

_____, F. Lancaster Jones, and Robert M. Hauser. 1975. "Assumptions of Mobility Research in the United States: The Case of Occupational Status." *Social Science Research* 4: 329–60.

Febvre, Lucien, and Henri-Jean Martin. 1976 [1958]. *The Coming of the Book: The Impact of Printing 1450–1800*, David Gerard, trans. London: NLB.

Feeley, Malcolm M., and Jonathan Simon. 1992. ""The New Penology: Notes on the Emerging Strategy of Corrections and Its Implications." *Criminology* 30: 449–74.

Fein, Helen. 1979. *Accounting for Genocide: National Responses and Jewish Victimization During the Holocaust*. New York: Free Press.

Felson, Richard B. 1996. "Mass Media Effects on Violent Behavior." *Annual Review of Sociology* 22: 103–28.

Feminist.com. "The Wage Gap." 1999. On the World Wide Web at http://www.feminist.com/wgot.htm (30 April 2000).

Fernandez-Dols, Jose-Miguel, Flor Sanchez, Pilar Carrera, and Maria-Angeles Ruiz-Belda. 1997. "Are Spontaneous Expressions and Emotions Linked? An Experimental Test of Coherence." *Journal of Nonverbal Behavior* 21: 163–77.

Ferrante, Joan, and Pierre Brown Jr., eds. 2001 [1998]. *The Social Construction of Race and Ethnicity in the United States*, 2nd ed. Upper Saddle River, NJ: Prentice Hall.

Feynman, Richard P. 1999. *The Pleasure of Finding Things Out*. Jeffrey Robbins, ed. Cambridge, MA: Perseus Books.

Figart, Deborah M., and June Lapidus. 1996. "The Impact of Comparable Worth on Earnings Inequality." *Work and Occupations* 23: 297–318.

Finke, Roger, and Rodney Starke. 1992. *The Churching of America, 1776–1990: Winners and Losers in Our Religious Economy*. New Brunswick, NJ: Rutgers University Press.

Firebaugh, Glenn, and Frank D. Beck. 1994. "Does Economic Growth Benefit the Masses? Growth, Dependence and Welfare in the Third World." *American Journal of Sociology* 59: 631–53.

Fischer, Claude S. 1981. "The Public and Private Worlds of City Life." *American Sociological Review* 46: 306–16.

_____. 1984 [1976]. *The Urban Experience*, 2nd ed. New York: Harcourt Brace Jovanovich.

_____, Michael Hout, Martín Sánchez Jankowski, Samuel R. Lucas, Ann Swidler, and Kim Voss. 1996. *Inequality by Design: Cracking the Bell Curve Myth*. Princeton, NJ: Princeton University Press.

Fitzgerald, Mark. 1997. "Media Perpetuate a Myth." *Editor and Publisher* 130, 33: 13.

Flexner, Eleanor. 1975. *Century of Struggle: The Woman's Rights Movement in the United States*, rev. ed. Cambridge, MA: Harvard University Press.

Fludd, Robert. 1617–19. *Utriusque Cosmi Maioris Scilicet et Minoris Metaphysica, Physica Atqve Technica Historia*. Oppenheim, Germany: Johan-Theodori de Bry.

Folbre, Nancy. 1995. *The New Field Guide to the U.S. Economy: A Compact and Irreverent Guide to Economic Life in America*. New York: New Press.

Forbes.com. 1999. "Forbes Top 40 Entertainers." On the World Wide Web at http://www.forbes.com/tool/toolbox/entertain (28 April 2000).

_____. 1999a. "Forbes 400 Richest People in America." On the World Wide Web at http://www.forbes.com/tool/toolbox/rich400/asp/WorthIndex.asp?year=1999&value2=1 (29 April 2000).

_____. 1999b. "Forbes: The World's Richest People: USA 98." On the World Wide Web at http://www.forbes.com/tool/toolbox/billnew/country98.asp?country=United%20States,98 (1 May 2000).

_____. 2000. "Forbes 500 Annual Directory." On the World Wide Web at http://www.forbes.com/tool/toolbox/forbes500s/asp/rankindex.asp (30 April 2000).

Foucault, Michel. 1977 [1975]. *Discipline and Punish: The Birth of the Prison*. Alan Sheridan, trans. New York: Pantheon.

_____. 1990 [1978]. *The History of Sexuality: An Introduction*, Vol. 1. Robert Hurley, trans. New York: Vintage.

Frank, André Gundar. 1967. *Capitalism and Underdevelopment in Latin America: Historical Studies of Chile and Brazil*. New York: Monthly Review Press.

Frank Porter Graham Child Development Center. 1999. "Early Learning, Later Success: The Abecedarian Study." On the World Wide Web at http://www.fpg.unc.edu/~abc/abcedarianWeb/index.htm (10 August 2000).

Frank, Robert H. 1988. *Passions Within Reason: The Strategic Role of the Emotions*. New York: W. W. Norton.

Frank, Thomas. 1997. *The Conquest of Cool*. Chicago: University of Chicago Press.

_____ and Matt Weiland, eds. 1997. *Commodify Your Dissent: Salvos from the Baffler*. New York: W. W. Norton.

Franke, Richard H., and James D. Kaul. 1978. "The Hawthorne Experiments: First Statistical Interpretation." *American Sociological Review* 43: 623–43.

Franke, Richard W., and Barbara H. Chasin. 1992. *Kerala: Development Through Radical Reform*. San Francisco: Institute for Food and Development Policy.

Franklin, Karen. 1998. "Psychosocial Motivations of Hate Crime Perpetrators." Paper presented at the annual meetings of the American Psychological Association (San Francisco: 16 August).

Franklin, Sara, and Helen Ragone, eds. 1999. *Reproducing Reproduction*. Philadelphia: University of Pennsylvania Press.

Freidson, Eliot. 1986. *Professional Powers: A Study of the Institutionalization of Formal Knowledge*. Chicago: University of Chicago Press.

Freire, Paolo. 1972. *The Pedagogy of the Oppressed*. New York: Herder and Herder.

Freud, Sigmund. 1962 [1930]. *Civilization and Its Discontents*. James Strachey, trans. New York: W. W. Norton.

_____. 1973 [1915–17]. *Introductory Lectures on Psychoanalysis*. James Strachey, trans. James Strachey and Angela Richards, eds. Harmondsworth, UK: Penguin.

_____. 1977 [1905]. *On Sexuality*. James Strachey, trans., Angela Richards, comp. and ed. Vol. 7 of the Pelican Freud Library. Harmondsworth, UK: Penguin.

Freudenburg, William R. 1997. "Contamination, Corrosion and the Social Order: An Overview." *Current Sociology* 45, 3: 19–39.

Frey, William H., with Cheryl First. 1997. *Investigating Change in American Society: Exploring Social Trends with U.S. Census Data and StudentChip*. Belmont, CA: Wadsworth.

Fried, Martha Nemes, and Morton H. Fried. 1980. *Transitions: Four Rituals in Eight Cultures*. New York: W. W. Norton.

Friedenberg, Edgar Z. 1959. *The Vanishing Adolescent*. Boston: Beacon Press.

Froissart, Jean. 1968 [c. 1365]. *Chronicles,* selected and translated by Geoffrey Brereton. Harmondsworth, UK: Penguin.

Frum, David. 2000. "Values Gulf Has Split U.S. in Two." *National Post* 9 November: A3.

Furstenberg, Frank F., Jr., and Andrew Cherlin. 1991. *Divided Families: What Happens to Children When Parents Part*. Cambridge, MA: Harvard University Press.

Gaines, Donna. 1990. *Teenage Wasteland: Suburbia's Dead End Kids*. New York: Pantheon.

Galinsky, Ellen, Stacy S. Kim, and James T. Bond. 2001. *Feeling Overworked: When Work Becomes Too Much*. New York: Families and Work Institute.

Galper, Joseph. 1998. "Schooling for Society." *American Demographics* 20, 3: 33–4.

Galt, Virginia. 1999. "Jack Falling Behind Jill in School, Especially in Reading." *Globe and Mail* 30 October: A10.

Gambetta, Diego, ed. 1988. *Trust: Making and Breaking Cooperative Relations*. Oxford, UK: Blackwell.

Gamson, William A. 1975. *The Strategy of Social Protest*. Homewood, IL: Dorsey Press.

_____, Bruce Fireman, and Steven Rytina. 1982. *Encounters with Unjust Authority*. Homewood, IL: Dorsey Press.

Gans, Herbert. 1962. *The Urban Villagers: Group and Class in the Life of Italian-Americans*. New York: Free Press.

_____. 1979a. *Deciding What's News: A Study of CBS Evening News, NBC Nightly News, Newsweek and Time*. New York: Pantheon.

Gans 1979b, "Symbolic Ethnicity: The Future of Ethnic Groups and Cultures in America." Pp. 193-220 in Herbert Gans et al., eds. *On the Making of Americans: Essays in Honor of David Reisman*. Philadelphia: University of Pennsylvania Press.

_____. 1995. *The War Against Poverty: The Underclass and Antipoverty Policy*. New York: Basic Books.

Ganz, Marshall. 1996. "Motor Voter or Motivated Voter?" *The American Prospect* 28. On the World Wide Web at http://www.prospect.org/archives/28/28ganz.html (21 November 2000).

Gap.com. "Gap." 1999. On the World Wide Web at http://www.gap.com/onlinestore/gap/advertising/khakitv.asp (28 April 2000).

"Garciaparra Explains his Superstitions." 2000. On the World Wide Web at http://www.geocities.com/Colosseum/Track/4242/nomar3.wav (28 April 2000).

Garfinkel, Harold. 1967. *Studies in Ethnomethodology*. Englewood Cliffs, NJ: Prentice-Hall.

Garfinkel, Simson. 2000. *Database Nation: The Death of Privacy in the 21st Century*. Cambridge, MA: O'Reilly and Associates.

Garkawe, Sam. 1995. "The Impact of the Doctrine of Cultural Relativism on the Australian Legal System." *E Law* 2, 1. On the World Wide Web at http://www.murdoch.edu.au/elaw/issues/v2n1/garkawe.txt (10 May 2000).

Garland, David. 1990. *Punishment and Modern Society: A Study in Social Theory*. Chicago: University of Chicago Press.

Garner, David M. 1997. "The 1997 Body Image Survey Results." *Psychology Today* 30, 1: 30–44.

Garrau, Joel. 1991. *Edge City: Life on the New Frontier*. New York: Doubleday.

Gates, Bill with Nathan Myhrvold and Peter Rinearson. 1996. *The Road Ahead*. New York: Penguin.

Gaubatz, Kathlyn Taylor. 1995. *Crime in the Public Mind*. Ann Arbor, MI: University of Michigan Press.

Gelbspan, Ross. 1997. *The Heat is On: The High Stakes Battle Over Earth's Threatened Climate*. Reading, MA: Addison-Wesley.

_____. 1999. "Trading Away Our Chances to End Global Warming." *Boston Globe* 16 May: E2.

Gelles, Richard J. 1997 [1985]. *Intimate Violence in Families*, 3rd ed. Thousand Oaks, CA: Sage.

Gellner, Ernest. 1988. *Plough, Sword and Book: The Structure of Human History*. Chicago: University of Chicago Press.

General Motors. 2000. "GM Energy and Environment Vehicle Strategy: Creating Products and Options." On the World Wide Web at http://www.gm.com/environment/products/chart/index.html (2 August 2000).

Gerber, Theodore P., and Michael Hout. 1998. "More Shock than Therapy: Market Transition, Employment, and Income in Russia, 1991–1995." *American Journal of Sociology* 104: 1–50.

Gerbner, George. 1998. "Casting the American Scene: A Look at the Characters on Prime Time and Daytime Television from 1994–1997." *The 1998 Screen Actors Guild Report*. On the World Wide Web at http://www.media-awareness.ca/eng/issues/minrep/resource/reports/gerbner.htm (5 August 2000).

Germani, Gino, and Kalman Silvert. 1961. "Politics, Social Structure and Military Intervention in Latin America." *European Journal of Sociology* 11: 62–81.

Giddens, Anthony. 1987. *Sociology: A Brief But Critical Introduction,* 2nd ed. New York: Harcourt Brace Jovanovich.

Gilligan, Carol. 1982. *In a Different Voice: Psychological Theory and Women's Development*. Cambridge, MA: Harvard University Press.

_____, Nona P. Lyons, and Trudy J. Hanmer, eds. 1990. *Making Connections: The Relational Worlds of Adolescent Girls at Emma Willard School.* Cambridge, MA: Harvard University Press.

Gilman, Sander L. 1991. *The Jew's Body.* New York: Routledge.

Gimbutas, Marija. 1982. *Goddesses and Gods of Old Europe.* Berkeley and Los Angeles: University of California Press.

Ginsberg, Benjamin. 1986. *The Captive Public: How Mass Opinion Promotes States Power.* New York: Basic Books.

Gitlin, Todd. 1983. *Inside Prime Time.* New York: Pantheon.

Glaser, Barney, and Anselm Straus. 1967. *The Discovery of Grounded Theory.* Chicago: Aldine.

Glazer, Nathan. 1997. *We Are All Multiculturalists Now.* Cambridge, MA: Harvard University Press.

_____ and Daniel Patrick Moynihan. 1963. *Beyond the Melting Pot.* Cambridge, MA: MIT Press.

Glick, Jennifer E. 1999. "Economic Support from and to Extended Kin: A Comparison of Mexican Americans and Mexican Immigrants." *The International Migration Review* 33: 745–65.

Global Reach. 2001. "Global Internet Statistics (by Language)." On the World Wide Web at http://www.glreach.com/globstats/index.php3 (23 September 2001).

"Global 1000." 1999. *Business Week Online.* On the World Wide Web at http://www.businessweek.com (12 July 1999).

Glock, Charles Y. 1962. "On the Study of Religious Commitment." *Religious Education* 62, 4: 98–110.

"Gnutella." 2000. On the World Wide Web at http://gnutella.wego.com (7 August 2000).

Goddard Institute for Space Studies. 2001. "Annual Mean Temperature Anomalies in .01 C: Selected Zonal Means." On the World Wide Web at http://www.giss.nasa.gov/data/update/gistemp/ZonAnn.Ts.txt (10 May).

Goffman, Erving. 1959 [1956]. *The Presentation of Self in Everyday Life.* Garden City, NY: Anchor.

_____. 1961. *Asylums: Essays on the Social Situation of Mental Patients and Other Inmates.* Garden City, NY: Anchor Books.

_____. 1963. *Stigma: Notes on the Management of Spoiled Identity.* Englewood Cliffs, NJ: Prentice-Hall.

_____. 1974. *Frame Analysis.* Cambridge, MA: Harvard University Press.

_____. 1981. *Forms of Talk.* Philadelphia: University of Pennsylvania Press.

Goldhagen, Daniel Jonah. 1996. *Hitler's Willing Executioners: Ordinary Germans and the Holocaust.* New York: Knopf.

Goldthorpe, John H. in collaboration with Catriona Llewellyn and Clive Payne. 1987 [1980]. *Social Mobility and Class Structure in Modern Britain,* 2nd ed. Oxford, UK: Clarendon Press.

Goode, Erich, and Nachman Ben-Yehuda. 1994. *Moral Panics: The Social Construction of Deviance.* Cambridge, MA: Blackwell.

Goodell, Jeff. 1999. "Down and Out in Silicon Valley." *Rolling Stone* 9 December: 64–71.

Gordon, David M. 1996. *Fat and Mean: The Corporate Squeeze of Working Americans and the Myth of Managerial "Downsizing."* New York: Free Press.

_____, Richard Edwards, and Michael Reich. 1982. *Segmented Work, Divided Workers: The Historical Transformation of Labor in the United States.* New York: Cambridge University Press.

"Gore-Bush Race Reflects Many Divisions in Nation." 2000. On the World Wide Web at http://www.cnn.com/2000/ALLPOLITICS/stories/11/08/exit.polls (8 November 2000).

Gorman, Christine. 1998. "Playing the HMO Game." *Time* 152, 2: 13 July. On the World Wide Web at http://www.time.com/time/magazine/1998/dom/980713/cover1.html (2 May 2000).

Gormley, Jr., William T. 1995. *Everybody's Children: Child Care as a Public Problem.* Washington, DC: Brookings Institution.

Gottdiener, Mark, and Ray Hutchison. 2000 [1994]. *The New Urban Sociology,* 2nd ed. Boston: McGraw-Hill.

Gottfredson, Michael, and Travis Hirschi. 1990. *A General Theory of Crime.* Stanford, CA: Stanford University Press.

Goubert, Jean-Pierre. 1989 [1986]. *The Conquest of Water,* Andrew Wilson, trans. Princeton, NJ: Princeton University Press.

Gould, Stephen Jay. 1988. "Kropotkin Was No Crackpot." *Natural History* 97, 7: 12–18.

_____. 1996 [1981]. *The Mismeasure of Man,* rev. ed. New York: W. W. Norton.

Gouldner, Alvin W. 1954. *Patterns of Industrial Bureaucracy: A Case Study of Modern Factory Administration.* New York: Free Press.

Granovetter, Mark. 1973. "The Strength of Weak Ties." *American Sociological Review* 78: 1360–80.

_____. 1984. "Small is Bountiful." *American Sociological Review* 49: 323–34.

_____. 1995. *Getting a Job: A Study of Contacts and Careers.* Chicago: University of Chicago Press.

Greeley, Andrew. 1974. *Ethnicity in the United States.* New York: John Wiley.

_____. 1989. *Religious Change in America.* Cambridge, MA: Harvard University Press.

Green, Richard. 1974. *Sexual Identity Conflict in Children and Adults.* Baltimore: Penguin.

Greenberg, David F. 1988. *The Construction of Homosexuality.* Chicago: University of Chicago Press.

Greenpeace. 2000. "Greenpeace Contacts Worldwide." 2000. On the World Wide Web at http://adam.greenpeace.org/information.shtml (2 May 2000).

Greenstein, Theodore N. 1996. "Husbands' Participation in Domestic Labor: Interactive Effects of Wives' and Husbands' Gender Ideologies." *Journal of Marriage and the Family* 58: 585–95.

Grescoe, P. 1996. *The Merchants of Venus: Inside Harlequin and the Empire of Romance.* Vancouver: Raincoast.

Grindstaff, Laura. 1997. "Producing Trash, Class, and the Money Shot: A Behind-the-Scenes Account of Daytime TV Talk Shows." Pp. 164–202 in James Lull and Stephen Hinerman, eds. *Media Scandals: Morality and Desire in the Popular Culture Marketplace.* Cambridge, UK: Polity Press.

Griswold, Wendy. 1992. "The Sociology of Culture: Four Good Arguments (And One Bad One)." *Acta Sociologica* 35: 322–28.

Grofman, Bernard. 2000. "Questions and Answers About Motor Voter." On the World Wide Web at http://www.fairvote.org/reports/1995/chp6/grofman.html (21 November 2000).

Grusky, David B., ed. 1994. *Social Stratification: Class, Race, and Gender in Sociological Perspective.* Boulder, CO: Westview.

_____, and Robert M. Hauser. 1984. "Comparative Social Mobility Revisited: Models of Convergence and Divergence in Sixteen Countries." *American Sociological Review* 49: 19–38.

Gurr, Ted Robert. 1970. *Why Men Rebel*. Princeton, NJ: Princeton University Press.

Gusfield, Joseph R. 1963. *Symbolic Crusade: Status Politics and the American Temperance Movement*. Urbana, IL: University of Illinois Press.

Gutiérrez, David G. 1995. *Walls and Mirrors: Mexican Americans, Mexican Immigrants, and the Politics of Ethnicity*. Berkeley, CA: University of California Press.

"GVU's WWW User Survey." 1999. On the World Wide Web at http://www.cc.gatech.edu/gvu/user_surveys/survey-1998-10/graphs/graphs.html#general (11 February 2001).

Haas, Jack, and William Shaffir. 1987. *Becoming Doctors: The Adoption of a Cloak of Competence*. Greenwich, CT: JAI Press.

Haas, Jennifer. 1998. "The Cost of Being a Woman." *New England Journal of Medicine* 338: 1694–5.

Habermas, Jürgen. 1989. *The Structural Transformation of the Public Sphere,* Thomas Burger, trans. Cambridge, MA: MIT Press.

Hacker, Andrew. 1992. *Two Nations: Black and White, Separate, Hostile, Unequal*. New York: Ballantine Books.

_____. 1997. *Money: Who Has How Much and Why*. New York: Scribner.

Hacker, Jacob S. 1997. *The Road to Nowhere: The Genesis of President Clinton's Plan for Health Security*. Princeton, NJ: Princeton University Press.

Hagan, John. 1989. *Structuralist Criminology*. New Brunswick, NJ: Rutgers University Press.

_____. 1994. *Crime and Disrepute*. Thousand Oaks, CA: Pine Forge Press.

_____, John Simpson, and A.R. Gillis. 1987. "Class in the Household: A Power-Control Theory of Gender and Delinquency." *American Journal of Sociology* 92: 788–816.

Haines, Herbert H. 1996. *Against Capital Punishment: The Anti-Death Penalty Movement in America, 1972–1994*. New York: Oxford University Press.

Hall, Edward, 1959. *The Silent Language*. New York: Doubleday.

_____. 1966. *The Hidden Dimension*. New York: Doubleday.

Hall, Stuart. 1980. "Encoding/Decoding." Pp. 128–38 in Stuart Hall, Dorothy Hobson, Andrew Lowe, and Paul Willis, eds. *Culture, Media, Language: Working Papers in Cultural Studies, 1972–79*. London: Hutchinson.

_____. 1992. "The Question of Cultural Identity." Pp. 274–313 in Stuart Hall, David Held, and Tim McGrew, eds. *Modernity and its Futures*. Cambridge, UK: Polity and Open University Press.

Hammer, Michael. 1999. "Is Work Bad for You?" *The Atlantic Monthly* August: 87–93.

Hampton, Janie, ed. 1998. *Internally Displaced People: A Global Survey*. London: Earthscan.

Hancock, Lynnell. 1994. "In Defiance of Darwin: How a Public School in the Bronx Turns Dropouts into Scholars." *Newsweek* 24 October: 61.

Handelman, Stephen. 1995. *Comrade Criminal: Russia's New Mafiya*. New Haven, CT: Yale University Press.

Haney, Craig, W. Curtis Banks, and Philip G. Zimbardo. 1973. "Interpersonal Dynamics in a Simulated Prison." *International Journal of Criminology and Penology* 1: 69–97.

Hanke, Robert. 1998. "'Yo Quiero Mi MTV!' Making Music Television for Latin America." Pp. 219–45 in Thomas Swiss, Andrew Herman, and John M. Sloop, eds. *Mapping the Beat: Popular Music and Contemporary Theory*. Oxford, UK: Blackwell.

Hannigan, John. 1995a. "The Postmodern City: A New Urbanization?" *Current Sociology* 43, 1: 151–217.

Hannigan, John. 1995b. *Environmental Sociology: A Social Constructionist Perspective*. London: Routledge.

_____. 1998a. *Fantasy City: Pleasure and Profit in the Postmodern Metropolis*. New York: Routledge.

_____. 1998b [1995]. "Urbanization." Pp. 337–59 in Robert J. Brym, ed. *New Society: Sociology for the 21st Century*, 2nd ed. Toronto: Harcourt Brace Canada.

Hannon, Roseann, David S. Hall, Todd Kuntz, Van Laar, and Jennifer Williams. 1995. "Dating Characteristics Leading to Unwanted vs. Wanted Sexual Behavior." *Sex Roles* 33: 767–83.

Hansen, James E., Makiko Sato, Reto Ruedy, Andrew Lacis, and Jay Glascoe. 1998. "Global Climate Data and Models: A Reconciliation." *Science* 281: 930–2.

Hao, Xiaoming. 1994. "Television Viewing Among American Adults in the 1990s." *Journal of Broadcasting and Electronic Media* 38: 353–60.

Harden, Blaine. 2001. "Two-Parent Families Rise After Changes in Welfare." *The New York Times Online.* On the World Wide Web at http://www.nytimes.com/2001/08/12/national/12FAMI.html?todaysheadlines=&pagewanted=print (12 August 2001).

Harrigan, Jinni A., and Kristy T. Tiang. 1997. "Fooled by a Smile: Detecting Anxiety in Others." *Journal of Nonverbal Behavior* 21: 203–21.

Harris, Chauncy D., and Edward L. Ullman. 1945. "The Nature of Cities." *Annals of the American Academy of Political and Social Science* 242: 7–17.

Harris, Kathleen Mullan. 1997. *Teen Mothers and the Revolving Welfare Door*. Philadelphia: Temple University Press.

Harris, Marvin. 1974. *Cows, Pigs, Wars and Witches: The Riddles of Culture*. New York, Random House.

Harrison, Bennett. 1994. *Lean and Mean: The Changing Landscape of Corporate Power in the Age of Flexibility*. New York: Basic Books.

Hart, David M., and David G. Victor. 1993. "Scientific Elites and the Making of U.S. Policy for Climate Change Research, 1957–74." *Social Studies of Science* 23:643–80.

Harvey, Andrew S., Katherine Marshall, and Judith A. Frederick. 1991. *Where Does the Time Go?* Ottawa: Statistics Canada.

Harvey, Elizabeth. 1999. "Sort-term and Long-term Effects of Early Parental Employment on Children of the National Longitudinal Survey of Youth." *Developmental Psychology* 35: 445–9.

Hastings, Arthur C., James Fadiman, and James C. Gordon, eds. 1980. *Health for the Whole Person: The Complete Guide to Holistic Medicine*. Boulder, CO: Westview Press.

Hawley, Amos. 1950. *Human Ecology: A Theory of Community Structure*. New York: Ronald Press.

Hechter, Michael. 1974. *Internal Colonialism: The Celtic Fringe in British National Development, 1536–1966*. Berkeley, CA: University of California Press.

_____. 1987. *Principles of Group Solidarity*. Berkeley, CA: University of California Press.

Hein, Simeon. 1992. "Trade Strategy and the Dependency Hypothesis: A Comparison of Policy, Foreign Investment, and Economic Growth in Latin America and East Asia." *Economic Development and Cultural Change* 40: 495–521.

Helsing, Knud J., Moyses Szklo, and George W. Comstock. 1981. "Factors Associated with Mortality After Widowhood." *American Journal of Public Health* 71: 802–9.

Henson, Kevin Daniel. 1996. *Just a Temp*. Philadelphia: Temple University Press.

Henwood, Doug. 1999. "The Material Rewards for Labor Over Time: Earnings." *Left Business Observer* 27 November. On the World Wide Web at http://www.panix.com/~dhenwood/Stats_earns.html (29 April 2000).

Herlihy, David. 1998. *The Black Death and the Transformation of the West*. Cambridge, MA: Harvard University Press.

Herman, Edward S., and Noam Chomsky. 1988. *Manufacturing Consent: The Political Economy of the Mass Media*. New York: Pantheon.

_____ and Gerry O'Sullivan. 1989. *The "Terrorism" Industry: The Experts and Institutions That Shape Our View of Terror*. New York: Pantheon.

Herrnstein, Richard J., and Charles Murray. 1994. *The Bell Curve: Intelligence and Class Structure in American Life*. New York: Free Press.

Hersch, Patricia. 1998. *A Tribe Apart: A Journey into the Heart of American Adolescence*. New York: Ballantine Books.

Hertzman, Clyde, 2000. "The Case for Early Childhood Development Strategy." *Isuma: Canadian Journal of Policy Research* 1, 2: 11–18.

Hesse-Biber, Sharlene. 1996. *Am I Thin Enough Yet? The Cult of Thinness and the Commercialization of Identity*. New York: Oxford University Press.

_____ and Gregg Lee Carter. 2000. *Working Women in America: Split Dreams*. New York: Oxford University Press.

Hilberg, Raoul. 1961. *The Destruction of the European Jews*. Chicago: Quadrangle Books.

Hirschi, Travis. 1969. *Causes of Delinquency*. Berkeley, CA: University of California Press.

_____ and Hanan C. Selvin. 1972. "Principles of Causal Analysis." Pp. 126–47 in Paul F. Lazarsfeld, Ann K. Pasanella, and Morris Rosenberg, eds. *Continuities in the Language of Social Research*. New York: Free Press.

Hobbes, Thomas. 1968 [1651]). *Leviathan*. Middlesex, UK: Penguin.

Hobsbawm, Eric. 1994. *Age of Extremes: The Short Twentieth Century, 1914–1991*. London: Abacus.

Hochschild, Arlie Russell. 1973. *The Unexpected Community: Portrait of an Old Age Subculture*. Berkeley, CA: University of California Press.

_____ with Anne Machung. 1989. *The Second Shift: Working Parents and the Revolution at Home*. New York: Viking.

Hochstetler, Andrew L., and Neal Shover. 1998. "Street Crime, Labor Surplus, and Criminal Punishment, 1980–1990." *Social Problems* 44: 358–68.

Hodge, Robert, and David Tripp. 1986. *Children and Television: A Semiotic Approach*. Cambridge UK: Polity.

Hodson, Randy, and Teresa Sullivan. 1995 [1990]. *The Social Organization of Work*, 2nd ed. Belmont, CA: Wadsworth.

Hoffman, Donna L., and Thomas P. Novak. 1998. "Bridging the Racial Divide on the Internet." *Science* 280: 390–1.

Hoggart, Richard. 1958. *The Uses of Literacy*. Harmondsworth, UK: Penguin.

Holloway, Marguerite. 1999. "The Aborted Crime Wave?" *Scientific American* 281, 6: 23–4.

"Hollywood Lights Up." *The Washington Post* 30 August 1997: A26.

Homans, George Caspar. 1950. *The Human Group*. New York: Harcourt, Brace.

_____. 1961. *Social Behavior: Its Elementary Forms*. New York: Harcourt, Brace and World.

Hooks, Bell. 1984. *Feminist Theory: From Margin to Center*. Boston: South End Press.

Hoover, Robert N. 2000. Cancer—Nature, Nurture, or Both." *New England Journal of Medicine* 343, 2. On the World Wide Web at http://www.nejm.org/content/2000/0343/0002/0135.asp (16 July 2000).

Horan, Patrick M. 1978. "Is Status Attainment Research Atheoretical?" *American Sociological Review* 43: 534–41.

Horkheimer, Max, and Theodor W. Adorno. 1986 [1944]. *Dialectic of Enlightenment*, John Cumming, trans. London: Verso.

Houseknecht, Sharon K., and Jaya Sastry. 1996. "Family 'Decline' and Child Well-Being: A Comparative Assessment." *Journal of Marriage and the Family* 58: 726–39.

Hout, Michael. 1988. "More Universalism, Less Structural Mobility: The American Occupational Structure in the 1980s." *American Journal of Sociology* 93: 1358–1400.

_____ and William R. Morgan. 1975. "Race and Sex Variations in the Causes of the Expected Attainments of High School Seniors." *American Journal of Sociology* 81: 364–94.

_____, Clem Brooks, and Jeff Manza. 1993. "The Persistence of Classes in Post-Industrial Societies." *International Sociology* 8: 259–77.

"How to Tell Your Friends from the Japs." 1941. *Time* 22 December: 33.

Hoyt, Homer. 1939. *The Structure and Growth of Residential Neighborhoods in American Cities*. Washington, DC: Federal Housing Authority.

Hughes, Fergus P. 1995 [1991]. *Children, Play and Development*, 2nd ed. Boston: Allyn and Bacon.

Hughes, H. Stuart. 1967. *Consciousness and Society: The Reorientation of European Social Thought, 1890–1930*. London: Macgibbon and Kee.

Human Rights Campaign. 1999. "The Hate Crime Prevention Act of 1999." On the World Wide Web at http://www.hrc.org/issues/leg/hcpa/index.html (30 April 2000).

Human Rights Watch. 1995. *The Human Rights Watch Global Report on Women's Human Rights*. New York: Human Rights Watch.

Hunter, James Davison. 1991. *Culture Wars: The Struggle to Define America*. New York: Basic Books.

"Hunting bin Laden." 1999. On the World Wide Web at http://www.pbs.org/wgbh/pages/frontline/shows/binladen (13 September 2001).

Huntington, Samuel. 1968. *Political Order in Changing Societies.* New Haven, CT: Yale University Press.

_____. 1991. *The Third Wave: Democratization in the Late Twentieth Century.* Norman, OK: University of Oklahoma Press.

IBM. 1997. *IBM Annual Report 1997.* On the World Wide Web at http://www.ibm.com/annualreport/1997/arbm.html (30 April 2000).

"ICQ.com." On the World Wide Web at http://web.icq.com (17 May).

Ignatiev, Noel. 1995. *How the Irish Became White.* New York: Routledge.

Illich, Ivan. 1976. *Limits to Medicine: Medical Nemesis: The Expropriation of Health.* New York: Penguin.

Inglehart, Ronald. 1997. *Modernization and Postmodernization: Cultural, Economic, and Political Change in 43 Societies.* Princeton, NJ: Princeton University Press.

Inkeles, Alex, and David H. Smith. 1976. *Becoming Modern: Individual Change in Six Developing Countries.* Cambridge, MA: Harvard University Press.

International Institute for Management Development. 1999. "The World Competitiveness Scoreboard." On the World Wide Web at http://www.imd.ch/wcy/ranking/ranking.cfm?CFID=5473&CFTOKEN=31174325 (30 April 2000).

Internet Software Consortium. 2000. "Distribution by Top-Level Domain Name by Host Count, July 1999." On the World Wide Web at http://www.isc.org/ds/WWW-9907/dist-bynum.html (2 May 2000).

Interuniversity Consortium for Political and Social Research. "Appendix F - 1980 Occupational Classification." 2000. On the World Wide Web at http://www.icpsr.umich.edu/GSS99/appendix/occu1980.htm (29 April 2000).

Iyengar, Shanto. 1991. *Is Anyone Responsible? How Television Frames Political Issues.* Chicago: University of Chicago Press.

Jacobs, Jane. 1961. *The Death and Life of Great American Cities.* New York: Random House.

_____. 1969. *The Economy of Cities.* New York: Random House.

Kimmerling, Baruch, ed. 1996. "Political Sociology at the Crossroads." *Current Sociology* 44, 3: 1–176.

"Internet Growth." 2000. On the World Wide Web at http://citywideguide.com/InternetGrowth.html (29 April 2000).

Isajiw, W. Wsevolod. 1978. "Olga in Wonderland: Ethnicity in a Technological Society." Pp. 29–39 in Leo Driedger, ed. *The Canadian Ethnic Mosaic: A Quest for Identity.* Toronto: McClelland and Stewart.

Jacobs, Jerry A. 1993. "Careers in the U.S. Service Economy." Pp. 195–224 in Gøsta Esping-Andersen, ed. *Changing Classes: Stratification and Mobility in Post–Industrial Societies.* London: Sage.

Jackman, Mary R., and Robert W. Jackman. 1983. *Class Awareness in the United States.* Berkeley CA: University of California Press.

James, William. 1976 [1902]. *The Varieties of Religious Experience: A Study in Human Nature.* New York: Collier Books.

Jencks, Christopher. 1994. *The Homeless.* Cambridge MA: Harvard University Press.

_____, Marshall Smith, Henry Acland, Mary Jo Bane, David Cohen, Herbert Gintis, Barbara Heyns, and Stephan Michelson. 1972. *Inequality: A Reassessment of the Effect of Family and Schooling in America.* New York: Basic Books.

Jeness, Valerie. 1995. "Social Movement Growth, Domain Expansion, and Framing Processes: The Gay/Lesbian Movement and Violence Against Gays and Lesbians as a Social Problem." *Social Problems* 42: 145–70.

Jenkins, J. Craig. 1983. "Resource Mobilization Theory and the Study of Social Movements." *Annual Review of Sociology* 9: 527–53.

Jensen, Margaret Ann. 1984. *Love's Sweet Return. The Harlequin Story.* Toronto: Women's Press.

Johns, Adrian. 1998. *The Nature of the Book: Print and Knowledge in the Making.* Chicago: University of Chicago Press.

Johnson, Neil F. and Sushil Jajodia. 1998. "Exploring Steganography: Seeing the Unseen." *IEEE Computer* February. On the World Wide Web at http://www.jjtc.com/pub/r2026a.htm (13 September 2001).

Johnson, Terence J. 1972. *Professions and Power.* London: Macmillan.

Jones, Christopher. 1999. "Chiapas' Well–Connected Rebels." *Wired News* 1 February. On the World Wide Web at http://www.wired.com/news/print/0,1294,17633,00.html (30 July 2000).

Jones, Jacqueline. 1986. *Labor of Love, Love of Sorrow: Black Women, Work and Slavery from Slavery to the Present.* New York: Random House.

Jordan, Penny. 1999. *A Treacherous Seduction.* Toronto: Harlequin.

Joy, Bill. 2000. "Why the Future Doesn't Need Us." *Wired* 8, 4. On the World Wide Web at http://wired.com/wired/archive/8.04/joy_pr.html (4 November 2000).

Juergensmeyer, Mark. 2000. *Terror in the Mind of God: The Global Rise of Religious Violence.* Berkeley CA: University of California Press.

Kalmijn, Matthijs. 1996. "The Socioeconomic Assimilation of Caribbean American Blacks." *Social Forces* 74: 911–30.

_____. 1998. "Intermarriage and Homogamy: Causes, Patterns, Trends." *Annual Review of Sociology* 24: 395–421.

Kanter, Rosabeth Moss. 1977. *Men and Women of the Corporation.* New York: Basic Books.

_____. 1983. *The Change Masters: Innovation and Entrepreneurship in the American Corporation.* New York: Simon & Schuster.

_____. 1989. *When Giants Learn to Dance: Mastering the Challenges of Strategy, Management, and Careers in the 1990s.* New York: Simon & Schuster.

Karklins, Rasma. 1986. *Ethnic Relations in the USSR: The Perspective from Below.* London: Unwin Hyman.

Karl, Thomas R., and Kevin E. Trenberth. 1999. "The Human Impact on Climate." *Scientific American* 281, 6: September: 100–5.

Kasindorf, Martin, Stephanie Armour, and Andrea Stone. 1998. "In Work World, Affairs Can Drag Down People at the Top." *USA Today* 24 August: 8A.

Katz, Elihu. 1957. "The Two-Step Flow of Communication: An Up-to-Date Report on an Hypothesis." *Public Opinion Quarterly* 21: 61–78.

Katznelson, Ira, and Margaret Weir. 1985. *Schooling for All: Class, Race, and the Decline of the Democratic Ideal.* New York: Basic Books.

Kaufman, Bruce E. 1982. "The Determinants of Strikes in the United States, 1900–1977." *Industrial and Labor Relations Review* 35: 473–90.

Keller, Larry. 2000. "Dual Earners: Double Trouble." On the World Wide Web at http://www.cnn.com/2000/CAREER/trends/11/13/dual.earners (13 November 2000).

Kelley, Jack. 2001. "Terror Groups Hide behind Web Encryption." *USA Today* (June 19). On the World Wide Web at http://www.usatoday.com/life/cyber/tech/2001-02-05-binladen.htm (13 September 2001).

Kellner, Douglas. 1995. *Media Culture: Cultural Studies, Identity and Politics Between the Modern and the Postmodern.* New York: Routledge.

Kennedy, Paul. 1993. *Preparing for the Twenty-First Century.* New York: Harper Collins.

Kepel, Gilles. 1994 [1991]. *The Revenge of God: The Resurgence of Islam, Christianity and Judaism in the Modern World*, Alan Braley, trans. University Park, PA: Pennsylvania State University Press.

Kerig, Patricia K., Philip A. Cowan, and Carolyn Pape Cowan. 1993. "Marital Quality and Gender Differences in Parent-Child Interaction." *Developmental Psychology* 29: 931–39.

Kett, Joseph F. 1977. *Rites of Passage: Adolescence in America, 1790 to the Present.* New York: Basic Books.

Kevles, Daniel J. 1999. "Cancer: What Do They Know?" *New York Review of Books* 46, 14: 14–21.

Kidron, Michael, and Ronald Segal. 1995. *The State of the World Atlas*, 5th ed. London: Penguin.

Kinder, Donald R., and Lynn M. Sanders. 1996. *Divided by Color: Racial Politics and Democratic Ideals.* Chicago: University of Chicago Press.

Kinsey, Alfred C., Wardell B. Pomeroy, and Clyde E. Martin. 1948. *Sexual Behavior in the Human Male.* Philadelphia: W. B. Saunders.

Kirp, David L. 1982. *Just Schools: The Idea of Racial Equality in American Education.* Berkeley: University of California Press.

Klee, Kenneth. 1999. "The Siege of Seattle." *Newsweek* 13 December. On the World Wide Web at http://server5.ezboard.com/fdrugpolicytalkwtoseattleriots1999.showMessage?topicID=12.topic (3 August 2000).

Klingemann, Hans-Dieter. 1999. "Mapping Political Support in the 1990s: A Global Analysis." In Pippa Norris, ed. *Critical Citizens: Global Support for Democratic Governance.* Oxford, UK: Oxford University Press. On the World Wide Web at http://ksgwww.harvard.edu/people/pnorris/Chapter_2.htm (20 October 1999).

Klockars, Carl B. 1974. *The Professional Fence.* New York: Free Press.

Kluegel, James R., and Eliot R. Smith. 1986. *Beliefs about Inequality: Americans' Views of What Is and What Ought to Be.* New York: Aldine de Gruyter.

Kluger, Richard. 1996. *Ashes to Ashes: America's Hundred-Year Cigarette War, the Public Health, and the Unabashed Triumph of Philip Morris.* New York: Knopf.

Knight, Graham. 1998 [1995]. "The Mass Media." Pp. 103–27 in Robert J. Brym, ed. *New Society: Sociology for the 21st Century,* 2nd ed. Toronto: Harcourt Brace Canada.

Koepke, Leslie, Jan Hare, and Patricia B. Moran. 1992. "Relationship Quality in a Sample of Lesbian Couples with Children and Child-Free Lesbian Couples." *Family Relations* 41: 224–9.

Kohlberg, Lawrence. 1981. *The Psychology of Moral Development: The Nature and Validity of Moral Stages.* New York: Harper and Row.

Kolbert, E. 1995. "Americans Despair of Popular Culture." *New York Times* 20 August: Section 2: 1, 23.

Kornblum, William. 1997 [1988]. *Sociology in a Changing World*, 4th ed. Fort Worth: Harcourt Brace College Publishers.

Korpi, Walter. 1983. *The Democratic Class Struggle.* London: Routledge and Kegan Paul.

Kosmin, Barry A. 1991. *Research Report of the National Survey of Religious Identification.* New York: CUNY Graduate Center.

Koss, Mary P., Christine A. Gidycz, and Nadine Wisniewski. 1987. "The Scope of Rape: Incidence and Prevalence of Sexual Aggression and Victimization in a National Sample of Higher Education Students." *Journal of Consulting and Clinical Psychology* 55: 162–70.

Kozol, Jonathan. 1991. *Savage Inequalities: Children in America's Schools.* New York: Crown.

Krahn, Harvey, and Graham Lowe. 1998. *Work, Industry, and Canadian Society*, 3rd ed. Toronto: ITP Nelson.

Kreisi, Hanspeter, et al. 1995. *New Social Movements in Western Europe.* Minneapolis, MN: University of Minnesota Press.

Kristof, Nicholas D. 1997. "With Stateside Lingo, Valley Girl Goes Japanese," *New York Times* 19 October: Section 1, 3.

Kropotkin, Petr. 1908 [1902]. *Mutual Aid: A Factor of Evolution*, revised ed. London: W. Heinemann.

Kuhn, Thomas. 1970 [1962]. *The Structure of Scientific Revolutions*, 2nd ed. Chicago: University of Chicago Press.

Kurzweil, Ray. 1999. *The Age of Spiritual Machines: When Computers Exceed Human Intelligence.* New York: Viking Penguin.

Kuttner, Robert. 1997. "The Limits of Markets." *The American Prospect* 31: 28–41. On the World Wide Web at http://www.prospect.org/archives/31/31kuttfs.html (30 April 2000).

_____. 1998a. "In this For-Profit Age, Preventive Medicine Means Avoiding Audits." *Boston Globe* 22 March: E7.

_____. 1998b. "Toward Universal Coverage." *The Washington Post* 14 July: A15.

LaFeber, Walter. 1993. *Inevitable Revolutions: The United States in Central America*, 2nd ed. New York: W. W. Norton.

LaFree, Gary D. 1980. "The Effect of Sexual Stratification by Race on Official Reactions to Rape." *American Sociological Review* 45: 842–54.

_____. 1998. "Social Institutions and the Crime 'Bust' of the 1990s." *Journal of Criminal Law and Criminology* 88: 1325–68.

Lamont, Michele. 1992. *Money, Morals, and Manners.* Chicago: University of Chicago Press.

Lantz, Herman, Martin Schultz, and Mary O'Hara. 1977. "The Changing American Family from the Preindustrial to the Industrial Period: A Final Report." *American Sociological Review* 42: 406–21.

Lapidus, Gail Warshofsky. 1978. *Women in Soviet Society: Equality, Development, and Social Change.* Berkeley, CA: University of California Press.

Laslett, Peter. 1991 [1989]. *A Fresh Map of Life: The Emergence of the Third Age.* Cambridge, MA: Harvard University Press.

Lasswell, Harold. 1936. *Politics: Who Gets What, When and How.* New York: McGraw-Hill.

Laumann, Edward O., John H. Gagnon, Robert T. Michael, and Stuart Michaels. 1994. *The Social Organization of Sexuality: Sexual Practices in the United States*. Chicago: University of Chicago Press.

Lazare, Daniel. 1999. "Your Constitution Is Killing You: A Reconsideration of the Right to Bear Arms." *Harper's* 299, 1793, October: 57–65.

Lazonick, William. 1991. *Business Organization and the Myth of the Market Economy*. Cambridge, UK: Cambridge University Press.

Leacock, Eleanor Burke. 1981. *Myths of Male Dominance: Collected Articles on Women Cross-Culturally*. New York: Monthly Review Press.

Lears, Jackson. 1994. *Fables of Abundance: A Cultural History of Advertising in America*. New York: Basic Books.

Le Bon, Gustave. 1969 [1895]. *The Crowd: A Study of the Popular Mind*. New York: Ballantine Books.

Lee, Richard B. 1979. *The !Kung San: Men, Women and Work in a Foraging Society*. Cambridge, UK: Cambridge University Press.

Leidner, Robin. 1993. *Fast Food, Fast Talk: Service Work and the Routinization of Everyday Life*. Berkeley, CA: University of California Press.

Leman, Christopher. 1977. "Patterns of Policy Development: Social Security in the United States and Canada." *Public Policy* 25: 26–291.

Leman, Nicholas. 1998. "I'd Walk a Mile for a Fee." *New York Review of Books* 45, 11: 33–5.

Lenski, Gerhard. 1966. *Power and Privilege: A Theory of Social Stratification*. New York: McGraw-Hill.

_____, Patrick Nolan, and Jean Lenski. 1995. *Human Societies: An Introduction to Macrosociology*, 7th ed. New York: McGraw-Hill.

Lenton, Rhonda. 2001. "Sex, Gender, and Sexuality." Pp. 68-88 in Robert J. Brym, ed. *New Society: Sociology for the 21st Century*, 3rd ed. Toronto: Harcourt Canada.

Lerner, Gerda. 1986. *The Creation of Patriarchy*. New York: Oxford University Press.

"The Lesson Nobody Learns." 1999. *The Economist* 24 April: 25–6.

Levine, David. 1987. *Reproducing Families: The Political Economy of English Population History*. Cambridge, UK: Cambridge University Press.

Levy, Frank. 1998. *The New Dollars and Dreams: American Incomes and Economic Change*. New York: Russell Sage Foundation.

Lewontin, R. C. 1991. *Biology as Ideology: The Doctrine of DNA*. New York: Harper Collins.

Lichtenstein, Paul, Niels V. Holm, Pia K. Verkasalo, Anastasia Iliadou, Jaakko Kaprio, Markku Koskenvuo, Eero Pukkala, Axel Skytthe, and Kari Hemminki. 2000. "Environment and Heritable Factors in the Causation of Cancer—Analyses of Cohorts of Twins from Sweden, Denmark, and Finland." *New England Journal of Medicine* 343, 2: On the World Wide Web at http://content.nejm.org/cgi/content/short/343/2/78 (12 July 2001).

Lie, John. 1992. "The Concept of Mode of Exchange." *American Sociological Review* 57: 508–23.

_____. 1998. *Han Unbound: The Political Economy of South Korea*. Stanford, CA: Stanford University Press.

_____. 2001. *Multiethnic Japan*. Cambridge, MA: Harvard University Press.

Lieberson, Stanley. 1980. *A Piece of the Pie: Blacks and White Immigrants Since 1880*. Berkeley, CA: University of California Press.

_____.1991. "A New Ethnic Group in the United States." Pp. 444–57 in Norman R. Yetman, ed. *Majority and Minority: The Dynamics of Race and Ethnicity in American Life*, 5th ed. Boston: Allyn and Bacon.

_____, _____ and Mary Waters. 1986. "Ethnic Groups in Flux: The Changing Ethnic Responses of American Whites." *Annals of the American Academy of Social and Political Science* 487: 79–91.

Liebow, Elliot. 1967. *Tally's Corner: A Study of Negro Street-Corner Men*. Boston: Little, Brown.

_____. 1993. *Tell Them Who I Am: The Lives of Homeless Women*. New York: Free Press.

Light, Ivan. 1991. "Immigrant and Ethnic Enterprise in North America." Pp. 307–18 in Norman R. Yetman, ed. *Majority and Minority: The Dynamics of Race and Ethnicity in American Life*, 5th ed. Boston: Allyn and Bacon.

Lijphart, Arend. 1997. "Unequal Participation: Democracy's Unresolved Dilemma." *American Political Science Review* 91: 1–14.

Lindner, Rolf. 1996 [1990]. *The Reportage of Urban Culture: Robert Park and the Chicago School*. Adrian Morris, trans. Cambridge, UK: Cambridge University Press.

Lipset, Seymour Martin. 1971 [1951]. *Agrarian Socialism: The Cooperative Commonwealth Federation in Saskatchewan*, revised ed. Berkeley, CA: University of California Press.

_____. 1977. "Why No Socialism in the United States?" Pp. 31–363 in Seweryn Bialer and Sophia Sluzar, eds. *Sources of Contemporary Radicalism*. Boulder, CO: Westview Press.

_____. 1981 [1960]. *Political Man: The Social Bases of Politics*, 2nd ed. Baltimore: Johns Hopkins University Press.

_____. 1994. "The Social Requisites of Democracy Revisited." *American Sociological Review* 59: 1–22.

_____ and Reinhard Bendix. 1963. *Social Mobility in Industrial Society*. Berkeley, CA: University of California Press.

_____ and Stein Rokkan. 1967. "Cleavage Structures, Party Systems, and Voter Alignments: An Introduction." Pp. 1–64 in Seymour Martin Lipset and Stein Rokkan, eds. *Party Systems and Voter Alignments: Cross-National Perspectives*. New York: Free Press.

Lisak, David. 1992. "Sexual Aggression, Masculinity, and Fathers." *Signs* 16: 238–62.

"Liszt's Usenet NewsGroups Directory." 2000. On the World Wide Web at http://www.liszt.com/news (2 May 2000).

Livernash, Robert, and Eric Rodenburg. 1998. "Population Change, Resources, and the Environment." *Population Bulletin* 53, 1. On the World Wide Web at http://www.prb.org/pubs/population_bulletin/bu53-1.htm (25 August 2000).

Livi-Bacci, Massimo. 1992. *A Concise History of World Population*. Cambridge, MA: Blackwell.

Lock, Margaret. 1993. *Encounters with Aging: Mythologies of Menopause in Japan and North America*. Berkeley, CA: University of California Press.

Lodhi, Abdul Qaiyum, and Charles Tilly. 1973. "Urbanization, Crime, and Collective Violence in 19th Century France." *American Journal of Sociology* 79: 296–318.

Lofland, John, and Lyn H. Lofland. 1995 [1971]. *Analyzing Social Settings: A Guide to Qualitative Observation and Analysis*, 3rd ed. Belmont, CA: Wadsworth.

Logan, John R., and Harvey L. Molotch. 1987. *Urban Fortunes: The Political Economy of Place*. Berkeley, CA: University of California Press.

London, Scott. 1994. "Electronic Democracy—A Literature Survey." On the World Wide Web at http:///www.west.net/~insight/london/ed.htm (15 August 1998).

Long, Elizabeth. 1997. *From Sociology to Cultural Studies*. Malden, MA: Blackwell.

Longwood College Library. 2000. "Doctor-Assisted Suicide—Chronology of the Issue." On the World Wide Web at http://web.lwc.edu/administrative/library/death.htm (2 May 2000).

Lowe, Graham. 2000. *The Quality of Work: A People-Centred Agenda*. Toronto: Oxford University Press.

Lurie, Alison. 1981. *The Language of Clothes*. New York: Random House.

Luxembourg Income Study. 1999a. "LIS Inequality Indices." On the World Wide Web at http://lissy.ceps.lu/ineq.htm (29 April 2000).

_____. 1999b. "LIS Low Income Measures." On the World Wide Web at http://lissy.ceps.lu/lim.htm (29 April 2000).

Lynch, Michael, and David Bogen. 1997. "Sociology's Asociological 'Core': An Examination of Textbook Sociology in Light of the Sociology of Scientific Knowledge." *American Sociological Review* 62: 481–93.

Lyon, David, and Elia Zureik, eds. 1996. *Computers, Surveillance, and Privacy*. Minneapolis: University of Minnesota Press.

MacCarthy, Fiona. 1999. "Skin Deep." *New York Review of Books* 46, 15: 19–21.

MacDonald, K., and R. D. Parke. 1986. "Parent-Child Physical Play: The Effects of Sex and Age on Children and Parents." *Sex Roles* 15: 367–78.

Machung, Anne. 1989. "Talking Career, Thinking Jobs: Gender Differences in Career and Family Expectations of Berkeley Seniors." *Feminist Studies* 15: 35–58.

Macionis, John J. 1997 [1987]. *Sociology*, 6th ed. Upper Saddle River, NJ: Prentice-Hall.

MacKinnon, Catharine A. 1979. *Sexual Harassment of Working Women*. New Haven, CT: Yale University Press.

Macklin, Eleanor D. 1980. "Nontraditional Family Forms: A Decade of Research." *Journal of Marriage and the Family* 42: 905–22.

Maguire, Kathleen, and Ann L. Pastore, eds. 1998. *Sourcebook of Criminal Justice Statistics 1997*. On the World Wide Web at http://www.albany.edu/sourcebook/1995/pdf/t256.pdf (29 April 2000).

Maguire, Mike, Rod Morgan, and Robert Reiner, eds. 1994. *The Oxford Handbook of Criminology*. Oxford, UK: Clarendon Press.

Mahony, Rhona. 1995. *Kidding Ourselves: Breadwinning, Babies, and Bargaining Power*. New York: Basic Books.

Maloney, Wendi A. 1999. "Brick and Mortar." *Academe* 85, 5: 19–24.

Malthus, Thomas Robert. 1966 [1798]. *An Essay on the Principle of Population*. J.R. Bodnar, ed. London: Macmillan.

Manga, Pran, Douglas E. Angus, and William R. Swan. 1993. "Effective Management of Low Back Pain: It's Time to Accept the Evidence." *Journal of the Canadian Chiropractic Association* 37: 221–9.

Mankiw, N. Gregory. 1998. *Principles of Macroeconomics*. Fort Worth, TX: The Dryden Press.

Mann, Susan A., Michael D. Grimes, Alice Abel Kemp, and Pamela J. Jenkins. 1997. "Paradigm Shifts in Family Sociology? Evidence From Three Decades of Family Textbooks." *Journal of Family Issues* 18: 315–49.

Mannheim, Karl. 1952. "The Problem of Generations." Pp. 276–320 in *Essays on the Sociology of Knowledge*, Paul Kecskemeti, ed. New York: Oxford University Press.

Manza, Jeff, Michael Hout, and Clem Brooks. 1995. "Class Voting in Capitalist Democracies Since World War II: Dealignment, Realignment, or Trendless Fluctuation?" *Annual Review of Sociology* 21: 137–62.

Markowitz, Fran. 1993. *A Community in Spite of Itself: Soviet Jewish Émigrés in New York*. Washington, DC: Smithsonian Institute Press.

_____. 2000. *Coming of Age in Post-Soviet Russia*. Urbana-Champaign, IL: University of Illinois Press.

Marmor, Theodore R. 1994. *Understanding Health Care Reform*. New Haven, CT: Yale University Press.

Marrus, Michael Robert. 1987. *The Holocaust in History*. Hanover, NH: University Press of New England for Brandeis University Press.

Marshall, Gordon, Howard Newby, David Rose, and Carolyn Vogler. 1988. *Social Class in Modern Britain*. London: Hutchinson.

_____, Adam Swift, and Stephen Roberts. 1997. *Against the Odds? Social Class and Social Justice in Industrial Societies*. Oxford, UK: Clarendon Press.

Marshall, T. H. 1965. "Citizenship and Social Class." Pp. 71–134 in T. H. Marshall, ed. *Class, Citizenship, and Social Development: Essays by T. H. Marshall*. Garden City, NY: Anchor.

Martin, Joanne. 1992. *Cultures in Organizations*. New York: Oxford University Press.

Martin, Patricia Yancey, and Robert A. Hummer. 1989. "Fraternities and Rape on Campus." *Gender and Society* 3: 457–73.

Martineau, Harriet. 1985. *Harriet Martineau on Women*, Gayle Graham Yates, ed. New Brunswick, NJ: Rutgers University Press.

Marx, Gary T. 1997. "Of Methods and Manners for Aspiring Sociologists: 36 Moral Imperatives." *The American Sociologist* 28: 102–25. On the World Wide Web at http://web.mit.edu/gt-marx/www/37moral.html (27 April 2000).

Marx, Karl. 1904 [1859]. *A Contribution to the Critique of Political Economy*, N. Stone, trans. Chicago: Charles H. Kerr.

_____. 1970 [1843]. *Critique of Hegel's 'Philosophy of Right'*, Annette Jolin and Joseph O'Malley, trans. Cambridge, MA: Harvard University Press.

_____ and Friedrich Engels. 1972 [1848]. "Manifesto of the Communist Party." Pp. 331–62 in R. Tucker, ed. *The Marx-Engels Reader*. New York: W. W. Norton.

Massey, Douglas S., and Nancy A. Denton. "Trends in the Residential Segregation of Blacks, Hispanics, and Asians: 1970–1980." *American Sociological Review* 52: 802–25.

_____ and _____. 1993. *American Apartheid: Segregation and the Making of the Underclass*. Cambridge, MA: Harvard University Press.

Massing, Michael. 1999. "The End of Welfare?" *New York Review of Books* 46, 15: 22–6.

_____ et al. 1999. "Beyond Legalization: New Ideas for Ending the War on Drugs." *The Nation* 20 September: 11–48.

Masters, W. H., and V. E. Johnson. 1966. *Human Sexual Response*. Boston: Little, Brown.

Matalon, Jean-Marc. 1997. "Jeanne Calment, World's Oldest Person, Dead at 122." *The Shawnee News-Star* 5 August. On the World Wide Web at http://www.news-star.com/stories/080597/life1.html (2 May 2000).

Matsueda, Ross L. 1988. "The Current State of Differential Association Theory." *Crime and Delinquency* 34: 277–306.

_____. 1992. "Reflected Appraisals, Parental Labeling, and Delinquency: Specifying a Symbolic Interactionist Theory." *American Journal of Sociology* 97: 1577–1611.

Mattern, Mark. 1998. *Acting in Concert: Music, Community, and Political Action*. New Brunswick, NJ: Rutgers University Press.

Mauer, Marc. 1994. "Americans Behind Bars: The International Use of Incarceration, 1992–1993." On the World Wide Web at http://www.druglibrary.org/schaffer/Other/sp/abb.htm (29 April 2000).

Maume, David J., Jr., A. Silvia Cancio, and T. David Evans. 1996. "Cognitive Skills and Racial Wage Inequity: Reply to Farkas and Vicknair." *American Sociological Review* 61: 561–4.

Mayer, J. P. 1944. *Max Weber and German Politics*. London: Faber and Faber.

Mazur, Allan. 1998. "Global Environmental Change in the News: 1987–90 vs. 1992–96." *International Sociology* 13: 457–72.

McAdam, Doug. 1982. *Political Process and the Development of Black Insurgency, 1930–1970*. Chicago: University of Chicago Press.

_____, John D. McCarthy, and Mayer N. Zald. 1996. "Introduction: Opportunities, Mobilizing Structures, and Framing Processes—Toward a Synthetic, Comparative Perspective on Social Movements." Pp. 1–20 in Doug McAdam, John D. McCarthy, and Mayer N. Zald, eds. *Comparative Perspectives on Social Movements: Political Opportunities, Mobilizing Structures, and Cultural Framing*. New York: Cambridge University Press.

_____ and Douglas A. Snow. 1997. *Social Movements: Readings on their Emergence, Mobilization and Dynamics*. Los Angeles: Roxbury.

McCarthy, John D., and Mayer N. Zald. 1977. "Resource Mobilization and Social Movements: A Partial Theory." *American Journal of Sociology* 82: 1212–41.

McCarthy, Shawn. 1999. "Entrepreneurs Find a Place in the Sun." *Globe and Mail* 31 July: B1, B5.

McChesney, Robert W. 1999. "Oligopoly: The Big Media Game Has Fewer and Fewer Players." *The Progressive* November: 20–4. On the World Wide Web at http://www.progressive.org/mcc1199.htm (7 August 2000).

McClendon, McKee J. 1976. "The Occupational Status Attainment Processes of Males and Females." *American Sociological Review* 41: 52–64.

McCrum, Robert, William Cran, and Robert MacNeil. 1992. *The Story of English*, new and rev. ed. London: Faber and Faber.

McCullagh, Declan. 2000a. "Bin Laden: Steganography Master?" *Wired* February 7. On the World Wide Web at http://www.wired.com/news/print/0.1294.41658.00.html (13 September 2001).

_____. 2000b. "Regulating Privacy: At What Cost?" *Wired* September 19. On the World Wide Web at http://www.wired.com/news/print/0.1294.38878.00.html (13 September 2001).

McDonald's Corporation. 1999. "McDonald's Nutrition Facts." On the World Wide Web at http://www.mcdonalds.com/food/nutrition/index.html (16 August 1999).

"McDonald's Testing E-Burgers." 1999. *Wall Street Journal Interactive Edition*. 11 August. On the World Wide Web at http://www.zdnet.com/zdnn/stories/news/0,4586,2312611,00.html (29 April 2000).

McGinn, Anne Platt. 1998. "Promoting Sustainable Fisheries." Pp. 59–78 in Lester R. Brown, Christopher Flavin, Hilary French, et al. *State of the World 1998*. New York: W. W. Norton.

McGovern, James R. 1982. *Anatomy of a Lynching: The Killing of Claude Neal*. Baton Rouge, LA: Louisiana State University Press.

McKelvey, Bill. 1982. *Organizational Systematics: Taxonomy, Evolution, Classification*. Berkeley, CA: University of California Press.

McKenna, Barrie. 1998. "Clinton Says, 'I'm Sorry' as Old Allies Desert Him." *Globe and Mail* 5 September: A1, A12.

_____. 1998. "How a U.S. Tobacco Settlement Was Stubbed Out." *Globe and Mail* 20 June: A15.

McLaughlin, Margaret L., Kerry K. Osborne, and Christine B. Smith. 1995. "Standards of Conduct on Usenet." Pp. 90-111 in Steven G. Jones, ed., *CyberSociety*. Thousand Oaks, CA: Sage.

McLuhan, Marshall. 1964. *Understanding Media: The Extensions of Man*. New York: McGraw-Hill.

McNeill, William H. 1976. *Plagues and Peoples*. Garden City, NY: Anchor Press.

_____ 1990. *Population and Politics since 1750*. Charlottesville, WV: University Press of Virginia.

McPhail, Clark. 1991. *The Myth of the Madding Crowd*. New York: Aldine de Gruyter.

_____. 1994. "The Dark Side of Purpose: Individual and Collective Violence in Riots." *The Sociological Quarterly* 35: 1–32.

_____ and Ronald T. Wohlstein. 1983. "Individual and Collective Behaviors Within Gatherings, Demonstrations, and Riots." *Annual Review of Sociology* 9: 579–600.

Mead, G. H. 1934. *Mind, Self and Society*. Chicago: University of Chicago Press.

Medawar, Peter. 1996. *The Strange Case of the Spotted Mice and Other Classic Essays on Science*. New York: Oxford University Press.

"Media Mergers, Consolidation, and Conglomeration." 2000. On the World Wide Web at http://fargo.itp.tsoa.nyu.edu/~walter/foi/index.htm (2 May 2000).

Meek, Ronald L., ed. 1971. *Marx and Engels on the Population Bomb: Selections from the Writings of Marx and Engels Dealing with the Theories of Thomas Robert Malthus*. Dorothea L. Meek and Ronald L. Meek, trans. Berkeley, CA: Ramparts Press.

Meier, Diane E., Carol-Ann Emmons, Sylvan Wallenstein, Timothy Quill, R. Sean Morrison, and Christine K. Cassell.

1998, "A National Survey of Physician Assisted Suicide and Euthanasia in the United States." *New England Journal of Medicine* 338: 1193–1201.

Meisenheimer II, Joseph R. 1998. "The Service Industry in the 'Good' Versus 'Bad' Jobs Debate." *Monthly Labor Review.* 121, 2: 22–47.

Melton, J. Gordon. 1996 [1978]. *Encyclopedia of American Religions,* 5th ed. Detroit: Gale.

Melucci, Alberto. 1980. "The New Social Movements: A Theoretical Approach." *Social Science Information* 19: 199–226.

_____. 1995. "The New Social Movements Revisited: Reflections on a Sociological Misunderstanding." Pp. 107–19 in Louis Maheu, ed. *Social Classes and Social Movements: The Future of Collective Action.* London: Sage.

Merisotis, Jamie P. 1999. "The 'What's-The-Difference?' Debate." *Academe* 85, 5: 47–51.

Merrick, Thomas W., et al. 1986. "World Population in Transition." *Population Bulletin* 41, 2.

Merton, Robert K. 1938. "Social Structure and Anomie." *American Sociological Review* 3: 672–82.

_____. 1968 [1949]. *Social Theory and Social Structure.* New York: Free Press.

Messner, Michael. 1995 [1989]. "Boyhood, Organized Sports, and the Construction of Masculinities." Pp. 102–14 in Michael S. Kimmel and Michael A. Messner *Men's Lives,* 3rd ed. Boston: Allyn and Bacon.

Metropolitan Museum of Art. 2000. "Mrs. Charles Dana Gibson (1873–1956)" On the World Wide Web at http:// costumeinstitute.org/gibson.htm (13 June 2000).

Meyer, John W., Francisco O. Ramirez, and Yasemin Nuhoglu Soysal. 1992. "World Expansion of Mass Education, 1870–1980." *Sociology of Education* 65: 128–49.

_____ and W. Richard Scott. 1983. *Organizational Environments: Ritual and Rationality.* Beverly Hills, CA: Sage.

Michael, Robert T., John H. Gagnon, Edward O. Laumann, and Gina Kolata. 1994. *Sex in America: A Definitive Survey.* Boston: Little, Brown and Company.

Michels, Robert. 1949 [1911]. *Political Parties: A Sociological Study of the Oligarchical Tendencies of Modern Democracy,* E. and C. Paul, trans. New York: Free Press.

"Mild Labor: The World at Work and Play." 1999. *Wired.* 7, 12: 144.

Milem, Jeffrey F. 1998. "Attitude Change in College Students: Examining the Effect of College Peer Groups and Faculty Normative Groups." *The Journal of Higher Education* 69: 117–140

Miles, Robert. 1989. *Racism.* London: Routledge.

Milgram, Stanley. 1974. *Obedience to Authority: An Experimental View.* New York: Harper.

Miller, Jerome G. 1996. *Search and Destroy: African-American Males in the Criminal Justice System.* New York: Cambridge University Press.

Miller, Mark Crispin. 1996. "Free the Media." *The Nation* 3 June 3: 9–15.

Miller, Robert L. 1998. "The Limited Concerns of Social Mobility Research." *Current Sociology* 46, 4: 145–70.

Mills, C. Wright. 1956. *The Power Elite.* New York: Oxford University Press.

_____. 1959. *The Sociological Imagination.* New York: Oxford University Press.

Mills, Janet Lee. 1985. "Body Language Speaks Louder Than Words." *Horizons* February: 8–12.

Mink-Cee. 2000. "Rap vs. Hip-Hop." On the World Wide Web at http://www.geocities.com/BourbonStreet/9459/rapvshiphop.htm (10 May 2000).

Minois, George. 1989 [1987]. *History of Old Age: From Antiquity to the Renaissance,* Sarah Hanbury Tenison, trans. Chicago: University of Chicago Press.

Mintz, Alexander. 1946. "A Re-Examination of Correlations Between Lynchings and Economic Indices." *Journal of Abnormal and Social Psychology* 41: 154–60.

Mintz, Beth. 1989. "United States of America." Pp. 207–36 in Tom Bottomore and Robert J. Brym, eds. *The Capitalist Class: An International Study.* New York: New York University Press.

_____ and Michael Schwartz. 1985. *The Power Structure of American Business.* Chicago: University of Chicago Press.

Mishel, Lawrence, Jared Bernstein, and John Schmitt. 1999. *The State of Working America, 1998–99.* Ithaca, NY: Cornell University Press.

Mitchell, Alison. 1999. "Vote on Campaign Finances Is Blocked by Senate GOP." *New York Times on the Web* 20 October. On the World Wide Web at http://www.nytimes.com/library/ politics/102099campaign-finance.html (1 May 2000).

Mittelstaedt, Martin. 2001. "When a Car's Tailpipe Is More Lethal Than a Car Crash." *Globe and Mail* 29 September: F9.

Mizruchi, Mark S. 1982. *The American Corporate Network, 1904–1974.* Beverly Hills, CA: Sage.

_____. 1992. *The Structure of Corporate Political Action: Interfirm Relations and Their Consequences.* Cambridge, MA: Harvard University Press.

Molloy, Beth L., and Sharon D. Herzberger. 1998. "Body Image and Self-Esteem: A Comparison of African-American and Caucasian Women." *Sex Roles* 38: 631–43.

Molm, Linda D. 1997. *Coercive Power in Social Exchange.* Cambridge, MA: Cambridge University Press.

"Monitoring the Future Study." 1998. On the World Wide Web at http://monitoringthefuture.org (1 May 2000).

Moore, Barrington, Jr. 1967. *Social Origins of Dictatorship and Democracy: Lord and Peasant in the Making of the Modern World.* Boston: Beacon Press.

Moore, David S. 1995. *The Basic Practice of Statistics.* New York: W. H. Freeman.

Morris, Aldon D. 1984. *The Origins of the Civil Rights Movement: Black Communities Organizing for Change.* New York: Free Press.

Morris, Charles R. 1996. *The AARP: America's Most Powerful Lobby and the Clash of Generations.* New York: Times Books.

Morris, Norval, and David J. Rothman, eds. 1995. *The Oxford History of the Prison: The Practice of Punishment in Western Society.* New York: Oxford University Press.

Mortimer, Jeylan T., and Roberta G. Simmons. 1978. "Adult Socialization." *Annual Review of Sociology* 4: 421–54.

Mosquera, Mary. 1999. "Yahoo Beats Estimates." *TechWeb* 7 April. On the World Wide Web at http://www.techweb.com/ wire/story/TWB19990407S0029 (2 May 2000).

"MP3stego." 2001. On the World Wide Web at http://www.cl. cam.ac.uk/~fapp2/steganography/mp3stego (13 September 2001).

"Mr. Showbiz." 1999. On the World Wide Web at http://mrshowbiz. go.com/reviews/tvreviews/numbers/index.html (8 June 1999).

"The MUD Connector." 2001. On the World Wide Web at http://www.mudconnect.com (25 June 2001).

Mumford, Lewis. 1961. *The City in History: Its Origins, Its Transformations, and Its Prospects*. New York: Harcourt, Brace, and World.

Mundell, Helen. 1993. "How the Color Mafia Chooses Your Clothes." *American Demographics* November. On the World Wide Web at http://www.demographics.com/publications/ad/93_ad/9311_ad/ad281.htm (2 May 2000).

Murdoch, Guy. 1995. "Child Care Centers." *Consumers' Research Magazine* 78, 10: 2.

Murdock, George Peter. 1937. "Comparative Data on the Division of Labor by Sex." *Social Forces* 15: 551–3.

_____. 1949. *Social Structure*. New York: Macmillan.

Mustard, Cameron A., Patricia Kaufert, Anita Kozyrskyj, and Teresa Mayer. 1998. "Sex Differences in the Use of Health Care Services." *New England Journal of Medicine* 338: 1678–83.

Myerhoff, Barbara. 1978. *Number Our Days*. New York: Dutton.

Myers, Daniel J. 1997. "Racial Rioting in the 1960s: An Event History Analysis of Local Conditions." *American Sociological Review* 62: 94–112.

Myles, John. 1988. "The Expanding Middle: Some Canadian Evidence on the Deskilling Debate." *Canadian Review of Sociology and Anthropology* 25: 335–64.

_____. 1989 [1984]. *Old Age in the Welfare State: The Political Economy of Public Pensions*, 2nd ed. Lawrence, KS: University Press of Kansas.

_____ and Adnan Turegun. 1994. "Comparative Studies in Class Structure." *Annual Review of Sociology* 20: 103–24.

Nagel, Joane. 1996. *American Indian Ethnic Renewal: Red Power and the Resurgence of Identity and Culture*. New York: Oxford University Press.

National Association for the Advancement of Colored People. 1934. *The Lynching of Claude Neal*. New York: National Association for the Advancement of Colored People.

National Basketball Association. 2000. "New York Knicks History." On the World Wide at http://nba.com/knicks/00400499.html#2 (29 May 2000).

National Center for Injury Prevention and Control. 2000. "Suicide in the United States." On the World Wide Web at http://www.cdc.gov/ncipc/factsheets/suifacts.htm (27 April 2000).

"The National Council of La Raza." 2000. On the World Wide Web at http://www.nclr.org/about (29 April 2000).

National Gay and Lesbian Task Force. 2000. "Anti-Same-Sex Marriage Laws in the U.S.—April 2000." On the World Wide Web at http://www.ngltf.org/downloads/marriagemap0400.gif (20 August 2000).

National Opinion Research Center. 1999. *General Social Survey, 1972–98*. Chicago: University of Chicago. Machine readable file.

National Organization for Men Against Sexism. 2000. On the World Wide Web at http://nomas.idea-net.com (15 June 2000).

Neal, Mark Anthony. 1999. *What the Music Said: Black Popular Music and Black Public Culture*. New York: Routledge.

Neugarten, Bernice. 1974. "Age Groups in American Society and the Rise of the Young Old." *Annals of the American Academy of Political and Social Science* 415: 187–98.

Nevitte, Neil. 1996. *The Decline of Deference*. Peterborough, Canada: Broadview Press.

Newman, Katherine S. 1988. *Falling From Grace: The Experience of Downward Mobility in the American Middle Class*. New York: Free Press.

_____. 1999. *No Shame in My Game: The Working Poor in the Inner City*. New York: Knopf and the Russell Sage Foundation.

Newport, Frank. 2000. "Support for Death Penalty Drops to Lowest Level in 19 Years, Although Still High at 66%." The Gallup Organization. On the World Wide Web at http://www.gallup.com/poll/releases/pr000224.asp (8 August 2000).

Nie, Norman H., Sidney Verba, and John R. Petrocik. 1979 [1976]. *The Changing American Voter*, revised ed.. Cambridge, MA: Harvard University Press.

"The Nike Campaign." 2000. On the World Wide Web at http://www.web.net/~msn/3nike.htm (23 June 2000).

Nisbett, Richard E., Kaiping Peng, Incheol Choi, and Ara Norenzayan. 2001. "Culture and Systems of Thought: Holistic Versus Analytic Cognition." *Psychological Review* 108: 291–310.

Nolen, Stephanie. 1999. "Gender: The Third Way." *Globe and Mail* September 25: D1, D4.

"Nomar Garciaparra Pictures." 1998. On the World Wide Web at http://www.geocities.com/Colosseum/Track/4242/photos.htm (15 June 1998).

Notestein, F.W. 1945. "Population—The Long View." Pp.36–57 in T.W. Schultz, ed., *Food for the World*. Chicago: University of Chicago Press.

Nowak, Martin A., Robert M. May, and Karl Sigmund. 1995. "The Arithmetics of Mutual Help." *Scientific American* 272, 6: 76–81.

Nowell, Amy, and Larry V. Hedges. 1998. "Trends in Gender Differences in Academic Achievement from 1960 to 1994: An Analysis of Differences in Mean, Variance, and Extreme Scores." *Sex Roles* 39: 21–43.

Nuland, Sherwin B. 1994. *How We Die: Reflections on Life's Final Chapter*. New York: Vintage.

Oates, Joyce Carol. 1999. "The Mystery of JonBenét Ramsey." *New York Review of Books* 24 June: 31–7.

Oberschall, Anthony. 1973. *Social Conflict and Social Movements*. Englewood Cliffs, NJ: Prentice-Hall.

O'Connor, Julia S., and Robert J. Brym. 1988. "Public Welfare Expenditure in OECD Countries: Towards a Reconciliation of Inconsistent Findings." *British Journal of Sociology* 39: 47–68.

_____ and Gregg M. Olsen, eds. 1998. *Power Resources Theory and the Welfare State: A Critical Approach*. Toronto: University of Toronto Press.

"Official Ballot, General Election, Palm Beach County, Florida, November 7, 2000." 2000. On the World Wide Web at http://www.cnn.com/2000/ALLPOLITICS/stories/11/10/election.president.02/large.ballot.ap.jpg (10 November 2000).

O'Hare, William P. 1996. "A New Look at Poverty in America." *Population Bulletin* 51, 2: 2–46.

Oliver, Melvin L., and Thomas M. Shapiro. 1995. *Black Wealth/White Wealth: A New Perspective on Racial Inequality*. New York: Routledge.

Olsen, Gregg, and Robert J. Brym. 1996. "Between American Exceptionalism and Swedish Social Democracy: Public and Private Pensions in Canada." Pp. 261–79 in Michael Shalev, ed. *The Privatization of Social Policy? Occupational Welfare and the Welfare State in America, Scandinavia and Japan.* London: Macmillan.

Olshansky, S. Jay. 1997. "Infectious Diseases: New and Ancient Threats to World Health." *Population Bulletin* 52: 2.

_____, Bruce A. Carnes, and Christine Cassel. 1990. "In Search of Methusaleh: Estimating the Upper Limits of Human Longevity." *Science* 250: 634–40.

Olzak, Susan, and Suzanne Shanahan. 1996. "Deprivation and Race Riots: An Extension of Spilerman's Analysis." *Social Forces* 74: 931–62.

_____, _____, and Elizabeth H. McEneaney. 1996. "Poverty, Segregation, and Race Riots: 1960 to 1993." *American Sociological Review* 61: 590–614.

Omega Foundation. 1998. "An Appraisal of the Technologies of Political Control: Summary and Options Report for the European Parliament." On the World Wide Web at http://home.icdc.com/~paulwolf/eu_stoa_2.htm (29 April 2000).

Omi, Michael, and Howard Winant. 1986. *Racial Formation in the United States.* New York: Routledge.

Optometrists Network. 2000. "Attention Deficit Disorder." On the World Wide Web at http://www.add-adhd.org/ADHD_attention-deficit.html (14 August 2000).

Orfield, Gary, and Susan E. Eaton. 1996. *Dismantling Desegregation: The Quiet Reversal of Brown v. Board of Education.* New York: New Press.

Ornstein, Michael. 1998. "Survey Research." *Current Sociology* 46, 4: 1–87.

Ossowski, Stanislaw. 1963. *Class Structure in the Social Consciousness,* S. Patterson, trans. London: Routledge and Kegan Paul.

Pacey, Arnold. 1983. *The Culture of Technology.* Cambridge, MA: MIT Press.

Pammett, Jon H. 1997. "Getting Ahead Around the World." Pp. 67–86 in Alan Frizzell and Jon H. Pammett, eds. *Social Inequality in Canada.* Ottawa: Carleton University Press.

Parillo, Vincent N., John Stimson, and Ardyth Stimson. 1999. *Contemporary Social Problems,* 4th ed. Boston: Allyn and Bacon.

Park, Robert Ezra. 1914. "Racial Assimilation in Secondary Groups." *Publications of the American Sociological Society* 8: 66–72.

_____ 1950. *Race and Culture.* New York: Free Press.

_____, Ernest W. Burgess, and Roderick D. McKenzie. 1967 [1925]. *The City.* Chicago: University of Chicago Press.

Parshall, Gerald. 1998. "Brotherhood of the Bomb." *US News & World Report* 125, 7 (17–24 August): 64–8.

Parson, E. A., and D. W. Keith. 1998. "Fossil Fuels Without CO_2 Emissions." *Science* 282: 1053–4.

Parsons, Talcott. 1951. *The Social System.* New York: Free Press.

_____. 1955. "The American Family: Its Relation to Personality and to the Social Structure." Pp. 3–33 in Talcott Parsons and Robert F. Bales, eds. *Family, Socialization and Interaction Process.* New York: Free Press.

Pastore, Michael. 2001. "The World's Online Populations." *Cyberatlas.* On the World Wide Web at http://cyberatlas.internet.com/big_picture/demographics/article/0.1323.5911_151151.00.html (October 7 2001).

Patterson, Orlando. 1982. *Slavery and Social Death.* Cambridge, MA: Harvard University Press.

_____. 1997. *The Ordeal of Integration: Progress and Resentment in America's "Racial" Crisis.* Washington, DC: Civitas.

Peacock, Mary. 2000. "The Cult of Thinness." On the World Wide Web at http://www.womenswire.com/image/toothin.html (13 June 2000).

Perrow, Charles B. 1984. *Normal Accidents.* New York: Basic Books.

Perry-Castaneda Library Map Collection. 2000. "Comparative Soviet Nationalities by Republic." On the World Wide Web at http://www.lib.utexas.edu/Libs/PCL/Map_collection/commonwealth/USSR_NatRep_89.jpg (23 November 2000).

Pessen, Edward. 1984. *The Log Cabin Myth: The Social Backgrounds of the Presidents.* New Haven, CT: Yale University Press.

Peters, John F. 1994. "Gender Socialization of Adolescents in the Home: Research and Discussion." *Adolescence* 29: 913–34.

Phillips, Kevin. 1990. *The Politics of Rich and Poor: Wealth and the American Electorate in the Reagan Aftermath.* New York: Random House.

Piaget, Jean, and Bärbel Inhelder. 1969. *The Psychology of the Child,* Helen Weaver, trans. New York: Basic Books.

Pinker, Steven. 1994. "Apes—Lost for Words." *New Statesman and Society* 15 April: 30–1.

Piven, Frances Fox, and Richard A. Cloward. 1977. *Poor People's Movements: Why They Succeed, How They Fail.* New York: Vintage.

_____ and _____. 1989 [1988]. *Why Americans Don't Vote.* New York: Pantheon.

_____ and _____. 1993 [1971]. *Regulating the Poor: The Functions of Public Welfare,* updated ed. New York: Vintage.

A Place Called Chiapas. 1998. Vancouver: Canada Wild Productions. (Movie).

Podolny, Joel M., and Karen L. Page. 1998. "Network Forms of Organization." *Annual Review of Sociology* 24: 57–76.

Polanyi, Karl. 1957 [1944]. *The Great Transformation: The Political and Economic Origins of Our Time.* Boston: Beacon Press.

Polsby, Nelson W. 1959. "Three Problems in the Analysis of Community Power." *American Sociological Review* 24: 796–803.

Pool, Robert. 1997. *Beyond Engineering: How Society Shapes Technology.* New York: Oxford University Press.

Popenoe, David. 1988. *Disturbing the Nest: Family Change and Decline in Modern Societies.* New York: Aldine de Gruyter.

_____. 1991. "Family Decline in the Swedish Welfare State." *Public Interest* 102: 65–78.

_____. 1992. "Family Decline: A Rejoinder." *Public Interest* 109: 116–18.

_____. 1993. "American Family Decline, 1960–1990: A Review and Appraisal." *Journal of Marriage and the Family* 55: 527–55.

_____. 1996. *Life Without Father: Compelling New Evidence that Fatherhood and Marriage are Indispensable for the Good of Children and Society.* New York: Martin Kessler Books.

Population Reference Bureau. 2000. "2000 World Population Data Sheet." On the World Wide Web at http://www.prb.org/pubs/wpds2000/wpds2000_Infant_Mortality-Life_Expectancy_At_Birth.html (16 August 2000).

Portes, Alejandro. 1996. "Global Villagers: The Rise of Transnational Communities." *The American Prospect* 25: 74–77. On the World Wide Web at http://www.prospect.org/archives/25/25port.html (29 April 2000).

_____ and Robert D. Manning. 1991. "The Immigrant Enclave: Theory and Empirical Examples." Pp. 319–32 in Norman R. Yetman, ed. *Majority and Minority: The Dynamics of Race and Ethnicity in American Life*, 5th ed. Boston: Allyn and Bacon.

_____ and Rubén G. Rumbaut. 1990. *Immigrant America: A Portrait*. Berkeley, CA: University of California Press.

_____ and Alex Stepick. 1993. *City on the Edge: The Transformation of Miami*. Berkeley, CA: University of California Press.

_____ and Cynthia G. Truelove. 1991. "Making Sense of Diversity: Recent Research on Hispanic Minorities in the United States." Pp. 402–19 in Norman R. Yetman, ed. *Majority and Minority: The Dynamics of Race and Ethnicity in American Life*, 5th ed. Boston: Allyn and Bacon.

Postel, Sandra. 1994. "Carrying Capacity: Earth's Bottom Line." Pp. 3–21 in Linda Starke, ed. *State of the World 1994*. New York: W. W. Norton.

Postman, Neil. 1982. *The Disappearance of Childhood*. New York: Delacorte.

_____. 1992. *Technopoly: The Surrender of Culture to Technology*. New York: Vintage.

Powers, Elizabeth T. 1994. "The Impact of AFDC on Birth Decisions and Program Participation." *Working Papers*. Cleveland: Federal Reserve Bank of Cleveland.

Pred, Allan R. 1973. *Urban Growth and the Circulation of Information*. Cambridge, MA: Harvard University Press.

Press, Andrea. 1991. *Women Watching Television: Gender, Class and Generation in the American Television Experience*. Philadelphia: University of Pennsylvania Press.

Press, Eyal. 1999. "A Nike Sneak." *The Nation* 5 April. On the World Wide Web at http://www.thenation.com/issue/990405/0405press.shtml (8 August 2000).

Prittie, Jennifer. 2000. "The Serious Business of Rubber Noses and Big Shoes." *National Post* 29 March: A1–A2.

Public Broadcasting System. 1997. "Double Talk?" On the World Wide Web at http://www.pbs.org/newshour/bb/education/july-dec97/bilingual_9-21.html (11 August 2000).

Quadagno, Jill. 1988. *The Transformation of Old Age Security: Class and Politics in the American Welfare State*. Chicago: University of Chicago Press.

_____. 1994. *The Color of Welfare: How Racism Undermined the War on Poverty*. New York: Oxford University Press.

Quinn, Tom. 2000. "The 'Motor Voter' Question." *Ford Foundation Report*. On the World Wide Web at http://www.fordfound.org/publications/ff_report/view_ff_report_detail.cfm?report_index=247 (21 November 2000).

Raag, Tarja, and Christine L. Rackliff. 1998. "Preschoolers' Awareness of Social Expectations of Gender: Relationships to Toy Choices." *Sex Roles* 38: 685–700.

Rainwater, Lee, and Timothy M. Smeeding. 1995. "Safety Nets for Children are Weakest in US." United National Children's Emergency Fund. On the World Wide Web at http://www.unicef.org:80/pon96/indust4.htm (1 May 2000).

Rank, Mark Robert. 1994. *Living on the Edge: The Politics of Welfare in America*. New York: Columbia University Press.

Rapp, R., and E. Ross. 1986. "The 1920s: Feminism, Consumerism and Political Backlash in the U.S." Pp. 52–62 in J. Friedlander, B. Cook, A. Kessler-Harris, and C. Smith-Rosenberg, eds. *Women in Culture and Politics*. Bloomington, IN: Indiana University Press.

Reed, Dan. 2000. "Janitors Get Pay Increase; Strike Averted." *San Jose Mercury News* 4 June. On the World Wide Web at http://www.mercurycenter.com/premium/local/docs/janitors04.htm (6 June 2000).

Reich, Robert B. 1991. *The Work of Nations: Preparing Ourselves for 21st-Century Capitalism*. New York: Knopf.

Reiman, Jeffrey H. 1995 [1979]. *The Rich Get Richer and the Poor Get Prison: Ideology, Class, and Criminal Justice*, 4th ed. Boston: Allyn and Bacon.

_____. 1996. *And the Poor Get Prison: Economic Bias in American Criminal Justice*. Boston: Allyn and Bacon.

Reinarman, Craig, and Harry G. Levine, eds. 1999. *Crack in America: Demon Drugs and Social Justice*. Berkeley, CA: University of California Press.

Reiter, Ester. 1991. *Making Fast Food: From the Frying Pan into the Fryer*. Montreal: McGill-Queen's University Press.

ReligiousTolerance.org. 2000. "Homosexual (Same-Sex) Marriages." On the World Wide Web at http://www.religioustolerance.org/hom_marr.htm (20 August 2000).

Remennick, Larissa I. 1998. "The Cancer Problem in the Context of Modernity: Sociology, Demography, Politics." *Current Sociology* 46, 1: 1–150.

Remnick, David. 1998. "How Russia Is Ruled." *New York Review of Books* 45, 6: 10–15.

Renner, Michael. 2000. "Creating Jobs, Preserving the Environment." Pp. 162–83 in Lester R. Brown et al., eds. *State of the World 2000*. New York: Norton.

Resnick, Mitchel and Natalie Rusk. 1996. "Access Is Not Enough: Computer Clubhouses in the Inner City." *The American Prospect* 27, July–August: 60–68. On the World Wide Web at http://www.prospect.org/archives/27/27resn.html (2 May 2000).

Ridgeway, Cecilia L. 1983. *The Dynamics of Small Groups*. New York: St. Martin's Press.

Riedmann, Agnes. 1993. *Science That Colonizes: A Critique of Fertility Studies in Africa*. Philadelphia: Temple University Press.

Rifkin, Jeremy. 1995. *The End of Work: The Decline of the Global Labor Force and the Dawn of the Post Market Era*. New York: G. P. Putnam's Sons.

_____. 1998. *The Biotech Century: Harnessing the Gene and Remaking the World*. New York: Jeremy P. Tarcher/Putnam.

Riley, Matilda White, Anne Foner, and Joan Waring. 1988. "Sociology of Age." Pp. 243–90 in Neil Smelser, ed., *Handbook of Sociology*. Newbury Park, CA: Sage.

Riley, Nancy. 1997. "Gender, Power, and Population Change." *Population Bulletin* 52, 1. On the World Wide Web at p://www.prb.org/pubs/population_bulletin/bu52-1.htm (25 August 2000).

Ritzer, George. 1993. *The McDonaldization of Society*. Thousand Oaks, CA: Pine Forge Press.

_____. 1996. "The McDonalidzation Thesis: Is Expansion Inevitable?" *International Sociology* 11: 291–307.

Roberts, Siobhan. 2000. "Web Could Be New STD Breeding Ground." *National Post* 17 June: A2.

Robinson, John P., and Suzanne Bianchi. 1997. "The Children's Hours." *American Demographics* December: 20–4.

Robinson, Robert V., and Wendell Bell. 1978. "Equality, Success, and Social Justice in England and the United States." *American Sociological Review* 43: 125–43.

Roche, Maurice. 1995. "Rethinking Citizenship and Social Movements: Themes in Contemporary Sociology and Neoconservative Ideology." Pp. 186–219 in Louis Maheu, ed. *Social Classes and Social Movements: The Future of Collective Action*. London: Sage.

Roediger, David R. 1991. *The Wages of Whiteness: Race and the Making of the American Working Class*. London: Verso.

Roethlisberger, Fritz J., and William J. Dickson. 1939. *Management and the Worker*. Cambridge, MA: Harvard University Press.

Rogers, Jackie Krasas, and Kevin D. Henson. 1997. "'Hey, Why Don't You Wear a Shorter Skirt?' Structural Vulnerability and the Organization of Sexual Harassment in Temporary Clerical Employment." *Gender and Society* 11: 215–37.

Rollins, Boyd C., and Kenneth L. Cannon. 1974. "Marital Satisfaction over the Family Life Cycle." *Journal of Marriage and the Family* 36: 271–84.

Rootes, Chris. 1995. "A New Class? The Higher Educated and the New Politics." Pp. 220–35 in Louis Maheu, ed. *Social Classes and Social Movements: The Future of Collective Action*. London: Sage.

Roscoe, Lori A., L. J. Dragovi , and Donna Cohen. 2000. "Dr. Jack Kevorkian and Cases of Euthanasia in Oakland County, Michigan, 1990–1998." *New England Jurnal of Medicine* 343, 23. On the World Wide Web at http://www.nejm.org/content/2000/0343/0023/1735.asp (12 July 2000).

Rosenberg, Charles E. 1962. *The Cholera Years: The United States in 1832, 1849, and 1866*. Chicago: University of Chicago Press.

_____. 1987. *The Care of Strangers: The Rise of America's Hospital System*. New York: Basic.

Rosenberg, Nathan. 1982. *Inside the Black Box: Technology and Economics*. Cambridge, UK: Cambridge University Press.

Roslin, Alex. 2000. "Black & Blue." *Saturday Night* 23 September: 44–9.

Rossi, Peter H. 1989. *Down and Out in America: The Origins of Homelessness*. Chicago: University of Chicago Press.

Rostow, W. W. 1960. *The Stages of Economic Growth: A Non-Communist Manifesto*. New York: Cambridge University Press.

Rothman, Barbara Katz. 1982. *In Labor: Women and Power in the Birthplace*. New York: W. W. Norton.

_____. 1989. *Recreating Motherhood: Ideology and Technology in a Patriarchal Society*. New York: W. W. Norton.

Rothman, David J. 1991. *Strangers at the Bedside: A History of How Law and Bioethics Transformed Medical Decision Making*. New York: Basic Books.

_____. 1998. "The International Organ Traffic." *New York Review of Books* 45, 5: 14–17.

Rothman, Stanley, and Amy E. Black. 1998. "Who Rules Now? American Elites in the 1990s." *Society* 35, 6: 17–20.

Rubin, J. Z., F. J. Provenzano, and Z. Lurra. 1974. "The Eye of the Beholder." *American Journal of Orthopsychiatry* 44: 512–19.

Rubin, Lillian B. 1994. *Families on the Fault Line*. New York: Harper Collins.

Rueschemeyer, Dietrich, Evelyne Huber Stephens, and John Stephens. 1992. *Capitalist Development and Democracy*. Chicago: University of Chicago Press.

Ruggles, Steven. 1997. "The Effects of AFDC on American Family Structure, 1940–1990." *Journal of Family History* 22: 307–25.

Rural Advancement Foundation International. 1999. "The Gene Giants." On the World Wide Web at http://www.rafi.org/web/allpub-one.shtml?dfl=allpub.db&tfl=allpub-one-frag.ptml&operation=display&ro1=recNo&rf1=34&rt1=34&usebrs=true (2 May 2000).

Rushton, J. Phillipe. 1995. *Race, Evolution, and Behavior: A Life History Perspective*. New Brunswick, NJ: Transaction Publishers.

Russett, Cynthia Eagle. 1966. *The Concept of Equilibrium in American Social Thought*. New Haven, CT: Yale University Press.

Ryan, Kathryn M., and Jeanne Kanjorski. 1998. "The Enjoyment of Sexist Humor, Rape Attitudes, and Relationship Aggression in College Students." *Sex Roles* 38: 743–56.

Rytina, Steve. 1992. "Scaling the Intergenerational Continuity of Occupation: Is Occupational Inheritance Ascriptive After All?" *American Journal of Sociology* 97: 1658–88.

Sampson, Robert. 1997. "The Embeddedness of Child and Adolescent Development: A Community-Level Perspective on Urban Violence." Pp. 31–77 in Joan McCord, ed. *Violence and Childhood in the Inner City*. Cambridge, UK: Cambridge University Press.

_____ and John H. Laub. 1993. *Crime in the Making: Pathways and Turning Points through Life*. Cambridge, MA: Harvard University Press.

_____ and William J. Wilson. 1995. "Toward a Theory of Race, Crime and Urban Inequality." Pp. 37–54 in John Hagan and Ruth D. Peterson, eds., *Crime and Inequality*. Stanford, CA: Stanford University Press.

Samuelsson, Kurt. 1961 [1957]. *Religion and Economic Action*, E. French, trans. Stockholm: Scandinavian University Books.

Sanday, Peggy Reeves. 1990. *Fraternity Gang Rape: Sex, Brotherhood, and Privilege on Campus*. New York: New York University Press.

Sanderson, Stephen K. 1995. *Macrosociology: An Introduction to Human Societies*, 3rd ed. New York: Harper Collins.

Sandqvist, Karin, and Bengt-Erik Andersson. 1992. "Thriving Families in the Swedish Welfare State." *Public Interest* 109: 114–16.

Sartre, Jean-Paul. 1965 [1948]. *Anti-Semite and Jew*, George J. Becker, trans. New York: Schocken.

Sassen, Saskia. 1991. *The Global City: New York, London, Tokyo*. Princeton, NJ: Princeton University Press.

Saxton, Alexander. 1900. *The Rise and Fall of the White Republic*. London: Verso.

Savelsberg, Joachim, with contributions by Peter Brühl. 1994. *Constructing White-Collar Crime: Rationalities, Communication, Power*. Philadelphia: University of Pennsylvania Press.

Scarr, Sandra, and Richard A. Weinberg. 1978. "The Influence of 'Family Background' on Intellectual Attainment." *American Sociological Review* 43: 674–92.

Schiebinger, Londa L. *Nature's Body: Gender in the Making of Modern Science*. Boston: Beacon Press.

Schiff, Michel, and Richard Lewontin. 1986. *Education and Class: The Irrelevance of IQ Genetic Studies*. Oxford, UK: Clarendon Press.

Schiller, Herbert I. 1969. *Mass Communications and American Empire*. New York: Kelley.

_____. 1989. *Culture Inc.: The Corporate Takeover of Public Expression*. New York: Oxford University Press.

_____. 1996. *Information Inequality: The Deepening Social Crisis in America*. New York: Routledge.

Schlesinger, Arthur. 1991. *The Disuniting of America: Reflections on a Multicultural Society*. New York: W. W. Norton.

Schlosser, Eric. 1998. "The Prison-Industrial Complex." *The Atlantic Monthly* December. On the World Wide Web at http://www.theatlantic.com/issues/98dec/prisons.htm (29 April 2000).

Schluchter, Wolfgang. 1981 [1980]. *The Rise of Western Rationalism: Max Weber's Developmental History*, Guenther Roth, trans. Berkeley, CA: University of California Press.

Schneider, Barbara, and David Stevenson. 1999. *The Ambitious Generation: America's Teenagers: Motivated But Directionless*. New Haven, CT: Yale University Press.

Schor, Juliet B. 1992. *The Overworked American: The Unexpected Decline of Leisure*. New York: Basic Books.

_____. 1999. *The Overspent American: Why We Want What We Don't Need*. New York: Harper.

_____ and Samuel Bowles. 1987. "Employment Rents and the Incidence of Strikes." *The Review of Economics and Statistics* 69: 584–92.

Schrag, Peter. 1997. "The Near-Myth of Our Failing Schools." *The Atlantic Monthly* October. On the World Wide Web at http://www.theatlantic.com/issues/97oct/fail.htm (2 May 2000).

Schudson, Michael. 1991. "National News Culture and the Rise of the Informational Citizen." Pp. 265–82 in Alan Wolfe, ed., *America at Century's End*. Berkeley, CA: University of California Press.

_____. 1995. *The Power of News*. Cambridge, MA: Harvard University Press.

Schwartz, Michael, ed. 1987. *The Structure of Power in America: The Corporate Elite as a Ruling Class*. New York: Holmes and Meier.

Schweingruber, David, and Clark McPhail. 1999. "A Method for Systematically Observing and Recording Collective Action." *Sociological Methods and Research* 27: 451–498.

Scoon-Rogers, Lydia, and Gordon H. Lester. 1995. *Child Support for Custodial Mothers and Fathers: 1991*. Washington, DC: U.S. Bureau of the Census, U.S. Department of Commerce, Economics and Statistics Administration. On the World Wide Web at http://www.census.gov/prod/2/pop/p60/p60-187.pdf (1 May 2000).

Scott, James C. 1998. *Seeing Like a State: How Certain Schemes to Improve the Human Condition Have Failed*. New Haven, CT: Yale University Press.

Scott, Janny. 2001. "Study Puts Census Errors at $4 Billion." *The New York Times Online* August 8, 2001. On the World Wide Web at http://www.nytimes.com/2001/08/08/nyregion/08CENS.html?searchpv=day04&pagewanted=print (12 August 2001).

Scott, Peter Dale, and Jonathan Marshall. 1991. *Cocaine Politics: Drugs, Armies, and the CIA in Central America*. Berkeley, CA: University of California Press.

Scott, Wilbur J. 1990. "PTSD in *DSM-III*: A Case in the Politics of Diagnosis and Disease." *Social Problems* 37: 294–310.

Scully, Diana. 1990. *Understanding Sexual Violence: A Study of Convicted Rapists*. Boston: Unwin Hyman.

Seccombe, Wally. 1992. *A Millennium of Family Change: Feudalism to Capitalism in Northwestern Europe*. London: Verso.

Seeman, Neil. 2000. "Capital Questions." *The National Post* 6 August: B1, B6.

Segundo, Juan Luis, S. J. 1976 [1975]. *The Liberation of Theology*, John Drury, trans. Maryknoll, NY: Orbis.

Seiter, Ellen. 1999. *Television and New Media Audiences*. Oxford, UK: Clarendon Press.

Selznick, Philip. 1957. *Leadership in Administration: A Sociological Interpretation*. New York: Harper and Row.

Sen, Amartya. 1981. *Poverty and Famines: An Essay on Entitlement and Deprivation*. Oxford, UK: Clarendon Press.

_____. 1994. "Population: Delusion and Reality." *New York Review of Books* 41, 15: 62–71.

The Sentencing Project. 1997. "Americans Behind Bars: U.S. and International Use of Incarceration, 1995." On the World Wide Web at http://www.sentencingproject.org/pubs/tsppubs/9030data.html (29 April 2000).

_____. 2001. "U.S. Surpasses Russia as World Leader in Rate of Incarceration." On the World Wide Web at http://www.sentencingproject.org/brief/usvsrus.pdf (27 June 2001).

Sewell, William H. 1958. "Infant Training and the Personality of the Child." *American Journal of Sociology* 64: 150–9.

"SexQuiz.org." 2000. On the World Wide Web at http://www.sexquiz.org (12 August 2000).

Shakur, Sanyika (a.k.a. Monster Kody Scott). 1993. *Monster: The Autobiography of an L.A. Gang Member*. New York: Penguin.

Shalev, Michael. 1983. "Class Politics and the Western Welfare State." Pp. 27–50 in S. E. Spiro and E. Yuchtman-Yaar, eds. *Evaluating the Welfare State: Social and Political Perspectives*. New York: Academic Press.

Shapiro, Andrew L. 1992. *We're Number One*. New York: Vintage.

Shattuck, Roger. 1980. *The Forbidden Experiment: The Story of the Wild Boy of Aveyron*. New York: Farrar, Straus, and Giroux.

Shea, Christopher. 1994. "'Gender Gap' on Examinations Shrank Again This Year." *Chronicle of Higher Education* 41, 2: A54.

Shea, Sarah E., Kevin Gordon, Ann Hawkins, Janet Kawchuk, and Donna Smith. 2000. "Pathology in the Hundred Acre Wood: A Neurodevelopmental Perspective on A. A. Milne." *Canadian Medical Association Journal* 163(12):1557–9. On the World Wide Web at http://www.cma.ca/cmaj/vol-163/issue-12/1557.htm (12 December 2000).

Shekelle, Paul G. 1998. "What Role for Chiropractic in Health Care?" *New England Journal of Medicine* 339: 1074–5.

Shelton, Beth Anne, and Daphne John. 1996. "The Division of Household Labor." *Annual Review of Sociology* 22: 299–322.

Sherrill, Robert. 1990. "The Looting Decade: S&Ls, Big Banks and Other Triumphs of Capitalism." *The Nation* 251, 17: 19 November: 589–623.

_____. 1997. "A Year in Corporate Crime." *The Nation* 7 April: 11–20.

Shipler, David K. 1997. *A Country of Strangers: Blacks and Whites in America*. New York: Knopf.

Short, James F., Jr., and Fred L. Strodtbeck. 1965. *Group Process and Gang Delinquency*. Chicago: University of Chicago Press.

Shorter, Edward. 1997. *A History of Psychiatry: From the Era of the Asylum to the Age of Prozac*. New York: John Wiley and Sons.

Shulman, Alix Kates. 1997 [1969]. *Memoirs of an Ex-Prom Queen*. New York: Penguin.

Siegel, Jacob. 1996. "Aging Into the 21st Century." Administration on Aging. On the World Wide Web at http://www.aoa.dhhs.gov/aoa/stats/aging21/default.htm (2 May 2000).

Silberman, Steve. 2000. "Talking to Strangers." *Wired* 8, 5: 225–33, 288–96. On the World Wide Web at http://www.wired.com/wired/archive/8.05/translation.html (23 May 2000).

"The Silent Boom." 1998. *Fortune* 7 July: 170–1.

Silverstein, Louise B., and Carl F. Auerbach. 1999. "Reconstructing the Essential Father." *American Psychologist* 54: 397–407.

Simmel, George. 1950. *The Sociology of Georg Simmel*, Kurt H. Wolff, trans and ed. New York: Free Press.

Simon, Jonathan. 1993. *Poor Discipline: Parole and the Social Control of the Underclass, 1890–1990*. Chicago: University of Chicago Press.

Simon, Julian. 1998. "The Five Greatest Years for Humanity." *Wired* 6, 1: 66–8.

Sjöberg, Gideon. 1960. *The Preindustrial City: Past and Present*. New York: Free Press.

Skinner, B. F. 1953. *Science and Human Behavior*. New York: Macmillan.

Skocpol, Theda. 1979. *States and Revolutions: A Comparative Analysis of France, Russia, and China*. Cambridge, UK: Cambridge University Press.

_____. 1996. *Boomerang: Clinton's Health Security Effort and the Turn Against Government in U.S. Politics*. New York: W. W. Norton.

Skolnick, Arlene. 1991. *Embattled Paradise: The American Family in an Age of Uncertainty*. New York: Basic Books.

Skolnick, Jerome K. 1997. "Tough Guys." *The American Prospect* 30: 86–91. On the World Wide Web at http://www.prospect.org/archives/30/fs30jsko.html (29 April 2000).

Smelser, Neil. 1963. *Theory of Collective Behavior*. New York: Free Press.

Smith, Adam. 1981 [1776]. *An Inquiry into the Nature and Causes of the Wealth of Nations*, 2 vols. Indianapolis, IN: Liberty Press.

Smith, Anthony. 1980. *Goodbye Gutenberg: The Newspaper Revolution of the 1980s*. New York: Oxford University Press.

Smith, Christian. 1991. *The Emergence of Liberation Theology: Radical Religion and Social Movement Theory*. Chicago: University of Chicago Press.

_____. 2000. *Christian America? What Evangelicals Really Want*. Berkeley, CA: University of California Press.

Smith, Jackie. 1998. "Global Civil Society? Transnational Social Movement Organizations and Social Capital." *American Behavioral Scientist* 42: 93–107.

Smith, Michael. 1990. "Patriarchal Ideology and Wife Beating: A Test of a Feminist Hypothesis." *Violence and Victims* 5: 257–73.

Smith, Tom W. 1992. "A Methodological Analysis of the Sexual Behavior Questions on the GSS." *Journal of Official Statistics* 8: 309–25.

Smith, Vicki. 1990. *Managing in the Corporate Interest: Control and Resistance in an American Bank*. Berkeley, CA: University of California Press.

Snow, David A., E. Burke Rochford, Jr., Steven K. Worden, and Robert D. Benford. 1986. "Frame Alignment Processes, Micromobilization, and Movement Participation." *American Sociological Review* 51: 464–81.

Snyder, David. 1979. "Collective Violence Processes: Implications for Disaggregated Theory and Research." Pp. 35–61 in Louis Kriesberg, ed. *Research in Social Movements, Conflict and Change: A Research Annual*. Greenwich, CT: JAI Press.

_____ and Charles Tilly. 1972. "Hardship and Collective Violence in France, 1830–1960." *American Sociological Review* 37: 520–32.

Sofsky, Wolfgang. 1997 [1993]. *The Order of Terror: The Concentration Camp*, William Templer, trans. Princeton, NJ: Princeton University Press.

Solomon, Charlene Marmer. 1999. "Stressed to the Limit." *Workforce* 78, 9: 48–54.

Sørenson, Aage B. 1992. "Women, Family and Class." *Annual Review of Sociology* 18: 39–61.

Sorenson, Elaine. 1994. *Comparable Worth: Is It a Worthy Policy?* Princeton, NJ: Princeton University Press.

Soule, Sarah A. 1992. "Populism and Black Lynching in Georgia, 1890–1900." *Social Forces* 71: 431–49.

"Southern Poverty Law Center. 2000. "Intelligence Report." On the World Wide Web at http://www.splcenter.org/intelligenceproject/ip-index.html (29 April 2000).

Spigel, Lynn. 1992. *Make Room for TV*. Chicago: University of Chicago Press.

Spilerman, Seymour. 1970. "The Causes of Racial Disturbances: A Comparison of Alternative Explanations." *American Sociological Review* 35: 627–49.

_____. 1976. "Structural Characteristics of Cities and the Severity of Racial Disorders." *American Sociological Review* 41: 771–93.

Spitz, René A. 1945. "Hospitalism: An Inquiry into the Genesis of Psychiatric Conditions in Early Childhood." Pp. 53–74 in *The Psychoanalytic Study of the Child*, Vol. 1. New York: International Universities Press.

_____. 1962. "Autoerotism Re-examined: The Role of Early Sexual Behavior Patterns in Personality Formation." Pp. 283–315 in *The Psychoanalytic Study of the Child*, Vol. 17. New York: International Universities Press.

Spitzer, Allan. 1973. "The Historical Problem of Generations." *American Historical Review* 78: 1353–85.

Spitzer, Steven. 1980. "Toward a Marxian Theory of Deviance." Pp. 175–91 in Delos H. Kelly, ed. *Criminal Behavior: Readings in Criminology*. New York: St. Martin's Press.

Springhall, John. 1998. *Youth, Popular Culture and Moral Panics: Penny Gaffs to Gangsta-Rap, 1830–1996*. New York: Routledge.

Srinivas, M. N. 1952. *Religion and Society among the Coorgs of South India*. Oxford, UK: Oxford University Press.

Stacey, Judith. 1991. *Brave New Families*. New York: Basic Books.

Stack, Carol. 1974. *All Our Kin: Strategies for Survival in a Black Community*. New York: Harper.

Stack, Stephen, and J. Ross Eshleman. 1998. "Marital Status and Happiness: A 17-Nation Study." *Journal of Marriage and the Family* 60: 527–36.

Stanfield, Rochelle L. 1997. "Blending of America." *National Journal* 13 September: 1780–2.

Stark, Rodney. 1985. *Sociology*. Belmont, CA: Wadsworth.

_____ and William Sims Bainbridge. 1979. "Of Churches, Sects, and Cults: Preliminary Concepts for a Theory of Religious Movements." *Journal for the Scientific Study of Religion* 18: 117–31.

Starr, Paul. 1982. *The Social Transformation of American Medicine*. New York: Basic Books.

_____. 1994 [1992]. *The Logic of Health Care Reform: Why and How the President's Plan Will Work*, rev. ed. New York: Penguin.

Steele, Claude M. 1992. "Race and the Schooling of Black Americans." *The Atlantic Monthly* April. On the World Wide Web at http://www.theatlantic.com/unbound/flashbks/blacked/steele.htm (2 May 2000).

Steinberg, Stephen. 1989 [1981]. *The Ethnic Myth: Race, Ethnicity, and Class in America*, updated ed. Boston: Beacon Press.

_____. 1995. *Turning Back: The Retreat from Racial Justice in American Thought and Policy*. Boston: Beacon Press.

Steinem, Gloria. 1994. *Moving Beyond Words*. New York: Simon & Schuster.

Steinhart, David. 2000. "Flint's Future: Meaner than a Junkyard Bond." *Financial Post* 12 August: D1, D6.

Stephens, W. Richard, Jr. 1999. *Careers in Sociology*, 2nd ed. Boston: Allyn and Bacon.

Sternberg, Robert J. 1986. "A Triangular Theory of Love." *Psychological Review* 93: 119–35.

Stewart, Abigail, Anne P. Copeland, Nia Lane Chester, Janet E. Malley, and Nicole B. Barenbaum. 1997. *Separating Together: How Divorce Transforms Families*. New York: The Guilford Press.

Stolzenberg, Ross. M. 1990. "Ethnicity, Geography, and Occupational Achievement of Hispanic Men in the United States." *American Sociological Review* 55: 143–54.

Stone, Lawrence. 1977. *The Family, Sex and Marriage in England, 1500–1800*. New York: Harper and Row.

Stotsky, Sandra. 1999. *Losing Our Language: How Multicultural Classroom Instruction Is Undermining Our Children's Ability to Read, Write, and Reason*. New York: Free Press.

Straus, Murray A. 1994. "State-to-State Differences in Social Inequality and Social Bonds in Relation to Assaults on Wives in the United States." *Journal of Comparative Family Studies* 25: 7–24.

_____. 1995. "Trends in Cultural Norms and Rates of Partner Violence: An Update to 1992." Pp. 30–33 in Sandra M. Stith and Murray A. Straus, eds. *Understanding Partner Violence: Prevalence, Causes, Consequences, and Solutions*. Minneapolis, MN: National Council on Family Relations.

Strauss, Anselm L. 1993. *Continual Permutations of Action*. New York: Aldine de Gruyter.

Stretesky, Paul, and Michael J. Hogan. 1998. "Environmental Justice: An Analysis of Superfund Sites in Florida." *Social Problems* 45: 268–87.

Sudweeks, Fay, Margaret McLaughlin, and Sheizaf Rafaeli, eds. 1999. *Network and Netplay: Virtual Groups on the Internet*. Menlo Park, CA: AAAI Press.

Sullivan, Amy, Katrina Hedberg, and David W. Fleming. 2000. "Legalized Physician-Assisted Suicde in Oregon—The Second Year." *New England Journal of Medicine* 342, 8. On the World Wide Web at http://www.nejm.org/content/2000/0342/0008/0598.asp (14 August 2000).

Sumner, William Graham. 1940 [1907]. *Folkways*. Boston: Ginn.

"Super Bowl TV Ratings Are Lowest Since 1990." 1999. On the World Wide Web at http://209.97.20.174/archives/9902/0202ratings.shtml (19 July 1999).

Supreme Court of the State of Hawaii. 1999. "Baehr v. Miike." On the World Wide Web at http://www.state.hi.us/jud/20371.htm (19 August 2000).

"Supreme Court Rejects Census Sampling." 1999. *Catalog Age Weekly*. On the World Wide Web at http://www.catalogagemag.com/content/Weekly/1999/1999012803.HTM (6 May 2000).

"A Survey of Human-Rights Law." 1998. *The Economist* 5 December.

Sutherland, Edwin H. 1939. *Principles of Criminology*. Philadelphia: Lippincott.

_____. 1949. *White Collar Crime*. New York: Dryden.

Suttles, Gerald D. 1968. *The Social Order of the Slum: Ethnicity and Territory in the Inner City*. Chicago: University of Chicago Press.

"Sweathshops.org." 2000. On the World Wide Web at http://www.sweatshops.org (23 June 2000).

Sweezy, Kate, and Jill Tiefenthaler. 1996. "Do State-Level Variables Affect Divorce Rates?" *Review of Social Economy* 54: 47–65.

Sykes, Gresham, and David Matza. 1957. "Techniques of Neutralization: A Theory of Delinquency." *American Sociological Review* 22: 664–70.

Szasz, Andrew, and Michael Meuser. 1997. "Environmental Inequalities: Literature Review and Proposals for New Directions in Research and Theory." *Current Sociology* 45, 3: 99–120.

Szreter, Simon. 1996. *Fertility, Class and Gender in Britain, 1860–1940*. Cambridge, UK: Cambridge University Press.

Tannen, Deborah. 1990. *You Just Don't Understand Me: Women and Men in Conversation*. New York: William Morrow.

_____. 1994a. *Talking from 9 to 5: How Women's and Men's Conversational Styles Affect Who Gets Heard, Who Gets Credit, and What Gets Done at Work*. New York: William Morrow.

_____ 1994b. *Gender and Discourse*. New York: Oxford University Press.

Tarrow, Sidney. 1994. *Power in Movement: Social Movements, Collective Action and Politics*. Cambridge, UK: Cambridge University Press.

Tasker, Fiona L. and Susan Golombok. 1997. *Growing Up in a Lesbian Family: Effects on Child Development*. New York: The Guilford Press.

Tavris, Carol. 1992. *The Mismeasure of Woman*. New York: Simon & Schuster.

Taylor, Ronald, ed. 1998 [1994]. *Minority Families in the United States: A Multicultural Perspective*. Upper Saddle River, NJ: Prentice-Hall.

Tec, Nechama. 1986. *When Light Pierced the Darkness: Christian Rescue of Jews in Nazi-Occupied Poland*. New York: Oxford University Press.

Television Bureau of Advertising. 2000. "Gross Domestic Product, Total Ad Volume, and Television Ad Volume 1960–1998." On the World Wide Web at http://www.tvb.org/tvfacts/trends/gdpvolume/gdp1.html (11 May 2000).

"That's AOL Folks . . .". 2000. *CNNfn*. On the World Wide Web at http://cnnfn.com/2000/01/10/deals/aol_warner (2 May).

Thelen, David. 1996. *Becoming Citizens in the Age of Television: How Americans Challenged the Media and Seized Political Initiative during the Iran-Contra Debate*. Chicago: University of Chicago Press.

Theodore, Peter S. 1998. "Heterosexual Masculinity and Homophobia: A Reaction to the Self?" Paper presented at the annual meetings of the American Psychological Association (San Francisco: 17 August).

Thernstrom, Stephan, and Abigail Thernstrom. 1997. *America in Black and White: One Nation, Indivisible*. New York: Simon & Schuster.

Thomas, Keith. 1971. *Religion and the Decline of Magic*. London: Weidenfeld and Nicholson.

Thomas, William Isaac. 1996 [1931]. "The Relation of Research to the Social Process." Pp. 289–305 in Morris Janowitz, ed. *W.I. Thomas on Social Organization and Social Personality*. Chicago: University of Chicago Press.

_____ and Florian Znaniecki. 1958 [1918–20]. *The Polish Peasant in Europe and America: Monograph of an Immigrant Group*, 2 vols., 2nd ed. New York: Dover Publications.

Thompson, E. P. 1967. "Time, Work Discipline, and Industrial Capitalism." *Past and Present* 38: 59–67.

_____. 1968. *The Making of the English Working Class*. Harmondsworth, UK: Penguin.

Thorne, Barrie. 1993. *Gender Play: Girls and Boys in School*. New Brunswick, NJ: Rutgers University Press.

Thurow, Lester. 1999. "Building Wealth." *The Atlantic Monthly* June. On the World Wide Web at http://www.theatlantic.com/issues/99jun/9906thurow.htm (2 May 2000).

Tienda, Marta, and Ding-Tzann Lii. 1987. "Minority Concentration and Earnings Inequality: Blacks, Hispanics, and Asians Compared." *American Journal of Sociology* 93: 141–65.

Tierney, John. 1997. "Our Oldest Computer, Upgraded." *New York Times Magazine* 28 September: 46–9, 97, 100, 104–5.

Tilly, Charles. 1978. *From Mobilization to Revolution*. Reading, MA: Addison-Wesley.

_____. 1979a. "Collective Violence in European Perspective." Pp. 83–118 in H. Graham and T. Gurr, eds. *Violence in America: Historical and Comparative Perspective*, 2nd ed. Beverly Hills, CA: Sage.

_____. 1979b. "Repertoires of Contention in America and Britain, 1750–1830." Pp. 126–55 in Mayer N. Zald and John D. McCarthy, eds. *The Dynamics of Social Movements: Resource Mobilization, Social Control, and Tactics*. Cambridge, MA: Winthrop Publishers.

_____, Louise Tilly, and Richard Tilly. 1975. *The Rebellious Century, 1830–1930*. Cambridge, MA: Harvard University Press.

Tilly, Chris. 1996. *Half a Job: Bad and Good Part-Time Jobs in a Changing Labor Market*. Philadelphia: Temple University Press.

Tillyard, E. M. W. 1943. *The Elizabethan World Picture*. London: Chatto and Windus.

Toffler, Alvin. 1990. *Powershift: Knowledge, Wealth, and Violence at the Edge of the 21st Century*. New York: Bantam.

Tokar, Brian, ed. 2001. *Redesigning Life: The Worldwide Challenge to Genetic Engineering*. Montreal: McGill-Queen's University Press.

Tolnay, Stewart E., and E. M. Beck. 1995. *A Festival of Violence: An Analysis of Southern Lynchings, 1882–1930*. Urbana: University of Illinois Press.

Tong, Rosemarie. 1989. *Feminist Thought: A Comprehensive Introduction*. Boulder, CO: Westview.

Tönnies, Ferdinand. 1957 [1887]. *Community and Society*. Charles P. Loomis, ed. and trans. East Lansing, MI: Michigan State University Press.

Tonry, Michael. 1995. *Malign Neglect: Race, Crime, and Punishment in America*. New York: Oxford University Press.

Tornquist, Cynthia. 1998. "Students Head Back to Decaying Classrooms." *CNN.COM* 30 August. On the World Wide Web http://www.cnn.co.uk/US/9808/30/hazardous.schools (11 August 2000).

Traub, James. 2000. "What No School Can Do." *New York Times Magazine* 16 January: 52–7, 68, 81, 90–1.

Troeltsch, Ernst. 1931 [1923]. *The Social Teaching of the Christian Churches*, Olive Wyon, trans. 2 vols. London: George Allen and Unwin.

Troy, Leo. 1986. "The Rise and Fall of American Trade Unions: The Labor Movement from FDR to RR." Pp. 75–109 in Seymour Martin Lipset, ed. *Unions in Transition: Entering the Second Century*. San Francisco: ICS Press.

Tschannen, Olivier. 1991. "The Secularization Paradigm: A Systematization." *Journal for the Scientific Study of Religion* 30: 395–415.

Tsutsui, William M. 1998. *Manufacturing Ideology: Scientific Management in Twentieth-Century Japan*. Princeton, NJ: Princeton University Press.

Tuljapurkar, Shripad, Nan Li, and Carl Boe. 2000. "A Universal Pattern of Mortality Decline in the G7 Countries." *Nature* 405: 789–92.

Tumin, Melvin. 1953. "Some Principles of Stratification: A Critical Analysis." *American Sociological Review* 18: 387–94.

Turkle, Sherry. 1995. *Life on the Screen: Identity in the Age of the Internet*. New York: Simon & Schuster.

Turnbull, Colin M. 1961. *The Forest People*. New York: Doubleday.

Turner, Bryan S. 1986. *Citizenship and Capitalism: The Debate over Reformism*. London: Allen and Unwin.

Turner, Ralph H., and Lewis M. Killian. 1987 [1957] *Collective Behavior*, 3rd ed. Englewood Cliffs, NJ: Prentice-Hall.

Tuxill, John, and Chris Bright. 1998. "Losing Strands in the Web of Life." Pp. 41–58 in Lester R. Brown, Christopher Flavin, Hilary French, et al. *State of the World 1998*. New York: W. W. Norton.

Twenge, Jean M. 1997. "Changes in Masculine and Feminine Traits Over Time: A Meta-analysis." *Sex Roles* 36: 305–25.

Twitchell, James B. 1999. *Lead Us Into Temptation: The Triumph of American Materialism*. New York: Columbia University Press.

Tyree, Andrea, Moshe Semyonov, and Robert W. Hodge. 1979. "Gaps and Glissandos: Inequality, Economic Development, and Social Mobility in 24 Countries." *American Sociological Review* 44: 410–24.

Ungar, Sheldon. 1992. "The Rise and (Relative) Decline of Global Warming as a Social Problem." *Sociological Quarterly* 33: 483–501.

_____. 1995. "Social Scares and Global Warming: Beyond the Rio Convention." *Society and Natural Resources* 8: 443–56.

_____. 1998. "Bringing the Issue Back In: Comparing the Marketability of the Ozone Hole and Global Warming." *Social Problems* 45: 510–27.

_____. 1999. "Is Strange Weather in the Air? A Study of U.S. National Network News Coverage of Extreme Weather Events." *Climatic Change* 41: 133–50.

Union of Concerned Scientists. 1993. "World Scientists' Warning to Humanity." On the World Wide Web at http://www.englib. cornell.edu/scitech/u95/warn.html (2 May 2000).

United Nations. 1993. *The Age and Sex Distribution of the World Population*. New York.

_____. 1994. *Human Development Report 1994*. New York: Oxford University Press.

_____. 1997a. "Total Fertility Rate." On the World Wide Web at http://www.undp.org/popin/wdtrends/fer/ffer.htm (2 May 2000).

_____. 1997b. "Percentage of Population Living in Urban Areas in 1996 and 2030." On the World Wide Web at http:// www.undp.org/popin/wdtrends/ura/uracht1.htm (2 May 2000).

_____. 1998a. "Below-Replacement Fertility." On the World Wide Web at http://www.popin.org/pop1998/7.htm (4 July 1999).

_____. 1998b. *Human Development Report 1998*. New York: Oxford University Press.

_____. 1998c. "World Population Growth from Year 0 to 2050." On the World Wide Web at http://www.popin.org/ pop1998/4.htm (3 July 1999).

_____. 1999a. "Gender Empowerment Measure." On the World Wide Web at http://www.undp.org/hdro/98gem.htm (28 April 2000).

_____. 1999b. "Human Development Index." On the World Wide Web at http://www.undp.org/hdro/98hdi1.htm (28 April 2000).

_____. 1999c. "Indicators on Income and Economic Activity." On the World Wide Web at http://www.un.org/ Depts/unsd/social/inc-eco.htm (28 April 2000).

_____. 2000. *Report on the Global HIV/AIDS Epidemic*. On the World Wide Web at http://www.unaids.org/ epidemic_update/report/Epi_report.htm (12 August 2000).

U.S. Administration on Aging. 1999. "Older Population by Age: 1900 to 2050." On the World Wide Web at http://www.aoa. dhhs.gov/aoa/stats/AgePop2050.html (2 May 2000).

U.S. Bureau of the Census. 1993. "We the American . . . Foreign Born." On the World Wide Web at http://www.census.gov/ apsd/wepeople/we-7.pdf (29 April 2000).

_____. 1997. "Country of Origin and Year of Entry into the US of the Foreign Born, By Citizenship Status: March 1997." On the World Wide Web at http://www.bls.census.gov/cps/ pub/1997/for_born.htm (29 April 2000).

_____. 1998a. "Historical Income Tables—Households." On the World Wide Web at http://www.census.gov/hhes/income/ histinc/h02.html (29 April 2000).

_____. 1998b. "Population of the 100 Largest Urban Places: 1950." On the World Wide Web at http://www.census.gov/ population/documentation/twps0027/tab18.txt (2 May 2000).

_____. 1998c. *Statistical Abstract of the United States: 1998*. On the World Wide Web at http://www.census.gov/ prod/www/statistical-abstract-us.html (29 April 2000).

_____. 1998d. "Total Midyear Population for the World: 1950–2050." On the World Wide Web at http://www.census. gov/ipc/www/worldpop.html (2 May 2000).

_____. 1999a. "Households, by Type: 1940 to Present." On the World Wide Web at http://www.census.gov/population/ socdemo/hh-fam/htabHH-1.txt (1 May 2000).

_____. 1999b. "Money Income in the United States." *Current Population Reports, 1998*. On the World Wide Web at http:// www.census.gov/prod/99pubs/p60-206.pdf (6 June 2000).

_____. 1999c. "Population Estimates for Cities with Populations of 100,000 and Greater." On the World Wide Web at http://www.census.gov/population/estimates/ metro–city/SC100K98-T1-DR.txt (2 May 2000).

_____. 1999d. *Poverty in the United States: 1998*. On the World Wide Web at http://www.census.gov/prod/99pubs/ p60-207.pdf (7 June 2000).

_____. 1999e. "Region and Country or Area of Birth of the Foreign-Born Population, With Geographic Detail Shown in Decennial Census Publications of 1930 or Earlier: 1850 to 1930 and 1960 to 1999." On the World Wide Web at http://www.census.gov/population/www/documentation/ twps0029/tab04.html (29 April 2000).

_____. 1999f. *Statistical Abstract of the United States: 1999*. On the World Wide Web at http://www.census. gov/prod/99pubs/99statab/sec04.pdf (15 January 2001).

_____. 1999g. "Table 2. Poverty Status of People, by Family Relationship, Race, and Hispanic Origin: 1959 to 1998." On the World Wide Web at http://www.census. gov/hhes/poverty/histpov/hstpov2.html (29 April 2000).

_____. 2000a. "Countries Ranked by Population: 2000." On the World Wide Web at http://www.census.gov/cgi-bin/ipc/ idbrank.pl (29 April 2000).

_____. 2000b. "Historical National Population Estimates: July 1, 1900 to July 1, 1999." On the World Wide Web at http://www.census.gov/population/estimates/nation/ popclockest.txt (29 April 2000).

_____. 2000c. "IDB Population Pyramids." On the World Wide Web at http://www.census.gov/ipc/www/idbpyr.html (4 August 2000).

_____. 2000d. "National Population Projections." On the World Wide Web at http://www.census.gov/population/www/ projections/natproj.html (28 April 2000).

_____. 2000e. "Percentage of Industry Statistics Accounted for by Largest Companies: 1992." On the World Wide Web at http://www.census.gov:80/mcd/mancen/download/mc92cr. sum (30 April 2000).

_____. 2000f. "Resident Population Estimates of the United States by Age and Sex: April 1, 1990 to July 1, 1999, with Short-Term Projection to March 1, 2000." On the World Wide Web at http://www.census.gov/population/estimates/nation/ intfile2-1.txt (29 April 2000)

_____. 2000g. "Resident Population Estimates of the United States by Sex, Race, and Hispanic Origin: April 1, 1990 to July 1, 1999, with Short-Term Projection to March 1, 2000." On the World Wide Web at http://www.census.gov/ population/estimates/nation/intfile3-1.txt (25 May 2000).

_____. 2000h. "Selected Characteristics of Households, by Total Money Income in 1998." On the World Wide Web at http://ferret.bls.census.gov/macro/031999/hhinc/new01_ 001.htm (7 June 2000).

_____. 2001a. "Census 2000 Shows Resident Population of 281,421,906; Apportionment Counts Delivered to President." On the World Wide Web at http://www.census.gov/ Press-Release/www/2000/cb00cn64.html (12 August 2000).

_____ 2001b. *Mapping Census 2000: The Geography of U.S. Diversity.* On the World Wide Web at http://www.census.gov/population/cen2000/atlas/censr01-1.pdf (August 12 2000).

U.S. Department of Education. 1997. *Pursuing Excellence: A Study of U.S. Fourth Grade Mathematics and Science Achievement in International Context.* On the World Wide Web at http://nces.ed.gov/timss/report/97255-01.html (2 May 2000).

_____. 1998. *Pursuing Excellence: A Study of U.S. Twelfth Grade Mathematics and Science Achievement in International Context.* On the World Wide Web at http://nces.ed.gov/timss/twelfth/index.html (2 May 2000).

_____. 1999. *The Condition of Education 1999.* On the World Wide Web at http://nces.ed.gov/pubs99/condition99 (2 May 2000).

_____. 2000. *The Digest of Education Statistics 1999.* Washington D.C.: National Center for Education Statistics. On the World Wide Web at http://www.nces.ed.gov/pubs2000/digest99 (15 January 2001).

U.S. Department of Health and Human Services. 1999. *Healthy People 2000: National Health Promotion and Disease Prevention Objectives.* Hyattsville, MD: Centers for Disease Control and Prevention, National Center for Health Statistics. On the World Wide Web at http://www.cdc.gov/nchswww/data/hp2k99.pdf (29 April 2000).

U.S. Department of Justice. 1999. "Criminal Victimization 1998: Changes 1997–98 with Trends 1993–98." On the World Wide Web at http://www.ojp.usdoj.gov/bjs/pub/ascii/cv98.txt (25 May 2000).

U.S. Department of Labor, Bureau of Labor Statistics. 1998a. "Median Usual Weekly Earnings of Full-Time Wage and Salary Workers by Detailed Occupation and Sex, 1997." On the World Wide Web at ftp://ftp.bls.gov.pub/specialrequests/lf/aat39.txt (20 April 1999).

_____. 1998b. "Table A-1. Employment Status of the Civilian Population by Sex and Age." On the World Wide Web at http://stats.bls.gov/webapps/legacy/cpsatab1.htm (30 April 2000).

_____. 1999a. "Comparative Civilian Labor Force Statistics: Ten Countries, 1959–1998." On the World Wide Web at ftp://ftp.bls.gov/pub/special.requests/ForeignLabor/flslforc.txt (30 April 2000).

_____ 1999b. "Employment Status of the Civilian Population by Sex and Age." On the World Wide Web at http://stats.bls.gov/news.release/empsit.t01.htm (30 April 2000).

_____. 1999c. "Median Usual Weekly Earnings of Full-Time Wage and Salary Workers by Detailed Occupation and Sex, 1998." On the World Wide Web at ftp://ftp/bls.gov/pub/special.requests/lf/aat39.txt (23 January 2000).

_____. 1999d."Table 5. Civilian Labor Force by Sex, Age, Race, and Hispanic Origin, 1978, 1988, 1998, and projected 2008." On the World Wide Web at http://stats.bls.gov/emplt985.htm (30 April 2000).

_____. 1999e. "Usual Weekly Earnings Summary." On the World Wide Web at http://stats.bls.gov/news.release/wkyeng.nws.htm (30 April 2000).

_____. 1999f. "Value of the Federal Minimum Wage." On the World Wide Web at http://www.dol.gov/dol/esa/public/minwage/chart2.htm (30 April 2000).

_____. 2000a. "Labor Force Statistics from the Current Population Survey." On the World Wide Web at http://146.142.4.24/cgi-bin/surveymost (31 December 2000).

_____. 2000b. "Median Usual Weekly Earnings of Full-Time Wage and Salary Workers by Selected Characteristics, Quarterly Averages, Not Seasonally Adjusted." On the World Wide Web at http://stats.bls.gov/news.release/wkyeng.t01.htm (15 June 2000).

_____. 2000c. *Occupational Outlook Handbook, 2000–01.* On the World Wide Web at http://stats.bls.gov/oco/ocos054.htm (25 November 2000).

_____. 2000d. "Union Members Summary." On the World Wide Web at http://stats.bls.gov/news.release/union2.nws.htm (30 April 2000).

_____. 2000e. "Union Membership Data from the National Directory Series." On the World Wide Web at ftp://146.142.4.23/pub/special.requests/collbarg/unmem.txt (1 August 2000).

_____. 2000f. "Work Stoppages Involving 1,000 Workers or More, 1947–2000." On the World Wide Web at http://stats.bls.gov/news.release/wkstp.t01.htm (28 April 2000).

_____. 2001. "Union Members Summary." On the World Wide Web at http://stats.bls.gov/news.release/union2.nws.htm (22 March 2000).

U.S. Environmental Protection Agency, Office of Air Quality Planning and Standards. 1998. *National Air Pollutant Emission Trends Report, 1900–1996.* On the World Wide Web at http://www.epa.gov/ttn/chief/trends96/chapter5.pdf (29 April 1999).

_____. 2000. *National Air Pollutant Emission Trends, 1900–1998.* On the World Wide Web at http://www.epa.gov/ttn/chief/trends98/emtrnd.html (3 August 2000).

U.S. Federal Bureau of Investigation. 1997. *Hate Crime Statistics 1997.* On the World Wide Web at http://www.fbi.gov/ucr/hc97all.pdf (29 April 2000).

_____. 1999a. *Uniform Crime Reports for the United States 1998.* On the World Wide Web at http://www.fbi.gov/ucr/98cius.htm (25 May 2000).

_____. 1999b. "Uniform Crime Reports: January–June 1999." On the World Wide Web at http://www.fbi.gov/ucr.htm (29 April 2000).

_____. 2000. *Crime in the United States, 1999.* On the World Wide Web at http://www.fbi.gov/ucr/Cius_99/99crime/99cius.pdf (21 May 2001).

U.S. Information Agency. 1998–99. *The People Have Spoken: Global Views of Democracy,* 2 vols. Washington, DC: Office of Research and Media Reaction.

U.S. Secretary of State. 1999. "Summary and Highlights: International Affairs (Function 150): Fiscal Year 2000 Budget Request." On the World Wide Web at http://www.state.gov/www/budget/2000_budget.html (2 May 2000).

Unschuld, Paul. 1985. *Medicine in China.* Berkeley, CA: University of California Press.

Urban Institute. 1998. "Policy Challenges Posed by the Aging of America." On the World Wide Web at http://www.urban.org/health/oldpol.html (2 May 2000).

Useem, Bert. 1998. "Breakdown Theories of Collective Action." *Annual Review of Sociology* 24: 215–38.

Useem, Michael. 1984. *The Inner Circle: Large Corporations and the Rise of Business Political Activity in the U.S. and U.K.* New York: Oxford University Press.

Valelly, Richard. 1999. "Voting Rights in Jeopardy." *The American Prospect* 46: 43–9. On the World Wide Web at http://www.prospect.org/archives/46/46valelly.html (14 January 2001).

Valocchi, Steve. 1996. "The Emergence of the Integrationist Ideology in the Civil Rights Movement." *Social Problems* 43: 116–30.

Van de Kaa, Dirk. 1987. "Europe's Second Demographic Transition." *Population Bulletin* 42, 1: 1–58.

Van der Laan, Luc J. W., Christopher Lockey, Bradley C. Griffeth, Francine S. Frasier, Carolyn A. Wilson, David E. Onions, Bernhard J. Hering, Zhifeng Long, Edward Otto, Bruce E. Torbett, and Daniel R. Salomon. 2000. "Infection by Porcine Endogenous Retrovirus After Islet Xenotransplantation in SCID Mice." *Nature* 407: 501–4. On the World Wide Web at http://www.nature.com/cgi-taf/DynaPage.taf?file=/nature/journal/v407/n6800/full/407501a0_fs.html (17 August 2000).

Vanneman, Reeve, and Lynn Weber Cannon. 1987. *The American Perception of Class*. Philadelphia: Temple University Press.

Veblen, Thorstein. 1899. *The Theory of the Leisure Class*. On the World Wide Web at http://socserv2.socsci.mcmaster.ca/~econ/ugcm/3ll3/veblen/leisure/index.html (29 April 2000).

Verba, Sidney, Kay Lehman Schlozman, and Henry E. Brady. 1997. "The Big Tilt: Participatory Inequality in America." *The American Prospect* 32: 74–80.

Vernarec, Emil. 2000. "Depression in the Work Force: Seeing the Cost in a Fuller Light." *Business and Health* 18, 4: 48–55.

Veugelers, John. 1997. "Social Cleavage and the Revival of Far Right Parties: The Case of France's National Front." *Acta Sociologica* 40: 31–49.

Vidal, John. 1999. "World Burdened by Wars and Pollution." *The Guardian* 6 June: 7.

Vitale, Ami. 2000. "The War Next Door." *Saturday Night* 3 June: 48–54.

Vogel, David J. 1996. "The Study of Business and Politics." *California Management Review* 38, 3: 146–65.

Vogel, Ezra F., ed. 1975. *Modern Japanese Organization and Decision-Making*. Berkeley, CA: University of California Press.

Vygotsky, Lev S. 1987. *The Collected Works of L. S. Vygotsky*, Vol. 1, N. Minick, trans. New York: Plenum.

Waldfogel, Jane. 1997. "The Effect of Children on Women's Wages." *American Sociological Review* 62: 209–17.

Wallace, Anthony F. C. 1993. *The Long, Bitter Trail: Andrew Jackson and the Indians*. New York: Hill and Wang.

Wallace, James. 1997. *Overdrive: Bill Gates and the Race to Control Cyberspace*. New York: John Wiley.

_____ and Jim Erickson. 1992. *Hard Drive: Bill Gates and the Making of the Microsoft Empire*. New York: John Wiley.

Wallerstein, Immanuel. 1974–89. *The Modern World-System*, 3 vols. New York: Academic Press.

_____, ed. 1998. "The Heritage of Sociology and the Future of the Social Sciences in the 21st Century." *Current Sociology* 46, 2.

Wallerstein, Judith S., and Sandra Blakeslee. 1989. *Second Chances: Men, Women, and Children a Decade After Divorce*. New York: Ticknor and Fields.

_____, Julia Lewis, and Sandra Blakeslee. 2000. *The Unexpected Legacy of Divorce: A 25 Year Landmark Study*. New York: Hyperion.

Waters, Mary C. 1990. *Ethnic Options: Choosing Identities in America*. Berkeley, CA: University of California Press.

_____. 2000. *Black Identities*. Cambridge, CA MA: Harvard University Press.

Weakliem, David L. 1991. "The Two Lefts? Occupation and Party Choice in France, Italy, and the Netherlands." *American Journal of Sociology* 96: 1327–61.

"Webcam Search.com." 2000. On the World Wide Web at http://www.webcamsearch.com (2 May 2000).

Webb, Eugene J., Donald T. Campbell, Richard D. Schwartz, and Lee Sechrest. 1966. *Unobtrusive Measures: Nonreactive Research in the Social Sciences*. Chicago: Rand McNally.

Webb, Stephen D., and John Collette. 1977. "Rural–Urban Differences in the Use of Stress-Alleviating Drugs." *American Journal of Sociology* 83: 700–7.

_____. 1979. "Reply to Comment on Rural–Urban Differences in the Use of Stress-Alleviating Drugs." *American Journal of Sociology* 84: 1446–52.

Weber, Max. 1946. *From Max Weber: Essays in Sociology*, Hans Gerth and C. Wright Mills, eds. and trans. New York: Oxford University Press.

_____. 1947. *The Theory of Social and Economic Organization*, T. Parsons, ed., A. M. Henderson and T. Parsons, trans. New York: Free Press.

_____. 1958 [1904–5]. *The Protestant Ethic and the Spirit of Capitalism*. New York: Charles Scribner's Sons.

_____. 1963. *The Sociology of Religion*, Ephraim Fischoff, trans. Boston: Beacon Press.

_____. 1964 [1949]. "'Objectivity' in Social Science and Social Policy." Pp. 49–112 in *The Methodology of the Social Sciences*, Edward A. Shils and Henry A. Finch, trans. and eds. New York: Free Press of Glencoe.

_____. 1978a [1968]. *Economy and Society*, Guenther Roth and Claus Wittich, eds. Berkeley, CA: University of California Press.

Weeks, Jeffrey. 1986. *Sexuality*. London: Routledge.

Weis, Joseph G. 1987. "Class and Crime." Pp. 71–90 in Michael Gottfredson and Travis Hirschi, eds. *Positive Criminology*. Beverly Hills, CA: Sage.

Welch, Michael. 1997. "Violence Against Women by Professional Football Players: A Gender Analysis of Hypermasculinity, Positional Status, Narcissism, and Entitlement." *Journal of Sport and Social Issues* 21: 392–411.

Weller, Jack M., and E. L. Quarantelli. 1973. "Neglected Characteristics of Collective Behavior." *American Journal of Sociology* 79: 665–85.

Wellman, Barry. 1979. "The Community Question: The Intimate Networks of East Yorkers." *American Journal of Sociology* 84: 201–31.

_____ and Stephen Berkowitz, eds. 1997 [1988]. *Social Structures: A Network Approach*, updated ed. Greenwich, CT: JAI Press.

_____ et al. 1996. "Computer Networks as Social Networks: Collaborative Work, Telework, and Virtual Community." *Annual Review of Sociology* 22: 213–38.

Welsh, Sandy. 1999. "Gender and Sexual Harassment." *Annual Review of Sociology* 25: 169–90.

Wente, Margaret. 2000. "How David Found His Manhood." *Globe and Mail* 29 January: A15–A16.

Wentz, Frank J. and Matthias Schabel. 1998. "Effects of Orbital Decay on Satellite-Derived Lower-Tropospheric Temperature Trends." *Nature* 394: 661–4.

West, Candace, and Don Zimmerman. 1987. "Doing Gender." *Gender and Society* 1: 125–51.

Westen, Tracy. 1998. "Can Technology Save Democracy?" *National Civic Review* 87: 47–56.

Wheeler, Stanton. 1961. "Socialization in Correctional Communities." *American Sociological Review* 26: 697–712.

Whitefield, S., and G. Evans. 1994. "The Russian Election of 1993: Public Opinion and the Transition Experience." *Post-Soviet Affairs* 10: 38–60.

Whitman, David. 2000. "When East Beats West Old Money Bests New." *Business Week* 8 May: 28.

Whittier, Nancy. 1995. *Feminist Generations: The Persistence of the Radical Women's Movement.* Philadelphia: Temple University Press.

Whyte, William Foote. 1981 [1943]. *Street Corner Society: The Social Structure of an Italian Slum*, 3rd revised and expanded ed. Chicago: University of Chicago Press.

Wilensky, Harold L. 1967. *Organizational Intelligence: Knowledge and Policy in Government and Industry.* New York: Basic Books.

_____. 1997. "Social Science and the Public Agenda: Reflections on the Relation of Knowledge to Policy in the United States and Abroad." *Journal of Health Politics, Policy and Law* 22: 1241–65.

Wiley, Norbert. 1994. *The Semiotic Self.* Chicago: University of Chicago Press.

Wilkinson, Richard G. 1996. *Unhealthy Societies: The Afflictions of Inequality.* London: Routledge.

Willardt, Kenneth. 2000. "The Gaze He'll Go Gaga For." *Cosmopolitan* April: 232–7.

Williams, Daniel T. 1970. "The Lynching Records at Tuskegee Institute." *Eight Negro Bibliographies.* New York: Kraus Reprint Co.

Williams, David R., and Chiquita Collins. 1995. "U.S. Socioeconomic and Racial Differences in Health: Patterns and Explanations." *Annual Review of Sociology* 21: 349–86.

Willis, Paul. 1984 [1977]. *Learning to Labour: How Working-Class Kids Get Working-Class Jobs.* New York: Columbia University Press.

Wilson, Edward O. 1975. *Sociobiology: The New Synthesis.* Cambridge, MA: Belknap Press of the Harvard University Press.

Wilson, William Julius. 1980 [1978]. *The Declining Significance of Race: Blacks and Changing American Institutions*, 2nd ed. Chicago: University of Chicago Press.

_____. 1987. *The Truly Disadvantaged: The Inner City, the Underclass, and Public Policy.* Chicago: University of Chicago Press.

_____. 1996. *When Work Disappears: The World of the New Urban Poor.* New York: Knopf.

Winch, Donald. 1987. *Malthus.* Oxford, UK: Oxford University Press.

Wirth, Louis. 1938. "Urbanism as a Way of Life." *American Journal of Sociology* 44:1–24.

Witt, Susan D. 1997. "Parental Influence on Children's Socialization to Gender Roles." *Adolescence* 32: 253–9.

Wolf, Diane Lauren. 1992. *Factory Daughters: Gender, Household Dynamics, and Rural Industrialization in Java.* Berkeley, CA: University of California Press.

Wolf, Eric R. 1982. *Europe and the People without History.* Berkeley, CA: University of California Press.

Wolf, Naomi. 1997. *Promiscuities: The Secret Struggle for Womanhood.* New York: Vintage.

Wolff, Edward N. 1996 [1995]. *Top Heavy: The Increasing Inequality of Wealth in America and What Can Be Done About It*, expanded ed. New York: New Press.

Wolff, Janet. 1999. "Cultural Studies and the Sociology of Culture." *Invisible Culture* 1. On the World Wide Web at http://www.rochester.edu/in_visible_culture/issue1/wolff/wolff.html (10 May 2000).

Wood, Julia. 1999 [1996]. *Everyday Encounters: An Introduction to Interpersonal Communication*, 2nd ed. Belmont, CA: Wadsworth.

Wood, W., F. Y. Wong, and J. G. Chachere. 1991. "Effects of Media Violence on Viewers' Aggression in Unconstrained Social Interaction." *Psychological Bulletin* 109: 371–83.

Woodrow Federal Reserve Bank of Minneapolis. 2000. "What's a Dollar Worth?" On the World Wide Web at http://woodrow.mpls.frb.fed.us/economy/calc/cpihome.html (2 May 2000).

Wooton, Barbara H. 1997. "Gender Differences in Occupational Employment." *Monthly Labor Review* 120, 4: 15–24. On the World Wide Web at http://stats.bls.gov/opub/mlr/1997/04/art2full.pdf (30 April 2000).

Wordsworth, Araminta. 2000. "Family Planning Officials Drown Baby in Rice Paddy." *National Post* 25 August: A10.

"Work-related Stress: A Condition Felt 'Round the World'." 1995. *HR Focus* 72, 4: 17.

World Bank. 1999a. "Aid Dependency." On the World Wide Web at http://www.worldbank.org/data/wdi/pdfs/tab6_10.pdf (27 April 2000).

_____. 1999b. *Global Development Finance 1999.* On the World Wide Web at http://www.worldbank.org/prospects/gdf99/tables.pdf (27 April 2000).

_____. 1999c. "GNP Per Capita 1997, Atlas Method and PPP." http://www.worldbank.org/data/databytopic/GNPPC97.pdf (10 July 1999).

World Health Organization. 1996. "Most Recent Values of Global Health-For-All Indicators." On the World Wide Web at http://www.who.int/whosis/hfa/countries/index.html (29 April 2000).

_____. 1999a. "WHO Estimates of Health Personnel." On the World Wide Web at http://www.who.int/whosis/healthpersonnel/index.html (19 November 1999).

_____. 1999b. "World Health Report 1999: Basic Indicators for all Member States." On the World Wide Web at http://www.who.int/whr/1999/en/indicators.htm (2 May 2000).

_____. 2000. "WHO Terminology Information System." On the World Wide Web at http://www.who.int/terminology/ter/wt001.html#health (2 May 2000).

World Values Survey, 1990–1993. 1994. Ann Arbor, MI: Inter-University Consortium for Political and Social Research.

"The World's Online Populations." 2001. On the World Wide Web at http://cyberatlas.internet.com/big_picture/geographic/article/0,1323,5911_151151,00.html (17 December 2001).

Worth, Robert. 1995. "A Model Prison." *The Atlantic Monthly* November. On the World Wide Web at http://www.theatlantic.com/issues/95nov/prisons/prisons.htm (28 May 2000).

_____. 2001. "The Deep Intellectual Roots of Islamic Terror." *The New York Times on the Web* (13 October 2001). On the World Wide Web at http://www.nytimes.com/2001/10/13/arts/13ROOT.html?searchpv=past7days (15 October 2001).

Wright, Charles Robert. 1975. *Mass Communication: A Sociological Perspective*. New York: Random House.

Wright, Erik Olin. 1985. *Classes*. London: Verso.

_____. 1997. *Class Counts: Comparative Studies in Class Analysis*. Cambridge, UK: Cambridge University Press.

Wright, John W., ed. 1998. *The New York Times Almanac 1999*. New York: Penguin.

X, Malcolm. 1965. *The Autobiography of Malcolm X*. New York: Grove.

Yamane, David. 1997. "Secularization on Trial: In Defense of a Neosecularization Paradigm." *Journal for the Scientific Study of Religion* 36: 109–22.

Yancey, William L., Eugene P. Ericksen, and George H. Leon. 1976. "Emergent Ethnicity: A Review and Reformulation." *American Sociological Review* 41: 391–403.

Yoo, Peter S. 1998. "Still Charging: The Growth of Credit Card Debt Between 1992 and 1995." *Review* (Federal Reserve Bank of St. Louis) January–February.

Zakaria, Fareed. 1997. "The Rise of Illiberal Democracy." *Foreign Affairs* 76, 6: 22–43.

Zald, Meyer N., and John D. McCarthy. 1979. *The Dynamics of Social Movements*. Cambridge, MA: Winthrop.

Zangwill, Israel. 1909. *The Melting Pot: Drama in Four Acts*. New York: Macmillan.

Zaslavsky, Victor, and Robert J. Brym. 1978. "The Functions of Elections in the USSR." *Soviet Studies* 30: 62–71.

_____ and _____. 1983. *Soviet Jewish Emigration and Soviet Nationality Policy*. London: Macmillan.

Zimbardo, Philip G. 1972. "Pathology of Imprisonment." *Society* 9, 6: 4–8.

Zimmermann, Francis. 1987 [1982]. *The Jungle and the Aroma of Meats: An Ecological Theme in Hindu Medicine*, Janet Lloyd, trans. Berkeley, CA: University of California Press.

Zimring, Franklin E., and Gordon Hawkins. 1995. *Incapacitation: Penal Confinement and the Restraint of Crime*. New York: Oxford University Press.

Zinsser, Hans. 1935. *Rats, Lice and History*. Boston: Little, Brown.

Zola, Irving Kenneth. 1982. *Missing Pieces: A Chronicle of Living with a Disability*. Philadelphia: Temple University Press.

Zuboff, Shoshana. 1988. *In the Age of the Smart Machine: The Future of Work and Power*. New York: Basic Books.

Zukin, Sharon. 1980. "A Decade of the New Urban Sociology." *Theory and Society* 9: 539–74.

Zurcher, Louis A., and David A. Snow. 1981. "Collective Behavior and Social Movements." Pp. 447–82 in Morris Rosenberg and Ralph Turner, eds. *Social Psychology: Sociological Perspectives*. New York: Basic Books.

Zussman, Robert. 1992. *Intensive Care: Medical Ethics and the Medical Profession*. Chicago: University of Chicago Press.

_____. 1997. "Sociological Perspectives on Medical Ethics and Decision-Making." *Annual Review of Sociology* 23: 171–89.

INDEX

CREDITS AND ACKNOWLEDGMENTS

Photo Credits

Chapter 1 © Bettmann/Corbis, 6 (left); Courtesy of A. C. Fine Art, Nova Scotia. Photographer: James Chambers, 6 (right); © SuperStock, 7; Brown Brothers, Sterling, PA, 10; © Phillip Caruso/The Everett Collection, 11; © Paul Almasy/CORBIS, 12; By permission of the Houghton Library, Harvard University, 13; © Musee du Louvre, Paris/Giraudon, Paris/Superstock, 14 (top); Photograph Copyright © 2001 The Detroit Institute of the Arts, 14 (center); © Associated Press/AP/Wide World Photos, 17; © Archivo Iconografico, S.A./CORBIS, 18 (top); Brown Brothers, Sterling, PA, 18 (bottom); © Bettmann/CORBIS, 19; © Hulton-Deutsch Collection/CORBIS, 20; Copyright © 2001 Time, Inc. Reprinted by permission, 24 & 25.

Chapter 2 © Archivo Iconografico, S.A./CORBIS, 34 (top); © Bettmann/CORBIS, 34 (bottom); Courtesy of Carol Wainio, London, Ontario, Canada, 36; Courtesy of Lillian B. Rubin, 37; © David Tumely/CORBIS, 42; © 2001 Joel Gordon, 45; The Everett Collection, 46; © Associated Press/AP/Wide World Photos, 47; © 1995 Antonio Rosario/The Image Bank, 49; Figure 2.4 © Silver Image Photo Agency, 51; Charles Tilly by John Sheretz. © 1997 CASAS, 54.

Chapter 3 Figure 3.1a © Chris Stanford/Allsport, 64 (left); Figure 3.1b © Allsport, 64 (right); © Mark Richards/PhotoEdit, 66; © Dinodia/V.H. Mishra, 70; © Phil Bray/The Everett Collection, 71; © 1992 Joel Gordon, 74; Figure 3.3 Courtesy of Kelloggs, 77 (top); © Owen Franklin/CORBIS, 79; The Everett Collection, 81; Courtesy of Carol Wainio, London, Ontario, Canada, 83 (top); © SuperStock, 83 (bottom); © Mitchell Gerber/CORBIS, 87.

Chapter 4 © Martin Rogers / Stock Boston, 92; © CORBIS, 94; The Everett Collection, 95; © Myrleen Cate/PhotoEdit, 96; Courtesy of Carol Gilligan, 98; © Spencer Grant/PhotoEdit, 99; © SuperStock, 100 (bottom); © Paul Conklin/PhotoEdit, 101; © Jonathan Blair/CORBIS, 106; © Spencer Grant/PhotoEdit, 111.

Chapter 5 The Everett Collection, 119; © Brian Leng/CORBIS, 121; © Bettmann/CORBIS, 124; Figure 5.2 Mills, Janet Lee. 1985. "Body Language Speaks Louder than Words," *Horizons*, February, pp. 6-12, 127; Appeared in *Leatherneck*, March, 1945, 129; © Erich Lessing/Art Resource, NY, 130; © Brandom Films/The Everett Collection, NY, 133; The Everett Collection, 136.

Chapter 6 © Bettmann/CORBIS, 144; National Library of Medicine, Washington, DC, 145; © WeeGEE/ICP/Getty Images, 146; New York Public Library, 148; © David James/The Everett Collection, 150; © Bob Marshak/The Everett Collection, 151; The Everett Collection, 155; The Everett Collection, 158; New York Public Library, 163; The Everett Collection, 165.

Chapter 7 © Christie's Images/CORBIS, 175; © Reuters NewMedia, Inc./CORBIS credit for both, 178; Figure 7.2a from *Saturday Night*, June 3, 2000, "The War Next Store, " p 48 story and photographs by Ami Vitale, 181 (left); Figure 7.2b from *Saturday Night*, June 3, 2000, "The War Next Store, " p 49 story and photographs by Ami Vitale, 181(right); © Art Resource, NY, 184; © 1995 Alex Webb/Magnum Photos, Inc., 190 (left); © Richard T. Nowitz/CORBIS, 190 (right); © The Granger Collection, New York, 193; © Robert Hepier/The Everett Collection, 197 (left); © Gerrit Greve/CORBIS, 197 (right); Copyright © the Dorothea Lange Collection, Oakland Museum of California, 200.

Chapter 8 From *Mismeasurement of Man* by Stephen Jay Gould (Norton), 211; © Duomo/CORBIS, 213 (left); © Annie Sachs/Archive Photo, 213 (right); Figure 8.1 Copyright © 1941 Time, Inc. Reprinted by permission, 214; The Everett Collection, NY, 216; © Heather Titus/Photo Source Hawaii, 218; © Bettmann/CORBIS, 224; Jacob A. Riis Collection, Museum of the City of New York. 90.13.4.160, 226; The Phillips Collection, Washington, DC, 229; © Lewis H. Hines/Archive Photos, 231 (left); © David & Peter Tumley/CORBIS, 231 (right).

Chapter 9 © Peter Carmichael/The Stock Market, 248; © Vivian Zink/The Everett Collection, NY, 250 (top); © Reuters News Media, Inc./CORBIS, 251; © David Young-Wolff/PhotoEdit, 252; Courtesy of the White Rock Beverage Company, 254 (left); Courtesy of the White Rock Beverage Company, 254 (right); © Michael Kooren/Archive Photos, NY, 259; © Joseph Sohm, ChromoSohm, Inc./CORBIS, 261; The Granger Collection, New York, 263; © Janette Beckman/CORBIS, 265; © Rachel Epstein/PhotoEdit, 267; © SuperStock, 272; © Joseph Sohm, ChromoSohm, Inc./CORBIS, 273.

Chapter 10 From *In the Age of the Smart Machine* by Shoshana Zuboff, © 1988 by Basic Books, Inc., 280; Cover of *Wired Magazine*, December, 1999. Photography by James Porto/Courtesy of Wired Magazine, Conde Nast Publications, 281; © Sebastiao Salgado/Contact Press Images, 282; The Everett Collection, 284; © Spencer Grant/PhotoEdit, 287; The Everett Collection, New York, 293; Figure 10.8 Copyright © San Jose Mercury News, 295; The Granger Collection, New York, 302; © Associated Press/AP/Wide World Photos, 304.

Chapter 11 © John Duricka/AP/Wide World Photos, NY, 312; © Archive Iconografico, S.A./Corbis, 315 (top); © SuperStock, 315 (center); © CORBIS, 315 (bottom); Courtesy of the Artist and the Nancy Poole Studios, Toronto, Ontario, 317; © Mark Richards/PhotoEdit, 318; © Reuters NewsMedia, Inc./Corbis, 322 (both); © SuperStock, 325 (left); © Associated Press/AP/Wide World Photos, 325 (right); © Associated Press/AP/Wide World Photos, NY, 326; © 1995 Lise Sarfati/Magnum Photos, Inc., 328 (left); © AFP/Corbis, 328 (right); © Marc Riboud/Magnum Photos, Inc., 330 (left); © Reuters/Jeremiah Kamau/Archive Photos, 330 (right); The Everett Collection, 332.

Chapter 12 The Everett Collection, New York, 341; © Peter Johnson/Corbis, 345; The Everett Collection, New York, 348; Courtesy of the artist and Bau-Xi Gallery, Toronto, ON, Canada, 349; The Everett Collection, New York, 351; Andrew Benjei, *Pink Couch*, 1993. Fiberglass, 24 x 15 x 19 inches. Photo: Ron Giddings, Reproduced with permission of the artist, 355 (top); © SuperStock, 355 (bottom); © Jacques M. Chenet/Corbis, 357; © Michael Newman/PhotoEdit, 359; © Associated Press/AP/Wide World Photos, 361; © Laura Dwight/CORBIS, 363; © Jonathan Blair/CORBIS, 366.

Chapter 13 © Michael Newman/PhotoEdit, 374; © 1994 Wendy Chan/The Image Bank, 376; © Bettmann/CORBIS, 377; © A. Ramey/PhotoEdit, 379; © Bill Varie/CORBIS, 383; © Myrleen Ferguson/PhotoEdit, 385; © CORBIS, 387; © Michael Newman/PhotoEdit, 388; © David Young-Wolf/PhotoEdit, 390 (left); © Cleve Bryant/PhotoEdit, 390 (right); The Everett Collection, New York, 392; © The Purcell Team/CORBIS, 393; © Diane Bondareff/AP/Wide World Photos, NY, 394; Figure 13.8 © Mike Derer/AP/Wide World Photos, NY, 397; © Michael Newman/PhotoEdit, 398.

Chapter 14 The Everett Collection, NY, 404; © Renée Zellweger/The Everett Collection, 405; © Spencer Grant Original/PhotoEdit, 407; © The Pierpont Morgan Library/Art Resource, NY, 409; © Myrleen Ferguson/PhotoEdit, 410; © AFP/CORBIS, 412; Figure 14.6 © Associated Press, Melbourne Age/AP/Wide World Photos, 421; © Reuters NewMedia, Inc./CORBIS, 422.

Chapter 15 © Figaro Magazine/N'gun Tien/Gamma Presse Images, 430; © Mike Hutchings/Reuters/Archive Photos, 432; © Mark Richards/PhotoEdit, 434; © Spencer Grant/PhotoEdit, 440; © Jeff Greenberg/PhotoEdit, 443; © Reuters/John Hillary/Archive Photos, NY, 445; © Ted Streshinsky/CORBIS, 446; © Melinda Sue Gordon/The Everett Collection, NY, 450; © Lisa M. McGeady/CORBIS, 452.

Chapter 16 © 1990 Christopher Morris/Black Star Publishing/PictureQuest, 462; © Scala/Art Resource, NY, 463; © The Purcell Group/CORBIS, 469; The Granger Collection, NY, 470; © 1982 The Ladd Company/The Everett Collection, NY, 475; © 1996 Joel Gordon, 476; Photo by Miro Cernetig, *Globe & Mail*, Toronto, Canada, 479; From Eric Hobsbawm's *The Age of Empire* (Vintage Books), Copyright © 1987 by E. J. Hobsbaum, 481; © Giraudon/Art Resource, NY, 483; © Brent Patterson/CORBIS, 484.

Chapter 17 National Association for the Advancement of Colored People, 1934, 492 (top); © AFP/CORBIS, 492 (bottom); © AFP/CORBIS, 498; © Associated Press/AP/Wide World Photos, NY, 500; © Joseph Sohm, ChromoSohm, Inc./CORBIS, 502; © Hintz

Diltz/CORBIS, 504; The Everett Collection, NY, 506; Bibliotheque Nationale de France, 508; © Associated Press/AP/Wide World Photos, 509; © Canadian Press Picture Archives/Sean White, 511 (center); Figure.17.07 © Reuters NewsMedia, Inc./CORBIS, 511 (bottom).

Chapter 18 © Bettmann/CORBIS, 518; © 1999 WB & Village Roadshow Film Limited/The Everett Collection, NY, 519; © SuperStock, 520 (top); © Gary Braasch/CORBIS, 520 (bottom); U. S. Army Photos, 523; © SuperStock, 525; © SuperStock, 526; Figure 18.6 Copyright © 2000 Time, Inc. Reprinted by permission, 527; The Everett Collection, NY, 529; © Philip Gould/CORBIS, 533; Figure 18.7 © Joe Polemi/Reuters/Archive Photos, NY, 535; Figure 18.10 Copyright © 1990 Bill Watterson. Reprinted with permission of Universal Press Syndicate. All rights reserved, 540.

Literary Acknowledgements

Chapter 3

78 Figure 3.5 National Opinion Research Center, 1999. *General Social Survey*, 1972-98. Copyright © 1999 NORC. Used with permission.

80 Figure 3.7 From *World Values Survey 1994*. Distributed by the Inter-University Consortium for Political and Social Research. Used with permission.

Chapter 4

100 Figure 4.1 Figure from THE AMBITIOUS GENERATION by Schneider and Stevenson. © 1999 Yale University Press. Reprinted by permission.

104 Figure 4.4 "The Children's Hours" by John P. Robinson and Suzanne Bianchi, *American Demographics*, Vol. 20, No. 4, December 1997. Copyright © 1997 American Demographics. Used with permission.

Chapter 5

120 Figure 5.1 Bar graph based on information in Chapter 4, Closeness of the Victim from OBEDIENCE TO AUTHORITY by Stanley Milgram. Copyright © 1974 by Stanley Milgram. Reprinted by permission of HarperCollins Publishers, Inc.

Chapter 6

147 Figure 6.1 Figure from CRIME AND DISREPUTE by John Hagan, © 1994 Pine Forge Press.

152 Figure 6.2 From "Criminal Victimization" *Juristat*, Vol. 18, No. 6, 1998. Copyright © 1998.

153 Figure 6.5 From "The Aborted Crime Wave?" Marguerite Holloway, *Scientific American*, Vol. 281, No. 6, pp. 23-24. Copyright © 1999 Sarah Donelson.

Chapter 8

233 Figure 8.6 From *The State of Working America, 1998-1999* by Mischel et al. and Economic Policy Institute. Copyright © 1999

Cornell University. Used by permission of the publisher, Cornell University Press.

234 Figure 8.7 National Opinion Research Center, 1999. *General Social Survey*, 1972-98. Copyright © 1999 NORC. Used with permission.

234 Figure 8.9 National Opinion Research Center, 1999. *General Social Survey*, 1972-98. Copyright © 1999 NORC. Used with permission.

237 Figure 8.10 Figures from SOCIOLOGY IN A CHANGING WORLD, Fourth Edition by William Kornblum. Copyright © 1997 by Harcourt, Inc. Adapted by permission of the publisher.

237 Figure 8.11 Data from *The Decline of Deference* by Neil Nevitte, Broadview Press.

Chapter 9

256 Figure 9.2 "The 1997 Body Image Survey Results" by David M. Garner, *Psychology Today*, Vol. 30, No. 1, pp. 30-44. Reprinted with permission from Psychology Today Magazine. Copyright © 1997 Sussex Publishers, Inc.

269 Figure 9.5 From UNITED NATION'S HUMAN DEVELOPMENT REPORT 1998. Copyright © 1998. Reprinted by permission of Oxford University Press.

Chapter 10

294 Figure 10.7 From WORKING WOMEN IN AMERICA: Split Dreams by Sharlene Hesse-Biber and Gregg Lee Carter. Copyright © 2000 Oxford University Press. Used with permission.

Chapter 11

319 Figure 11.2 From Sidney Verba et al., "The Big Tilt" *The American Prospect*, Vol. 32, 1997, pp. 74-80. Copyright © 1997 American Prospect. All rights reserved.

320 Figure 11.3 From MONEY TALKS: CORPORATE PACS AND POLITICAL INFLUENCE by Dan Clawson et al. Copyright © 1992 Basic Books. Reprinted by permission of the author.

334 Figure 11.5 From "Social Cleavages and Political Alignments" Brooks & Manza, *ASR*, vol. 62, pp. 937-46, 1997. Copyright © 1997 American Sociological Association.

Chapter 12

350 Figure 12.3 Figure from *Psychological Review*, 1986, Vol. 93, pp. 119-135. Copyright © 1986 by the American Psychological Association. Adapted with permission.

358 Figure 12.6 From "Family Diversity and the Division of Domestic Labor: How Have Things Really Changed" by David H. Demo and Alan C. Acock, *Family Relations*, Issue 9307, Vol. 42:3,

1993. Copyright © 1993 by the National Council on Family Relations, 3989 Central Ave., NE, Suite 550, Minneapolis, MN 55421.

359 Figure 12.7 Based on material in UNDERSTANDING PARTNER VIOLENCE: Prevalence, Causes, Consequences and Solutions by Murray A. Straus and Sandra M. Stith, editors.

Chapter 13

378 Figure 13.1 National Opinion Research Center, 1999. *General Social Survey*, 1972-98. Copyright © 1999 NORC. Used with permission.

382 Figure 13.2 National Opinion Research Center, 1999. *General Social Survey*, 1972-98. Copyright © 1999 NORC. Used with permission.

389 Figure 13.6 Figures from SOCIOLOGY IN A CHANGING WORLD, Fourth Edition by William Kornblum. Copyright © 1997 by Harcourt, Inc. Adapted by permission of the publisher.

399 Figure 13.9 Campbell et al., in press. APPLIED DEVELOPMENTAL SCIENCE.

Reference: Campbell, F.A. Ramey, C.T., Pungello, E.P., Miller-Johnson, S., & Sparling, J. (2002), EARLY CHILDHOOD EDUCATION: Young Adult Outcomes from the Abecedarian Project, *Applied Developmental Science*.

Chapter 14

417 Figure 14.4 From George Gerbner, "Casting the American Scene" in *The 1998 Screen Actors Guild Report*. Copyright © 1998 SAG.

Chapter 15

435 Figure 15.3 National Opinion Research Center, 1999. *General Social Survey*, 1972-98. Copyright © 1999 NORC. Used with permission.

Chapter 16

473 Figure 16.5 From "The Nature of Cities" by Chauncy D. Harris and Edward L. Ullman in *Annals of the American Academy of Political and Social Science*, Vol. 242, 1945.

471 Figure 16.4 From "The Growth of the City: An Introduction to a Research Project" Ernest W. Burgess, pp. 47-62 in THE CITY by Robert E. Park et al. Copyright © 1967 University of Chicago Press. Used with permission.

Chapter 18

525 Figure 18.3 From "The Silent Boom" *Forbes*, July 7, 1998. Reprinted by permisison of Forbes Magazine © 2001 Forbes Inc.